Industry 4.1

Industry 4.1

Intelligent Manufacturing with Zero Defects

Edited by

Fan-Tien Cheng

IEEE Press Series on Systems Science and Engineering
MengChu Zhou, Series Editor

IEEE PRESS
WILEY

Published by John Wiley & Sons, Inc., Hoboken, New Jersey.
Published simultaneously in Canada.

Library of Congress Cataloging-in-Publication Data is applied for

Hardback: 9781119739890

Cover Design: Wiley
Cover Image: © Fan-Tien Cheng

Set in 9.5/12.5pt STIXTwoText by Straive, Pondicherry, India

10 9 8 7 6 5 4 3 2 1

MIX
Paper from responsible sources
FSC
www.fsc.org
FSC® C013604

Contents

Editor Biography

Soon after Fan-Tien Cheng graduated from the department of Electrical Engineering of National Cheng Kung University (NCKU) in 1976, he got in the Chung Shan Institute of Science and Technology (CSIST), Taiwan, ROC, serving as Research Assistant at the most basic level and then got promoted to Senior Scientist in 19 years. Then he went back to NCKU to start his teaching career and devoted the knowledge and practices he had learned in CSIST to the research domains of production improvement, manufacturing automation, and e-manufacturing for industries such as semiconductor, TFT–LCD, solar cell, machine tool, and aerospace to help achieve the goal of enhancing the industry competitiveness by successfully improving manufacturing processes and lowering production cost.

Professor Cheng has devoted himself to the academic research and industrial applications of the Intelligent Manufacturing and Industry 4.0 for the past decades and his accomplishments are eminent. Among them, the academic and applied research of Automatic Virtual Metrology (AVM) are especially unmatched worldwide. More than 40 journal papers related to VM had won him dozens of patents from Taiwan ROC, USA, Japan, China, Germany, and Korea; and 54 technology transfers had been successfully executed on several high-tech industries such as semiconductor (TSMC, UMC, ASE, and SUMCO), TFT–LCD (Innolux and CPT), solar cell (Motech); and traditional industries like aerospace (AIDC), machine tool industry (FEMCO/FATEK), blow molding machine (ChumPower), and carbon fiber (FPC), as well as foundations, constituted as a juristic person (ITRI and MIRDC).

Some of Professor Cheng's honors and awards include **2011 Award for Outstanding Contributions in Science and Technology, Executive Yuan, Taiwan, ROC, three times of Ministry of Science and Technology (MoST) Outstanding Research Award (2006, 2009, 2013), three times of the National Invention and Creation Award of Ministry of Economic Affairs (MoEA) (2011, 2012, 2018)**, University-Industry Economic Contribution Award from MoEA, Industry-University Cooperation Award for College Teachers, Ministry of Education (MoE), **NCKU Chair Professor since January 2009**, 17th TECO Award from TECO Technology Foundation, 2010, **2013 IEEE Inaba Technical Award for Innovation Leading to Production (for contributions to the development of the AVM System)**, 2014 Outstanding Research Award of Pan Wen Yuan Foundation, and 2015 20th Outstanding Achievement Award of The Phi Tau Phi Scholastic Honor Society. Moreover, Professor Cheng won **IEEE Fellow since January 2008, two times of IEEE ICRA Best Automation Paper Award (1999 and 2013)** as well as **CASE 2017 Best Application Paper Award**. He is currently in his second-term of President of Chinese Institute of Automation Engineers (CIAE) since 2017. Besides, he is the **Senior Editor of the IEEE T-ASE since October 2017**. Furthermore, he is honored to be the **IEEE CASE Conference Steering Committee Chair since September 2020**.

List of Contributors

1) **Fan-Tien Cheng**, Director/Chair Professor, Intelligent Manufacturing Research Center/ Institute of Manufacturing Information and Systems, National Cheng Kung University, Tainan, Taiwan, ROC
2) **Min-Hsiung Hung**, Professor, Department of Computer Science and Information Engineering, Chinese Culture University, Taipei, Taiwan, ROC
3) **Yu-Chen Chiu**, Specialist, Intelligent Manufacturing Research Center, National Cheng Kung University, Tainan, Taiwan, ROC
4) **Yu-Ming Hsieh**, Associate Research Fellow, Intelligent Manufacturing Research Center, National Cheng Kung University, Tainan, Taiwan, ROC
5) **Hao Tieng**, Associate Research Fellow, Intelligent Manufacturing Research Center, National Cheng Kung University, Tainan, Taiwan, ROC
6) **Haw-Ching Yang**, Professor, Department of Electrical Engineering, National Kaohsiung University of Science and Technology, Kaohsiung, Taiwan, ROC
7) **Yu-Chuan Lin**, Secretary General, Intelligent Manufacturing Research Center, National Cheng Kung University, Tainan, Taiwan, ROC
8) **Chin-Yi Lin**, Postdoctoral Research Fellow, Intelligent Manufacturing Research Center, National Cheng Kung University, Tainan, Taiwan, ROC
9) **Chao-Chun Chen**, Professor, Institute of Manufacturing Information and Systems, National Cheng Kung University, Tainan, Taiwan, ROC
10) **Hung-Chang Hsiao**, Professor, Department of Computer Science and Information Engineering, National Cheng Kung University, Tainan, Taiwan, ROC
11) **Kuan-Chou Lai**, Professor, Department of Computer Science, National Taichung University of Education, Taichung, Taiwan, ROC
12) **Hsien-Cheng Huang**, Deputy Director, e-Manufacturing Research Center, National Cheng Kung University, Tainan, Taiwan, ROC
13) **Yu-Yong Li**, Postdoctoral Research Fellow, Intelligent Manufacturing Research Center, National Cheng Kung University, Tainan, Taiwan, ROC

Preface

In the era of global competition, improving productivity and increasing yield of the manufacturing industries through information and communication as well as cloud computing technologies, big data analytics, and cyber-physical systems (CPS) are the common goals for the manufacturers worldwide. For instance, Germany proposed Industry 4.0 in hopes to construct Smart Factory so as to enhance its global competitiveness and continue to take the lead in manufacturing. The Advanced Manufacturing Partnership issued by the United States aimed to regain the leadership in international manufacturing competitiveness for attracting the manufacturing industries back to the United States. Chinese government brought out the plan called "Made in China 2025" in 2015, where the guidelines and strategies of becoming one of the countries with strong manufacturing power by 2025 were clearly stated. Facing the United States-China trade war that began in 2018, even though Chinese government stopped to mention Made in China 2025 in public to not intensify the conflict, they claimed that their goals of enhancing Intelligent Manufacturing have not been changed. Adjustments on the strategies in accordance with the international trend would be made, which is considered as the new version of Made in China 2025.

Overview and Goals

The current Industry 4.0 related technologies emphasize on productivity improvement but not on quality enhancement; in other words, they can only keep the faith of achieving nearly Zero-Defects (ZD) state without realizing this goal. The key reason for this inability is the lack of an affordable online and real-time Total Inspection technology. By adopting the Automatic Virtual Metrology (AVM) that has been certified with the invention patents from six countries (Taiwan ROC, USA, Japan, Germany, China, and Korea) developed by the research team of Fan-Tien Cheng, the Editor and main author of this book, ZD can be achieved as AVM can provide the Total Inspection data of all products online and in real time. A defective product will be discarded once it is detected by AVM; in this way, all of the deliverables will be ZD. Further, the Key-variable Search Algorithm (KSA) of the Intelligent Yield Management (IYM) system developed by our research team can be utilized to find out the root causes of the defects for continuous improvement on those defective products. As such, ZD of all products can be achieved. Therefore, once AVM and IYM are integrated into the successfully developed Industry 4.0 platform constituting of Internet of things (IoT), CPS, big data analytics, and cloud computing, the state of ZD can be realized, which is defined as Industry 4.1 by Fan-Tien Cheng. The concepts of Industry 4.1 were disclosed in *IEEE Robotics and Automation Letters* in January 2016.

To realize and promote Intelligent Manufacturing, National Cheng Kung University (NCKU) established the Intelligent Manufacturing Research Center (iMRC) with Professor Fan-Tien Cheng being its Director. Based on the platform of Advanced Manufacturing Cloud of Things (AMCoT) that won IEEE Conference on Automation Science and Engineering (CASE) 2017 Best Application Paper Award, iMRC integrates cross-disciplinary research resources, utilizes various Intelligent-Manufacturing related technologies, and implements Intelligent Manufacturing services [including AVM, IYM, Intelligent Predictive Maintenance (IPM), . . ., etc.] to develop the so-called Intelligent Factory Automation (iFA) System Platform. Through implementing the iFA System Platform to the manufacturing tools and production lines of high-tech (e.g. semiconductor, TFT-LCD, and solar cell) and traditional industries (e.g. machine tool, aerospace, blow molding machine, and carbon fiber), ZD of all products as well as highly efficient and flexible intelligence capabilities (single-machine intelligence, production-line intelligence, and global-fab intelligence) can be accomplished for improving the competitiveness and profits of all Intelligent-Manufacturing related industries.

Organization and Features

To promote Intelligent Manufacturing and carry out the vision of Industry 4.1, our research team decided to include the survey and introduction of Intelligent Manufacturing, as well as the intact concept and the core technologies of Industry 4.1, along with the successful cases of iFA implementation for Industry 4.1 in different industries all in this book. This book contains 11 chapters in total. Chapters 1–5 describe the evolution of automation and the development strategy of Intelligent Manufacturing and introduce the mandatory components and fundamental technologies for constructing Intelligent Manufacturing. Chapter 6 introduces the overall concept of iFA integrated by the AMCoT framework with pluggable modules of AVM, IPM, and IYM intelligent services. Two versions of the iFA System Platform are provided for different business models. Chapter 7 illustrates the AMCoT framework for constructing the advanced cloud manufacturing platform. Chapters 8–10 address the principles and implementation of AVM, IPM, and IYM, respectively. Finally, the actual Intelligent-Manufacturing implementation and application cases adopting all the techniques mentioned above in seven industries, including flat panel display, solar cell, semiconductor, automobile, aerospace, carbon fiber, and blow molding, are presented in Chapter 11. Moreover, all the major patents related to AMCoT, AVM, IPM, and IYM are listed in Appendices 7.C, 8.C, 9.C, and 10.C, respectively.

Fan-Tien Cheng
Life Fellow, IEEE
Chair Professor, NCKU
Director of Intelligent Manufacturing Research Center (iMRC), NCKU
July 2021

Acknowledgments

Firstly, I would like to thank all the contributors for sharing their precious knowledge and experience in their professional fields. Throughout numerous discussions, the outline of this book is gradually shaped. I really appreciate their time and effort devoted in writing this book.

In addition to the authors listed, I would also like to express my deepest gratitude to my secretaries and research assistants: Pei-Ying Du, Ken-Ying Liao, Yan-Yu Shih, and Benny Suryajaya, for their dedication in completing the book. I want to sincerely thank them for helping with the translation, layout arrangement, proofreading, artwork illustration, and contribution in any manuscript preparation tasks. Without their effort, I, the editor and main contributor, as well as the other contributors couldn't have made it this far.

Fan-Tien Cheng
Life Fellow, IEEE
Chair Professor, NCKU
Director of Intelligent Manufacturing Research Center (iMRC), NCKU
July 2021

Foreword

Since the term "Industry 4.0" was coined in Germany in 2011, industries worldwide have been investing in the development of smart factories that are more efficient and better adaptive to digital transformation to enhance their service-oriented and customized-supply capabilities.

To take Industry 4.0 a step further, Professor Fan-Tien Cheng proposed the upgraded version of Industry 4.1, core to which is the realization of Zero Defects, a solution taking advantage of the newly developed Intelligent Factory Automation (iFA) System Platform to address the production quality issue that has received relatively scant attention in Industry 4.0. To put into practice Zero Defects as well as in response to the Intelligent Manufacturing Industry Innovation policy of the Taiwan, ROC government, he further established the Intelligent Manufacturing Research Center (iMRC) at National Cheng Kung University (NCKU) in 2018.

As NCKU's President, I always take great pride in the achievements of all my colleagues and students. In the latest Times Higher Education (THE) Impact Rankings 2020, NCKU is ranked first in Taiwan, ROC, second in the Asia region, and 38th globally. It excels especially in "Industry, Innovations, and Infrastructure," one of United Nation's 17 Sustainable Development Goals (SDGs), and earns the 10th place worldwide. Professor Cheng's innovative research has played a critical role in this intense competition because, as I understand it, his Industry 4.1 and Intelligent Manufacturing are key to NCKU's success in the SDG of "Industry, Innovations, and Infrastructure" and NCKU's continuous leadership in the Engineering field.

As NCKU is accelerating its research momentum, especially in disciplines of traditional strengths like Intelligent Manufacturing and Engineering, I am glad that Professor Cheng is willing to share his valuable research and industry-university cooperation experiences in this book, one that will become an important reference not only for students but professors and researchers alike, and not only at NCKU but in industries and higher education worldwide whose focus is Intelligent Manufacturing.

This book not only details the essentials that have paved the way from Industry 4.0 to Industry 4.1 but also provides numerous practical industrial application cases in different manufacturing industries. It thus offers readers a comprehensive perspective of what they are and will be facing in the industry. I am sure this book is fundamental – a must-have indeed – for researchers, engineers, and focused students in the fields of, among others, Intelligent Manufacturing and Industry 4.1.

Huey-Jen Su
President, National Cheng Kung University (NCKU)

Industry 4.0 is a confluence of trends and technologies for the fourth industrial revolution. It has been "pushed" by the digital revolution over the past many decades and the recent Internet of Things (IoT); and "pulled" by demand from customers for high quality and customized products at reasonable prices and lead times. With (i) the ubiquitous connection and interaction of machines, things, and people; (ii) the integration of cyber and physical systems; and (iii) the emerging of disruptive technologies such as big data, machine learning, artificial intelligence, 3D printing and robotics, the ways we design and manufacture products and provide services are undergoing fundamental changes.

Although much R&D progress has been made, industries have been slow to develop effective holistic Industry 4.0 strategies. From a recent survey of 2000 C-suite executives by Deloitte (https://www2.deloitte.com/content/dam/insights/us/articles/us32959-industry-4-0/DI_Industry4.0.pdf), only 10% of the executives surveyed indicated they had long-range strategies to leverage new technologies that reach across their organizations. This is not surprising since creating and implementing holistic Industry 4.0 strategies are complicated, and require deep understanding, sharp vision, inspirational leadership, and resolute persistence. Among those with comprehensive Industry 4.0 strategies, the results have been impressive: 73% of those with a strategy report success in protecting their businesses from disruption, versus 12% of those with more scattershot approaches; 61% of those with Industry 4.0 strategies report that they have developed innovative products and services, versus 12% of those lacking strategies; and 60% of those with Industry 4.0 strategies report that they have found growth opportunities for existing products and services, versus 8% of those lacking strategies. Those companies with strategies also are growing more financially, and making more progress investing in technologies that have a positive societal impact.

Consider specifically a key area of Industry 4.0, the quality of products and processes. It is well-known that a host of methods and processes such as Statistical Process Control (SPC), Zero Defect Manufacturing (ZDM), Six Sigma Methodologies, Preventive Maintenance (PM), Continuous Improvement (Kaizen), Total Quality Management (TQM), etc., have been around for years and are contributing to the quality of products and processes. Integrating the digital revolution, the Internet of Things, big data, machine learning, and artificial intelligence to raise the quality of products and processes to a new level and with practical and scalable implementations, however, remains a major challenge for scholars, practitioners, and C-suite executives alike.

This book "Industry 4.1 – Intelligent Manufacturing with Zero Defects" focuses on improving the quality of products and processes, and is the culmination of the brilliant but down-to-the-earth efforts of the team led by Professor Fan-Tien Cheng over the past many years. The efforts started with Virtual Metrology. In view of the incompatible paces of fast production and slow metrology, 100% inspection is impossible, and sample inspection has been the practice. With the advancements in sensing, metrology, analytics and Industry 4.0 technologies, the team innovatively integrated physical metrology with its cyber counterparts, Virtual Metrology (VM). The resulting Automatic Virtual Metrology (AVM) system presented in this book is capable of predicting the quality of a product based on machine parameters, sensor data in the production equipment, and off-line sampling measurements. It also provides on-line and real-time total inspection of all work pieces to timely detect abnormalities during production. As a result, the sampling rate of real measurements can be cut down, the production costs can be reduced, and the goal of nearly zero defects of deliverables can be achieved.

Effective implementation of Automatic Virtual Metrology, however, is not easy, especially if we want it to be scalable to large factories and transferrable to other companies and other industries. Major infrastructure needs to be established efficiently and flexibly. Based on the team's successful research, development, implementation, and redeployment at many factories and across multiple industries, this book methodically presents the essential infrastructure components. The content includes data collection and management and feature extraction; communication standards; computation infrastructure of cloud, edge, Internet of Things and big data; container-related software development, deployment, and management technologies of Docker and Kubernetes; the overall architecture of the advanced manufacturing "Cloud of Things" framework, and the specific design and implementation of key components such as cyber-physical agents, big data analytics application platform, the automated construction scheme for manufacturing services, and AVM and other servers.

Extending the ideas, methods, and infrastructure presented above, the book then focuses on Intelligent Predictive Maintenance (IPM). Predictive maintenance, sometimes known as "condition-based maintenance," is to monitor the performance and conditions of equipment during operations to predict when equipment performance is deteriorating and when equipment is going to fail, followed by scheduled or corrective maintenance. Intelligent Predictive Maintenance presented in this book detects the abnormality of key components of manufacturing tools based on advanced fault detection and classification techniques and predicts their Remaining Useful Lives (RUL) using time series prediction algorithms. Factory-wide implementation is then discussed to improve tool availability and prevent unscheduled down of manufacturing tools.

Since modern manufacturing facilities are generally capital intensive, it is critical to have consistently high yields to justify the investment and to have a positive bottom line. Intelligent Yield Management (IYM) presented in this book is a closely related cousin of Intelligent Predictive Maintenance, with the purpose to effectively detect root causes that affect the yield. It consists of data collection and management; statistical, big data, and machine learning tools for defect and yield analysis; and timely resolution of issues discovered while maintaining the requisite quality and reliability standards. The kernel of the above is the "Key-variable Search Algorithm" (KSA), which includes new root-cause search methods for solving the high-dimensional variable selection problem, and modules for checking the quality of input data and for evaluating the reliability of search results.

The current Industry 4.0-related technologies emphasize productivity improvement but not on quality enhancement. They can have the faith of achieving nearly Zero-Defect Manufacturing but without effective methods to achieve it. By developing and implementing the novel methods, technologies, and infrastructure presented above, zero defects of products can be effectively achieved. This is what is defined as Industry 4.1 in the book. The actual deployment cases in seven industries, including flat panel display, semiconductor, solar cell, automobile, aerospace, carbon fiber, and blow molding, are presented in the final Chapter 11. The ingenuity is outstanding, the effort is tremendous, and the impact is far-reaching and long-lasting.

Since many acronyms are used throughout the book, readers are advised to have Abbreviation Lists handy when reading the book. Beyond this point, I sincerely hope that you enjoy reading the book, and delightfully discover the wonderful world of Industry 4.1.

Peter B. Luh
Board of Trustees Distinguished Professor
SNET Professor of Communications & Information Technologies
Dept. of Electrical & Computer Engineering
University of Connecticut

Since Germany brought up Industry 4.0 in 2012, the trend of Intelligent Manufacturing has boomed globally. By integrating the innovative information-and-communication technologies such as IoT, Cloud, Big Data, AI, etc., various Cyber-Physical Systems are developed to promote factory process optimization, yield improvement, efficiency enhancement, and cost reduction. Besides, in response to changes in consumers' habits, Zero Defects, High Variety Low Volume, and Rapid Change have become mandatory indicators for Intelligent Manufacturing.

Advanced Semiconductor Engineering Inc. (ASE), is the leading provider of independent semiconductor manufacturing services in assembly and test. ASE develops and offers complete turnkey solutions in IC packaging, design and production of interconnect materials, front-end engineering test, wafer probing, and final test. In 2011, ASE started to vigorously promote Intelligent Manufacturing and established over 15 lights-out factories in response to changes in the global industrial environment. Moreover, ASE also collaborated with various top universities in Taiwan, ROC for R&D of IoT, Cloud, Big Data, and AI technologies, which have cultivated more than 400 professionals in the automation field via co-hosting educational trainings and industry programs to improve the automation capability within ASE.

ASE began the industry-university collaboration with Prof. Fan-Tien Cheng in 2014. Initially, we implemented Automatic Virtual Metrology (AVM) to achieve total inspection in an efficient and economic way so as to reduce the measurement cost. The project was a great success, and ever since then Prof. Cheng has become one of our major collaborators. The subsequent cooperation includes Intelligent Yield Management (IYM), Intelligent Predictive Maintenance (IPM), Advanced Manufacturing Cloud of Things (AMCoT), and Scheduling, which can be said to be the practical applications of all the research essence of Prof. Cheng on the production line.

The Industry 4.1 proposed by Prof. Cheng aims at Zero Defects, it applies AVM to accomplish total inspection and utilizes IYM to find the root causes of a yield loss. In addition to enhancing production efficiency, it also improves product yield and makes products close to Zero Defects, which is a great step forward in the realm of Industry 4.0.

Although Intelligent Manufacturing is a hot subject nowadays, it is challenging for the enterprises to actually carry it out; many enterprises still struggle to realize the vision of Intelligent Manufacturing. The implementation of novel technologies isn't the only core for Intelligent Manufacturing, the shaping of the ecological chain of the automation industry and the cultivation of talents are also important factors.

As the development of hardware like sensor, microcontroller, Automatic Material Handling System (AMHS), and robot is coming to a mature state gradually, the focus of Intelligent Manufacturing has shifted to the software. The cloud-based technologies such as Big Data and AI application modules draw more attention to the researchers and professionals at present. The technologies introduced in this book are a series of automation technologies developed upon IoT, Cloud, Big Data, and AI. Aside from explaining through the theories in detail, it also includes hands-on application cases in various industries. This is a book worth reading for both industrial professionals and scholars, and I highly recommend these materials for Intelligent Manufacturing education.

Michael Lee

Michael Lee
Vice GM of ASEKH MIS Center
Former Plant Manager of ASE Testing and Wafer Bumping Plants
Former Executive Secretary of ASE Security Committee
Former Committee Member of ASE Automation Committee

By the time we established the Precision Machinery Research & Development Center (PMC) in 1993, the board of directors agreed to my suggestion of focusing our efforts on two fields of expertise, IT and total quality control, to speed up our competitiveness on machine tools made in Taiwan, ROC.

Back then, we were totally unaware that IT could even be developed outside our expertise realm to missions such as Apollo 13 by NASA through digital twins.

However, we began to appreciate our choice of focusing on IT when the U.S. National Science Foundation announced the development of Cyber-Physical Systems in 2006. I am glad to report that PMC was the first organization in Taiwan, ROC to join the IMS Center founded by Prof. Jay Lee while he was a professor at the University of Wisconsin-Milwaukee before he moved to Cincinnati. Our affiliation with the IMS center guided us to recognize the worth of Industry 4.0 initiated by Germany later in 2013.

In the meantime, virtual metrology (VM) has emerged as a key tool for controlling complex process such as semiconductor device manufacturing. VM utilizes mathematical models to estimate quality variables that may be difficult or expensive to measure using readily available process information.

Professor Fan-Tien Cheng, the Editor and leading author of this book, realized that if VM can be fully automated, the quality of a process can be monitored without processing interruption. His team applied their Automatic Virtual Metrology (AVM) to the chemical vapor deposition for a thin film transistor liquid display manufacturing process in Taiwan, ROC. Since AVM allows the possibility of acquiring Zero-Defects production, he claimed that AVM should be coined into Industry 4.1, i.e., one step ahead of the original Industry 4.0.

In 2013, his team began to expand AVM into the semiconductor packaging process in cooperation with the ASE group. The success of AVM implementation was then followed by the integration of Intelligent Predictive Maintenance (IPM) and Intelligent Yield Management (IYM) into their production lines through the Intelligent Factory Automation (iFA) platform Professor Cheng developed.

In the meantime, the iFA platform was applied to the machining of aluminum alloy wheels at FEMCO Machine Tool Manufacturing Co., Ltd. in Chiayi, Taiwan, ROC. Their success has helped FEMCO to export numerous similar systems to worldwide automobile wheel manufactures. In addition to the semiconductor and automotive industries, his team has deployed these systems constituting Industry 4.1 to many other manufacturing enterprises such as TFT-LCD, solar cell, jet engine case machining, plastic bottle blow molding, machine tools, 3D metal printing, and thermal process for making carbon fibers.

Professor Cheng and his team aim to upgrade the manufacturing industries to achieve Zero Defects through the implementation of Industry 4.1. This book is the embodiment of their dedication on the advanced technologies that pave the way from Industry 4.0 to Industry 4.1. I highly recommend this practical book to those who are interested in or preparing themselves to take parts in the manufacturing industries, they can see a whole picture of the industry evolvement with actual on-site application cases.

Kuo-Chin Chuang
Ph.D. of Materials Sci. and Eng., Massachusetts Institute of Technology (MIT)
Honorary Chairman, Taiwan Association of Machinery Industry (TAMI)
Former Chairman, Far East Machinery Co, Ltd. (FEMCO)
Chairman, LOGICOM, Inc.

1

Evolution of Automation and Development Strategy of Intelligent Manufacturing with Zero Defects

Fan-Tien Cheng

Director/Chair Professor, Intelligent Manufacturing Research Center/Institute of Manufacturing Information and Systems, National Cheng Kung University, Tainan, Taiwan, ROC

1.1 Introduction

The evolution of automation from Industry 1.0 to 3.0 as well as e-Manufacturing, which is the predecessor of Industry 4.0 is described in this chapter. Then, the core technologies of Industry 4.0 and the concept of mass customization are presented. After that, the concept of Zero Defects (ZD), which is the vision of Industry 4.1, is introduced. Finally, the five-stage strategy of yield enhancement and ZD assurance is proposed in this chapter.

1.2 Evolution of Automation

While the first industrial revolution (Industry 1.0) introducing the steam engine, the second (Industry 2.0) carrying out the assembly line mass-production, and the third (Industry 3.0) framing the automated manufacturing with electronic controllers, industrial production requirements need further changes nowadays. There is an increasing demand for manufacturing to satisfy customer expectations precisely; at the same time, companies face growing pressure to manufacture at more competitive prices. To adapt to this evolution, the tools of systems engineering, information and communication technology (ICT), artificial intelligence, and business strategies will be applied to achieve a higher level than before for developing new scenarios of the automated production. Thus, the so-called Industry 4.0, which aims to increase productivity of the traditional manufacturing scenario, was proposed. In fact, e-Manufacturing presented by the semiconductor industry is the predecessor of Industry 4.0. Therefore, prior to introducing Industry 4.0, the concept and key components of e-Manufacturing are described as follows.

1.2.1 e-Manufacturing

Since market competition in the consumer electronics industry has intensified, short product lifecycle becomes essential. A company that generates innovative research and development can garner market share. The rapid development of the information and Internet technologies facilitates

Industry 4.1: Intelligent Manufacturing with Zero Defects, First Edition. Edited by Fan-Tien Cheng.

the computerization of the intra-company manufacturing execution system (MES) [1–3] and equipment engineering system (EES) [4–5], as well as expedites the networking of the inter-company supply chain (SC) [6–8] and engineering chain (EC) [9–11] to move toward a global business model of e-Manufacturing [12–13].

National Coalition for Advanced Manufacturing (NACFAM) [12] stated in 2001 that in the e-Manufacturing era, companies will be able to exchange information of all types with their suppliers at the speed of light. Also, design cycle times and intercompany costs of manufacturing complex products will implode. Information on design flows will be instantly transmitted from repair shops to manufacturers and their supply chains.

Figure 1.1 shows the e-Manufacturing hierarchy created by the international SEMATECH (ISMT) [4]. This hierarchy can be divided into the manufacturing portion and the engineering portion. In Figure 1.1, MES is a core system in the manufacturing portion that connects its upper factory-to-factory modules and lower equipment modules to dominate the overall manufacturing management. The highest (company-to-company) layer in the manufacturing portion is mainly for the purpose of SC. On the other hand, EES takes charge of the engineering portion that deals with equipment health monitoring, real-time quality control, and maintenance scheduling (e.g. e-diagnostics [15, 16]).

In the semiconductor manufacturing industry, Tag and Zhang [13] defined e-Manufacturing as the complete electronic integration of all factory components using industry standards. This e-Manufacturing model extends from equipment-to-equipment automation systems to the manufacturing execution system/yield management system/equipment engineering system (MES/YMS/EES) and to the enterprise resource planning (ERP).

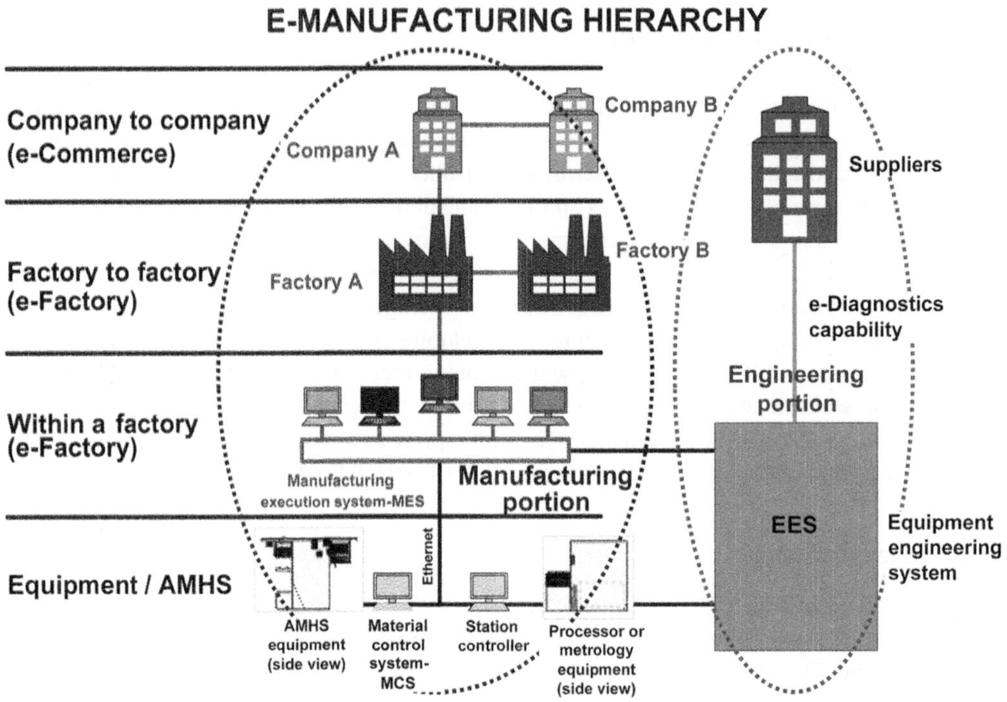

Figure 1.1 ISMT e-Manufacturing hierarchy. *Source:* Reprinted with permission from Ref. [14]; © 2010 IEEE.

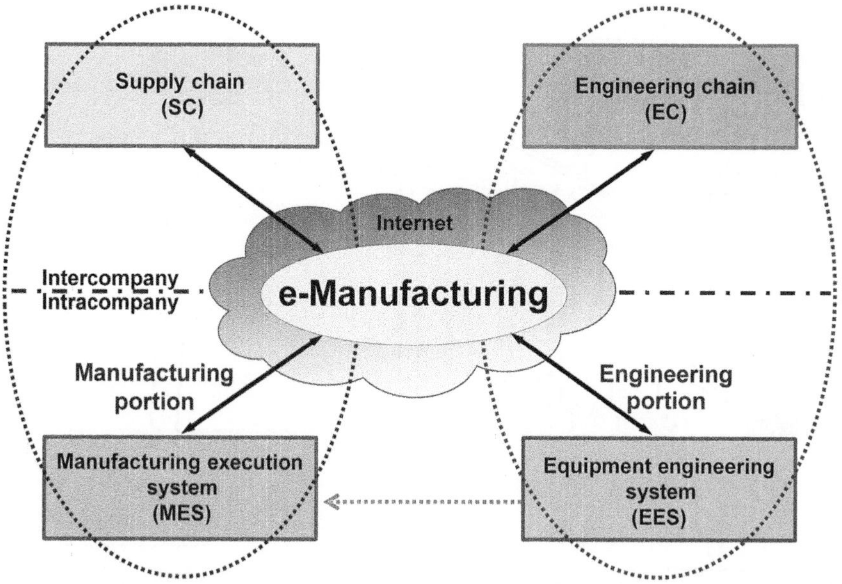

Figure 1.2 Four key components for the advanced e-Manufacturing model. *Source:* Reprinted with permission from Ref. [14]; © 2010 IEEE.

The ISMT e-Manufacturing hierarchy shown in Figure 1.1 [4] merely takes care of the functions of MES, EES, and SC without EC. Another model defined in [13] also takes the related functions of MES, EES, and SC into consideration only.

To consider all of the functions and applications of MES, SC, EES, and EC simultaneously, and enhance the integrity of e-Manufacturing as shown in Figure 1.2, Cheng et al. [14] proposed an advanced e-Manufacturing model that takes advantage of the information and Internet technologies to efficiently integrate the MES and EES within a company (intra-company integration), and the SC and EC among member companies (inter-company integration). With this advanced e-Manufacturing model, the productivity and yield of a complete production platform can be improved (by MES), the overall equipment effectiveness (OEE) can be enhanced (by EES), the order-to-delivery (O2D) period can be reduced (by SC), and the time-to-market (T2M) can be shortened (by EC). Furthermore, the goal of improving agility, efficiency, and decision-making for the entire semiconductor manufacturing processes can be reached.

In the advanced e-Manufacturing model, both the MES and SC belong to the manufacturing portion, whereas the EES and EC are closely related to the engineering portion. The proposed e-Manufacturing model fully integrates the four key components (MES, EES, SC, and EC) to enhance the globalization and competitiveness of the semiconductor industry. The definitions, missions, primary issues, and feasible implementation frameworks of the four key components of e-Manufacturing are discussed in the following sections.

1.2.1.1 Manufacturing Execution System (MES)

The MES is a shop floor control system which includes either manual or automatic labor and production reporting as well as on-line inquiries and links to tasks that take place on the production floor. The MES provides links to work orders, receipt of goods, shipping, quality control, maintenance, scheduling, and other related tasks [17]. The mission of MES is to increase productivity and yield.

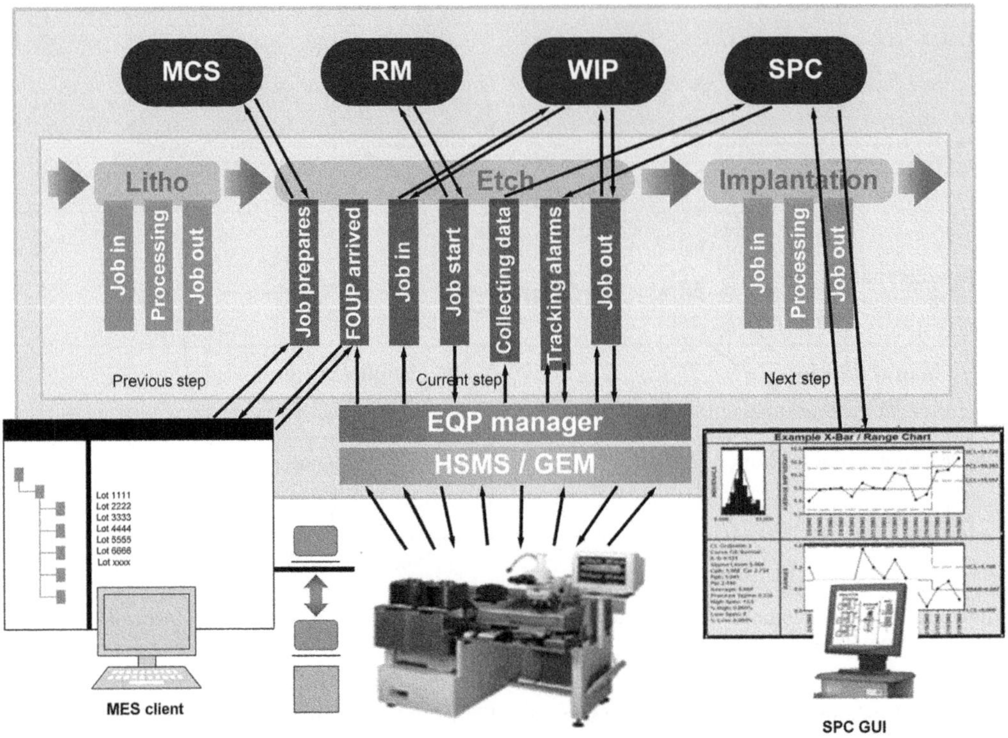

Figure 1.3 MES operation procedures. *Source:* Reprinted with permission from Ref. [14]; © 2010 IEEE.

Figure 1.3 presents the MES operation procedures in semiconductor manufacturing. In Figure 1.3, a front opening unified pod (FOUP), containing 25 wafers, is processed via lithography, etching, and implantation. After finishing its procedures in the lithography process, the FOUP is prepared for the etching process by the MES. First, the MES client requires a material control system (MCS) to move the FOUP to the process equipment. When the FOUP arrives at the etching equipment, the equipment manager sends a message to notify the MES, reads the information of work in process (WIP), acquires a recipe for this FOUP from the recipe management (RM) system, and initiates fabrication. Next, the equipment manager sends the process data of each wafer under fabrication to the statistical process control (SPC) server for quality monitoring. Eventually, the equipment manager updates the WIP information when the etching process completes and asks the MCS to move the FOUP from the etching equipment to the implantation equipment.

Notably, ISMT developed a SEMATECH computer-integrated manufacturing (CIM) framework (Figure 1.4) [1] to specify the common MES infrastructure and the software functions of MES applications, and incorporate those MES applications into a coherent system. By specifying the standard interfaces and behaviors of the common MES components, manufacturers can collect system components from multiple suppliers. Thus, manufacturers can develop systems by extending the common components and substituting old components with improved ones of the same interfaces and behaviors.

The SEMATECH CIM framework is an abstract model for typical semiconductor manufacturing systems. This CIM framework is developed based on open-distributed system and object technologies. However, the fragility and security problems are not considered in this framework. For these reasons, Cheng et al. [3] adopted the concepts of holon and holarchy to propose a holonic

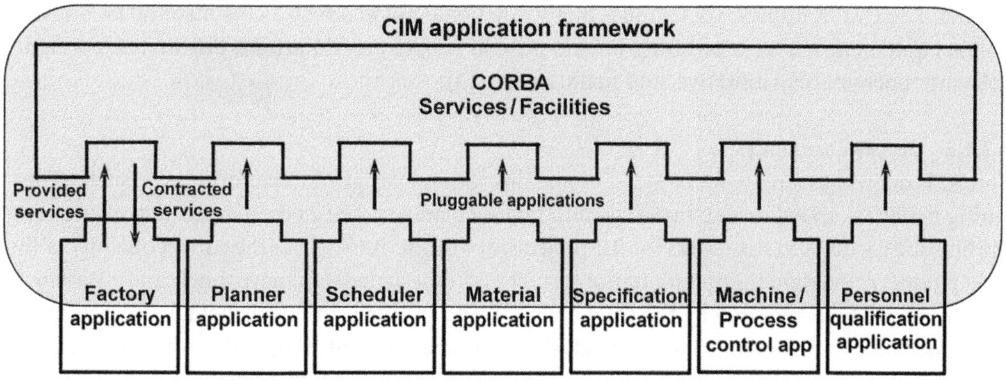

Figure 1.4 Functional architecture of the ISMT CIM framework. *Source:* Reprinted with permission from Ref. [14]; © 2010 IEEE.

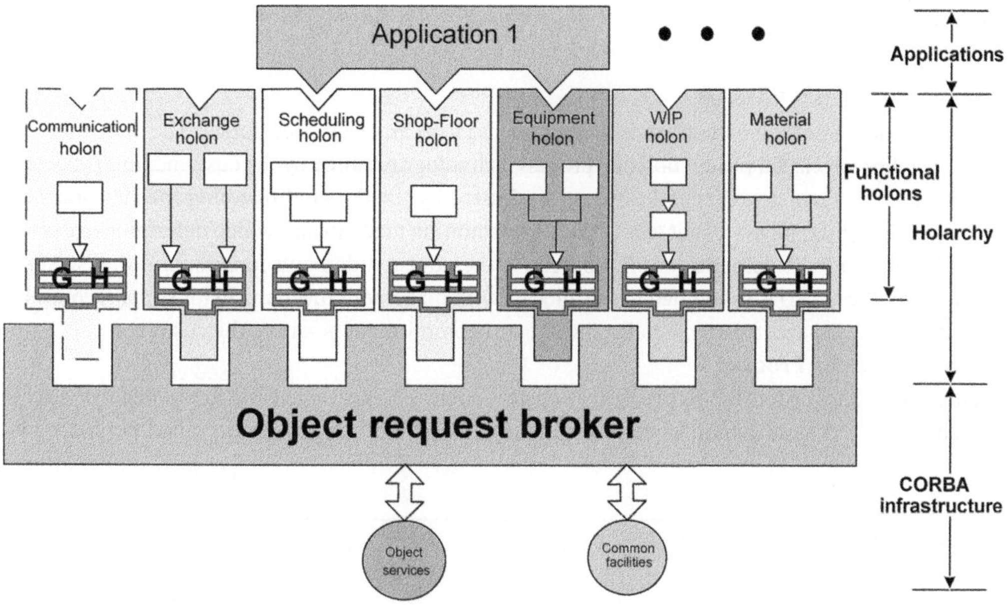

Figure 1.5 The HMES framework. *Source:* Reprinted with permission from Ref. [14]; © 2010 IEEE.

manufacturing execution system (HMES) framework that is also suitable for industrial application (Figure 1.5). The HMES framework not only owns the characters of open-distributed system and object technologies but also has the properties of failure recovery and security certification.

A systematic approach was proposed for developing the HMES framework of the semiconductor industry (Figure 1.5) [3]. This systematic approach starts with a system analysis by collecting domain requirements and analyzing domain knowledge. The HMES holarchy is designed through the following processes: (i) constructing an abstract object model based on the domain knowledge; (ii) partitioning the application domain into functional holons; (iii) identifying the generic parts among functional holons; (iv) developing the generic holon (GH); (v) defining the holarchy messages and holarchy framework of HMES; and finally, (vi) designing the functional holons based on the GH. The HMES framework includes many functional holons, such as the material holon, WIP holon, equipment holon, scheduling holon, etc. and is open, modularized, distributed, configurable, interoperable, collaborative, and maintainable [3].

1.2.1.2 Supply Chain (SC)

The SC is defined as a network of facilities and distribution designed to perform tasks, such as procuring materials, transforming materials into intermediate and finished products, and distributing the finished products to customers [6]. The objective of the SC is to deliver the correct quantity of the right product at the right time at minimum cost. The SC is designed to achieve timely and economical delivery of products required by the O2D cycle [7], and to support the collaborative computing of distributed orders in the semiconductor industry to ensure coherent IC manufacturing operations.

Figure 1.6 presents the architecture of an electronic supply chain management (ESCM) and its key processes [18]. This ESCM has been deployed in Taiwan Semiconductor Manufacturing Company (tsmc) [19]. The ESCM architecture comprises demand-planning, allocation-planning, capacity-modeling, allocation-management, order-management, available-to-promise (ATP), and output-planning mechanisms. The demand-forecast process and purchase-order process of ESCM are presented in the following paragraphs.

- **Demand-Forecast Process**
  The demand-planning mechanism receives demand forecasts from a customer. The demand forecast specifies forecasted production of a process technology required by the customer in a predetermined period. Then, the demand forecast is adjusted by the demand-planning mechanism. The adjusted demand forecast is sent to an allocation-planning mechanism, which determines a capacity-allocated-support demand (CASD) based on the adjusted demand forecast and the capacity plan. Next, the CASD is forwarded to the allocation-management mechanism for support commitment is generated accordingly. Finally, the support commitment is sent to the customer.
- **Purchase-Order Process**
  When a purchase order (PO) is placed by a customer, the PO is received and forwarded to the ATP mechanism by the order-management mechanism. After receiving the information pertaining to

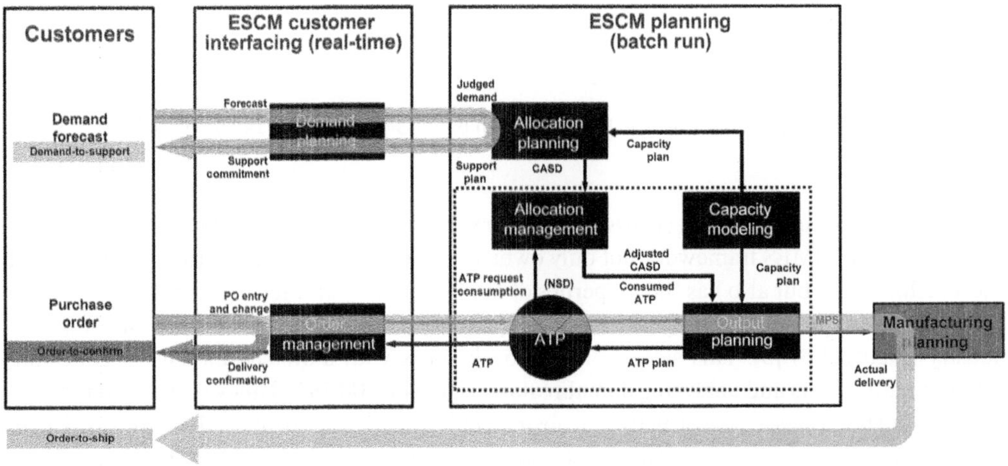

Figure 1.6 ESCM architecture and key processes. *Source:* Reprinted with permission from Ref. [14]; © 2010 IEEE.

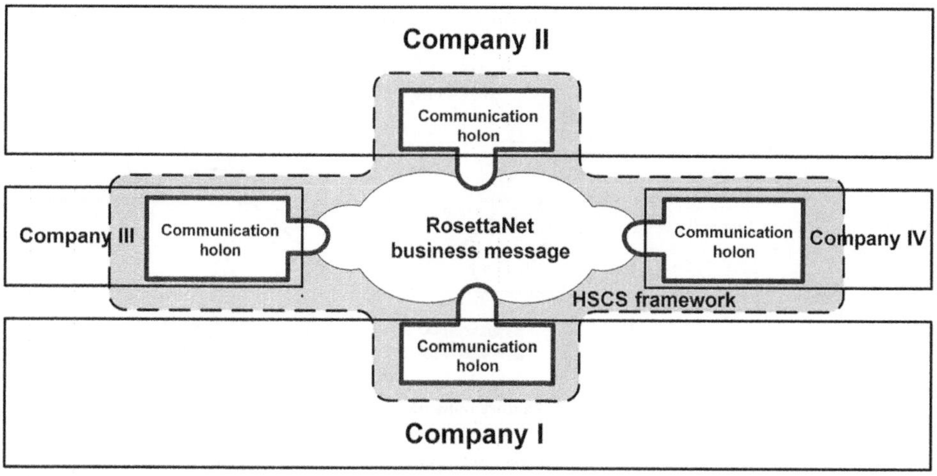

Figure 1.7 Functional-block diagram of the holonic supply-chain system. *Source:* Reprinted with permission from Ref. [14]; © 2010 IEEE.

the PO, the ATP mechanism determines the amount of CASD to be booked and the ATP production to be consumed. Next, the ATP mechanism sends the information pertaining to the booked CASD to the allocation-management mechanism. Once the booked CASD is received, the allocation-management mechanism adjusts the initial CASD accordingly. Meanwhile, the ATP mechanism also sends information pertaining to the consumed ATP production to the output-planning mechanism. With the consumed ATP production received from the ATP mechanism and the capacity plan from the capacity-modeling mechanism, the output-planning mechanism can generate a master production schedule (MPS) accordingly, which is sent to the manufacturing planning subsystem for shipping the product to the customer.

Cheng et al. [8] have also developed a holonic supply-chain system (HSCS) as shown in Figure 1.7. The HSCS consists of several communication holons. Each company in the SC should possess a communication holon. The HSCS employs distributed object and mobile object technologies, RosettaNet implementation framework, and holon and holarchy concepts. The systematic approach applied to develop the HMES is also utilized in constructing the HSCS. The GH is first developed. Next, the communication holon is generated by inheriting the GH. As shown in Figure 1.7, each company in the SC, such as Company I, requires a communication holon as the communication component for correspondence with other companies in the SC. The communication holon exhibits basic holonic attributes, such as intelligence, autonomy, and cooperation. Furthermore, the communication holon can handle partner interface processes and data exchange of various data formats by following the standards of RosettaNet business messages. As a result, the HSCS can meet the future requirements of the SC information integration of virtual enterprises [8].

1.2.1.3 Equipment Engineering System (EES)

The EES is defined as the physical implementation of the equipment engineering capabilities (EECs), which are applications that address specific areas of equipment engineering (EE), such as fault detection and classification (FDC), predictive maintenance (PdM), virtual metrology (VM), run-to-run (R2R) control, etc. [4, 5].

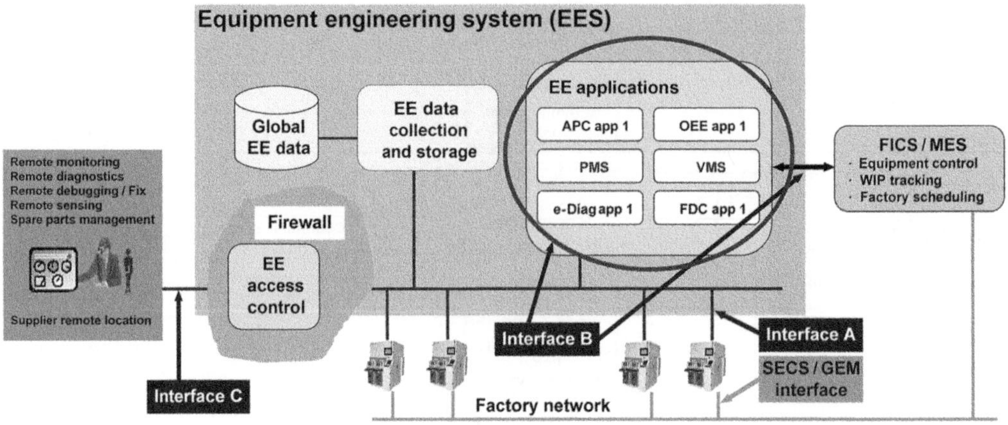

Figure 1.8 The ISMT EES framework. *Source:* Reprinted with permission from Ref. [14]; © 2010 IEEE.

An EES framework is required to support the EECs [4]. Therefore, ISMT proposed an EES conceptual framework as shown in Figure 1.8 [4]. In the ISMT EES framework, three interfaces (Interface A, Interface B, and Interface C) are defined for different purposes. Interface A is an equipment data acquisition interface for getting more and better data from the equipment [20]. Interface B defines interfaces among EE applications and creates a connection between the MES and EES [24]. Interface C describes the external access to e-diagnostics [16, 25].

As displayed in Figure 1.8, the ISMT EES framework posits all the EE applications (such as advanced process control (APC), OEE, FDC, PdM, VM, and others) outside the equipment. Those architectures are suitable for the applications of R2R-type controls involving more than one piece of equipment. However, for self-related equipment applications (e.g. FDC, PdM, and VM), such architectures heavily consume factory network bandwidth. Another disadvantage of those architectures is that all the data are sent to the same remote client for processing and monitoring, which may result in data overloading to the remote client and further impact the real-time analysis efficiency. Additionally, if the remote client breaks down and lacks backup, the entire prognostics system is paralyzed [5].

To resolve the problems mentioned above, Su et al. [5] proposed another EES framework as shown in Figure 1.9. The proposed EES framework divides all the EE applications into three categories. The R2R-type applications (e.g. APC and RM) are installed in the remote-client side; the self-related-type applications (e.g. FDC, PdM, and VM) are plugged in the generic embedded devices (GEDs) [5] and distributed in each individual tools; the e-diagnostics-type applications are implemented in a remote client via Interface C for security considerations.

Among the abovementioned EES applications, VM is an emerging technology [26]. VM is a method to conjecture the manufacturing quality of a process tool based on the data sensed from the process tool and without physical metrology operation [27].

Fab-wide R2R control [28] is one of essential EES applications for semiconductor manufacturing. In general, a run can be a batch, lot, or an individual wafer. When lot-to-lot (L2L) control is applied, the promptness of each wafer's VM result for the feedback and feedforward purposes will not be necessary. However, when wafer-to-wafer (W2W) control is adopted, obtaining the real-time and on-line VM result of each wafer in the feedback loop is essential.

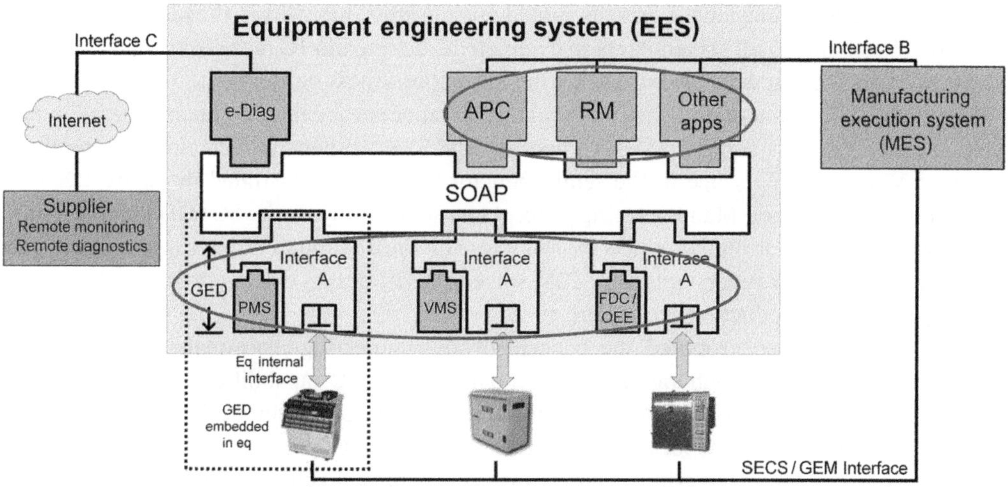

Figure 1.9 The proposed EES framework. *Source:* Reprinted with permission from Ref. [14]; © 2010 IEEE.

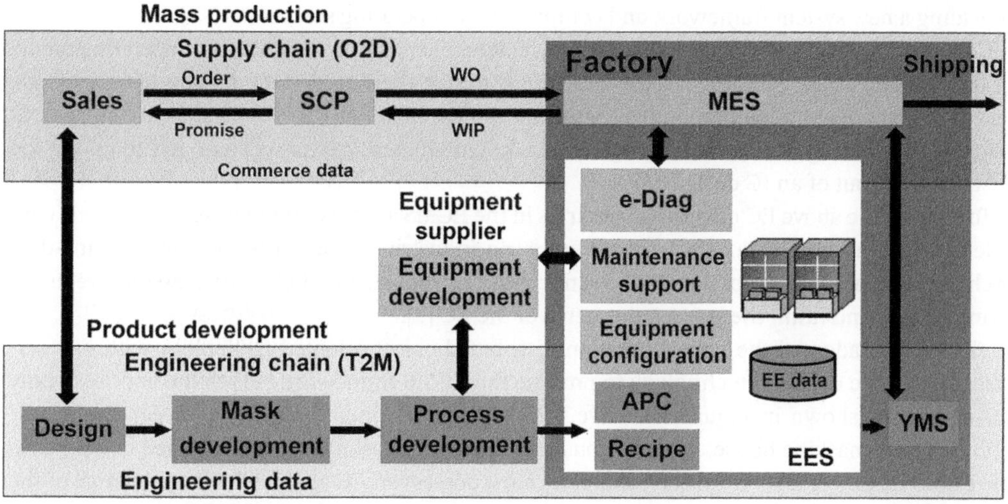

Figure 1.10 Comparison of SC and EC. *Source:* Reprinted with permission from Ref. [14]; © 2010 IEEE.

1.2.1.4 Engineering Chain (EC)

International technology roadmap of semiconductor (ITRS) proposed the concept of EC to cope with the problems aroused from design collaboration in the semiconductor industry in 2003. In the semiconductor industry, the EC was defined as a network of facilities and distributed services that performs device design, verification of design, manufacturing pilot run, assembly and test operations, yield improvements, and final release for mass production [11]. Figure 1.10 compares the SC and the EC in the semiconductor industry [14].

In the mass-production phase after a successful IC design, the SC manages the entire operation from order input to wafer delivery. On the other hand, in the product-development phase, the EC plays the role of managing IC design operation from IC design to the release of mass production.

Both the SC and EC management systems should operate efficiently within the new collaborative operation model among all stakeholders to complete IC design and IC manufacturing. However, the EC supports product development, while the SC supports mass production.

The novel EC component and the traditional SC component for inter-company operation are combined with the intra-company MES component and EES component to create a new comprehensive e-Manufacturing scope in the semiconductor industry as illustrated in Figure 1.2. The proposed semiconductor e-Manufacturing concept focuses not only on the SC O2D for timely and economical delivery of desired products [6, 7] but also on the e-Manufacturing support to achieve a faster design cycle for reducing the EC T2M since some IC design cycles are longer than their corresponding mass-production cycles [29].

Cheng et al. [11] also proposed the concept of an engineering-chain-management system (ECMS) to supervise the collaborations with EC partners and EC capabilities for shortening the T2M in the IC production process. An EC environment consists of numerous design partners that are allocated among different locations but working together to produce advanced IC design. Therefore, considerable engineering data exchange is inevitable. Each professional partner of the EC focuses on its professional work. The ECMS can support the operating efficiency of the collaborative team, including first design success rate enhancement, design cycle time subtraction, and design cost reduction. The ECMS supports the EC operation and engineering data exchange by providing a new system framework and comprehensive operating scenarios.

Coherent IC design operations among many heterogeneous companies are the operating model of the EC. Enormous quantities of data, including design files, mask data, process specifications, and yield data, need to be exchanged among all the members of the EC. Therefore, transparent information sharing is essential for significant design efficiency improvement and to assure the first-pass tape out of an IC design.

To support the above EC operating scenarios in the ECMS architecture requires exchanging considerable engineering data. Although, there is no industrial standard for EC engineering-data exchange as that for logistics data in the semiconductor industry, an ECMS framework is required to implement and fulfill the key requirements of the EC [9].

The authors adopted the new generation distributed object-oriented technology with web services [15] as the enabling technology to propose the ECMS framework [11] that comprises many EC agents. As shown in Figure 1.11, five main companies, including the IC design house, IP/library house, masking house, foundry house, and assembly/test house, are consisted in the example ECMS framework. Notably, each company must possess an EC agent, which is in charge of the data exchange operations, to communicate with the other servers.

1.2.2 Industry 4.0

Germany proposed Industry 4.0 to takes first steps toward the next industrial revolution in 2012. The definition and core technologies of Industry 4.0 as well as the requirement of mass customization are introduced below.

1.2.2.1 Definition and Core Technologies of Industry 4.0

Industry 4.0 is a collective term for technologies and paves the way to and for visions of a smart factory and smart manufacturing [31], which can be achieved by the integration of both IoT [33, 35] and CPS [34]. A smart factory possesses smart-manufacturing scenarios; while smart manufacturing emphasizes man-machine cooperation and production logistics management by applying IoT, CPS, cloud-based approaches, big data, and communicating technologies [36, 37]. Briefly

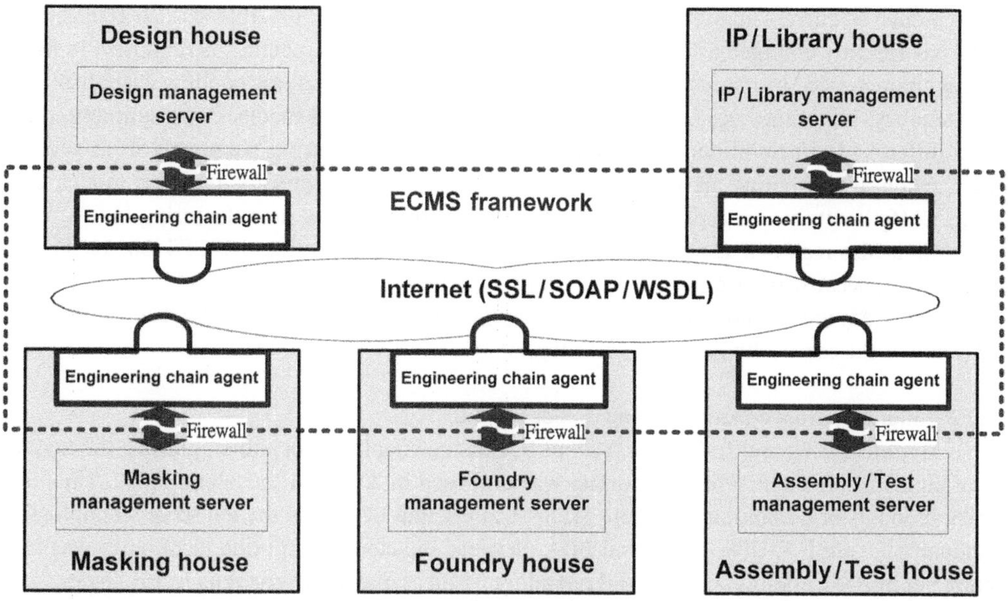

Figure 1.11 Engineering-chain-management system framework. *Source:* Reprinted with permission from Ref. [14]; © 2010 IEEE.

speaking, in order to step forward to the next advanced manufacturing level, Industry 4.0 focuses on enabling people, equipment, and products to communicate with each other independently; and allows venders and their customers to stay closer in the production processes and to react faster upon the changing market requirements.

IoT is a communication network for connecting every physical object (or "thing") in the real world which has naming, sensing, and processing abilities [38]. As a production factor with ubiquitous connections, it has been considered the promising technology of IT infrastructure for seamlessly integrating classical networks and networked objects, for data acquisition and sharing great effects of the performance for many enterprise systems in modern manufacturing [33, 35]. On the other hand, CPS is a term to describe the interconnection between the physical and cyber world. By integrating analog/digital hardware, middleware, and highly flexible software behind cyberspace, CPS achieve the creation of a link between virtual elements and real entities [34, 36, 39]. In this way, physical entities can be controlled by the intelligence from cyber elements. Currently, embedded systems only focus on the stand-alone computation rather than interaction with physical elements. Thus, CPS are usually referred as advanced embedded systems because of being intrinsically connected with internet-connected objects [38] for performing desired functions that are frequently accompanied with real-time computing capability and are able to link each embedded system to digital networks for independently facilitating data processing. A large number of recent studies [37, 38, 40, 41] also emphasized that IoT and CPS are supposed to have intelligence because they are assumed capable of being identified, sensing events, interacting with others, and making decisions by themselves. In summary, IoT provides a basic platform for connecting all CPS, and CPS cooperate seamlessly with real and virtual spaces to make Industry 4.0 possible. Therefore, we can definitely say that there is no CPS without IoT; no Industry 4.0 without CPS and IoT.

Cloud computing has emerged as a new trend of internet application in recent years [42]. By leveraging and extending the characteristics of cloud computing to meet the global and distributed requirements of current manufacturing industry, cloud-based manufacturing, also referred as CMfg, has recently emerged as a next-generation manufacturing paradigm. As remarked in [42], CMfg is characterized by many factors (such as scalability, agility, resource pooling, virtualization, multi-tenancy, ubiquitous access, self-service, search engine, social media, crowdsourcing, etc.) and is different from traditional web- and agent-based manufacturing paradigms from several aspects, such as computing architecture, data storage, operational process, business model, etc. Hence, CMfg is surely a new paradigm which will revolutionize the manufacturing industry. In fact, CMfg is also regarded as one of the best solutions for implementing IoT/CPS [30, 32, 34, 36, 37, 42, 43, 44] because of its powerful computing capability.

To realize CPS, the technology of Big Data Analytics (BDA) are adopted widely. Therefore, BDA is also one of the core technologies of Industry 4.0.

1.2.2.2 Migration from e-Manufacturing to Industry 4.0

Both e-Manufacturing and Industry 4.0 adopt ICT as the enabling tool and emphasize the necessity of big data collection; while the former was proposed in 2000 and the later in 2012. The four key components of e-Manufacturing are MES, SC, EES, and EC; while the four core technologies of Industry 4.0 are IoT, CPS, CMfg, and BDA. Because the cloud-computing technology was not mature yet in 2000, e-Manufacturing did not adopt CMfg as one of the enabling technologies.

e-Manufacturing utilizes equipment managers in MES to collect all the process and metrology data; while Industry 4.0 applies IoT devices to collect all the data required. The technologies of IoT, CPS, and CMfg of Industry 4.0 can be applied to implement various EES functions (such as VM, PdM, and APC) of e-Manufacturing with a more systematic and efficient fashion. The functions of SC in e-Manufacturing can be accomplished by the CPS technology of Industry 4.0 as well. Also, BDA of Industry 4.0 can be applied to find the root causes of a yield loss for yield enhancement and yield management. Therefore, as mentioned previously, e-Manufacturing is the predecessor of Industry 4.0. However, the function of EC in e-Manufacturing is not considered in Industry 4.0 because EC is specific for the semiconductor industry but not the machinery industry.

1.2.2.3 Mass Customization

With the upcoming age of IoT [35, 43, 45] and CPS [46], Industry 4.0 redefines the industrial manufacturing system in a completely automated scenario. The characteristics of "digitization, intelligentization, and customization" of this industrial evolution advance the traditional manufacturing techniques from mass production towards a deep-rooted mass-customization (MC) [47].

Although MC is not a new concept, it is emphasized again in Industry 4.0 for the fact that customers are returning to the center of the core value [48, 49]. One of the core values of Industry 4.0 targets to integrate people's demand into manufacturing for enhanced products, systems, and services for a wider variety of increasingly personalized customization of products [49]. Thus, a further change will happen to the manufacturing industries with Industry 4.0 that the customers can benefit from [50].

Frankly, it is the birth of IoT/CPS that lifts data-collection and communication technologies to a new level so as to allow a faster response to customers' needs. Industrial manufacturers can efficiently build relationships with the end-customers by combining the flexibility and personalization of "custom-made" in real-time. MC is also known as the concept of "made to order" or "build to order" [51]. The production only happens after manufacturers know what customers' demands are. Customers or end-users can easily decide the certain functionalities or personal attributes of a unique product or service what they exactly want just via a web portal. In other words, customers, manufacturers, and equipment closely interact with one another through seamless connections via IoT/CPS – a win-win situation for all participants in modern manufacturing relationships.

MC aims to provide customers with varieties of increasing customized products and a near mass-production efficiency without the corresponding increase in cost and lead time. Since MC first coined by Davis [52], it has attracted a large number of researchers to take their great efforts to make MC possible for decades. So far there has been a significant progress, such as Gilmore and Pine [53] defined four approaches: collaborative, adaptive, cosmetic, and transparent customizations for targeting different mass consumer groups in MC markets depending on degrees of customization in the product itself, and representation of the product. Collaborative customization seeks to help clients who struggle to spot exactly what they want and helps to understand the needs of the customers and strives to make it clear to them. Adaptive customization allows customers to handle customized products themselves without manufacturer's assistance. Cosmetic customization presents a standard product with various representation to different customers. Transparent customization means that manufacturers provide unique products without needing to inform customers. Silveira et al. [54] surveyed the earlier studies on MC to point out the visionary and practical conceptualizations of MC theory; also, fundamental requirements for developing a basic MC framework composed from eight generic levels of MC were thoroughly discussed in [54]. Further, as information technologies evolves, Fogliatto et al. [55] updated the latest successful MC applications among various fields, including the food industry, electronics, large engineered products, mobile phones, and personalized nutrition; or special MC applications such as homebuilding and the production of foot orthoses. They clearly identified required conditions in different fields and situations of implementing a suitable MC platform from the view of economics, success factors, enablers, and customer-manufacturer interactions.

For manufacturers, two mandatory factors of agility and quick responsiveness to manufacturing changes are expected to minimize the escalating costs [51, 53, 55]. They have to ensure the production facility must be flexible enough for switching between complex variants with some delay and be agile enough to adapt to changes in customized products at a low cost, thereby retaining economic benefits [55, 56]. For customers, after the emergence of Industry 4.0, the state-of-the-art of IoT/CPS replaces traditional MC scenarios, and gives customers more chances to actively participate in a collaborative design of customized products.

However, no matter how production technologies are improved in the era of Industry 4.0, the ultimate aim for manufacturing has not changed, which is the manufacturing quality of products. Manufacturers are imperative to ensure that the manufacturing quality of deliverables conforms to the design specifications before delivering them to customers. Thus, "quality control" is also listed as one of the promising areas to be achieved for future research in MC [55]. Namely, how to effectively minimize the defective product cost is still the biggest challenge of MC. As such, a fully automated and real-time total-inspection method is needed to withstand a global requirement on increasing product quality and reducing production cost.

1.2.3 Zero Defects – Vision of Industry 4.1

Since the late 1960s, ZD has been one of the quality-improvement objectives for accomplishing manufacturing quality [57]. Through prevention methods, ZD aims to boost production and minimize waste. ZD is based on the concept that the amount of mistakes a worker makes doesn't matter since inspectors will catch them before they reach customers [57].

Industry 4.0, since its first presentation at the Hannover Messe 2014, is set to be one of the new manufacturing objectives and, most of all, keep the faith of achieving nearly ZD state in the manufacturing industry [58, 59]. The current Industry 4.0 related technologies emphasize on productivity improvement but not on quality enhancement; in other words, they can only keep the faith of achieving nearly ZD state without realizing this goal. The key reason for this inability is the lack of an affordable online and real-time Total Inspection technology. By adopting the Automatic Virtual Metrology (AVM) technology that has been certified with the invention patents from six countries (Taiwan ROC, USA, Japan, Germany, China, and Korea) developed by the research team of Fan-Tien Cheng, the Editor and main author of this book, ZD can be achieved as AVM can provide the Total Inspection data of all products online and in real time. A defective product will be discarded once it is detected by AVM; in this way, all of the *deliverables* will be ZD. Further, the Key-variable Search Algorithm (KSA) of the Intelligent Yield Management (IYM) system developed by Fan-Tien Cheng's research team can be utilized to find out the root causes of the defects for continuous improvement on those defective products. As such, ZD of all *products* can be achieved. Therefore, once AVM and IYM are integrated into the successfully developed Industry 4.0 platform, the state of ZD can be realized, which is defined as Industry 4.1 by Fan-Tien Cheng. The concepts of Industry 4.1 were disclosed in *IEEE Robotics and Automation Letters* in January 2016 [60]. The technical details of AVM and IYM will be elaborated in Chapters 8 and 10, respectively.

1.2.3.1 Two Stages of Achieving Zero Defects

Generally speaking, two stages are involved for achieving Zero Defects in Industry 4.1:

- Stage I: accomplish Zero Defects of all the *deliverables* by applying efficient and economical total-quality-inspection techniques.
- Stage II: ensure Zero Defects of all the *products* gradually by improving the yield with big data analytics and continuous improvement.

Stage I can be accomplished by directly applying AVM to perform Total Inspection on all the possible deliverables. If any defects are found in a possible deliverable, then this one should not be delivered to customers. As a result, the goal of ZD for all the deliverables is achieved.

The manufacturing-related data of all the defective products found in Stage I should be collected such that the KSA in IYM can be performed on those data to find out the root causes that result in the defects. Then, those root causes should be fixed for reducing possible defects that may occur in the subsequent production run. As such, the goal of nearly ZD for all the products can be accomplished by continuous improvements. The process mentioned in this paragraph is the so-called Stage II.

1.3 Development Strategy of Intelligent Manufacturing with Zero Defects

As semiconductor manufacturing technologies advance, semiconductor manufacturing processes are becoming more and more sophisticated. Thus, how to maintain their feasible production yield becomes an important issue. As shown in Figure 1.12, during the product life cycle, the product yield (blue solid line) gradually rises up in the research-and-development (RD) phase and ramp-up phase and then keeps steady in the mass-production (MP) phase. On the contrary, the product cost (red solid line) continuously decreases during the production life cycle. If a company can improve its changing curves of yield and cost from the solid lines into their corresponding segmented lines, the company's competitiveness would be enhanced effectively. This implies that rapidly increasing the yield in the RD phase to transfer products into the MP phase, and then assuring the yield in the MP phase while promptly finding out and resolving the root causes of yield losses is a feasible

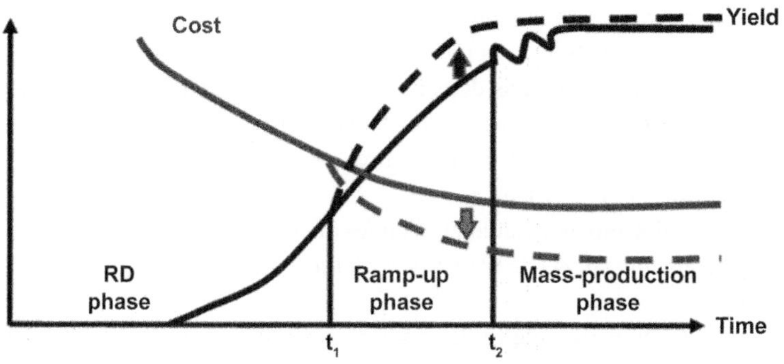

Figure 1.12 Changing curves of yield and cost during the product life cycle. *Source:* Reprinted with permission from Ref. [61]; © 2017 IEEE.

strategy for increasing the company's competitiveness. However, no literature has proposed a systematic approach of enhancing and assuring production yield, which targets both the RD phase and the MP phase of the product life cycle.

In the following, a five-stage approach as shown in Figure 1.13 for enhancing production yield and assuring nearly ZD, taking a semiconductor bumping process as an illustrative example, is proposed.

1.3.1 Five-Stage Strategy of Yield Enhancement and Zero-Defects Assurance

As shown in Figure 1.13, a five-stage approach for enhancing yield and assuring ZD of a manufacturing process is proposed. This five-stage approach involves RD, ramp-up, and MP phases. Observing the left portion of Figure 1.13, the RD and ramp-up phases cover the first two stages; while the right portion of Figure 1.13 contains the last three stages for the MP phase.

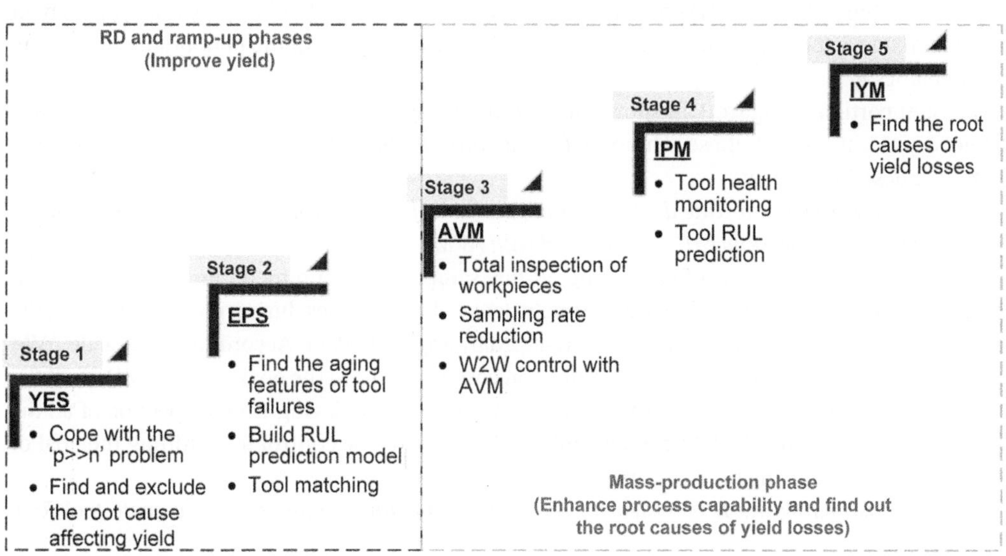

Figure 1.13 Five-stage strategy for increasing yield in RD/ramp-up and MP phases of a manufacturing process.

The production line of the bumping process shown in Figure 1.14 consists of two sub-processes, i.e. Re-Distribution Layer (RDL) and Under Bump Metallurgy (UBM). In the following, UBM is selected as the illustrative manufacturing process, as depicted at the bottom of Figure 1.14. The UBM bumping process contains the following stations: Sputtering Deposition, Photoresist (including Positive Photoresist Coating, Edge Bead Remover, Exposure, and Developing), Cu Plating, Striping, Etching, Ball Mount, Reflow, and Flux Clean.

The proposed actions at Stages 1 and 2 and associated challenges are described below.

Stage 1: developing a yield enhancement system (YES) to cope with the 'p>>n' problem to find and exclude the root cause affecting yield

In the illustrative bumping process, there are roughly 25 production stations, each station has 10 tools, each tool has 4 chambers, and each chamber has 100 sensors. Thus, there are totally about 100,000 parameters affecting the yield of the bumping process. If the information of tool maintenance and different material sources is also considered, the number of yield-affecting parameters, i.e. p, is even higher. In the RD phase of the product life cycle, the number of samples, i.e. n, is relatively small, thereby leading to a challenge of finding the root causes of poor yield under the condition '*p>>n*.' This is the so-called high dimensional regression problem [61]. Thus,the developed YES should be able to promptly find the root causes affecting yield from the enormous number of parameters (*p*) under the constraint of small number of samples (*n*) and exclude them so as to effectively enhance the yield in the RD phase.

Stage 2: developing an equipment prognosis system (EPS) to find the aging features of tool failures and perform tool matching

While the YES at Stage 1 can be used to identify the problematic tool affecting yield, an equipment prognosis system (EPS) shall be developed to facilitate assuring the capability of the tool. Specifically, by creating the cause-effect relationship of failure and prognosis model of equipment, the developed EPS should be able to observe the variation trend of key aging features and further estimate the remaining useful life (RUL) of equipment. Accordingly, the problematic tool can be maintained at proper time before it fails. Consequently, the possibility of tool's abnormality can be reduced, and the yield in the RD phase can be enhanced. Moreover, after building a successful pilot production line, a tool matching scheme shall be applied for rolling out the pilot production line to multiple lines.

The right portion of Figure 1.13 shows the last three-stage (Stages 3, 4, and 5) actions for assuring good yield in the MP phase. The proposed actions at Stages 3, 4, and 5 and associated challenges are described below.

Stage 3: conducting a fab-wide deployment of AVM to achieve the goal of Total Inspection and to perform workpiece-to-workpiece (W2W) control with AVM

The AVM system is capable of converting off-line sampling measurement into on-line and real-time Total Inspection of all workpieces to timely detect abnormalities during production. Also, the sampling rate of real measurements can be reduced by applying AVM. Accordingly, fab-wide AVM applications can effectively reduce the production cost and achieve the goal of nearly ZD of all the deliverables in the MP Phase. In addition, due to the ability of achieving Total Inspection of all the workpieces, the outputs of AVM can be applied to support W2W control for fulfilling the goal of enhancing manufacturing process capability.

Stage 4: constructing the IPM system to perform tool health monitoring and execute tool RUL prediction

At this stage, the IPM system is constructed to detect abnormality on key components of all the manufacturing tools and to predict the RULs of all the key components so as to improve the tool availability and prevent unscheduled down of all the manufacturing tools.

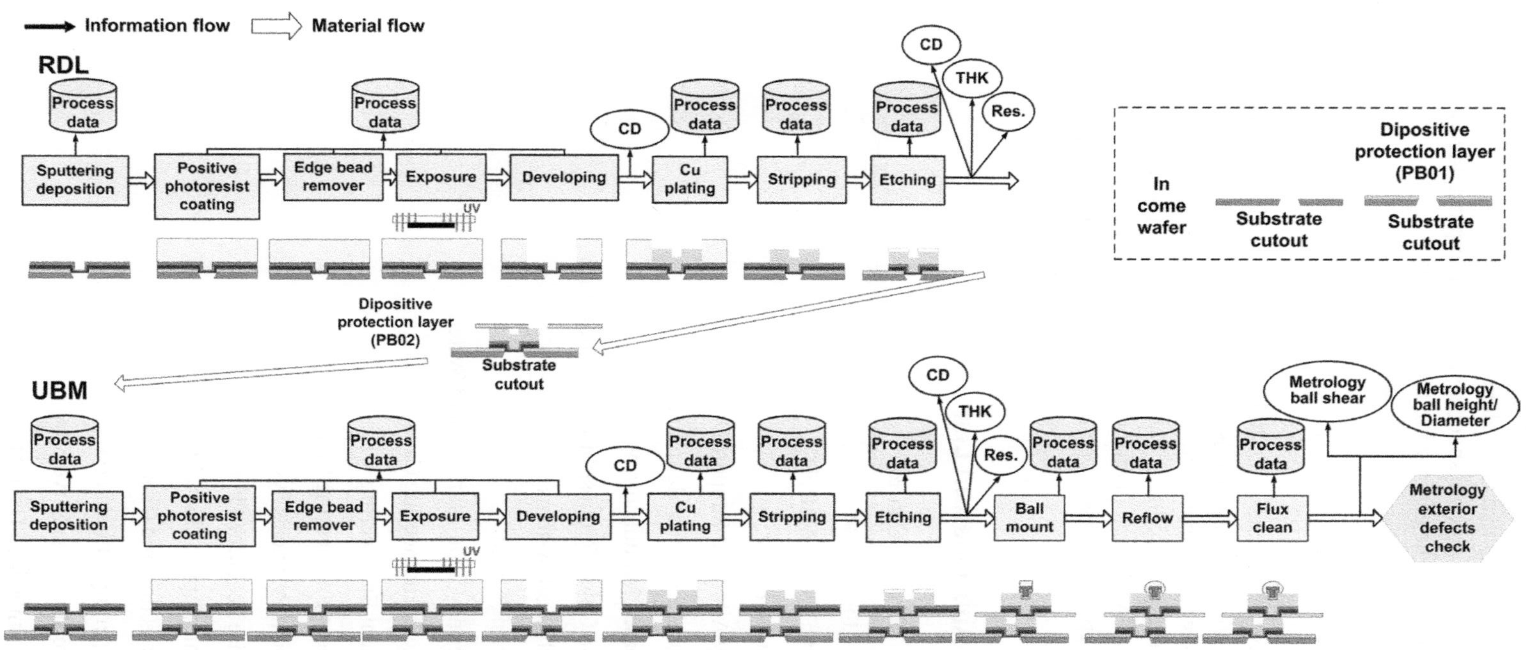

Figure 1.14 Production line of the bumping process.

Stage 5: developing the IYM system to promptly find the root causes of yield losses

In the MP phase, the IYM system is developed to promptly find out the root causes which affect the yield so as to reduce the trouble shooting time and improve the yield. As such, the goal of nearly ZD of all products can be achieved in the MP phase.

1.4 Conclusion

The evolution of automation is surveyed in this chapter, including e-Manufacturing and Industry 4.0. Then, the importance of ZD, which is the vision of Industry 4.1 is presented. Finally, the five-stage strategy of yield enhancement and ZD assurance is proposed. This five-stage strategy is the guideline for developing the Intelligent Manufacturing System with ZD. As a result, an Intelligent Factory Automation (iFA) System Platform is designed and elaborated in Chapter 6 to realize the proposed five-stage approach of yield enhancement and ZD assurance.

Appendix 1.A – Abbreviation List

APC	Advanced Process Control
ATP	Available-to-Promise
AVM	Automatic Virtual Metrology
BDA	Big Data Analytics
CASD	Capacity-Allocated-Support Demand
CIM	Computer-Integrated Manufacturing
CMfg	Cloud-based Manufacturing
CORBA	Common Object Request Broker Architecture
CPS	Cyber-Physical Systems
EC	Engineering Chain
ECMS	Engineering-Chain-Management System
EECs	Equipment Engineering Capabilities
EE	Equipment Engineering
EES	Equipment Engineering System
EPS	Equipment Prognosis System
ERP	Enterprise Resource Planning
ESCM	Electronic Supply Chain Management
FDC	Fault Detection and Classification
FOUP	Front Opening Unified Pod
FICS	Free Internet Chess Server
GEDs	Generic Embedded Devices

GEM	Generic Equipment Model
GH	Generic Holon
GUI	Graphical User Interface
HMES	Holonic Manufacturing Execution System
HSCS	Holonic Supply-Chain System
HSMS	High-Speed SECS Message Services
IC	Integrated Circuit
ICT	Information and Communication Technology
iFA	Intelligent Factory Automation
IoT	Internet of Things
IP	Internet Protocol Address
IPM	Intelligent Predictive Maintenance
ISMT	International SEMATECH
IT	Information Technology
ITRS	International Technology Roadmap of Semiconductor
IYM	Intelligent Yield Management
KSA	Key-variable Search Algorithm
L2L	Lot-to-Lot
MC	Mass Customization
MCS	Material Control System
MES	Manufacturing Execution System
MP	Mass Production
MPS	Master Production Schedule
NACFAM	National Coalition for Advanced Manufacturing
O2D	Order-to-Delivery
OEE	Overall Equipment Effectiveness
PdM	Predictive Maintenance
PO	Purchase Order
R2R	Run-to-Run
RD	Research and Development
RDL	Re-Distribution Layer
RM	Recipe Management
RUL	Remaining Useful Life
SC	Supply Chain
SECS	SEMI Equipment Communications Standard

SOAP	Simple Object Access Protocol
SPC	Statistical Process Control
SSL	Secure Sockets Layer
T2M	Time-to-Market
tsmc	Taiwan Semiconductor Manufacturing Company
UBM	Under Bump Metallurgy
VM	Virtual Metrology
W2W	Wafer-to-Wafer
W2W	Workpiece-to-Workpiece
WIP	Work in Process
WSDL	Web Services Description Language
YES	Yield Enhancement System
YMS	Yield Management System
ZD	Zero Defects

References

1 SEMATECH (1998). Computer Integrated Manufacturing (CIM) Framework Specification Version 2.0. *SEMATECH Technology Transfer # # 93061697J-ENG*. https://bit.ly/3hDCvJH (accessed 27 Aug 2020).

2 Cheng, F.-T., Shen, E., Deng, J.-Y. et al. (1999). Development of a system framework for the computer-integrated manufacturing execution system: a distributed object-oriented approach. *International Journal of Computer Integrated Manufacturing* 12 (5): 384–402. https://doi.org/10.1080/095119299130137.

3 Cheng, F.-T., Chang, C.-F., and Wu, S.-L. (2004). Development of holonic manufacturing execution systems. *Journal of Intelligent Manufacturing* 15 (2): 253–267. https://doi.org/10.1023/B:JIMS.0000018037.63935.a1.

4 International SEMATECH (2002). Equipment Engineering Capabilities (EEC) guidelines, v2.5. https://reurl.cc/Q3dR2o (accessed 17 Aug 2020).

5 Su, Y.-C., Cheng, F.-T., Hung, M.-H. et al. (2006). Intelligent prognostics system design and implementation. *IEEE Transactions on Semiconductor Manufacturing* 19 (2): 195–207. https://doi.org/10.1109/TSM.2006.873512.

6 Harrison, P. (2015). An introduction to supply chain management. https://reurl.cc/Y61Yv0 (accessed 17 Aug 2020).

7 NEVEM-workgroup (1989). *Performance Indicators in Logistics*. Bedford: IFS.

8 Cheng, F.-T., Yang, H.-C., and Lin, J.-Y. (2004). Development of holonic information coordination systems with failure-recovery considerations. *IEEE Transactions on Automation Science and Engineering* 1 (1): 58–72. https://doi.org/10.1109/TASE.2004.829350.

9 Chang, J. and Cheng, F.-T. (2005). Engineering-chain requirements for semiconductor industry. *Proceedings of the 2005 IEEE Conference on Automation Science and Engineering*, Edmonton, Canada (1-2 Aug 2005). USA: IEEE.

10 Chang, J. and Cheng, F.-T. (2005). Framework development of an Engineering-Chain-Management-System for the semiconductor industry. *Proceedings of the Thirteenth International Symposium on Temporal Representation and Reasoning,* Perth, Australia (10-12 Aug 2005). USA: IEEE.

11 Cheng, F.-T., Chen, Y.-L., and Chang, J.Y.-C. (2012). Engineering chain: a novel semiconductor engineering collaboration model. *IEEE Transactions on Semiconductor Manufacturing* 25 (3): 394–407. https://doi.org/10.1109/TSM.2012.2191626.

12 National Coalition for Advanced Manufacturing (NACFAM) (2001). Exploiting e-Manufacturing: Interoperability of Software Systems Used by U.S. Manufacturers.

13 Tag, P.-H. and Zhang, M.-T. (2006). e-Manufacturing in the semiconductor industry. *IEEE Robotics and Automation Magazine* 13 (4): 25–32. https://doi.org/10.1109/MRA.2006.250570.

14 Cheng, F.-T., Tsai, W.-H., Wang, T.-L. et al. (2010). Advanced e-Manufacturing model. *IEEE Robotics and Automation Magazine* 17 (1): 71–84. https://doi.org/10.1109/MRA.2010.935796.

15 Hung, M.-H., Cheng, F.-T., and Yeh, S.-C. (2005). Development of a web-services-based e-diagnostics framework for the semiconductor manufacturing industry. *IEEE Transactions on Semiconductor Manufacturing* 18 (1): 122–135. https://doi.org/10.1109/TSM.2004.836664.

16 Hung, M.-H., Wang, T.-L., Hsu, F.-Y. et al. (2008). Development of an interface C framework for semiconductor e-diagnostics systems. *Robotics and Computer-Integration Manufacturing* 24 (3): 370–383. https://doi.org/10.1016/j.rcim.2007.02.020.

17 TechTarget (2007). Manufacturing Execution Systems. http://www.bitpipe.com/rlist/term/Manufacturing-Execution-Systems.html ().

18 Lin, W.Y., Hsu, R., and Chiu, Y.T. (2005). Systems and methods for determining production availability. US Patent 20,070,016,318, filed 15 July 2005 and issued 18 Jan 2007.

19 Lee, H. and Whang, S. (2006). Taiwan semiconductor manufacturing company: the semiconductor services company. *Stanford Graduate School of Business case: GS-40.* https://www.gsb.stanford.edu/faculty-research/case-studies/taiwan-semiconductor-manufacturing-company-semiconductor-services (accessed 28 August 2020).

20 SEMI (2020). SEMI EDA standards: E120, E125, E132, E134. www.semi.org ().

21 SEMI E125 (2014). *SEMI E125 - Specification for Equipment Self Description (EqSD).* USA: SEMI https://bit.ly/3j5C2Bj (accessed 1 September 2020).

22 SEMI E132 (2019). *SEMI E132 - Specification for Equipment Client Authentication and Authorization.* USA: SEMI https://bit.ly/3haoJgS (accessed 1 September 2020).

23 SEMI E134 (2019). *SEMI E134 - Specification for Data Collection Management.* USA: SEMI https://bit.ly/3fCBWPq (accessed 1 September 2020).

24 SEMI E133 (2018). *SEMI E133 - Specification for Automated Process Control Systems Interface.* USA: SEMI https://bit.ly/2ESLuIF (accessed 17 Aug 2020).

25 International SEMATECH Manufacturing Initiative (2005). Interface C Requirements of e-Diagnostics Guidebook (version 2.1).

26 Weber, A. (2007). Virtual metrology and your technology watch list: ten things you should know about this emerging technology. *Future Fab International* 22 (4): 52–54.

27 Chang, J. and Cheng, F.-T. (2005). Application development of virtual metrology in semiconductor industry. *Proc. 31st Annual Conference of the IEEE Industrial Electronics (IECON 2005),* Raleigh, U.S.A. (6-10 Nov 2005). Raleigh, USA: IEEE.

28 Moyne, J., del Castillo, E., and Hurwitz, A.M. (2001). *Run-to-Run Control in Semiconductor Manufacturing.* Boca Raton, FL: CRC.

29 Madhavan, R.(2004). Changing economics of chip design. *FSA Presentation.*

30 Werr, P. (2015). How Industry 4.0 and the internet of things are connected. https://bit.ly/2YSX8dr (accessed 17 Aug 2020).

31 Ferber, S. (2012). Industry 4.0 – Germany takes first steps toward the next industrial revolution. https://bit.ly/3lEQe5B (accessed 17 Aug 2020).

32 Jazdi, N. (2014). Cyber physical systems in the context of Industry 4.0. *IEEE International Conference on Automation, Quality and Testing, Robotics*, Cluj-Napoca, Romania (22-24 May 2014). Cluj-Napoca, Romania: IEEE.

33 Perera, C., Liu, C.H., Jayawardena, S. et al. (2015). A survey on internet of things from industrial market perspective. *Access, IEEE* 2: 1660–1679. https://doi.org/10.1109/ACCESS.2015.2389854.

34 Wan, J., Chen, M., Xia, F. et al. (2013). From machine-to-machine communications towards cyber-physical systems. *Computer Science and Information Systems* 10 (3): 1105–1128. https://doi.org/10.2298/CSIS120326018W.

35 Gubbi, J., Buyya, R., Marusic, S. et al. (2013). Internet of things (IoT): a vision, architectural elements, and future directions. *Future Generation Computer Systems* 29 (7): 1645–1660. https://doi.org/10.1016/j.future.2013.01.010.

36 Colombo, A.W., Bangemann, T., Karnouskos, S. et al. (2014). *Industrial cloud-based cyber-physical systems, The IMC-AESOP Approach.* Switzerland: Springer International Publishing.

37 Tsai, C.-W., Lai, C.-F., Chiang, M.-C. et al. (2014). Data mining for internet of things: a survey. *Communications Surveys & Tutorials, IEEE* 16 (1): 77–97. https://doi.org/10.1109/SURV.2013.103013.00206.

38 Perera, C., Zaslavsky, A., Liu, C.-H. et al. (2014). Sensor search techniques for sensing as a service architecture for the internet of things. *IEEE Sensors Journal* 14 (2): 406–420. https://doi.org/10.1109/JSEN.2013.2282292.

39 Xia, F., Vinel, A., Gao, R. et al. (2011). Evaluating IEEE 802.15. 4 for cyber-physical systems. *EURASIP Journal on Wireless Communications and Networking 2011: 596397.* https://doi.org/10.1155/2011/596397.

40 Sánchez López, T., Ranasinghe, D.C., Harrison, M. et al. (2012). Adding sense to the internet of things. *Personal and Ubiquitous Computing* 16 (3): 291–308. https://doi.org/10.1007/s00779-011-0399-8.

41 Miorandi, D., Sicari, S., De Pellegrini, F. et al. (2012). Internet of things: vision, applications and research challenges. *Ad Hoc Networks* 10 (7): 1497–1516. https://doi.org/10.1016/j.adhoc.2012.02.016.

42 Huang, H.-C., Lin, Y.-C., Hung, M.-H. et al. (2015). Development of cloud-based automatic virtual metrology system for semiconductor industry. *Robotics and Computer-Integrated Manufacturing* 34: 30–43. https://doi.org/10.1016/j.rcim.2015.01.005.

43 Bi, Z., Xu, L.-D., and Wang, C. (2014). Internet of things for enterprise systems of modern manufacturing. *IEEE Transactions on Industrial Informatics* 10 (2): 1537–1546. https://doi.org/10.1109/TII.2014.2300338.

44 Mattern, F. and Floerkemeier, C. (2010). From the internet of computers to the internet of things. *From Active Data Management to Event-Based Systems and More. Lecture Notes in Computer Science* 6462: 242–259. https://doi.org/10.1007/978-3-642-17226-7_15.

45 Li, S., Li, D.-X., and Zhao, S. (2015). The internet of things: a survey. *Information Systems Frontiers* 17: 243–259. https://doi.org/10.1007/s10796-014-9492-7.

46 Brettel, M., Friederichsen, Keller, N.M. et al. (2014). How virtualization, decentralization and network building change the manufacturing landscape: an Industry 4.0 perspective. *International Journal of Mechanical, Industrial Science and Engineering* 8 (1): 37–44. https://doi.org/10.5281/zenodo.1336426.

47 Shen, J., Majid, B.N., Xie, L. et al. (2017). Interactive UHF/UWB RFID tag for mass customization. *Information Systems Frontiers* 19: 1177–1190. https://doi.org/10.1007/s10796-016-9653-y.

48 Nirmala, J. (2016). Japan embracing Industry 4.0 and IoT to leap into next industrial automation. https://bit.ly/2YXoBe4 (accessed 17 Aug 2020).

49 The Boston Consulting Group (2015). Industry 4.0 lifts automation and mass customization to new levels. http://goo.gl/ilYMVD (accessed 17 Aug 2020).

50 Gross, D. (2016). Siemens CEO Joe Kaeser on the next industrial revolution. http://goo.gl/ZSGgqo (accessed 17 Aug 2020).

51 Pollard, D., Chuo, S., and Lee, B. (2016). Strategies for mass customization. *Journal of Business & Economics Research* 14 (3): 101–110. https://doi.org/10.19030/jber.v14i3.9751.

52 Davis, S.M. (1989). From future perfect: mass customizing. *Planning Review* 17 (2): 16–21. https://doi.org/10.1108/eb054249.

53 Gilmore, J.H. and Pine, B.J. 2nd (1997). The four faces of mass customization. *Harvard Business Review* 75 (1): 91–101.

54 Da Silveira, G.J., Borenstein, D., and Fogliatto, F.S. (2001). Mass customization: literature review and research directions. *International Journal of Production Economics* 72 (1): 1–13. https://doi.org/10.1016/S0925-5273(00)00079-7.

55 Fogliatto, F.S., Da Silveira, G.J.C., and Borenstein, D. (2012). The mass customization decade: an updated review of the literature. *International Journal of Production Economics* 138 (1): 14–25. https://doi.org/10.1016/j.ijpe.2012.03.002.

56 Peng, D.X., Liu, G., and Heim, G.R. (2011). Impacts of information technology on mass customization capability of manufacturing plants. *International Journal of Operations & Production Management* 31 (10): 1022–1047. https://doi.org/10.1108/01443571111182173.

57 Halpin, J.F. (1966). *Zero Defects: A New Dimension in Quality Assurance*. New York: McGraw-Hill.

58 Weisenberger, S. (2015). Hannover Messe Day 1 - will Industry 4.0 enable zero defects? how are business models impacted by Industry 4.0. https://bit.ly/3331HDB (accessed 17 Aug 2020).

59 Somers, D. (2014). Enter the world of 'Industrial 4.0' at Hannover Messe 2014. https://goo.gl/47yfdw (accessed 17 Aug 2020).

60 Cheng, F.-T., Tieng, H., Yang, H.-C. et al. (2016). Industry 4.1 for wheel machining automation. *IEEE Robotics and Automation Letters* 1 (1): 332–339. https://doi.org/10.1109/LRA.2016.2517208.

61 Cheng, F.-T., Hsieh, Y.-S., Zheng, J.-W. et al. (2017). A scheme of high-dimensional key-variable search algorithms for yield improvement. *IEEE Robotics and Automation Letters* 2 (1): 179–186. https://doi.org/10.1109/LRA.2016.2584143.

2

Data Acquisition and Preprocessing

Hao Tieng[1], *Haw-Ching Yang*[2], *and Yu-Yong Li*[3]

[1]*Associate Research Fellow, Intelligent Manufacturing Research Center, National Cheng Kung University, Tainan, Taiwan, ROC*
[2]*Professor, Department of Electrical Engineering, National Kaohsiung University of Science and Technology, Kaohsiung City, Taiwan, ROC*
[3]*Postdoctoral Research Fellow, Intelligent Manufacturing Research Center, National Cheng Kung University, Tainan, Taiwan, ROC*

2.1 Introduction

Various intelligent applications (such as predictive maintenance, virtual metrology, etc.) should be developed for achieving the goals of Intelligent Manufacturing. Taking predictive-maintenance-related applications as the illustrative examples, Chen et al. [1] installed one accelerometer, one acoustic emission (AE) sensor and two current sensors on a lathe to estimate the reliability, and remaining useful life (RUL) for cutting tools based on the logistic regression model using vibration signals. Suprock et al. [2] installed one strain gauge and one instrumentation amplifier with the Bluetooth transmitter on a cutting tool to calculate dynamic torque values, which are as accurate as the real measurements by the dynamometer. Ghosh et al. [3] developed an artificial neural network (ANN)-based sensor fusion model for tool condition monitoring using cutting force, spindle vibration, spindle current, and sound pressure. Abuthakeer et al. [4] analyzed vibration signals based on the full factorial design and utilized ANN to validate the effect of cutting parameters on cutting tools during machining.

The fundamental steps for developing intelligent applications are depicted in Figure 2.1. As shown in Figure 2.1, before developing an intelligent application, the associated process/metrology data source needs to be acquired followed by appropriate data preprocessing. The main purposes of the aforementioned steps are briefly introduced in this subsection and more details can be found in the remaining subsections of Chapter 2.

- **Data Acquisition**
  Data acquisition is a basic and vitally important procedure to prepare and collect all needed process and metrology data for developing intelligent applications. In general, process data consists of sensing signals and manufacturing parameters. More details are introduced in Section 2.2.
- **Data Preprocessing**
  Without reliable data acquired from sensors and/or controllers, an intelligent application will not be feasible. To correctly interpret the acquired process and metrology data for deriving useful information, data preprocessing must be performed in advance.

  Data preprocessing focuses on improving data quality through techniques such as de-noising, synchronization, modification, and signal compression to enhance the transmission speed and

Industry 4.1: Intelligent Manufacturing with Zero Defects, First Edition. Edited by Fan-Tien Cheng.

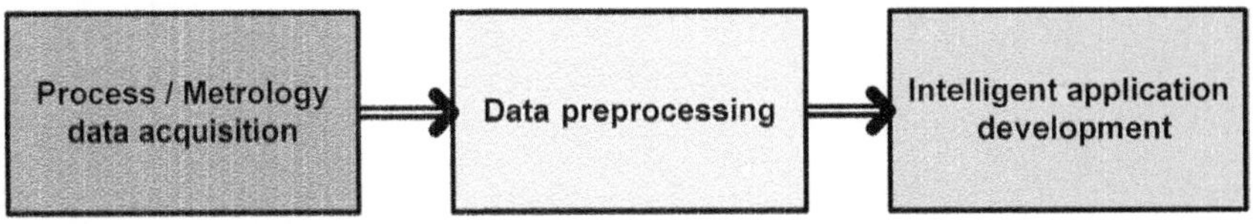

Figure 2.1 Fundamental steps for developing an intelligent application.

storage efficiency as well as emphasize the major components and key features of signals. More commonly adopted techniques are introduced in Section 2.3 in three steps of data preprocessing: segmentation, cleaning, and feature extraction.

- **Intelligent Application Development**
  Based on the extracted features, final decisions and actions for the current situation can be carried out through learned functions by artificial intelligence (AI) approaches, including machine learning and/or deep learning techniques [5]. The ultimate purpose of the intelligent application is to extract useful knowledge and explanation from the AI models, so that correct decisions and actions can be made.

This chapter presents the existing techniques for data acquisition and data preprocessing in general; while the adoption of selected AI models for solving the problems in different industries, such as Thin-Film-Transistor Liquid-Crystal Display (TFT-LCD), solar cell, semiconductor, automotive, aerospace, chemical, and bottle industries, will be illustrated in Chapter 11 respectively.

2.2 Data Acquisition

Figure 2.2 illustrates the connection of an equipment and an external data acquisition system with a basic hardware architecture meeting the aforementioned requirements for implementing intelligent applications. The analog-to-digital converter (ADC) connects to various sensors installed on the equipment side to convert analog sensing signals into digital signals via its analog input/output (AIO) ports. The external data acquisition system is connected to the controller of the equipment through an Ethernet card for retrieving the manufacturing parameters. The corresponding metrology data by the measurement tool for training and tuning the AI models can also be acquired through the Ethernet card, which interconnects network devices based on the Transmission Control Protocol/Internet Protocol (TCP/IP) protocol. In this way, required process and metrology data are collected and sent to an industrial personal computer (PC) for further processing.

2.2.1 Process Data Acquisition

By acquiring the process data, including sensing signals and manufacturing parameters, the machining stability can then be evaluated and the tool health status can be monitored. Details are introduced as below.

2.2.1.1 Sensing Signals Acquisition

A sensor is a device that detects and measures a physical quantity from the real-world environment and converts it into signals. Almost an infinite number of parameters can be acquired, such as light, temperature, location, displacement, movement, sound, pressure, moisture, voltage, current, and a great number of other environmental phenomena. Sensors are the key enabling devices for improving manufacturing capability and productivity.

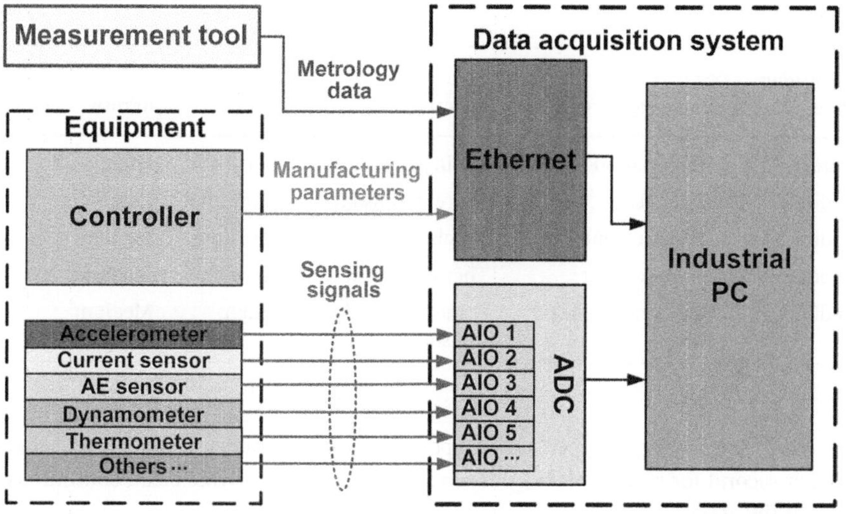

Figure 2.2 An external data acquisition system for acquiring process and metrology data from the equipment and measurement tool.

With the emergence of the Fourth Industrial Revolution (Industry 4.0), cyber-physical systems (CPSs), and industrial Internet of Things (IIoT), the number of sensors on equipment is increasing rapidly. Currently, various sensors can be seamlessly connected to equipment through generic platforms for integrating numerous devices, various tools, and shop floor information into a smart factory. In ubiquitous sensing, all possible sensors are connected to provide essential data for both IoT and big data in a given environment. These pervasive sensors enhance awareness of time-varying information for physical manufacturing processes, thus they bridge the cyber and physical worlds. The output of a sensor is generally a converted digital format that is human-readable at the sensor location or electronically transmitted to an external system to serve as the input for further processing and application. In the following, various sensing techniques and the issues of sensor selection and installation are presented.

Sensing Techniques

To measure process accuracy or production quality of a process tool, direct and indirect techniques may be applied. For direct techniques, the process accuracy or production quality can be measured in the machine by various sensors such as touch sensor, charge-coupled device (CCD), laser detector, and ultrasonic sensors. However, direct techniques are limited in practice due to extreme environment of machine workspace (such as affected by cutting fluid and chips). Furthermore, valuable production cycle time is reduced along with the measurement of device accuracy or workpiece quality. Also, these sensors are usually very expensive and difficult to apply to the in-line production due to the increased cycle time.

Relatively, indirect techniques for measuring process accuracy or production quality are less-accurate. However, using the sensors to sense indirect factors (such as force, vibration, temperature, and power consumption) is more economical and feasible for achieving the purpose of on-line and real-time diagnosis and prognosis.

Sensor Selection and Installation

When it comes to selecting appropriate sensors for monitoring a specific machining process, the cutting force is regarded as the best indicator to describe the performance of cutting processes and determine product quality. Five commonly used sensors for capturing cutting-force information are listed in Table 2.1.

Table 2.1 Sensor comparison.

Physical quantity	Measuring type	Detecting principle	Typical device	Cost	Intrusive nature
Force	Direct	Deformation	Strain gauge	High	High
Current	Indirect	Hall effect	Current transducer	Low	Low
Speed	Indirect	Displacement	Accelerometer	Medium	Medium
Temperature	Indirect	Seebeck effect	Thermal couple	Low	Medium
Sonic	Indirect	Radiation of elastic wave	Acoustic emission transducer	High	Medium

The other critical issue for selecting a proper sensor type is the sampling rate, which represents the scanned frequency per second for reading data from sensors that have to provide clear, continuous enough, and distortion-free data.

To make signals distortion-free, the Nyquist–Shannon sampling theorem states that the sampling rate for signal F_s is at least more than twice the highest frequency F_B component contained in the acquired signals during machining processes so that no information gets lost by discretization [6–8]. The relationship can be defined as $F_s > 2F_B$.

In the past, the sampling rate is restricted by the hardware specification and cost. As hardware becomes faster and cheaper, in a modern monitoring system, the sampling rate can be set at a very high frequency (>10 k Hz) to ensure that captured signals are nondistorting and can represent any situation during production since the sensor can serve as an independent unit from the manufacturing tool or device, which means that they do not decrease the production performance or disturb the machining process. However, a high F_s generates a high-dimensionality of data size, which makes raw data hard to be used directly. Some signal processing techniques for simplifying the data will be introduced in Section 2.3. In the following, several types of sensors, such as force: strain gauge, loading: current transducer, vibration: accelerometer, temperature: thermal couple, and AE waves: AE transducer, are introduced and followed by the description of sensor fusion.

Force: Strain Gauge Cutting force is an intuitional physic quantity that affects the machining quality and product quality the most. This force generated from the relative motions between the cutting tool and workpiece is necessary to form the shape of the workpiece, this relationship is illustrated in Figure 2.3. Any change occurred in the cutting-tool paths, heat, weight, tension, or structure of the material during the cutting operation can be directly reflected on the cutting force. Thus, the cutting force is considered to be the best indicator to perform the monitoring of cutting-tool/equipment condition and prediction of the workpiece accuracy.

Observing Figure 2.3, the electrical strain gauge, or the so-called piezoresistive sensor, detects forces or strains on the target device and converts the amount of the deformation into an electrical signal. As shown in the right side of Figure 2.3, the electrical resistance element changes its resistance length once the strain gauge is elongated or contracted under the external force. Through this direct measurement, the strain gauge provides extremely accurate and rapid response of the forces and strains that happened in the movement direction of cutting-tool. The excellent durability and stability also enable them to be operated in harsh environment. Figure 2.3 demonstrates an example of the adhesive-backed strain-gauge installation, which can be directly attached to the tool holder to monitor the torque and force on the cutting tool during the operations.

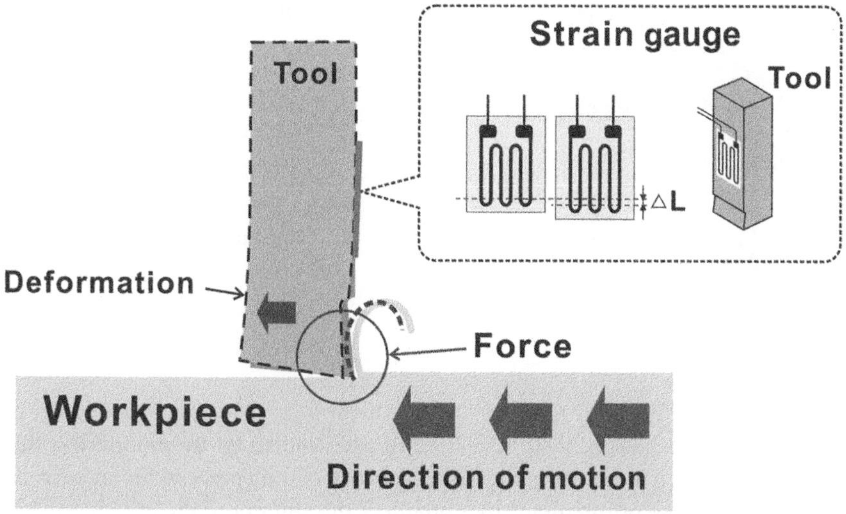

Figure 2.3 Relative motion between a cutting tool and a workpiece.

Many similar sensor types derived from the same working principle of strain-gauge technology, such as strain gauge-based pressure transducer and dynamometer are also commonly used in the machining factories. The tool dynamometer or force dynamometer is also frequently adopted as a sensor for monitoring cutting force signals. The sampling rate usually ranges from 1 to 100 kHz.

The installation of a force dynamometer is illustrated in Figure 2.4, which has to be fixed on the machining table. The performance is also sensitive to the installation direction, humidity, temperature, or long-time usage. Yet the installation process of the dynamometer is tedious because of its intrusive nature, which may deteriorate the machining performance. In addition, a dynamometer may reduce the working space and available machining conditions so that it is not acceptable and practical in the industry.

Figure 2.4 Installation of a Dynamometer.

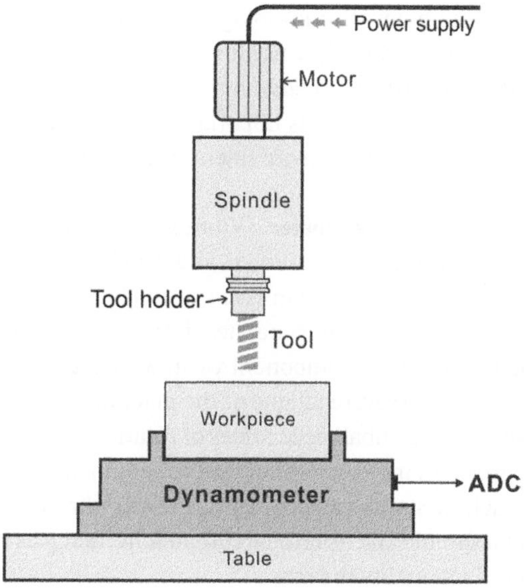

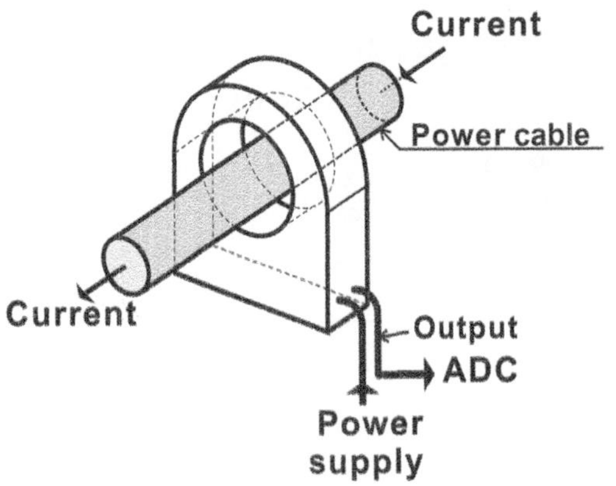

Figure 2.5 Installation of a CT.

Loading: Current Transducer The cutting force can be estimated indirectly by monitoring the change of the motor current since it is related to the cutting force of cutting process. Motor current is proportional to the output torque of the motor that produces the amount of the cutting forces, which provide the needed mechanical force to remove material from the workpiece. Thus, the motor current is very useful for the tool wear, breakage/failure detection, and quality prediction of workpiece.

Current measurement sensors can be used to monitor the cutting process in a manner similar to a tool dynamometer. The Hall effect current sensor, or the so-called current transducer (CT), converts motor current into an output voltage without any physical contact through detecting changes in the electromagnetic field, which is caused by the motion electric charges. As illustrated in Figure 2.5, a CT detects the electromagnetic field in the air surrounding when the electric current passes through a power cable.

Compared to traditional sensing techniques based on Ohm's Law, which generates a voltage that is proportional to current by directly connecting a resistor in series with the circuits, the Hall-effect CT sensor does not disturb the machining since it has no direct connection to the circuit that carries the current.

The installation location can be attached directly on the motor cables without fixed costs. This isolated or noninvasive advantage makes the Hall current sensor become the most popular measurement method to indirectly sense the motor power or spindle currents with the sampling rate ranging from 100 Hz to 10 kHz. In addition, a Hall current sensor is available at a very low cost and is extremely durable. However, it is important to note that the unrelated magnetic objects have to be kept away from the operating environment since the Hall current sensor is vulnerable to magnetic fields.

Vibration: Accelerometer Vibration describes the state of an object moving repetitively back/forward, right/left, or up/down, and usually can be expressed by the physical quantity of acceleration, which is the changes in velocity divided by time. Machining vibration and noise, or the so-called chatter, are usually generated from reciprocating motions of the cutting processes, mechanical malfunction, or component wear, which are undesirable since the irregular and random behavior may lead to severe fatigue of the machine structure and further deteriorate machine performance, such as the unbalanced forces of rotating parts. These failure incidents sometimes may not directly happen in motor power or current but in vibration. Therefore, the vibration is especially crucial to machine status monitoring. In a word, vibration data are useful for monitoring the rotation quality of a spindle and its critical components (such as bearings inside the spindle) that affect cutting and product quality the most.

A piezoelectric accelerometer is the most widely used electromechanical device converting a dynamically mechanical change (strain, force, vibration, acceleration, ...) into an electric signal. This electric signal is proportional to the piezoelectric effect occurred through mechanical changes during machine operations. The piezoelectric accelerometer is designed in a small size and has rugged construction that mounts it on the surface of specific axes and positions close to the vibration source.

In addition, the accelerometer provides good data quality with low-loss signal in forms of high frequency and transient response. Some critical characteristics of high-frequency can be detected and outputted in linear waveforms within microseconds. However, this high-sensitivity property is also prone to obtaining erroneous data that needs to be de-noised. For example, the installation position and the use of cutting fluid may affect the accuracy of the vibration signal.

Figure 2.6 illustrates how the accelerometer is mounted on the metal shell that surrounds the machinery spindle to collect rotation vibration. Vibration is very useful to monitor the rotation quality for the spindle and its critical components (such as bearings inside the spindle) which also affect the cutting and product quality. Thus, the proper installation position should be as closer to the rotor as possible. Typically, the sampling rate of an accelerometer is ranged from 100 Hz to 100 kHz.

Note that, there are various installation types for the accelerometer: probe tips, direct adhesive, adhesive pad, magnetic base, stud mounting, or insulating flange. The most recommended approach is stud mounting, which possesses the best relative sensitivity and highest frequency response among all these fixing types. Various types and sizes for studs and captive screws with mounting threads are all available. By mounting the stud or screw to fix the accelerometer on a specific location can improve repeatability of signal and reduce collection errors. Thus, the instructive nature is another critical issue to be taken into deep consideration once the desired vibration source is on the cutting tool.

The installation orientation of the accelerometer is as important as the attaching location and types. The accurate signal means that it reflects real situation in a straightforward and noncomplex form. The collected vibration should be clear and easy-to-understand strained conditions so as to notify where the force is from, since vibration independently occurs in the X, Y, or Z axes. Figure 2.7 demonstrates a vibration data collection of the Z-axis from a machinery spindle from Figure 2.6.

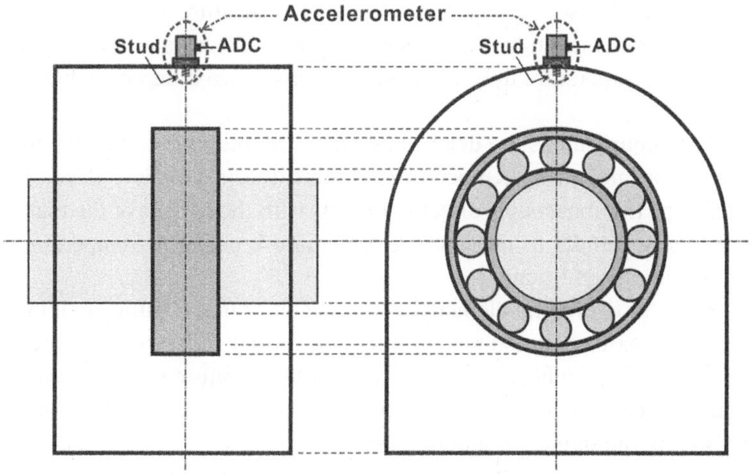

Figure 2.6 Installation of accelerometer by stud mounting.

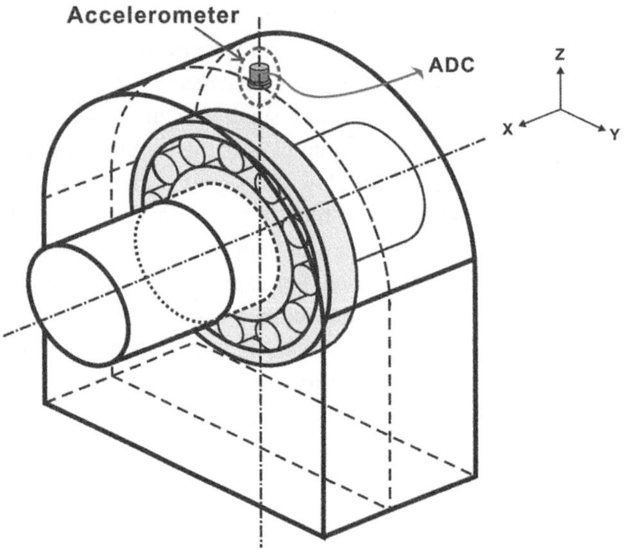

Figure 2.7 Vibration data collection of Z-axis.

Once the direction of the accelerometer's receiving surface is not orthogonal with one of the specific axes, the collected signals may contain multi-axis characteristics, and that will increase the difficulty of analyzation.

Temperature: Thermal Couple The machining temperature changes not only affect equipment operation and machining performance but can also reflect on product quality and component status during processing.

For example, the measurement of temperature in the cutting zone has high correlation with machining quality. In general, cutting temperature gets higher along with the increase of cutting speed, feed rates, and depths due to the frictional heat generated on the cutting tool-workpiece interface. This increased temperature can soften the workpiece so that material can be removed from the workpiece easily; however, higher temperature might also accelerate tool wears.

Thermal couple is the most commonly used sensor to measure the temperature of target objects by directly converting heat into electricity through thermoelectric effect, which creates the temperature-dependent voltage when the temperature difference exists between two different semiconductors inside the sensor.

Once the equipment or specific components are determined to be the monitoring targets, the measurement can be completed by measuring temperatures around the installation place. As shown in Figure 2.8, one patch-type thermal couple directly contacts with the surface of the metal shell that surrounds the machinery spindle to monitor the temperature changes. Temperature information here reflects the operation conditions.

When the quality of oil or grease for the rolling-element bearings deteriorates, insufficient lubricant quantity and viscosity may increase operation temperature and cause bearing or the spindle malfunctions. Temperature information can be a very wide-range of temperature up to thousand degrees at a low sampling rate compared to other sensing techniques. Thus, more storage space and de-noising methods for filtering the signals are not necessary.

However, sensing distance is one challenge that has to be taken into consideration. Figure 2.9 indicates that the linear distance between the thermal couple and the spindle can also lead to the inaccuracy of the obtained temperature.

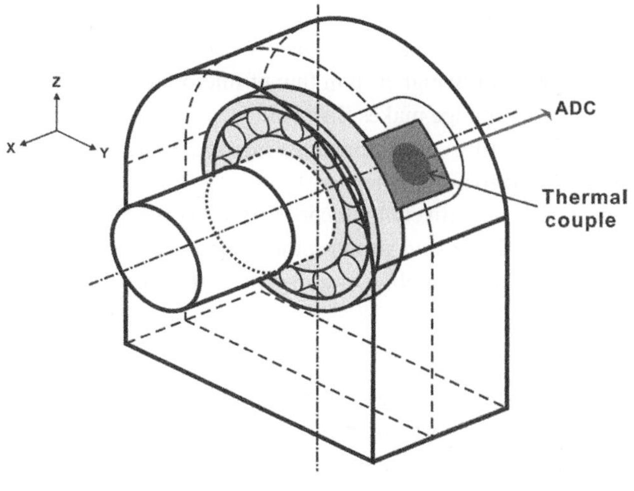

Figure 2.8 Installation of thermal couple.

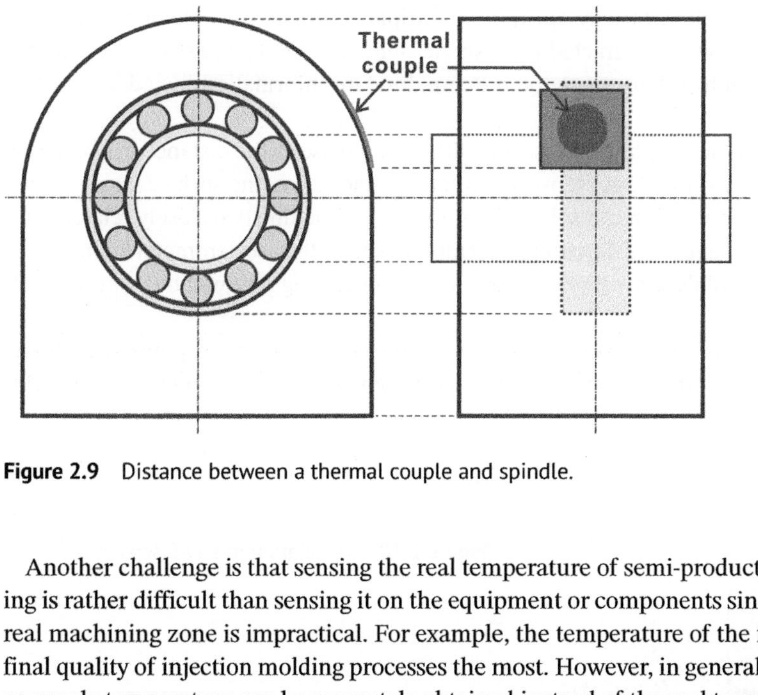

Figure 2.9 Distance between a thermal couple and spindle.

Another challenge is that sensing the real temperature of semi-product directly during machining is rather difficult than sensing it on the equipment or components since the installation on the real machining zone is impractical. For example, the temperature of the melted plastic affects the final quality of injection molding processes the most. However, in general, only the injection mold or nozzle temperature can be accurately obtained instead of the real temperature from the melted plastic. In this case, indirect optical temperature sensor is probably the only solution to this issue, yet the cost is relatively high and it is hard to be fixed inside the machine or the mold.

AE Waves: AE Transducer When the material deformation or defect occurs, the AE signal comes with the suddenly redistributed stress and transient elastic waves from a specific source within a material. This material's stress can be generated from mechanical energies (e.g. cutting forces, component wears, thermal variation, machine structures, or tool breakages) and released in the form of the acoustic radiation into an electric signal. These materials' voice can be received and interpreted by the AE sensor.

The most widely used AE sensor is the piezoelectric transducer that converts the mechanical energy into an electrical voltage signal. The AE sensor has the highest sensitivity in frequency response from 30 k Hz to 1 M Hz, which is significantly higher than that of microphone (20–100 k Hz), accelerometer, or strain gauge. Thus, this advantage makes the obtained signals less likely to be disturbed or attenuated by the mechanical structures or components. Figure 2.10 demonstrates the installation of an AE sensor on a tool and next to a strain gauge.

However, to acquire such high-frequency data, a relatively high corresponding sampling rate is required. Therefore, a sensing system with large storage and strong computing power for the large volumes of data is indispensable. In addition, the high sampling rate also increases the difficulty of data preprocessing for raw AE signals.

Sensor Fusion There is not a single sensor that can perfectly capture all signs released from the equipment during production. Every type of sensors has its own limitations in different aspects; therefore, it is difficult to obtain a comprehensive result by using only one sensory source.

To meet the requirement of enhanced accuracy with greater robustness under the varying environment, the sensor fusion technique that combines sensory data from individual and multiple sensory sources in a complementary manner is extensively proposed to improve the resolution and reliability of the acquired signals. Strictly speaking, multiple sensors with noncomplementary measurements and various features extracted from single signal can only be regarded as a multisensor but not a sensor fusion system, since too much redundant information may decrease the accuracy [6, 7].

Note that, the sensor accuracy might be worsened with the increase of the distance between the sensor and the cutting zone. If the direct sensors cannot be attached on the surface closer to the cutting zone, the sensor fusion system can solve this problem by attaching two closely related sensors such as the combination of the accelerometer/current sensor, or the accelerometer/dynamometer, to provide the cross-validation scheme between dependent signals without any loss of important information.

Generally speaking, AE sensors have very sensitive frequency and transient response, which are very adequate for combining with the current sensor, accelerometer, or dynamometer. Figure 2.11

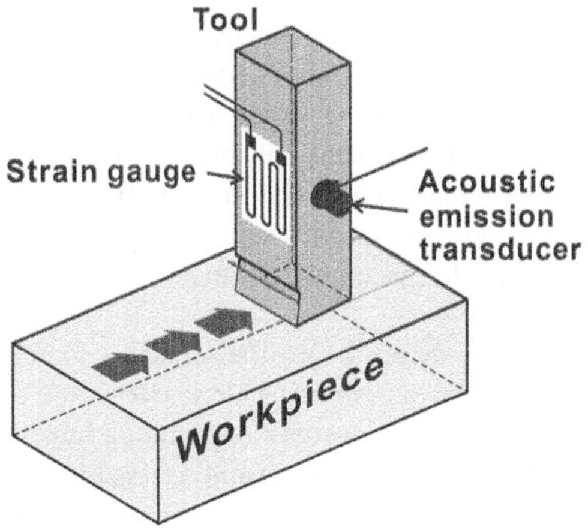

Figure 2.10 Installation of an AE sensor.

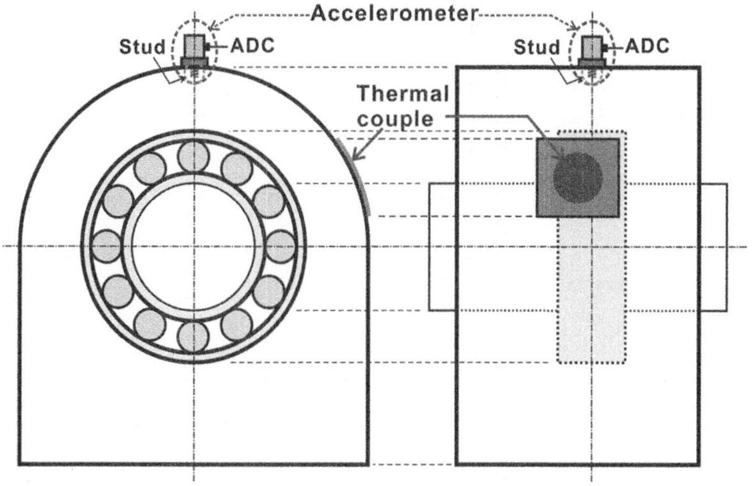

Figure 2.11 Sensor fusion system comprising an accelerometer and a thermal couple.

illustrates a machinery spindle monitored by a sensor fusion system using an accelerometer and one thermal couple.

Figure 2.12 demonstrates a sensor fusion system with five built-in sensors near the cutting zone on a computer numerical control (CNC) machine to capture critical information. The accelerometer and the thermal couple are mounted on the spindle, CT is attached on the power cable, AE is mounted and dynamometer is fixed on the machine table.

2.2.1.2 Manufacturing Parameters Acquisition

Manufacturing parameters are actual values retrieved from specific devices (e.g. machine controllers) during the production process. The actual values of manufacturing parameters are based on manually set target values and usually affected by numerous variables in the real-world environment.

For example, two commonly measured parameters that affect the injection molding process are the actual pressure value and temperature value. The manufacturing parameters are periodically updated under the supported refreshing rate of the machine controller.

Nowadays, with many easy-to-use Ethernet-based communication standards for industrial environments, such as Ethernet for Control Automation Technology (EtherCAT), MTConnect, and Open Platform Communication Unified Architecture (OPC-UA), accessing and retrieving dynamic manufacturing parameters seem not as difficult compared to installing external sensors. This advantage enables manufacturing parameters to be widely applied to record stable data that is not likely to change, so that lower sampling rate can suffice the demand (e.g. temperature, tool number, feed rate, etc).

However, high-frequency data are not available via applying communication standards. The machine controller is designed to continuously guarantee the machining quality of all time and it executes the machining processes with a high computational complexity. Since frequent requests for accessing parameters have a high risk to interrupt normal operations, the sampling rate is usually set to be within 300–500 μs for each cycle. Once the machine controller overloads, returned values of process parameters might be just default values rather than actual values. Thus, the mainstream methods of machining condition monitoring usually adopt the combination of both sensor signals and process parameters. More details of communication standards can be found in Chapter 3.

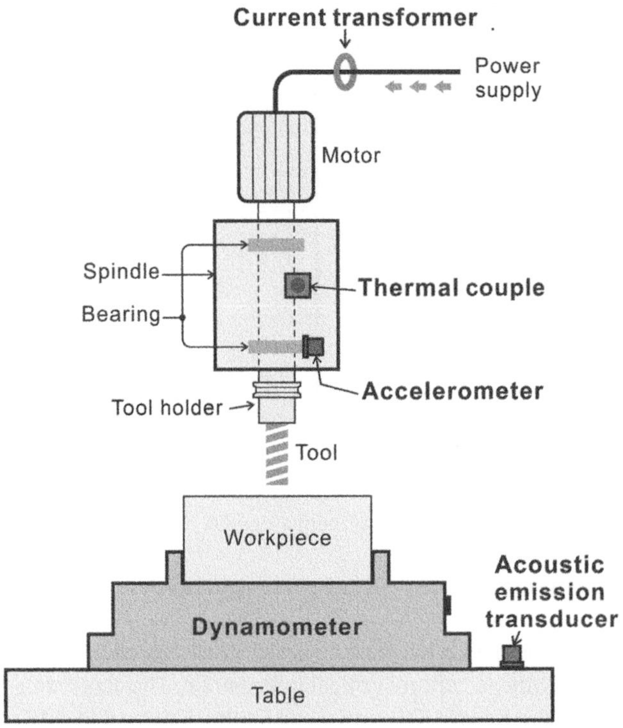

Figure 2.12 Sensor fusion system using five types of sensors.

2.2.2 Metrology Data Acquisition

The metrology data of the product quality aim to conform to customer requirements and minimize defect costs via quality control. Product quality means the overall quality of a product, which is what really matters for the customers.

A quality report that records measurement results for a specific product (e.g. dimensional accuracy, surface roughness, and position tolerance, etc.) can be completed via the inspection of various measurement equipment, such as the coordinate measuring machine (CMM) or the automated optical inspection (AOI) device, so that it shows whether the status of the product conforms to customer's requirements, specifications, and expectations. For any case, the best situation is to measure the practical quality of all end-products.

Note that, metrology delay varying with the inspection frequency and measurement place would restrict the feasibility of total inspection. Therefore, intelligent applications [1–8] should come in place to solve this issue.

In short, process data are the major contributors to metrology data. They are used as the predictor variables; while the metrology data of products can be used as a label variable or category for typical supervised learning or reinforcement learning methods. In other words, given that the data acquisition quality is reliable, the causation between process data and metrology data would be a strong correlation that efficiently indicates if the manufacturing process performs well or not.

Thus, the root cause analysis can be conducted to narrow down to the specific process steps that affect product quality. Manufactures, therefore, have a chance to make their production line free from deficiencies or defects via continuous improvement and corrective actions.

2.3 Data Preprocessing

The goal of data preprocessing is to eliminate the noise imbedded in the signal and extract key-feature-related information. In general, data preprocessing consists of three steps: segmentation, cleaning, and feature extraction, they are introduced as follows.

2.3.1 Segmentation

This step accurately segments essential parts of the raw process data from the original manufacturing process, such as from a numerical control (NC) file. Because a typical machining process could take several minutes or even hours, the entire collected process data may exceed 100 k bytes per second. The premise of building a specific decision-making model is to find the cause-and-effect relationship between the machining/process data and the corresponding precision/metrology data. As such, only the final machining process data are required for purposes like precision prediction or condition monitoring. In general, the final machining process takes only a few seconds.

Hence, a segmentation method is required for acquiring the essential segments from the NC file. These essential segments are directly related to the final precision items. Therefore, how to automatically specify the essential ranges from the original NC file as the signal segments and correlate the segments with precision items is the first step.

To acquire the desired sensor data, updated machining status and data from the controller of a machine tool are required. Conventionally, a polling method is used to retrieve the updated machining status and data from the controller through a TCP/IP based communication library provided by the vendor of the controller. However, this polling method has three issues as shown below.

1) The minimal time interval between two consecutive demands for querying machining status is only 100 μs.
2) The response time for a demand may vary from 50 to 300 μs depending on the loading of the controller.
3) A demand could be ignored by the controller due to flow exceptions or message errors.

Due to these three problems, the machining statuses retrieved by using the polling method may have random time delay, which will result in synchronization errors during machining. These synchronization errors may degrade the accuracy of predicting machining precision.

To avoid synchronization errors, a segmentation method by applying the pre-defined miscellaneous function codes, denoted as M codes, is proposed in this chapter. Originally, the M codes are used to control the miscellaneous devices of machine tools; while, in this work, the M codes are applied to specify essential blocks of an NC program. When executing the pre-defined M codes, the programmable logic controller (PLC) in the machine tool's controller can turn on or off the specified devices. For instance, M08 and M09 are used to turn on and off a specific coolant valve, respectively.

An illustrative example for applying M codes to segment the "X44Y50" block is shown in Table 2.2. The "X44Y50" block is displayed in Line l. To segment this block, the M codes, M_i and M_j, are inserted in lines $l-1$ and $l+1$, respectively. When M_i is executed by the controller, the controller will turn on Relays 2 and 4. Therefore, the combination of Relays 1–4 becomes the relay code: "0101." These four digits will then be sent through the digital input/output (DIO) ports into the external data acquisition system to acquire the sensor data via the ADC. Then, the "X44Y50" block is executed under the condition "0101." As a result, all the machining process data of block "X44Y50" can be collected. After finishing the "X44Y50" block, M_j is executed such that the relay code becomes "0000," which will notify the external data collection system to stop collecting the sensor data.

Table 2.2 Illustrative example of applying M codes to segment the "X44Y50" block in an NC code for machine tool.

Line	Block	Relay 1	Relay 2	Relay 3	Relay 4
	...	0	0	0	0
$l-1$	M_i	0	1	0	1
l	X44Y50	0	1	0	1
$l+1$	M_j	0	0	0	0
	...	0	0	0	0

Source: Reprinted with permission from Ref. [8]; © 2015 JCIE.

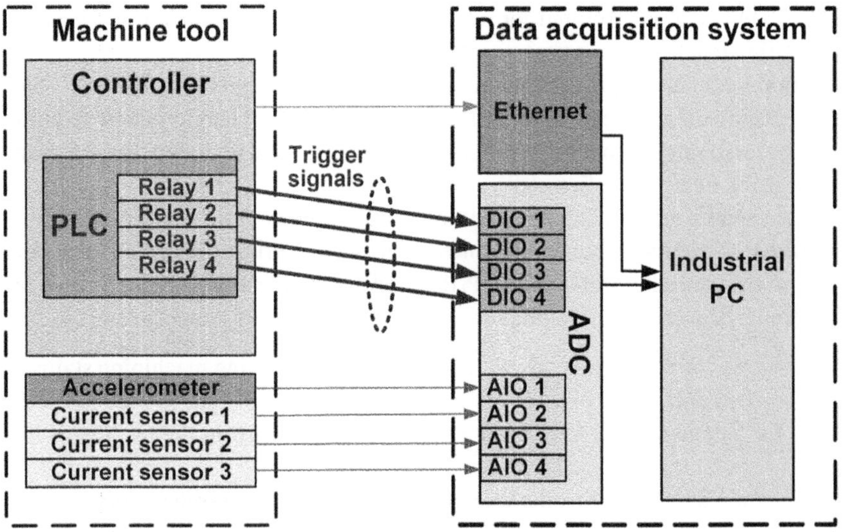

Figure 2.13 An external data acquisition system triggered by electronic relays. *Source:* Reprinted with permission from Ref. [8]; © 2015 JCIE.

Figure 2.13 demonstrates an example of an external data acquisition system triggered by four electronic relays (i.e. DIO) to acquire manufacturing parameters from the controller and sensing signals from one accelerometer and three current sensors.

Because the operation status can be updated in 0.5 μs with a sampling rate of 2 kHz and the time delay between the controller and the external data collection system is fixed to be less than 100 μs, synchronization errors can be effectively reduced. Thus, the segmented sensor data collected can be accurately synchronized with the final precision items of the machining process. Note that, with the four relays adopted in Figure 2.13 and Table 2.2, 15 M-codes plus a reset code are allowed in this case to specify 15 pre-defined M codes.

2.3.2 Cleaning

The second step, data cleaning, emerges after acquiring the segmented signals. Data cleaning effectively handles raw sensing signals with noises. Basically, a process observed and recorded via signals might contain a series of stationary and non-stationary components, especially in a machinery manufacturing environment. If the properties of the process that generate the events do not change in time, then the process is defined as stationary.

The nonstationary process (such as edge, peak, break, and noise embedded in sensing signals) includes anything that does not satisfy the conditions for stationary. These result in low signal-to-noise (S/N) ratios of the raw data collected from sensors, and thereby affect the diagnosis or prediction accuracy.

Data cleaning attempts to cancel the noise in signal and improve the S/N ratio to prepare for post-processing. Generally, data cleaning can be done by using various time-domain or frequency-domain techniques. Trend removal and wavelet thresholding methods are respectively introduced below.

2.3.2.1 Trend Removal

A signal with a trend is called nonstationary. A trend is a long-term continuous increase or decrease embedded in signals over time, which is not a good situation for designing a stable process and achieving the high product quality. To identify the pattern of a trend helps to determine whether the changes resulted from equipment or environment are normal or not. By removing or correcting the unmeaning trend information from signals (de-trending), problems can be simplified and model efficiency can be improved.

For example, Figure 2.14a demonstrates a direct current (DC) signal with 10 Hz in time-domain, and Figure 2.14b shows its frequency spectrum, where the DC component, or the so-called mean value, hardly contains any useful information to reflect mechanical faults due to an unchanged constant. In addition, the amplitude of the frequency spectrum is much larger than real characteristic frequency so that the critical information might be out of focus.

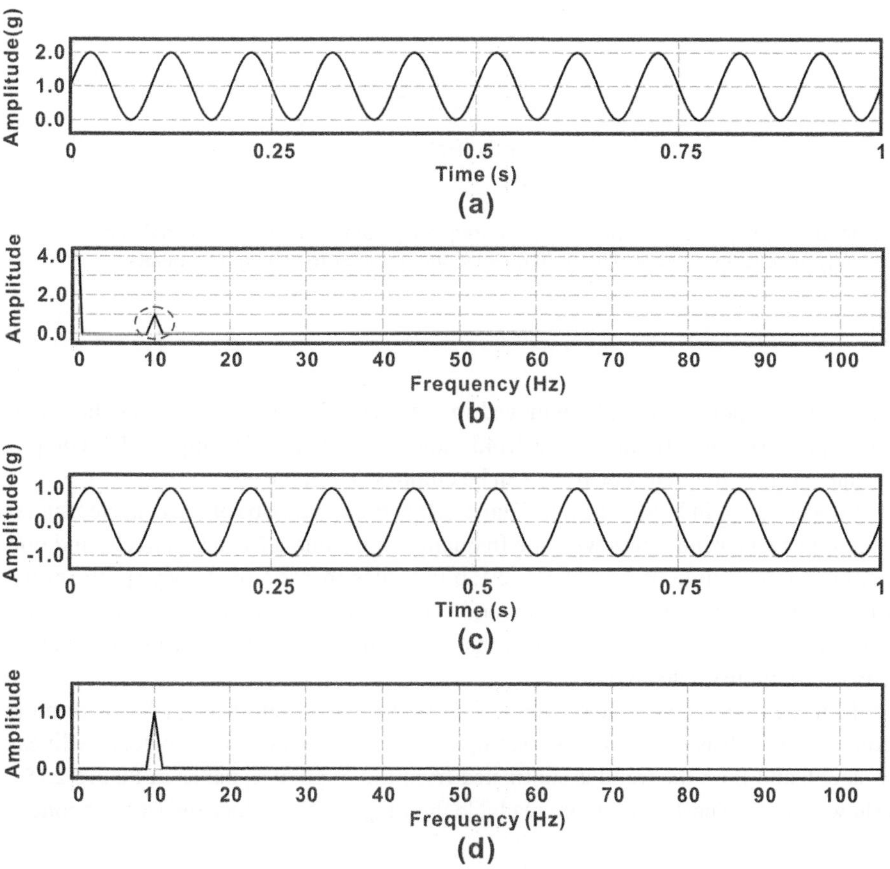

Figure 2.14 An DC signal: (a) in time-domain; (b) in frequency-domain; (c) in time-domain after removing DC trend; and (d) in frequency-domain after removing DC trend.

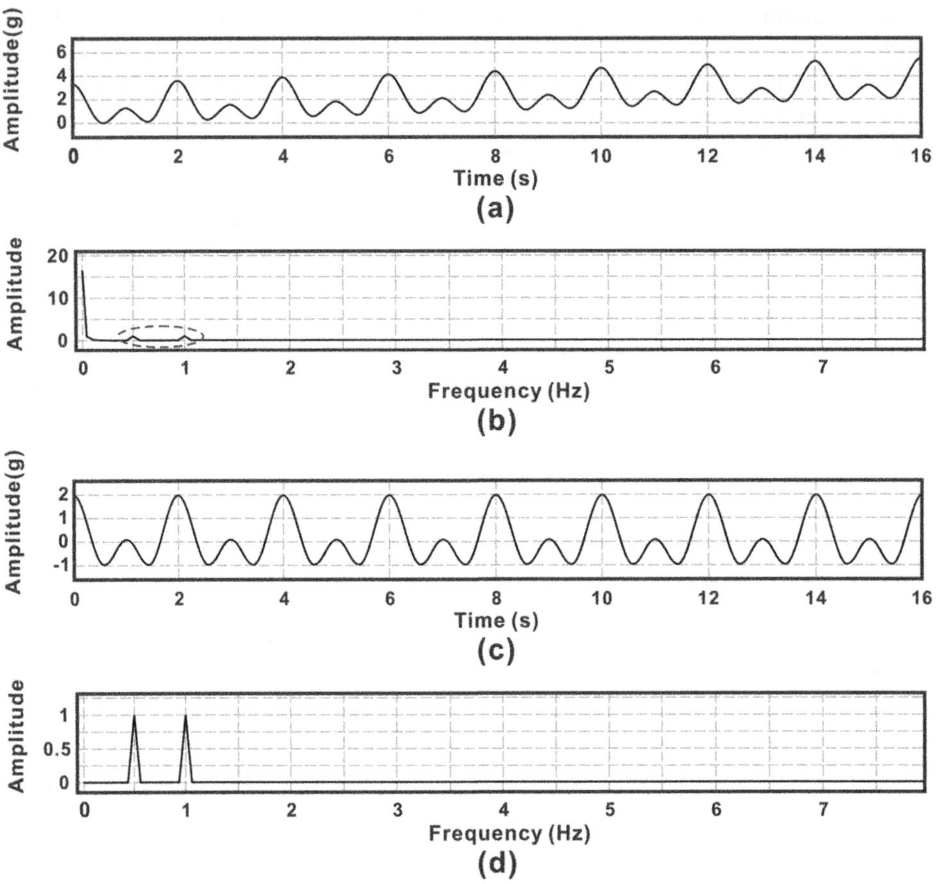

Figure 2.15 A random signal: (a) in time-domain; (b) in frequency-domain; (c) in time-domain after removing the linearly increasing trend; and (d) in frequency-domain after removing the linearly increasing trend.

Therefore, this DC component must be removed in advance. Figure 2.14c removes the mean value from the original DC signals and Figure 2.14d indicates that after deleting the DC component, the real characteristic frequency of 10 Hz can be emphasized in the spectrum.

The other example is given in Figure 2.15. In Figure 2.15a, the waveshape of the collected signal is the composition of two cosine waves with the frequency of 0.5 and 1 Hz, respectively, and one linear function increases as time passes. Figure 2.15b indicates that the frequency spectrum for recognizing the characteristic frequencies at 0, 0.5, and 1 Hz. As the example shown in Figure 2.14b, the amplitude of the DC component is too large so that real characteristic frequencies are not obvious, which may result in miss detection.

Therefore, this linear trend is unexpected and should be removed from the signal. The trend is obtained by computing the least-squares fit of a straight line (or composite line for piecewise linear trends) to the signal, and then the trend is subtracted from the original signal as in Figure 2.15c. Figure 2.15d shows that the spectrum is more readable than Figure 2.15b since the DC component is removed.

2.3.2.2 Wavelet Thresholding

Wavelet thresholding, or the so-called wavelet de-noising, mainly adopts the discrete wavelet transform (DWT) technique [8–10] to filter noises in signals. DWT provides a multi-resolution representation using wavelets, which can discretely capture rich information both in time and frequency domains.

The discrete wavelet coefficient capability of spare distribution and auto-zooming in the time and frequency domains provided by DWT can be applied to deal with the non-stationary signals for enhancing the S/N ratios. Thus, critical information behind signals with noises can be accurately obtained. More details can be referred to Sections 2.3.3.2 and 2.3.3.3.

Suppose that M sets of machining signals related to machining precision are collected and each set has data length N, denoted as $s[i]$, $i = 1, \ldots, N$. Let r and c represent the raw and cleaned data, respectively. Based on DWT, data cleaning adopts the wavelet de-noising algorithm to purify the discrete raw sensor data of precision item p, denoted as $\mathbf{S}_p^r$, to become the cleaned discrete sensor data, $\mathbf{S}_p^c$, by using the function:

$$\mathbf{S}_p^c = \text{Denosing}\left(\mathbf{S}_p^r\right) \tag{2.1}$$

where $\mathbf{S}_p^r = \left[s[1], s[2], \ldots, s[N]\right]_p, p = 1, \ldots, M.$

The general wavelet de-noising process consists of three steps: decomposition, thresholding, and reconstruction. They are described as follows.

Decomposition

The wavelet coefficients are calculated by passing $\mathbf{S}_p^r$ through a series of filters. DWT can be implemented efficiently by using a multi-resolution analysis (MRA) of fast wavelet transform invented by Mallat [9].

MRA computationally decomposes $\mathbf{S}_p^r$ into a two-scale relation with various time and frequency resolutions by DWT, which is composed of a series of low-pass and high-pass filters.

An L-level wavelet decomposition is illustrated in detail. The process starts with inputting $\mathbf{S}_p^r$ of length $N=2^L$ into a low-pass filter $g[k]$ and a high-pass filter $h[k]$, then $\mathbf{S}_p^r$ is convolved with $g[k]$ and $h[k]$ for generating two vectors $c_1^a[n]$ and $c_1^d[n]$ of length $N/2$, respectively.

The contents of $c_1^a[n]$ and $c_1^d[n]$ are approximation and detail coefficients of DWT at the first level, respectively. Further, $c_1^a[n]$ is used as an input for obtaining wavelet coefficients $c_2^a[n]$ and $c_2^d[n]$ at the second level of resolution with the length of coefficients being $N/4$. In other words, a recursive relationship exists between the approximation and detail coefficients at successive levels of resolution. Therefore, the general decomposition form of wavelet coefficients of length N at the j level can be expressed as follows:

$$c_j^a[n] = \sum_{k=0}^{\left(\frac{N}{2^{j-1}}\right)-1} c_{j-1}^a[2n-k]g[k], \tag{2.2}$$

$$c_j^d[n] = \sum_{k=0}^{\left(\frac{N}{2^{j-1}}\right)-1} c_{j-1}^a[2n-k]h[k], \tag{2.3}$$

where

j scale parameters $j = 1, 2, ..., L$;
L number of decomposition levels;
N data length of a discrete signal;
n translation parameter $n = 0, 1, 2, \ldots, \frac{N}{2^j}$;
$c_j^a[n]$ nth approximation coefficient at level j;
$c_j^d[n]$ nth detail coefficient at level j;
$g[k]$ DWT low-pass filters; and
$h[k]$ DWT high-pass filters.

Figure 2.16 illustrates an example of three-level DWT decomposition. A signal with available bandwidth 2,000 Hz is decomposed into the sets $c_j^a[n]$ and $c_j^d[n]$, where $\{\mathbf{A_1}, \mathbf{D_1}\}$, $\{\mathbf{A_2}, \mathbf{D_2}\}$, and $\{\mathbf{A_3}, \mathbf{D_3}\}$ represent the approximate (gray (A_i) blocks) and detail (white (D_i) blocks) coefficients in levels 1–3, respectively.

Thresholding

The performance of wavelet de-noising depends on the determination of two factors, the threshold value λ and the threshold function. In this section, the wavelet de-noising algorithm adopts the soft thresholding method to adaptively filter the specific spectrums of the noisy signals to obtain modified wavelet coefficients $\hat{c}_j[n]$. Let $c_j[n] = \{c_j^a[n], c_j^d[n]\}$ at level j.

The soft thresholding method sets every wavelet coefficient $c_j[n]$ to zero if $|c_j[n]|$ is less than or equal to a chosen threshold λ; otherwise, the threshold is subtracted from any $c_j[n]$. Then, all modified coefficients $\hat{c}_j[n] = \left(\left\{\hat{c}_j^a[n], \hat{c}_j^d[n]\right\}\right)$ at level j can be defined as:

$$\hat{c}_j[n] = \begin{cases} sign(c_j[n]) \cdot (|c_j[n]| - \lambda), & \text{if } |c_j[n]| > \lambda \\ 0, & \text{if } |c_j[n]| \leq \lambda \end{cases}, \tag{2.4}$$

$$\lambda = 1.483 \cdot \text{MAD}(c_{L-1}[n]) \cdot \sqrt{2\log(N)}; n = 1, 2, \ldots, 2^{L-1}, \tag{2.5}$$

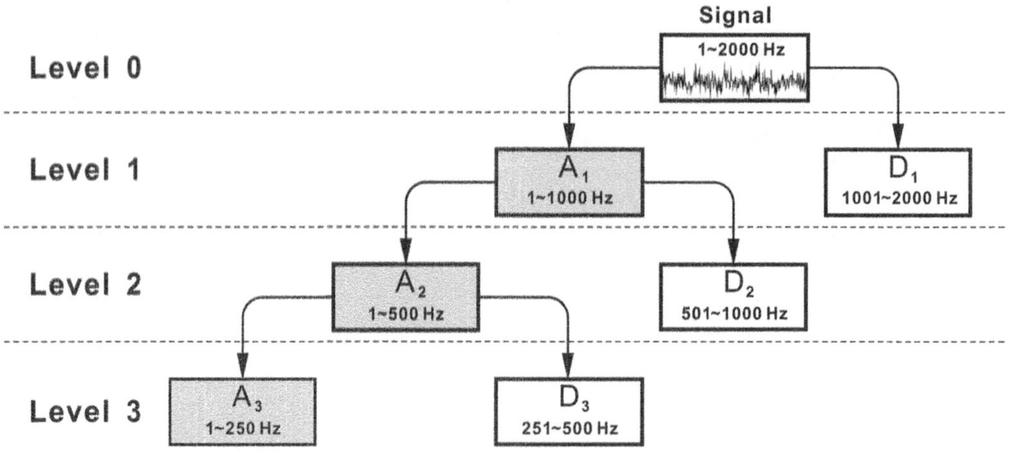

Figure 2.16 Three-level decomposition tree of the DWT.

where

$c_j[n]$ nth wavelet coefficients at level j;
$\hat{c}_j\left[n\right]$ nth modified wavelet coefficients at level j;
λ threshold of $c_j[n]$; and
$\text{MAD}(c_{L-1}[n])$ mean absolute deviation of $c_{L-1}[n]$.

Reconstruction

The de-noised sensor data $\mathbf{S}_p^r$ can be reconstructed from all modified wavelet coefficients at the level of resolution L. The reconstruction process is in the opposite direction to the decomposition process; that is, the process proceeds with level $j=L, L-1, \ldots,$ till 1. As such, intermediate modified approximation coefficients $\hat{c}_{j-1}^{a}\left[n\right], \left(n=1,2,\ldots,N/2^j\right)$ at each level of resolution, $j-1$ can be recovered by up-sampling modified approximation and detail coefficients at the level of resolution j. Finally, $\mathbf{S}_p^r$ is obtained with $j = 1$ as follows:

$$\mathbf{S}_p^c = \sum_{k=0}^{\left(\frac{N}{2}\right)-1} \hat{c}_1^{a}\left[2n-k\right]g\left(k\right) + \sum_{k=0}^{\left(\frac{N}{2}\right)-1} \hat{c}_1^{d}\left[2n-k\right]h\left(k\right); n=1,2,\ldots,\frac{N}{2}. \tag{2.6}$$

2.3.3 Feature Extraction

Feature extraction [6–8] is the process to generate a smaller linear or nonlinear combination set to represent the original high-dimensional data set. Thus, the de-noised sensing signals need to be transformed into meaningful signal features (SFs), which can adequately describe the physical meaning of the signal and maintain relevant information of the machining operations [6]. However, monitoring machining conditions based on a single SF is not enough [7]. To properly describe machining precisions, a set of multiple SFs is required to provide further insight into coordination [8].

This section introduces feature extraction approaches that are commonly used in the time, frequency, and time–frequency domains. Feature selection is a process used to define a small and concise feature subset through the removal of redundant features from a feature set and it is introduced in Section 2.3.3.1(A.1). In addition, an Autoencoder (AEN) as a popular ANN-based feature extraction method [11, 12] is introduced in Section 2.3.3.4.

2.3.3.1 Time Domain

SFs extracted from time-domain signals are very intuitive to understand how the signals change in the past or at a specific time. Statistical SFs and correlation-based SFs are two main methods used to describe these changes in practical manufacturing applications.

In signal processing, SFs of the statistical description can essentially identify changes from the shape of the waveform; while SFs of cross-correlation and autocorrelation, based on the Pearson product-moment correlation coefficient [13], investigate similarity relationship between time-varying signals. The feature extraction methods of statistical SFs, cross-correlation SFs, and autocorrelation SFs are presented as follows.

Statistical SFs

If the machining parameters (such as feed rate, spindle speed, depth of cut, etc.) are fixed and the precision after machining is within specifications, then the machining operations are a kind of a quasi-static condition [8]. Under this condition, Yang et al. [8] summarized that the nine most common SFs for various types of sensor signals are derived from all the elements of $\mathbf{S}_p^c$.

Table 2.3 Definition of time-domain SFs.

SF	Formula	Description
avg	$\frac{1}{N}\sum_{i=1}^{N}\left(\mathbf{S}_p^c(i)\right)$	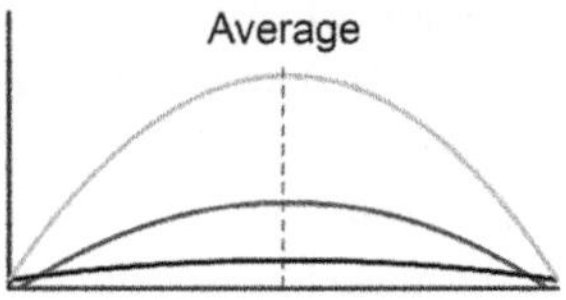 Average
std	$\sqrt{\frac{1}{N}\sum_{i=1}^{N}\left(\mathbf{S}_p^c(i)-avg\right)^2}$	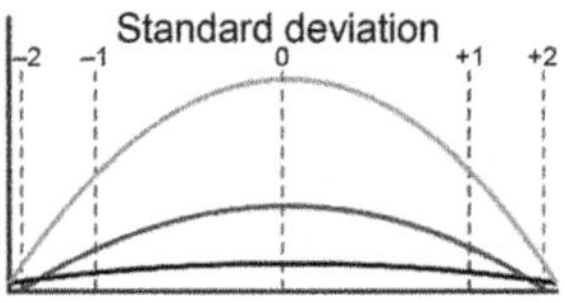
max	$\max\left(\mathbf{S}_p^c\right)$	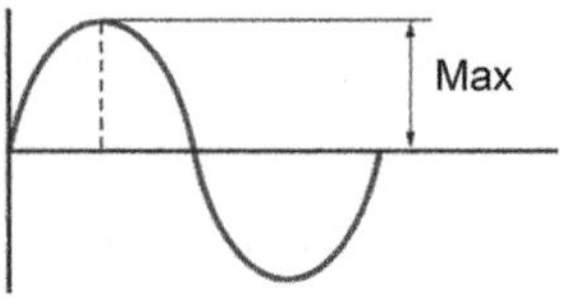 Max
min	$\min\left(\mathbf{S}_p^c\right)$	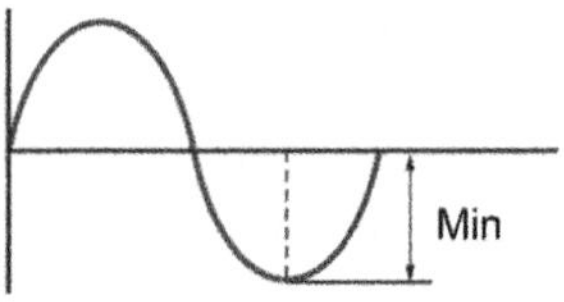 Min
ptp	$\max\left(\mathbf{S}_p^c\right)-\min\left(\mathbf{S}_p^c\right)$	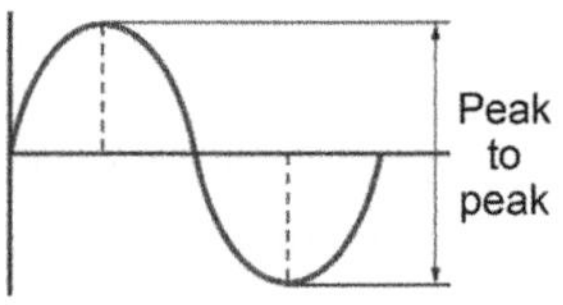 Peak to peak
kurt	$\frac{\frac{1}{N}\sum_{i=1}^{N}\left(\mathbf{S}_p^c(i)-avg\right)^4}{std^4}$	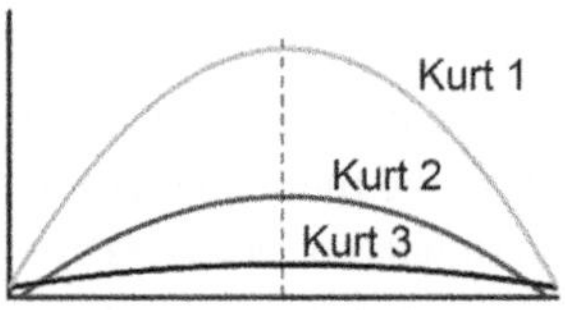 Kurt 1 Kurt 2 Kurt 3
skew	$\frac{\frac{1}{N}\sum_{i=1}^{N}\left(\mathbf{S}_p^c-avg\right)^3}{std^3}$	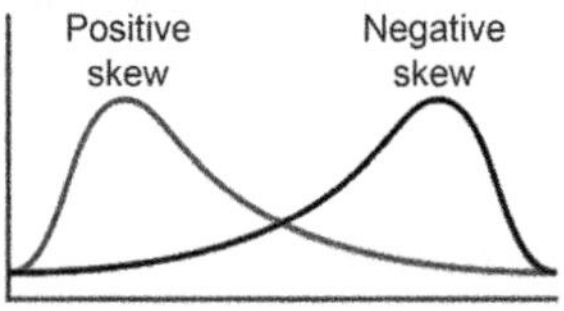 Positive skew Negative skew

Table 2.3 (Continued)

SF	Formula	Description
RMS	$\sqrt{\frac{1}{N}\sum_{i=1}^{N}\mathbf{S}_p^c(i)^2}$	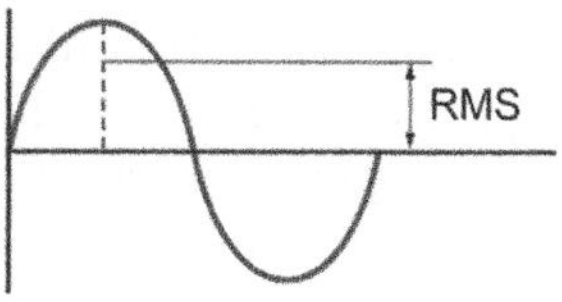
CF	$\frac{\max}{RMS}$	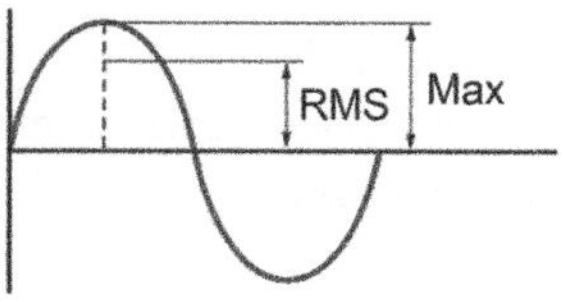

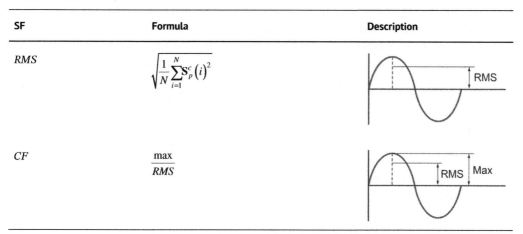

For the de-noised time-domain signals $\mathbf{S}_p^c$ that contain N elements (data length = N), these nine SFs are represented and defined as in Table 2.3, including:

- Arithmetic mean or average (*avg*)
- Standard derivation (*std*)
- Maximal magnitude (*max*)
- Minimal magnitude (*min*)
- Peak-to-peak amplitude for presenting difference of peaks (*ptp*)
- Kurtosis for measuring the peakedness of signals (*kurt*)
- Skewness for measuring the asymmetry of signals (*skew*)
- Root mean square value for indicating the weighting effect of variances (*RMS*)
- Crest factor for representing how extreme the peaks are in a waveform (*CF*)

As such, these nine SFs can be used as a feature set based on expert knowledge. Suppose that one vibration sensor and three current sensors are installed as the sensor fusion example illustrated in Figure 2.13, then there are 36 SFs in total because each sensor has nine SFs.

These 36 SFs may be adopted as the input variables of any intelligent system. However, redundancy, irrelevancy, and/or dependency may exist among these 36 SFs, which may deteriorate the model accuracy; and the more SFs, the more training samples are needed [4]. Conventional feature selection methods [6–8] can be applied to automatically search for key SFs to reduce the number of SFs during the model-building and model-refreshing processes so as to improve the model accuracy. However, due to the dynamic nature of these methods, the content of key SFs could vary after applying automatic search in each model refreshing, which might not be appropriate for implementation considerations.

For easy implementation, a fixed and concise set of SFs is required to represent the significance of the entire manufacturing process. Therefore, an expert-knowledge-based (EK-based) selection procedure to find a fixed and concise set of SFs is illustrated below.

EK-based Selection Procedure In view of selecting the SFs of a vibration sensor (i.e. accelerometer), since the machining quality is affected directly by the tool states, SFs that can accurately monitor tool status should be selected. In high-speed machining operations, a serious increase in cutting energy generated due to tool breakage or flank wear will amplify vibration magnitude that can be detected by the *max*, *RMS*, and *avg* of the vibration signal. These three SFs are crucial to the detection of vibration amplitude and energy variance between workpieces and tools.

A rolling bearing is one of the most important components widely embedded in machine tools. Therefore, abnormal statuses of rolling bearings may also cause breakdowns in rotating systems and result in serious machining failures. *skew* and *kurt* are two useful SFs to detect rolling bearing faults at an early stage. Besides, *kurt*, which works very well in the whole range from slow to very fast rolling speed, is sensitive enough to provide rich information of incipient faults for characterizing the impact existing in the rolling bearings. Further, *std*, a kind of precision-related SF responsible for investigating any small changes during machining, can indicate good correlation with real precision values of workpieces.

Then, the essential and concise SFs of electric-current signals are investigated. *RMS* of spindle current can correctly represent dynamic cutting-force variation for monitoring tool fracture and precision prediction. When dealing with the alternating current (AC), *CF* is applied for detecting whether an electrical system has the ability to generate a particular current output. In addition, *avg* can also be used as an SF of tool flute breakage or tool-wear estimations.

Finally, with the same reason as for vibration signals, *max* is used for detecting any abnormal current peaks during machining. Accelerometer and current sensors are selected here to demonstrate how to choose essential and concise SFs. By the same token, the essential and concise SFs of other sensors, such as dynamometers, acoustic-emission sensors, and thermometers, can also be identified. In summary, the six vibration SFs selected include: *max*, *RMS*, *avg*, *skew*, *kurt*, and *std*; while the four current SFs chosen are *RMS*, *avg*, *max*, and *CF*.

Cross-Correlation SFs

Generally, the Pearson product-moment correlation coefficient [13] measures the similarity degree between two signals in the same time without any time lag. Given that $\mathbf{x}(t)$ and $\mathbf{y}(t)$ are two continuous-time signals with time T, and their correlation coefficient value cr_{xy}, ranging from negative one to positive one, is defined in (2.7).

$$cr_{\mathrm{xy}} = \frac{\sum_t^T \left[\mathbf{x}(t) - avg(\mathbf{x})\right]\left[\mathbf{y}(t) - avg(\mathbf{y})\right]}{\sqrt{\sum_t^T \left(\mathbf{x}(t) - avg(\mathbf{x})\right)^2}\sqrt{\sum_t^T \left(\mathbf{y}(t) - avg(\mathbf{y})\right)^2}} \tag{2.7}$$

The cross-correlation is similar to correlation coefficient, but it takes time lag into consideration. One signal is allowed to be time-shifted and slide over the other to compare the similarity of two independent signals at each stride. It helps to find out where the two waveforms match the best at a certain time.

For random signals, the cross-correlation CR_{xy} between $\mathbf{x}(t)$ and $\mathbf{y}(t)$ is expressed in (2.8):

$$CR_{\mathrm{xy}} = \frac{\gamma_{\mathrm{xy}}}{\sqrt{std(\mathbf{x})^2\, std(\mathbf{y})^2}} \tag{2.8}$$

where γ_{xy} is the covariance expressed in (2.9):

$$\gamma_{\mathrm{xy}} = \frac{1}{T}\sum_{t=1}^{T}\left[\mathbf{x}(t) - avg(\mathbf{x})\right]\left[\mathbf{y}(t) - avg(\mathbf{y})\right]. \tag{2.9}$$

Thus, cross-correlation can be simplified by the ratio in (2.10):

$$CR_{\mathrm{xy}} = \frac{\sum_t^T \left[\mathbf{x}(t) - avg(\mathbf{x})\right]\left[\mathbf{y}(t) - avg(\mathbf{y})\right]}{\sqrt{\sum_t^T \left[\mathbf{x}(t) - avg(\mathbf{x})\right]^2}\sqrt{\sum_t^T \left[\mathbf{y}(t) - avg(y)\right]^2}}. \tag{2.10}$$

For deterministic signals, the cross-correlation of two continuous and periodic signals $\mathbf{x}(t)$ and $\mathbf{y}(t)$ can be defined by the integration from $+\infty$ to $-\infty$ as in (2.11), where notation * denotes the complex conjugate; $\mathbf{x}(t)$ is fixed and $\mathbf{y}(t)$ is shifted forward/backward by m, which is the displacement, or the so-called lag.

$$CR_{\mathrm{xy}}(m) = \int_{-\infty}^{\infty} \mathbf{x}(t)\mathbf{y}^*(t-m)dt \tag{2.11}$$

For discrete signals $\mathbf{x}[n]$ and $\mathbf{y}[n]$ with length N, the cross-correlation is defined as in (2.12), the products of two signals and the integration are replaced by any interval of period T at each point.

$$CR_{\mathrm{xy}}[m] = \sum_{n=-\infty}^{\infty} \mathbf{x}[n]\mathbf{y}^*[n-m] \tag{2.12}$$

Cross-correlation repeats to successively slide one signal along the x-axis and compare with the other until the maximized correlation value is found. The reason is that two signals with the same sign (both positive or negative) tend to have a large correlation. Especially, when both peaks or troughs are aligned, it must be the best correlation. On the other hand, when signals have opposing signs at a certain time, its correlation or integral area must be small.

Cross-correlation is very useful in the pattern recognition within a signal or between two signals. It is widely used to check the stability of sensor data and remove noise in a mass production environment. Note that, each CR_{xy} can serve as a critical SF in a set, which can be expressed as $\mathbf{SF}_{\mathrm{CR(xy)}} = CR_{\mathrm{xy}}$.

Autocorrelation SFs

Autocorrelation, or the so-called serial correlation, performs the same cross-correlation procedure of a signal with the time-shifted form of itself. Thus, all autocorrelation has to do is to replace $\mathbf{y}(t)$ with $\mathbf{x}(t)$ from (2.8) to (2.12). Note that, the maximized value of the autocorrelated signal always exists at the displacement zero, which means that two signals are totally overlapped. Autocorrelation is widely used in signal processing for recognizing some repeating patterns, such as detecting the missing frequencies or presence of critical information in a periodic signal.

2.3.3.2 Frequency Domain

Frequency-domain SFs can reflect the signal's power distribution over a range of frequencies. Theoretically, periodic signals are composed of many sinusoidal signals with different frequencies, such as the triangle signal, which is actually composed of infinite sinusoidal signal (fundamental and odd harmonics frequencies).

Thus, the frequency spectrum of the periodic signal can be obtained by the projection of these sine and cosine signals in the frequency axis by the Fourier transform (FT) technique [10], which is probably the most widely used method for signal processing. Since then, a signal can be represented by the spectrum of frequency components in the frequency domain.

As the conversion of time and frequency domain shown in Figure 2.17, one time-domain signal composed of two different waveforms with frequency is converted into the frequency domain. Two magnitudes of corresponding sine or cosine signals are represented at the specific location on the frequency spectrum.

One drawback is that the calculation and execution are very time-consuming when dealing with a large amount of datasets. Thus, fast Fourier transform (FFT) based on FT is implemented to deal with nonperiodic functions and discrete time-domain signals [10]. FFT can reduce the complexity of computing FT and rapidly compute the global information of the frequency distribution from

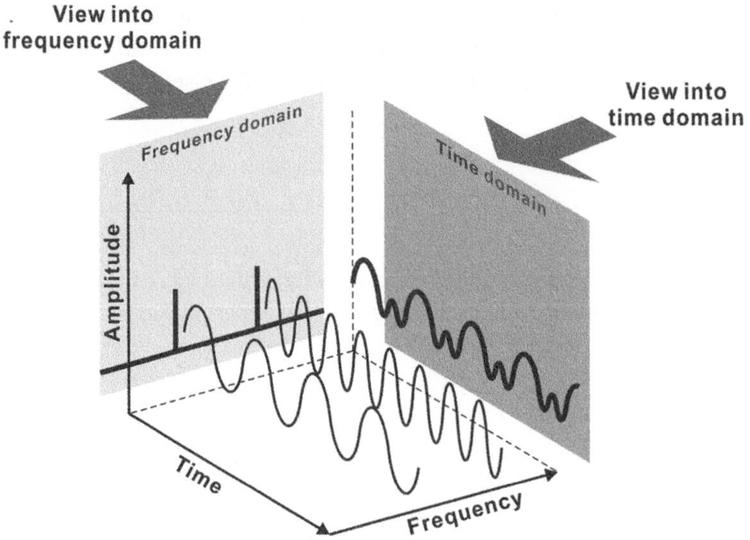

Figure 2.17 View of the time and frequency domains.

any signal. The famous mathematician Gilbert Strang also described that FFT is "the most important numerical algorithm of our lifetime" in 1994 [14].

FFT directly decomposes any discrete signal $\mathbf{x}[n]$ into the frequency spectrum by the orthogonal trigonometric basis functions as in (2.13), where $l = 1, 2, \ldots, N$.

$$\mathbf{FFT}[n] = \sum_{n=1}^{N} e^{-i\frac{2\pi}{T}nl}\mathbf{x}[n] \tag{2.13}$$

Figure 2.18a shows a vibration signal collected from a practical rotary spindle under the speed of 2,000 rotations per minute (rpm) with the sampling rate being 2,048 Hz. Figure 2.18b illustrates that there are three major peaks at 33.3, 66.6, and 99.9 Hz on the frequency spectrum, which correctly represent the fundamental frequency, the second harmonic frequency, and the third harmonic frequency, respectively. Some unimportant frequency components with relatively small amplitude (usually the noises) among three peaks are hard to observe in the time domain, but they are very clear to be detected and can be ignored.

The set of FFT-based SFs $\mathbf{SF}_{\mathrm{FFT}(q)}$ can be extracted from the summation of $\mathbf{FFT}[n]$ values close to the qth certain frequency band delimited by a lower frequency and an upper frequency of critical characteristics, as expressed in (2.14).

$$\mathbf{SF}_{\mathrm{FFT}(q)} = \sum_{n=lf_q}^{uf_q} \mathbf{FFT}[n] \tag{2.14}$$

where q=1, 2, ..., Q and

uf_q qth upper frequency of the critical characteristics; and
lf_q qth lower frequency of the critical characteristics.

For stationary signals, FFT provides a good description in global frequency bandwidth without indicating the happening time of a particular frequency component and whether the resolution scale in both time and frequency domains are enough or not.

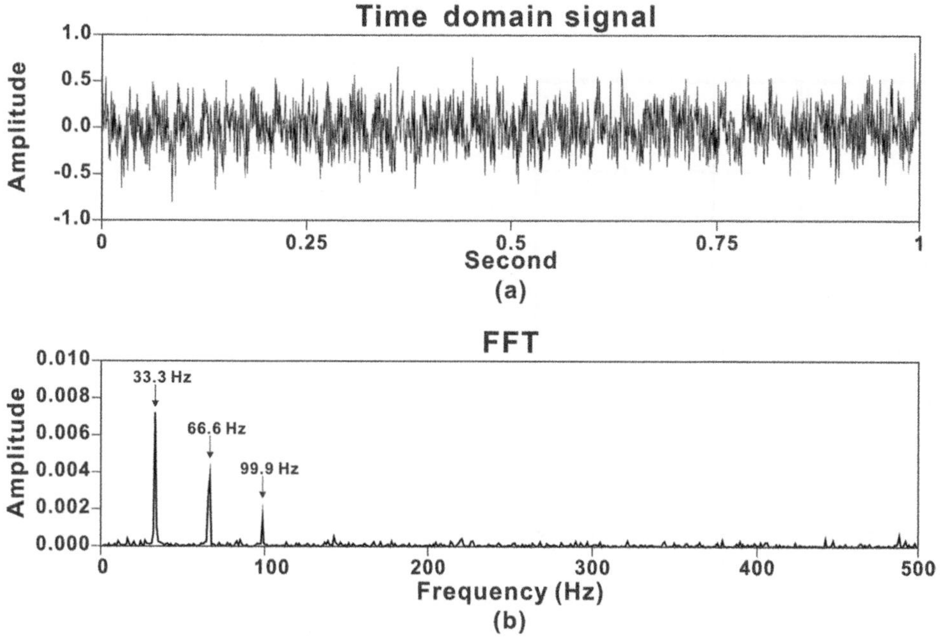

Figure 2.18 A vibration signal: (a) in time-domain; and (b) in FFT spectrum.

However, FFT might be limited to processing stationary signals. A highly non-stationary signal cannot be adequately described in the frequency domain by FFT, since its frequency characteristics dynamically change over time. Thus, extracting other SFs in the time–frequency domain is necessary.

2.3.3.3 Time–Frequency Domain

The time-frequency analysis describes a nonstationary signal in both the time and frequency domains simultaneously, using various time-frequency representations. The advantage is the ability to focus on local details compared to other traditional frequency-domain techniques.

Although short-time Fourier transform (STFT) method is proposed to retrieve both frequency and time information from a signal afterward, the deficiency is still yet to be overcome completely. STFT calculates FT components of a fixed time-length window, which slides over the original signal along the time axis.

STFT adopts an unchanged resolution in both time and frequency domains, as shown in Figure 2.19. Heisenberg uncertainty principle [15] states that it is impractical to use good resolution in both time and frequency axes since the product of the two axes is a constant. A longer window has better time resolution but worse frequency resolution, and vice versa. In general, nonstationary components often appear in high frequency and only happen in a very short period of time, but this unchanged window length makes resolution in high frequency unclear.

One representative technique to solve the FT-related issues is the wavelet packet transform (WPT) decomposition [10, 11, 16]. WPT not only dynamically changes resolutions both in time and frequency scales but also has more options to change its convolution function depending on characteristics of the signal.

In regards to the resolution of Figure 2.20, it is assumed that low frequencies last for the entire duration of the signal, whereas high frequencies appear from time to time as short bursts. This is often the case in practical applications.

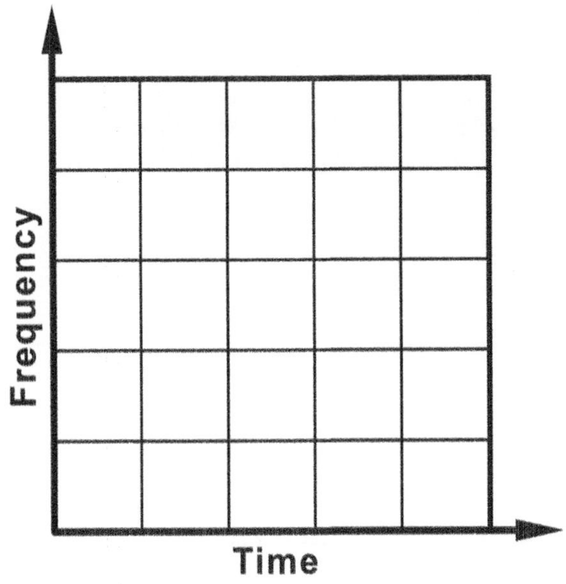

Figure 2.19 Unchanged resolution of STFT time-frequency plane.

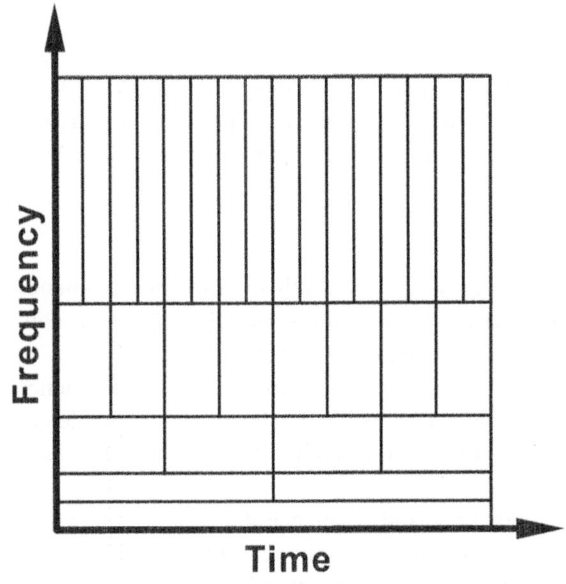

Figure 2.20 Dynamic window of WPT time-frequency plane.

In this section, WPT serves as the major time-frequency analysis method to extract useful SFs for various machinery applications. WPT is a generalization of DWT to provide a richer information and it can be implemented by DWT-based MRA as introduced in Section 2.3.2.2.

As illustrated in Figure 2.16, although DWT provides flexible time-frequency resolution, it suffers from a relatively low resolution in the high-frequency region since only the approximation coefficients $c_j^a\left[n\right]$ can be sent to the next level and split into approximation and detail coefficients $c_j^d\left[n\right]$ repeatedly. Thus, some transient elements existing in the high-frequency region are difficult to be captured and differentiated. By these procedures, any detail and approximation of the signal can be obtained at each resolution level depending on the analysis requirements.

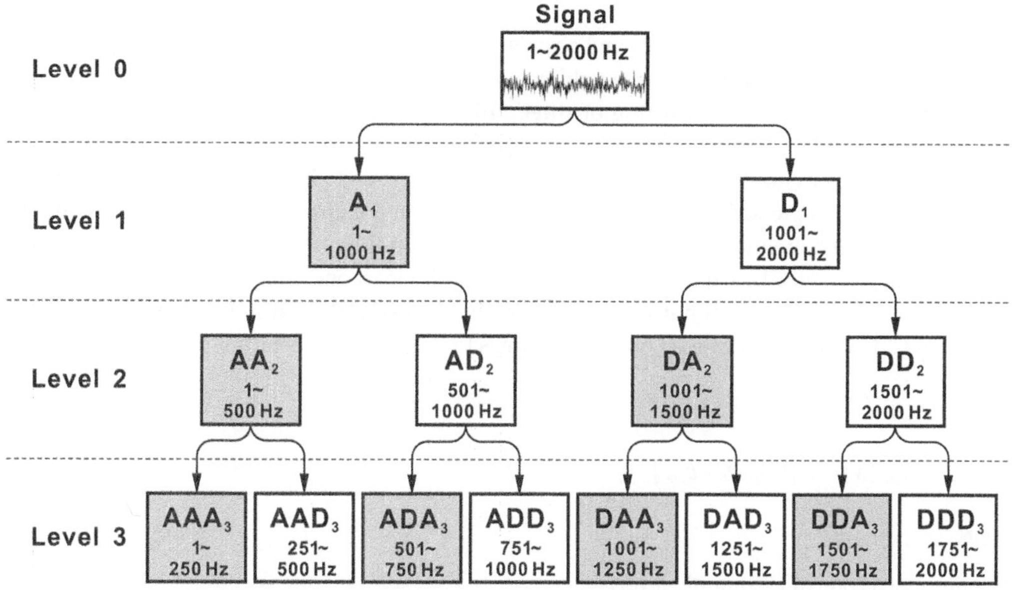

Figure 2.21 WPT decomposition binary tree.

Figure 2.21 illustrates a WPT-based fully binary decomposition tree. In WPT, the decomposition occurs in both approximation and detail coefficients. Then the same signal as illustrated in Figure 2.16 can be successively decomposed into different levels using a series low-pass $g[k]$ (scaling function) and high-pass $h[k]$ (wavelet function) filters that divide spectrums into one low-frequency band and one high-frequency band, which can be represented by approximation $c_j^a[n]$ (gray blocks) and detail $c_j^d[n]$ (white blocks) coefficients, respectively.

Note that, even detail coefficients in the high-frequency region can be decomposed into higher level with a better resolution. Finally, a three-level WPT produces a total of eight frequency sub-bands in the third level, with each frequency sub-band covering one-eighth of the signal frequency spectrum.

Thus, for a discrete signal with length N, and given WPT coefficients $c_L[n] = \{c_L^a[n], c_L^d[n]\}$ at the final level L as defined in Section 2.3.2.2, the set composed of the uth WPT-based node energy SF $\mathbf{SF}_{\mathrm{WTP}(u)}$ [16] can then be expressed in (2.15).

$$\mathbf{SF}_{\mathrm{WPT}(u)} = \sum_{k=(u-1)*v}^{u*v-1} \left(c_L[k]\right)^2 \tag{2.15}$$

where

u uth wavelet packet node at level L, $u = 1, 2, ..., L$;
v subband length for each wavelet packet node at level L, $v = N/2^L$.

The signal's energy distribution contained in a specific frequency band is calculated based on all $c_L[n]$ in each wavelet packet node using (2.15) and can be used as a SF [16], which provides more useful information than directly using $c_L[n]$.

In this way, the WPT technique precisely localizes information behind the non-stationary signals in both time and frequency domains and thus it is widely applied to mechanical fault diagnosis.

2.3.3.4 Autoencoder

Recently, AEN becomes an important and popular technique to efficiently reduce the dimensionality and generate the abstract of large volumes of data [11, 12]. AEN is an unsupervised back-propagation neural-network consisting of three fully-connected layers of encoder (input), code (middle), and decoder (output).

The encoder layer encodes and compresses the data to the code layer, and then decoder layer reconstructs the compressed internal representation of input data from the code layer into output data as closer to the original input as possible. As depicted in Figure 2.22, the architecture of the encoder, code, and decoder can be designed to constitute at least one layer each.

Let x be one variable of the input set, then the mathematical relationships between layers can be defined as (2.16) and (2.17), and its output $\hat{x}$ belongs to the same space with x.

$$h = f_{\mathrm{EN}}(x) = f_{\mathrm{a}}(W_{\mathrm{EN}} \cdot x + b_{\mathrm{EN}}) \tag{2.16}$$

$$\hat{x} = f_{\mathrm{DE}}(x) = f_{\mathrm{a}}(W_{\mathrm{DE}} \cdot h + b_{\mathrm{DE}}) \tag{2.17}$$

where

h compressed code of the middle layer;
$\hat{x}$ output reconstructed from c in the middle layer;
f_{EN} encoder layer;
f_{DE} decoder layer;
f_{a} activation function;
W_{EN} network weight for node in the encoder;
W_{DE} network weight for node in the decoder;
b_{EN} bias for node in the encoder layer;
b_{DE} bias for node in the decoder layer.

The number of input and output nodes depends on the size of raw data, while the number of nodes in the code layer is a hyperparameter that varies according to the AEN architecture and input data format as other hyperparameters do.

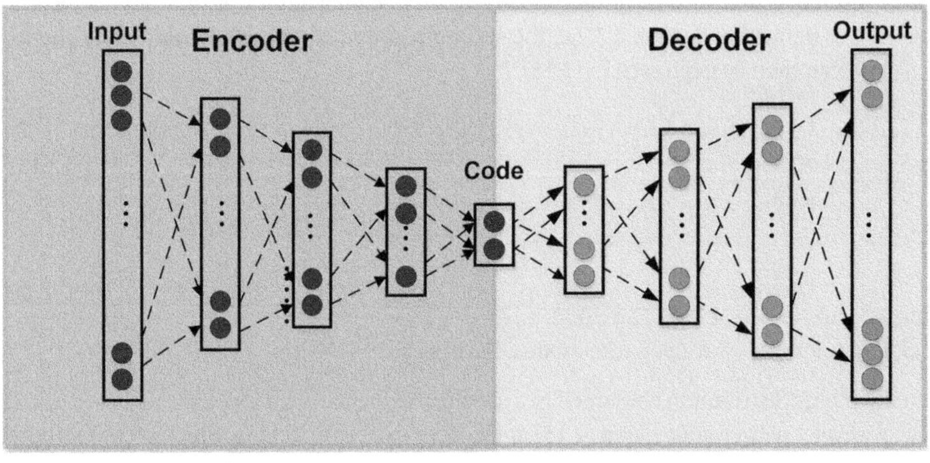

Figure 2.22 Architecture of the AEN.

All weights and biases are usually initialized randomly, and then the learning procedure starts to iteratively update weights through back-propagation algorithm, which minimizes the reconstruction errors between x and $\hat{x}$. This training technique can force each layer to learn a higher-level representation of its previous layer.

Instead of adopting the entire AEN, the compressed code h is widely used as condensed SFs to represent the original input set. If there are cp components in the code layer, then the SF set $\mathbf{SF}_{AEN}$ can be defined as $\{h_1, h_2, ...h_{cp}\}$. This feature extraction method is very similar to adopting the other well-known dimensionality reduction technique: principle component analysis (PCA).

2.4 Case Studies

Four practical examples using real-world data are respectively demonstrated to validate techniques of data acquisition and data preprocessing addressed in the previous sections. Details are described as below.

2.4.1 Detrending of the Thermal Effect in Strain Gauge Data

To detect force and torque during machining, a smart tool holder is developed and used in an CNC milling machine. When several corresponding gauges are attached to the holder, a strain gauge can be used to detect variation in the bending and torsion of the tool holder based on the proportional ratio of the resistance to the length of the stain gauge. As illustrated in Figure 2.23, the values of strain on the tool holder are sensed and detected by Wheatstone bridges, digitized using an ADC, processed via a microprocessor, and transmitted to an edge computer by using the message queuing telemetry transport (MQTT) protocol through a Wi-Fi module.

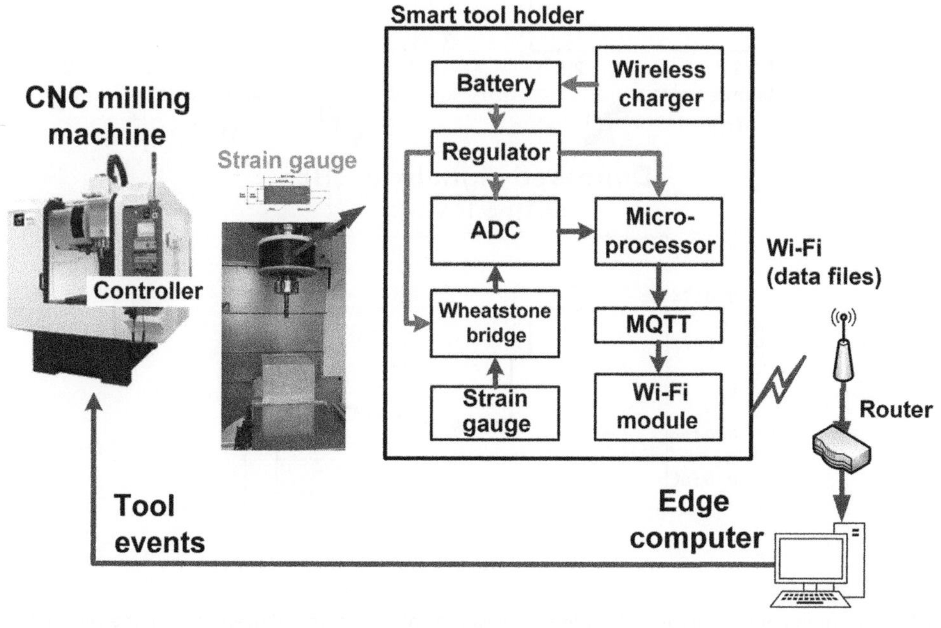

Figure 2.23 Using a smart tool holder to detect tool state.

The edge computer located near the CNC machine receives and processes strain values and issues tool events to the controller when tool breakage is detected or tool's RUL is short. A tool holder is stiff enough to enable clamping of a tool under various machining conditions and lead to tiny machining variation in the length and resistance of a strain gauge. Although a high-gauge-factor sensor is employed, a length difference (<1 μm) in a tool holder can be detected during machining. However, the strain gauge appears to have considerable thermal variations even in a stationary state. Thus, one challenge is how the thermal effect in strain-gauge data can be removed to derive effective strain values; the details are described in [17] via IEEE DataPort.

For example, Figure 2.24a depicts raw signals collected during machining for 2.8 s (sampling rate of 10 kHz). Because of the high heat capacity of a real machine, the thermal variation can be assumed to be constant within a short period such as 5 s. After applying the wavelet de-noising method with five levels (denoted DB5), the thermal trend can be derived as illustrated in Figure 2.24b. Then, the de-trended data can be obtained by subtracting the thermal trend from the raw data; the result is depicted in Figure 2.24c.

The thermal trend in the raw signal can be removed and normalized to the same criterion by subtracting the mean value. As shown in Figure 2.25, the darker and lighter lines represent the

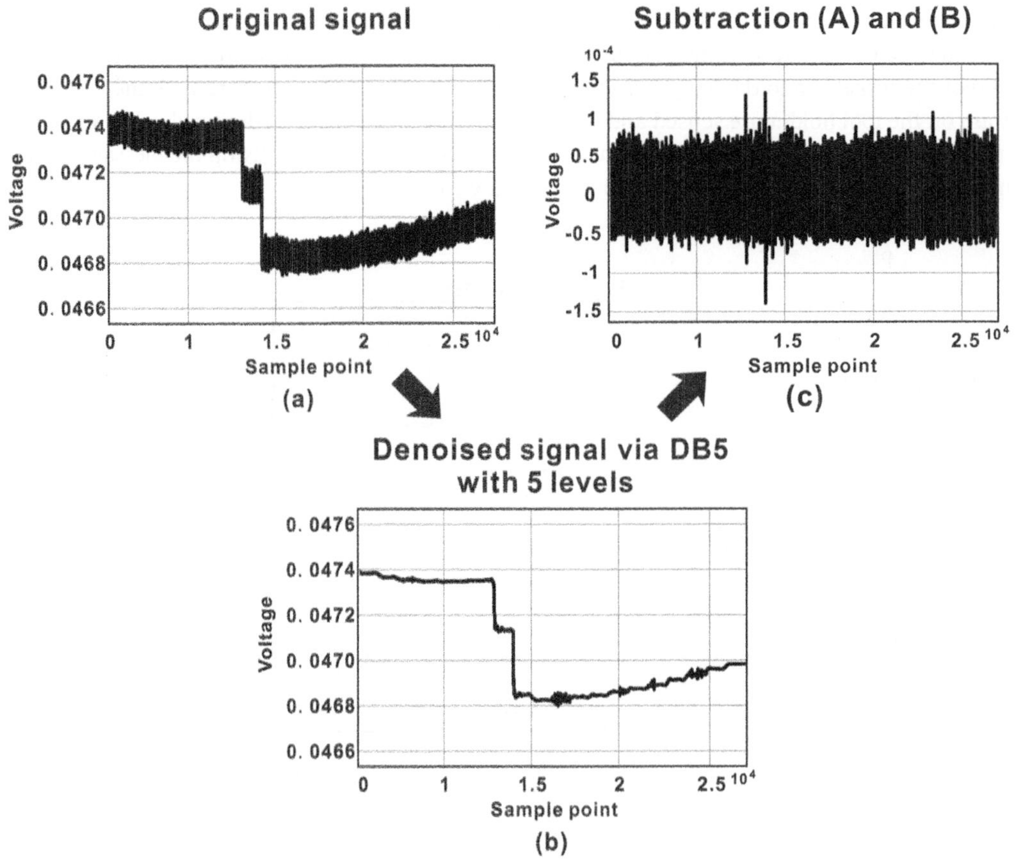

Figure 2.24 Detrending of the thermal effect in strain-gauge data: (a) before detrending; (b) the thermal trend; (c) after detrending.

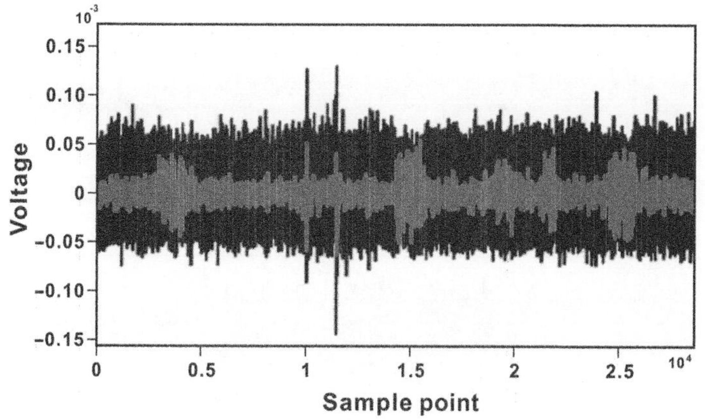

Figure 2.25 De-noising signals to highlight differences between dry-run and tool-use periods.

signals before and after de-noising using the wavelet mother function DB3, respectively. The signals collected during machining imply that the signals contain the dry-run and tool-use periods.

By comparing with the values measured during the dry-run period (range $\pm$ 0.035 mV), the values measured during tool-use can be filtered from the background noise and derived in the range of $\pm$0.083 mV. After de-trending and de-noising the raw data, the ratio between the dry-run and tool-use signals can be improved from 1 to 2.37, which makes the extraction of effective features for modeling considerably easier.

2.4.2 Automated Segmentation of Signal Data

To detect tool wear during machining, a vibration-based evaluation method is developed and used on a CNC milling machine. During any machining operation, a tool holder chunked by the spindle holds the cutting tool in place as precisely and firmly as possible. The spindle stability affects the quality of tool holder and cutting tool. Thus, a high-resolution accelerometer attached close to the spindle is adopted to monitor the cutting-tool wear based on the changes of spindle vibration.

However, when M codes are not available in some cases, for example, the required number of DIO is too large to achieve some complicated operations, so that the raw data cannot be timely divided into several critical parts of final machining process by M Codes during the machining time. When dealing with the segmentation issue under the condition of an insufficient number of M codes, the feasible solution is to decrease the usage of the pairs of M codes by extending the duration of each data acquisition. In this way, however, not only the specified final machining parts that affect product quality most are included but various types of machining operations would be involved during each data acquisition.

In this manner, the acquired vibration data may be a long signal that contains the process during the idling (dry-run) and the real machining (tool-workpiece contact) periods. Thus, the challenge is how to automatically segment the collected data so as to estimate the tool-wear status.

Three-axis vibration data sampled at 2,048 Hz during the drilling operation of seven holes on a medal plate with material FDAC are respectively illustrated in Figure 2.26. To increase tool availability, a 0.5 mm hole is pecked by a two-flute tool in each cycle until a total depth of 4 mm is achieved. The feed rate and spindle speed are 100 mm/min and 4,500 rpm, respectively. Obviously, the main loading happens in the Z-axis vibration, and the bottom of Figure 2.26 shows real machining periods

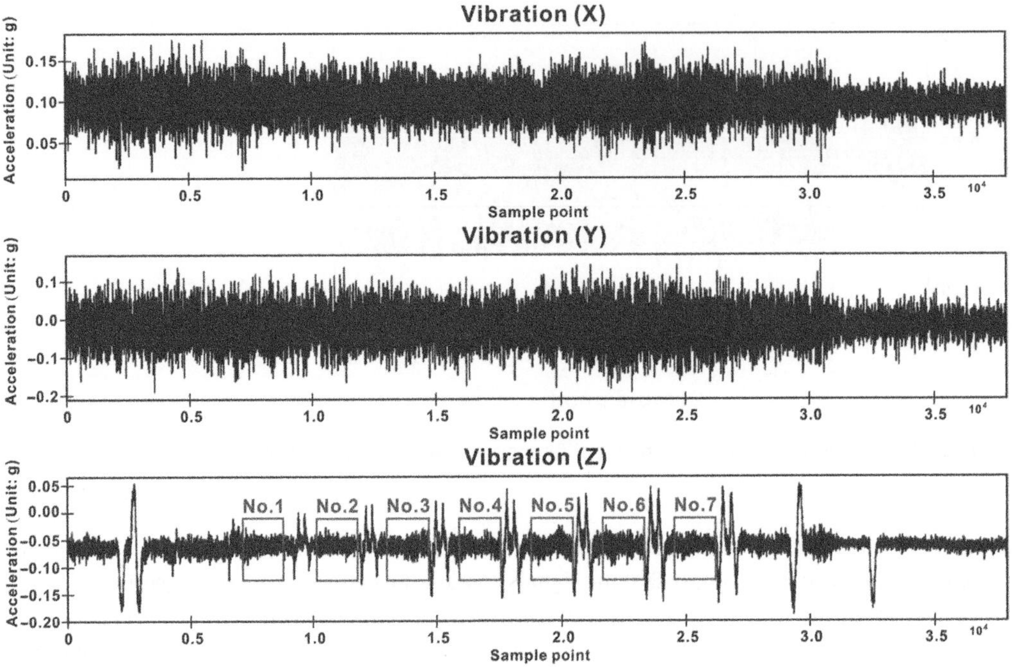

Figure 2.26 Collected vibration signals (including idling and machining periods).

of seven parts, which are manually segmented and numbered from 1 to 7. The details of this application case can also be found in [18] via IEEE DataPort.

To automatically segment the aforementioned Z-axis data for identifying the actual drilling periods, an AEN model integrated with an encoder, code, and decoder is used to learn the idling characteristics under specific conditions. In this segmentation case, the encoder (four-layer structure with 32, 16, 8, and 4 nodes, respectively) compresses the inputs into code in the middle layer, and the decoder (the inverse structure with 4, 8, 16, and 32 nodes) decompresses code into the outputs. Only one node in the code layer is used to evaluate the distance between the modeling and testing features.

First, 32 features are derived from vibration data collected during the idling periods by using the five-level WPT. Figure 2.27 compares the two trends from 32 WPT features extracted from Z-axis idling vibration (original data) and AEN training results using the same 32 WPT features (decoder data), under four different spindle speeds (3,000, 3,500, 4,500, and 5,000 rpm), where the node number starts counting from 0. Note that, the high similarity of both trends indicates that AEN is reliable.

The reason why Z-axis vibration is chosen to be a learning criterion for training the AEN model is that the main drilling loading occurs in Z-axis but not in X-axis or Y-axis. The AEN accuracies would be worse if X-axis or Y-axis vibration is adopted to train AEN since loading difference between the idling section and the real drilling section is not significant enough.

Thus, once AEN learns the feature patterns of idling sections, it is able to achieve the period detection of tool–workpiece contact by comparing the certain distance between the idling section and the real machining section. Figure 2.28 illustrates the automated segmentation results of AEN. Figure 2.28a compares the distances between modeling features (Z-axis vibration during idling) of the AEN model and testing features of real machining (X-axis vibration during drilling) every 0.2 s from the beginning to the end of the data acquisition.

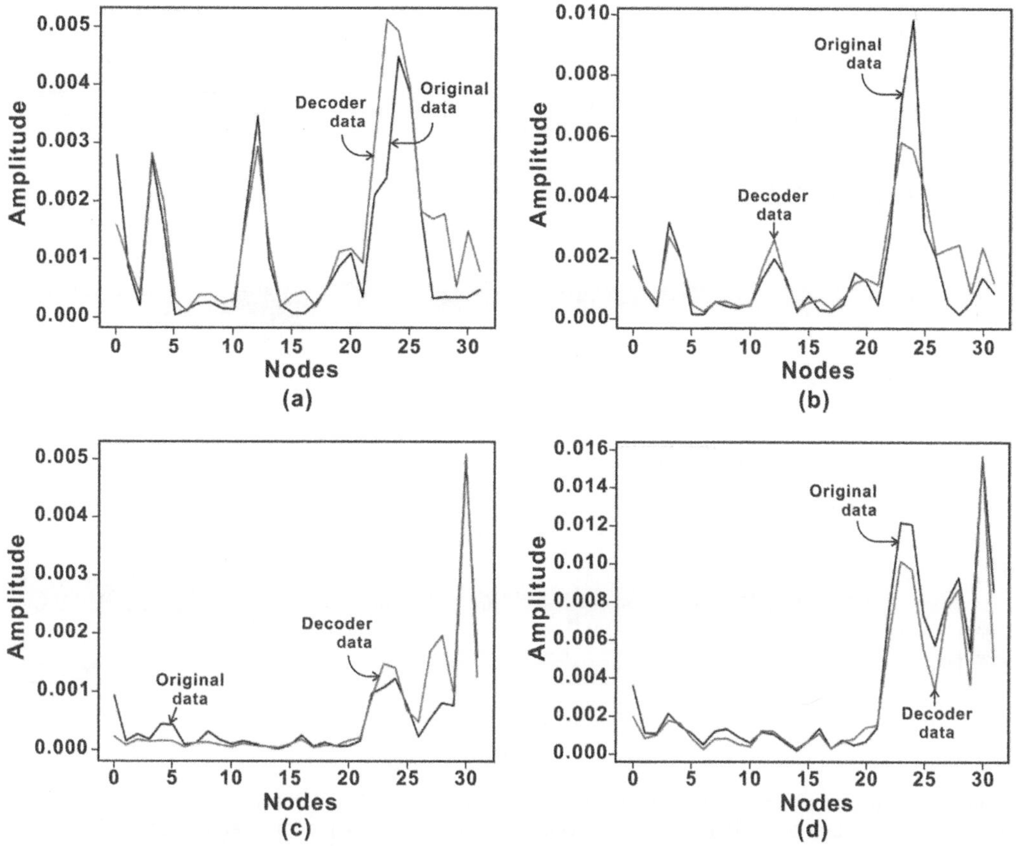

Figure 2.27 Comparison of the original and decoded features under four idling conditions of spindle speeds: (a) 3,000 rpm; (b) 3,500 rpm; (c) 4,500 rpm; and (d) 5,000 rpm.

As depicted in Figure 2.28a, a stable maximum distance, which means a high dissimilarity between the modeling and testing features, can be used to recognize a real drilling section. A certain duration of X-axis vibration data can be segmented into the real drilling section according to Z-axis vibration, as highlighted within the two red dotted lines in Figure 2.28a. In this manner, seven real drilling sections of Z-axis vibration as in the bottom of Figure 2.26 can also be automatically segmented using AEN. To sum up, raw data can be segmented using the vibration characteristics of various rotor components such as motors and spindles to automatically reduce time-consuming manual segmentation.

2.4.3 Tool State Diagnosis

Generally speaking, tool wear is concomitant to vibration and it gradually increases due to long-term usage. In this case, the vibration is acquired from the cutting tool used in the side milling at a sampling rate of 2,000 Hz and the data of the cutting tool records from new to worn status. The entire data set is available in [19] via IEEE DataPort.

Features are extracted from time-domain signals into the number of 2^4 WPT nodes based on the 4-level WPT manner. Although differences in time-domain signals between new and worn statuses are small as shown in the upper portion of Figure 2.29, the values of the 13, 14, and 15 WPTs of

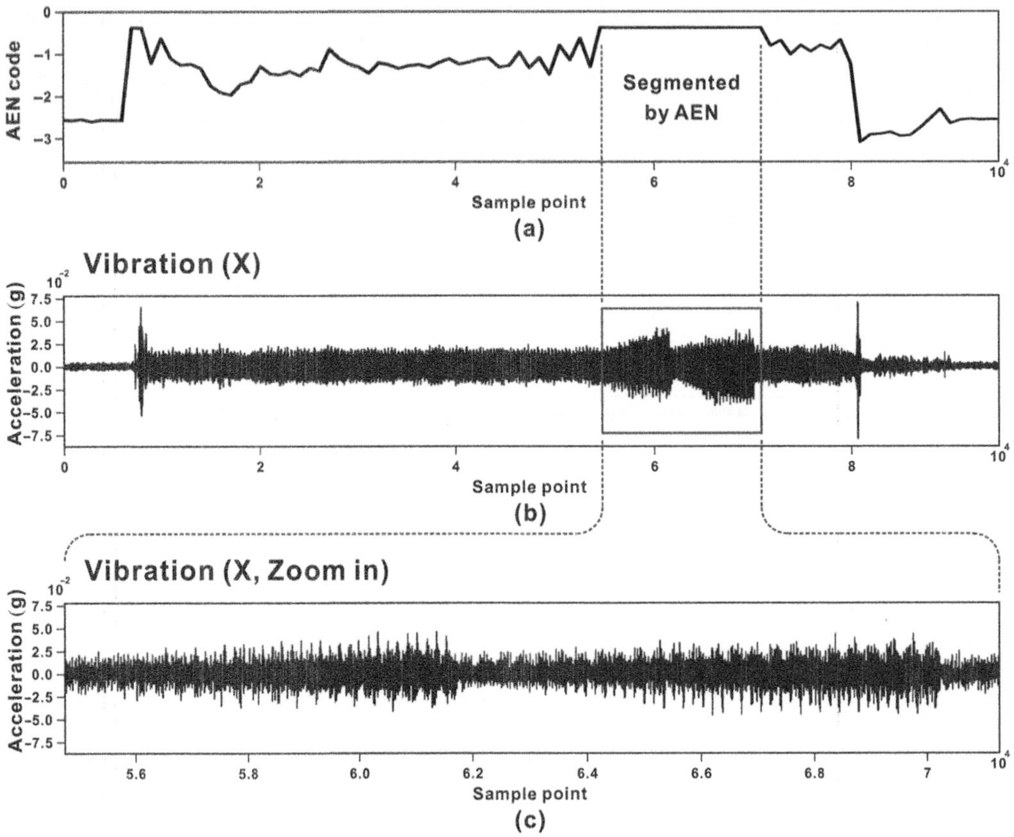

Figure 2.28 Automated segmentation of machining signals using an AEN: (a) distance derived by AEN based on idling vibration features of Z-axis; (b) collected X-axis vibration signal; and (c) zoom in segmented X-axis signal.

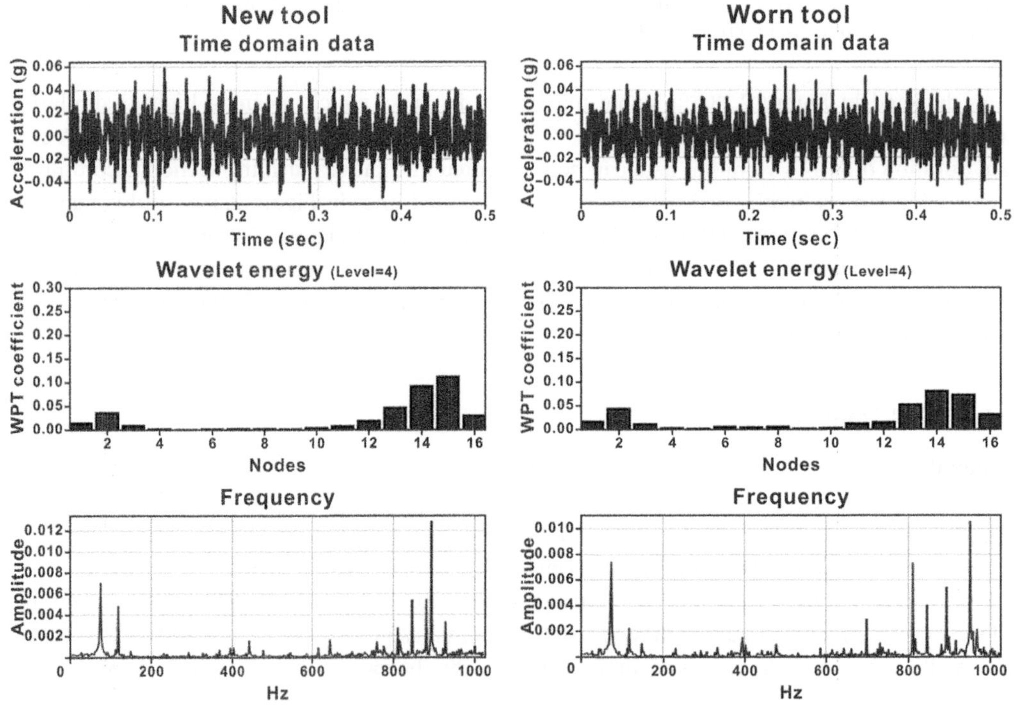

Figure 2.29 Comparison of time-domain signals (upper portion), WPT features (middle portion), and frequency-domain spectrums (lower portion) between new and worn statuses.

worn tool signals are clearly different from those of a new tool as portrayed in the middle portion of Figure 2.29. Here, the bandwidths of the 13, 14, and 15 WPTs are about 750–812.5, 812.5–875, and 875–937.5 Hz, respectively. The detailed amplitudes of each frequency band are illustrated in the lower portion of Figure 2.29.

Figure 2.30 illustrates four energy distributions of the 32 WPT features extracted from the X-axis and Y-axis vibrations under four cutting depths (from 4 to 7 mm). Note that, main differences of amplitudes exist among high-frequency bands (especially from 13 to 15) between new and worn statuses. These features provide the AEN model with useful data source to extract information for the tool state diagnosis.

Hence, 32 WPT-based features of the X-axis and Y-axis serve as the inputs to the encoder in an AEN model. As shown in Figure 2.31, four SFs (fAE1, fAE2, fAE3, and fAE4) extracted from the fourth layer in encoder can be used as a compressed representation of the original feature set to reduce the number of feature dimensions; the left side (sample nos. 1–133) and right side (sample nos. 134–266) represent the new and worn cutting tools, respectively. Finally, the four SFs show their capability in classifying the new or the worn tool.

The accuracy of four feature sets are compared using a random forest (RF) model and evaluated in a cross-validation scenario. As shown in Table 2.4, the average accuracies of tool state diagnosis when using 32 WPT-based features and 4 fAE features are 89.5 and 69.1%, respectively. Furthermore, when the cutting depth is added as one of the inputs, the average accuracies are improved to 90.9 and

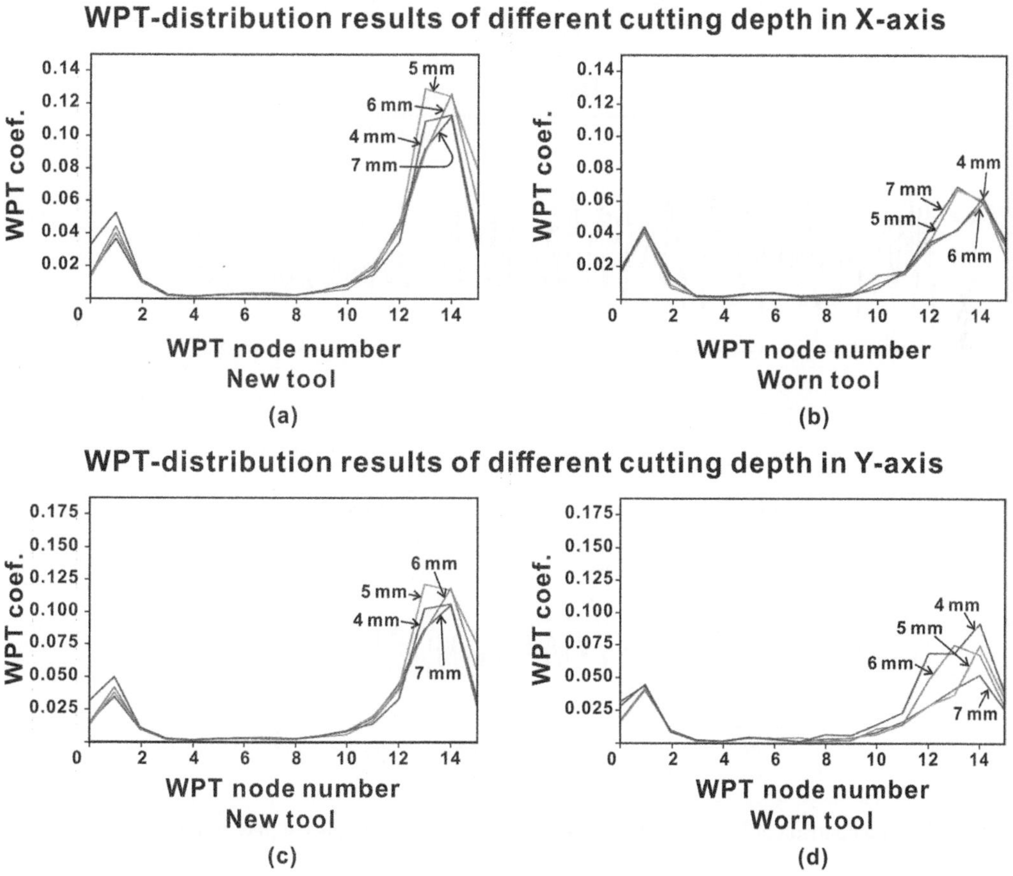

Figure 2.30 WPT distribution results for different cutting depths in the X and Y axis (node number counted from 0): (a) the new tool in X-axis; (b) the worn tool in X-axis; (c) the new tool in Y-axis (d) the worn tool in Y-axis.

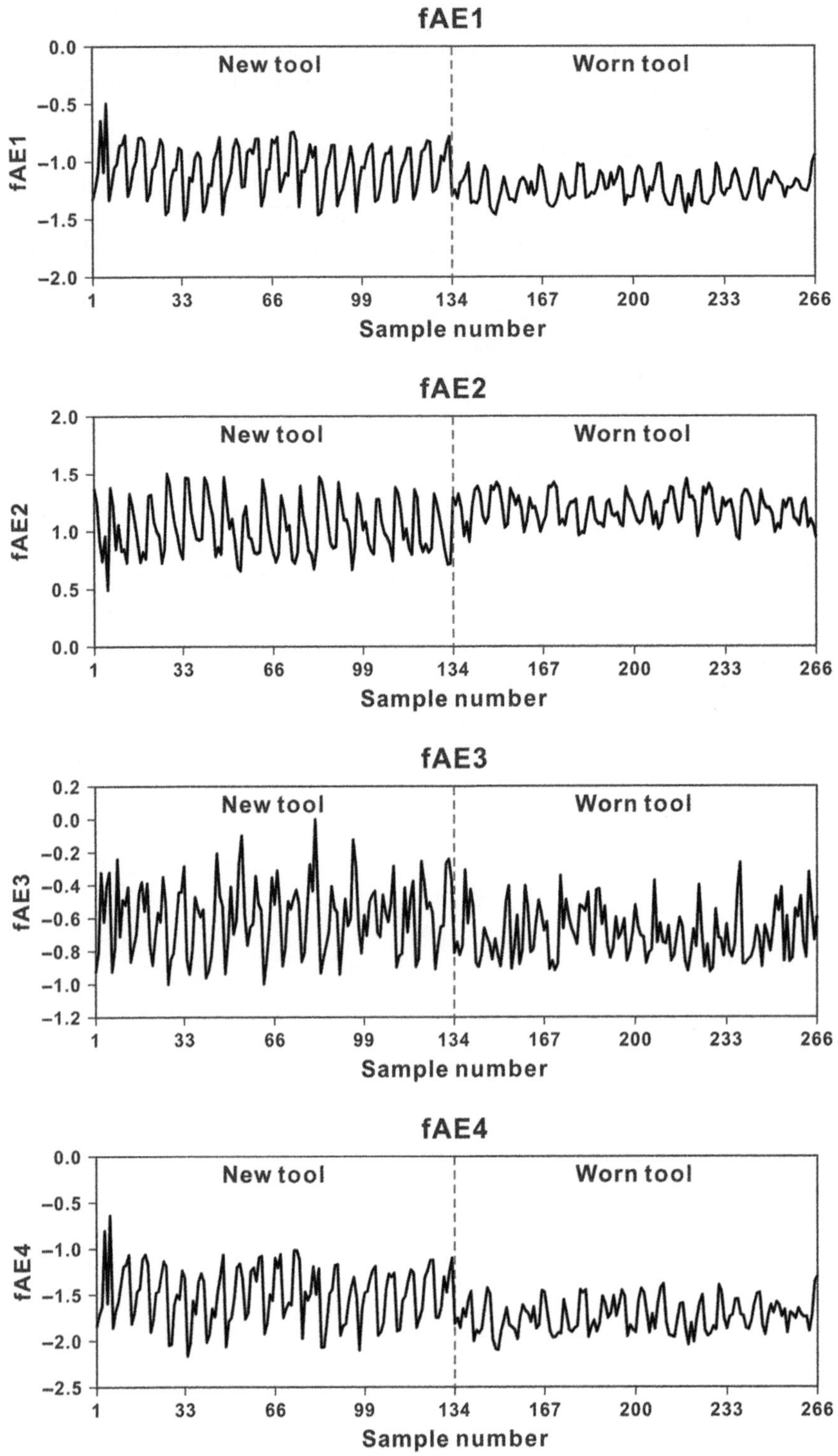

Figure 2.31 Comparison of four SFs extracted by using an AEN for samples of new and worn tools.

Table 2.4 Tool diagnosis example: results of using an RF model.

Model inputs	Average (%)	Best (%)	Worst (%)
32 WPT SFs (X/Y axis with level = 4)	89.5	95.0	81.2
32 WPT SFs (X/Y axis with level = 4) + Cutting depth	90.9	96.2	85.0
fAE1~fAE4	69.1	77.5	60.0
fAE1~fAE4 + Cutting depth	81.7	90.0	76.2

81.7%, respectively. This indicates that the accuracy by applying WPT-based SFs is better than that by utilizing AE-based SFs regardless of whether the cutting depth is added to the feature set or not.

2.4.4 Tool Diagnosis using Loading Data

Retaining experienced machine operators is difficult because of poor manufacturing environment. Some forming machines are now equipped with pressure sensors to indicate operators with machine status for compensating their inexperience. To detect failures in a forming process, a pressure sensor (load cell) is attached to a forging die to detect variation in the forging signals of a bolt-forging machine. Forging failures and loads are generally strongly correlated, but the load distribution may vary with numerous failure modes.

Further, an issue regarding big data exists in identifying failures after long-term data collection. The cycle time for forming a bolt is only 0.3 s; thus, the signal length is approximately 300 points under the 1 kHz sampling rate of the sensor. The amount of data collected daily is almost 10 MB for one forming machine with data collection performed for 22 hours/day when using eight pressure sensors of four stages (i.e. 8 channels × 1000 data samples/second × 3600 seconds/hour × 22 hours/day). Thus, how to automatically diagnose failure modes from loading data in a forging process becomes a challenge.

Observing Figure 2.32, a forging load (pressure)-stroke curve demonstrates two intervals: T1 and T2, which can be defined to indicate the characteristics of fastener forming. T1 is an interval representing the time from the die contacting the workpiece to forming after the material has exceeded its yield strength and shown plastic deformation; whereas T2 is the time taken for the cavity to be completely filled after T1.

The forging energy during T2 is mainly related to the geometric variation of the die, as illustrated in Figure 2.32. A feature engineering method is used to extract features, including those in time and frequency domains, from the load-stroke signal in interval T2. For example, *avg*, *std*, *kurt*, *RMS*, *skew*, and *max* are extracted from the time domain and six frequency bands (5 Hz in each bandwidth) are extracted from the frequency domain as defined in Sections 2.3.3.1 and 2.3.3.2.

Additionally, an AEN model is used for reducing the number of forging features; the inputs and outputs are the original and encoded forging load signals, respectively. In this case, the AEN monitors the stability of the bolt-forming processes and identifies invalid samples, which are mainly affected by the forging pressure.

For example, three failure modes of the bolt-forming processes and their end products are shown in the upper part of Figure 2.33; these failures usually result from three failure modes including length over-specification, die notching, and die adhesion. The pressure patterns of the valid process and three failure modes are depicted in the solid and segmented curves in the lower part of Figure 2.33, respectively. Although longer (+0.3 mm) material does not strongly affect the forming

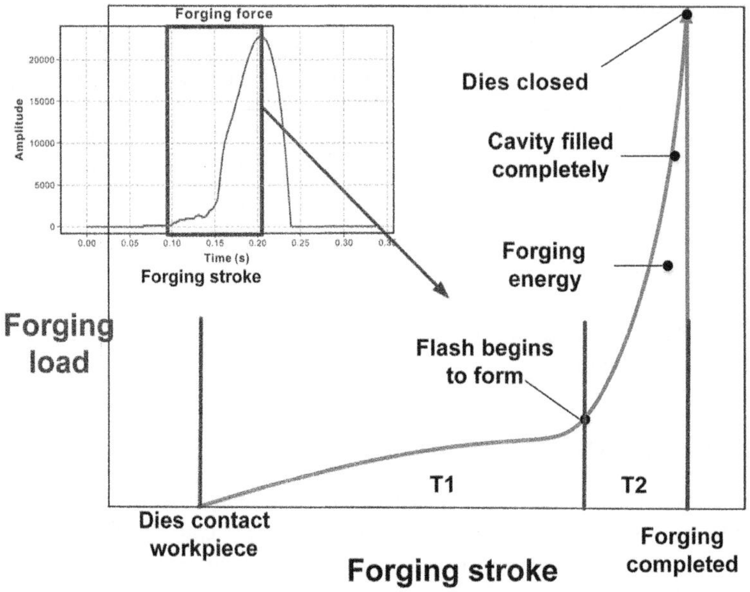

Figure 2.32 A forging load (pressure)-stroke curve.

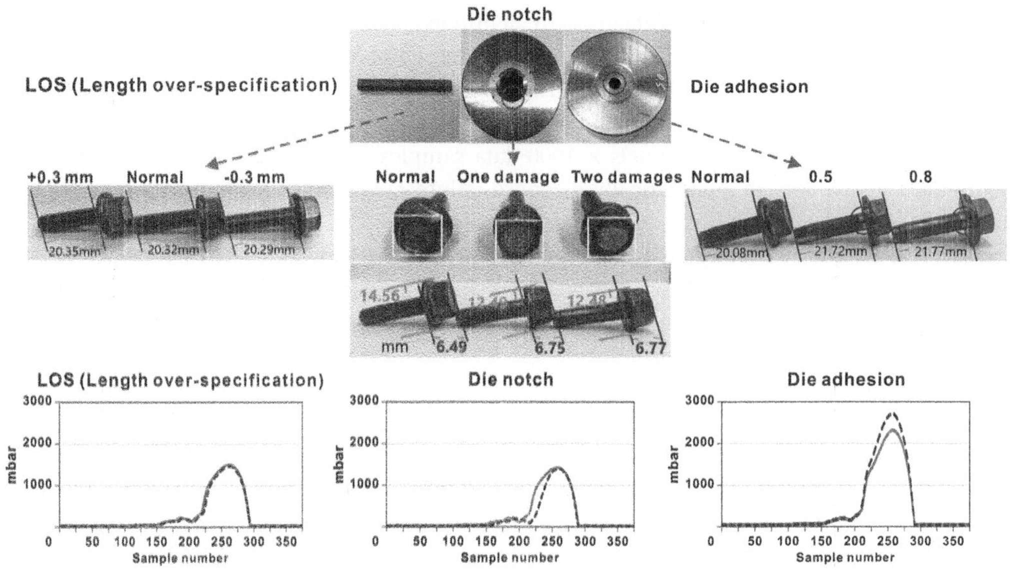

Figure 2.33 Failure diagnosis in a forming process.

process, both die notching and adhesion result in extremely different pressure patterns compared with the original patterns.

The single dimension feature in the code of AEN can be used to determine whether invalid bolts have been formed. Because the actual curve of the forging stroke varies, validating the received signal is difficult. Fortunately, the pressure pattern for invalid bolts can be observed using reconstruction of the compressed code by the decoder of AEN with correct results. As shown in the upper part of Figure 2.34, the value for a valid sample is approximately 2.1, whereas that for an invalid sample is approximately −10. The raw data of valid and invalid samples, shown as the

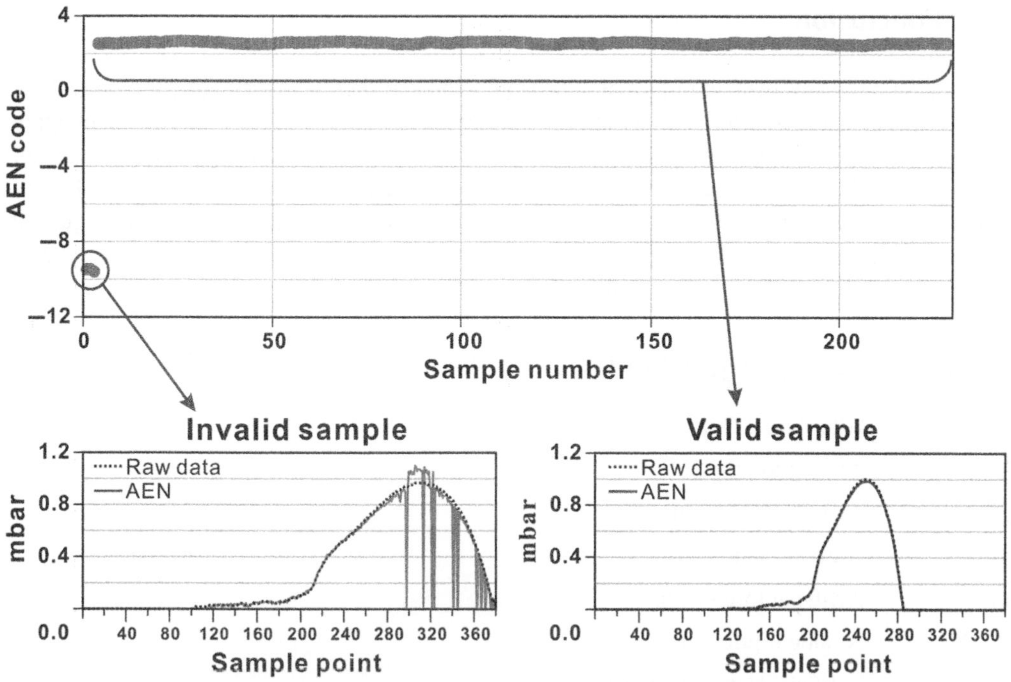

Figure 2.34 Sample validation using the single dimension feature of the middle layer in the AEN model.

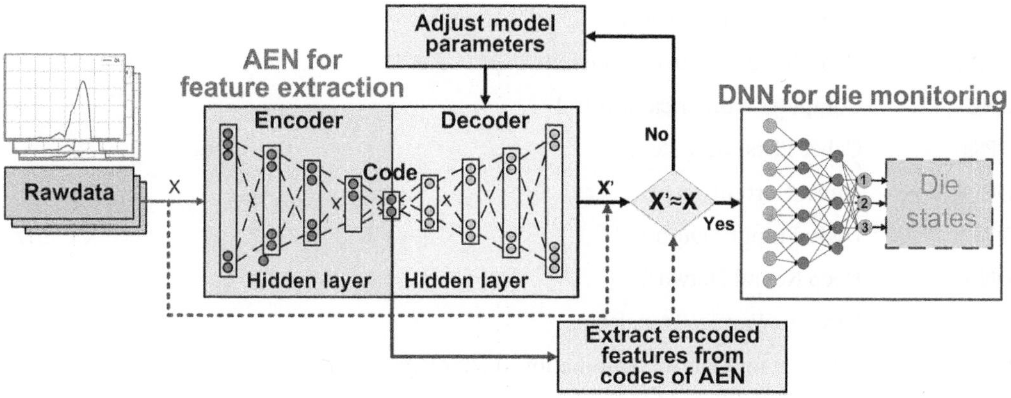

Figure 2.35 AEN-DNN architecture for failure diagnosis.

dotted curves in the lower part of Figure 2.34, are so similar that the difference could not be identified using a rule or a threshold system.

As illustrated in Figure 2.35, an AEN–deep neural network (DNN) is employed to diagnose failures in the forming process. The features extracted by the codes of AEN serve as the DNN model inputs when the reconstructed output X' of AEN is similar enough to the original input X, which validates that the extracted features of codes are reliable. Consequently, the AEN-DNN model can not only accurately distinguish valid samples (positive detection rate > 99%) but also correctly diagnose various failure modes (accuracy > 95%). The details of this application case are described in [20] via IEEE DataPort.

2.5 Conclusion

This chapter addresses the techniques of data acquisition and preprocessing. For data acquisition, both process data and metrology data have to be collected for developing various intelligent applications. In general, process data consists of sensing signals and manufacturing parameters. As for data preprocessing, the key steps are segmentation, cleaning, and feature extraction. Finally, four practical examples using real-world data are respectively demonstrated to validate techniques of data acquisition and data preprocessing addressed in this chapter; the detailed experimental data are uploaded in the IEEE DataPort for references.

Appendix 2.A – Abbreviation List

AC	Alternating Current
ADC	Analog-to-Digital Converter
AE	Acoustic Emission
AEN	Autoencoder
AI	Artificial Intelligence
AIO	Analog Input / Output
ANN	Artificial Neural network
AOI	Automated Optical Inspection
CbM	Condition-based Maintenance
CCD	Charge-Coupled Device
CMM	Coordinate Measuring Machine
CNC	Computer Numerical Control
CPSs	Cyber-Physical Systems
DC	Direct Current
DIO	Digital Input / Output
DNN	Deep Neural Network
DWT	Discrete Wavelet Transform
EtherCAT	Ethernet for Control Automation Technology
FFT	Fast Fourier Transform
FT	Fourier Transform
Industry 4.0	Fourth Industrial Revolution
IIoT	Industrial Internet of Things
MRA	Multi-Resolution Analysis
MQTT	Message Queuing Telemetry Transport
NC	Numerical Control
OPC-UA	Open Platform Communication Unified Architecture
PC	Personal Computer

PLC	Programmable Logic Controller
PM	Preventive Maintenance
RM	Reactive Maintenance
RF	Random Forest
rpm	rotation per minute
RUL	Remaining Useful Life
SFs	Signal Features
S/N	Signal-to-Noise
STFT	Short-Time Fourier Transform
TCP/IP	Transmission Control Protocol/Internet Protocol
TFT-LCD	Thin-Film-Transistor Liquid-Crystal Display
WPT	Wavelet Packet Transform

Appendix 2.B – List of Symbols in Equations

F_s	sampling rate for acquiring a signal
F_B	highest frequency component contained in a signal
M	number of precision items
N	data length of discrete signals
$s[i]$	ith data point of a discrete signal
p	pth precision item
$\mathbf{S}_p^r$	raw sensor data of pth precision item
$\mathbf{S}_p^c$	cleaned sensor data of pth precision item
$g[k]$	DWT low-pass filters
$h[k]$	DWT high-pass filters
L	number of decomposition levels;
j	scale parameters
n	translation parameter
$c_j[n]$	nth discrete wavelet coefficient of $\mathbf{S}_p^r$ at level j
$c_j^a[n]$	nth approximation coefficient at level j
$c_j^d[n]$	nth detail coefficient at level j
$\hat{c}_j[n]$	nth modified wavelet coefficients at level j
$\hat{c}_j^a[n]$	nth modified approximation coefficient at level j
$\hat{c}_j^d[n]$	nth modified detail coefficient at level j
λ	threshold of $c_j[n]$
$\mathrm{MAD}(c_{L-1}[n])$	mean absolute deviation of $c_{L-1}[n]$
$\hat{c}_{j-1}^a[n]$	modified approximation coefficients

avg	arithmetic mean or average of $\mathbf{S}_p^c$
std	standard derivation of $\mathbf{S}_p^c$
max	maximal magnitude of $\mathbf{S}_p^c$
min	minimal magnitude of $\mathbf{S}_p^c$
ptp	peak-to-peak amplitude of $\mathbf{S}_p^c$
$kurt$	kurtosis for measuring the peakedness of $\mathbf{S}_p^c$
$skew$	skewness for measuring the asymmetry of $\mathbf{S}_p^c$
RMS	root mean square value of $\mathbf{S}_p^c$
CF	crest factor of $\mathbf{S}_p^c$
T	duration of a continuous-time signal
t	specific time t of a continuous-time signals
cr_{xy}	correlation coefficient value of two signals: $\mathbf{x}(t)$ and $\mathbf{y}(t)$
CR_{xy}	cross-correlation value of two signals: $\mathbf{x}(t)$ and $\mathbf{y}(t)$
γ_{xy}	covariance value of two signals: $\mathbf{x}(t)$ and $\mathbf{y}(t)$
m	displacement unit of a signal
*	complex conjugate of a signal
$\mathbf{SF}_{\mathrm{CR(xy)}}$	a feature set of CR_{xy}
$\mathbf{FFT}[n]$	frequency amplitude at the nth Hz by FFT
$\mathbf{SF}_{\mathrm{FFT}(q)}$	qth certain frequency band delimited by a lower frequency and an upper frequency
Q	number of frequency band
uf_q	qth upper frequency of the critical characteristics; and
lf_q	qth lower frequency of the critical characteristics.
$\mathbf{SF}_{\mathrm{FFT}(q)}$	qth FFT-based SF
u	uth wavelet packet at level L, u= 1, 2, ..., L;
v	subband length for each wavelet packet at level L, $v = N/2^L$
$\mathbf{SF}_{\mathrm{WPT}(u)}$	uth WPT-based SF
h	compressed code of the middle layer in AEN;
$\hat{x}$	output reconstructed from c in the middle layer of AEN;
f_{EN}	encoder layer of AEN;
f_{DE}	decoder layer of AEN;
f_{a}	activation function of AEN;
W_{EN}	network weight for node in the encoder;
W_{DE}	network weight for node decoder;
b_{EN}	bias for node in the encoder layer;
b_{DE}	bias for node in the decoder layer
$\mathbf{SF}_{\mathrm{AEN}}$	SFs extracted from AEN

References

1 Chen, B., Chen, X., Li, B. et al. (2011). Reliability estimation for cutting tools based on logistic regression model using vibration signals. *Mechanical System and Signal Process* 25 (7): 2526–2537. https://doi.org/10.1016/j.ymssp.2011.03.001.

2 Suprock, C.A. and Nichols, J.S. (2009). A low cost wireless high bandwidth transmitter for sensor-integrated metal cutting tools and process monitoring. *International Journal of Mechatronics and Manufacturing Systems* 2 (4): 441–454. https://doi.org/10.1504/IJMMS.2009.027128.

3 Ghosh, N., Ravi, Y.B., Patra, A. et al. (2007). Estimation of tool wear during CNC milling using neural network-based sensor fusion. *Mechanical Systems and Signal Processing* 21 (1): 466–479. https://doi.org/10.1016/j.ymssp.2005.10.010.

4 Abuthakeer, S.S., Mohanram, P.V., and Kumar, G.M. (2011). Prediction and control of cutting tool vibration CNC lathe with ANOVA and ANN. *International Journal of Lean Thinking* 2 (1): 1–23.

5 Tieng, H., Li, Y.Y., Tseng, K.P. et al. (2020). An Automated Dynamic-Balancing-Inspection Scheme for Wheel Machining. *IEEE Robotics and Automation Letters* 5 (4): 2224–2231. https://doi.org/10.1109/LRA.2020.2970953.

6 Abellan-Nebot, J.V. and Subirón, F.R. (2010). A review of machining monitoring systems based on artificial intelligence process models. *The International Journal of Advanced Manufacturing Technology* 47 (14): 237–257. https://doi.org/10.1007/s00170-009-2191-8.

7 Teti, R., Jemielniak, K., O'Donnell, G. et al. (2010). Advanced monitoring of machining operations. *CIRP Annals* 59 (2): 717–739. https://doi.org/10.1016/j.cirp.2010.05.010.

8 Yang, H.C., Tieng, H., and Cheng, F.T. (2015). Total precision inspection of machine tools with virtual metrology. *Journal of the Chinese Institute of Engineers* 39 (2): 221–235. https://doi.org/10.1109/CoASE.2015.7294301.

9 Mallat, S.G. (1989). A theory for multiresolution signal decomposition: the wavelet representation. *IEEE Transactions on Pattern Analysis and Machine Intelligence* 11 (7): 674–693. https://doi.org/10.1109/34.192463.

10 Zhang, Z., Wang, Y., and Wang, K. (2013). Fault diagnosis and prognosis using wavelet packet decomposition, Fourier transform and artificial neural network. *Journal of Intelligent Manufacturing* 24: 1213–1227. https://doi.org/10.1007/s10845-012-0657-2.

11 Kim, J., Lee, H., Jeon, J.W. et al. (2020). Stacked auto-encoder based CNC tool diagnosis using discrete wavelet transform feature extraction. *Processes* 8 (4): 456. https://doi.org/10.3390/pr8040456.

12 Jiang, G., He, H., Xie, P. et al. (2017). Stacked multilevel-denoising autoencoders: a new representation learning approach for wind turbine gearbox fault diagnosis. *IEEE Transactions on Instrumentation and Measurement* 66 (9): 2391–2402. https://doi.org/10.1109/TIM.2017.2698738.

13 Lee, R.J. and Nicewander, W.A. (1988). Thirteen ways to look at the correlation coefficient. *The American Statistician* 42 (1): 59–66. https://doi.org/10.1080/00031305.1988.10475524.

14 Strang, G. (1994). Wavelets. *American Scientist* 82 (3): 250–255.

15 Belinsky, A.V.E. and Lapshin, V.B. (2017). The uncertainty principle and measurement accuracy. *Physics-Uspekhi* 60 (3): 325–326. https://doi.org/10.3367/UFNe.2017.02.038069.

16 Yen, G.G. and Lin, K.C. (2000). Wavelet packet feature extraction for vibration monitoring. *IEEE Transactions on Industrial Electronics* 47 (3): 650–667. https://ieeexplore.ieee.org/document/847906.

17 Yang, H.C. (2020). Strain data collected from machine tool. https://ieee-dataport.org/documents/strain-data-collected-machine-tool (accessed 26 October 2020).

18 Yang, H.C. (2020). Signal segmentation for milling process. https://ieee-dataport.org/open-access/signal-segmentation-milling-process (accessed 6 November 2020).

19 Yang, H.C. (2020). Roughness of milling process. https://ieee-dataport.org/open-access/roughness-milling-process (accessed 6 November 2020).

20 Yang, H.C. (2020). Cold forging process failures. https://ieee-dataport.org/open-access/cold-forging-process-failures (accessed 5 November 2020).

3

Communication Standards

Fan-Tien Cheng[1], Hao Tieng[2], and Yu-Chen Chiu[3]

[1] *Director/Chair Professor, Intelligent Manufacturing Research Center/Institute of Manufacturing Information and Systems, National Cheng Kung University, Tainan, Taiwan, ROC*
[2] *Associate Research Fellow, Intelligent Manufacturing Research Center, National Cheng Kung University, Tainan, Taiwan, ROC*
[3] *Specialist,Intelligent Manufacturing Research Center, National Cheng Kung University, Tainan, Taiwan, ROC*

3.1 Introduction

Communication standards support the entire data-exchange requirements of Intelligent Manufacturing. Since e-Manufacturing of the semiconductor industry is the predecessor of Industry 4.0, the communication standards are organized into two categories: semiconductor equipment as well as industrial devices and systems.

3.2 Communication Standards of the Semiconductor Equipment

The communication standards for the semiconductor industry were emphasized prior to other industries due to its industrial characteristics. As shown in Figures 1.1 and 1.2, the e-Manufacturing model is divided into the manufacturing portion and the engineering portion. The standards for the manufacturing portion including SECS-I (Semiconductor Equipment Communication Standard I), SECS-II, GEM (Generic Equipment Model), and HSMS (High Speed SECS Message Services) as well as the standards for the engineering portion, Interface A, will be introduced below, respectively.

3.2.1 Manufacturing Portion

Communication standards for the semiconductor equipment have been evolved in the past decades, SECS-I, SECS-II, GEM, and HSMS are derived progressively to adapt to the advancement of the semiconductor industry.

SECS-I (defined in SEMI E4) [1], officially published in 1980, is a data transfer protocol in charge of ensuring the data being transmitted correctly from Host to Equipment or vice versa via RS-232C protocol (one start bit, one stop bit, full-duplex line, half-duplex communication). The Host can be any type of computer on the other side of the Equipment, such as a personal computer, workstation,

Industry 4.1: Intelligent Manufacturing with Zero Defects, First Edition. Edited by Fan-Tien Cheng.

mainframe, robot, or even another piece of Equipment. SECS-I is a low technology with low cost and its slowness has been a limiting factor for implementing SECS in the factories.

SECS-II (SEMI E5) [2], a message protocol, was published in 1982. SECS-II defines the messages and related data items exchanged between Host and Equipment. It serves as the core specification of defining specific messages and message formats for manufacturing information exchange.

In 1992, GEM (SEMI E30) [3] was published. GEM defines Equipment behavior that indicates which SECS-II messages should be used, in what situations, and what the results would be. GEM is categorized as both a specification and a design guide, which sets minimum capabilities for the SECS communication on the Equipment and also determines that this communication is closely tied to Equipment activities through a state diagram. The specification was developed by the semiconductor manufacturers as a directive to Equipment manufacturers upon request.

HSMS (SEMI E37) [4] is also a data transfer protocol and was published in 1994 for defining a communication interface suitable for message exchange in a specific environment of the semiconductor industry known as TCP/IP (Transmission Control Protocol/Internet Protocol). TCP/IP provides reliable and connection-oriented message exchanges between computers within a network.

Generally speaking, the most common applications of SECS include data collection, alarm collection, recipe management (RM), wafer mapping, remote control, and material control. SECS is also sometimes used for internal communication to a piece of Equipment. SECS-I, SECS-II, GEM, and HSMS will be introduced with more details in the sub-sections below.

3.2.1.1 SEMI Equipment Communication Standard I (SECS-I) (SEMI E4)

SECS-I [1] provides the communication formats between the Host and the Equipment via RS-232C protocol. The Host can be any type of computer on the other side of the Equipment, such as personal computer, workstation, mainframe, robot, or even another piece of Equipment.

Block Protocol

In SECS-I, a message is sent in blocks. As shown in Figure 3.1, a block contains a 10-byte header and a 2-byte checksum, starting with a length byte that describes the length of the block with the checksum excluded. A block's maximum length is 254 bytes and the minimum is 10 bytes (header only), which means that a block can contain up to 244 bytes of data. Therefore, any message with more than 244 bytes of data should have multiple blocks. A message can contain up to 32 767 blocks (approximately 8 million bytes long) at the most.

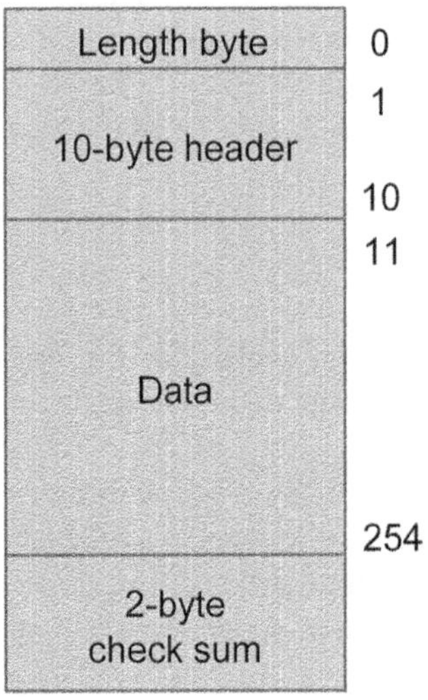

Figure 3.1 SECS block.

For the block protocol of SECS-I, the Sender can be either the Host or the Equipment. Referring to the handshake codes shown in Table 3.1 and Figure 3.2 for explanation of block transfer, the Sender of a SECS message first sends an ENQ (Request to Send); if the Receiver of the ENQ is able to accept a message, the Receiver will send an EOT (Ready to Receive). Then, after receiving the EOT, the Sender will send one block of data. Once the block of data is received correctly, the Receiver will acknowledge with an ACK (Correct Reception); if data are not correctly received, the Receiver will issue a NAK (Incorrect Reception). The example of a multi-block message (677 data bytes) is shown in Figure 3.3 with the numbers of bytes of the first, second, and third blocks being 244, 244, and 189, respectively.

Table 3.1 Handshake codes (bytes).

Name	Binary code	Hex	Function
ENQ	00000101	05	Request to send
EOT	00000100	04	Ready to receive
ACK	00000110	06	Correct reception
NAK	00010101	15	Incorrect reception

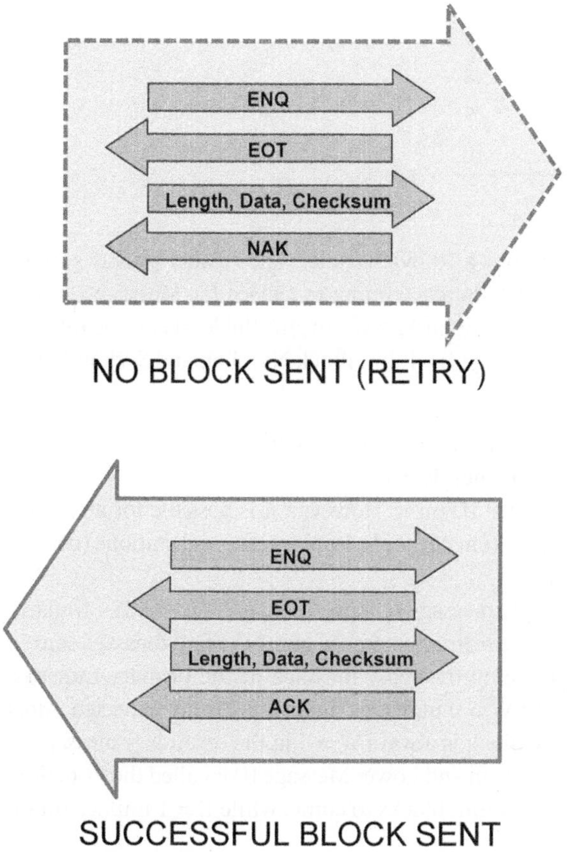

Figure 3.2 Block transfer.

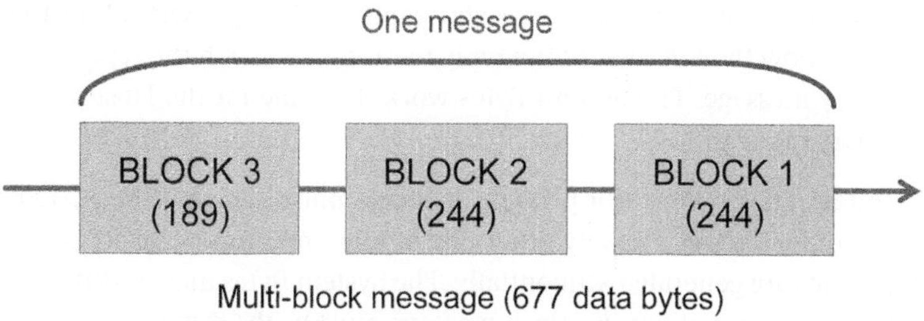

Figure 3.3 Multi-block message (677 data bytes).

Table 3.2 Header composition.

8	7 6 5 4 3 2 1	byte
R	Upper Device ID	1
	Lower Device ID	2
W	Upper Message ID	3
	Lower Message ID	4
E	Upper Block No.	5
	Lower Block No.	6
	System Bytes	7
	System Bytes	8
	System Bytes	9
	System Bytes	10

Header

As depicted in Table 3.2, every SECS block contains a 10-byte Header. The Header provides information so that the Receiver can identify the block data according to Device ID, Message ID, and Block Number with Upper means the most significant and Lower means the least significant.

A Header contains R-bit, Device ID, W-bit, Message ID, E-bit, Block Number, and System Bytes. They are described as follows:

- **R-bit:** Reverse bit, which signifies the direction of the message. When R = 0, the message goes from the Host to the Equipment and vice versa when R = 1.
- **Device ID:** Each equipment must have a Device ID on it. However, it is possible for an equipment with multiple tubes (diffusion), chambers (cluster tools, implanters), and stations (coater/developer) to have more than one Device ID.
- **W-bit:** Wait bit, which indicates whether a reply message is expected or not. The SECS-I transaction consists of a pair of SECS messages with the primary message being the first message sent in the transaction and the secondary message being the reply message of the primary message. When W = 1, the message expects a reply, and W = 0 indicates that a reply is not expected. Most primary messages expect replies but not all, while it is always W = 0 in the secondary messages.
- **Message ID**: Upper Message ID is called the Stream and Lower Message ID is called the Function.
- **E-bit**: End bit. When E = 0, it means there are more blocks to come, while E = 1 implies that it is the final block.
- **Block Number**: A two-byte number that counts the number of blocks in a message. Block numbers start with 1.
- **System Bytes**: When a Sender of a primary message expects a reply, the System Bytes allow the Sender of the primary message (now the Receiver of the reply message) to match the reply message to the appropriate primary message. The System Bytes work the same for the Host or the Equipment sending the primary message.

Responsibility of the Sender: The Sender of a primary message must generate the System Bytes. The System Bytes must not be the same as any other outstanding messages. This is called "Distinction." Usually, System Bytes are generated sequentially. The System Bytes may be different for different Device IDs. All System Bytes for a multi-block message must be the same.

Responsibility of the Receiver: The Receiver should copy the System Bytes from the primary message to the reply message.

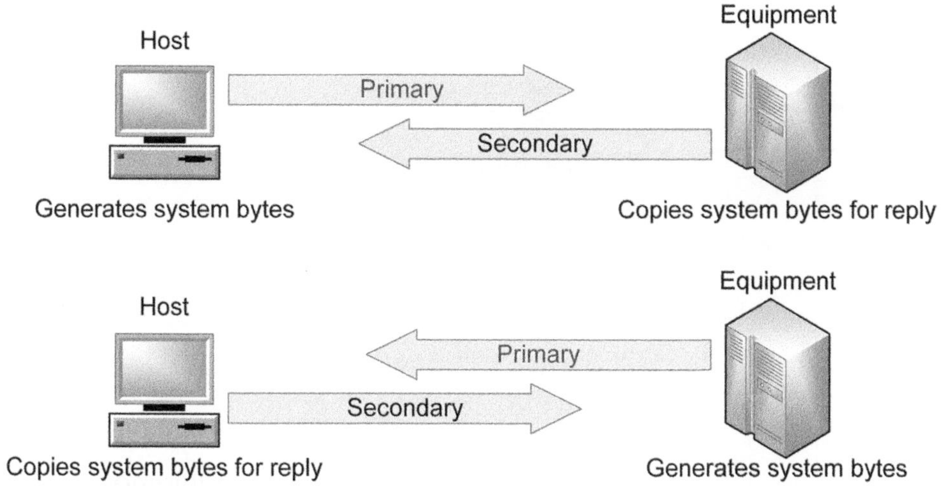

Figure 3.4 Illustration of system bytes.

Figure 3.4 portraits the example of System Bytes. In the upper portion, the Host generates the System Bytes in the primary message and the Equipment copies the System Bytes into the secondary message replying to the Host. As for the lower portion, the reverse happens.

A message more than 254 bytes long must be sent as a multi-block message and each sent block must contain the block transfer protocol. Figure 3.5 illustrates an example of a three block multi-block message sent from the Host to the Equipment with a single block reply.

Timeouts

If the Receiver responds with a NAK instead of an ACK, the block transfer protocol will retry sending the block. There are four types of timeouts that result in NAK, including

- **T1**: Inter-Character Timeout limits the time between receipts of characters.
- **T2**: Protocol Timeout constrains the time between handshake bytes and between data and handshake bytes.
- **T3**: Replay Timeout is the time that the sender of a primary message will wait until the receipt of the secondary message with W-bit being equal to 1. The Sender starts counting time whenever the last block of the primary message is successfully sent until the first block of the secondary message is received. If the T3 timeout is exceeded, which means the message protocol fails at the application level, and there is no retry.
- **T4**: Inter-block Timeout is the time interval that the Receiver of a multi-block message will wait between blocks. The Sender starts courting time whenever each block of the primary message is successfully received and the E-bit is equal to 0. If the T4 timeout is exceeded, which means the message protocol fails at the application level and there is no retry.

Among them, T1, T2, and T3 are illustrated in Figure 3.6. The typical values, range, and resolution of T1–T4 timeouts are tabulated in Table 3.3.

Protocol Parameters Summary

- Retry Limit

  During a block transfer send attempt, the Retry Limit is the maximum number of times. The Sender will attempt to retry sending data before declaring a block send failure. For example, as

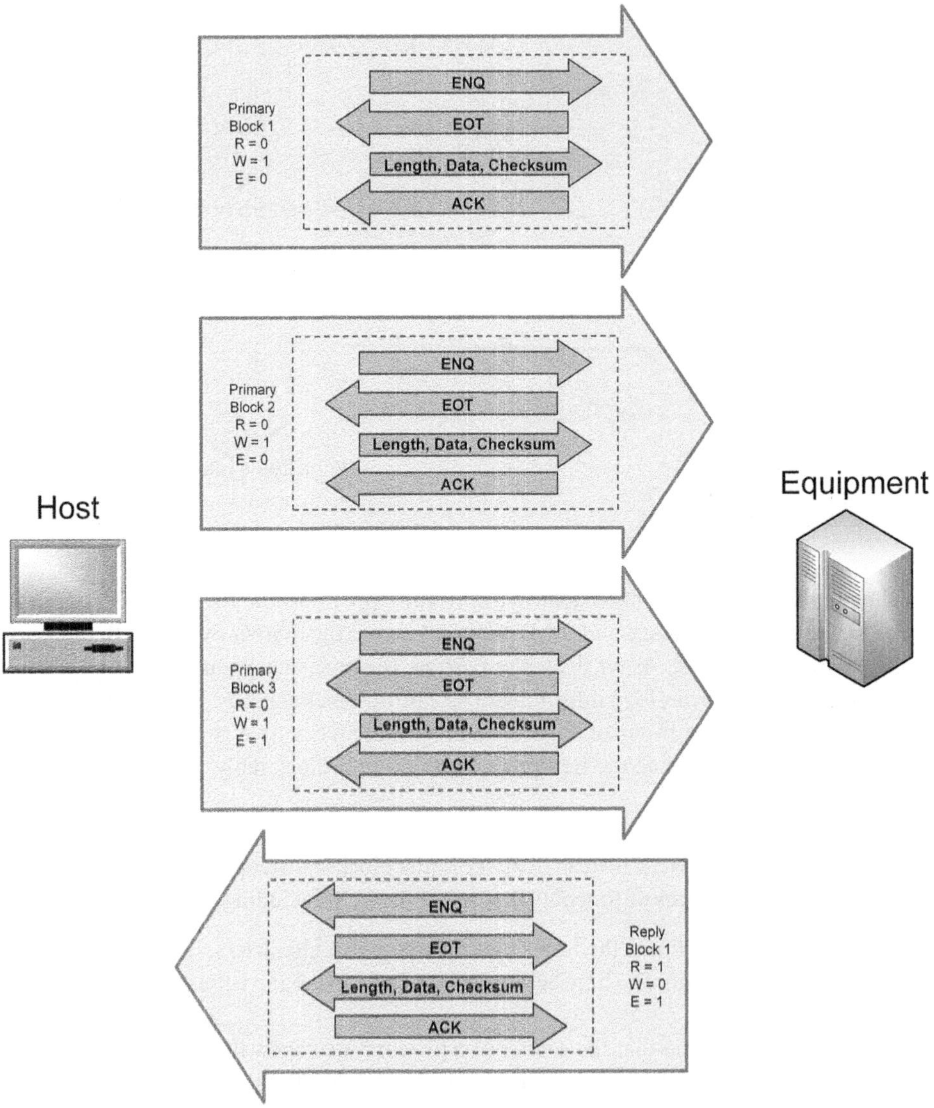

Figure 3.5 Block transfer protocol in a multi-block message.

shown in Figure 3.7, ENQ is retried repeatedly waiting for an EOT response before declaring a failure. Note that the T2 timeout and Retry count are different between the Host send case and the Equipment send case. Retry count is typically set to be 3. Increase if the equipment fails to complete block protocol. The normal case for this is failure of EOT response to ENQ.

- Baud Rate
  Baud rate is usually 9600. Reduce baud rate for T1 timeouts or general communication problems.
- Master/Slave
  SECS operates at a full-duplex line. Both Sender and Receiver can be sending ENQs to each other simultaneously. The Equipment is the Master and the Host is the Slave, as shown in Figure 3.8. The Host will give up control of the "line" to the Equipment and reply to the Equipment's ENQ with an EOT.

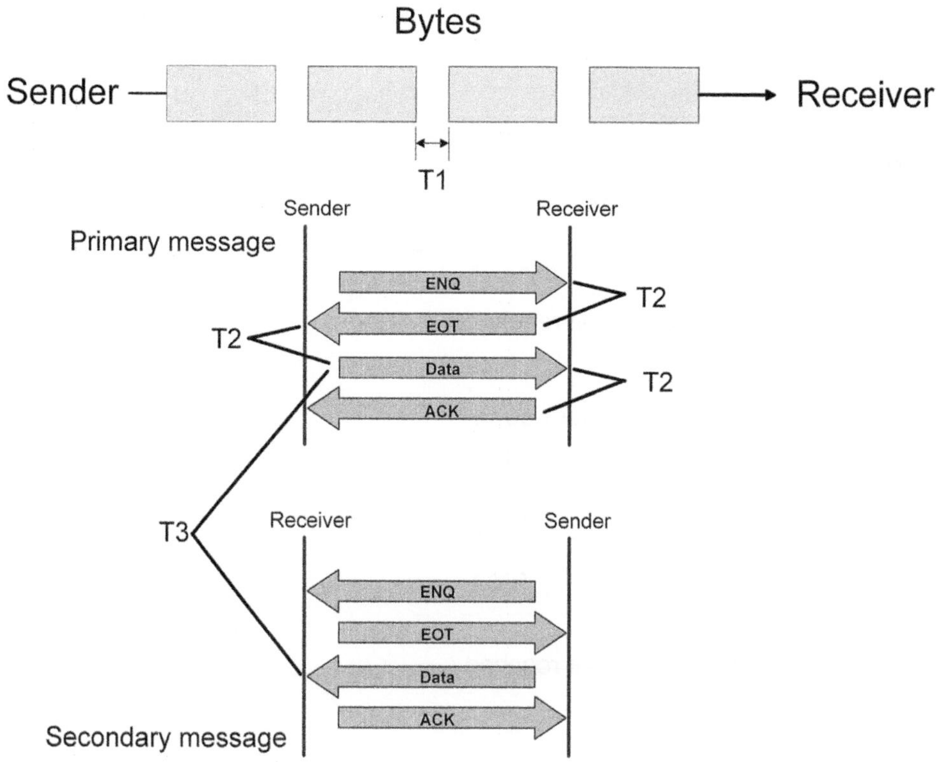

Figure 3.6 T1, T2, and T3 timeouts.

Table 3.3 Timeout parameters (in seconds).

T	Name	Typical values	Range	Resolution
T1	Inter-Character	0.5	0.1–10	0.1
T2	Protocol	2	0.2–25	0.2
T3	Reply	45	1–120	1
T4	Inter-Block	10	1–120	1

3.2.1.2 SEMI Equipment Communication Standard II (SECS-II) (SEMI E5)

SECS-II [2] provides a set of data structure rules and message structure guidelines that allow an equipment manufacturer to provide communication services to its equipment. The data structure (individual items) rules are quite strict with rare deviations. The message structure (lists and items) guidelines are a set of recommendations for particular messages that can be used to allow specific kinds of message communication. These guidelines allow for tremendous flexibility for the equipment manufacturer to allow communication with their equipment.

Message Transfer Protocol

The message interchange is managed by the transaction protocol, which is composed of a set of rules for message handling. And this protocol has some minimum requirements on the SECS-II implementation.

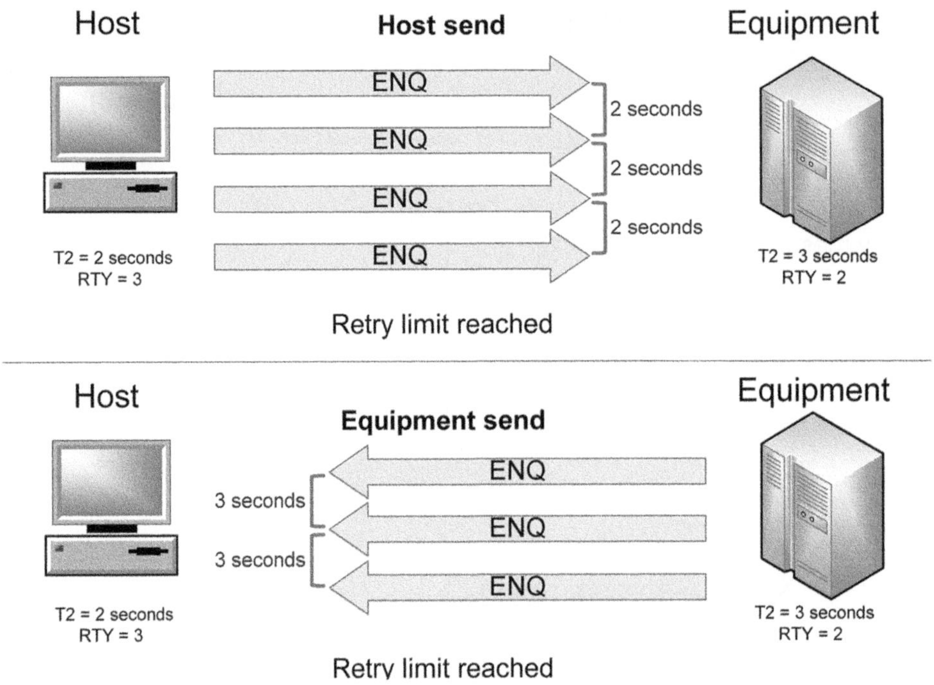

Figure 3.7 Example of T2 timeout and retry limit.

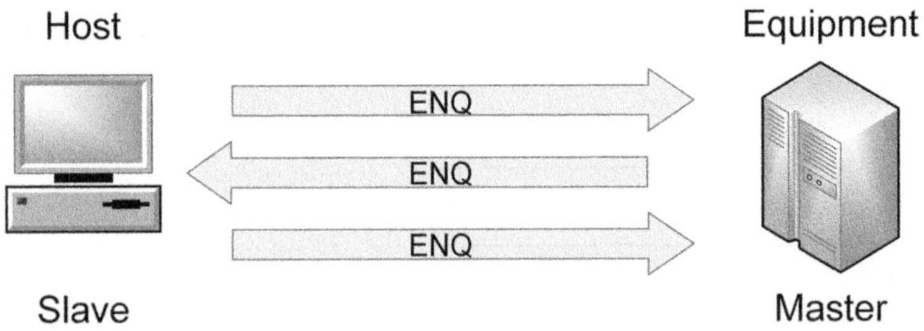

Figure 3.8 Equipment is Master and Host is Slave.

The message transfer protocol is applied to deliver messages between the Equipment and Host. It sends a primary message to indicate if a reply is demanded. If yes, it should be able to relate the corresponding secondary message or reply message to the original primary message. The Sender creates the original primary message, and the Receiver deciphers the primary message at its destination and then generates a reply upon demand.

As in SECS-I, the maximum length for a single-block message in SECS-II is 244 bytes. Messages longer than 244 bytes are multi-block messages. The message transfer protocol should contain message header of every message.

Streams and Functions

Messages delivered in SECS-II are organized into streams, which mean the categories of activities; and a stream contains specific messages denoted as functions. The transmission request between information and its corresponding data, for instance, can be considered as an activity. The categories of activities of various streams are tabulated in Table 3.4.

Table 3.4 Streams.

Stream 1	Equipment status
Stream 2	Equipment control and diagnostics
Stream 3	Materials status
Stream 4	Material control
Stream 5	Exception handling
Stream 6	Data collection
Stream 7	Process program management
Stream 8	Control program transfer
Stream 9	System errors
Stream 10	Terminal services
Stream 11	(Deleted from the 1989 edition)
Stream 12	Wafer mapping
Stream 13	Data set transfers
Stream 14	Object services
Stream 15	Recipe management
Stream 16	Processing management
Stream 17	Equipment control and diagnostics
Stream 18	Subsystem control and data

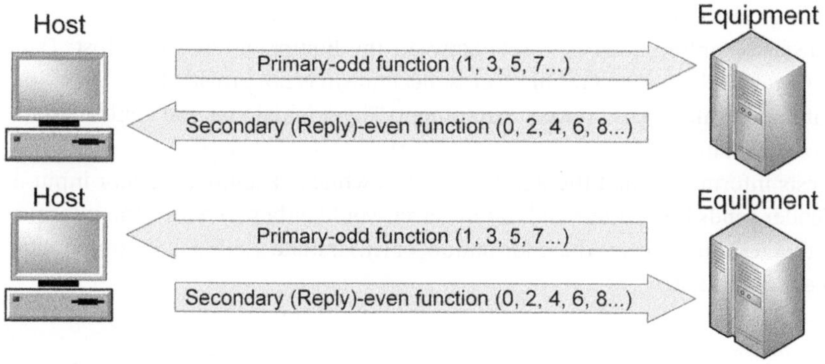

Figure 3.9 Primary and secondary messages.

A stream is a category of messages supposed to assist similar or associated activities, and a function is a specific message within a stream for a particular purpose. Both streams and functions are identified by an integer between 0 and 255. A numbering convention relating to the primary and secondary message pairs will cover all functions used in SECS-II. As shown in Figure 3.9, all primary messages will be assigned an odd-numbered function code, while the corresponding reply secondary message (even-numbered) function code is defined by adding one to the primary message function code. Each SECS-II message exchange is assigned a unique transaction ID number.

Equipment manufacturers are allowed to create their very own streams under SECS-II standard if none of the existing streams suits their requirement. The newly created stream will begin at number 64.

The equipment manufacturer decides the most appropriate streams and functions for their own models. Note that, two equipment might use the same stream and function but contain different data.

Transaction and Conversation Protocols

A transaction is constituted of a primary message requesting no reply as well as a primary message requesting a reply along with its corresponding secondary message. Yet unlike the primary message, a secondary message cannot request a reply.

A conversation is composed by one or more related transactions for accomplishing a particular task. An uncompleted conversation causes a conversation timeout, which would halt any further action on the conversation.

The seven types of conversations in SECS-II are introduced as follows:

1) A primary message with no reply: The simplest single-block SECS-II message. The Sender expects that the Receiver replies to the message; and there is nothing the Sender can do if the message is rejected.
2) The Sender requests data from the Receiver with a primary message: The Sender expects to receive the data returned as a reply message. This is the request/data conversation.
3) The Sender sends a single-block SECS-II message: The Sender wants an acknowledgement from the Receiver. This is the send/acknowledgement conversation.
4) The Sender wants to send a multi-block SECS-II message: The Sender needs to receive the permission from the Receiver first. The Receiver can accept or reject this request. If the Receiver permits, the Sender sends the data and the Receiver replies accordingly. This is the inquire/grant/send/acknowledge conversation. If a conversation timeout happens due to the preparation by the Receiver, the Receiver would send an error message to the Sender, and the Equipment needs to issue an error message to the Host.
5) Conversation transfer of unformatted data sets between the Equipment and the Host: Data transfer between systems where a general file access mechanism is not provided.
6) Conversation that handles materials between equipment: The original material control protocol is included in the stream.
7) The Sender requests information that the Receiver takes a while to acquire (operator input is required): The Sender sends the request and the Receiver would either (i) return the information; (ii) indicate that it cannot receive the information; or (iii) initiate a subsequent transaction after receiving the information.

Data Structure Rules

Information is formatted in two structures, items and lists. An item is one or more data values of the same data format. All data in a SECS message are contained in items, and a SECS message contains at least one item. There are currently 15 item data formats recognized by SECS-II.

A list can be composed by either a group of items or items with other lists. The data structures define the logical divisions of the message to distinguish them from the physical divisions of the message transfer protocol. They provide a self-describing internal structure to the messages transmitted between the Equipment and Host.

The length and format of an item are determined by its first 2, 3, or 4 bytes, and these first bytes are called the item header (IH), where the first two bits indicate how many of the following bytes consist the length of the item, so that long items do not require the byte overhead for shorter items. The item length is decided by the item body, which refers to the number of bytes after the item header; item body is the actual data of the item. As item length only counts the item body and

Table 3.5 Item header.

Byte Number	Bit Number 8 7 6 5 4 3 2 1	Description
1	Format code (octal) I NLB	Format byte
2	Most significant length byte	Length byte
3		1, 2, or 3
4	Least significant length byte	

excludes the item header, the actual number of bytes for an item is the item length plus 2, 3, or 4 bytes for the item header. All bytes in the item body follow the specified format byte.

Observing Table 3.5, the format of an item is set as the octal numbers (6 bits). The number of length bytes (NLB, 2 bits) informs the receiver whether the next 1, 2, or 3 bytes would be used for indicating the length of the item. For instance, if the length of a message is 258 bytes, the length bytes of the message would be 0102 Hex, which takes 2 bytes to describe the message length. Therefore, NLB = 2. In general, with NLB being 1, 2, or 3, the maximal length of bytes of the message will be 255, 64K, or 7.99 million, respectively.

A list is a group of items and/or lists. The format code for a list is 0 octal with one length byte. The format code for a list in hex is therefore 01. And the length byte in a list (following the format code 01) is the number of items in the list. The length bytes indicate the number of items in a list instead of the number of bytes. Related items with different formats can also be organized into a list owing to its character.

The item format codes are tabulated in Table 3.6. The Hex column of Table 3.6 shows the code if the NLB = 1 (01 binary). Except for recipes and data files, all items are generally less than 256 bytes long and NLB = 1. For integers, data are sent with most significant byte first.

Example Messages

The examples of the SECS-II (S1F3) message sent from the Host to the Equipment and vice versa (S1F4) are depicted as Figures 3.10 and 3.11.

Notations of SECS-II Messages

One of the popular abbreviations for SECS-II messages comes from SEMI. You will find this one in most equipment documents.

- **SEMI Data Item Dictionary**
  The Data Item Dictionary of the SEMI E5 document contains a list of recommended variable names. These names are for equipment companies to use in their equipment SECS documentation. Two of the most common variable names used in SECS are SV (status variable value) and SVID (status variable ID). Status variables may include any parameter that can be sampled by time such as temperature or quantity of consumable.
- **SEMI SECS Notation**
  Every message has a style that is used as displayed in Figure 3.12. The entries in the form of Figure 3.12 are described in Table 3.7.
- **Message Structure**
  The following abbreviations are used:

L, n	List of n elements
< variable name>	Item
< var name 1, . . ., var name n>	Item with more than one variable

Table 3.6 Item format codes.

Binary Bit 876543	Octal	Hex NLB = 1	Meaning
000000	00	01	List
001000	10	21	Binary
001001	11	25	Boolean
010000	20	41	ASCII
010001	21	45	JIS-8
011000	30	61	8-byte integer (signed)
011001	31	65	1-byte integer (signed)
011010	32	69	2-byte integer (signed)
011100	34	71	4-byte integer (signed)
100000	40	81	8-byte floating point
100100	44	91	4-byte floating point
101000	50	A1	8-byte integer (unsigned)
101001	51	A5	1-byte integer (unsigned)
101010	52	A9	2-byte integer (unsigned)
101100	54	B1	4-byte integer (unsigned)

		S1F3	Host to equipment
byte	Hex	Value	Description
1	00		R-bit=0; Host->Equip
2	00	Device ID=0	
3	81	Stream 1	W-bit=1; Reply expected
4	03	Function3	
5	80		E-bit=1; Last block
6	01	Block # =1	
7	XX		
8	XX		
9	XX	System bytes	
10	XX		
11	69	Item 1 format	2-byte signed integers
12	0A	Item 1 length	10 bytes; 5 SVIDs
13	00		
14	02	SVID1 = 2	Gas flow 1
15	00		
16	10	SVID2 = 16	Hexcode DC bias
17	00		
18	11	SVID3 = 17	RF forward power
19	00		
20	60	SVID4 = 96	Current recipe
21	00		
22	01	SVID5 = 1	Turbo purge

Figure 3.10 SECS-II message detail of host sending S1F3 to equipment.

		S1F4	Equipment to host
byte	Hex	Value	Description
1	80		R-bit = 1 Host <- Equipment
2	00	Device ID=0	
3	01	Stream 1	W-bit = 0 No reply
4	04	Function 4	
5	80		E-bit = 1 Last block
6	01	Block #	First block
7	XX		
8	XX		
9	XX	System bytes	same as primary
10	XX		
11	01	List	
12	05	5 Items	
13	69	2-byte signed	Item 1 format
14	02		Item 1 length
15	00		SV1
16	40	64 sccm	Gas flow 1
17	69		Item 2 format
18	02		Item 2 length
19	01		SV2
20	00	256 volts	Hexode DC bias
21	69		Item 3 format
22	02		Item 3 length
23	02		SV 3

		S1F4	Equipment to host (continued)
byte	Hex	Value	Description
24	00	512 watts	RF forward power
25	41		Item 4 format
26	10	16 bytes	Item 4 length
27	41	A	
28	42	B	
29	43	C	SV 4
30	44	D	Current recipe =" ABCDEF"
31	45	E	
32	46	F	
33	20	blank	
34	20	.	
35	20	.	
36	20	.	
37	20	.	
38	20	.	
39	20	.	
40	20	.	
41	20	.	
42	20	blank	
43	69	Item 5 format	
44	02	Item 5 length	
45	00		SV 5
46	53	83 cc	Turbo purge

Figure 3.11 SECS-II message detail of Equipment sending S1F4 to Host.

Sn, Fm name of function (xxx)   S, M direction reply

Description:
Structure:
Exception:

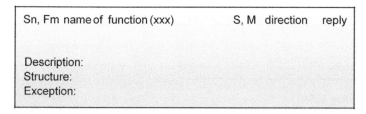

Figure 3.12 Message detail style.

Table 3.7 SECS notations.

Abbreviated entry	Expanded entry and description	Example
Sn	Stream n where n = the stream number	S1 (Stream 1)
Fm	Function m where m = the function number	F3 (Function 3)
name	Every function has a name	Selected equipment status request
(xxx)	Mnemonic-abbreviated notation for message	SSR
S, M	Single Block or Multi-Block	S
Direction	Host to Equipment, Equipment to Host, or both	H->E
Reply	Reply expected, [reply] = optional	reply
Description:	Explanation of purpose of message	
Structure:	List and item structure	
Exception:	To the structure	

An element can be either an item or another list and most items contain one variable. It is legal in some messages to have more than one variable as long as every variable has the same format. The examples of the message structures receiving S1F3 shown in Figure 3.10 and sending S1F4 shown in Figure 3.11 are displayed in Figures 3.13 and 3.14, respectively.

3.2.1.3 Generic Model for Communications and Control of Manufacturing Equipment (GEM) (SEMI E30)

The Generic Equipment Model (GEM) Standard (SEMI E30) [3] was published in 1992. A large number of semiconductor companies and equipment supplier companies had implemented the

```
S1F3 Selected equipment status request (SSR)          S, H → E, reply
        L, 5
            1. <SVID1>
            .
            .
            5. <SVID5>
Exceptions: Zero length list means report all SVIDs.
```

Figure 3.13 Message structure receiving S1F3.

```
S1F4 Selected equipment status data (SSD)             M, H ← E
        L, 5
            1. <SV1>
            .
            .
            5. <SV5>
Exceptions: Zero length item means SVID doesn't exist.
```

Figure 3.14 Message structure sending S1F4.

SECS-1 and SECS-II standards for factory automation by the year of 1987. SECS-I and SECS-II were important steps towards simplifying factory automation. However, at that time, some problems as listed below had been discovered. As such, there was a need for additional standards to work with SECS-II.

1) Divergence among Equipment Implementations of SECS-II: As most capabilities specified by SECS-II standard are optional, various equipment types would choose to implement different capabilities from others.
2) Divergence of User Requirements: Different purchasers demand different SECS-II implementations even from the same equipment suppliers. As a result, the equipment supplier has to implement several different versions of the SECS-II communications on the same equipment.
3) Limited Minimum Set: The minimum requirements that SECS-II imposes are too limited to develop equipment capabilities necessary for effective factory automation.

GEM Requirements and Capabilities

As depicted in Figure 3.15, the GEM standard was developed based on SECS-II. While SECS-II defines a generic message protocol for data transmission, GEM defines the Equipment behavior via SECS-II communications.

Being both a specification and a guide, GEM not only determines the minimum capabilities for the SECS communication on the Equipment but also delimits that this communication is highly dependent on the Equipment activities through a state diagram. The GEM is an interface between the Host's software and the Equipment's software, which can be utilized for remote equipment monitoring.

GEM is composed of two types of specifications: the fundamental GEM requirements and requirements for additional GEM capabilities. Equipment that implements GEM is expected to support all the fundamental requirements. The additional GEM capabilities provide functionality required for some types of factory automation or functionality applicable to specific types of equipment. Thus, the GEM standard's flexibility enables both simple and complex equipment to implement GEM.

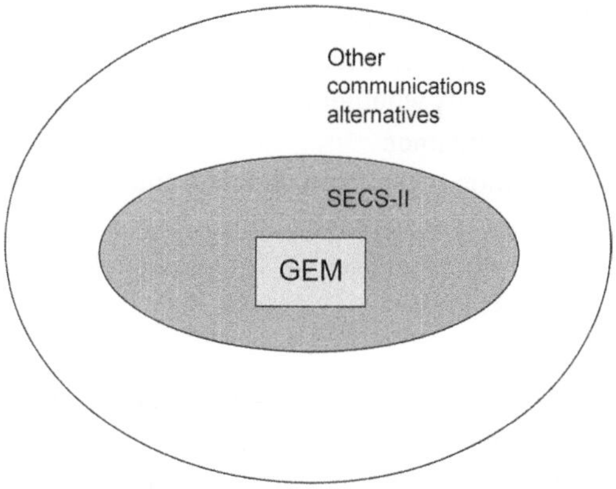

Figure 3.15 Scopes of GEM, SECS-II, and other communications alternatives.

Fundamental GEM requirements include

- State Models
- Equipment Processing States
- Host-Initiated S1F13/F14 Scenario
- Event Notification
- On-line Identification
- Error Messages
- Control (Operator Initiated)
- Documentation

The additional GEM capabilities provide functionality required for some types of factory automation or functionality applicable to specific types of equipment. GEM capabilities can be divided into three broad categories as described below.

- **Data Collection:** Variable Data Collection, Status Data Collection, Event Notification, Dynamic Event Report Configuration, Trace Data Collection, Alarm Management, Clock, Limits Monitoring.
- **Equipment Control:** Remote Control, Equipment Constants, Process Program Management, Control State (On-Line/Off-Line), Equipment Terminal Services.
- **Protocol:** Establish Communications, Spooling.

It is worth noticing that the requirements in the GEM standard apply only to the Equipment but not the Host. In other words, the Equipment behavior is more predictable, while the Host can be more flexible on the decision of selecting which features from the Equipment's GEM interface would be implemented.

GEM Compliant

An Equipment is GEM Compliant for a specific GEM capability if and only if (i) the fundamental GEM requirements are satisfied; (ii) the capability is implemented to conform with all applicable definitions, descriptions, and requirements defined for the capability in this standard; and (iii) the Equipment does not exhibit behavior related to this capability that conflicts with the GEM behavior defined for the capability.

Further, an Equipment is Fully GEM Capable if and only if it meets the following two criteria: (i) the Equipment supplies all the GEM capabilities; and (ii) every implemented GEM capability is GEM Compliant.

The SECS-II interface documentation provided by an equipment supplier should emphasize GEM compliance. It has to contain a GEM Compliance Statement which accurately indicates for each capability whether it has been implemented and whether it has been implemented in a GEM compliant manner.

In addition to the semiconductor industry, GEM is also widely adopted by other industries for its adaptability and maturity. Quite a few SEMI standards were developed to improve the applicability of GEM, which enables GEM to be even better accommodated to various kinds of manufacturing industries globally.

3.2.1.4 High-Speed SECS Message Services (HSMS) (SEMI E37)

For the semiconductor industry, SECS-I has been primarily applied since 1980. Until mid-1980s, there was a demand to develop a higher speed alternative as the semiconductor industry and network continued to advance and the existing protocol cannot satisfy the growing requirements.

HSMS [4] was designed by merits of network technology advancement to substitute SECS-I. Since TCP/IP protocol was widely used and applicable for most computer platforms, HSMS was planned to define a communication interface favorable for message exchange between computers utilizing TCP/IP.

HSMS-Generic Services

The HSMS document [4] provides the fundamental components for developing an HSMS-compliant communication interface. The message exchange procedures using the TCP/IP protocol are stated below.

1) Establish a communication link between entities via TCP/IP connection.
2) Create and maintain the protocol conventions required for SECS message exchanges.
3) Exchange data through TCP/IP.
4) Recognize error conditions.
5) End the communication formally to ensure both parties don't need the TCP/IP connection anymore.
6) Break the communication links without physical disconnection from the network medium.
7) Test the communication links for connection integrity.
8) Reject connection attempts from incompatible subsidiary standards.

HSMS can be further divided into two subsidiary standards: HSMS-SS (single session) and HSMS-GS (general session), as shown in Figure 3.16.

HSMS-SS and HSMS-GS

HSMS-GS [5] is the other subsidiary standard of HSMS. HSMS-GS defines the support for complex systems consisting of multiple and accessible subsystems, including cluster tool or track systems. HSMS-GS extends the HSMS state machine by adding state transition definitions and state information. The additional capabilities allow the individual sub-entities of complex systems to be accessed during the HSMS procedures. These extensions include (i) a list of all session entities that are accessible through TCP/IP connections from outside entities; (ii) a list of selected entities that are currently accessible on a given TCP/IP connection; and (iii) a count of the currently selected entity IDs. In addition, HSMS-GS also indicates the application notes of simultaneously supporting both HSMS-GS and HSMS-SS.

HSMS-SS [6] is a subsidiary standard of HSMS, which includes the minimum services necessary for a direct SECS-I substitution. HSMS-SS was defined with limited capabilities, and it (i) reduces the HSMS-GS procedures adopted for implementations that support multiple TCP/IP connections;

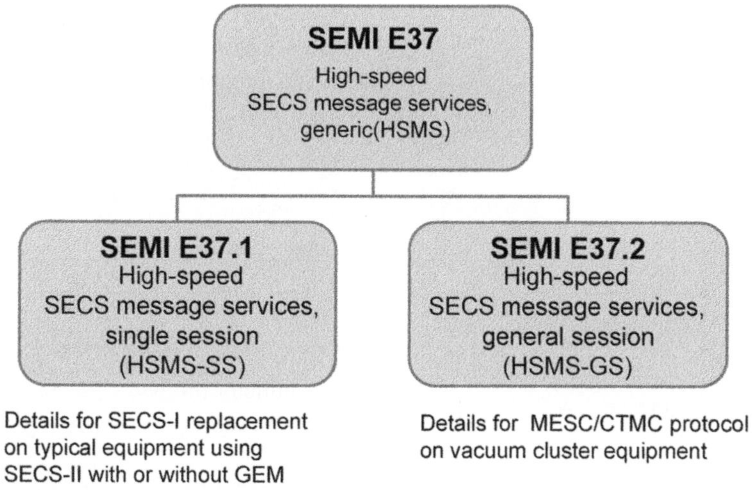

Figure 3.16 Subsidiary standards of HSMS.

(ii) constrains other HSMS-GS procedures to simplify the operations for certain SECS-I replacement cases; and (iii) provides the application notes of supporting multiple hosts.

HSMS Adoption

The first adopters include equipment suppliers in the metrology and inspection area, who have had problems with transferring a large amount of measurement data through RS-232. Under the constraints of RS-232, some metrology and inspection manufacturers had to develop better network environment to support information delivery via the Equipment's interface.

The adoption of HSMS by different types of equipment manufacturers usually follows the requirement of the IC manufacturers. The availability of TCP/IP environment in these factories enables them to adopt HSMS in an easier manner. Also, utilizing HSMS-compliant equipment in the factories provides more flexibility on physical equipment placement. By adopting HSMS, the constraint of maximum cable length is released. In addition, it only needs to configure a network address for equipment moving in the factory without needing to supply a dedicated RS-232 cable as in Figure 3.17.

HSMS Design Principles

Due to the fact that a SECS-I message is sent in blocks, observing Table 3.2 for header composition of SECS-1, bytes 5 and 6 of the Header specify the Block Count of the message. On the other hand, HSMS is intended as an alternative to SECS-I which utilizes RS-232 for applications. And, HSMS defines a communication interface suitable for the exchange of messages between computers in a semiconductor factory using a TCP/IP environment, whose messages are sent by TCP/IP stream transmission and need not specify the Block Count. Therefore, the design principle is to substitute the bytes 5 and 6 (for specifying the Block Count) of the SECS-1 Header into session-type (SType) and presentation-type (PType) bytes of the HSMS Header. The functions of SType and PType will be elaborated in the following subsection.

HSMS Message Format

The composition of the HSMS-SS message is shown in Figure 3.18. A HSMS-SS message begins with a Message Length field of 4 bytes, which is followed by a Message Header of 10 bytes. Then comes the useful Message Text of the HSMS-SS message formatted as specified in SECS-II and the Message

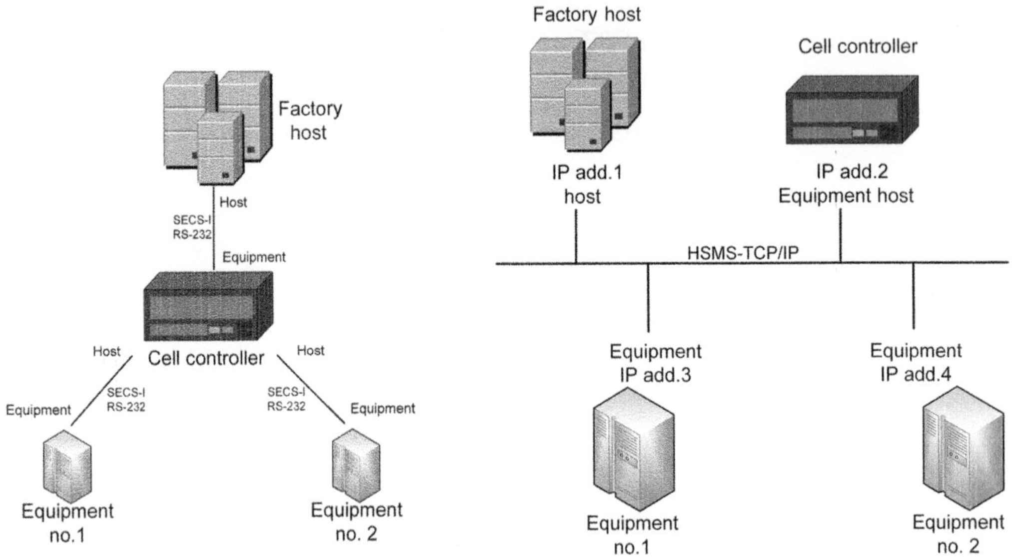

Figure 3.17 SECS-I RS-232 connections versus HSMS TCP/IP Ethernet connections.

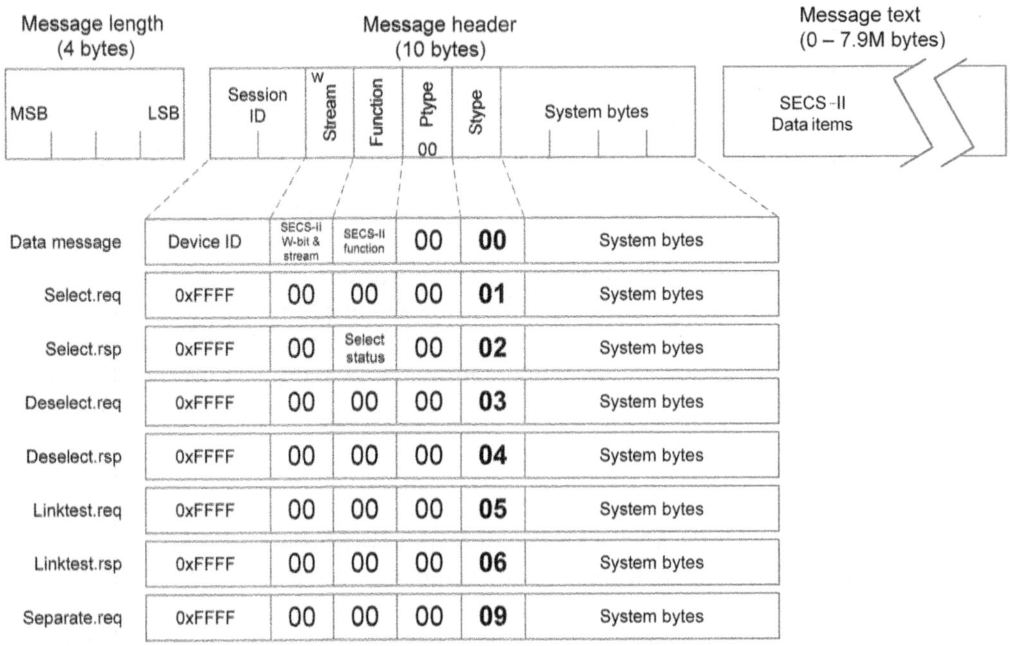

Figure 3.18 HSMS-SS message format.

Text can reach to several megabytes in size. The number of bytes specified in Message Length contains the Message Header and the Message Text, with the four bytes of the Message Length excluded.

Within the ten-byte Message Header shown in Figure 3.18, the first two bytes contain a Device ID, useful in complex equipment to identify a major subsystem. The third and fourth bytes of the Message Header have different uses depending on Stype, as described below. The fifth byte of the Message Header (Ptype) is always zero. The sixth byte of the Message Header (Stype) contains a code indicating whether this message is a Data Message (containing useful application data) or one of the five or six HSMS-SS Control Messages used for link management.

The Control Messages Select.req and Select.rsp are used to establish a connection between the Host and the Equipment. Linktest.req and Linktest.rsp are used to verify that the connection is still active. Separate.req is used to terminate the connection.

The seventh through tenth bytes of the Message Header contain the System Bytes, which are used logically to associate a Primary Message with the corresponding Reply Message.

For a Data Message (Stype 00), bytes three and four of the Message Header contain SECS-II Stream and Function codes, which identify the topic of the message and which are described previously in the SECS-II standard. An odd-numbered Function (least significant bit of Function is "1") signifies a Primary Data Message, and an even-numbered Function (with value one greater than the corresponding Primary Data Message) signifies a Reply Data Message.

The ten-byte Data Message (Stype 00) of HSMS-SS Message Header shown in Figure 3.18 looks a lot like the older SECS-I Block Header depicted in Table 3.2. In SECS-I, we had a one-byte Block Length, and a ten-byte header for each block. For HSMS-SS, the TCP/IP layer provides "hidden" logic for blocking the TCP/IP Stream transmission, in HSMS-SS we don't need to worry about blocks. Instead, we deal with complete SECS-II messages. Only one ten-byte header is needed for the entire HSMS-SS message. In SECS-I, the fifth and sixth bytes of the Block Header contained a Block Count and E-Bit. In HSMS-SS, we don't worry about blocks, so these bytes of the header are used for Ptype and Stype, as described above.

HSMS-SS Procedures

The HSMS-SS contains the following procedures: Connect, Data, Disconnect, Linktest, etc. They are described below.

- Connect Procedure

  As shown in Figure 3.19, The HSMS Connect procedure establishes a logical connection between the Host and the Equipment. One end is the Active entity that initiates the connection establishment by means of the TCP "connect" function; and the other end, the Passive entity, accepts the connection via TCP "accept" function. To avoid the Active entity from sending the connections to inappropriate partners (e.g. printer server), the HSMS-SS standard requires the Active entity to send the HSMS-SS control message Select.req, and the Passive entity to reply with Select.rsp, which is more effective to ensure both ends are HSMS entities.

  There are several error types in the Connect procedure. For example, a connection might fail if there is no matching "accept." In such case, the Active entity waits for the specified interval (T5 Separation Timeout) and tries to connect again. The Passive entity initiates T7 NOT SELECTED Timeout after the TCP/IP connect/accept completes. The connect

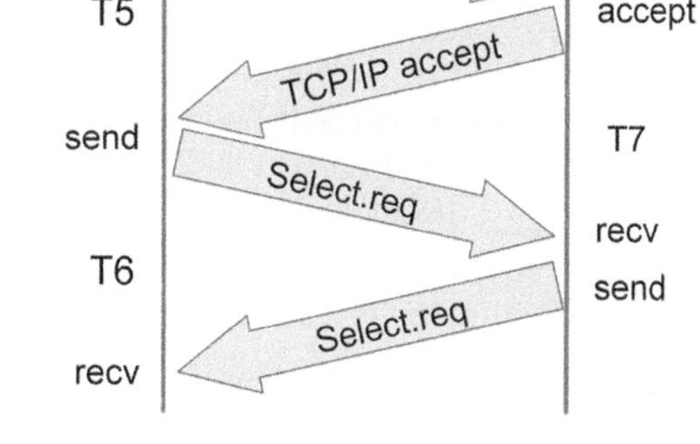

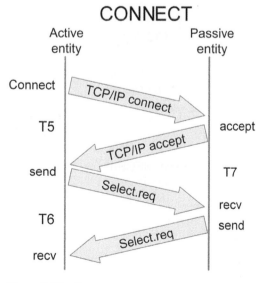

Figure 3.19 Connect.

procedure fails if the Passive entity cannot receive Select.req within T7 Timeout. Similarly, the Active entity starts T6 Control Transaction Timeout after sending Select.req; and the connect procedure fails if the Active entity cannot receive Select.rsp within T6 Timeout.

In most conventional TCP/IP environment, a second Active entity can, even after the Passive entity accepts the TCP/IP connect, still try to connect to the same Passive entity. On the contrary to the non-HSMS TCP/IP protocols which usually allow several Active entities to connect to a single Passive entity simultaneously, HSMS does not give such permission in general. The Passive entity sends Select.rsp with Select Status being zero in a successful connect procedure; on the other hand, it sends Select.rsp with Select Status nonzero to reject a simultaneous connect from a second Active entity's attempt. This logic has been verified on various TCP/IP environments for providing effective connect failure detection from both ends.

- Data
  After a logical connection between the Host and the Equipment is established, the most important procedure is the Data procedure depicted in Figure 3.20, which consists of sending a Primary SECS-II message in one direction and starts the T3 Timeout, and waiting for the appropriate SECS-II reply message back. Like SECS-I, either end of the HSMS-SS connection can initiate a transaction, several transactions can be in progress simultaneously, and HSMS-SS associates each Reply Message to the appropriate Primary Message. Unlike SECS-I, HSMS-SS also defines several additional procedures which are used to manage the TCP/IP connection.
- Disconnect Procedure
  Either side can disconnect an HSMS-SS connection by sending the HSMS-SS Control Message Deselect.req; the other end then replies with Deselect.rsp. After sending or receiving Deselect.req, the HSMS-SS entity issues TCP/IP "close" function to stop the connection, as shown in Figure 3.21.
- Linktest Procedure
  There are chances that one end of a TCP/IP connection dies or terminates the link abruptly. Some implementations can report the disconnection to the other end immediately, while certain implementations might take up to 15 min to do so. To solve this problem, the Linktest procedure can be utilized to judge if an HSMS-SS connection is still active. As shown in Figure 3.22, one entity

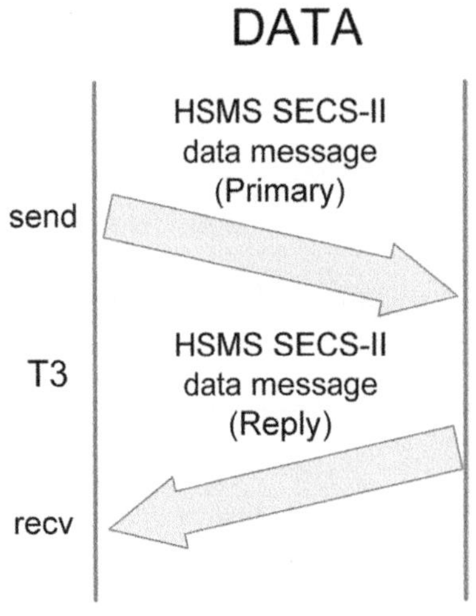

Figure 3.20 Data.

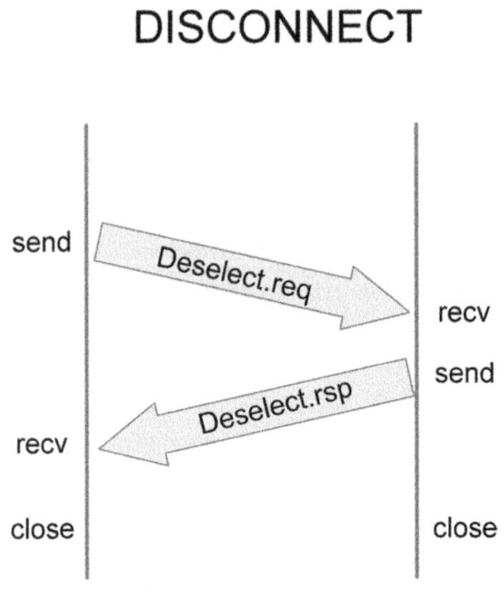

Figure 3.21 Disconnect.

initiates the Linktest by sending the HSMS-SS Control Message Linktest.req, and starts the T6 Timeout; the other end then replies with Linktest.rsp. If the initiating entity cannot receive Linktest.rsp. within the T6 Timeout, it is assumed that the connection has failed and TCP/IP Close function will be used to terminate this connection.

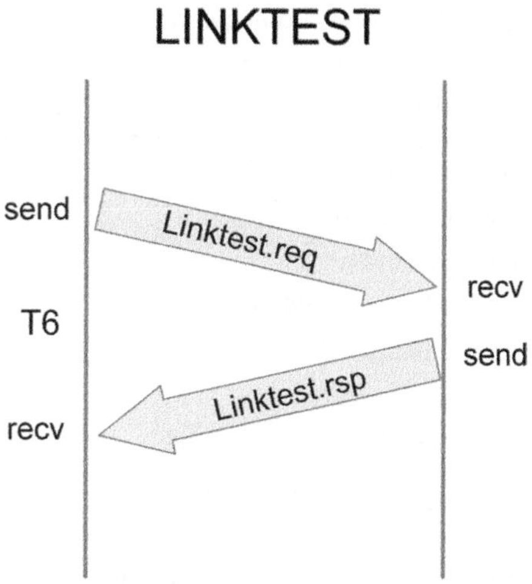

Figure 3.22 Linktest.

HSMS Timeouts

There are five types of HSMS timeouts, including

- **T3**: Reply Timeout is the same as for SECS-I.
- **T5**: Connect Separation Timeout controls how often the Active entity will re-try its attempts to establish the TCP/IP connection.
- **T6**: Control Transaction Timeout is a reply timeout for HSMS Control Transactions (Select and Linktest).
- **T7**: Not Selected Timeout controls how long the Passive entity will wait to receive Select.req during a Connect procedure.
- **T8**: Network Inter-Character Timeout controls how long an application waits for an unresponsive TCP/IP layer.

All the HSMS-SS timeouts and tabulated in Table 3.8.

Configuring HSMS-SS Parameters

Like SECS-I, one end of the link is configured as the Host and the other end as Equipment, with the Equipment's SECS Device ID indicated. Either the Host or the Equipment should be assigned

Table 3.8 HSMS timeouts.

Parameter name	Value range	Resolution	Typical value	Description
T3 Reply Timeout	1–120 s	1 s	45 s	The maximum amount of time an entity expecting a reply message will wait for that reply
T5 Connect Separation Timeout	1–240 s	1 s	10 s	The amount of time which must elapse between successive attempts to connect to a given remote entity
T6 Control Transaction Timeout	1–240 s	1 s	5 s	The time which a control transaction may remain open before it is considered a communication failure
T7 Not Selected Timeout	1–240 s	1 s	10 s	The time which a TCP/IP connection can remain in NOT SELECTED state (i.e. no HSMS activity) before it is considered a communication failure
T8 Network Inter-Character Timeout	1–120 s	1 s	5 s	The maximum time between successive bytes of a single HSMS message which may expire before it is considered a communication failure

Note: Parameter defaults shown above are for small networks (10 nodes or less). Settings may need to be adjusted for larger network configurations.

Table 3.9 HSMS-SS configuration.

Parameter	Typical value	
Host or Equipment		a
SECS Device ID		a
Active or Passive	Host is Active	
Passive Entity IP Address	140.116.86.150	
Passive Entity TCP Port	5000	
T3 Reply Timeout	45 s	a
T5 Connect Separation Timeout	10 s	
T6 Control Transaction Timeout	5 s	
T7 NOT SELECTED Timeout	10 s	
T8 Network Inter-Character Timeout	5 s	

[a] Same as SECS-I (SECS-I parameters not needed for HSMS-SS: Baud Rate, T1, T2, T4, RTY).

as Active, while the other end as Passive so as to form a connection. On the perspective of the factory, configuring the Host as Active is preferable since it is easier to switch the Host computers. The IP address and TCP port number of the Passive entity should be configured. The typical values of HSMS-SS configuration are depicted in Table 3.9.

Sharing Network with Other TCP/IP Protocols

As HSMS is a communication interface suitable for the exchange of messages between computers using a TCP/IP environment, it can share network with other TCP/IP protocols, as shown in Figure 3.23.

Comparison and Migration of SECS-I and HSMS

Figure 3.24 displays the SECS-I and HSMS-SS protocol stacks.

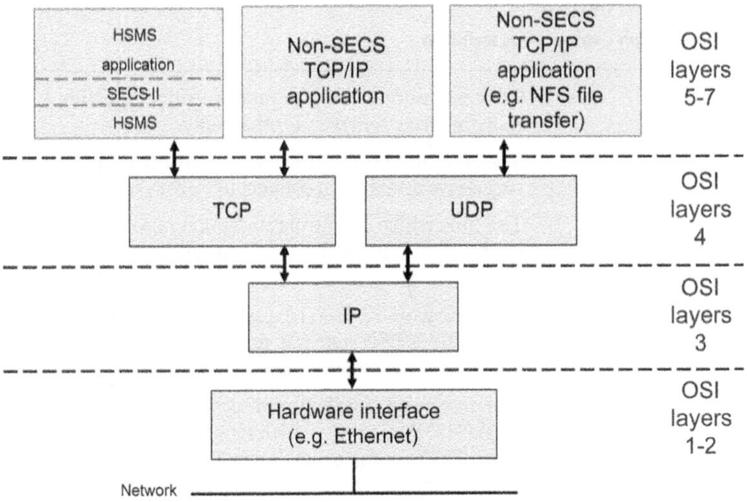

Figure 3.23 HSMS can share network with other TCP/IP protocols.

Figure 3.24 SECS-I and HSMS-SS protocol stacks.

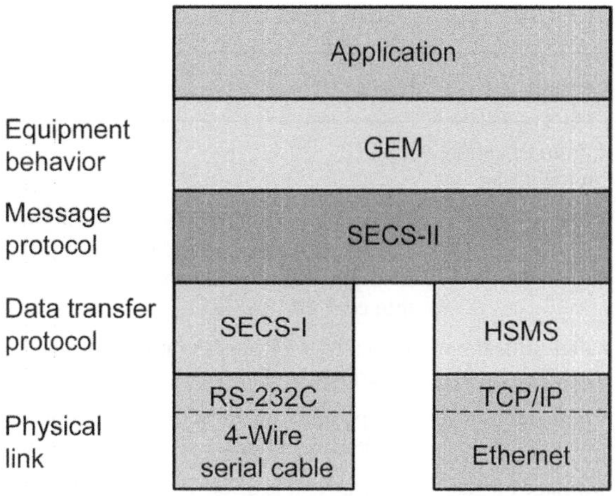

Some old existing factories are set up to use SECS-I, while a few newer factories are ready to use HSMS-SS. A gradual shift to HSMS-SS is anticipated, but for several years, equipment suppliers will need to provide both protocols for different customers. Equipment can be designed with "plug and play" software components to make this straightforward. Figure 3.24 shows how substituting SECS-I for HSMS changes only the lower levels of the protocols. SECS-II, GEM, and (most importantly) the equipment or line control computer application software does not need to change. Converting an equipment or line control computer is easy — one simply swaps models of the low-level commercial SECS Driver software (from SECS-I Driver to HSMS Driver). Except for speed, most other aspects of the equipment remain unchanged. The feature comparisons of HSMS-SS and SECS-I are tabulated in Table 3.10.

3.2.2 Engineering Portion (Interface A)

Interface A (also called Equipment Data Acquisition, EDA) is a group of standards developed by SEMI which allows high volume and high rate data from Equipment to be collected via a common client/host data collection system across tool types. Different from SECS/GEM, Interface A focuses on the engineering equipment system (EES) part (engineering portion) of e-Manufacturing hierarchy as shown in Figure 1.1. It is designed to support SECS/GEM standards as it features more flexible approaches such as offering multiple independent client access, restricted access based on client credentials, the compatibility with common internet technology like simple object access protocol (SOAP)/extensible markup language (XML) over the HTTP/1.1 infrastructure, and self-describing interface. It is a well-defined XML/SOAP interface which enables the implementation to help the communication between the Equipment and its Interface A clients. Clients are granted access to data of interest after authentication and are able to browse available data by querying the tools. As the Equipment type may vary, we need a solution to ensure the Interface A clients can gather data from any type of Equipment.

The Common Equipment Model (CEM) is defined for each Equipment providing a structured view of the Equipment, which includes the physical hardware, logical software components, and all the data can be shared: objects, attributes (variables), events, and exceptions (alarms). Typically, these data requirements align with SEMI standards supported by the Equipment, and also the extra hardware or process-specific data like sensor readings and processing results. Customized Data Collection Plans (DCPs) can be defined by Interface A clients to retrieve the relevant data.

Table 3.10 Comparison of HSMS and SECS-I.

Feature	HSMS	SECS-I
Communication protocol base	TCP/IP	RS-232
Physical layer	Physical layer not defined. HSMS allows any TCP/IP supported physical medium. Typical examples include Ethernet (IEEE 802.3) and thin coax (10-BASE-2)	25-pin connector and 4-wire serial cable
Communication speed	Typically 10 M Bits/second (assuming typical Ethernet)	Typically about 1000 bytes/second (assuming 9600 baud)
Connections	One physical network cable can support many HSMS connections	One physical RS-232 cable per SECS-I connection
Message format	Message text is SECS-II Data Items. Transmits a SECS-II message as a TCP/IP byte stream. The message has a four-byte Message Length, a ten-byte Message Header, and text. The TCP/IP layer may impose blocking limits depending on the physical layer used, but the blocking is transparent to the TCP/IP API and is outside the scope of HSMS	Message text is SECS-II Data Items. Transmits a SECS-II message as a series of transmitted blocks, with each block containing approximately 256 bytes in size. Each block has a one-byte block length, a ten-byte Block Header, text, and a two-byte Checksum
Header	One ten-byte Header for the entire message. Header bytes 4–5 contain PType and SType. Header bytes 2–3 are W-bit, Stream, and Function when Stype = 0 (Data Message). For SType not equal to 0 (Control Message), bytes 2–3 have other uses. No R-bit	Ten-byte header on each block of a message. Header bytes 4–5 contain E-bit and Block Number
Maximum message size	Message size limited by 4-byte message length (approximately 4 GBytes). Local implementation of TCP/IP and HSMS may further limit this in practice	Limited to approximately 7.9 million bytes (32 767 blocks times 244 text bytes per block)
Protocol parameters (common)	T3 Reply Timeout Session ID (analogous to Device ID)	T3 Reply Timeout Device ID
Protocol parameters (SECS-I only)	Not used in HSMS. Corresponding issues addressed by TCP/IP layers	Baud Rate T1 Inter-Character Timeout T2 Block Protocol Timeout T4 Inter-Block Timeout RTY Retry Count Host/Equipment
Protocol parameters (HSMS only)	IP Address and Port of Passive Entity T5 Connect Separation Timeout T6 Control Transaction Timeout T7 NOT SELECTED Timeout T8 Network Inter-Character Timeout	Not needed by SECS-I

The implementation of Interface A has grown since 2005 and been widely integrated in semiconductor equipment. Primary Interface A standards include specification for Equipment Client Authentication and Authorization (A&A, SEMI E132) [7], specification for the Common Equipment Model (CEM, SEMI E120) [8], specification for Equipment Self-Description (EqSD, SEMI E125) [9], specification for Data Collection Management (DCM, SEMI E134) [10] and the latest launched specification for EDA Common Metadata (ECM, SEMI E164) [11].

Briefly speaking, as shown in Figure 3.25, E132 defines the way to create sessions for equipment client authentication and authorization. E120 provides a consistent vocabulary for describing equipment structure; while E125 provides a comprehensive set of data structures to describe information available from the equipment. E125 and E120 together simplify the naming and identification of equipment data items. E164 then further defines the specification of metadata which makes the standards even more universal. Moreover, E125 and E134 together create a more manageable and flexible data collection environment.

The Interface A integrated scenario is illustrated in Figure 3.26, which also shows how the individual standards (including E120, E125, E132, E134, and E164) fit together. To begin with, the Client requests to create a session with the Equipment for the purpose of authentication and authorization. After the authorization being granted and the session being created, the Client then sends the Metadata request to the Equipment. The Equipment, then, replies the Metadata to the Client. Based on the Metadata, the Client creates the desired DCP and sends this DCP Create Request followed by the DCP Activate Request to the Equipment. After receiving this DCP request, the Equipment then starts the Data Collection Activity to generate the corresponding Data Collection Report (DCR) so as to reply to the Client. The DCR may contain Event, Exception, and/or Trace Reports. Finally, the Client closes the session. All the Interface A related standards will be elaborated as follows respectively.

3.2.2.1 Authentication & Authorization (A&A) (SEMI E132)

The purpose of E132 [7] is to provide means for factories to control, in software, which applications are permitted to communicate with the Equipment and the Equipment services application are permitted to use. E132 defines two security features for Interface A: client authentication and client authorization. Client authentication decides the way the client starts the session before it takes any action, while client authorization manages the accessible data for the Client after the session is established. A security admin which is a utility in charge of administrative configuration provided by the Equipment is necessary for setting up the client authentication and authorization after installation in the fab. For implementation in SOAP/XML, a subordinated standard E132.1 – Specification for SOAP Binding of Equipment Client Authentication and Authorization is published by SEMI to instruct.

The E132 authentication model can be divided into three parts: central Security Admin service, Client service, and Equipment service as Figure 3.27 shown. The central Security Admin service is implemented by factory to assign credentials to applications; the Client service establishes session with Equipment and provides credentials; the Equipment service challenges Client for credentials, and it will deny the session if not accepted. To establish a session from an Interface A Client, the Client must insert valid credentials and be authenticated by the Security Admin, also the Equipment must be in the allowed session establishment state. The credentials are composed by a Client ID, an encrypted session key, and an encrypted client ID proof-of-identity key. Without authentication, any attempts to access the services will be rejected. After the session is established, the Client authorization is effectively verified based upon the Client's credentials. The Client can then access to the services.

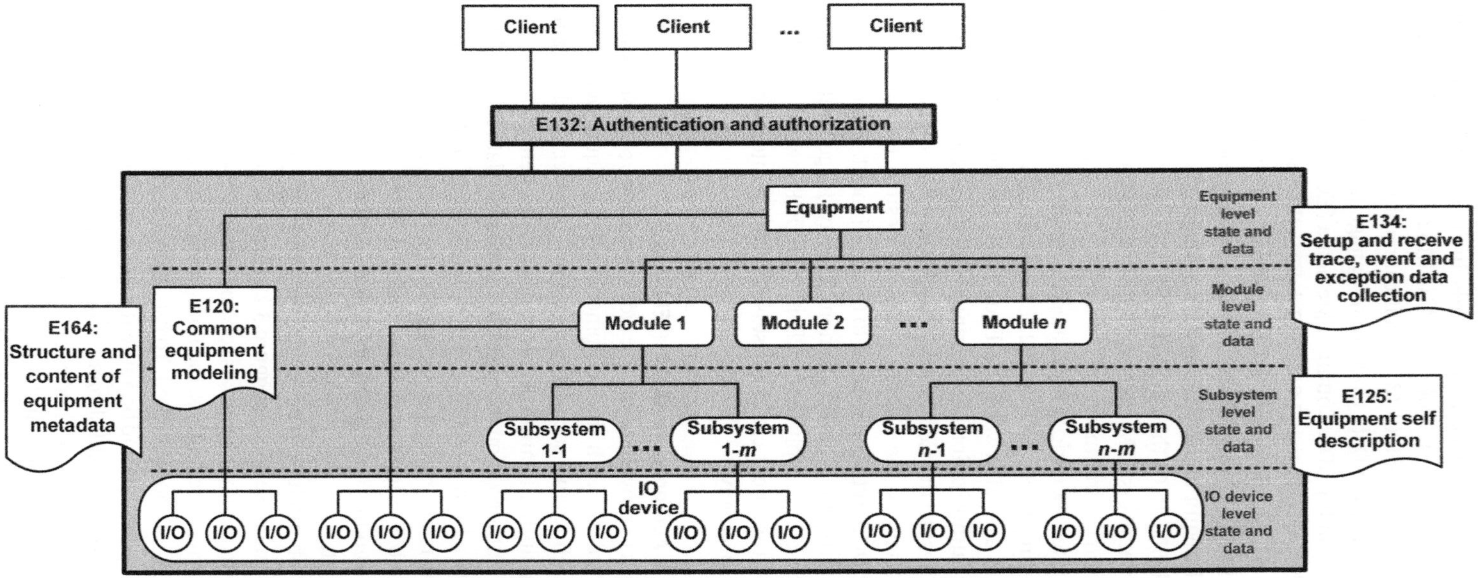

Figure 3.25 Interface A Standards on equipment.

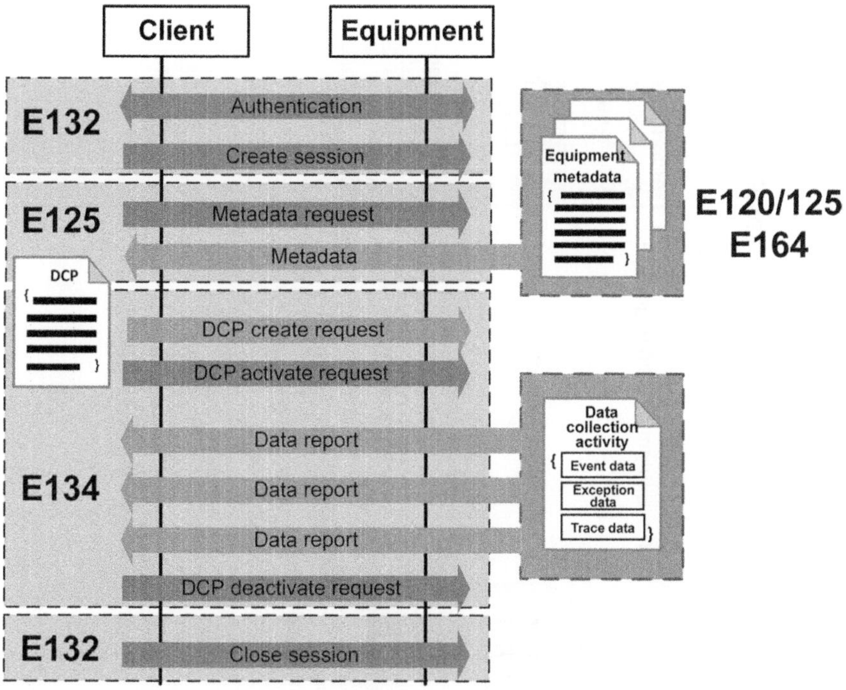

Figure 3.26 Interface A Integrated Scenario.

3.2.2.2 Common Equipment Model (CEM) (SEMI E120)

E120 [8] is the specification for CEM. The purpose is to provide means for suppliers to describe the physical structure of their Equipment using common attributes and terminology and for SEMI standards that depend on information about Equipment structure to have a basis for doing so using common attributes and terminology. E120 defines a general object model which depicts the Equipment from an external view. The model consists of various classes organized in a logical hierarchy, which summarizes all the Equipment's major hardware and software components. Moreover, SEMI also published as subordinate Standard E120.1 – XML Schema for the CEM to support how to map CEM into a specific XML implementation.

The major classes in E120 include Equipment, Module, Subsystem, I/O Device, and material location. Take a conveyor system for delivering glass as shown in Figures 3.28 and 3.29 for example, the whole conveyor system is composed by two conveyors CV-A & CV-B and two turntables TT-A and TT-B. To describe this system in CEM, the Equipment class is the whole conveyor system. The conveyors and the turn tables are Subsystems. Underneath the Subsystem, there are several IO devices, such as motors (MO1 to MO3) and conveyor sensors (X22-28) to slow or stop it. The conveyor CEM description diagram is depicted in Figure 3.30. E120 provides a consistent vocabulary for describing an Equipment structure which allows E125 to describe a comprehensive set of data structures.

3.2.2.3 Equipment Self-Description (EqSD) (SEMI E125)

E125 [9] provides means for applications to discover the physical Equipment structure, available data items, events, and exceptions via software. This standard permits the Client to request useful descriptions about all available information for gathering data, including the parameters (specific data, units, and types), events, exceptions, state machines, SEMI E39 data, and physical configuration.

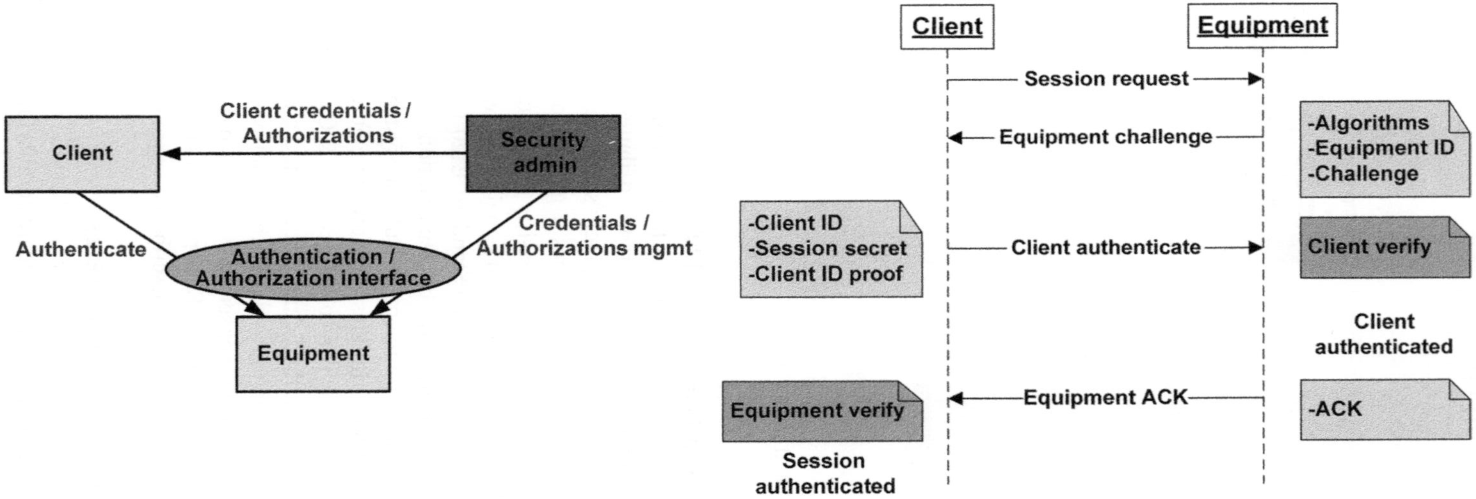

Figure 3.27 E132 Authentication model.

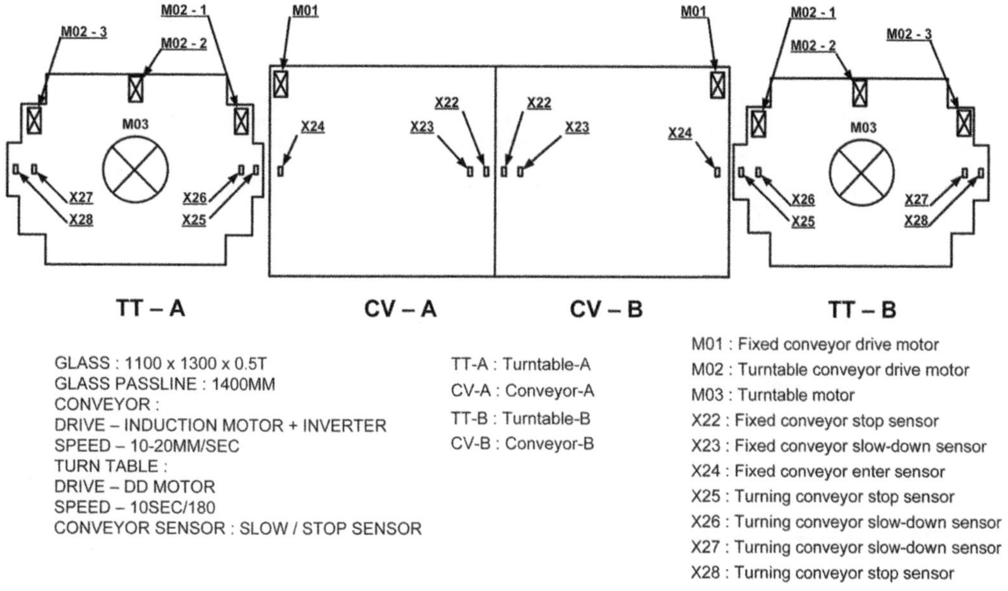

Figure 3.28 CEM example – diagram of a conveyor system.

Figure 3.29 CEM example – photo of a conveyor system.

All the information is then mapped into E120 CEM object hierarchy. The EqSD is defined in E125 with client-initiated operations to query the available metadata information of the Equipment. This interface is obligatory for the Equipment compatible with E125. To further explain how to implement E125 in SOAP/XML environment, SEMI published a subordinate document SEMI E125.1-0414 – Specification for SOAP Binding for EqSD as an instruction. With E125 and E120 working together, it simplifies the naming and identification of Equipment data items.

The E125 Equipment metadata structure is shown in Figure 3.31. In E125, metadata is modularized. The units, types, Equipment structure, and exceptions, etc. are defined independently of one another. The associations with the Equipment structure are centralized. Data produced by each

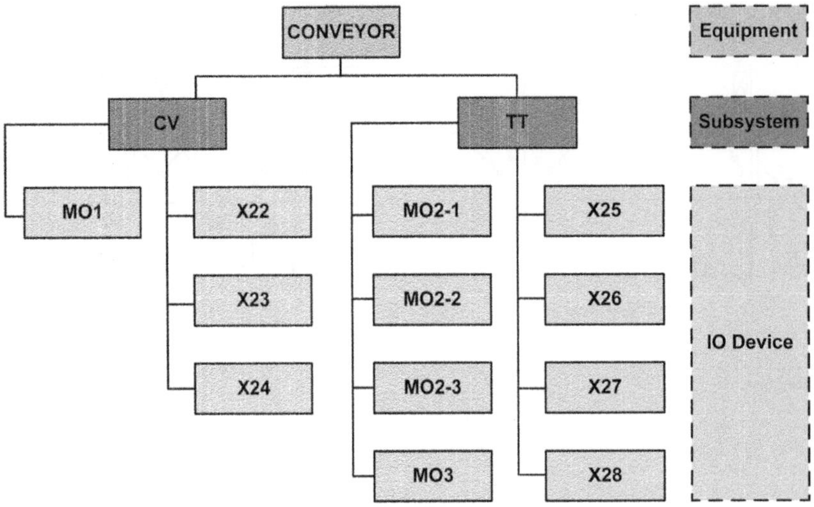

Figure 3.30 CEM example-conveyor CEM description diagram.

Equipment node with a unique ID (UID) are referenced by name, data types, events, exceptions, etc. With E125, end-users can release the dependence on Equipment documentation as they are able to directly acquire information of what data can be monitored on the Equipment and the context (i.e. process chamber #1 or #2). Moreover, Interface A clients can be equipped with plug-and-play methodology which will refresh information automatically right after any revision on the Equipment interface. If an unexpected error occurs, data recovery is also available.

The same conveyor system example mentioned above in E120 is adopted for explanation. As Figure 3.32 shown, CEM is on the left and the EqSD is on the right. Both of them coordinates with each other.

3.2.2.4 Equipment Data Acquisition (EDA) Common Metadata (ECM) (SEMI E164)

Launched in July 2012, E164 [11] intends to motivate companies developing a more generic method to implement equipment metadata upon Interface A connections. Although E120 and E125 have defined the standard for Equipment models and the resulting metadata sets, respectively, the types can vary a lot. To create a guidance to complement E120 and E125, E164 was born. According to E164 standard, a greater consistency from tool-to-tool and from fab-to-fab is achieved for Equipment suppliers and fabs. This consistency helps both ends in providing a universal interface from the Equipment suppliers and developing Interface A client applications from fabs. Deriving from E125 EqSD, E164 defines common metadata set to support the consistent implementation of Interface A Freeze Version II. E164 also supports SEMI E30, E40, E87, E90, E94, E148, and E157 standards. In particular between Original Equipment Manufacturers (OEMs), Equipment modeling is more clearly defined to ensure the consistency and help the users to easily and quickly locate the models and find the required data.

The benefit of E164 not only has standards for data content but can build an application beyond just visualization. E164 ensures the automation of the data collection process and reduces the work to interpret the collected data. With E164, we can achieve true plug-and-play applications. All features above significantly increase the engineering efficiency. E164 specifies the structure and content of Equipment metadata such as the naming of event, parameter type, and hierarchy in a generic way as shown in Figure 3.33. The role of E164 to Interface A is just like GEM to SECS-II.

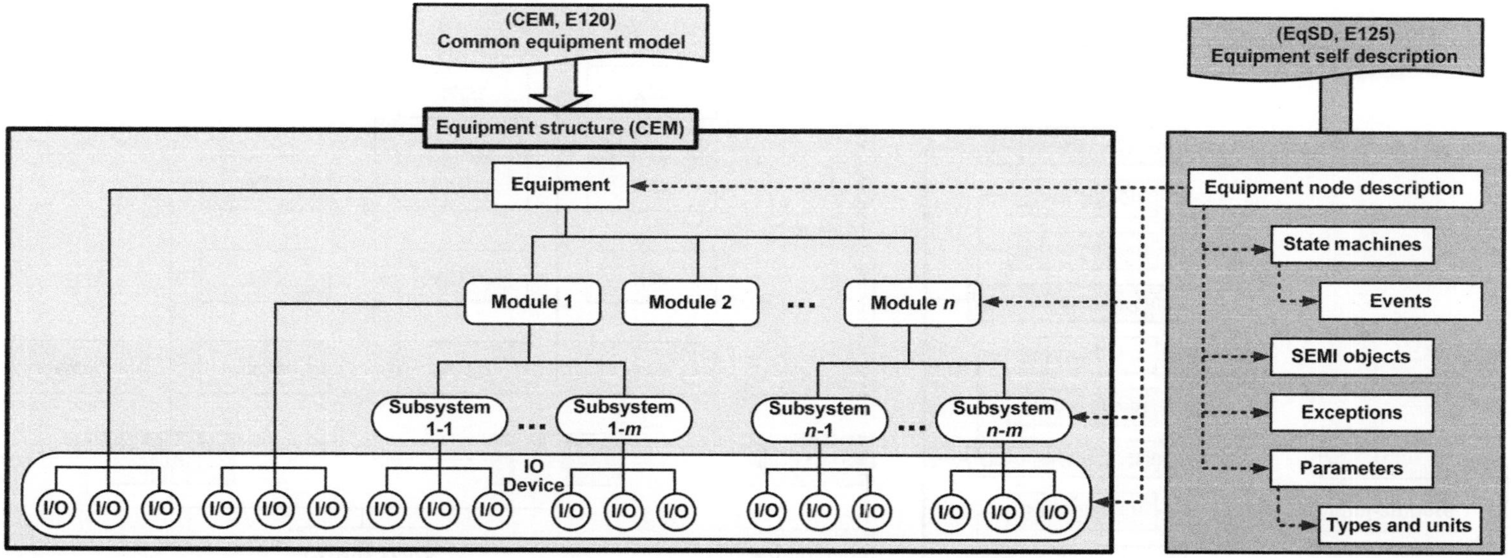

Figure 3.31 E125 Equipment metadata.

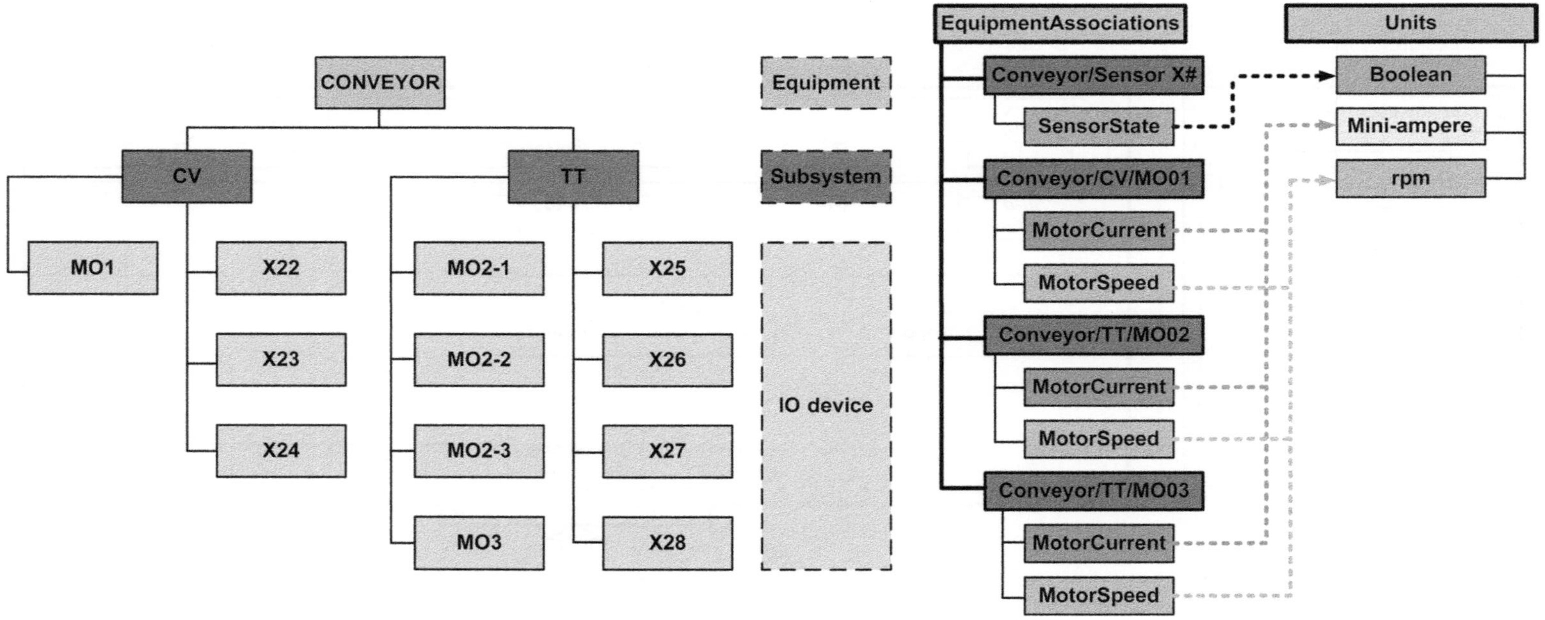

Figure 3.32 Example of CEM and EqSD – conveyor system.

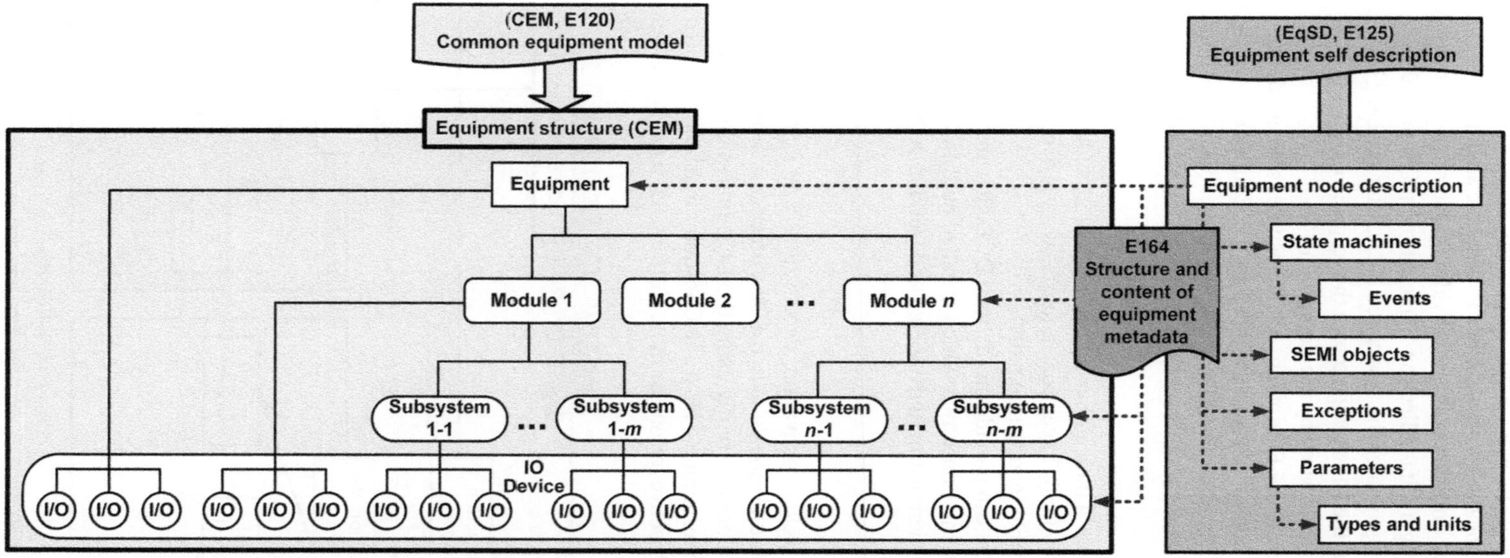

Figure 3.33 E164 in CEM and EqSD.

3.2.2.5 Data Collection Management (DCM) (SEMI E134)

E134 [10] aims to offer a means for applications to organize all data needs (Event, Exception, and Trace) into logical, named units that can be individually activated and deactivated. As depicted in Figure 3.34, after receiving a data-collection request from a Client, the Data Collection Manager (DCM) of the Equipment generates the associated DCP. After the DCP being successfully created, it must be activated. Then the Client will receive the respective Data Collection Reports (DCRs) as configured in the DCP and prepared by the Equipment.

In E134, both interfaces of the Equipment and Consumer side are defined. The DCM interface includes client-initiated operations implemented in the Equipment, while the DCP Consumer interface which includes equipment-initiated fire-and-forget operations implemented in the client for the Consumer side. These operations are as shown in Figure 3.35.

DCP Definition

The DCP defined in E134 is a data gathering request that includes a set of Event requests (with a configurable set of parameters), Exception requests (with a fixed set of parameters), and Trace requests (with a configurable set of parameters).

The equipment suppliers can provide pre-defined DCPs that are included in the Equipment, and no definition required by the Consumer, which is called built-in DCP. Built-in DCPs can be activated and viewed by any client using ActivatePlan and GetPlanDefinition but cannot be deleted by any Consumer.

DCP Operation

The operations of DCP include activation, deactivation, and deletion. Multiple Consumers can activate the same DCP simultaneously. When a Consumer deactivates a DCP, the DCP will remain active for all other Consumers currently using the DCP, unless the Consumer indicates the DCP should be terminated. To terminate a DCP, the Consumer must have granted sufficient user privileges. When a DCP is still active for one or more Consumers, it cannot be deleted.

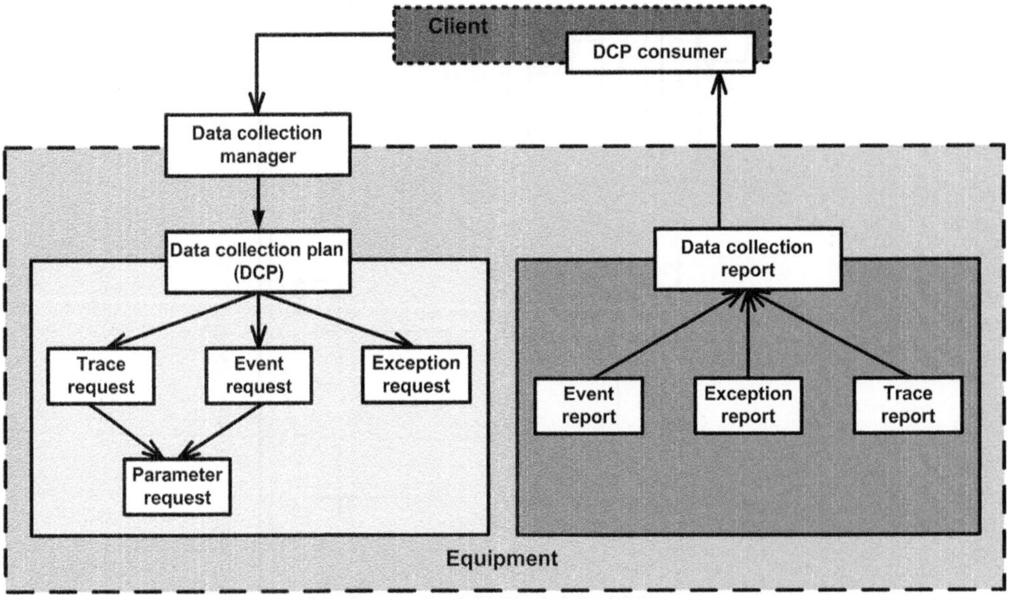

Figure 3.34 Overview of data collection.

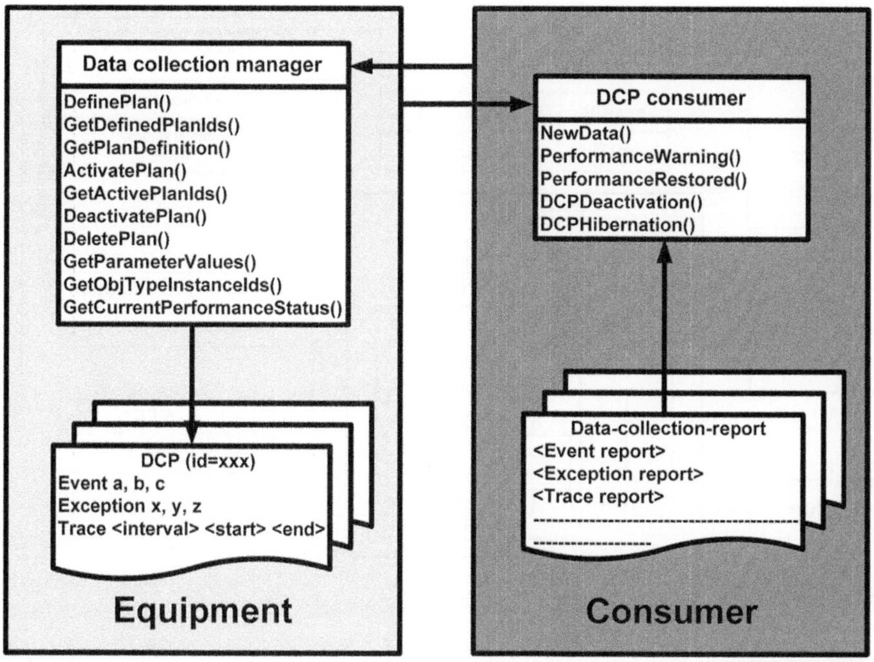

Figure 3.35 Data collection manager and consumer interfaces.

DCP is uneditable. If the change of DCP is required, the Consumer has to delete the defined DCP and define a new DCP instead.

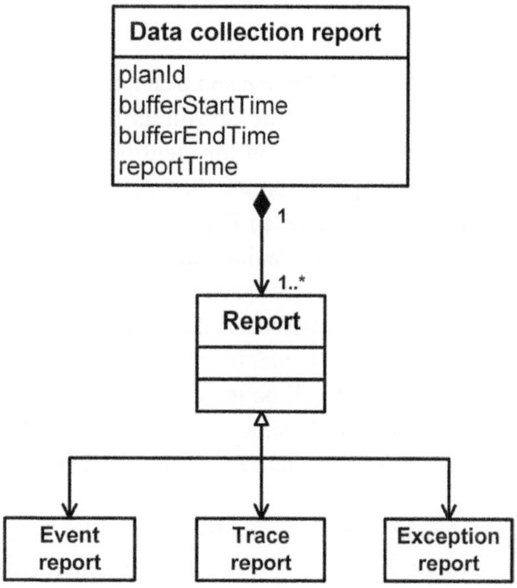

Figure 3.36 Definition of data collection report.

DCR Definition

The DCR consists of Event Report, Exception Report, and Trace Report as shown in Figure 3.36. The following paragraphs elaborate the definitions and examples of each part.

Event Report Definition and Example When an Event occurs and upon the Consumer's request, the Event Report is generated. An Event Report is identified by sourceId, eventId, and eventTime. The structure is as the left part Figure 3.37 shown. Every Report may have zero or more ParameterValues. The ParameterValues are the actual data value and appear in the same order as they were listed in the DCP. There may be more than one ParameterValue or zero in an Event Report. While AbstractValue appears as a value in the data type, and its formats vary according to the data types.

An example of the Event Report is shown in the right part of Figure 3.37. The sourceId "PM-1" is a part in the Equipment, while the eventId "ProcessEnd" indicates this event report is generated

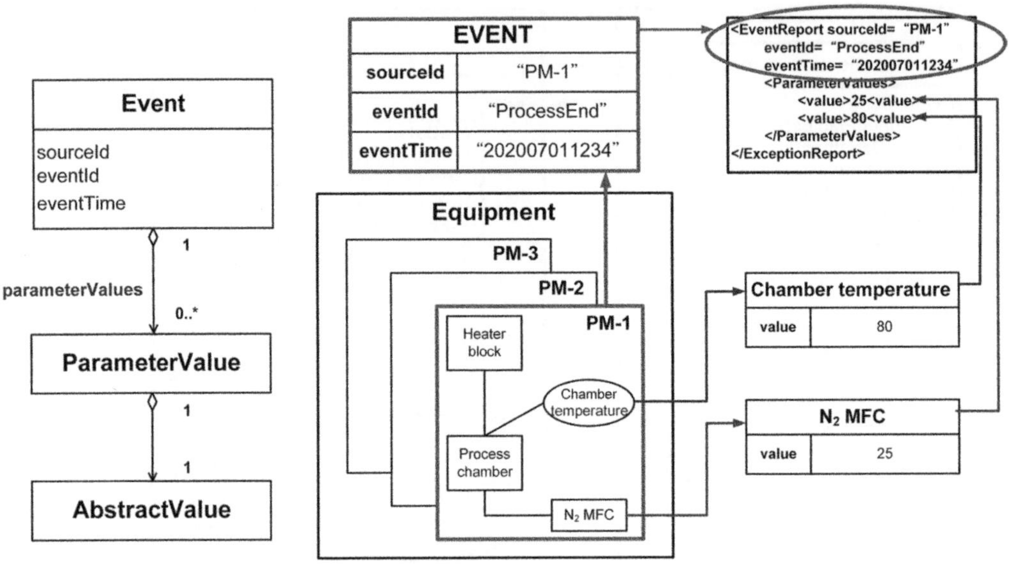

Figure 3.37 Definition and example of event report.

at the end of the process. The digits in eventTime further explain the year, month, date, and time of the specific point. In this report, we can see two different ParameterValues from different sections of the Equipment. One is for chamber temperature, while the other is the flow of N2 mass flow controller.

Exception Report Definition and Example If the Consumer requested, an Exception Report will be generated and sent to the Consumer once an Exception occurs. An Exception Report consists of the following information for identification: sourceId, exceptionId, exceptionTime, severity, and state. The state can be either "Set" or "Clear" and must be set if the exception is applicable for this option. Same as the Event Report, the ParameterValue is the actual data value and it may be zero. There may also be more than one ParameterValues in one Exception Report and they are listed in E125. The AbstractValue may be a value in a data type, such as °C or °F in temperature. The formats are different based on data types. Take Figure 3.38 as an example, the scourceId "PM-1. Heater" means the heater block in the PM-1 part of the equipment; while the eventId "OverHeat" means the report will be generated when the heater block is overheating, then the eventTime, severity, state, and ParameteValues further define the details of this exception: When the voltage reaches 30, and the chamber temperature reached 95°.

Trace Report Definition and Example The Trace Report is the collection of collected data. The structure is as Figure 3.39 shown. It is generated when an Event or Exception happens, and either of them can be the Trigger. There will be a specific startTriggerTime and stopTriggerTime, between these two points is the specific sampling period defined within DCP. Depending on whether the DCP is buffering, two possible scenarios may happen. When DCP is buffering, the CollectedData will be on hold and send out to the Consumer one time at the point of reaching the stopTrigger. If the DCP is not buffering, then the content (CollectedData) of Trace Report will keep growing until it reaches the maximum group size defined in the DCP and send to the Consumer. The Consumer may receive more than one Trace Report before reaching the

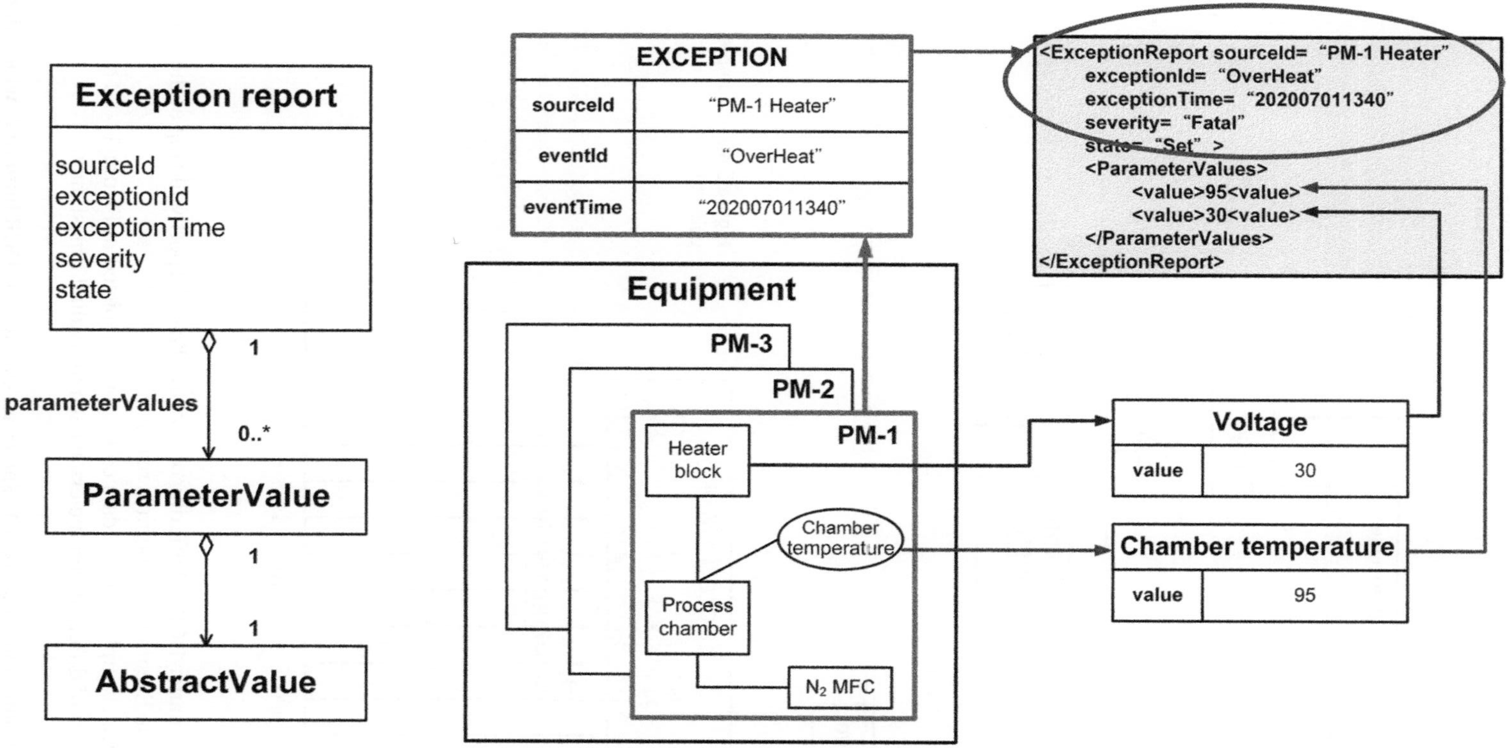

Figure 3.38 Definition and example of exception report.

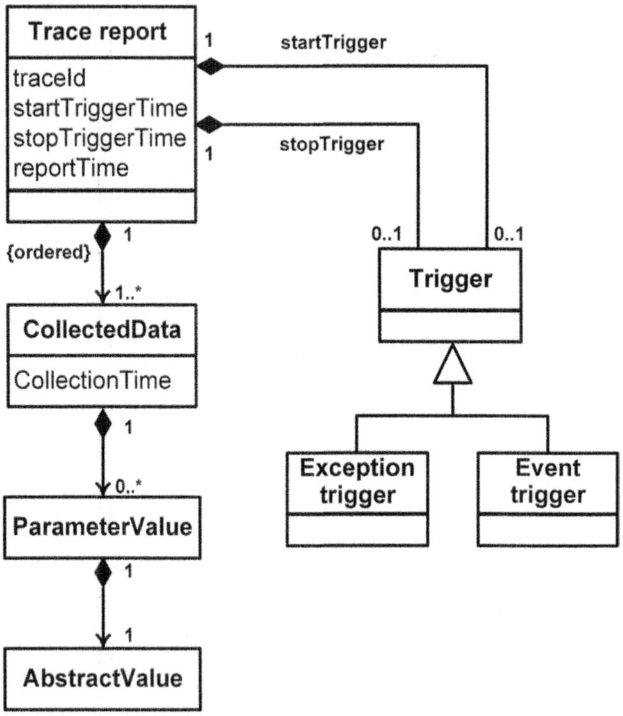

Figure 3.39 Definition of trace report.

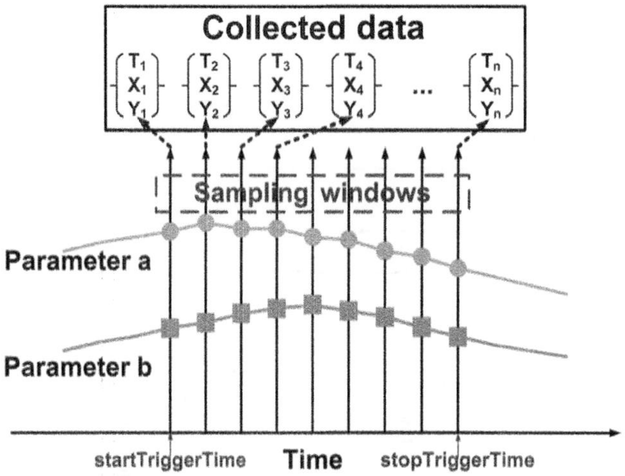

Figure 3.40 Example of trace data collection.

stopTrigger. No matter DCP is buffering or not, the Trace Report will stop at the point when the stopTrigger of the Event or Exception is reached.

The illustrative example of Trace data collection is depicted in Figure 3.40. The Collected Data contain the values of different parameters at each individual timestamp.

Example of DCR Buffering Figure 3.41 shows an example of DCR buffering, which demonstrates that two Event Reports, one Trace Report, and one Exception Report are mixed in the DCR.

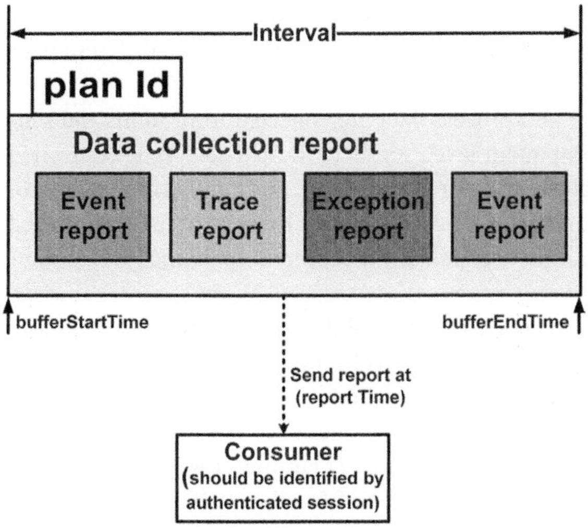

Figure 3.41 Example of DCR buffering.

3.3 Communication Standards of the Industrial Devices and Systems

As depicted in Figure 3.42, the industrial automation pyramid consists of component, device, control, plant, and company levels. A Reference Architecture Model for Industry 4.0, (RAMI 4.0) [12] was presented at the Hannover fair in April, 2015. RAMI 4.0 integrates elements in Industry 4.0 into a three-dimensional layer's model so as to classify and develop those structure technologies based on Industry 4.0. Melo and Godoy [13] developed an open source controller interface for Industry 4.0 based on RAMI 4.0 and Open Platform Communication Unified

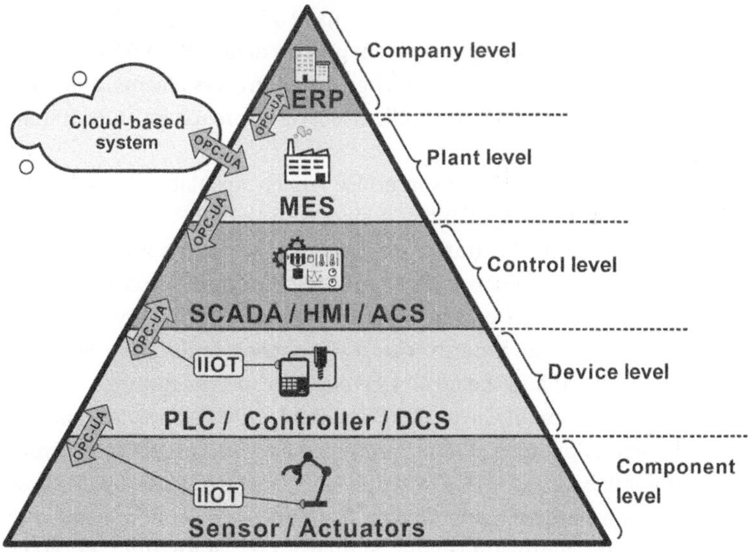

Figure 3.42 Automation pyramid.

Architecture (OPC-UA) protocol due to the fact that OPC allows industrial devices at all levels of the automation pyramid shown in Figure 3.42 to operate with different protocols on different platforms to communicate with each other.

By using OPC-UA, production information and process data are collected from industrial sensors, machine actuators, or motors in the component level. As such, these data can be controlled or monitored by machine tools, programmable logic controllers (PLCs), and/or distributed control system (DCS) in the device level. A large number of field devices can be horizontally connected among the device level of the automation pyramid to enable the machine-to-machine (M2M) communication without human intervention.

Further, OPC-UA is not only targeted at the sensors, actuators, PLC, and DCS interfaces but also as a way to provide greater interoperability between higher level units. Vertical integration with higher levels, including supervisory control and data acquisition systems (SCADA), human-machine interfaces (HMI), and advanced control systems (ACS) in the control level, manufacturing execution systems (MES) in the plant level, and enterprise resource planning (ERP) systems in the company level, should also be established for the requirements of Intelligent Manufacturing. In addition, OPU-UA also brings out information integration with the industrial Internet of things (IIoT) and cloud-based services.

In the following, the historical roadmaps of classic OPC and OPC-UA are described. Then, the specifications and target applications of OPC-UA are introduced. Finally, an Intelligent Manufacturing hierarchy example of Flexible Copper Clad Laminate (FCCL) industry applying OPC-UA protocols is presented.

3.3.1 Historical Roadmaps of Classic Open Platform Communications (OPC) and OPC Unified Architecture (OPC-UA) Protocols

The historical roadmaps of Classic OPC and OPC-UA are described as below.

3.3.1.1 Classic OPC

According to different requirements within industrial applications, three major OPC specifications have been developed [14]: Data Access (DA) in 1996, Alarm & Events (A&E) in 1999, and Historical Data Access (HDA) in 2001. OPC DA defines the access to current process data, OPC A&E describes an interface for event-based information, including acknowledgment of process alarms, and OPC HDA presents various functions to access archived data. A client–server approach for information exchange is utilized by OPC applications.

An OPC server makes the information available via its interface and encapsulates the source of process information like a device. As for an OPC client, it connects to the OPC server and can access and consume the offered data [14]. An illustrative use case of OPC clients and servers is depicted in Figure 3.43. An OPC DA server connects with a CNC machine and is accessed by an OPC DA client built in the SCADA, HMI, or ACS system. Similarly, the OPC HDA server and the OPC A&E server can send historical reports and alarm records to external units from SCADA, HMI, or ACS system via the OPC HDA client and the OPC A&E client, respectively.

OPC DA is the first and still the most successful Classic OPC standard. OPC DA was designed as an interface to communication drivers, allowing a standardized read and write access to current data in automation devices. The major use case of OPC DA is that HMI and SCADA systems access data from different types of automation hardware and devices from different vendors using one defined software interface supplied by the hardware vendor. Similarly, OPC A&E and OPC HDA are also designed to access information provided by the SCADA systems.

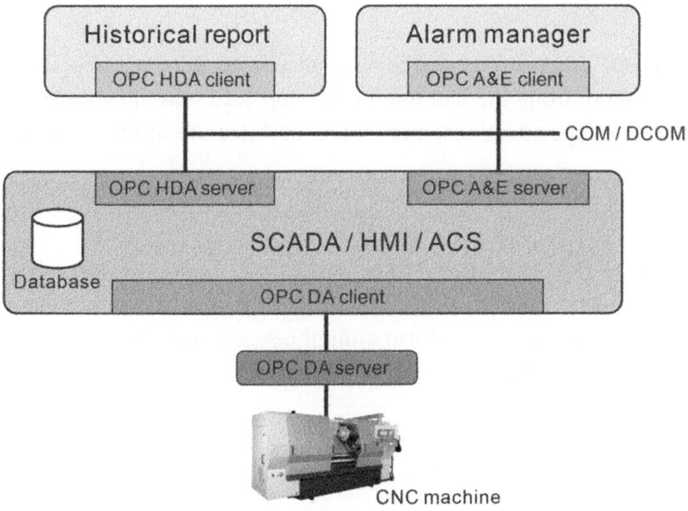

Figure 3.43 Illustrative use case of OPC clients and servers.

The Component Object Model (COM) and Distributed COM (DCOM) technologies from Microsoft are adopted for implementing Classic OPC interfaces so as to reduce the specification work to the definition of different APIs for different specialized needs without the requirement to define a network protocol or a mechanism for inter-process communication. Nevertheless, DCOM is difficult to configure due to its long and nonconfigurable timeouts, and may not be used for Internet communication. As a result, there are two main disadvantages: the DCOM issues when using remote communication with OPC and the Windows-platform-dependency of OPC.

3.3.1.2 OPC-UA

Classic OPC is used today as a standardized interface between automation systems in different levels of the automation pyramid. Manufacturers wish to use a standard like Classic OPC in many more areas without success because of the COM dependency of OPC or because of the limitations for remote access using DCOM.

The first approach of the OPC Foundation to maintain successful features of OPC but to use a vendor and platform neutral communication infrastructure is to define OPC XML-DA. However, there are several reasons why a created web-services version of the successful OPC specification cannot fulfill the requirements of a new OPC generation. One reason is the poor performance of XML Web Service (WS) compared with the original COM version. Moreover, interoperability problems occur when applying different XML Web Service stacks. To remove the limitations of Classic OPC and deal with the issue of platform independence, the OPC member companies brought forward the requirements to expose complex data and complex systems.

In 2006, OPC-UA version 1.0 was born out of the desire to create a true replacement for all existing COM-based specifications without losing any features or performance. Additionally, it must cover all requirements for platform-independent system interfaces with rich and extensible modeling capabilities being also able to describe complex systems. The wide range of applications where OPC is used requires scalability from embedded systems across SCADA and DCS up to the MES and ERP systems [14].

3.3.2 Fundamentals of OPC-UA

As the aforementioned introduction, OPC-UA is proposed to be applicable to all components in a standard industrial automation-pyramid from the bottom to the top levels as illustrated in Figure 3.42. Requirements, foundations, specifications, and system architecture of OPC-UA are addressed in this subsection.

3.3.2.1 Requirements

To overcome the limitations of Classic OPC discussed in Section 3.3.1.2, the requirements for OPC-UA are specified: Data Modeling and Communication. The first requirement is the Data Modeling rules describing a system and the available information; the second requirement is the establishment of the basic Communication between distributed systems.

Data Modeling

Items for Data-Modeling requirement are listed below [14]:

- Common model for all OPC data
- Object-oriented design
- Abstract base model
- Meta information
- Complex data and methods
- Extensible type system
- Scalability from simple to complex models
- Base for other standard data models

A common descriptive model is definitely the most important for all data that have to be modeled via OPC-UA. In the past, Classic OPC was very limited by its COM/DCOM-based technologies developed by Microsoft. Moreover, the common model should be able to describe relationships between underlying systems and applications in object-oriented properties with the abstract-based models and meta-information structures. An extensible architecture design of the common model is also required to provide the ability to rapidly extend the system without major changes in its basic architecture. The methods need to be reusable and callable by distributed systems in various field levels to make OPC-UA flexible.

Communication

Items for the requirement of Communication between distributed systems are listed below [14]:

- Robustness design
- Fault-tolerant ability
- Redundancy
- Platform-independence
- Scalability
- High performance
- Internet and firewalls
- Security and access control
- Interoperability

To create an effective communication platform, the reliability and availability of exchange information among distributed systems have to be improved first. The redundancy, platform-independence, and scalability enable OPC-UA to be rapidly executed on a variety of hardware and software. The performance in Intranet and Internet communication environments is another issue to be concerned. Moreover, network security and access control issues must be taken care of; and the interoperability between different systems should be fulfilled as well.

OPC-UA applies two roles: client and server to achieve its basic concept. This basic concept enables an OPC-UA client to access the smallest pieces of data without needing to understand the whole model exposed by the OPC-UA server from complex systems.

3.3.2.2 Foundations

To satisfy aforementioned requirements addressed in Section 3.3.2.1, two fundamental components: Meta Model and Transport, are defined in OPC-UA foundations and depicted in the basement of Figure 3.44.

Meta Model defines the modeling rules and base modeling constructs (or the so-called building blocks) which are necessary to expose an information model with OPC-UA formats. It defines the entry points of OPC-UA into a specific memory structure (the so-called Address Space) and basic types used to build an OPC-UA hierarchy. These basic building blocks can be extended by Base Information Models and Information Models as shown at the top of Figure 3.44.

Transport defines two mechanisms for different use cases: web services and optimized binary TCP protocol. Web services are used for firewall-friendly Internet communication; while optimized binary TCP protocol, or TCP UA binary, is used for high-performance Intranet communication.

Services, shown in the middle of Figure 3.44, are the interfaces defined in an abstract manner between OPC-UA servers and OPC-UA clients. OPC-UA servers play the role of supplier to provide Information Model; while OPC-UA clients act the role of consumers to use the Information Model. All Services are executed to exchange Information Models based on Meta models between OPC-UA clients and OPC-UA servers using the Transport mechanisms.

Based on these foundations, OPC-UA defines rules to describe and transport data; while collaborating vendors define data structures they want to describe and transport depending on their Information Models.

Figure 3.44 Foundations of OPC-UA

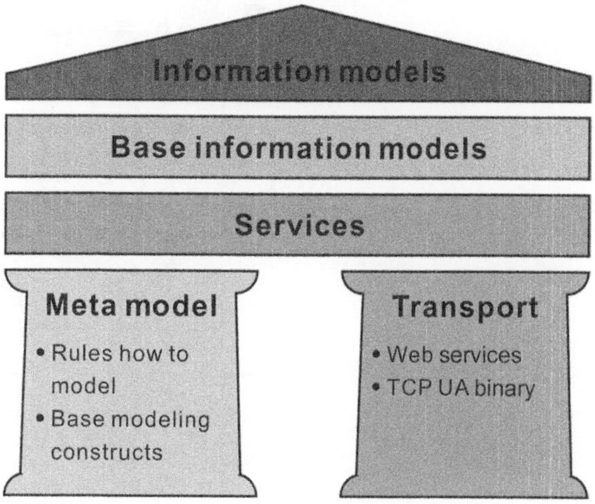

3.3.2.3 Specifications

OPC-UA specifications address an overview of multiple specification parts as below [15].

- **Core Specifications**
  - Part 1: Concepts
  - Part 2: Security Model
  - Part 3: Address Space Model
  - Part 4: Services
  - Part 5: Information Model
  - Part 6: Service Mappings
  - Part 7: Profiles
- **Access Type Specifications**
  - Part 8: Data Access
  - Part 9: Alarms & Conditions
  - Part 10: Programs
  - Part 11: Historical Access
- **Utility Specifications**
  - Part 12: Discovery
  - Part 13: Aggregates

The Core Specifications (Part 1–Part 7) split into the basic features of generic parts (Part 1–Part 5) and the web-service parts (Part 6 and Part 7) for an OPC-UA system; the Access Type Specifications (Part 8–Part 11) mainly specify the OPC-UA information models defined in Classic OPC; and the Utility Specifications (Part 12 and Part 13) provide additional tools designed to help supporting the OPC-UA infrastructure. More detailed contents of these specifications can be referred to the OPC-UA Online Reference provided by the OPC Foundation [16].

3.3.2.4 System Architecture

An OPC-UA system architecture refers to a client-server communication infrastructure that concurrently connects all distributed devices. Figure 3.45 illustrates the platform-independent and service-oriented features of OPC-UA. Each OPU-UA-enabled device exchanges secure and reliable

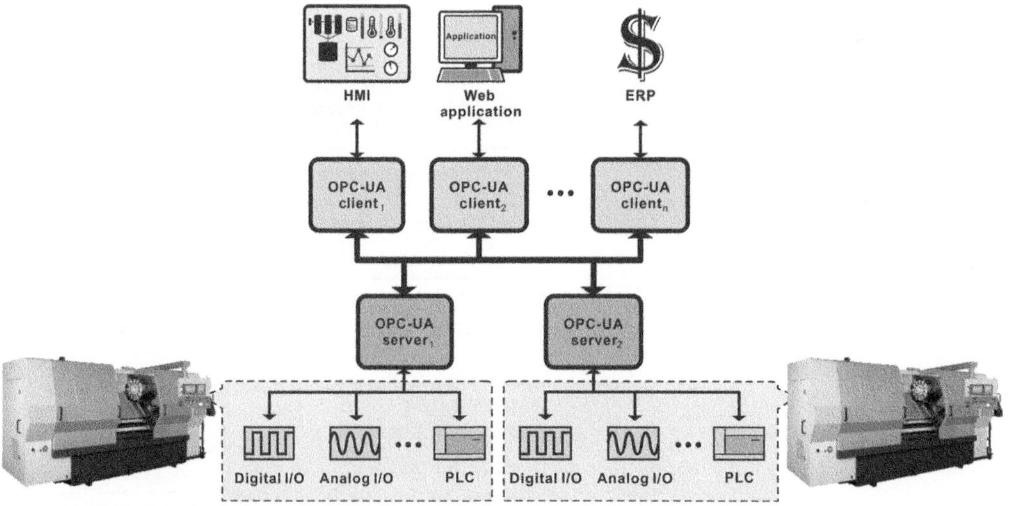

Figure 3.45 OPU-UA architecture.

data between multiple OPC-UA clients and servers over various networks. An OPC-UA client may interact concurrently with one or more OPC-UA servers and an OPC-UA server may also interact concurrently with one or more OPC-UA clients [15].

OPC-UA servers are able to access specific data sources like digital/analog/PLC signals from devices, sensors, databases, and/or other applications. OPC-UA clients are integrated into various application systems, such as an MES or an ERP.

End-users have to define sets of services and attributes that the OPC-UA server can provide. Accordingly, the OPC-UA server provides current and historical data to the OPC-UA client, or OPC A&E to notify the OPC-UA client of important changes.

Thus, there are large-scale interactions among the platform-independent OPC-UA clients and servers. For example, as shown in Figure 3.45, OPC-UA Client_1 may send request messages to OPC-UA Server_1 and Server_2; then the two OPC-UA servers send corresponding response messages to OPC-UA Client_1 and transmit required information to OPC-UA Client_1. Moreover, when OPC-UA Client_1 receives any event notification, it can also directly read more information from the specific OPC-UA server or write values into the specified data item in the OPC-UA server.

The security issue of aforementioned interactions should also be taken into considerations. Brief concepts of services, architecture overview, and security model of OPC-UA, as well as an illustrative sequence diagram, are described in the following subsections.

OPC-UA Services

One core concept of OPC-UA specifications is the service-oriented architecture. OPC-UA Services are the collection of remote procedure calls, which clearly defines several service-sets with parameters and behaviors to describe various standard interactions and operations occurring between OPC-UA clients and servers.

All service-sets are designed in an abstract and generic type so as to enable the implementation to be achieved in platform-independency and scalability. For example, Read is the one and only method used to read attribute values in OPC-UA standards even when there are many OPC-UA-enabled systems exist in different platforms. Required service-sets will be briefly stated using the italic font along with the overview and concepts description in following contents of this subsection. Detailed functionalities of different service-sets can be found in [17].

Architecture Overview

Three key layers of Application, Software Development Kit (SDK), and Stack are respectively designed to fulfill demands of developing an OPC-UA-based system [15]. As shown in Figure 3.46, both the OPC-UA client and server are composed of the three layers. This architecture separates the design logic into two parts: (i) the common functionality that can be reusable by OPC-UA developers, corresponding to the SDK layer and the Stack layer; while (ii) the use-case-specific requirements that can be defined by end-users, corresponding to the Application layer.

To save developing time and cost, the SDK layer and Stack layer aim to build a maintainable system, which defines reusable plug-and-play functions, application programming interface (API), services, or modules to efficiently deal with some common and routine procedures.

Lower-level functions defined in the Stack layer perform message processing including encoding, securing, and transmitting for each OPC-UA message; while higher-level functions defined in the SDK layer provide various interfaces and methods so that OPC-UA clients can access OPC-UA servers over networks. Both OPC-UA clients and servers are located on the Application layer since their design logic covers well-tailored functionalities strictly specified by end-users.

Note that, any message transmitted in the OPC-UA client and server is passed via the SDK layer and the Stack layer in the special APIs or functions. Both the OPC-UA client and server use the

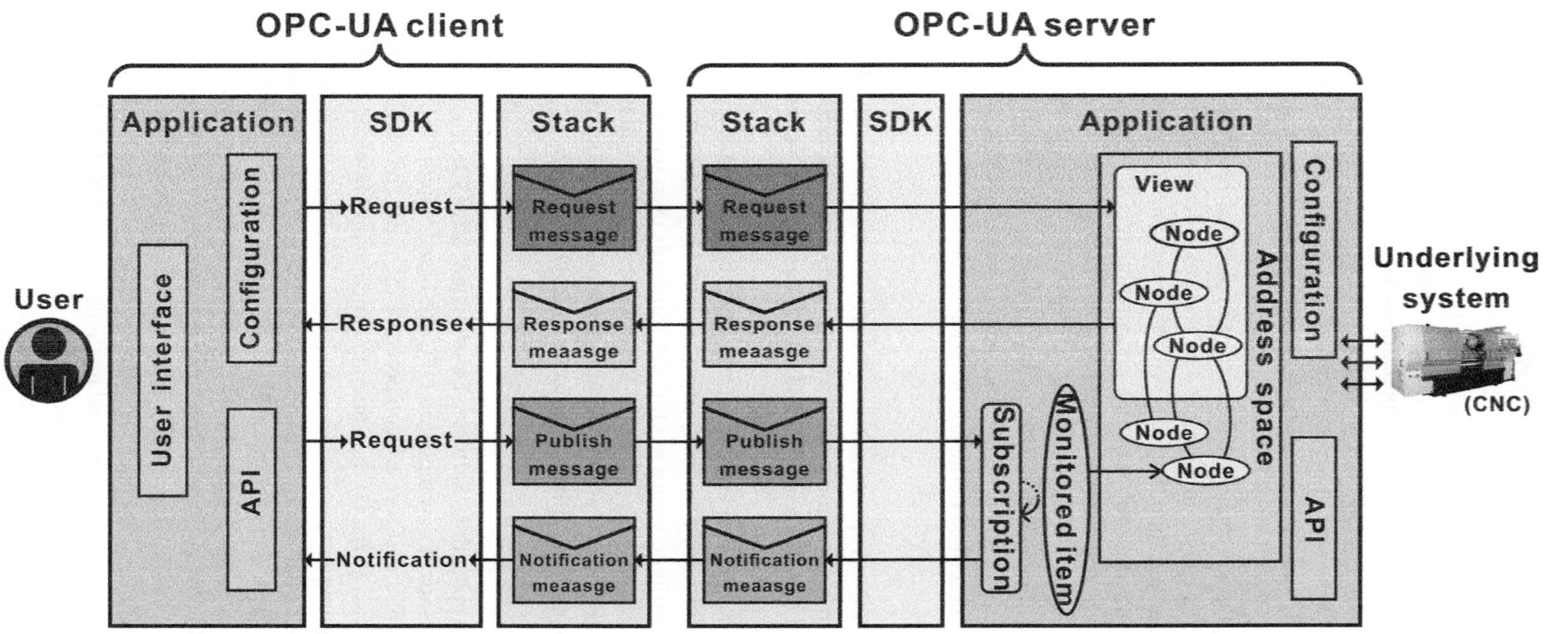

Figure 3.46 Architecture overview of an OPC-UA client and an OPC-UA server.

same stack infrastructure, which provides a lot of common functionality that can be reused for both sides such as encoding or securing messages.

OPC Foundation not only enables OPC-UA in any operating system but also provides three available programing language of implementing the complete Stack deliverables with C/C++, .NET, or JAVA. Detailed architectures of the OPC-UA can be referred to Part 1 of [15].

OPC-UA Client The OPC-UA client is an application that can access and consume the offered data that are temporarily stored within the memory resided on the OPC-UA server-side. The OPC-UA client connects with the OPC-UA server, to send request messages to the OPC-UA server and receive responses from the OPC-UA server and then go back into wait state.

Since OPC-UA is a two-way (read/write) communication standard, the OPC-UA client can also instruct the OPC-UA server to send updates when they come in the OPC-UA server. This benefit provides the connectivity advantage for all levels of the automation architecture. OPC-UA is well-suited for connecting the factory floor to the enterprise.

As shown on left-hand side of Figure 3.46, the OPC-UA client needs a specific user interface to display available data items on it and allow users to navigate the OPC-UA server's memory. Thus, three types of OPC-UA Services: Data Search, Data Monitoring, and Memory Management of the huge OPC-UA server system are required to provide the OPC-UA client with the ability of defining an elaborate information-rich data-collection structure.

- **Data Search**
  View service-sets allow OPC-UA clients to browse through a specific subset of data items that users interest in within the OPC-UA server's memory since underlying systems are often large and complex. *Query* service-sets allow OPC-UA clients to perform a directly filtered search of subset data from the OPC-UA server's memory without knowing the logical schema represented for internal storage structures.
- **Data Monitoring**
  MonitoredItem service-sets allow OPC-UA clients to define a set of data items that have to be monitored by the OPC-UA server. *Subscription* service-sets allow OPC-UA clients to create/modify/delete a subscription. Once a *Subscription* is created, the OPC-UA server periodically returns *Notifications* generated from *MonitoredItem* to the OPC-UA clients in response to *Publish* at every publishing interval.
- **Memory Management**
  NodeManagement service-sets allow the OPC-UA clients to modify data items and references in the OPC-UA server's memory hierarchy. *Attribute* service-sets allow OPC-UA clients to read, write, or update current/historical values of data items.

As the Stack layer of the OPC-UA client in Figure 3.46, four messages are queued from top to bottom. The first request is sent to the OPC-UA server and lately the corresponding response is then received from the OPC-UA server through the OPC-UA client APIs. After that, a *Publish* request command is sent to the OPC-UA server for updating the monitored items, and a *Notification* message is returned to the OPC-UA client from the OPC-UA server in response to that *Publish* request, which contains *notifications* that have not yet been reported to the OPC-UA client.

OPC-UA Server The OPC-UA server is an application that can access underlying systems to retrieve data and make the information available via its interface to OPC-UA clients. The main responsibilities are to communicate with the underlying system according to requested commands sent from OPC-UA clients.

As illustrated on the right-hand side of Figure 3.46, the OPC-UA server has to directly convert the underlying system's signals and parameters into formats that OPC-UA can read. Four messages corresponding to the Stack of OPC-UA client are also queued in the Stack layer of the OPC-UA server from top to bottom.

Note that, one of the OPC-UA server's feature is to utilize a specific memory space called the Address Space to interchange and share data with the OPC-UA clients in a standard way. All the underlying real-time process data and the entire enterprise information are integrated into the OPC-UA server's Address Space so that they are available for further processing. As the Address Space depicted in Figure 3.46, the OPC-UA server allows the OPC-UA clients to access and manipulate data items via a set of OPC-UA Nodes within its Address Space, which are visible and free to be organized by *View*. These Nodes can be expanded/collapsed to check its relationships between other children Nodes in a tree structure. Each OPC-UA Node can be modified by the OPC-UA client through various OPC-UA Services introduced in Section 0, such as data monitoring or memory management.

Thus, an OPC-UA client might firstly check Nodes via *View* and send a command request to the OPC-UA server. When passively receiving a request sent from the OPC-UA client, the OPC-UA server starts to retrieve Node values and send response messages to the OPC-UA client. In contrast, when the OPC-UA server detects the data changes of selected Nodes or an event/alarm occurrence, a *Notification* message is actively generated and transferred to the OPC-UA client by *Subscription*, which contains a set of *MonitoredItem*.

Security Model

OPC-UA connects industrial controllers in factories with the Internet. Thus, there are many different aspects of security that have to be considered when developing applications, including the security threats of the physical, hardware, and software environments when OPC-UA is expected to run. Possible damages caused by industrial devices in critical facilities can be severe. Since OPC-UA specifies a communication protocol, the focus is on the security of the data exchanged between applications.

OPC-UA utilizes a sophisticated security model to reduce internal and external threats and attacks, such as message flooding, unauthorized disclosure of message data, message spoofing, message alteration, and message replay [18]. It supports robust, secure communication that assures the identity of OPC-UA applications and resists attacks. Detailed descriptions of the security model can be found in Part 2 of [18].

Figure 3.47 shows that the OPC-UA security model secures all the incoming and outgoing messages between a pair of OPC-UA server and client. Processes of security check are performed in the three layers of the OSI Model [19] above the Network Layer: Transport Layer, Communication Layer, and Application Layer.

- Transport Layer
  The Transport Layer creates the TCP socket connection to transmit and receive the encrypted/decrypted data between OPC-UA clients and servers. For any OPC-UA client that connecting with an OPC-UA server at the transport layer, it firstly starts with a handshaking process. This layer has to ensure the connection state is available and recover any broken connection so that it can prevent data loss.
- Communication Layer
  On the top of the Transport Layer, a secure channel is built at the Communication Layer to meet the confidentiality and achieve integrity via *SecureChannel* service-sets. *SecureChannel* allows the OPC-UA client to establish a secure communication channel to ensure the confidentiality

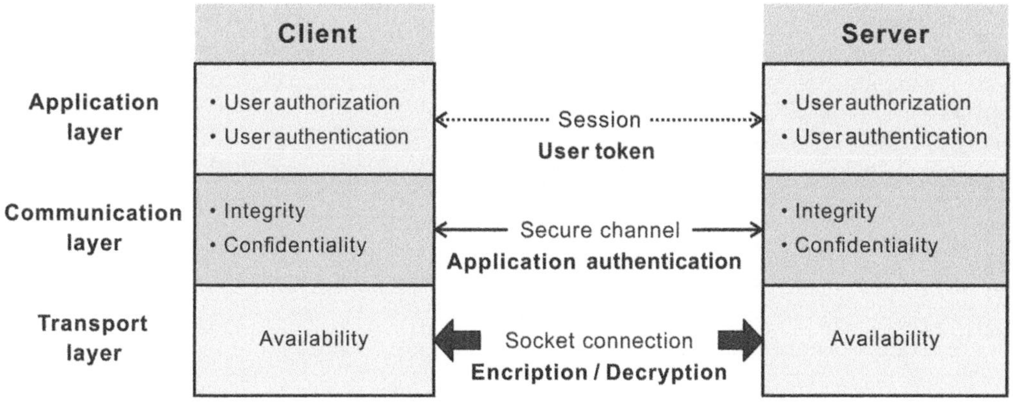

Figure 3.47 OPC-UA security model.

and integrity of all incoming messages exchanged with the OPC-UA server. Different from most service-sets designed in the applications, *SecureChannel* is directly implemented in the OPC-UA Stack; while other service sets are built in the OPC-UA applications.

Communication Layer prevents all messages from unauthorized changes and eavesdropping by the encryption and the digital signature between OPC-UA clients and servers. Furthermore, the digital certificate is used to authenticate the communication of individual instances between OPC-UA applications.

- Application Layer
  A session is built to secure the Application Layer via *Session* service-sets. Session allows the OPC-UA client to establish and administrate user-specific connection with the single OPC-UA server, and authenticate the process of verifying the user.

Application Layer is used to exchange the plant information, parameters, settings, or real-time process data. For any user who intends to establish a Session with the specified OPC-UA server on the OPC-UA client side, it has to be allowed to become a trusted user at first by the authentication and authorization processes of the OPC-UA server. The OPC-UA server verifies the user identity token passed from the OPC-UA client so that both of the OPC-UA client and server are able to recognize who the user is and what resources can be authorized to use. Note that, the Session has to be executed based on a SecureChannel connection, which is built in advance at the Communication Layer. Thus, the Session also can be secured by the SecureChannel.

More basic concepts for OPC-UA security are defined in OPC-UA Specification Part 2 [15]. In addition, OPC-UA Specification Part 6 describes how the SecureChannel is implemented.

Sequence Diagram of OPC-UA

To make a concise summary of Section 3.3.2.4, a sequence diagram of the connection between a pair of OPC-UA client and OPC-UA server is depicted in Figure 3.48.

Application and Stack introduced in Section 0 are demonstrated in Figure 3.48 to illustrate the interactions between OPC-UA client and OPC-UA server. Two crucial roles, Address Space and Nodes inside the OPC-UA server, interact with others. SDK is skipped in Figure 3.48 since it is a middleware that manages and passes services to each other. Most of the callable APIs are implemented in SDK. A Local Discovery Server is responsible to ensure the very first interoperability step between an OPC-UA client and all available OPC-UA servers by providing a list of available OPC-UA servers to any application that queries it. The sequence diagram can be divided into seven

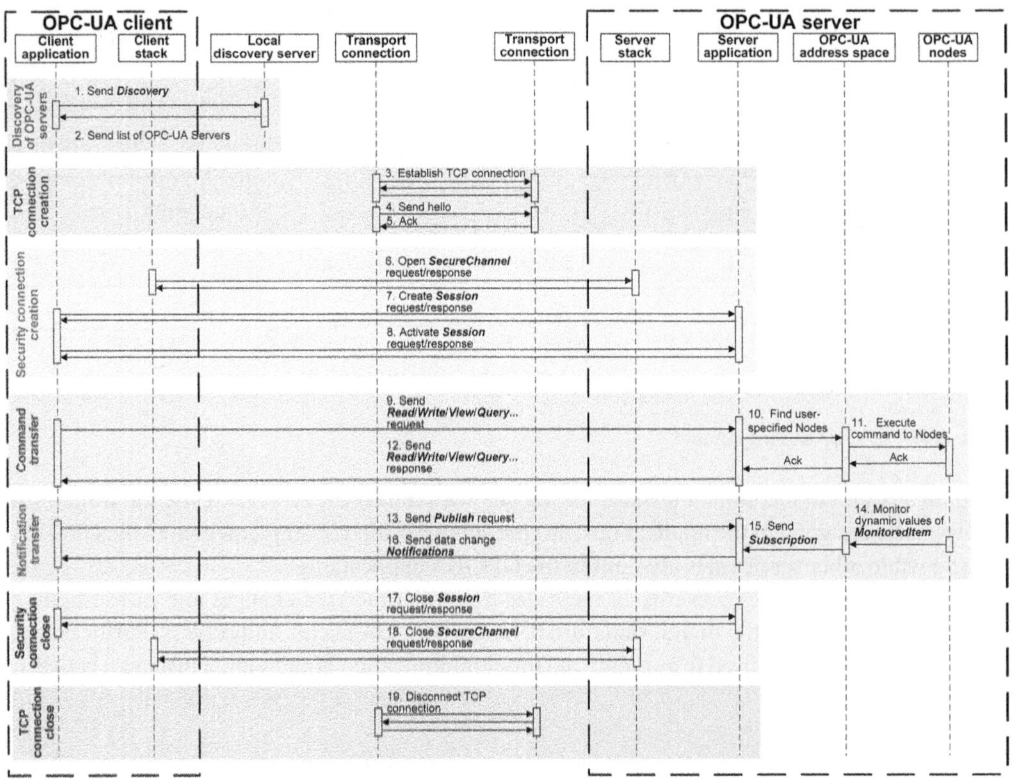

Figure 3.48 Sequence diagram of operating procedure between a pair of OPC-UA client and OPC-UA server.

parts of interactions as highlighted in different color blocks in Figure 3.48, and procedures are described as below.

- **Discovery of OPC-UA Servers**
  Step 1: The OPC-UA client sends *Discovery* to a free-access Local Discovery Server to find available OPC-UA servers in the network.
  Step 2: Local Discovery Server provides required information including the network addresses, corresponding protocols, and security configurations for the OPC-UA client to establish a connection with a specified OPC-UA server.
- **TCP Connection Creation**
  Steps 3–5: The TCP connection between the OPC-UA client and server is built at the Transport Layer. Before executing the security models, the OPC-UA client sends a "hello" message and then the OPC-UA server replies an "acknowledge" message.
- **Security Connection Creation**
  Step 6: *SecureChannel* is built to secure messages between the Stacks of both the OPC-UA client and server at the lower level.
  Step 7: *Session* is built between the Applications of both the OPC-UA client and server at the higher level.
  Step 8: *Session* is activated to authenticate user between the Applications of both the OPC-UA client and server.

- **Command Transfer**
  Step 9: Application of the OPC-UA client sends request commands to the OPC-UA server, such as *Read*, *Write*, *View*, *Query*, or *NodeManagement*.
  Steps 10–12: Application of the OPC-UA server handles request commands via its Address Space and Nodes. Then it sends the corresponding response messages to the requested commands. The OPC-UA security model is also applied to secure all messages interacted between both the OPC-UA client and server.
- **Notification Transfer**
  Step 13: Application of the OPC-UA client sends *Publish* request to periodically update the monitored items in the OPC-UA server.
  Steps 14–16: Application of the OPC-UA server sends the *Notifications* of *Subscription*, which is generated by *MonitoredItem*.
- **Security Connection Close**
  Steps 17–18: The connection between the OPC-UA client and server is disconnected.
- **TCP Connection Close**
  Step 19: The connection between the OPC-UA client and server is disconnected.

In Section 3.3.3, a practical OPC-UA example is illustrated based on the above procedures to demonstrate the interaction relationships among users, devices, OPC-UA clients, and OPC-UA servers.

3.3.3 Example of Intelligent Manufacturing Hierarchy Applying OPC-UA Protocol

The Intelligent Manufacturing communication realization applying OPC-UA in the FCCL industry is taken as the illustrative example. The illustrative Intelligent Manufacturing architecture is depicted in Figure 3.49. There are three factories in this example, each factory has one or more Equipment Application Program (EAP) Servers to connect to all the production tools. The MES and other factory-wide external information control systems including Recipe Management System (RMS), Statistical Process Control (SPC), Engineering Data Analysis, Fault Detection and Classification (FDC), and Equipment Management System (EMS) communicate with these EAP Servers for engineering data exchange. After the extract-transform-load processes of raw data, the processed engineering data are sent to the Reporting System and Data Warehouse, then operators in the control room can monitor the dashboards of the factory and factory managers can access the reports via mobile devices. Moreover, logistics-related systems including ERP, Customer Relationship Management (CRM), and Business Intelligence (BI) can get the needed data for business operations via Data Warehouse.

In this example, OPC-UA is adopted for two-way communication with production tools. The most important production tool of FCCL processes is Coater. To enhance the yield and OEE of Coater, the monitoring and control systems are required. Originally, there is no communication protocol in Coater, so that the equipment data were collected manually. However, the quality of manually collected data is poor and the volume of the manually collected data is not efficient for the purpose of big data analytics. Therefore, a task force is set up to improve the situation.

To begin with, the required parameter list is defined after the discussion with domain experts. As shown in Table 3.11, besides the sensor physical values (PVs) of key parameters, the corresponding setting values (SVs) of those key parameters are also included for data collection.

Then, referring to RAMI 4.0, the task force decides to adopt OPC-UA as the communication protocol [13]. The master controller of Coater is PLC, and the brands of PLCs include Siemens, Mitsubishi, and Yaskawa. To avoid the interference between PLC and HMI, an additional Ethernet

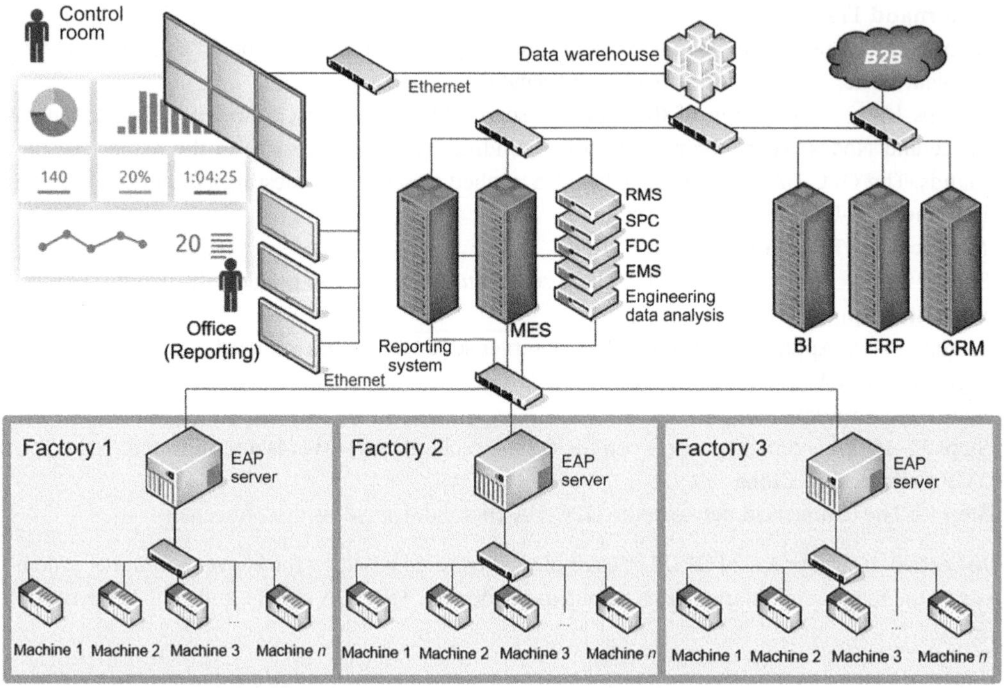

Figure 3.49 Intelligent Manufacturing architecture in the FCCL industry.

Table 3.11 Required parameter list of Coater.

No	Data item	No	Data item
1	Machine speed setting value (SV)	10	Oven intake system physical value (PV)
2	Machine speed physical value (PV)	11	Oven exhaust system setting value (SV)
3	Tension setting value (SV)	12	Oven exhaust system physical value (PV)
4	Tension physical value (PV)	13	Gas density physical value (PV)
5	Lamination temperature setting value (SV)	14	Lamination force (right side)
6	Lamination temperature physical value (PV)	15	Lamination force (left side)
7	Oven temperature setting value (SV)	16	Pump RPM (3-Layer)
8	Oven temperature physical value (PV)	17	Pump RPM (2-Layer)
9	Oven intake system setting value (SV)		

module is installed and configured to each master PLC for individual connection tunnel with OPC-UA. At the same time, the Application modules of OPC-UA servers are installed into the backend servers in the server room, and the Ethernet cables are deployed between PLCs and OPC-UA servers. Afterward, the memory addresses of each individual parameters in PLC listed in Table 3.11 are provided by the equipment engineer; and then Nodes with memory addresses are configured via the OPC-UA operation interface.

The integration tests are performed via OPC-UA testing tools after a Coater is connected to the OPC-UA server. The memory addresses are then inputted with Nodes to the OPC-UA testing tools installed with the OPC-UA client in the EAP Server. Then, the OPC-UA testing tool will start to request the values of parameters in Coater via the OPC-UA client and show the values after receiving the reply from the OPC-UA server that is connected to Coater.

3.3.3.1 Equipment Application Program (EAP) Server

In fact, the functions of an EAP Server are similar as those of an Equipment Manager of the semiconductor industry introduced in Section 1.2.1.1. Observing Figure 3.49, the EAP Server serves as a dual-way communication bridge between Coaters and external systems such as MES, RMS, and FDC, etc.

As depicted in Figure 3.50, Data Access Manager (DAM) is the kernel of EAP. DAM manages the dual-way data exchange between tools (Coaters) and external systems. As for Data Access Plan (DAP) and Data Access Report (DAR), the former defines the data items, types, and update rates and the latter prepares the report according to the corresponding DAP and the buffer-start-time and buffer-end-time. As a result, when a request defined in DAP for DAM to receive/write the data from/to Coater occurs, DAM will send/report the statuses/results via DAR.

DAM can also communicate with the PLC in Coater via OPC-UA protocol. For example, the Dynamic-Link Library (DLL) of the OPC-UA client in the EAP Server can be activated if the development environment is C# of Visual Studio. As such, DAM can read/write the parameters of PLCs via node ID, which is the register address located inside the PLC. The node mentioned here can be the same with the OPC-UA Node defined in Section 0 if the OPC-UA server is directly installed into a device which is PC-based. OPC-UA server used in this section is an external and independent third system from the device. For example, if DAM wants to read the value of a specific node ID, DAM can send the requested command to the OPC-UA server of Coater via the DLL of the OPC-UA client in the EAP Server, and then the OPC-UA server will send the request of

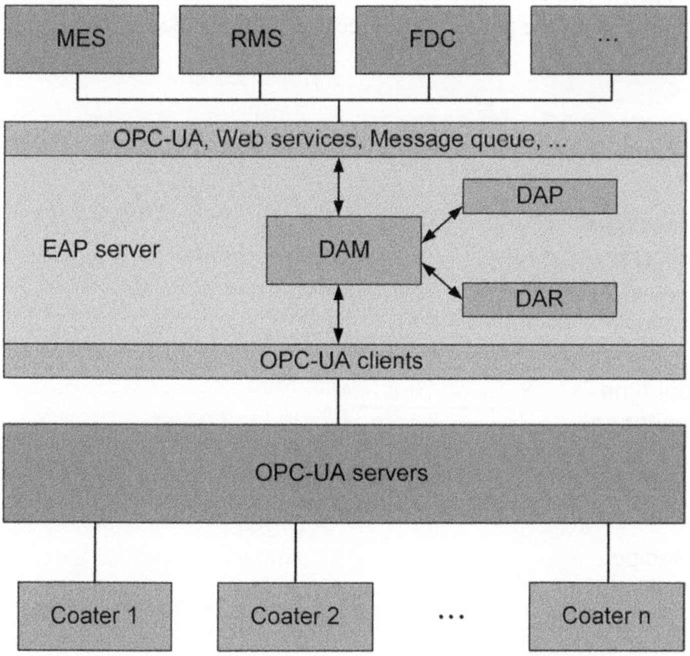

Figure 3.50 Monitoring and control system architecture of Coater by applying the EAP Server.

updating the OPC-UA address space to the device to get the desired data according to the address designated by the OPC-UA client.

OPC-UA also can be served as the communication protocol between EAP and the external systems such as MES, RMS, FDC, etc. [13]. However, some other communication protocols such as web services and message queue may also be adopted by the external systems as shown in Figure 3.50.

3.3.3.2 Use Cases of Data Manipulation

Three use cases shown in Figure 3.51 are planned for the communications between EAP and Coater. The first use case is Data Collection. Hundreds of sensors are deployed in a single Coater for monitoring the status of FCCL production. The EAP Server is deployed to collect values of key parameters of Coater via OPC-UA. The purpose of sensor data collection is to collect big data and then analyze the data for possible productivity and/or quality improvements.

The second use case is Real-Time Monitoring. The purpose of Real-Time Monitoring is to make sure that the production environment is healthy and stable for product quality assurance. Essential and key parameters are selected, and the control limits of those parameters are defined by process engineers after data analysis. Then, the FDC system is applied to perform the anomaly detection beyond control limits. The purpose of Real-Time Monitoring is to make sure that the production environment is healthy and stable for product quality assurance.

Finally, the third use case is Recipe Download. When the Work-In-Process (WIP) arrives the production equipment, the process parameters need to be configured according to the features of this product. Recipe Download is a software function of RMS that can accomplish the purpose of recipe configurations automatically. The contents of various recipes are stored in the database of RMS. When RMS receives the information of batch ID and product ID, RMS will search the suitable recipe name and body (parameter settings), and send them to production equipment. The production equipment will configure the contents of recipe after receiving the recipe name and body from RMS; and then the production process will commence by using this designated recipe.

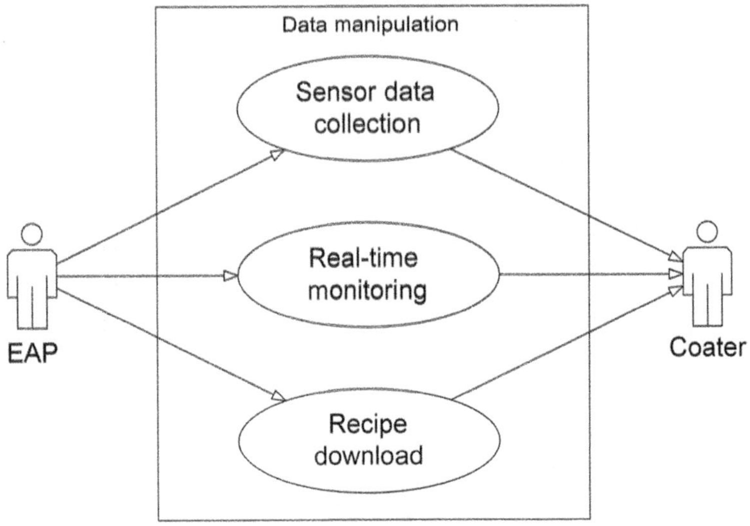

Figure 3.51 Use cases of data manipulation.

3.3.3.3 Sequence Diagrams of Data Manipulation

The sequence diagrams of data manipulation corresponding to Data Collection, Real-Time Monitoring, and Recipe Down are presented as follows.

Sequence Diagram of Data Collection

The sequence diagram of Data Collection is shown in Figure 3.52 and is described below.

Step 1: DAM in EAP loads DAP. DAP is specified by process engineers beforehand.

Step 2: EAP defines DAR based on DAP.

Steps 3–4: EAP sends the data-collection request of *Read* with node ID via the OPC-UA client in EAP to the OPC-UA server of the Coater.

Steps 5–8: The OPC-UA server connects to the PLC in Coater and then collects the data with the corresponding addresses specified in the node.

Steps 9–10: The OPC-UA server returns the collected data via the OPC-UA client to DAM in EAP and then DAM stores the data in DAR.

Step 11: EAP writes the collected data in DAR to Database.

In fact, Steps 1–10 can be grouped as a "Read data" macro procedure via OPC-UA. This macro procedure may also be re-used for other use cases that need the operation of "Read Data" such as Real-Time Monitoring and Recipe Down, etc.

Sequence Diagram of Real-Time Monitoring

Basically, the sequence diagram of Real-Time Monitoring is composed of (i) adopting the "Read Data" macro procedure shown in Figure 3.52 to read the target data in real time via OPC-UA; (ii) sending data to the FDC Server to check if the data is within the allowable specifications or not; and (iii) if it is out of specification, an alarm will be sent to the Alarm Manager.

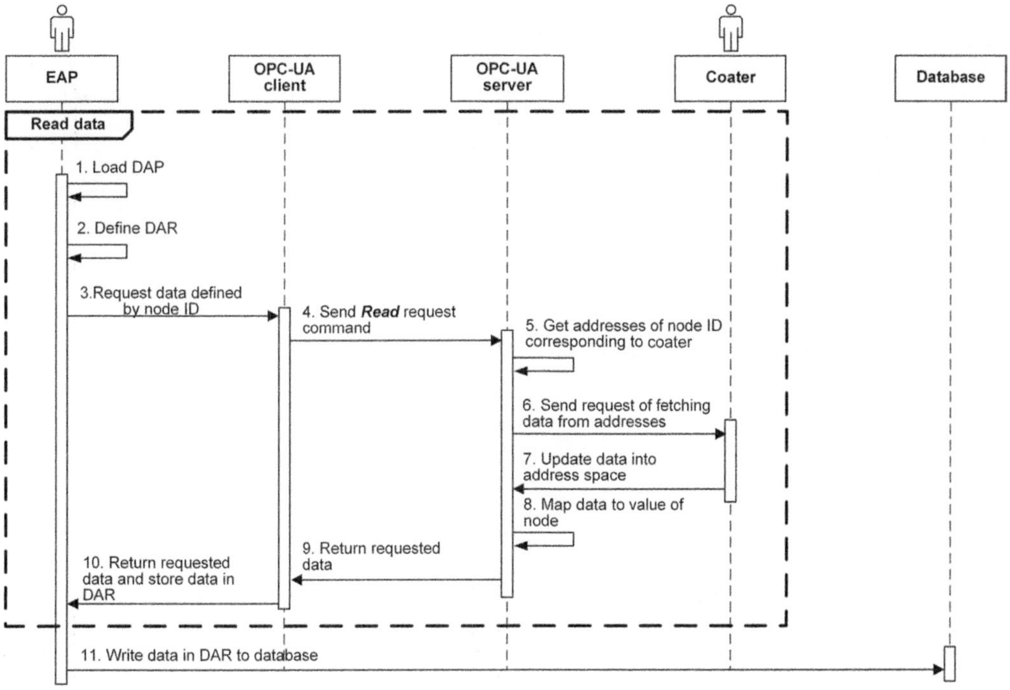

Figure 3.52 Sequence diagram of Data Collection.

Sequence Diagram of Recipe Download

Originally, the Coater in this example needs to set the recipe parameters manually before executing the production process. As a result, the operator needs to input more than 20 parameter settings each time, which is highly risky. Therefore, it is proposed to adopt RMS for avoiding wrong parameter settings on production equipment. The sequence diagram of Recipe Download is shown in Figure 3.53.

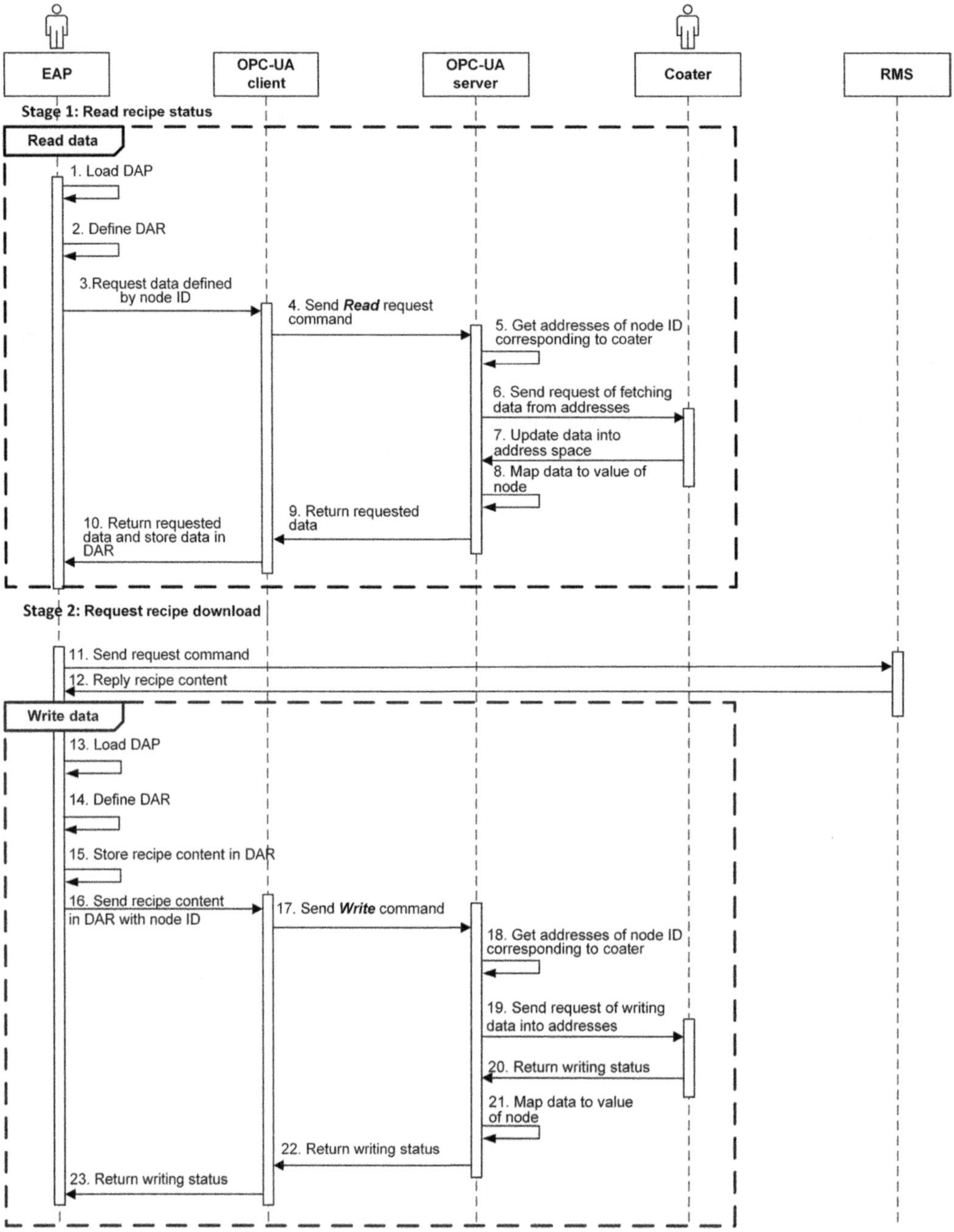

Figure 3.53 Sequence diagram of recipe download.

A two-stage process is designed as depicted in Figure 3.53. To begin with, the recipe status is checked in Stage-1: Read Recipe Status by adopting the “Read Data” macro procedure shown in Figure 3.52. After checking the recipe, if a new recipe is required, then Stage-2: Request Recipe Download is activated. They are elaborated as follows.

Stage-1: Read Recipe Status

Steps 1–10: Adopt the “Read data” macro procedure shown in Figure 3.52 to read recipe status.

Stage-2: Request Recipe Download

Steps 11–12: EAP sends request recipe command to RMS and RMS replies the selected recipe content to EAP.

Steps 13–15: EAP loads the proper DAP for this selected recipe and then defines DAR based on this DAP; finally, the recipe content is stored in DAR.

Steps 16–17: EAP sends the recipe content with node ID via the OPC-UA client in EAP to OPC-UA server of Coater.

Steps 17–21: The OPC-UA server connects to the PLC in Coater and then uses the *Write* service to write the data (recipe content) with the corresponding addresses specified in the node.

Steps 22–23: The OPC-UA server returns the writing status via the OPC-UA client to EAP. If the recipe writing status shows normal on HMI, then the operator will press the button to start the coating process.

3.4 Conclusion

This chapter introduces the communication standards that support the entire data-exchange requirements of Intelligent Manufacturing. The communication standards of the semiconductor equipment including SECS-I, SECS-II, GEM, and HSMS of the manufacturing portion as well as Interface A of the engineering portion are presented first. Then, the classic OPC and OPC-UA communication standards for connecting industrial devices and systems are illustrated.

Appendix 3.A – Abbreviation List

A&A	Authentication and Authorization
ACK	Correct Reception
ACS	Advanced Control System
API	Application Programming Interface
ASCII	American Standard Code for Information Interchange
BI	Business Intelligence
CEM	Common Equipment Model
COM	Component Object Model
CRM	Customer Relationship Management
DAR	Data Access Report
DAP	Data Access Plan
DCM	Data Collection Management

DCOM	Distributed COM
DCP	Data Collection Plan
DCR	Data Collection Report
DCS	Distributed Control System
DLL	Dynamic-Link Library
E-bit	End Bit
EAP	Equipment Application Program
ECM	EDA Common Metadata
EDA	Equipment Data Acquisition
EES	Equipment Engineering System
EMS	Equipment Management System
ENQ	Request to Send
ERP	Enterprise resource planning
EOT	Ready to Receive
EqSD	Equipment Self-Description
FCCL	Flexible Copper Clad Laminate
FDC	Fault Detection and Classification
Fm	Function m, where m = the function number
GEM	Generic Equipment Model
HMI	Human-Machine Interface
HSMS	High Speed SECS Message Services
HSMS-GS	A Subsidiary standard of HSMS, GS stands for General Session
HSMS-SS	A Subsidiary standard of HSMS, SS stands for Single Session
I/O Device	Input/Output Device
IC	Integrated Circuit
ID	Identification
IH	Item Header
IIoT	Industrial Internet of Things
JIS	Japanese Industrial Standards
M2M	Machine-to-Machine
MES	Manufacturing Execution System
NAK	Incorrect Reception
NLB	Number of Length Bytes
OEE	Overall Equipment Effectiveness

OEM	Original Equipment Manufacturer
OPC	Open Platform Communication
OPC A&E	OPC Alarms and Events
OPC DA	OPC Data Access
OPC HDA	OPC Historical Data Access
OPC-UA	OPC Unified Architecture
OSI	Open System Interconnection
PLC	Programmable Logic Controller
PType	Presentation Type
PVs	Physical Values
R-bit	Reserve Bit
RS-232	An Electronic Industry Association Recommended Standard
RAMI 4.0	Reference Architecture Model for Industry 4.0
RM	Recipe Management
RMS	Recipe Management System
RTY	Retry Limit
SCADA	Supervisory Control and Data Acquisition
SECS	Semiconductor Equipment Communication Standard
Sn	Stream n, where n = the stream number
SOAP	Simple Object Access Protocol
SSD	Selected Equipment Status Data
SSR	Selected Equipment Status Request
SType	Session Type
SV	Status Variable Value
SVID	Status Variable ID
SVs	Setting Values
TCP	Transmission Control Protocol
TCP/IP	Transmission Control Protocol/Internet Protocol
UDP	User Datagram Protocol
Unique ID	UID
W-bit	Wait Bit
WIP	Work-In-Process
WS	Web Service
XML	Extensible Markup Language

References

1 SEMI E4 (2018). *SEMI E4 – Specification for SEMI Equipment Communications Standard 1 Message Transfer (SECS-I)*. USA: SEMI. https://bit.ly/3iXrEvc (accessed 12 August 2020).

2 SEMI E5 (2020). *SEMI E5 – Specification for SEMI Equipment Communications Standard 2 Message Content (SECS-II)*. USA: SEMI. https://bit.ly/2Oralog (accessed 12 August 2020).

3 SEMI E30 (2020). *SEMI E30 – Specification for the Generic Model for Communications and Control of Manufacturing Equipment (GEM)*. USA: SEMI. https://bit.ly/2C295pf (accessed 12 August 2020).

4 SEMI E37 (2019). *SEMI E37 – Specification for High-Speed SECS Message Services (HSMS) Generic Services*. USA: SEMI. https://bit.ly/3jehYwG (accessed 12 August 2020).

5 SEMI E37.2-95 (Withdrawn 1109) (2019). *SEMI E37.2-95 (Withdrawn 1109) – High-Speed SECS Message Services General Session (HSMS-GS)*. USA: SEMI. https://bit.ly/3jehYwG (accessed 12 August 2020).

6 SEMI E37.1-0819 (2019). *SEMI E37.1-0819 – Specification for High-Speed SECS Message Services Single Selected-Session Mode (HSMS-SS)*. USA: SEMI. https://bit.ly/3jehYwG (accessed 12 August 2020).

7 SEMI E132 (2019). *SEMI E132 – Specification for Equipment Client Authentication and Authorization*. USA: SEMI. https://bit.ly/3haoJgS (accessed 12 August 2020).

8 SEMI E120 (2019). *SEMI E120 – Specification for the Common Equipment Model (CEM)*. USA: SEMI. https://bit.ly/3h97Kfa (accessed 12 August 2020).

9 SEMI E125 (2014). *SEMI E125 – Specification for Equipment Self Description (EqSD)*. USA: SEMI. https://bit.ly/3j5C2Bj (accessed 12 August 2020).

10 SEMI E134 (2019). *SEMI E134 – Specification for Data Collection Management*. USA: SEMI. https://bit.ly/3fCBWPq (accessed 12 August 2020).

11 SEMI E164 (2014). *SEMI E164 – Specification for EDA Common Metadata*. USA: SEMI. https://bit.ly/2Ow9yT7 (accessed 12 August 2020).

12 Adolph, L., Anlahr, T., Bedenbenderm, H. et al. (2016). *German Standardization Roadmap Industry – Version 3*. Germany: DKE Deutsche Kommission Elektrotechnik Elektronik Informationstechnik in DIN und VDE.

13 Soares de Melo, P. and Godoy, E.P. (2019). Controller interface for Industry 4.0 based on RAMI 4.0 and OPC UA. *Proceedings of the 2019 II Workshop on Metrology for Industry 4.0 and IoT (MetroInd4.0&IoT)*, Naples, Italy (4–6 June 2019). USA: IEEE.

14 Wolfgang, M., Leitner, S., and Damm, M. (2009). *OPC Unified Architecture*. Germany: Springer https://doi.org/10.1007/978-3-540-68899-0.

15 OPC 10000-1 Release 1.04 (2017). *OPC 10000-1: OPC Unified Architecture Part 1: Overview and Concepts*. USA: OPC Foundation. https://bit.ly/2H7TcPY (accessed 15 October 2020).

16 OPC Foundation (2020). *OPC UA Online Reference*. https://reference.opcfoundation.org/v104/ (accessed 15 October 2020).

17 OPC 10000-4 Release 1.04 (2017). *OPC 10000-4: OPC Unified Architecture Part 4: Services*. USA: OPC Foundation. https://bit.ly/3k0LvKb (accessed 15 October 2020).

18 OPC 10000-2 Release 1.04 (2018). *OPC 10000-2 OPC Unified Architecture Part 2: Security Model*. USA: OPC Foundation. https://bit.ly/3lGlaS6 (accessed 15 October 2020).

19 Boait P., Neville G., Norris R., Pickman M., Tolhurst M., Walmsley J. (1988). The OSI Model. In: Tolhurst M. (eds) *Open Systems Interconnection. Macmillan Computer Science Series*. London, UK: Palgrave. https://doi.org/10.1007/978-1-349-10306-5_3.

4

Cloud Computing, Internet of Things (IoT), Edge Computing, and Big Data Infrastructure

Hung-Chang Hsiao[1], *Min-Hsiung Hung*[2], *Chao-Chun Chen*[3], *and Yu-Chuan Lin*[4]

[1] *Professor, Department of Computer Science and Information Engineering, National Cheng Kung University, Tainan, Taiwan, ROC*
[2] *Professor, Department of Computer Science and Information Engineering, Chinese Culture University, Taipei, Taiwan, ROC*
[3] *Professor, Institute of Manufacturing Information and Systems, National Cheng Kung University, Tainan, Taiwan, ROC*
[4] *Secretary General, Intelligent Manufacturing Research Center, National Cheng Kung University, Tainan, Taiwan, ROC*

4.1 Introduction

Over the past few years, due to the vigorous development of the Internet and information and communication technology (ICT), major manufacturing countries in the world competed to propose projects and strategies [1–3] that leveraged advanced ICTs to promote the intelligence and automation of their manufacturing industry in response to the issues, such as rapid market changes, insufficient manpower, and limited production resources, to enhance the global competitiveness of their products and manufacturing technologies. Among them, Industry 4.0 proposed by Germany is the most well-known [2, 4].

The concept and vision of Industry 4.0 [2, 4, 5] aimed at creating Intelligent Manufacturing and smart factories [6] that can enable more responsive, adaptive, flexible, efficient, and connected processes to achieve mass production at reduced costs, compared to traditional factories. The underlying core technologies of Industry 4.0 and Intelligent Manufacturing include several advanced ICTs, such as Internet of Things (IoT) [7, 8], Cloud Computing [9–12], Edge Computing [13, 14], Big Data Analytics [15, 16], and Cyber-Physical Systems (CPS) [17, 18].

A smart factory could adopt IoT as its data collection and communication platform and use cloud computing which has abundant computing resources or edge computing which is near the data sources to perform big data analytics on a large amount of production-related data. Also, CPS with predictive algorithms could be constructed on IoT combined with cloud computing or edge computing to cooperate and interact seamlessly with real and virtual worlds to provide intelligent capabilities, such as equipment prognosis and predictive maintenance. The values and benefits of cloud computing, edge computing, IoT, and big data analytics for the manufacturing sector are briefly introduced as follows.

Cloud computing is an Internet-based computing that a pool of computing resources (e.g. computing power, data storage, network, software development and deployment environments, applications, etc.) on datacenters are virtualized as services which can be accessed by many users on-demand over the Internet. A datacenter refers to a well-constructed and safe room or building

Industry 4.1: Intelligent Manufacturing with Zero Defects, First Edition. Edited by Fan-Tien Cheng.

that contains a large number of computer servers and their associated components, including storage devices, communication networks, software, infrastructure, environmental controls, cabling, power, cooling, and more. Cloud computing can be regarded as a utility computing where the users can use the abundant computing resources through different-level services on a pay-as-you-go model. Via cloud computing, organizations and individuals could easily get enterprise-grade computing resources on demand and just pay what they have used. Also, they can cancel the rented computing resources at will when the demand is no longer needed.

Through the infrastructure-level service, the users can build virtual machines and install desired operating systems (OSs) and applications on them. By using the platform-level service, the users can develop and deploy applications that may access databases and data storage on the cloud. Via the software-level service, the users can just use their desired software applications without needing to install the applications locally. Further, the manufacturing industry can leverage cloud computing to encapsulate and virtualize distributed manufacturing resources into cloud services that can then be used on demand over networks to support manufacturing activities, including product design, simulation, manufacturing, testing, management, and all other tasks in the product life cycle. The above concept is referred to as cloud-based manufacturing (also known as cloud manufacturing, CMfg [9, 10]). CMfg has potential to satisfy the growing demands of manufacturing for producing individualized products more flexibly, supporting global cooperation and collaboration widely, facilitating knowledge-intensive innovation, and increasing the agility of responding to the changing market [11, 12].

Although cloud computing offers the abundance, flexibility, and scalability of computing resources, it may encounter the issue of latency (a delay) in transferring data from a factory to a remote datacenter for processing data and getting the resulting information back. Such a latency issue could fail to meet the demand of some time-critical applications in a factory, such as those needing to detect a machine's fault or the production quality of workpiece instantly for taking control actions. During the production, sensors equipped on production equipment and workpieces will generate a large amount of production-related raw data. Sending a large amount of raw data to a remote datacenter is a big burden for computers on the plant floor, particularly for resource-constrained IoT devices. It also leads to inefficient uses of their network bandwidth and energy. Thus, the concept of edge computing [13, 14] was proposed to tackle these real-time data processing challenges. Edge computing extends cloud computing to the network edge. The cloud services on the network edge (i.e. edge services) provide computing power and data storage in a small scale to support traditional cloud datacenters in dealing with real-time data processing and storage. Moving data processing closer to data sources can reduce communication delays and the overall size of the data that needs to be transferred across the Internet and datacenters, thereby saving energy and avoiding unnecessary network bandwidth consumption for resource-constrained edge devices.

IoT for manufacturing is a new Internet evolution. A smart factory can adopt IoT to integrate a large number of things, such as sensors, machines, devices, cameras, displays, and smartphones, with cloud datacenters and to process their data on virtualized computing resources in a timely manner to support manufacturing activities [19]. IoT also supports bidirectional machine-to-machine (M2M) communication in the factory or on the Internet. IoT can connect machines, sensors, and handhelds for collecting data from any factory machine or device, while the cloud services can perform real-time analysis on the collected data. Thus, the IoT and cloud services could bring benefits and insights in productivity and quality to the manufacturing industry, which could drive business and operational innovation.

In a smart factory, production equipment and workpieces are equipped with various sensors. During the production, a large amount of production-related data will be generated. If we use some appropriate tools to mine these production big data (i.e. perform big data analytics [20]) to extract useful information related to products or production, this information could be converted into an appropriate form to become the wisdom of the manufacturing system for decision making. For example, if advanced forecasting tools are used, these production-related data can be systematically processed and transformed into useful information that can explain uncertain conditions. Then, this high-value information could be applied in production-related applications, such as the quality control of workpieces and the fault diagnosis and predictive maintenance of production equipment [21]. Thus, for processing the big data generated in the factory, the development of various data exploration methods and tools is an important topic; meanwhile, the big data infrastructure [22] that consists of big data analysis software and hardware to support big data analytics is essential as well.

This chapter will introduce some underlying core technologies of Industry 4.0 and Intelligent Manufacturing. In Section 4.2, we will introduce cloud computing, including its essentials, service models, and deployment models, as well as the concept of CMfg. In Section 4.3, we will depict edge computing, including its essentials and platforms, together with its applications in manufacturing. Besides, Section 4.3 will present IoT, including its essentials and architectures, as well as its applications in manufacturing. Section 4.4 will introduce big data infrastructure, including the application demands, core software stack components, and the methods to bridge the gap between core software stack components and applications. Finally, Section 4.5 is the conclusion of this chapter.

4.2 Cloud Computing

In this section, we will introduce cloud computing, including its essentials, the cloud service models, and the cloud deployment models. We also present some cloud computing applications and benefits for the manufacturing industry.

4.2.1 Essentials of Cloud Computing

Cloud computing is a computing paradigm that IT applications in datacenters are delivered as services over the Internet. The users can easily use these Internet computing services on demand. Cloud computing moves computing and data away from desktop PCs and laptop computers into large datacenters, which contain the actual cloud infrastructure, i.e. the hardware and systems software that provide these services [23]. Cloud computing provides on-demand services, ubiquitous network access, location-independent computing resource pooling, ease of utilization, quality of service (QoS), and reliability for computing [24]. Thus, cloud computing could bring new models of IT utilization to enterprises for gaining commercial benefits.

The rise of cloud computing is driven by many factors. Among them, economics and simplification of software delivery and operation are two main drivers. Cloud computing could reduce the cost of IT to enterprises and free them from the expenditure and nuisance of needing to install and maintain applications locally. Cloud computing provides cloud services of IT to organizations and individuals in a pay-as-you-go pricing model, which is commercially viable for many small and medium enterprises (SMEs) [25].

Scalability is the most important idea behind cloud computing, and virtualization is the key technology that makes it possible. By using virtualization technology, multiple virtual servers (called virtual machines), each of which is installed with an operating system (OS) as well as applications, can be aggregated to run on a single physical server. Thus, virtualization could increase the utilization of computing resource of a physical server. Also, if a physical server becomes

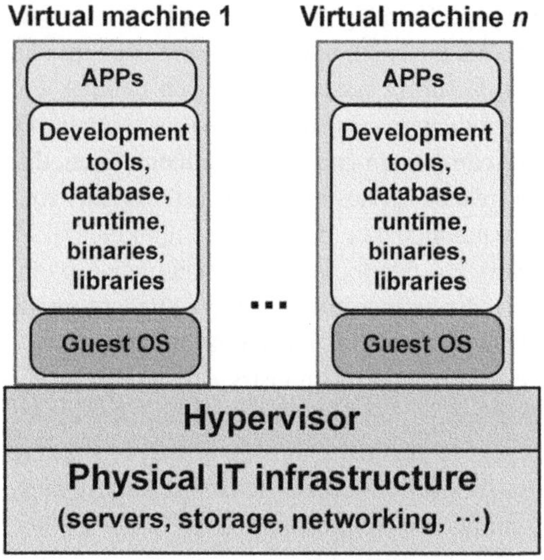

Figure 4.1 Architecture of virtual machines on top of hypervisor and physical IT infrastructure.

overloaded, virtualization easily allows some virtual machines to be on-line migrated to another physical server which is less loaded.

Figure 4.1 shows an architecture of virtual machines running on top of hypervisor and physical IT infrastructure. The virtual machine is a hardware-level virtualization technology using a hypervisor to abstract the underlying hardware. A hypervisor, such as VMware vSphere, Microsoft Hyper-V, and Citrix Xen, is a program that allows multiple OSs to share a single hardware host. On top of the hypervisor, we can create several virtual machines, each of which can install a guest OS, like Linux or Windows. Then, on top of the guest OS, we install the development tools, various applications (Apps), databases, and the dependencies required to run the applications, such as frameworks, runtimes, binary codes, and libraries.

National Institute of Standards and Technology (NIST) defined cloud computing as "a model for enabling convenient, on-demand network access to a shared pool of configurable computing resources (e.g. networks, servers, storage, applications, and services) that can be rapidly provisioned and released with minimal management effort or service provider interaction [26]." This cloud model contains five essential characteristics [26]: (i) On-demand Self-service: Consumers can use cloud services on their own without interacting with cloud providers; (ii) Ubiquitous Network Access: Cloud services can be accessed over the internet at anytime from anywhere; (iii) Resource Pooling: Cloud providers provide consumers a large pool of virtualized resources through a multi-tenant model and can assign or reassign virtualized resources according to consumers' demands; (iv) Rapid Elasticity: The scale of resources can be adjusted rapidly and flexibly according to different requirements; and (v) Measured Service: All levels of cloud services are measured and supervised by the cloud provider for the usage of the pay-as-you-go pricing model. Key enabling technologies of cloud computing include fast wide-area networks, powerful and inexpensive server computers, and high-performance virtualization for commodity hardware [26].

Although cloud computing could bring many advantages to enterprises, it faces challenges and concerns as well, just to mention some below [24]. The first challenge with cloud computing is privacy. As data are accessed from anywhere over the Internet, users' privacy may be compromised. The second challenge is security. Information security is about keeping data secret. Transferring confidential information over the Internet can be unsecure and risky. For some applications that require responsive data processing, the latency of cloud computing is another critical issue. This latency refers to the delay from the moment of sending data to remote datacenters to the time of getting the data-processing result back.

4.2.2 Cloud Service Models

As cloud computing services became mature, several service models were introduced, including Infrastructure as a Service (IaaS), Platform as a Service (PaaS), and Software as a Service (SaaS) [26]. These cloud service models can be used for categorizing not only cloud computing services but also specific cloud providers' offerings, products, and services.

- **IaaS**
  IaaS involves the delivery of huge computing resources, such as computing power (virtual machines), storage, and networking. IaaS allows the users to remotely access computing resources over the Internet. The major advantages of IaaS are pay per use, security, scalability, and high availability. IaaS is also known as a "hardware as a service" cloud infrastructure. The most notable vendors under this model include Amazon providing Elastic Compute Cloud (EC2), Microsoft providing Azure IaaS, and Google providing Google Cloud Trusted Infrastructure.
- **PaaS**
  PaaS provides users with a development and hosting platform that can integrate a variety of different application software and tools (e.g. databases, middleware, Web servers, and other software) under the same interface. Thus, PaaS allows the users to deploy their developed software on the cloud, becoming cloud services. The most notable PaaS platforms include Amazon Web Services (AWS), Microsoft Azure, and Google App Engine.
- **SaaS**
  SaaS provides software deployed on the cloud as cloud services that are directly consumable by end users over the Internet. Under this model, instead of installing and maintaining software locally, one can simply access the software functionalities via the Internet without needing to managing complex software and hardware. The cloud service of this type could offer a variety of application functionalities that range from productivity applications (e.g. word processing, spreadsheets, etc.) to enterprise programs for Customer Relationship Management (CRM) and Enterprise Resource Management (ERM). Products under this category include Google Apps (e.g. Google Maps, Google Docs, Gmail, Google Drive, etc.), Salesforce CRM, Dropbox, and Microsoft Office 365.

Figure 4.2 shows the comparison of cloud computing service models [27]. In the IaaS model, the underlying IT infrastructure (e.g. servers, storage, networks) and the hypervisor are managed by the cloud provider. The users can install their desired OSs, runtimes, middleware, software development kits (SDKs), and applications with data. In the PaaS model, the underlying IT infrastructure, the hypervisor, OSs, runtimes, and middleware are managed by the cloud provider. The users can

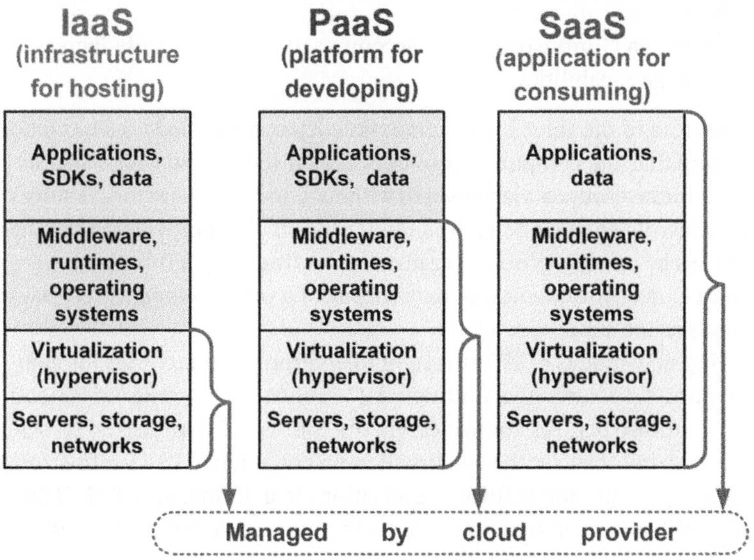

Figure 4.2 Comparison of cloud computing service models.

develop and deploy their software application on the hosted environment of the cloud. In the SaaS model, everything is managed by the cloud provider. The users can just use the software functionalities on the cloud.

4.2.3 Cloud Deployment Models

Regarding to the deployment of cloud computing, three cloud deployment models, i.e. public cloud, private cloud, and hybrid cloud, are introduced as follows [26]:

- **Public Cloud**
  A public cloud refers to a cloud computing environment provided by an independent cloud provider. The cloud services on a public cloud can be accessed by many enterprises and users in a multi-tenancy mode. The public cloud is suitable for small and medium-sized enterprises to extend their on-premises computing and storage capabilities in a pay-per-use manner because it can save the big cloud infrastructure investment. The most notable public clouds include AWS, Microsoft Azure, and Google Cloud Platform (GCP).
- **Private Cloud**
  A private cloud refers to a cloud computing environment that is constructed and used solely by an enterprise or organization. All users of a private cloud are internal members of an enterprise or organization and share the private cloud's resources. Thus, the private cloud is more secure and has less privacy concerns than the public cloud. However, the scale and resources of a private cloud are much smaller than a public cloud that has virtually unlimited resources for the users. The private cloud is suitable for large enterprises, such as financial institutions, big manufacturing companies, and government agencies, because they have already had heavy IT investments and can further reduce the usage rate of their IT resources via the private cloud.
- **Hybrid Cloud**
  The hybrid cloud refers to the cloud computing environment that uses both the private cloud and public cloud. Generally, large organizations, such as financial institutions, big manufacturing companies, government agencies, and other large enterprises, already have a certain scale of IT infrastructure in a private cloud where their vital applications are housed. Meanwhile, they can offload some workloads and applications to public clouds during the peak time of workloads for reducing the burden of their own IT infrastructure or accommodating the suddenly-increased demand of computing and storage capabilities.

Figure 4.3 shows the architecture of the three cloud computing deployment models. The cloud symbol is a metaphor indicating that the computing resources are on remote datacenters somewhere, and the users can access these resources via networks without knowing where the resources are located. A public cloud is constructed and offered by an independent cloud provider. A private cloud is constructed on-premises by an enterprise or organization or hosted by a third-party vendor. A hybrid cloud is a multi-cloud environment that may consist of a private cloud as the main cloud and parts of public clouds as the supporting cloud.

Cloud computing could bring new modes of IT utilization to enterprises or factories for gaining commercial benefits. Figure 4.4 shows the traditional IT utilization of enterprises and factories using on-premises IT Infrastructures connected via the Internet. Regardless of its size and scale, every company must spend costs on the construction and upgrade of its IT infrastructure, such as servers, storages, networks, and software applications (e.g. Database, CRM, HRM, SCM, Web Portal, MES, and ERP), and need manpower to maintain its IT systems. These traditional companies or factories use the Internet to exchange information to achieve e-commerce

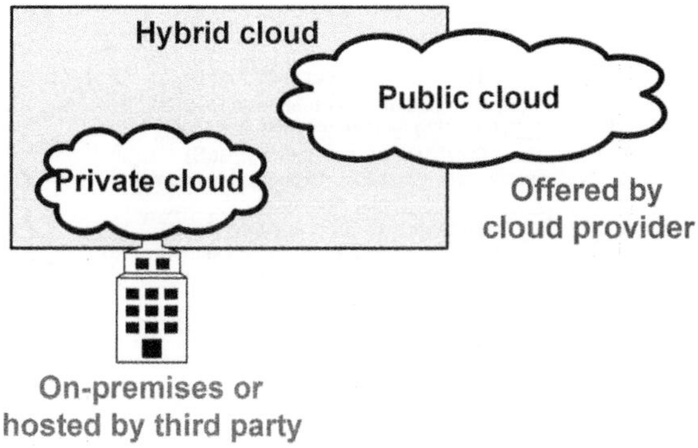

Figure 4.3 Architecture of the three cloud deployment models.

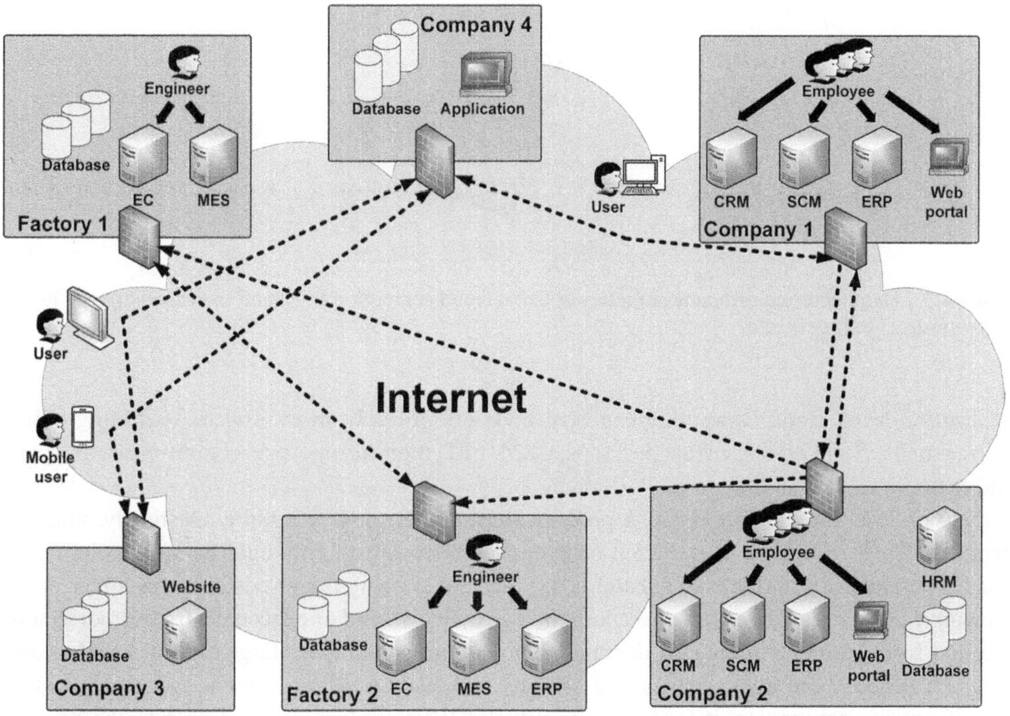

Figure 4.4 Traditional IT utilization of enterprises and factories using on-premises IT Infrastructures connected via the Internet.

and engineering collaboration of business to business (B2B), business to customer (B2C), and customer to customer (C2C).

By contrast, cloud computing can be leveraged to extend the computing and storage capabilities of the enterprises' on-promises IT infrastructure (e.g. servers, PCs, laptops, mobile devices, Web browsers, mobile Apps, thin clients, terminals, and more) [27], as shown in Figure 4.5. At the infrastructure level, cloud computing could provide services like virtual machines, storages, networks, and load balancers, etc. At the platform level, cloud computing offers services like Web servers, databases, runtimes, development tools, deployment environments, and more. At the

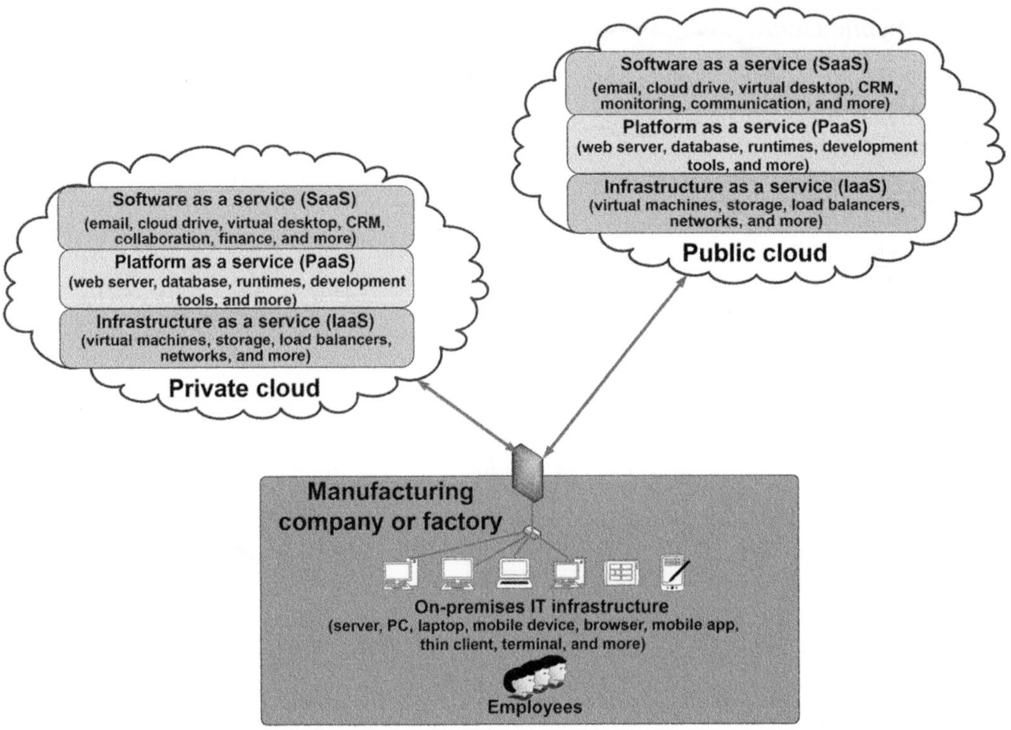

Figure 4.5 Manufacturing company or factory utilizing cloud computing to extend the capabilities of its on-premises IT infrastructure.

application level, cloud computing can host a variety of application services, including office Apps, emails, cloud drives, virtual desktops, CRM, ERP, monitoring Apps, communication Apps, and so on.

The public cloud can provision a pool of abundant resources rapidly, elastically, and on demand. Thus, compared to the fixed on-premises IT resources, it would be more suitable to handle on-and-off workloads (e.g. batch jobs) to avoid wasting the provisioned resources, fast-growing workloads to keep up with a company's growth fast, and unexpected bursting workloads resulted from suddenly-increased demands. On the other hand, for large enterprises that are very concerned about data security and privacy, such as banks, government agencies, and big manufacturing companies, it is suitable to migrate their applications and data to the private cloud. From the perspective of return on investment (ROI), building a private cloud may require a large investment at the beginning; however, in the long term, it will be beneficial for a company in terms of increasing the usage rate of the IT resources and reducing the maintenance cost and manpower.

Nowadays, smartphones are widely used in our daily lives. They can run a wide range of applications (Apps), some of which would demand huge computational power. This poses a challenge because smartphones are resource-constrained mobile devices with limited computation power, memory, storage, and energy. Thus, cloud computing services with virtually unlimited dynamic resources for computation and storage can be utilized to support smartphones to overcome the resource constraints of mobile devices [28]. Notable kinds of such cloud computing services include maps, navigation, and big-data-analytics cloud services.

Artificial intelligence (AI) technology is capable of learning, reasoning, and problem solving. It can be applied in many fields, such as e-commerce, healthcare, cybersecurity, supply chain, machine monitoring, manufacturing, inventory management, and more. However, AI technology could be costly in terms of hardware (e.g. high-end servers with GPUs and storage) and software required to train and run AI programs with big data. Thus, building an AI platform with sufficient capabilities on-premises and hiring relevant talents may incur a big IT budget that small to medium enterprises cannot afford it. Now, cloud providers offer AI services in their public clouds, AI is within the reach of most enterprises' budgets [29]. Amazon AWS artificial intelligence, Microsoft Azure artificial intelligence, and Google Cloud artificial intelligence are three examples of such public cloud AI services. Public clouds also provide cheap data storage which can be used as the data input for the AI-related applications. They all provide SDKs and APIs that allow developers to embed AI functionalities directly into the applications. Thus, enterprises can leverage these affordable AI services and other resources on public clouds to develop AI-enabled applications that could bring actual values and benefits for their businesses.

However, it is worth to note that using public clouds for business applications can face some limitations [30], such as interoperability, portability, and migration. For example, all the major cloud providers offer unique and proprietary data storage, so cloud users can find difficulties in moving their data from one cloud provider to another. Also, business applications based on a cloud provider's APIs are locked into that cloud provider.

According to the report of the 2020 cloud computing usage trends mentioned in [31], most enterprises participating in this survey embrace a multi-cloud or hybrid-cloud strategy, and their public cloud adoption continues to accelerate. Also, public cloud adoption is evolving, and the top three public cloud providers for these enterprises are AWS, Azure, and Google. Besides, these enterprises' cloud expenditure is growing, and the use of public cloud PaaS services is rising. The top three PaaS services from cloud providers are relational database as a service, container as a service, and data warehouse. In particular, the enterprises grow their interests in container as a service to leverage containers to expedite deployment, scale operations, and increase efficiency of their workloads and applications running in the cloud.

4.2.4 Cloud Computing Applications in Manufacturing

Like other businesses, the manufacturing enterprises can leverage cloud computing to extend the computing and storage capabilities and other functionalities of their on-premises IT infrastructure on demand based on a pay-as-you-go pricing model. For example, by leveraging public cloud platforms, such as AWS [32], Microsoft Azure [33], Google Cloud Platform (GCP) [34], an manufacturing enterprise can rent virtual machines for getting extra computing capabilities, building and hosting Web sites for auto-scaling to accommodate the growing traffics, and building and deploying scalable cloud services (i.e. Web APIs) that support rich multi-tier architectures and automated management. The public cloud platforms also provide relational databases, each of which is a fully managed service and is clustered and fault-tolerance for high availability. Besides, they provide highly available, scalable, and secure file storage systems that could keep continuous geo-replication across datacenters. A manufacturing enterprise can also utilize service buses provided by public clouds to set up messaging that connects applications and services across on-premises and cloud environments [35].

With abundant computing and storage resources, the public clouds offer big data analytics platforms for processing and storing factory big data. For example, Amazon EMR [36] is a cloud big data platform for processing vast amounts of data using open source tools such as Apache Spark,

Apache Hive, Apache HBase, Apache Flink, Apache Hudi, and Presto. By using EMR, you can run Petabyte-scale analysis at less cost and faster speed, compared to using traditional on-premises solutions and standard Apache Spark. Microsoft Azure provides robust services for analyzing big data [37]. You can store data in Azure Data Lake Storage Gen2 and then process it using Spark on Azure Databricks. Google also provides a cloud platform-based big data analytics service – BigQuery [38]. Google BigQuery was designed to perform analytics of read-only data from billions of data source rows using an SQL-like syntax. It runs on the Google Cloud Storage platform and can be invoked through REST APIs. There are good reasons for manufacturing enterprises to conduct big data analytics on cloud [39]. First, cloud big data analytics solutions allow them to focus on solving business problems without needing to worry about the associated IT infrastructure. Second, they can keep up on innovation with access to new releases and features of open source and managed services and achieve faster time to market to generate values on data applications. Third, they can leverage hybrid cloud to expand private data-processing power to resource-abundant cloud on usage spikes.

Another cloud computing application in manufacturing is CMfg [9, 10]. By leveraging and extending the characteristics of cloud computing to meet the global and distributed requirements of the manufacturing industry, CMfg was proposed to be a new manufacturing paradigm. It has potential to satisfy the growing demands of the manufacturing enterprises for producing individualized products more flexibly, increasing the agility of responding to the changing market, supporting global cooperation and collaboration widely, and facilitating knowledge-intensive innovation [10, 11].

The work in [9] introduced CMfg as the manufacturing version of cloud computing, which encapsulates and virtualizes distributed manufacturing resources into cloud services that can then be used on demand over networks to support manufacturing activities ranging from product design, simulation, manufacturing, testing, and management to all other tasks in the product life cycle. A four-layered architecture of a CMfg system was also presented, consisting of a manufacturing resource layer, a virtual service layer, a global service layer, and an application layer. The manufacturing resource layer contains the resources required during the product life cycle, including manufacturing equipment, manufacturing software, and manufacturing capabilities. At the virtual service layer, manufacturing resources are identified, virtualized, and then packaged into CMfg services. The global service layer can operate in a complete-service mode where the entire cloud operational activities are wholly handled by this layer. The application layer allows the user to define and construct a manufacturing application through the virtualized manufacturing resource and services.

The work in [40] further regarded CMfg as the next-generation manufacturing model and proposed a service-oriented, interoperable CMfg system, containing three layers, i.e. Manufacturing Cloud, Smart Cloud Manager (SCM), and Cloud Users. Manufacturing resources and capabilities were virtualized as CMfg services on the manufacturing cloud, which could be reused for different purposes. The service provider, cloud user, and SCM were the three key roles of this cloud-based Service-Oriented Architecture (SOA), where SCM played the role of service broker, handling the interaction between the cloud users and service providers. SCM could take queries from the cloud user and then searched and provided the desired interoperable CMfg services.

The work in [41] defined CMfg as "a customer-centric manufacturing model that exploits on-demand access to a shared collection of diversified and distributed manufacturing resources to form temporary, reconfigurable production lines which enhance efficiency, reduce product lifecycle costs, and allow for optimal resource loading in response to variable-demand customer generated tasking." This work presented the drivers of CMfg, including reducing cost brought by the outsourcing and crowdsourcing models, agility for accommodating the rapid changing customer needs, ability to rapidly scale up/down manufacturing resources to quickly respond to changing requirements, capability to pool manufacturing resources to serve consumers in a pay-per-use manner and facilitate resource sharing between service providers, allowing information sharing across organizations, and fitting the trend of embracing social networks as a means of interaction and communication with users and customers. This work further provided a strategic vision for CMfg. It described a provider–consumer interaction model that the users' needs will be matched with the resource providers' capabilities through the application layer. Several key characteristics of CMfg were also depicted, such as (i) having customer-centric manufacturing supply chain relationships which can enhance efficiency, reduce cost, increase flexibility, and improve capabilities for the user, (ii) enabling temporary, reconfigurable, and dynamic production lines which allow for high efficiency, minimized down time, and instant response to demand, (iii) bringing Intelligent Manufacturing environments which could assure even load sharing across interchangeable manufacturing resources, and (iv) shifting to share-to-gain based business models which share manufacturing burdens and then share generated benefits among stakeholders.

Aiming at applying cloud computing to the field of product design and manufacturing, the work in [42] proposed a definition of cloud-based design and manufacturing (CBDM) and discussed its key characteristics. This work also presented a CBDM model which consisted of four key roles, i.e. Cloud Provider, Cloud Consumer, Cloud Broker, and Cloud Carrier. The cloud consumers can utilize the cloud services provided by the cloud providers to complete product design and manufacturing tasks. The cloud broker can manage the use (e.g. intermediation, aggregation, and arbitrage), performance, and delivery of cloud services between the cloud consumers and providers. The cloud carriers can provide connectivity and transport of cloud services between cloud providers and consumers. There are four service-delivery types in the proposed CBDM model: Hardware-as-a-Service (HaaS), Software-as-a-Service (SaaS), Platform-as-a-Service (PaaS), and Infrastructure-as-a-Service (IaaS), whose major activities are as follows. HaaS consumers rent hardware and manufacturing processes from HaaS providers via the CBDM environment. SaaS consumers use engineering software packages, installed and maintained in the cloud by SaaS providers, for design, manufacturing, and analysis. PaaS consumers use the CBDM platform to access necessary tools and communicate with other users for product development process. IaaS consumers utilize computing resources in the cloud without needing to purchase or maintain them.

Figure 4.6 shows a simplified generic system architecture of CMfg, which consists of two parts: cloud and factory. On the factory side, many edge computers serve as data collectors to acquire process data from production equipment and metrology data from metrology equipment. Each data collector can store the acquired raw data and further process the data to extract features related to production. The resulting feature data can then be used in the applications that need to run on the edge computer and very rely on instant data processing to take decisions locally, such as manufacturing-specific data analyses for conducting immediate fault detection and predictive maintenance on equipment's key components. In this case, only the most important data such as status and alarm messages are transmitted over the network. On the other hand, the feature data can also be periodically sent to the cloud for the applications that require more complex data processing and analyses, such as factory big data analytics for finding the root causes of yield loss. Each edge computer could also accept commands from the cloud to take actions accordingly on the equipment connected with it, such as starting and stopping the machining and performing a data collection plan.

With the advancement of manufacturing technologies, production equipment are more and more sophisticated. Therefore, creating intelligent capabilities, such as fault detection and

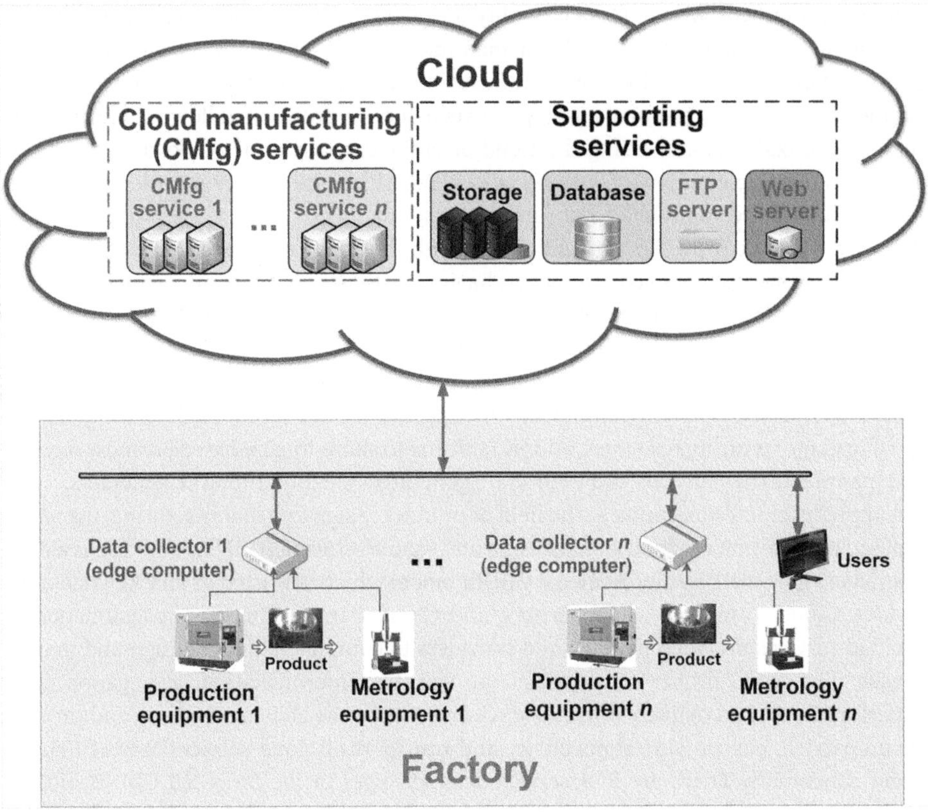

Figure 4.6 Simplified generic system architecture of cloud manufacturing (CMfg).

classification (FDC), manufacturing precision conjecture, and remaining useful life (RUL) prediction, for equipment to ensure their reliability and production quality has become an important issue for the manufacturing industry. Also, adding such kinds of intelligence to equipment fits the trend toward smarter machines and manufacturing systems. One feasible approach to adding intelligence to equipment is to use edge computers to connect to equipment or machine tools for providing functions, such as machine configuration, data collection and storage, fault detection and notification, communication, product quality monitoring, and machine prognosis, and so on.

On the cloud side, many CMfg services are built to encapsulate and virtualize distributed manufacturing resources into cloud services (REST or SOAP-based). The functions of these CMfg services could be utilized to support manufacturing activities in the product life cycle ranging from inception, through engineering design and manufacture, to service and disposal of manufactured products. Also, these CMfg services can be used on demand over networks. Besides, there are many supporting services on the cloud, including database services for storing and managing data, storage services for storing files and data, FTP Server for transferring files, and Web Server for hosting cloud services and Web GUIs. Because the Internet-based-access nature of the cloud, these CMfg services can be leveraged by the manufacturing enterprises over the globe and support global cooperation and collaboration for the manufacturing industry.

To illustrate the application of CMfg to manufacturing, the cloud-based Automatic Virtual Metrology (AVM) system developed in [43, 44] is presented in Figure 4.7 and described as follows. The design philosophy of the system architecture of the cloud-based AVM system contains two

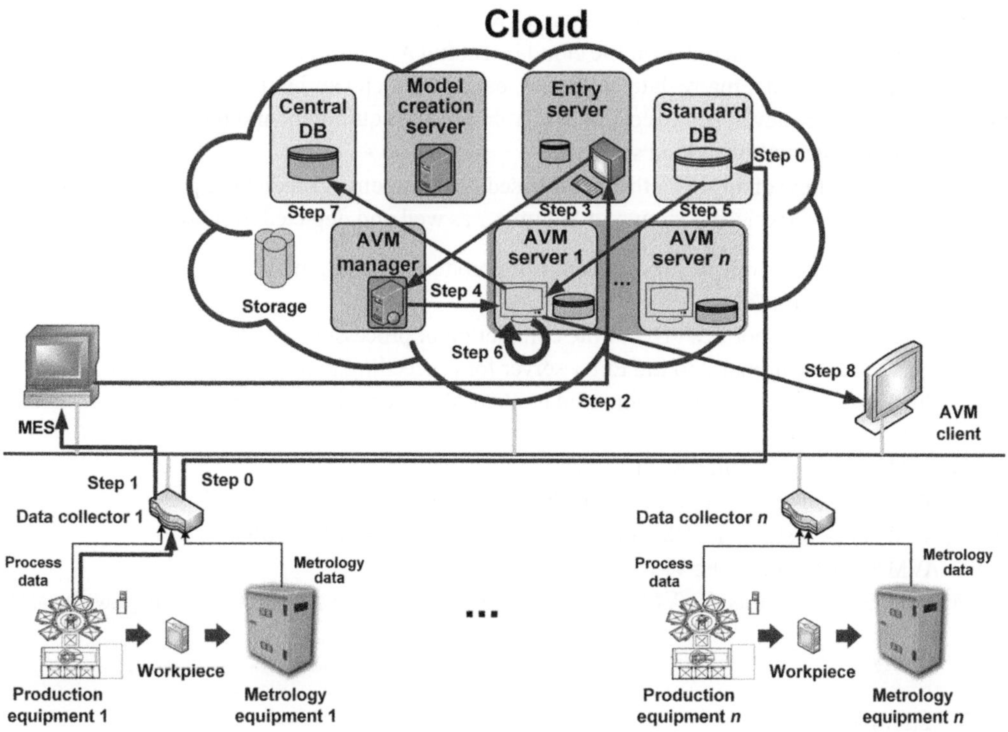

Figure 4.7 Architecture and operational flow of the cloud-based AVM system.

parts: (1) virtualizing all servers of the original AVM system while keeping the implementation of each server unchanged; and (2) designing an extra server to act as the system's operational core for hosting and performing the designed functional mechanisms so as to reduce the efforts of migrating the AVM system to the cloud. The architecture of the cloud-based AVM system is based on the simplified generic system architecture of CMfg in Figure 4.6.

To achieve the first part, the virtualization software (i.e. VMware vSphere and its associated software) is used to create a private cloud environment on top of several networked high-end physical servers, a common storage, and networking facilities. Then, many virtual machines are created in the private cloud environment to host all servers (i.e. the AVM servers for performing on-line virtual metrology (VM), the AVM manager for managing the AVM system, and the Model Creation server for creating VM models) and the required databases (i.e. the Standard DB for storing process and metrology data and the Central DB for storing VM results and the information of models) of the original AVM system, while leaving the original implementation codes of each server unchanged.

To achieve the second part, an Entry server is designed to host and perform the major designed functional mechanisms in the form of cloud services (i.e. the automatic-deployment mechanism, automatic-scaling mechanism, and automatic-serving mechanism). In the original AVM system, once a piece of equipment completes processing of a workpiece, it will trigger the manufacturing execution system (MES) to send a command to the AVM manager which will parse the command and then send an associated command to a target AVM server which is then activated to perform VM on that processed workpiece. By contrast, in the cloud-based AVM system, the MES will send commands to the Entry server which will then send associated commands to the AVM manager for activating AVM servers to perform the assigned VM tasks. Also, the Entry server is responsible for managing all the virtual machines on the cloud by calling APIs of the virtualization software.

On the equipment side, many data collectors, each of which can be installed in an embedded computer or a PC, are developed to collect data from process equipment and metrology equipment and to communicate with the cloud services.

The operational flow of applying the cloud-based AVM system to predict the production quality of a manufactured workpiece is shown in Figure 4.7 as well and depicted as follows.

Step 0: Data Collector 1 collects process data from production equipment 1 and stores the process data in Standard DB.
Step 1: Data Collector 1 notifies MES of the completion of processing a wafer.
Step 2: MES sends a command to the Entry server for proceeding to perform the VM task.
Step 3: The Entry server sends the command to AVM Manager, which in turn sends the command to AVM Server 1.
Step 4: AVM Manager activates AVM Server 1.
Step 5: AVM Server 1 retrieves process data from Standard DB.
Step 6: AVM Server 1 starts the VM computation to generate the VM results.
Step 7: AVM Server 1 stores the VM results into Central DB.
Step 8: AVM Server 1 sends the VM results to the browser of the AVM client to display the VM results.

4.2.5 Summary

This section has introduced the essentials of cloud computing. We describe what cloud computing is and the driving forces and core technologies of cloud computing. We also present the definition, characteristics, benefits, potential applications, and adoption trends of cloud computing. Besides, the cloud service models (i.e. IaaS, PaaS, and SaaS) and the cloud deployment models (i.e. public cloud, private cloud, and hybrid cloud) are depicted. Also, we portray some cloud computing applications in manufacturing, including big data analytics, AI-enabled applications, and CMfg.

4.3 IoT and Edge Computing

In this section, we will introduce the essentials and working principles of IoT and edge computing. Their applications and benefits in manufacturing are also presented.

4.3.1 Essentials of IoT

With the advancement of technology, the Internet has become one of the most important communication channels between people (e.g. Facebook, Line, WeChat, etc.), realizing the connection between people. Currently, the Internet is further developing toward IoT [45], which refers to the connection of many things (e.g. sensors, home appliances, devices, machines, etc.) to each other and to humans via the Internet. In addition to allowing people to communicate with each other, IoT can enable the dialogues between people and things, and between things and things. IoT can promote the convenience of human life and would thus bring great business opportunities. Many countries have regarded IoT as an enabling technology that could enhance the competitiveness of the nation's economy and industry, thereby developing various IoT-related strategies and projects.

Besides, large enterprises have also actively developed and deployed IoT-based solutions to seize the market and obtain benefits. IoT can be used in various fields, such as transportation, healthcare, energy, aerospace, smart home appliances, smart buildings, environment monitoring, logistics, and more [46, 47]. As the German Industry 4.0 Program and the American Advanced Manufacturing Program both intend to leverage IoT technology to build smart factories, IoT will also be widely used in the manufacturing industry.

IoT is a network based on information carriers, such as the Internet, broadband cellular networks (e.g. 3G, 4G, and 5G), and wireless networks (e.g. WiFi, Bluetooth, ZigBee, and 6LoWPAN), which enables all independently addressable objects to achieve interconnection. Objects on IoT rely on electronic devices that can communicate through different network interfaces so that they can be discovered, monitored, controlled, or interacted. Broadly speaking, the IoT technology includes Radio Frequency Identification (RFID) technology, Wireless Sensor Network (WSN) technology, M2M Communication technology, and Supervisory Control and Data Acquisition (SCADA) technology [7].

RFID is a technology [48] that can use radio waves to identify tagged objects. It can be used in several commercial and industrial applications, such as logistics and supply chain visibility, product inventory tracking, attendee tracking, materials management, access control, IT asset tracking, and tool tracking. WSN refers to a network that connects a group of spatially-dispersed sensors for monitoring and recording the conditions of the environment. WSNs can measure environmental conditions and the status of things, such as temperature, humidity, pollution level, sound, vibration, voltage, current, and so on [49]. M2M technology refers to the technology that could connect distributed machines using the Internet. It enables network-connected devices and machines to autonomously exchange information [50]. SCADA is a system of software and hardware that allows industrial enterprises to gather and process real-time data for monitoring and controlling industrial processes and assets locally or remotely [51].

As IoT is increasingly becoming a ubiquitous IT infrastructure, it requires a great amount of resources for data storage and processing. However, because each IoT device usually has a limited resource, it would need to delegate or offload highly-complex computation to the cloud, which contains abundant computing and storage capacities. Thus, IoT mainly adopts a cloud-based architecture [52]. Figure 4.8 shows a generic three-layer cloud-based IoT architecture: the IoT or Industrial IoT (IIoT) layer at the bottom, the communication layer in the middle, and the cloud layer at the top.

The IoT layer consists of a variety of things (e.g. sensors, actuators, RFID tags, machines, devices, mobile phones, tablets, laptops, PCs, etc.) connected via various types of networks, such as wire IP network, Wi-Fi, 4G, 5G, ZigBee, Bluetooth, 6LoWPAN, and so on. The wire IP network can be IPv4- or IPv6-based Ethernet. ZigBee [53] is a low-power short-distance wireless communication technology that has the characteristics of low cost, low power consumption, high reliability, and low complexity. The number of nodes in a ZigBee wireless sensor network can be many, say 100 for a reliable and low-latency network. The connection speed of join a node in a ZigBee network is only 30 milliseconds. Although ZigBee's transmission speed is about 20K~250K bps, it is very suitable for transmitting texts and commands, which is sufficient to many IIoT applications. ZigBee can be applied to many areas, including factory automation, home control, building automation, industrial control, healthcare, environmental sensing, and indoor positioning, etc. IP version 6 (IPv6) provides a massive address space for resource-rich networking environments such as Ethernet. However, typical IoT networking environments based on low-power wireless personal area network (LPWAN) offer constrained energy, computation, and communication capabilities. Thus, 6LoWPAN (i.e. IPv6 over LPWAN) [54] was developed as an adaptation layer to enable LPWAN devices to be able to run IPv6.

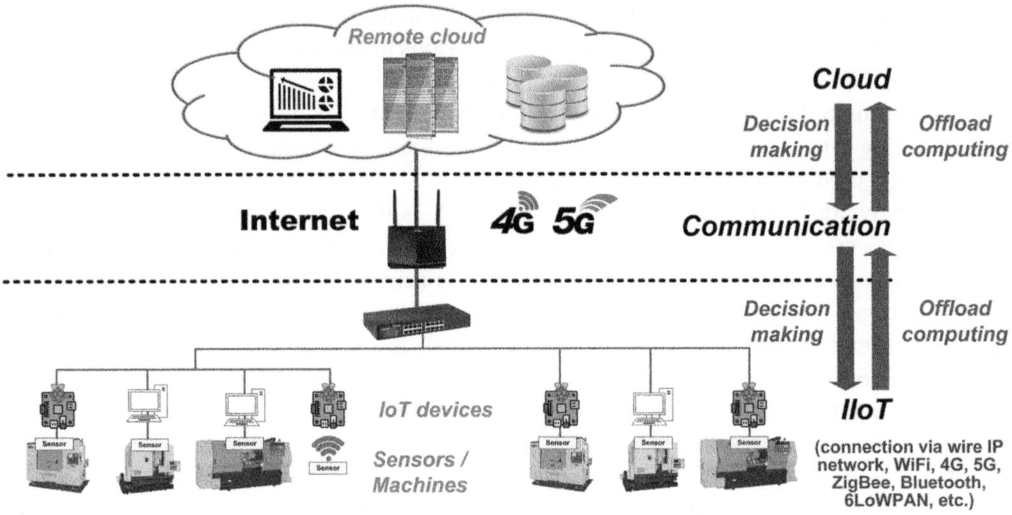

Figure 4.8 Generic three-layer cloud-based IoT architecture.

The communication layer comprises the Internet and mobile broadband cellular networks (e.g. 4G and 5G), which is responsible for transferring data between the IoT layer and the cloud layer. The cloud layer consists of remote datacenters that contain abundant virtualized computing and storage resources, which can be used to perform data analyses for decision making. Each IoT device typically has a limited resource and can only conduct simple data-processing tasks. Thus, IoT devices could offload computing-intensive tasks to the cloud.

IoT devices can be resource-constrained or resource-rich. The interoperability among diverse IoT devices is an importance issue. Several messaging protocols have been developed for the communication and integration among IoT devices, solutions or services [55], including Message Queueing Telemetry Transport (MQTT) [56], Advanced Message Queuing Protocol (AMQP) [57], Data Distribution Service (DDS) [58], and others. Among them, MQTT and AMQP are introduced below.

- **Message Queue Telemetry Transport Protocol (MQTT)**
  MQTT is a messaging protocol for IoT. It is designed as an extremely lightweight publish/subscribe messaging transport protocol that is ideal for connecting remote devices with a small code footprint and using minimal network bandwidth. Currently, MQTT has been applied in a variety of industries, such as automotive, telecommunications, oil and gas, and manufacturing, and so on [56].

  Figure 4.9 shows the MQTT Publish/Subscribe architecture for the communication of IoT. There are three roles in this architecture: MQTT Publisher, MQTT Broker, and MQTT Subscriber. The MQTT publisher can be any IoT device or object that would like to send out a message, such as a Raspberry Pi with a temperature sensor. The MQTT broker is the middleman that hosts a lot of topics for the publishers to publish messages and for the subscribers to subscribe them. The MQTT subscriber can be a laptop, a mobile phone, a tablet, or other IoT devices. The working principle of the communication between IoT devices using MQTT is illustrated as follows:

  Step 1: A publisher (e.g. a Raspberry Pi with a temperature sensor) publishes a message (e.g. 30 °C) to topic 1 (e.g. temperature).

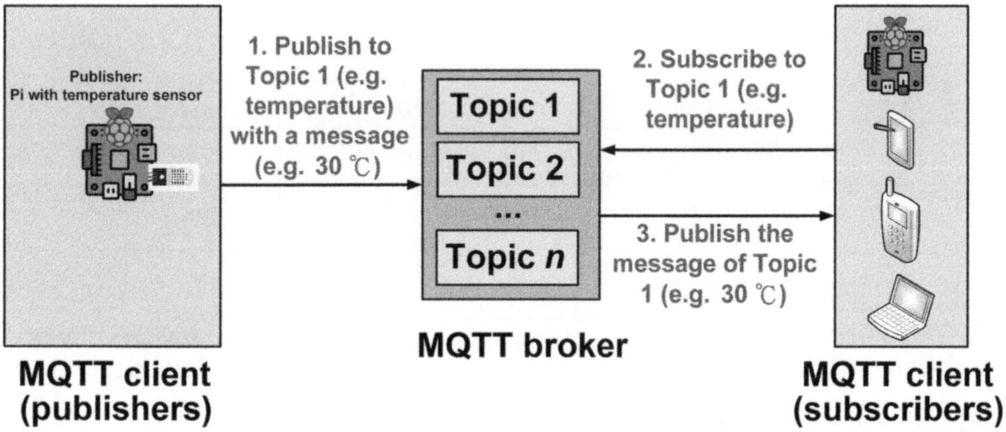

Figure 4.9 MQTT Publish/Subscribe architecture for the communication of IoT.

Step 2: A subscriber (e.g. a Raspberry Pi, a tablet, a smartphone, or a laptop) subscribes topic 1 (e.g. temperature).
Step 3: The MQTT broker publishes the message of topic 1 to the subscriber.

By this way, all MQTT clients only connect with a broker without needing to directly interact with each other, while achieving the interoperability among them.

- **Advanced Message Queuing Protocol (AMQP)**
AMQP is an open Internet Protocol standard for message-queuing communications between business applications or organizations [57], as well as among IoT devices or systems. AMQP runs over Transmission Control Protocol (TCP) to do asynchronous transfer of messages. It is based on a publish/subscribe architecture similar to the MQTT protocol architecture. The differences between AMQP and MQTT are briefly mentioned as follows [59]. MQTT is a message queuing telemetry transport protocol, which provides a simple way of message queuing services. MQTT uses TCP for transferring messages in an asynchronous manner. MQTT is specially designed for small devices with a constrained bandwidth of networking and is thus implemented mostly in embedded systems. By contrast, AMQP is an advanced message queuing protocol, which supports a wide range of messaging scenarios. AMQP also uses TCP for the asynchronous transfer of messages, regardless of the running OSs. Besides, the AMQP broker consists of two components (namely, exchange and queues) that are working together.
Figure 4.10 shows the AMQP Publish/Subscribe architecture for the communication of IoT. There are three roles in this architecture: AMQP Publisher, AMQP Broker, and AMQP Subscriber. The AMQP publisher can be any IoT device or object that would like to send out a message. The AMQP broker comprises an Exchange component and queues of message topics. The Exchange is the location to which publishers deliver messages. The published messages contain routing information which is used by "Exchange" to route them. The queues are the locations where messages are stored until they are delivered to or read by the subscribers. The AMQP subscriber can be a laptop, a mobile phone, a tablet, or other IoT devices. The working principle of the communication between IoT devices using AMQP is depicted as follows:

Step 1: A publisher publishes a message to the Exchange.
Step 2: The Exchange routes the message to a topic (e.g. Topic 1) stored in a message queue.

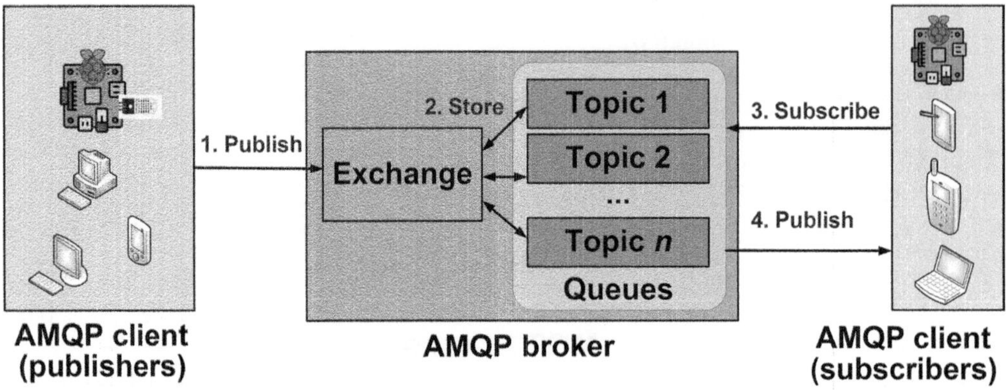

Figure 4.10 AMQP Publish/Subscribe architecture for the communication of IoT.

Step 3: A subscriber subscribes a topic (e.g. Topic 1).
Step 4: The AMQP broker publishes the message of topic 1 to the subscriber.

Although there are a lot of standards and technologies developed for IoT, there is no single protocol that can enable all devices, services, and applications to talk to each other seamlessly. This leads to that the current IoT systems or solutions are either in a small scale or are constrained to connect a few specific types of objects. On the other hand, existing Web standards and tools provide the ideal layer for connecting devices and applications to exchange data. This vision is called the Web of Things (WoT) [60], which refers to using a set of Web standards for solving the interoperability issues of different IoT platforms and application domains.

The Web technologies can bring many benefits to IoT, some of which are described below [60]:

- **Easier to Program**
  Whether reading data from a device or writing data to it, we can use the Web protocol, instead of using the complex IoT protocol. Thus, once having the basic code needed to build a web application, we can effortlessly start to code a program that can get into conversation with the device.
- **Offer Open and Extensible Standards**
  The Web standards are so popular because they are completely open and free. So, adopting them is of no risk. Furthermore, data can be easily moved between Web systems using Web protocols and interfaces, such as HyperText Transfer Protocol (HTTP) and Representational State Transfer (REST) APIs.
- **Vast and Easy in Deployment, Maintenance, and Integration**
  Stopping a Web application for an upgrade is fast. Also, the ability of the Web is continuously being enhanced according to people's demands. For example, the Web can now grab photos from the camera, share maps, exchange messages, play games, etc. Its transformation is never stopped. Besides, it is easy to integrate different systems by the Web. By contrast, there are always new devices in the IoT world. When a communication protocol is released or changed, some devices related to it must be updated accordingly.

4.3.2 Essentials of Edge Computing

IoT can connect a great number of objects or things via various networking technologies. Unconnected objects can also join an IoT. For example, an object attached with a bar code or an RFID tag can be sensed by a reader that would in turn transmit the sensed data via the IoT. When an IoT contains increasing numbers of objects and devices, it could generate a huge amount of diverse data. However, an IoT device typically has a limited resource and can only perform simple data-processing tasks. Thus, how to process and store the generated data efficiently is a big challenge for a standalone IoT. One feasible solution to this issue is to allow an IoT to offload computing-intensive tasks and big data storage to the cloud, as indicated in Figure 4.8.

Currently, data in the form of images, videos, and texts generated by user devices is usually transmitted to the cloud for processing. The traditional cloud refers to remote datacenters that can offer abundant computational resources through various services. However, mission-critical and latency-sensitive IoT services and applications require a very quick response and processing. For example, in an augmented reality (AR) application that a smart glass worn by a person provides instructions for navigation, there is always a steady stream of images that needs to be processed in the cloud [61]. In this case, it may not be feasible or sustainable to communicate huge data to the remote cloud over the Internet constantly. Also, when tens of billions of user devices, such as smartphones, tablets, appliances, and wearables, are connected to the Internet, it would be impossible to transmit the data generated by all devices to the remote cloud for processing [62].

To address the above-mentioned responsiveness and latency issues in offloading tasks to remote cloud, the edge computing [61–63] was developed. Edge computing refers to moving provisioned services from the cloud to the edge of the network. Edge computing becomes a promising solution for IoT applications to meet the latency demands, enhance the scalability and energy efficiency of resource-constrained IoT devices, and reduce the traffics of the IoT backbone network [64]. Also, using edge computing to process user requests will reduce end-to-end latency, decrease the traffic beyond the first hop of the network, and improve overall QoS [61].

Although both edge computing and cloud computing provide computation and storage resources, they are different in nature [62]. The edge computing is in proximity to the underlying accessing nodes, IoT devices, and WSNs. The edge computing is localized to the edge of the network, while the cloud computing is at remote datacenters.

Figure 4.11 shows a four-layer cloud-based IoT architecture with an edge computing layer for offloading computation: IoT or IIoT layer, edge computing layer, communication layer, and cloud layer. The descriptions about the IoT layer, the communication layer, and the cloud layer are the same as those for Figure 4.8. The edge computing layer is depicted in detail as follows.

In Figure 4.11, the IIoT is in the context of a manufacturing factory. The IIoT connects various objects (e.g. sensors, machines, equipment, IoT devices, etc.) via networking. In this case, the IoT devices are regarded as edge computers because they are beside the production equipment and machines for data acquisition and processing. IoT devices have different specifications and could be tiny devices (e.g. Arduino, Raspberry Pi, and Up Board), industrial PCs, servers, mobile devices, and smart gateways. IoT devices are typically resource-constrained and cannot perform data- or computing-intensive tasks. Thus, they can offload their workloads to other types of edge computing resources: Edge Server and Edge Cloud.

Edge servers refer to those high-end servers at the edge of the network about one hop away from IoT devices or user devices. These edge servers have much more resources than IoT devices or user devices. Thus, they can be used to perform data- or computing-intensive workloads, such as image processing, video streaming, online gaming, AR-related computing, and so on. On the other hand, some workloads that an edge server cannot afford to process can be offloaded to Edge Cloud. An edge cloud refers to a micro cloud at the edge, which could offer many virtualized resources provided by a cluster of several high-end servers. Although an edge cloud has a smaller scale compared to the remote cloud, it can store a great amount of data and perform highly-complex workloads to offload the IoT.

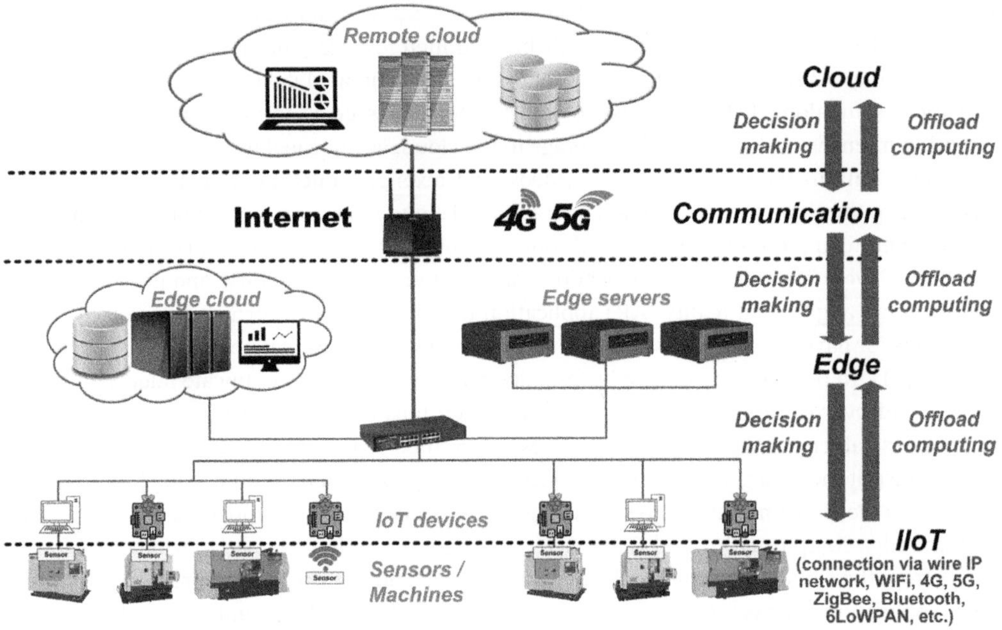

Figure 4.11 Cloud-based IoT architecture with an edge computing layer.

Industrial Internet of Things (IIoT) can be regarded as a subset of IoT because it specially interconnects a lot of industrial sensors, devices, actuators, processes, and machines, together with people at work. In a recent trend, IIoT is increasingly leveraging AI technologies to process and analyze data from diverse sources for conducting advanced predictive analytics, such as equipment FDC, production quality conjecture of workpieces, predictive maintenance of key devices, and demand forecasting [64]. Because the data-driven AI algorithms are able to extract critical features of an interested system without needing to understand the system's physics, they could play a significant role in IIoT applications. For example, predictive maintenance relies on machine learning to detect anomalies in industrial systems and then predict the RUL of their key devices. Further, integrating AI into edge computing to provide the intelligent edge for IoT or IIoT analytics and AI-enhanced offloading is a promising development trend [14, 64, 65].

4.3.3 Applications of IoT and Edge Computing in Manufacturing

IIoT can be regarded as a subset of IoT because it is applied in industrial environments. The IIoT for manufacturing can connect a great number of sensors, production machines, computing devices, and smart mobile devices for collecting and processing data from various sources. IIoT supports bidirectional M2M communication in the factory or anywhere on the Internet. When combining with the computing and storage resources of edge computing and cloud computing, IIoT allows real-time computation and delivery of high-value information for manufacturing activities. The manufacturing factories can leverage IIoT with edge and cloud computing solutions to be more efficient and effective in many production areas, including (i) capturing, storing, and analyzing data, (ii) understanding the production process better, (iii) enhancing worker performance by providing easy-to-use information relevant to the current task, and (iv) connecting things (e.g. sensors or machines) to collect their data [13].

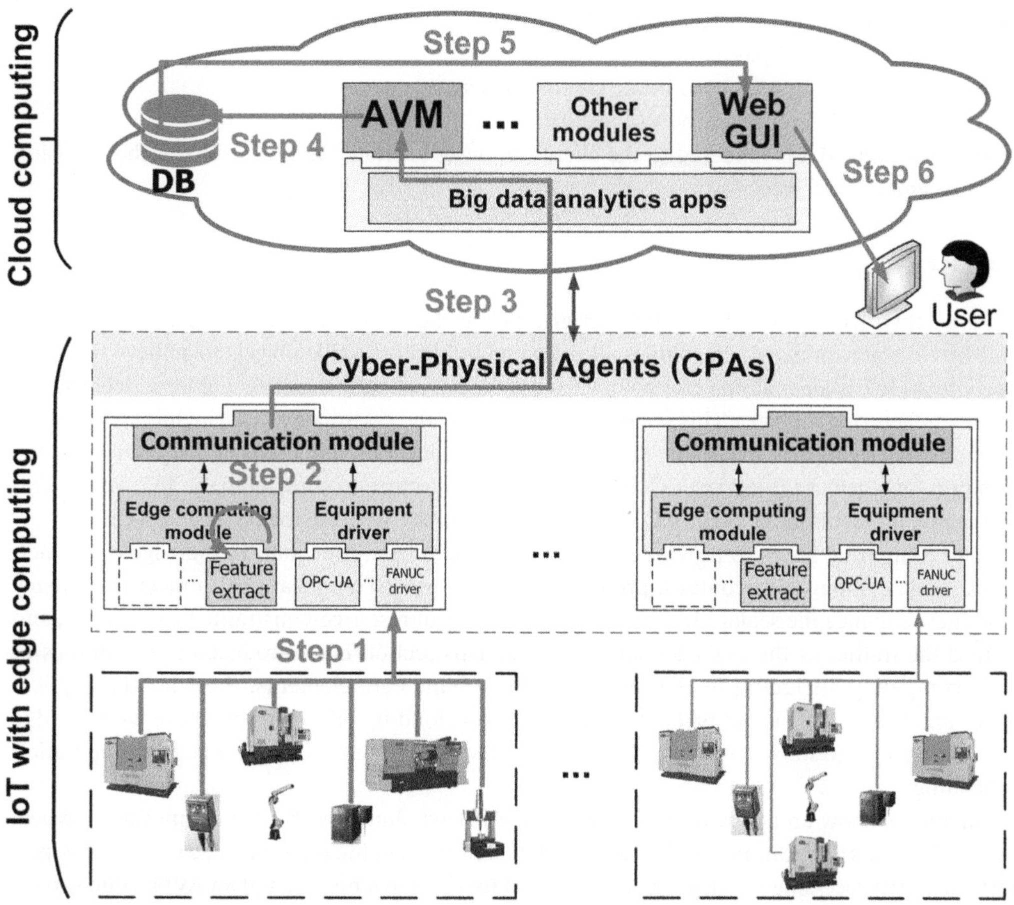

Figure 4.12 Application of IoT and edge computing in wheel machining.

Another valuable application of combining cloud-based IIoT with edge computing is to use the cloud to train models using data from IIoT devices. Then, these models can be executed at the edge to allow IIoT devices to respond more responsively to changes. Also, many companies have already used edge computing and industrial analytics to optimize their production processes to achieve accurate and efficient production [66].

Figure 4.12 shows an application of IoT and edge computing in wheel machining, which is developed by our team. The system architecture of this application consists of two parts: many IoT devices with edge-computing capability, and cloud computing. The IoT devices, denoted as Cyber-Physical Agents (CPAs), are used to connect machining tools, metrology machines, and other related devices for collecting and transferring data. The edge-computing module embedded in CPA is used to store and process the collected raw data for extracting features related to machining. The cloud is used to host and run AVM service, database, and Web GUIs. The functions of CPA and AVM will be elaborated in Chapters 7 and 8, respectively.

The workflow of conducting AVM for wheel machining is described as follows:

Step 1: CPA collects data from machine tools OP1, OP2, and OP3.

Step 2: CPA performs edge computing for data preprocessing to extract features related to wheel machining.

Step 3: The feature data are sent to the AVM service on the cloud to conjecture the machining quality of a workpiece.
Step 4: The VM results are stored in the cloud database.
Step 5: The Web GUI retrieves the VM results.
Step 6: The Web GUI is rendered to the client's browser to show the VM results of the precision items of workpieces.

4.3.4 Summary

IoT is a network based on information carriers, such as the Internet, broadband cellular networks, and wireless networks, which enables all independently addressable objects to achieve interconnection. As IoT is an enabling technology for building smart factories, IoT will be widely used in the manufacturing industry. This section has introduced the essentials of IoT, including its definition, the underlying core technologies, the potential applications, and the benefits it can bring. We also present a generic three-layer cloud-based IoT architecture to show how the IoT works. Two messaging protocols for IoT (i.e. MQTT and AMQP) and the WoT are introduced as well.

Edge computing refers to moving provisioned services from the cloud to the edge of the network. Edge computing becomes a promising solution for IoT applications to meet the latency demands, enhance the scalability and energy efficiency of resource-constrained IoT devices, and reduce the traffics of the IoT backbone network. This section has introduced the essentials of edge computing, including its definition, the underlying core elements, the potential applications, and the benefits it can bring. We also present a cloud-based IoT architecture with an edge computing layer to show how the IoT devices can leverage resources at the edge for computation offloading.

Finally, we show an application of cloud-based IoT combined with edge computing in wheel machining. This application uses IoT to collect data from machine tools, an edge computer to perform data preprocessing to extract features related to wheel machining, and an AVM cloud service to conjecture the machining quality of a workpiece.

4.4 Big Data Infrastructure

In this section, we will introduce big data infrastructure. First, the big-data application demands for semiconductor manufacturing are described. Then, the core software stack components of big data infrastructure are depicted. Then, how to bridge the gap between the core software stack components and the big-data-related applications is illustrated.

4.4.1 Application Demands

Semiconductor manufacturing is a high-end technology industry that not only improves its manufacturing technology over time but also relies on state-of-the-art information technology for production automation. In the semiconductor manufacturing industry, data volume increases exponentially during the manufacturing process, which greatly helps in monitoring and improving production quality. To accommodate excess data, Apache Hadoop [67] is often adopted. Hadoop is essentially a computing and storage facility for big data. That is, generated data are stored in Hadoop and then analyzed in a batch and/or real-time manner. Hadoop has been developed and used for a decade and its ecosystem remains prosperous.

Semiconductor wafer fabrication foundries have embraced Hadoop for big data applications, including FDC and yield analysis (YA). We discuss in this chapter two underlying infrastructural services, namely, Hadoop data service (HDS) and distributed R language computing service (DRS), based on Hadoop for a semiconductor manufacturing foundry in Taiwan, ROC. HDS and DRS have been designed, implemented, and operated since 2015. Both HDS and DRS are 7×24 operational services in production systems presently. Specifically, HDS and DRS are developed to meet the following requirements:

- **Transparency**
  Users are likely to access data objects stored in a big data system by using existing tools with which they are familiar, preventing the reduction of productivity due to the lengthy learning curve of users. Particularly, for accessing data objects in Hadoop, an essential step for users is to realize the architecture of Apache Hadoop distributed file systems (HDFS) and the practice of HDFS APIs [68–70]. Users need to be familiar with these APIs so they can fully utilize HDFS, thus maximizing the HDFS's performance. In addition, users have to be well trained to manipulate distributed databases in Hadoop (e.g. Apache HBase [71–73]). Moreover, existing applications require sophisticated efforts to migrate to Hadoop. Users have been familiar with traditional statistical and/or machine learning analytics tools such as R [74] and Python [75]. Specifically, earlier developments based on these analytical tools have been associated with production systems for quite some time. An unlikely approach is to reinvent the wheel to embrace novel technologies such as Hadoop. Thus, the use of existing efforts and experiences as a basis to transparently leverage emerging solutions has become a critical design consideration.
- **Autonomy**
  Big data storage and computation technologies are typically based on a server farm (or a cluster of commodity off-the-shelf servers). By an "off-the-shelf server," we mean a normal storage and computation device that is equipped with abundant computational resources (e.g. processor cycles, main memory, and secondary hard drive spaces) and software stack (e.g. Linux OS, Java virtual machine, and Hadoop) available in the public domain. The capacities of these technologies increase linearly if the number of serving nodes is expanded monotonically. (Servers and serving nodes are interchangeable in this chapter.) Each participating server may fail anytime; that is, failure is the norm. Consequently, a large-scale system shall self-organize, self-manage, and self-heal. In addition, deploying system-wide services should be as effortless as possible.
- **Efficiency**
  Given a set of resources, we will efficiently and effectively utilize these resources to provide services to users. While HDFS is mainly designed for sequential data access patterns and large data objects, HBase deals with random accesses for small ones. Accordingly, our developed services will intelligently select the underlying storage facilities to optimize data access requests issued by users. The optimization will not only maximize the utilization of resources but also introduce minimally operational overheads. To serve R language computational tasks, we will allocate computational resources to efficiently and effectively perform the R tasks. Specifically, the management includes adaptive resources allocation/deallocation and task scheduling. In addition to the optimizations for storage and computation facilities, we will further monitor the runtime usage for the resources to optimize system-wide services.

Our HDS and DRS services bear the above design considerations in mind. HDS is designed to reduce the gap for users accessing Hadoop storage facilities. It is a web-based service that allows users to replicate data objects from existing storage devices such as Samba [76] and FTP to Hadoop. Users can also transparently access data objects in HDFS and HBase in Hadoop through HTTP. In addition, existing analytically computational platforms (e.g. Apache Spark [77, 78]) and data query tools (e.g. Apache Hive [79, 80]) in the Hadoop ecosystem are oblivious to HDS. Moreover, HDS has the capability to self-scale, self-manage, and self-heal. Furthermore, HDS has been optimized for the efficient and effective use of resources to access data objects in Hadoop; it introduces more modest overheads compared with native Hadoop.

DRS is developed to serve R language computational tasks [81] transparently. Existing R tasks previously realized in centralized computing environments need minimal efforts to enjoy data-parallel computations provided by our proposed DRS. The DRS service intelligently allocates and deallocates resources to an R computational job. It also optimizes the resources allocated to an R job, considering distinct resource requirements for tasks assembling the job. Furthermore, DRS reliably commits the computation of an entire R job by detecting and resolving potential failures during runtime. Moreover, our DRS service details and reports statistical data measured and gathered during the runtime for a submitted R job.

4.4.2 Core Software Stack Components

We briefly discuss in this chapter the fundamental building blocks for our HDS and DRS services. The building blocks underlying our systems are ecosystems in Hadoop, including HDFS, HBase, YARN, and ZooKeeper, where HDFS and HBase serve as our storage layers of HD. YARN is a resource management framework employed by DRS, and ZooKeeper helps coordinate cooperative operations inside HDS and DRS.

- **HDFS**
  Apache Hadoop HDFS is a large-scale, distributed file system for sequential data accesses [68]. An HDFS system consists of a number of off-the-shelf commodity server machines (or a cluster). Among the machine nodes in the cluster, while a namenode server is selected for managing the naming space for files stored in HDFS, the remaining nodes are datanodes for storing files. A file published in HDFS may be segmented into a number of identical-sized data blocks. Given the fact that the namenode server may become the performance bottleneck in case excess data blocks need to be managed [82], the size of a data block is normally large (say, 64 or 128 MBytes). For a file with >1 data blocks, these blocks may be distributed to >1 datanodes, depending on the size of a HDFS cluster and the load status of each participating datanode [68]. In addition, each data block of a file may be replicated to ≥1 distinct datanodes for reliability. Clients requesting and accessing data files (or objects, informally) in HDFS need to first consult the namenode to resolve the locations of data blocks assembling the requested file. With the locations of data blocks, the client then directly manipulates the data blocks in the corresponding datanodes.
- **HBase**
  While HDFS mainly deals with sequential and bulk accesses for data, Apache HBase mainly supports random access for small data objects [71, 72]. HBase is a large-scale, distributed, and column-based NoSQL database storage engine, and it is essentially a key-value store. Clients in HBase can create a number of tables. Conceptually, each table manages any number of rows, and each row consists of a unique row key and a number of columnar values, where the value in a column of a row is simply a byte string.

  For scalability, a (large) table in the HBase can be horizontally partitioned into a number of regions, where the regions are the basic entities accommodating consecutive rows of data. Regions are disjoint, and each row can be hosted by a single region only. HBase operates in a cluster computing environment that consists of a number of region servers (a typical commodity server machine). If the number of regions of a table is c and the number of region servers available to the table is n, then the HBase ideally allocates $\frac{c}{n}$ regions to each region server. Here each region can only be assigned to one of n region servers.

For a typical setting, HBase is layered over HDFS. Precisely, each region server in HBase is also a HDFS datanode server. Each table populated in HBase is stored as a number of region files in HDFS. Through the replication mechanism provided by HDFS, each region file has a number of replicas stored in distinct HDFS datanode servers.

- **YARN**

  Apache YARN is a global resource management framework for Hadoop [83, 84]. Resources include processor cores and main memory space available in servers joining the Hadoop system. YARN is a unified resource manager in the sense that users (e.g. MapReduce jobs [85]) require computing resources to consult YARN for allocating the resources that meet the users' requirements. Resources shall be returned to YARN if their associated jobs have completed their computations. For example, a MapReduce (MR in short) job may request 8 containers, each with 2 processor cores and 4 GBytes main memory space. Here, containers in Hadoop are essentially akin to virtual machines, which is a process in an OS. Once YARN grants a user the requested resources, the user needs to manage those resources themselves. The management includes resource deallocation and allocation. Also included is task scheduling, for which data items should be performed given a designated container. Moreover, users need to ensure the reliability of their computations. Whether the allocated resources are utilized efficiently or not depends on how a user monitors and exploits the resources.

- **ZooKeeper**

  Apache ZooKeeper is a coordination service for cooperative entities that need to be coordinated and synchronized [86, 87]. By registering shared variables with ZooKeeper, entities can concurrently read and write those variables. Particularly, by publishing and subscribing the shared variables, the interested entities observe the values of shared variables and then determine their computational logic and flow.

  ZooKeeper is reliable and fault-tolerant because it is assembled by a cluster of computing nodes. Nodes in the ZooKeeper cluster farm synchronize themselves by replicating data stored from any one node to each other.

  Our HDS and DRS services, which will be discussed later, strongly depend on ZooKeeper for coordination and synchronization. The coordination and synchronization include performance data synchronization for load balancing in HDS servers, membership management for participating servers in the HDS cluster, and DRS job-level scheduling.

4.4.3 Bridging the Gap between Core Software Stack Components and Applications

The HDS and DRS infrastructural services for bridging the gap between software stack components and applications are presented as follows.

4.4.3.1 Hadoop Data Service (HDS)

HDS is a web-based distributed data repository. HDS is built over HDFS [68] and HBase [71, 72]. HDS allows application developers (or users) to replicate data objects (files, typically) from existing commodity storages such as Samba [76] and FTP servers to the HDS storage using the HTTP protocol while also supporting replicating data objects between any two existing heterogeneous storage servers. Precisely, users simply specify in HDS: http://HDS/copy?from=S:f_s&to=D:f_d where f_s (f_d) composites the storage server name, file path, and data object name; S and D respectively represent the source and destination storage entities; HDS is an HDS server arbitrarily selected from the HDS cluster. S and D may implement distinct storage communication protocols, including FTP, HTTP, Samba, the local file system, and even HDS itself.

For example, an invocation "http://earth/copy?from=SMB://venus/log/XYZ.csv&to=FTP://mars/fdc/ABC.csv" indicates a user replicates the file /log/XYZ.csv in the Samba file server named venus to the file /fdc/ABC.csv in the FTP file server called mars through the help of the specified HDS server earth. Notably, S and D may be designated as our HDS storage cluster. In particular, our operational HDS storage in the semiconductor fabrication foundry is a unified data lake that stores data files sent from existing FTP, HTTP, and Samba storage pools.

HDS is web-based, thereby ensuring ease of use for data scientists. The HDS web-based service is also scalable, reliable, and self-managed. Figure 4.13 shows the software stack architecture and modules for HDS. We discuss in the following the key design and implementation features of HDS.

- **Scalability and Reliability**
  Each data node in Hadoop joins our HDS serving cluster, that is, each data node (namely, an HDS server or an HDS serving node) is also a web server that accepts HTTP requests for accessing the HDS cluster. A user may issue his Hadoop data transfer request to any HDS server selected from the HDS cluster. Given k HDS serving nodes, our HDS cluster provides k times service capacity. Essentially, each Hadoop data transfer request utilizes resources of connections and memory space in an HDS server. For bulk transfer for a data object from S and D in http://HDS/copy?from=S:f_s&to=D:f_d, we pipeline the data blocks that assemble the object from S to D.

  The HDS server responsible for the transfer can be used to buffer the data blocks in order. Depending on the request and each HDS server hardware specification, our HDS service allows system managers to configure the buffer memory space for data delivery to exploit each HDS server's physical hardware resources. Note that our HDS cluster relies on ZooKeeper [86] that provides the coordination service. Precisely, each HDS server registers itself in the ZooKeeper by using heartbeat messages to realize our HDS membership service. An HDS server can simply inquire the status of any other HDS serving node member, such as its aliveness and loading

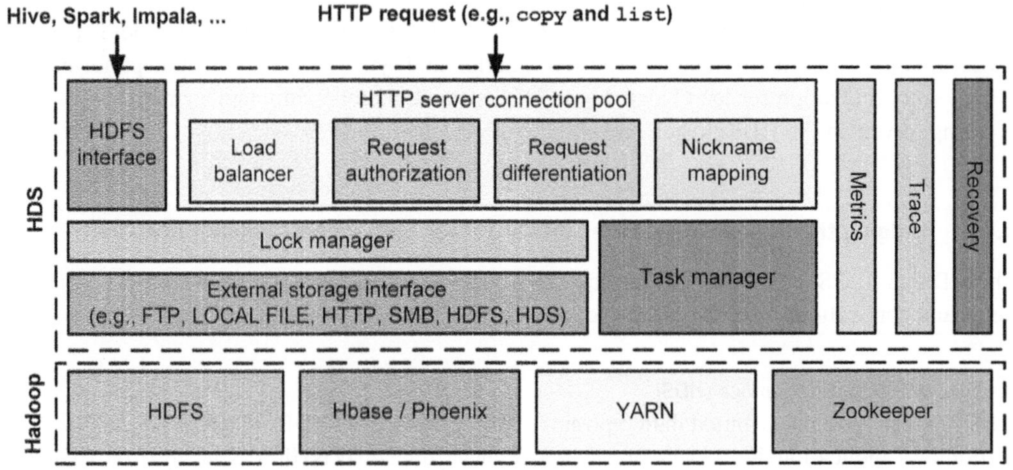

Figure 4.13 Software stack for a HDS server. *Source:* Reprinted with Permission from Ref. [88].

metrics (will be discussed in the following). Our HDS service is thus operated in a reliable manner without a single point of failure because our HDS membership management maintains up-to-date HDS cluster messages, and any failed HDS server can be promptly identified and then removed. While our HDS service depends on the ZooKeeper for membership management, each HDS serving node additionally acquires ZooKeeper for data consistency. Specifically, once a data delivery request is issued and accepted, its responsible HDS server locks the file using the ZooKeeper coordination service such that the file cannot be read and written during its delivery. Our HDS service currently implements the sequential consistency memory model [89].

- **Service Differentiation**
Our HDS service consists of control and data planes. The control plane handles real-time requests such as list and status, where list enumerates all data objects stored in HDS and status reports the progress of an ongoing data delivery request. Each HDS server statically allocates connections to control and data planes. This is because data delivery requests may take time to perform; if such requests dominate the connections available from an HDS server without service differentiation, then the server may not perform control messages promptly.
- **Load Balancing**
Our HDS cluster self-balances the loads of HDS servers joining the service.
We measure the load of an HDS serving node n_i as $L(n_i) = \alpha C_D(n_i) + \beta M(n_i)$, where $C_D(n_i) = \frac{C_{D_{occupied}}(n_i)}{C(n_i)}$ represents the ratio of the number of connections currently used by ongoing data delivery requests (that is, $C_{D_{occupied}}(n_i)$) to the total number of connections predefined for control and data planes ($C(n_i)$), $M(n_i)$ denotes the ratio of memory space used by the data plane to the total memory space allocated to the control and data planes; $0 \le \alpha \le 1$ and $0 \le \beta \le 1$ are two predefined parameters. We implement the ball-and-bin model [82, 90–92] to balance loads of participating HDS servers. More specifically, when an HDS server n_i accepts a data delivery request r_i, n_i randomly selects two alive HDS servers (say, n_j and n_k) from ZooKeeper (as mentioned previously, ZooKeeper maintains up-to-date HDS servers' aliveness status). n_i redirects r_i to n, where $n \in \{n_i, n_j, n_k\}$ and $L(n)$ is the smallest among $L(n_i)$, $L(n_j)$, and $L(n_k)$. The ball-and-bin model for balancing the loads of HDS servers operates explicitly in a lightweight manner as each HDS server forwards its load to another HDS serving node independent of the size of the HDS cluster (that is, the total number of HDS server participants) and without global knowledge.
- **Efficient Storage Management**
We observe that data objects stored in our HDS cluster are heavily tailed, that is, most data objects are small, whereas a few of them are large. To this end, the underlying storage infrastructure of our design for HDS thus mixes HDFS and HBase. When a data object is small (typically, ≤10 MBytes), the object is stored in HBase; otherwise, it is stored in HDFS. Given that the namenode server of HDFS strongly depends on its main memory storage for file naming space and metadata management, a large number of small files will rapidly exhaust the valuable main memory space in the namenode [82]. In addition, small data objects introduce the internal file block fragmentation issue because the size of a file data block in HDFS is typically large. (A typical data block is 64 MBytes or 128 MBytes.) By contrast, HBase is capable of providing random data access for small data objects stored in HDS. Although solutions such as Hadoop archives [93] and sequence files [94] exist in Hadoop to pack small data objects into a single large HDFS file, such solutions do not offer the indexing capability, which allows users to arbitrarily inquire any data object. Note that HDS manages HDFS and HBase transparently. Users simply issue their data access requests as defined in http://HDS/copy?from=S:f_s&to=D:f_d.

- **Transparency to Existing Data Query Languages and Analytic Frameworks**
  HDS is another distributed storage infrastructure over Hadoop. A number of data query language tools support SQL, such as Apache Hive SQL [79], which ensure ease of data manipulation in Hadoop. HDS is transparent to these language tools because it is compliant with HDFS APIs standard. Consequently, Hive, for example, may access data objects stored in HDS, thus leveraging applications built upon Hive. Analytical computing frameworks such as Apache Spark [77] can also transparently manipulate data objects stored in HDS without extra efforts for migration to HDFS. In compliance with the HDFS APIs standard, a critical issue for our HDS service is to support hierarchical naming space because existing applications may have organized their data in distinct directory folders of Hadoop. To support the hierarchical naming space, we take advantage of the HBase storage layout structure in indexing data objects. Precisely, the names of data objects in alphabetical order will be stored in increasingly consecutive rows in an HBase table.
- **Performance Dashboard**
  Our HDS service provides a performance metric framework to measure and break down the performance of service runtime. The performance monitoring framework allows application developers to trace each runtime stage of dealing with a data delivery request, thus optimizing their applications' data access behavior. Given that our HDS service is operational under production environments, facilitating system administrators in monitoring the HDS service is also crucial for the performance framework.
- **Security and Service Deployment**
  Users issuing their HDS data delivery requests need to be authenticated. Our authentication is currently role-based, that is, distinct users may be allowed to invoke different HDS RESTful APIs. For example, the user u_i may be granted authorization for copy, list, delete, while another user u_j can only invoke copy and list without being permitted to invoke delete. We also offer nickname mapping for ease of authentication. With regard to the deployment of HDS service, our implementation presently depends on the *HBase* coprocessor framework [95]. More precisely, when HBase bootstraps itself, HBase additionally initializes the booting procedure of HDS in the meantime, thus simplifying the management of HDS in a large-scale computing environment.

4.4.3.2 Distributed R Language Computing Service (DRS)

DRS is a data-parallel computing framework service for analytical users in R language, that is, an R program specifies the input data set and then processes the data objects in the set in parallel. Notably, the computational tasks for distinct data objects are independent. Consequently, DRS is a massively data-parallel computing framework for R.

Here, we define a DRS "job" as a single R program performing a set of independent data objects, and each data object computed by the R program is said to be a "task", thereby resulting in a computing job that consists of a set of independent tasks and these tasks can be performed simultaneously.

Our DRS framework service is based on Apache YARN [83], and it is akin to the Hadoop MapReduce [85] batch processing. YARN formulates a master/slave computing paradigm, where the master handles the workflow inside a computational job, and slaves are responsible for the computation. Particularly, the master in our DRS framework manages to schedule computational tasks by dispatching different data objects given by a user to slaves. In addition, the master in DRS allocates and reallocates computational resources to tasks for boosting performance. Moreover, it detects potential hardware and software failures during a job computation such that our DRS service can reliably perform the computation. By contrast, each slave computes with data objects

```
Configuration:
http://host/hds/v1/run?code="..."&data="..."&config="..."&codeout="..."&consoleto ="..." //fork a DRS job
code = ... //R program path
data = ... //input data file, containing a list of paths of data files
container_spec = ... //specification for a set of containers
codeout = ... //aggregation of output data files for R programs running in each container
consoleto = ... //R program console output
...

R program body:
tempData <- read.csv(DRS_INPUT_FILE_PATH , header = TRUE) //read data files locally
data <- ...(tempData) //filter and load data
...
sumResult <- sum(data) //compute sum
sdResult <- sd(data) //compute standard deviation
medianResult <- median(data) //find median
...
```

Figure 4.14 Programming in DRS. *Source:* Reprinted with permission from Ref. [88].

designated by the master, and it also periodically reports the status of on-the-fly computation so that the master can monitor whether a computational task allocated to a slave is ongoing or not. When a slave does not receive any task to be performed, the master then reallocates resources occupied by the slave (this subject will be discussed later) to another and/or even returns the resources to YARN. We note that each job in our DRS framework associates a single master and a set of cooperating slaves individually. Distinct jobs do not share their masters and slaves.

Figure 4.14 illustrates an example of programming in DRS, where a user needs to specify a configuration file that describes the location of the user's R program and the input data file paths for the R program. In addition, an R-job user needs to define the specification of computational resources. Given a configuration description, the requested computing nodes (i.e. containers in Hadoop, formally) are then allocated if successful. HDS is seamlessly enabled to help replicate specified data objects to containers. We note that while containers perform their computations, each data object demanded by a container is concurrently loaded in a pipeline fashion. Here, in addition to defining the configuration description, an R language program merely requires modest modification for its input data objects. We further discuss our design and implementation features for DRS as follows.

- **Scalability and Reliability**
  A user acquires the DRS service by specifying his job's metadata messages. The metadata messages primarily include the description for resources demanded by performing the job and locations of data objects for the job's input. By resources, we mean the number of computing nodes (i.e. containers) joining a computational job. In addition, the main memory space for each container node needs to be specified. The number of containers for a job is defined arbitrarily if such resources are available in Hadoop. In our DRS framework, each container node periodically heartbeats its status to its corresponding job master, preventing the job master from introducing considerable overheads in monitoring its slaves. If any slave fails, then the job master replaces the failed slave with healthy one by requesting YARN. Our design particularly detects software failure for slaves that have exhausted their allocated memory space. Such failures occur because users may underestimate their demanded memory space. We will discuss our solution to such failures later such that a job can be successfully performed.

- **Adaptive Resource Management**
  In a computational job submitted to DRS, the job's corresponding master manages resources, namely, slave computing nodes allocated from YARN, and dispatches tasks to the slaves. A slave may not be able to perform a computational task due to insufficient main memory space, that is, the specification of slaves is not defined well. According to our experience in operating DRS in a semiconductor fabrication foundry, the size of the main memory space required by slaves for a computational job is heavily tailed. Among the set of tasks, only a few of them require large memory space, while a small memory space is sufficient for most tasks. Designating the maximum size of main memory space required among tasks for each slave is clearly inefficient, considering that main memory space is one of the scarce resources in a big data computing system. By contrast, if we specify each slave's memory working space as the minimum size, then slaves for large tasks may not complete their computations, thereby failing to perform the entire job. To address this issue, our DRS service adaptively detects the resource demand for a task. Precisely, if the job master observes that a slave fails its computation due to shortage of memory space, then the master creates a slave assembled by a relatively large main memory space. This approach may remove existing slaves to aggregate resources. Our DRS framework guarantees that the total amount of resources is the same as that defined by the job's user, however. Consequently, we encourage users to modestly specify the resources for their tasks such that the entire system can accommodate as many jobs as possible.
- **Task Scheduling**
  DRS presently schedules tasks for a job in a per-batch manner. By scheduling a task to a slave computing node, DRS indicates the location of data object associated with a task; once a slave node receives the data object's processing request and thus the location of the object, the slave fetches the data object and then performs its computation with the input of the object. Notably, the location of a data object represents the data object's name in HDS. DRS can also manipulate data objects stored in another storage system (e.g. HDFS) in addition to HDS. Given a number of tasks, m, for a computational job, our DRS service evenly segments the task list into k disjoint bags, that is, each task bag contains m/k tasks. With k bags, our DRS framework overlaps the task scheduling and the task computation. Precisely, while the DRS service computes a schedule for the tasks in the ith bag (among the m/k bags), it also dispatches tasks in the $(i-1)$th bag that has been slotted to slave computing nodes. In DRS, the default task scheduler for a bag of tasks is longest-task first, that is, the tasks are scheduled in decreasing order in terms of the sizes of the input data objects in a bag, where distinct tasks in a job are uniformly distributed to different bags. We note that the task scheduler in DRS is replaceable if a DRS job developer intends to come up with his proprietary scheduler. However, for any scheduler, the metadata of a task is stored in HBase in our current design. The metadata attributes for a task include the data location of the task, the size of the data object, and access control list. A scheduler may be designed with distinct attributes in mind. Inquiring these attributes for a (large) set of tasks takes time in HBase, thus concluding our pipeline design and implementation for task scheduling and computation.
- **Performance Dashboard**
  DRS supports the performance dashboard to monitor the runtime metrics for an entire DRS framework. The dashboard details the execution time that spans from a job submission to a job computation commitment. In addition to resource allocation, the execution stages of our DRS framework consist of scheduling, dispatching, and computing for tasks in a job. In DRS, resources consumed by different execution stages are also measured. The resources include the time spent by allocated containers and the size of main memory spaces occupied. Our experience in

operating DRS shows that computational job developers are particularly interested in time elapsed by their R tasks.

- **Multi-Tenancy**
  YARN provides the hierarchical queue mechanism [83], allowing distinct users to allocate resources from Hadoop independently. For example, Hadoop may allocate 20% of resources to a Spark-job user, 30% of resources to an MR-job client, and the remaining 50% of resources to our DRS framework. To accommodate multiple DRS jobs simultaneously, our DRS service further partitions the allocated resources for different DRS jobs. For example, a total of 50% of resources allocated to DRS is evenly partitioned into five DRS job users, each with 10% such that the five DRS jobs can concurrently and fairly perform their computations. Multiple tenants supported by DRS would maximize the utilization of resources occupied by DRS and minimize the queuing delay of any pending DRS job. Notably, to manage ongoing and pending DRS jobs, DRS associates each job's status in ZooKeeper. When the computation of a job is completed, the job is removed from the ongoing job queue in ZooKeeper and the resources allocated to the job are released to YARN. Then, some jobs in the pending list are selected and migrated to the ongoing queue for execution. DRS delegates to allocate resources from YARN for these jobs. However, the total amount of resources allocated to the entire DRS service remains under its initial contract, such as 50% of resources as mentioned in the above example.

4.4.4 Summary

We presented in this section the HDS and DRS infrastructural services, which have been operated in a Taiwan, ROC's semiconductor wafer fabrication foundry since 2015. We discuss our design requirements and provide our implementations for HDS and DRS. Overall, HDS is designed for ease of use, thereby reducing the learning curve of a user for the sophisticated Hadoop software stack. Specifically, HDS is transparent to data accessing tools (e.g. Hive SQL) and computational frameworks (e.g. Spark) in Hadoop ecosystems. Compared with native HDFS, HDS performs well in terms of throughput and delay of accessing data objects stored in Hadoop.

In addition to HDS, DRS is designed and implemented to leverage existing efforts and experiences for conventional R language users. DRS not only exploits resources allocated to an R job for improving the job execution time but also intelligently and adaptively reallocates resources to boost a job's success rate. Particularly, DRS is transparent to R users by enjoying data-parallel computations in the Hadoop distributed computing environment.

4.5 Conclusion

This chapter has introduced some underlying core technologies of Industry 4.0 and Intelligent Manufacturing, including cloud computing, edge computing, IoT, and big data infrastructure.

In Section 4.2, we introduce cloud computing. First, the essentials of cloud computing are described, including its driving forces, core relevant technologies, main characteristics, and advantages. Besides, the cloud service models (i.e. IaaS, PaaS, and SaaS) are presented for categorizing not only cloud computing services but also specific cloud providers' offerings, products, and services. Also, the cloud deployment models (i.e. public cloud, private cloud, and hybrid cloud) are depicted for delineating how to leverage cloud computing. Some cloud computing applications in manufacturing, including big data analytics, AI-enabled applications, and CMfg, are portrayed as well.

In Section 4.3, we introduce the essentials of IoT, including its definition, the underlying core technologies, the potential applications, and the benefits it can bring. We also present a generic three-layer cloud-based IoT architecture to show how the IoT works. Two messaging protocols for IoT (i.e. MQTT and AMQP) and the WoT are introduced as well. This section also introduces the essentials of edge computing, including its definition, the underlying core elements, the potential applications, and the benefits it can bring. A cloud-based IoT architecture with an edge computing layer is presented to show how the IoT devices can leverage resources at the edge for computation offloading. Finally, we show an application of cloud-based IoT combined with edge computing in wheel machining.

In Section 4.4, we introduce big data infrastructure, including the application demands, core software stack components, and the methods to bridge the gap between core software stack components and applications. Specifically, we present two big data infrastructural services, i.e., HDS and DRS, based on Hadoop for a semiconductor manufacturing foundry in Taiwan, ROC. We describe the design requirements and depict the implementations of HDS and DRS. HDS is designed for ease of use, thereby reducing the learning curve of a user for the sophisticated Hadoop software stack. DRS is designed and implemented to leverage existing efforts and experiences for conventional R language users. Based on this chapter, the reader should be able to comprehend the contents related to these technologies in this book.

Appendix 4.A – Abbreviation List

6LoWPAN	IPv6 over LPWAN
AI	Artificial Intelligence
AMQP	Advanced Message Queuing Protocol
Apps	Applications
AR	Augmented Reality
AVM	Automatic Virtual Metrology
AWS	Amazon Web Services
B2B	Business to Business
B2C	Business to Customer
CBDM	Cloud-Based Design and Manufacturing
CMfg	Cloud Manufacturing
CPA	Cyber-Physical Agent
CPS	Cyber-Physical Systems
CRM	Customer Relationship Management
C2C	Customer to Customer
DDS	Data Distribution Service
DRS	Distributed R Language Computing Service
EC2	Elastic Compute Cloud
ERM	Enterprise Resource Management

FDC	Fault Detection and Classification
GCP	Google Cloud Platform
HaaS	Hardware-as-a-Service
HDFS	Hadoop Distributed File System
HDS	Hadoop Data Service
HTTP	HyperText Transfer Protocol
IaaS	Infrastructure as a Service
ICT	Information and Communication Technology
IIoT	Industrial Internet of Things
IoT	Internet of Things
IPv6	IP version 6
LPWAN	Low-Power Wireless Personal Area Network
M2M	Machine-to-Machine
MES	Manufacturing Execution System
MQTT	Message Queue Telemetry Transport Protocol
NIST	National Institute of Standards and Technology
PaaS	Platform as a Service
QoS	Quality of Service
OS	Operating System
OSs	Operating Systems
REST	Representational State Transfer
RFID	Radio Frequency Identification
ROI	Return on Investment
RUL	Remaining Useful Life
SaaS	Software as a Service
SCADA	Supervisory Control and Data Acquisition
SCM	Smart Cloud Manager
SDKs	Software Development Kits
SMEs	Small and Medium Enterprises
SOA	Service-Oriented Architecture
TCP	Transmission Control Protocol
VM	Virtual Metrology
WoT	Web of Things
WSN	Wireless Sensor Network
YA	Yield Analysis

Appendix 4.B – List of Symbols in Equations

c	number of regions of a table
n	number of region servers available to the table
f_s	storage server name, file path, and data object name of the source
f_d	storage server name, file path, and data object name of the destination
S	representation of source storage entities
D	representation of destination storage entities
n_i	HDS serving node
$L(n_i)$	load of an HDS serving node
$C_D(n_i)$	ratio of $C_{D_{\text{occupied}}}(n_i)$ to $C(n_i)$
$C_{D_{\text{occupied}}}(n_i)$	number of connections currently used by ongoing data delivery requests
$C(n_i)$	total number of connections predefined for control and data planes
$M(n_i)$	ratio of memory space used by the data plane to the total memory space allocated to the control and data planes
α	predefined parameters
β	predefined parameters
r_i	data delivery request
u_i	user issuing HDS data delivery request
m	number of tasks of a computational job
k	number of disjoint bags
i	sequential number of disjoint bags

References

1 Advanced Manufacturing National Program Office (AMNPO) (2020). Highlighting manufacturing USA. www.manufacturing.gov (accessed 2 November, 2020).

2 Kagermann, H., Wahlster, W., and Helbig, J. (2013) *Recommendations for Implementing the Strategic Initiative INDUSTRIE 4.0*. Germany: Forschungsunion, acatech.

3 Lee, X.E. (2015). Made in china 2025: a new era for Chinese manufacturing. http://knowledge.ckgsb.edu.cn/2015/09/02/technology/made-in-china-2025-a-new-era-for-chinese-manufacturing/ (accessed 2 November, 2020).

4 Drath, R. and Horch, A. (2014). Industrie 4.0: hit or hype? *IEEE Industrial Electronics Magazine* 8 (2): 56–58. https://doi.org/10.1109/MIE.2014.2312079.

5 Lee, J., Bagheri, B., and Kao, H.A. (2015). A cyber-physical systems architecture for industry 4.0-based manufacturing systems. *Manufacturing Letters* 3: 8–23. https://doi.org/10.1016/j.mfglet.2014.12.001.

6 Burke, R., Mussomeli, A., Laaper, S. et al. (2017). The smart factory: responsive, adaptive, connected manufacturing. https://bit.ly/34NWnG3 (accessed 2 November, 2020).

7 Zhou, H. (2013). *The Internet of Things in the Cloud: A Middleware Perspective*. UK: CRC Press.

8 Bi, Z., Xu, L.D., and Wang, C. (2014). Internet of Things for enterprise systems of modern manufacturing. *IEEE Transactions on Industrial Informatics* 10 (2): 1537–1546. https://doi.org/10.1109/TII.2014.2300338.

9 Xu, X. (2012). From cloud computing to cloud manufacturing. *Robotics and Computer-Integrated Manufacturing* 28 (1): 75–86. https://doi.org/10.1016/j.rcim.2011.07.002.

10 Tao, F., Cheng, Y., Xu, L.D. et al. (2014). CCIoT-CMfg: cloud computing and Internet of Things-based cloud manufacturing service system. *IEEE Transactions on Industrial Informatics* 10 (2): 1435–1442. https://doi.org/10.1109/TII.2014.2306383.

11 Ren, L., Zhang, L., Wang, L. et al. (2017). Cloud manufacturing: key characteristics and applications. *International Journal of Computer Integrated Manufacturing* 30 (6): 501–515. https://doi.org/10.1080/0951192X.2014.902105.

12 He, W. and Xu, L. (2015). A state-of-the-art survey of cloud manufacturing. *International Journal of Computer Integrated Manufacturing* 28 (3): 239–250. https://doi.org/10.1080/0951192X.2013.874595.

13 Georgakopoulos, D., Jayaraman, P.P., Fazia, M. et al. (2016). Internet of Things and edge cloud computing roadmap for manufacturing. *IEEE Cloud Computing* 3 (4): 66–73. https://doi.org/10.1109/MCC.2016.91.

14 Chen, B., Wan, J., Celesti, A. et al. (2018). Edge computing in IoT-based manufacturing. *IEEE Communications Magazine* 56 (9): 103–109. https://doi.org/10.1109/MCOM.2018.1701231.

15 Chen, X.W. and Lin, X. (2014). Big data deep learning: challenges and perspectives. *IEEE Access* 2: 514–525. https://doi.org/10.1109/ACCESS.2014.2325029.

16 Babiceanu, R.F. and Seker, R. (2015). Manufacturing cyber-physical systems enabled by complex event processing and big data environments: a framework for development. *Service Orientation in Holonic and Multi-agent Manufacturing, Studies in Computational Intelligence Series* 594: 165–173.

17 Bradley, J.M. and Atkins, E.M. (2015). Coupled cyber-physical system modeling and coregulation of a CubeSat. *IEEE Transactions on Robotics* 31 (2): 443–456. https://doi.org/10.1109/TRO.2015.2409431.

18 Thramboulidis, K. (2015). A cyber-physical system-based approach for industrial automation systems. *Computers in Industry* 72: 92–102. https://doi.org/10.1016/j.compind.2015.04.006.

19 Yang, C., Shen, W., and Wang, X. (2018). The Internet of Things in manufacturing: key issues and potential applications. *IEEE Systems, Man, and Cybernetics Magazine* 4 (1): 6–15. https://doi.org/10.1109/MSMC.2017.2702391.

20 Hu, H., Wen, Y., Chua, T.S. et al. (2014). Toward scalable systems for big data analytics: a technology tutorial. *IEEE Access* 2: 652–687. https://doi.org/10.1109/ACCESS.2014.2332453.

21 Lee, J., Kao, H.A., and Yang, S. (2014). Service innovation and smart analytics for Industry 4.0 and big data environment. *Procedia CIRP* 16: 3–8. https://doi.org/10.1016/j.procir.2014.02.001.

22 Fekete, J.D. (2013). Visual analytics infrastructures: from data management to exploration. *Computer* 46 (7): 22–29. https://doi.org/10.1109/MC.2013.120.

23 Dikaiakos, M.D., Katsaros, D., Mehra, P. et al. (2009). Cloud computing: distributed Internet computing for IT scientific research. *IEEE Internet Computing* 13 (5): 10–13. https://doi.org/10.1109/MIC.2009.103.

24 Sadiku, M.N.O., Musa, S.M., and Momoh, O.D. (2014). Cloud computing: opportunities and challenges. *IEEE Potentials* 33 (1): 34–36. https://doi.org/10.1109/MPOT.2013.2279684.

25 Sultan, N.A. (2011). Reaching for the “Cloud”: how SMEs can manage. *International Journal of Information Management* 31 (3): 272–278. https://doi.org/10.1016/j.ijinfomgt.2010.08.001.

26 National Institute of Standards and Technology (NIST) (2011). The NIST definition of cloud computing. *NIST Special Publication 800-145*. https://bit.ly/2TN5cK1 (accessed 2 November, 2020).

27 Wright, N. (2019). IaaS vs SaaS vs PaaS: a guide to Azure cloud service types. https://bit.ly/3kO0JTi (accessed 2 November, 2020).

28 Khan, A.R., Othman, M., Madani, S.A. et al. (2014). A survey of mobile cloud computing application models. *IEEE Communication Surveys & Tutorials* 16 (1): 393–413. https://doi.org/10.1109/SURV.2013.062613.00160.

29 Linthicum, D.S. (2017). Making sense of AI in public clouds. *IEEE Cloud Computing* 4 (6): 70–72. https://doi.org/10.1109/MCC.2018.1081067.

30 Hofmann, P. and Woods, D. (2010). Cloud computing: the limits of public clouds for business applications. *IEEE Internet Computing* 14 (6): 90–93. https://doi.org/10.1109/MIC.2010.136.

31 Weins, K. (2020). Cloud computing trends: 2020 state of the cloud report. https://www.flexera.com/blog/industry-trends/trend-of-cloud-computing-2020/ (accessed 2 November, 2020).

32 Amazon Web Services, Inc (2020). Amazon Web Services (AWS). aws.amazon.com (accessed 2 November, 2020).

33 Microsoft (2020). Microsoft Azure. azure.microsoft.com/en-us/ (accessed 2 November, 2020).

34 Google (2020). Google. cloud cloud.google.com/gcp/ (accessed 2 November, 2020).

35 Microsoft (2020). Azure service bus messaging documentation. https://docs.microsoft.com/en-us/azure/service-bus-messaging/ (accessed 2 November, 2020).

36 Amazon Web Services, Inc (2020). Data lakes and analytics on AWS. https://aws.amazon.com/big-data/datalakes-and-analytics/ (accessed 2 November, 2020).

37 Cloud Academy Inc. (2020). Big data analytics on Azure. https://cloudacademy.com/learning-paths/big-data-analytics-on-azure-200/ (accessed 2 November, 2020).

38 Google (2020). Google BigQuery. cloud.google.com/bigquery/ (accessed 2 November, 2020).

39 Srinivasan, S. (2019). Data and analytics on Google cloud platform. https://bit.ly/3egBd6s (accessed 2 November, 2020).

40 Wang, X.V. and Xu, X.W. (2013). An interoperable solution for cloud manufacturing. *Robotics and Computer-Integrated Manufacturing* 29 (4): 232–247. https://doi.org/10.1016/j.rcim.2013.01.005.

41 Wu, D., Greer, M.J., Rosen, D.W. et al. (2013). Cloud manufacturing: strategic vision and state-of-the-art. *Journal of Manufacturing Systems* 32 (4): 564–579. https://doi.org/10.1016/j.jmsy.2013.04.008.

42 Wu, D., Thames, J.L., Rosen, D.W. et al. (2013). Enhancing the product realization process with cloud-based design and manufacturing systems. *Journal of Computing and Information Science in Engineering* 13 (4): 041004. https://doi.org/10.1115/1.4025257.

43 Huang, H.C., Lin, Y.C., Hung, M.H. et al. (2015). Development of cloud-based automatic virtual metrology system for semiconductor industry. *Robotics and Computer-Integrated Manufacturing* 34: 30–43. https://doi.org/10.1016/j.rcim.2015.01.005.

44 Hung, M.H., Li, Y.Y., Lin, Y.C. et al. (2017). Development of a novel cloud-based multi-tenant model creation service for automatic virtual metrology. *Robotics and Computer-Integrated Manufacturing* 44: 174–189. https://doi.org/10.1016/j.rcim.2016.09.003.

45 Feki, M.A., Kawsar, F., Boussard, M. et al. (2013). The Internet of Things: the next technological revolution. *Computer* 46 (2): 24–25. https://doi.org/10.1109/MC.2013.63.

46 Jin, J., Gubbi, J., Marusic, S. et al. (2014). An information framework for creating a smart city through Internet of Things. *IEEE Internet of Things Journal* 1 (2): 112–121. https://doi.org/10.1109/JIOT.2013.2296516.

47 Kelly, S.D.T., Suryadevara, N.K., and Mukhopadhyay, S.C. (2013). Towards the implementation of IoT for environmental condition monitoring in homes. *IEEE Sensors Journal* 13 (10): 3846–3853. https://doi.org/10.1109/JSEN.2013.2263379.

48 Michahelles, F., Thiesse, F., Schmidt, A. et al. (2007). Pervasive RFID and Near Field Communication technology. *IEEE Pervasive Computing* 6 (3): 94–97. https://doi.org/10.1109/MPRV.2007.64.

49 Ullo, S.L. and Sinha, G.R. (2020). Advances in smart environment monitoring systems using IoT and sensors. *Sensors* 20 (11): 3113. https://doi.org/10.3390/s20113113.

50 Kulitski, A. (2017). What you need to know about machine-to-machine communication. https://smart-it.io/blog/machine-to-machine-m2m-how-does-it-work/ (accessed 2 November 2020).

51 Samtani, S., Yu, S., Zhu, H. et al. (2018). Identifying SCADA systems and their vulnerabilities on the Internet of Things: a text-mining approach. *IEEE Intelligent Systems* 33 (2): 63–73. https://doi.org/10.1109/MIS.2018.111145022.

52 Zhou, J., Cao, Z., Dong, X. et al. (2017). Security and privacy for cloud-based IoT: challenges, countermeasures, and future directions. *IEEE Communications Magazine* 55 (1): 26–35. https://doi.org/10.1109/MCOM.2017.1600363CM.

53 Wheeler, A. (2007). Commercial applications of wireless sensor networks using ZigBee. *IEEE Communications Magazine* 45 (4): 70–77. https://doi.org/10.1109/MCOM.2007.343615.

54 Gomez, C., Minaburo, A., Toutain, L. et al. (2020). IPv6 over LPWANs: connecting Low Power Wide Area Networks to the Internet (of Things). *IEEE Wireless Communications* 27 (1): 206–213. https://doi.org/10.1109/MWC.001.1900215.

55 Al-Fuqaha, A., Khreishah, A., Guizani, M. et al. (2015). Toward better horizontal integration among IoT services. *IEEE Communications Magazine* 53 (9): 72–79. https://doi.org/10.1109/MCOM.2015.7263375.

56 MQTT.org (2020). MQTT: the standard for IoT messaging. mqtt.org (accessed 2 November 2020).

57 OASIS® (2020). Advanced Message Queuing Protocol (AMQP). www.amqp.org ().

58 Object Management Group, Inc (2020). Data Distribution Service (DDS). www.dds-foundation.org (accessed 2 November 2020).

59 EDUCBA (2020). AMQP vs MQTT. https://www.educba.com/amqp-vsmqtt/ (accessed 2 November 2020).

60 Guinard, D. and Trifa, V. (2016). *Building the Web of Things: With examples in Node.js and Raspberry Pi*. New York: Manning Publications.

61 Varghese, B., Reaño, C., and Silla, F. (2018). Accelerator virtualization in fog computing: moving from the cloud to the edge. *IEEE Cloud Computing* 5: 28–37. https://doi.org/10.1109/MCC.2018.064181118.

62 Aazam, M. and Huh, E.N. (2016). Fog computing: the cloud-IoT/IoE middleware paradigm. *IEEE Potentials* 35 (3): 40–44. https://doi.org/10.1109/MPOT.2015.2456213.

63 Ren, J., Guo, H., Xu, C. et al. (2017). Serving at the edge: a scalable IoT architecture based on transparent computing. *IEEE Network* 31 (5): 96–105. https://doi.org/10.1109/MNET.2017.1700030.

64 Sun, W., Liu, J., and Yue, Y. (2019). AI-Enhanced offloading in edge computing: when machine learning meets industrial IoT. *IEEE Network* 33 (5): 68–74. https://doi.org/10.1109/MNET.001.1800510.

65 Patel, P., Ali, M.I., and Sheth, A. (2017). On using the intelligent edge for IoT analytics. *IEEE Intelligent Systems* 32 (5): 64–69. https://doi.org/10.1109/MIS.2017.3711653.

66 Johansson, C. (2020). How edge computing will unleash the potential of IIoT. https://www.controleng.com/articles/how-edge-computing-will-unleash-the-potential-of-iiot/ (accessed 2 November 2020).

67 The Apache Software Foundation (2020). Apache Hadoop. hadoop.apache.org (accessed 2 November 2020).

68 The Apache Software Foundation. (2008). HDFS architecture guide. http://hadoop.apache.org/docs/r1.2.1/hdfs_design.html (accessed 2 November 2020).

69 McKusick, K. and Quinlan, S. (2010). GFS: evolution on fast-forward. *Communications of the ACM* 53 (3): 42–49. https://doi.org/10.1145/1666420.1666439.

70 Ghemawat, S., Gobioff, H.B. and Leung, S.T. (2003). The Google file system. *Proceedings of the 19th ACM Symposium on Operating Systems Principles (SOSP'03)*, Bolton Landing, NY, USA (19-22 Oct. 2003). New York, USA: ACM.

71 The Apache Software Foundation (2020). Welcome to Apache HBase™. hbase.apache.org (accessed 2 November 2020).

72 Leavitt, N. (2010). Will NoSQL databases live up to their promise? *Computer* 43 (2): 12–14. https://doi.org/10.1109/MC.2010.58.

73 Chang, F., Dean, J., Ghemawat, S. et al. (2006). Bigtable: a distributed storage system for structured data. *Proceedings of the 7th USENIX Symposium on Operating Systems Design and Implementation (OSDI'06)*, Seattle, WA, USA (6-8 Nov. 2006). USA:USENIX Association.

74 The R Foundation (2020). The R project for statistical computing. r-project.org (accessed 2 November 2020).

75 Python Software Foundation (2020). Python™. www.python.org (accessed 2 November 2020).

76 The Samba team (2020). Samba. www.samba.org (accessed 2 November 2020).

77 The Apache Software Foundation (2018). Apache Spark™. spark.apache.org (accessed 2 November 2020).

78 Zaharia, M., Xin, R.S., Wendell, P. et al. (2016). Apache Spark: a unified engine for big data processing. *Communications of the ACM* 59 (11): 56–65. https://doi.org/10.1145/2934664.

79 The Apache Software Foundation (2014). Apache Hive™. hive.apache.org (accessed 2 November 2020).

80 Thusoo, A., Sarma, J.S., Jain, N. et al. (2010). Hive - a petabyte scale data warehouse using Hadoop. *Proceeding of the 2010 IEEE 26th International Conference on Data Engineering (ICDE 2010)*, Long Beach, CA, USA (1-6 Mar. 2010). USA: IEEE.

81 Ihaka, R. and Gentleman, R. (1996). R: a language for data analysis and graphics. *Journal of Computational and Graphical Statistics* 3 (5): 299–314.

82 Hsiao, H.C., Chung, H.Y., Shen, H. et al. (2013). Load rebalancing for distributed file systems in clouds. *IEEE Transactions on Parallel and Distributed Systems* 24 (5): 951–962. https://doi.org/10.1109/TPDS.2012.196.

83 The Apache Software Foundation (2020). Apache Hadoop YARN. https://hadoop.apache.org/docs/current/hadoop-yarn/hadoop-yarn-site/YARN.html (accessed 2 November 2020).

84 Vavilapalli, V. K., Murthy, A. C., Douglas, C. et al. (2013). Apache Hadoop YARN: yet another resource negotiator. *Proceedings of the 4th ACM Symposium on Cloud Computing (SOCC'13)*, Santa Clara, CA, USA (1-3 Oct. 2013). New York, USA: ACM.

85 Dean, J. and Ghemawat, S. (2004). MapReduce: simplified data processing on large clusters. *Proceedings of the 6th USENIX Symposium on Operating System Design and Implementation (OSDI'04)*, San Francisco, CA, USA (6-8 Dec. 2004). USA:USENIX Association.

86 The Apache Software Foundation (2020). Apache ZooKeeper™. zookeeper.apache.org (accessed 2 November 2020).

87 Hunt, P., Konar, M., Junqueira, F. P. et al. (2010). ZooKeeper: wait-free coordination for Internet-scale systems. *Proceedings of the 2010 USENIX Annual Technical Conference (USENIX ATC'10)*, Boston, MA, USA (23-25 Jun. 2010). USA:USENIX Association.

88 Tsai, C.P., Hsiao, H.C., Chao, Y.C. et al. (2018). Bridging the gap between big data system software stack and applications: the case of semiconductor wafer fabrication foundries. *Proceeding of the 2018 IEEE International Conference on Big Data*, Seattle, WA, USA (10-13 Dec. 2018). USA: IEEE.

89 Adve, S.V. and Gharachorloo, K. (1996). Shared memory consistency models: a tutorial. *Computer* 29 (12): 66–76. https://doi.org/10.1109/2.546611.

90 Hsiao, H.C. and Chang, C.W. (2013). A symmetric load balancing algorithm with performance guarantees for distributed hash tables. *IEEE Transactions on Computers* 62 (4): 662–675. https://doi.org/10.1109/TC.2012.13.

91 Hsiao, H.C., Liao, H., Chen, S.T. et al. (2011). Load balance with imperfect information in structured peer-to-peer systems. *IEEE Transactions on Parallel and Distributed Systems* 22 (4): 634–649. https://doi.org/10.1109/TPDS.2010.105.

92 Mitzenmacher, M. and Upfal, E. (2005). *Probability and Computing: Randomized Algorithms and Probabilistic Analysis*. Cambridge: Cambridge University Press.

93 The Apache Software Foundation (2020). Hadoop archives guide. https://hadoop.apache.org/docs/current/hadoop-archives/HadoopArchives.html (accessed 2 November 2020).

94 ASF Infrabot (2019). Hadoop SequenceFile. https://cwiki.apache.org/confluence/display/HADOOP2/SequenceFile (accessed 2 November 2020).

95 The Apache Software Foundation (2012). Apache HBase coprocessor introduction. https://blogs.apache.org/hbase/entry/coprocessor_introduction (accessed 2 November 2020).

5

Docker and Kubernetes

Chao-Chun Chen[1], *Min-Hsiung Hung*[2], *Kuan-Chou Lai*[3], *and Yu-Chuan Lin*[4]

[1] *Professor, Institute of Manufacturing Information and Systems & Department of Computer Science and Information Engineering, National Cheng Kung University, Tainan, Taiwan, ROC*
[2] *Professor, Department of Computer Science and Information Engineering, Chinese Culture University, Taipei City, Taiwan, ROC*
[3] *Professor, Department of Computer Science, National Taichung University of Education, Taichung City, Taiwan, ROC*
[4] *Secretary General, Intelligent Manufacturing Research Center, National Cheng Kung University, Tainan City, Taiwan, ROC*

5.1 Introduction

In recent years, container technologies have become a hot topic for the software industry. Among them, Docker [1] is a software platform that allows developers to package an application with all its dependencies into an image and run the application in a standardized unit, called a container, from the image. As such, Docker makes it very easy to build, ship, and run software containers across different environments, for instance, the development, testing, and production environments. Although Docker did not invent software containers, it changed how software is built, shared, and deployed during recent years.

The software containers play the same role in the software industry as the shipping containers do in the transportation industry [2]. With standardized shipping containers, the packaging, loading, shipping, and unloading of various goods could be standardized. Using containers to ship goods, the shippers regard the containers as black boxes and need not know what is inside a container. They can focus on transporting containers using suitable means, such as trucks, trains, and ships, according to their demands. Thus, the invention of shipping containers made transporting goods more efficient than it was before the time of using containers and thereby revolutionized the transporting industry.

Similarly, the Docker containers (software containers) allow the developers to package their applications with the required dependencies, such as configurations, frameworks, libraries, and runtimes, into them. Then, the developers can ship these containers to operations engineers in a standardized way. The operations engineers regard each Docker container as a black box. They can deploy and run containers on their servers without having to know what is inside a container. The Docker containers run on the same server (a virtual or physical one) are isolated from each other. Also, if a Docker container can run on a Docker host, it should be able to run on another Docker host. A Docker host refers to a server that has Docker installed on it because it can host and run Docker containers. Thus, with Docker, operations engineers can avoid wasting time in solving the

Industry 4.1: Intelligent Manufacturing with Zero Defects, First Edition. Edited by Fan-Tien Cheng.

inconsistency between development and operations (DevOps) environments and can focus on their expertise - providing IT infrastructure, applications, and services.

Besides, Docker can bring many advantages to the so-called software supply chain [2], which covers the development, testing, shipping, deployment, and execution of software. Some benefits of Docker are mentioned below. Docker can simplify the software development life cycle and, in turn, significantly shorten the release times of applications. With Docker, a developer can build an application locally, package the application with the required dependencies into one or more images, and then push the images to a testing environment for conducting automated or manual tests on the application running in the containers that are created from those images. When finding bugs, the developer can fix the application and update the images that package the application accordingly in the development environment. Then, the developer can redeploy the updated images to the test environment for testing and validation. When testing is complete, the developer can rapidly push the updated images to the production environment for running the application.

Also, Docker enhances the portability and scalability of applications. Docker container-based applications can run on various Docker hosts, such as laptop, desktop, physical server, virtual machine, and cloud. Using lightweight virtualization, Docker can rapidly and dynamically scale up or down the number of replicas of the containerized applications on demand. Besides, a Docker host can run many containerized applications simultaneously, each of which is isolated from the others. And, we can limit the resources each container can use, such as CPUs and RAMs. A containerized application's overhead is negligible, so its performance is close to the native application [3]. Hence, using Docker to run applications can increase computing resources' usage rate compared with traditional virtual machines.

Figure 5.1 illustrates the comparison of virtual machines (shown on the left-hand side) and Docker containers (shown on the right-hand side) running on a physical machine [4]. The virtual machine is a hardware-level virtualization technology using a hypervisor to abstract the underlying hardware. A hypervisor, such as VMware vSphere, Microsoft Hyper-V, and Citrix Xen, is a program that allows multiple operating systems to share a single hardware host. On top of the hypervisor, we can create several virtual machines, each of which can install a guest operating system (OS), like Linux or Windows. Then, on top of the guest OS, we install our applications (APPs) with their dependencies, such as binary codes and libraries (Bins/Libs). By contrast, the Docker container is an OS-level virtualization technology using Docker to abstract the underlying host OS. On top of Docker, we can create and run many containers, and each container can package an application with its dependencies.

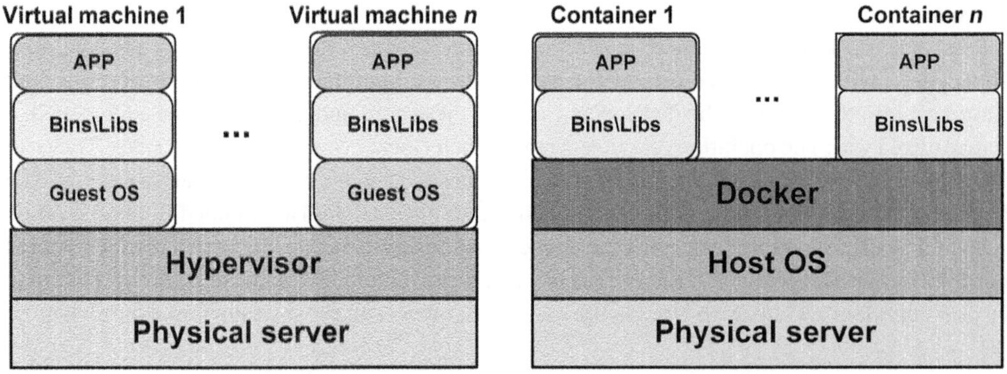

Figure 5.1 Comparison of virtual machines and Docker containers.

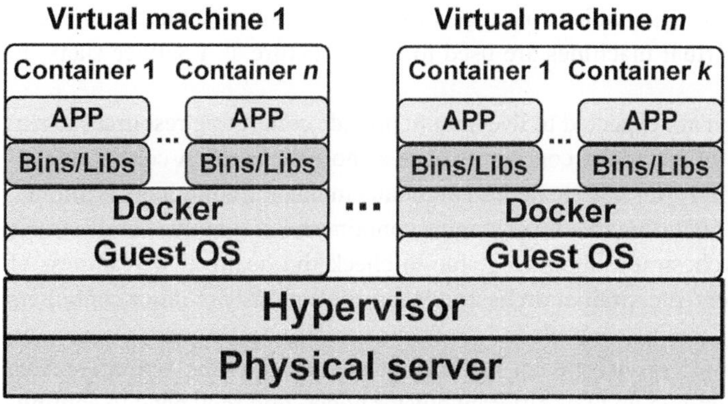

Figure 5.2 Illustration of Docker containers running on virtual machines.

Docker can also be installed on a virtual machine, as shown in Figure 5.2. In other words, Docker containers can co-work with virtual machines. Thus, a Docker host may refer to a physical or virtual machine.

Table 5.1 summarizes some distinctions between Docker containers and virtual machines [4]. Firstly, the size of a Linux Docker container image ranges from several MB (e.g. Alpine OS image: about 5MB) to a few hundred MB (e.g. Ubuntu OS image: about 188MB). However, a Linux virtual machine image is much bigger than a container image and is usually about several GB. For example, the Ubuntu 18.04 VMware image is about 3.7GB. Secondly, a Docker container's booting time is in seconds, while a virtual machine needs to take several minutes to boot. Thirdly, Docker achieves OS-level virtualization and isolates applications using containers. Virtual machines, oppositely, realize hardware-level virtualization and OS isolation. Fourthly, a Docker container requires fewer resources than a virtual machine and can run on resource-limited edge devices. On the contrary, a virtual machine requires intensive resources, it cannot run on resource-limited edge

Table 5.1 Some distinctions between Docker containers and virtual machines.

	Docker container	Traditional virtual machine
Deployment unit	container	virtual machine
Image size	the size of a Linux Docker container image ranges from several MB (e.g. Alpine OS image: about 5MB) to a few hundred MB (e.g. Ubuntu OS image: about 188MB)	the size of a Linux virtual machine image is usually about several GB (e.g. Ubuntu 18.04 VMware Image: about 3.7GB)
Booting time	seconds	minutes
Virtualization and isolation	OS-level virtualization with application isolation	hardware-level virtualization with OS isolation
Computing resource demand	• require less resources • can run on resource-limited edge devices	• require intensive resources • cannot run on resource-limited edge devices
Suitable use case	package, ship, run applications with their dependencies in an isolated manner	virtualize machines in Infrastructure as a Service (IaaS)

devices. Fifthly, Docker containers are suitable to package and ship applications with their dependencies in an isolated manner. Virtual machines are used in Infrastructure as a Service (IaaS) to realize machine virtualization.

Unlike virtual machines, which are expected to live long to provide computing resources to run various processes, applications, and services, containers have a short life cycle. A container running an application may crash soon after it is created, so manually managing containers is impracticable. Thus, a system designed for automatically managing containers, i.e. container orchestrator [5], is desirable. A container orchestrator should be able to check the health and readiness of containers and self-heal containers. A container orchestrator can continuously monitor containers and automatically create a new container to replace a crashed or malfunctioning one. Besides, an application may comprise multiple services, each running in a container. These multi-service application containers may run on one node or different cluster nodes, and they need to communicate with each other. Therefore, a container orchestrator should provide the service discovery functionality for containers within a cluster to communicate with each other and the outside world to access the containers.

According to the workload change, the desired number of a specific application container may need to increase or decrease. Thus, a container orchestrator should contain auto-scaling functionality to scale out or in the number of a container's replicas. Furthermore, for improving application responsiveness and increasing the availability of an application for users, load balancing is needed for a container orchestrator to distribute incoming requests across a group of replicas of a container to spread the workload evenly so that no single container replica undertakes too many requests. For example, a load balancer based on the round-robin algorithm can forward an incoming request to each container replica in turn from the first one to the last one. Then, the load balancer goes back to the first one of the list and repeats.

Moreover, failover is also an essential safeguard for running containerized applications. Failover across a group of container replicas can be achieved by the capabilities of self-healing and load balancing. When a replica of a container running on a node crashes, the container orchestrator must be able to detect this event and then create a new container replica instantly on another node to replace it. After that, the newly-created container replica should join the original group of container replicas to undertake the workloads according to the load balancing rule. Additionally, a node can run many containers simultaneously, and a container may predominantly consume most of the available resources, such as CPU and Memory, thereby leading to the starving of the rest containers. Thus, a container orchestrator must be able to prevent this problem from happening by limiting the consuming resources of each container. Besides, security is vital to applications, and containerized applications are no exception. A container orchestrator must protect the security of the containerized applications running on it. The security protection includes kernel-level security, host-level security, container images' security, the security of running containers, the security of container engine, and more.

In short, the management of containerized applications involves a lot of tasks, including abovementioned health check, readiness check, self-healing, communication, service discovery, autoscaling, failover, resource limitation, and security measures. Thus, a container orchestrator is a must tool [6] for automatically managing and orchestrating containerized applications.

There are two ways for a container orchestrator to work: imperative and declarative. The imperative working style means that the users need to provide the orchestrator with exact step-by-step instructions on completing a management task of containers. The declarative working style, nonetheless, means that the users only need to tell the orchestrator the desired state of the running containers. Then, the orchestrator does its best to achieve that desired state. A declarative container orchestrator is more potent than an imperative one.

The desired state of running containers includes the image used to create the target container, the number of replicas a container should have, the limitation of consumable resources for a container, the data volume attached to a container, the port a container exposes, just to name a few. The orchestrator gets containers up and running and then continuously monitor their state. If the running containers' current state deviates from the desired one, the orchestrator will automatically reconcile the desired state. For example, a container has four replicas as its desired state. Later on, one replica crashes for some reason. The orchestrator will instantly create a new replica to replace the crashed one to achieve the desired state. On the other hand, when the desired number of replicas is adjusted to three, the orchestrator will delete one replica to satisfy the desired state.

Currently, Kubernetes [7] and Docker Swarm [8] are the two most popular container orchestrators in the market. Kubernetes is an open-source container orchestration tool. Kubernetes was originally developed by Google and later donated to the Cloud Native Computing Foundation (CNCF) [9]. In terms of flexibility and capability, Kubernetes surpasses Docker Swarm [10, 11]. Kubernetes was designed for massive scalabilities and can support different container runtimes, such as Docker, CRI-O [12], and containerd [13], through Container Runtime Interface (CRI). The container management capabilities of Kubernetes include the deployment, running, monitoring, scaling, descaling, and load balancing of containers. Kubernetes supports both Linux and Windows containers. Kubernetes is now the leading container orchestrator. However, Kubernetes is complex to set up and manage. The newcomers often find difficulties in learning and using Kubernetes, yet Docker Swarm (SwarmKit) is an integral part of the Docker engine. Thus, after a Docker host is installed, it automatically has Docker Swarm. Docker designs SwarmKit with simplicity and security in mind. A Docker Swarm is easy to set up by just using the "docker swarm init" command to create the leader node of the cluster, and then using the "docker swarm join" command to join a worker node. Docker Swarm also supports both Linux and Windows containers. Docker Swarm is currently the second most popular container orchestration tool.

This chapter will introduce the fundamentals of Docker and Kubernetes to serve as the foundation for Chapters 6, 7, and 9, where Docker, Kubernetes, and various containerized applications are mentioned. We will introduce Docker in Section 5.2, including its architecture, operational principles, and applications. Then, Section 5.3 introduces Kubernetes, including its architecture, operational principles, and applications. Finally, Section 5.4 is the conclusion of this chapter.

5.2 Fundamentals of Docker

This section will introduce the fundamentals of Docker. In Section 5.2.1, we depict the Docker architecture. Section 5.2.2 describes the Docker operational principles. Section 5.2.3 presents some illustrative applications of Docker.

5.2.1 Docker Architecture

Docker can run Linux and Windows containers. In this section, we will describe essential concepts related to Docker from the perspective of architecture, including Docker Engine, a high-level Docker architecture, the architecture of a Linux Docker host, the architecture of a Windows Docker host, the architecture of Windows Server containers, and the architecture of Hyper-V containers.

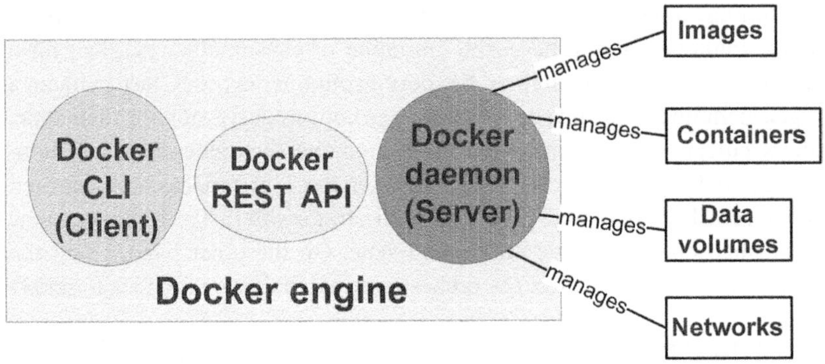

Figure 5.3 Constituent components of Docker Engine.

5.2.1.1 Docker Engine

After installing Docker on a physical or virtual machine, we create a Docker host, which contains Docker Engine [14]. As shown in Figure 5.3, Docker Engine consists of three components: Docker Daemon, Docker REST API, and Docker CLI, each of which is described as follows:

- Docker Daemon is the server-side component of Docker Engine. It can manage Docker images, containers, data volumes, and networks. Docker Daemon is a background process that continually listens for the requests from Docker REST API and processes them.
- Docker REST API is a RESTful application programming interface used by clients to interact with Docker Daemon. It can be accessed by Docker CLI or hypertext transfer protocol (HTTP) clients.
- Docker CLI is a command-line interface that serves as a client for interacting with Docker Daemon through Docker REST API. It is the interface for users to enter Docker commands to talk to Docker Daemon for managing Docker objects, such as images, containers, data volumes, and networks.

5.2.1.2 High-Level Docker Architecture

Docker adopts a client-server architecture. Figure 5.4 shows a high-level Docker architecture [14] with some of its workflows. A Docker host contains Docker CLI, Docker REST API, and Docker Daemon. Docker CLI serves as the client side that accepts docker commands (e.g. docker build, docker run, or docker pull) keyed in by the user and then sends the docker commands to the server side (i.e. Docker Daemon) through Docker REST API. Docker CLI interacts with the local Docker daemon on the same Docker host by default. It can also be configured to interact with a Docker daemon on a remote Docker host. Docker Daemon is responsible for managing Docker objects, such as images, containers, data volumes, networks, and plugins.

An image is a template for creating Docker containers. It is usually composed by many read-only layers, which can be used to package our desired application and its dependencies. An image is often built based on another image and then adding some customization. For example, we may build an image that packages an application using an official Ubuntu image as the base image and then installing the Apache web server, the application, and the dependencies needed to run the application on top of it. A container is an executable instance of an image. We can create, run, stop, or delete a container by sending a command using Docker CLI. A container can connect to one or more networks for communicating with other containers. When a container is deleted, any changes to its state will disappear. Thus, one or more data volumes can be mounted to a container to store persistent data.

An image registry is used to store and manage Docker images. We can create an image repository on an image registry to store different versions of an application image using different tags. The

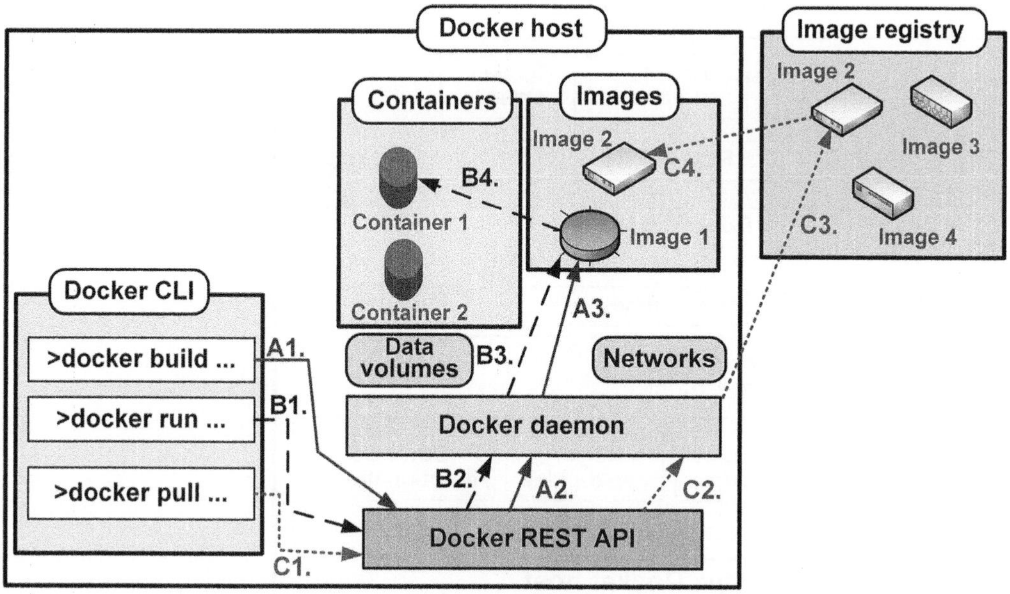

Figure 5.4 A high-level view of Docker architecture with some of its workflows.

docker run or docker pull command can pull an image from an image registry to the Docker host, yet the docker push command can push an image from a Docker host to an image registry. Docker Hub [15] is a public registry that Docker is configured to look for images by default. Other public image registries include Microsoft Container Registry (MCR) [16], Google Container Registry (GCR) [17], and Amazon Elastic Container Registry (ECR) [18]. We can even set up and run a private image registry.

Figure 5.4 also shows three workflows under this architecture:

- **Workflow of building an image**

  Step A1: A docker build command is keyed in Docker CLI, where the build context described in a Dockerfile is then sent to Docker REST API.

  Step A2: Docker REST API sends the build context to Docker Daemon.

  Step A3: Docker Daemon creates an image (e.g. Image 1) accordingly and stores it in the local cache.
- **Workflow of running a container from a local image**

  Step B1: A docker run command is keyed in Docker CLI, where the docker command is then sent to Docker REST API.

  Step B2: Docker REST API sends the docker command to Docker Daemon.

  Step B3 and B4: Docker Daemon creates and runs a container (e.g. Container 1) from the specified local image (e.g. Image 1). If the specified image does not exist locally, it will be pulled from the image registry to the local cache.
- **Workflow of pulling an image from an image registry**

  Step C1: A docker pull command is keyed in Docker CLI, where the docker command is then sent to Docker REST API.

  Step C2: Docker REST API sends the docker command to Docker Daemon.

  Step C3 and C4: Docker Daemon sends a request to the image registry to pull the specified image (e.g. Image 2) to the local cache.

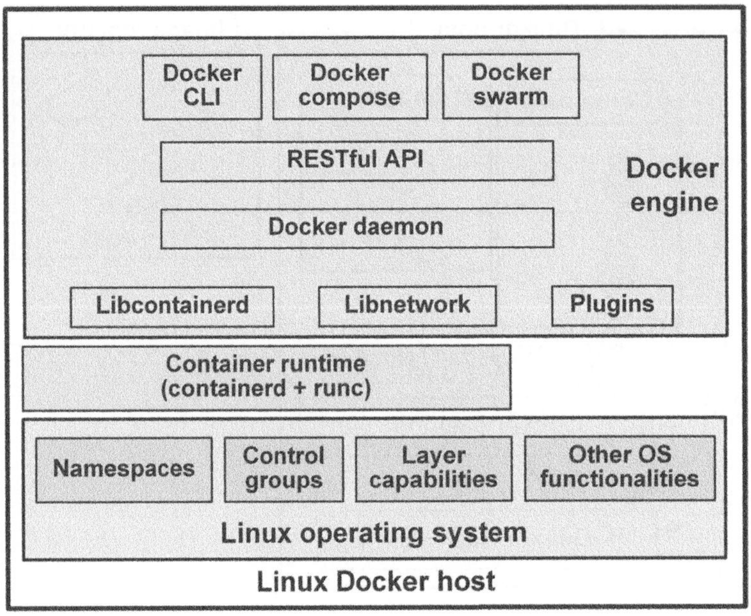

Figure 5.5 Architecture of a Linux Docker host.

5.2.1.3 Architecture of Linux Docker Host

Figure 5.5 shows the architecture of a Linux Docker host [19], which contains three parts: the Linux operating system (OS) on the bottom, the container runtime in the middle, and the Docker engine on the top.

The Linux OS provides some functionalities, such as namespaces, control groups, layer capabilities, and more, to make containers possible. The Linux kernel namespaces, such as process ID namespaces and network namespaces, allow Docker to isolate processes running inside each container. The control groups allow Docker to limit the computing resources, such as CPU and RAM, which each container can consume. The layer capabilities allow Docker to compose a virtual file system inside a container, consisting of different layers. The container runtime utilizes the above-mentioned functionalities in a specific way.

The container runtime is responsible for managing the life cycle of each container. It can pull an image from an image registry, create a container from that image, and then run the container. It can also stop a running container and delete the container from the Docker host. The container runtime on a Linux Docker host is composed by containerd and runc. The runc provides low-level functionalities of the container runtime, whereas the containerd is based on runc and provides higher-level functionalities.

The Docker engine provides additional functionalities on top of the container runtime, such as libraries for communicating with the container runtime (containerd), network libraries, and support for plugins. It also contains Docker Daemon and Docker REST API, which allow users to communicate and interact with the Docker Engine, in turn, communicating with containerd to rotate runc to run containers. Sitting on top of Docker Engine are Docker CLI and some Docker working tools, such as Docker Compose and Docker Swarm.

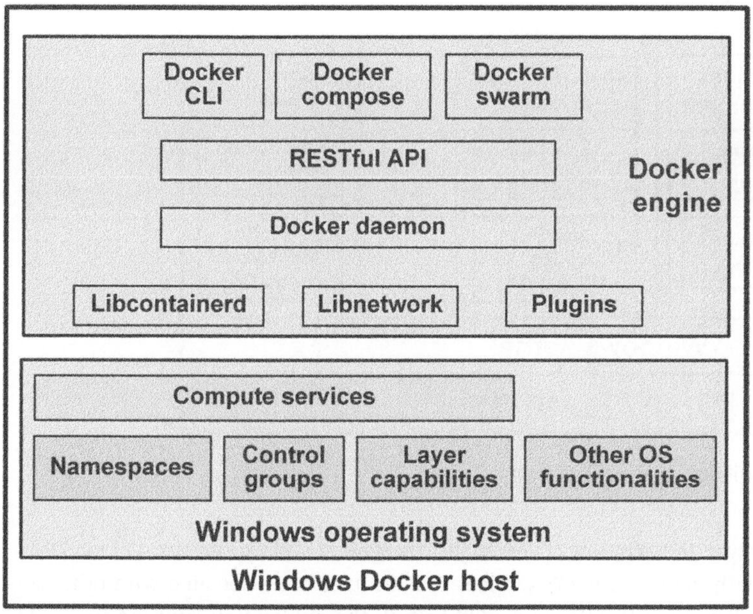

Figure 5.6 Architecture of a Windows Docker host.

5.2.1.4 Architecture of Windows Docker Host

Figure 5.6 shows a high-level architecture of a Windows Docker host [19]. It looks the same as a Linux Docker host's architecture in terms of the Docker Engine's components. Nevertheless, there is no containerd and runc concept available on the Windows Platform. From Windows Server 2016, Microsoft introduces the "Compute Services" layer at the OS level to provide a public interface for managing the containers, such as starting and stopping containers. The Compute Services abstracts the low-level capabilities, such as namespaces, control groups, and layer capabilities, provided by the Windows OS. In this way, Windows can change the low-level APIs for enhancing capabilities in managing containers without changing the public APIs of Compute Services, which are called by the Docker Engine.

5.2.1.5 Architecture of Windows Server Containers

Windows is very different from Linux. Within the Windows OS, there are two modes: user mode and kernel mode. The processor will continuously switch between these two modes according to the codes it runs. The code running in the kernel mode can directly access the underlying hardware and share the same memory address space as the operating system and other kernel drivers. Thus, running codes in the kernel mode is risky because it may cause the entire OS to crash if something is wrong. Nonetheless, an application running in the user mode has its own private virtual address space. An application cannot alter the data that belongs to another application, so the crash is limited to itself if an application crashes. Also, a processor running in the user mode cannot directly access the virtual memory addresses that are reserved for the OS.

Figure 5.7 depicts the architecture of Windows Server Containers (WSCs) [20]. The container management components such as Docker Engine and Compute Services run in the user mode on a Windows Docker host. Windows Server containers also run in the user mode and share the

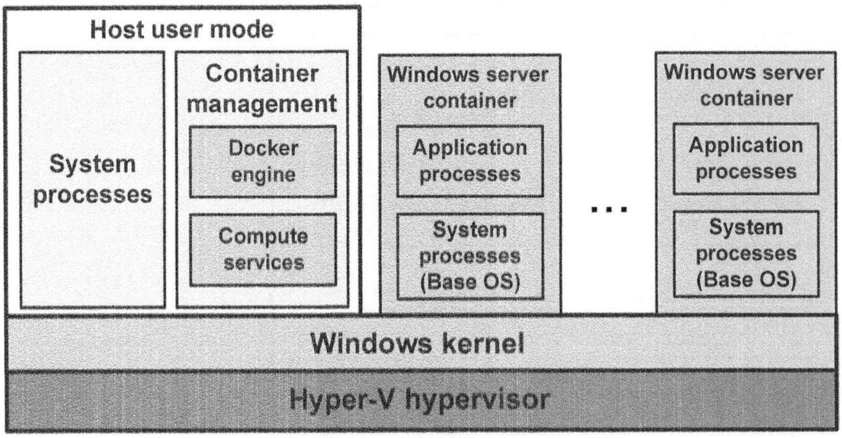

Figure 5.7 Architecture of Windows Server Containers.

Windows OS kernel. However, the Windows OS is highly integrated and exposes its API via DLLs, which is tightly coupled with the running OS services that, in turn, are associated with other services and DLLs. While it can share the Windows OS kernel, the application running in a WSC cannot be completely isolated from the system services and DLLs. Thus, each WSC inescapably needs a copy of essential system services and OS DLLs required to call Windows APIs. These are the reasons why each WSC contains a set of system processes (called Base OS), in addition to the application processes, as shown in Figure 5.7.

The above-mentioned design of WSCs has adverse effects on their portability. Windows Server Containers cannot run on different versions of Windows. For example, if we want to run a WSC on Windows Server 2016, we must build the container image based on a Windows Server 2016 base OS image. If the Windows host OS is Nano Server, we can only run WSCs based on Nano Server base OS images on that host. Because each WSC must contain a base OS, a Windows application image is, on average, much larger than a Linux application image.

5.2.1.6 Architecture of Hyper-V Containers

To address the limitations of WSCs, Microsoft introduced a new type of container isolation that uses Hyper-V Virtualization to run the container images. Figure 5.8 shows the architecture of Hyper-V Containers [20]. Each Hyper-V container runs in its Virtual Machine, which has its own Windows Kernel. Because Hyper-V containers do not need a shared OS kernel, they no longer need to match the Windows build versions between the host OS and the container. Nevertheless, the Hyper-V containers require nested virtualization support.

5.2.2 Docker Operational Principles

This section will introduce the operational principles of Docker, including Docker Image, Dockerfile, Docker Container, Container Network Model, and Docker Networking.

5.2.2.1 Docker Image

A Docker image [21] is the template for creating a Docker container. It is similar to a virtual machine image that can be used to create a virtual machine. However, a virtual machine image is

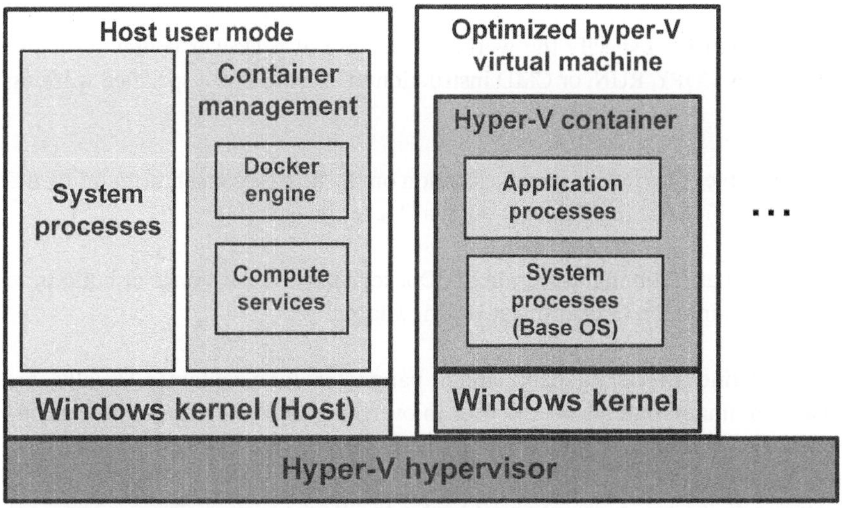

Figure 5.8 Architecture of Hyper-V Containers.

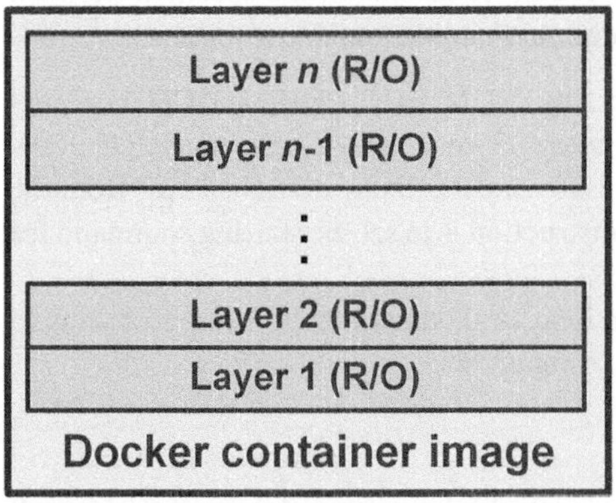

Figure 5.9 Anatomy of a Docker container image.

a huge single file, on the other hand, a Docker image consists of multiple read-only layers [22]. Usually, we use a Dockerfile to build an image, and each instruction in the Dockerfile will generate a layer in the image. Figure 5.9 depicts the anatomy of a Docker container image containing n layers. Docker uses storage drivers [23], such as aufs and overlay2, to build an image. The layers are stacked on top of each other. Each layer is only a set of differences from the layer below it.

5.2.2.2 Dockerfile

A Dockerfile is a text file containing instructions with specific keywords that instruct Docker to create a custom image. The following introduces some instructions frequently used in a Dockerfile [24]:

- **FROM**
  The FROM instruction initializes a new build stage and sets the base image for subsequent instructions. Thus, a valid Dockerfile must start with the FROM instruction.

- **WORKDIR**
  The WORKDIR instruction is to specify the working directory of a Docker container across instructions. It allows any COPY, RUN, or CMD instruction to execute in the specified working directory.
- **COPY**
  The COPY instruction copies files from a specific location on the host (the machine building the Docker image) into a specified destination location in a Docker image.
- **RUN**
  The RUN instruction executes commands inside the Docker image. It runs once at build time and gets the result written into the Docker image as a new layer.
- **CMD**
  The CMD instruction defines the default command or parameters for a container. Usually, it is used to set a default command that allows the user to override it when starting to run a container. If a Dockerfile has multiple CMD instructions, it only applies the last one.
- **ENV**
  The ENV instruction sets the environment variables for the Docker image. The environment variables which are set using ENV will persist when a container is run from the resulting image.
- **EXPOSE**
  The EXPOSE instruction informs Docker that the container listens on the specified network ports at runtime for communication between the container and the outside world.

Figure 5.10 shows an example of Dockerfile that contains FROM, RUN, COPY, and CMD instructions. The FROM instruction is for using the ubuntu:latest OS image as the base image. The RUN instruction is to install python. The COPY instruction is to copy the application (hello.py) from the host's current directory into the container. The CMD instruction is to set the starting command for the container (i.e. python /hello.py).

The following docker command can build a custom image called pythonapp:latest according to this Dockerfile: (Section 5.2.3 will show more docker commands.)

$$docker\ image\ build - t\ pythonapp : latest$$

where the option "-t" is to tag a name pythonapp:latest for the built image, and the path "." is to tell Docker to use the Dockfile in the current directory as the build context. Figure 5.11 shows the process of building an image. In this example, the current directory is build_pythonapp, which contains the Dockerfile and a python application hello.py. It is noted that there are four instructions in the Dockerfile, and thus there are four corresponding steps in the building process. Also,

```
# use ubuntu:latest as the base OS image
FROM ubuntu:latest
# install python
RUN apt-get update && apt-get install -y python
# copy the app from the host into the container
COPY hello.py /hello.py
# set starting command for the container
CMD ["python","/hello.py"]
```

Figure 5.10 An example of Dockerfile.

```
mhhung@imrc-host1:~/build_pythonapp$ ls
Dockerfile  hello.py
mhhung@imrc-host1:~/build_pythonapp$ docker image build -t pythonapp:latest .
Sending build context to Docker daemon  3.072kB
Step 1/4 : FROM ubuntu:latest
 ---> 9140108b62dc
Step 2/4 : RUN apt-get update && apt-get install -y python
 ---> Using cache
 ---> 4d54c3e75a3b
Step 3/4 : COPY hello.py /hello.py
 ---> Using cache
 ---> 8be88281b298
Step 4/4 : CMD ["python","/hello.py"]
 ---> Using cache
 ---> 08bed6b197a4
Successfully built 08bed6b197a4
Successfully tagged pythonapp:latest
```

Figure 5.11 The process of building an image.

```
mhhung@imrc-host1:~/build_pythonapp$ docker image history pythonapp
IMAGE          CREATED       CREATED BY                                       SIZE
08bed6b197a4   2 days ago    /bin/sh -c #(nop)  CMD ["python" "/hello.py"]    0B
8be88281b298   2 days ago    /bin/sh -c #(nop) COPY file:3f8a14a336efe96f…    21B
4d54c3e75a3b   2 days ago    /bin/sh -c apt-get update && apt-get install…    60.1MB
9140108b62dc   10 days ago   /bin/sh -c #(nop)  CMD ["/bin/bash"]             0B
<missing>      10 days ago   /bin/sh -c mkdir -p /run/systemd && echo 'do…    7B
<missing>      10 days ago   /bin/sh -c [ -z "$(apt-get indextargets)" ]      0B
<missing>      10 days ago   /bin/sh -c set -xe   && echo '#!/bin/sh' > /…    811B
<missing>      10 days ago   /bin/sh -c #(nop) ADD file:da80f59399481ffc3…    72.9MB
```

layer from the CMD instruction
layer from the COPY instruction
layer from the RUN instruction
layers from ubuntu base OS image

Figure 5.12 The history of the pythonapp:latest image.

if the resulting layer has existed in the local cache in each step, Docker just gets and uses it. For example, Step 2 uses the local cache result, so it does not need to create this layer again.

The following docker command can check the history of an image:

docker image history < image name >

Figure 5.12 shows the history of the pythonapp:latest image created using the Dockerfile in Figure 5.10. The five layers at the bottom result from the FROM instruction that instructs Docker to use the ubuntu:latest OS image as the base image. On top of these layers is the layer created by the RUN instruction. On top of it is the layer created by the COPY instruction. Finally, the top layer is created by the CMD instruction. These verify that a Docker image is made up of multiple layers.

The following docker command can examine the information of existing images on the host:

docker images

Figure 5.13 shows that the size of the pythonapp:latest image is 133 MB, whereas the ubuntu:latest OS image has a size of 72.9 MB. It is noted that the pythonapp:latest image contains or shares the ubuntu:latest image. The increased size of the pythonapp:latest image relative to the ubuntu:latest image is 60.1MB.

- **Example of building a Linux application image**

  The following illustrates another Dockerfile for building a Linux application image and the stacked layers of the resulting image. Figure 5.14 shows an example of Dockerfile for creating a

```
mhhung@imrc-host1:~/build_pythonapp$ docker images
REPOSITORY          TAG                 IMAGE ID            CREATED             SIZE
pythonapp           latest              08bed6b197a4        2 days ago          133MB
ubuntu              latest              9140108b62dc        10 days ago         72.9MB
```

Figure 5.13 The sizes of two Linux container images.

```
# use ubuntu:18.04 as the base OS image
FROM ubuntu:18.04
# Instructions for installing nginx web server
RUN apt update && apt install -y nginx ...

# Instructions for coping files of the web app

# Expose Ports for the web app
EXPOSE 80 443
# define the starting command for the container
CMD ["./start.sh"]
```

Figure 5.14 A shorthand Dockerfile for building a Linux container image.

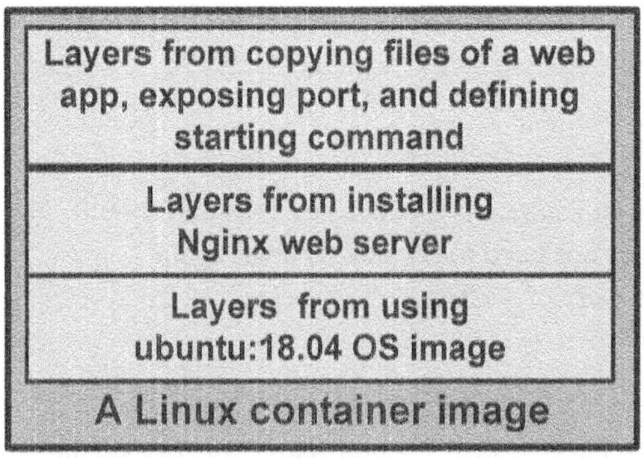

Figure 5.15 The stacked layers of the Linux container image built by the Dockerfile in Figure 5.14.

Linux web application image. The meanings of the instructions in the Dockerfile are explained in the comments. Figure 5.15 depicts the anatomy of the resulting image built by this Dockerfile. The image comprises some layers generated by using the ubuntu:18.04 base OS image at the bottom, some layers resulting from installing the Nginx web server in the middle, and some layers on the top, which are generated by copying the web application's files, exposing ports, and defining the starting command.

- **Example of building a Windows application image**
  Figure 5.16 shows an example of Dockerfile for building a Windows web application image. The meanings of the instructions in the Dockerfile are explained in the comments. Figure 5.17 depicts the anatomy of the resulting image built by this Dockerfile. The image consists of some layers generated by using the servercore:ltsc2019 base OS image at the bottom, some layers from installing the IIS web server and the .NET Framework in the middle, and some layers from copying the web application's files, exposing the port, and defining the starting command on the top.

```
# escape=`
# use servercore:ltsc2019 as the base OS image
FROM mcr.microsoft.com/windows/servercore:ltsc2019
# Instructions for installing IIS web server and .NET Framework
RUN powershell -Command `
    Add-WindowsFeature Web-Server; `
    ...
# Instructions for coping files of the web app

# Expose Port for the web app
EXPOSE 80
# define the starting command for the container
CMD ["./start.ps1"]
```

Figure 5.16 A shorthand Dockerfile for building a Windows container image.

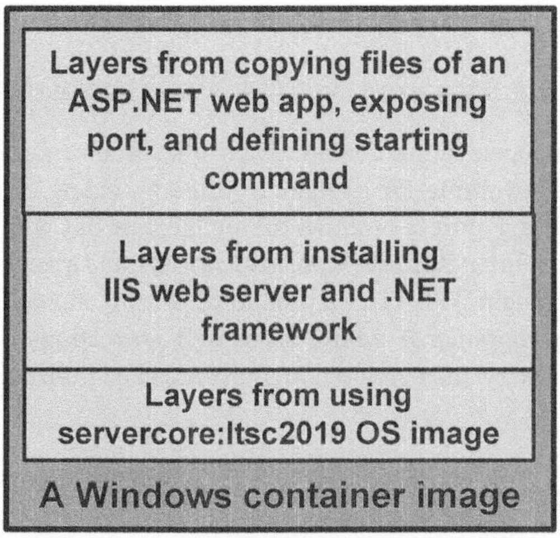

Figure 5.17 The stacked layers of the Windows container image built by the Dockerfile in Figure 5.16.

5.2.2.3 Docker Container

Docker Image and Docker Container are closely related to each other. The major difference between them is that an image packages an application and its dependencies required to run the application and serves as a template for creating containers. On the contrary, a container is a running instance of the image. That is to say, a container is created from an image, and an image can be used to create many containers that share the same image. Figure 5.18 is the illustration of many containers sharing a custom read-only image. When a container is created from an image, a thin writable container layer is added on the top of the shared image layers to form the container [25]. As such, each container has its own writable container layer, and all changes are stored in this container layer. Thus, while multiple containers share the same underlying image, they can have their own data state.

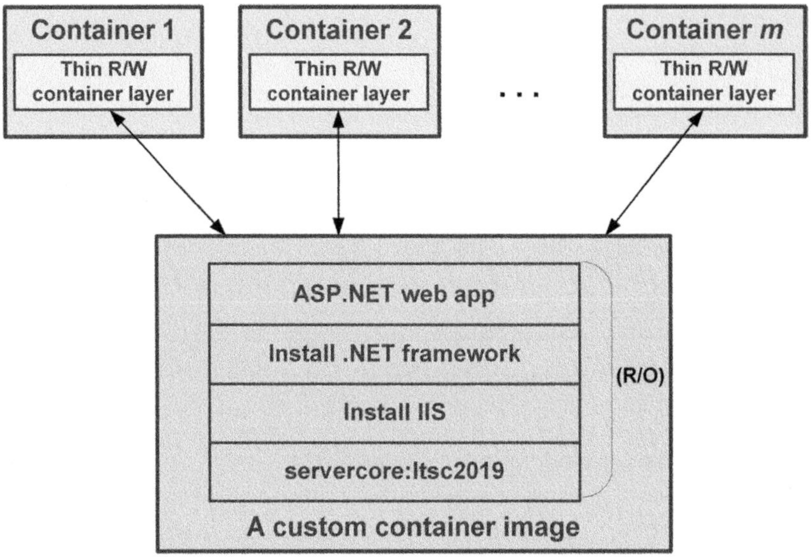

Figure 5.18 Illustration of many containers sharing the same image layers.

5.2.2.4 Container Network Model

The Container Network Model (CNM) [26, 27] is an open-source container network specification, which formalizes the steps required to provide networking for containers while providing an abstraction that can support multiple network drivers. Figure 5.19 shows the architecture of CNM for Linux containers [28]. The CNM has interfaces for IP Address Management (IPAM) Drivers plugins and Network Drivers plugins. The IPAM plugin APIs can be used to perform container IP-related tasks, such as allocating or deallocating container IP addresses, as well as creating or deleting IP address pools for containers. On the contrary, the network plugin APIs can be utilized

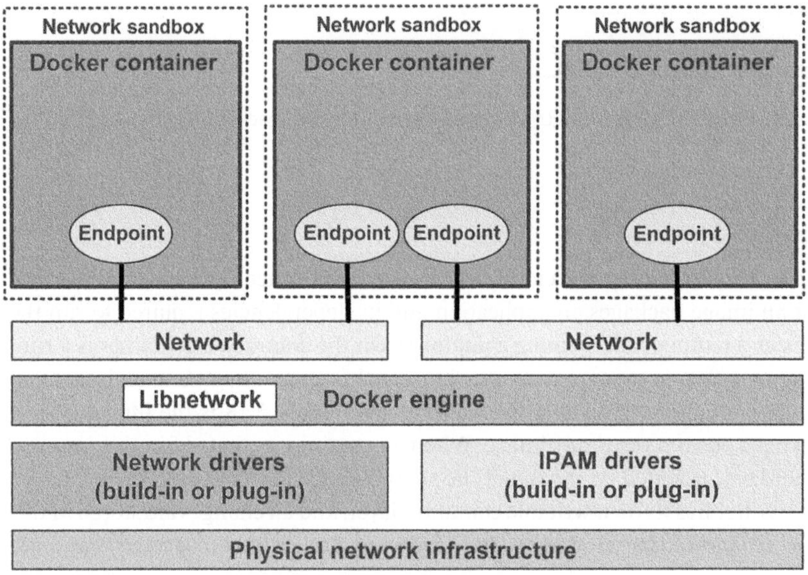

Figure 5.19 Architecture of the container network model (CNM) for Linux containers.

to conduct container networking-related tasks, such as creating or deleting networks, and adding containers to or removing containers from networks. Libnetwork [26] is an open-source library that Docker implements all of the key concepts in the CNM.

A CNM is based on the following five objects [26, 27]: NetworkController, Network Driver, Network, Sandbox, and Endpoint:

- The NetworkController provides the entry-point into Libnetwork that exposes simple APIs for Docker Engine to manage networks. Because Libnetwork supports multiple network drivers, the NetworkController allows users to attach a particular driver to a target network.
- The Driver is responsible for managing networks. Docker can include multiple network drivers to satisfy various container networking scenarios.
- A Network provides connectivity between a group of endpoints, which belong to the same network and can communicate with each other directly while isolating the endpoints on other networks.
- A Sandbox contains a container's network stack configuration, which includes the container's interfaces (e.g. IP-address and MAC-address), routing table, and DNS settings. A sandbox may contain many endpoints attached to different networks.
- An Endpoint provides the connectivity between the services exposed by a container and other containers' services in the same network. An endpoint connects a sandbox to a network. An endpoint can belong to only one network.

5.2.2.5 Docker Networking

Docker can use several network drivers for container networking, such as Bridge, Host, None, and Overlay drivers [28-30]. The first three network drivers are for single-host networking, whereas the last one is for multiple-host networking. The architecture of each type of container networking is described as follows.

- **Bridge Networking**

  Bridge networking uses software bridges on the Docker Host to allow containers to connect to the outside. The containers attached to the same bridge can communicate with each other. The docker bridge driver will automatically set the corresponding networking rules (e.g. iptables, network namespace, etc.) on the host to make the container network work correctly. The bridge network is to handle the communication between containers running on a single Docker host. The communication between containers on multiple Docker hosts should use the overlay network.

  Figure 5.20 depicts the architecture of bridge networking. After being installed on a host, Docker will create a software bridge, called docker0, as the default container bridge network. We can also create a user-defined bridge. As such, Containers 1 and 2 in the figure can communicate with each other through docker0, but Container 1 cannot communicate with Container 3 because they are attached to different bridge networks. The outside can access the containers on a Docker host using port mapping. For example, by mapping the host port 8080 to the port 80 of Container 2, we can use the Docker host's IP address and the port 8080 to access the web application running in Container 2.
- **Host Networking**

  The host network driver uses the Docker host's networking directly. It removes the network isolation between the Docker host and the Docker containers on it. As such, by using a host network, we cannot run multiple web containers on the same port of the Docker host because the port is common to all containers in the host network. If we use the host network mode for a container, that container's network stack is not isolated from the Docker host. The container shares the host's networking namespace and does not get its own IP address.

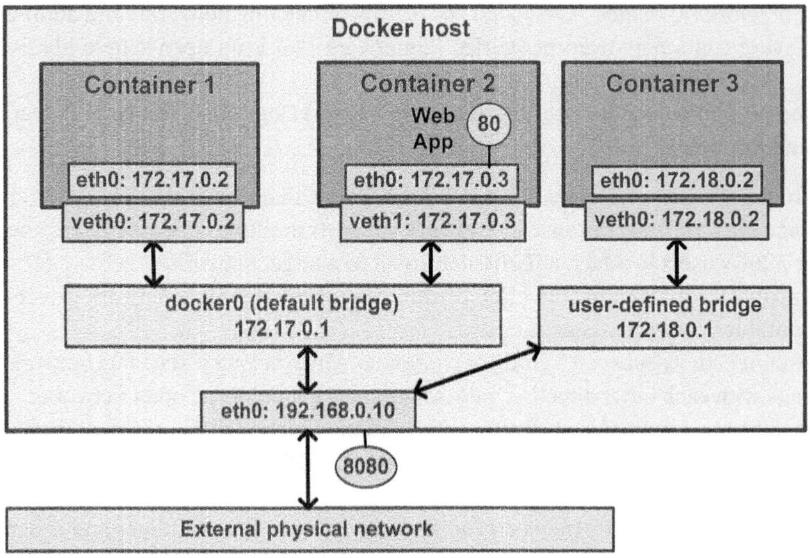

Figure 5.20 Architecture of bridge networking.

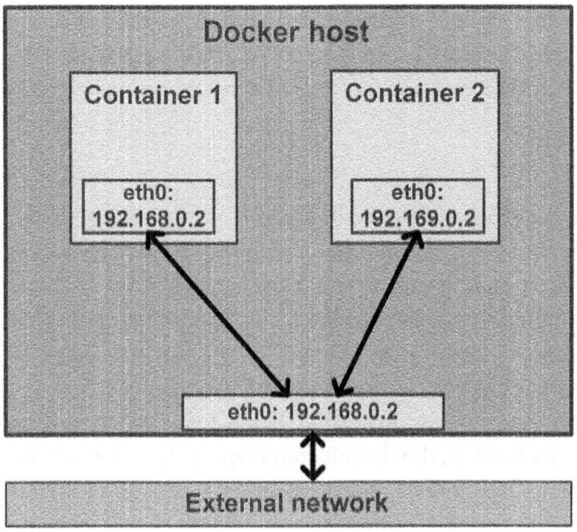

Figure 5.21 Architecture of host networking.

Figure 5.21 depicts the architecture of host networking. As shown in the figure, the IP address of Containers 1 and 2 is the same as the Docker host's IP. If we use host networking and run a container that binds to port 80, the container's application can be accessed by using the port 80 on the Docker host's IP address. Host networking can be useful to optimize the performance of containers in case that a container needs to handle a large range of ports, as it does not require network address translation (NAT) for each port. The host networking driver only works on Linux Docker hosts.

- **None Networking**
  In the mode of none networking, containers are not attached to any network and do not have any access to the external network or other containers, as shown in Figure 5.22. Thus, the none networking mode can be used when completely disabling the networking stack on a container is wanted.

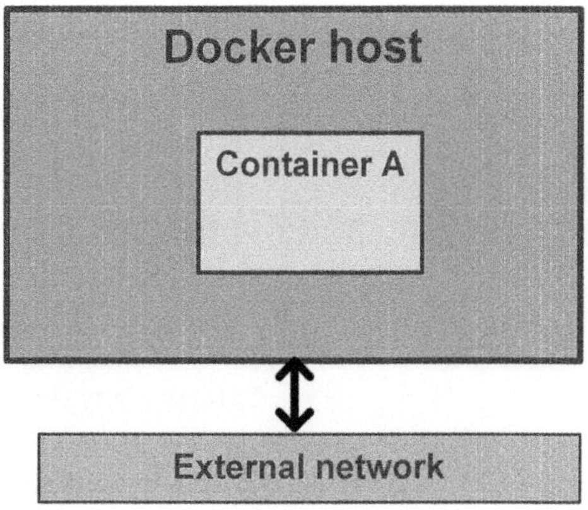

Figure 5.22 Architecture of none networking.

- **Overlay Networking**
  An overlay network can create an internal network that spans across all the nodes of a cluster. An overlay network enables the communication between a swarm service and a container, or between two containers on different Docker hosts of the cluster. Figure 5.23 shows the overlay networking architecture, where Containers 1, 2, and 3 on different Docker hosts communicate with one another through an overlay network that spanning three Docker hosts. The overlay networking does not need to do OS-level routing between these containers.

  The overlay network driver of Docker significantly simplifies the complexities in multi-host networking. With the overlay driver, multi-host networks are available inside Docker without needing to use external components. The overlay driver contains built-in functionalities of IPAM, service discovery, multi-host connectivity, and encryption. The overlay driver utilizes Virtual Extensible LAN (VXLAN) to encapsulate data, decoupling the container network from the underlying physical network.
- **CNM for Windows Containers**
  Windows containers function similarly to virtual machines in terms of networking. Each container has a virtual network adapter (vNIC) that connects to a Hyper-V virtual switch (vSwitch). Windows supports several different networking modes for Docker containers, such as NAT, Overlay, Transparent, and L2bridge. The 'NAT' network is the default network for Windows containers. When the Docker engine on a host runs for the first time, it will create a default NAT network, 'nat', which uses an internal vSwitch and a Windows component named WinNAT. Any container that runs on a Windows Docker host without setting specific network configurations will be attached to the default 'nat' network.

  The networking for Windows containers is very different from the networking for Linux containers. For easy management of Windows container networks, the Host Network Service (HNS) was invented to abstract the underlying container networking functionalities provided by Windows. Figure 5.24 shows the architecture of CNM for Windows containers [31] to illustrate this concept. The Docker engine calls Libnetwork, which, in turn, calls the APIs provided by the HNS, for managing Windows container networks. The HNS and the Compute Services mentioned in the architecture of a Windows Docker host in Figure 5.6 work together to create Windows containers and attach their endpoints to networks.

5.2.3 Illustrative Applications of Docker

This section illustrates some applications of Docker. First, the workflow of building, shipping, and deploying a Docker container is presented. Next, the deployment of a Docker container running a Linux application following the workflow is depicted. Then, we demonstrate the deployment of a Docker container running a Windows application according to the workflow.

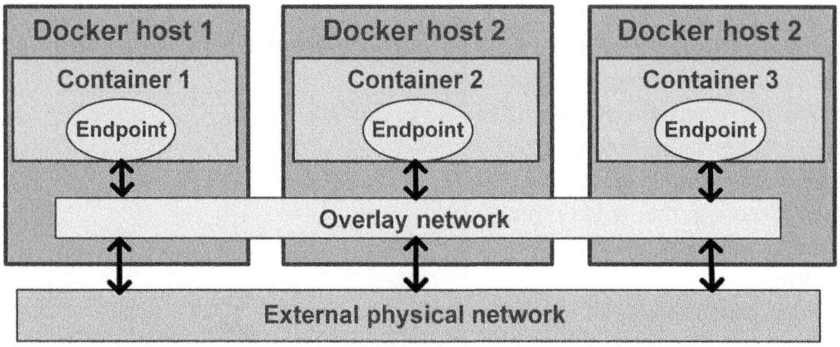

Figure 5.23 Architecture of overlay networking.

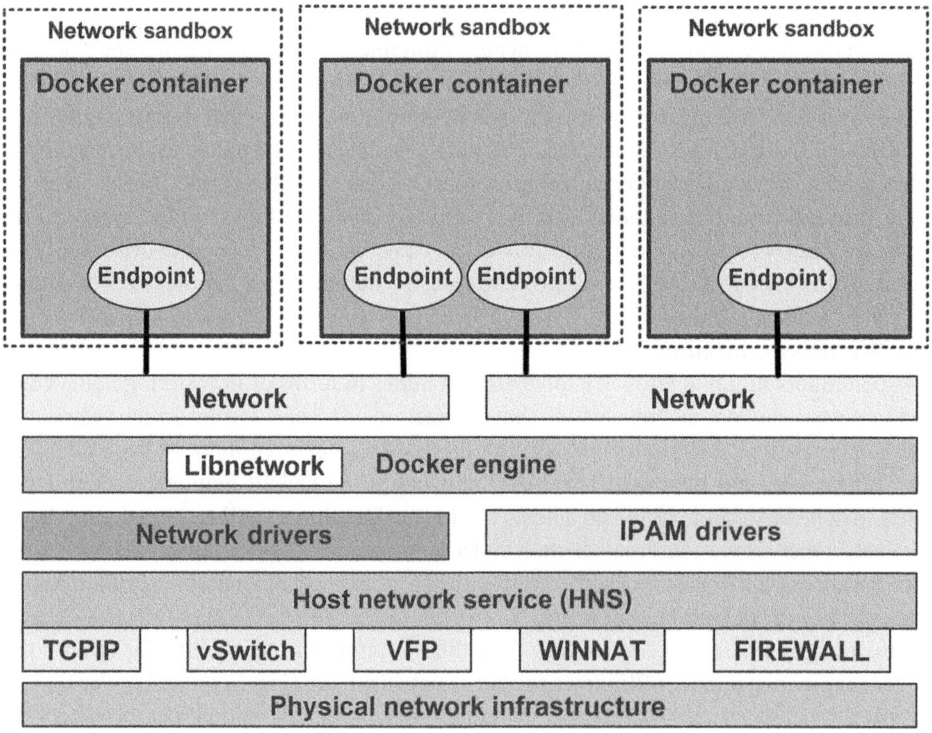

Figure 5.24 Architecture of CNM for Windows containers.

5.2.3.1 Workflow of Building, Shipping, and Deploying a Containerized Application

Figure 5.25 shows the workflow of building, shipping, and deploying a containerized application. This figure's architecture includes three parts: a Docker host on the development side, an image registry (Docker Hub in this example) on the cloud, and a Docker host on the production side. The working procedure is described as follows:

Step 1: A Base OS image is pulled from the image registry to the Docker host's local cache on the development side.

Steps 2 to 5: The developer builds and tags a custom application image (e.g. image n) using a Dockerfile that uses the Base OS image as the base and then adds the application and its dependencies (e.g. Configuration, Framework, and Libraries) on top of it.

Step 6: The resulting image n is pushed to the image registry on the cloud.

Step 7: The image n is pulled from the image registry on the cloud to the Docker host's local cache on the production side.

Step 8: The container n running the application is created and run from the image n on the production side.

5.2.3.2 Deployment of a Docker Container Running a Linux Application

In this illustrative example, we first prepare a Dockerfile shown in Figure 5.26 for building a Linux web application image. The Dockerfile instructs Docker to use an Nginx web server image as the base image, then copy the web application's files into the image, and finally expose the port 80 on the container.

Then, we build the Linux web application image by using the following docker build command:

docker build. −t example-linux

where the "." indicates that building the image according to the Dockerfile in the host's current directory, and the option "-t" is used to tag the name "example-linux" for the image. Figure 5.27 shows the building steps of the Linux web application image. It is noted that the Dockerfile has three instructions, so there are three building steps.

Next, we re-tag the example-linux image to include a Docker hub account named "imrc" in the image name by using the following docker tag command:

docker tag example-linux imrc/example-linux

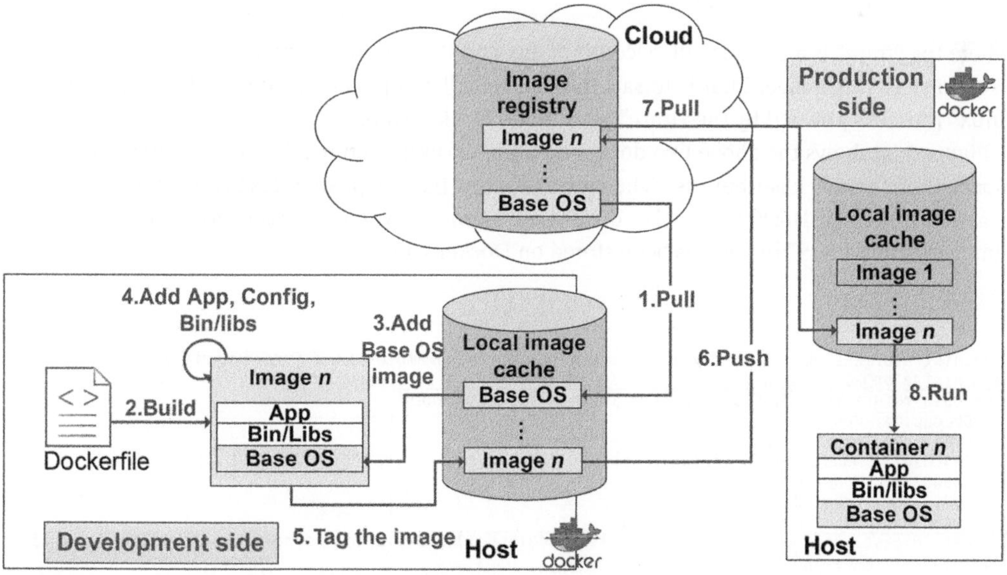

Figure 5.25 Workflow of building, shipping, and deploying a containerized application.

```
# use nginx as the base OS image, which includes Ubuntu OS and nginx environment
FROM nginx

# copy the web app from the host into the container
COPY . /usr/share/nginx/html/

# expose the port for web app
EXPOSE 80
```

Figure 5.26 An example Dockerfile for building a Linux web application image.

```
administrator@k8s-ubuntu-2:~/example-linux$ docker build  . -t example-linux
Sending build context to Docker daemon  3.072kB
Step 1/3 : FROM nginx
 ---> 7e4d58f0e5f3
Step 2/3 : COPY . /usr/share/nginx/html/
 ---> Using cache
 ---> 7df05e2370ef
Step 3/3 : EXPOSE 80
 ---> Using cache
 ---> 5c0a243dd242
Successfully built 5c0a243dd242
Successfully tagged example-linux:latest
administrator@k8s-ubuntu-2:~/example-linux$
```

Steps of building image according to Dockerfile

Image has been successfully built and tagged.

Figure 5.27 Building steps of the Linux web application image.

After doing this, we can then push the image from the local cache to the image registry by using the following docker push command:

docker push imrc/example-linux

where the "imrc" is a Docker hub account name, and the "example-linux" is the repository name used to store this image. That is to say, this image will be uploaded and stored in the "example-linux" repository owned by the "imrc" account on Docker Hub.

Figure 5.28 shows the above two docker commands' execution results in a terminal on a Linux Docker host, named k8s-ubuntu-2. The messages show that the push refers to the repository "imrc/example-linux" on Docker Hub (i.e. docker.io). Figure 5.29 is the screenshot showing that the "imrc/example-linux" image has been stored on Docker Hub.

```
administrator@k8s-ubuntu-2:~/example-linux$ docker tag example-linux imrc/example-linux
administrator@k8s-ubuntu-2:~/example-linux$ docker push imrc/example-linux
The push refers to repository [docker.io/imrc/example-linux]
bda8becdd1d4: Pushed
908cf8238301: Mounted from library/nginx
eabfa4cd2d12: Mounted from library/nginx
60c688e8765e: Mounted from library/nginx
f431d0917d41: Mounted from library/nginx
07cab4339852: Mounted from library/nginx
latest: digest: sha256:ec982dd01326b49819508c42f5c6d19d251fb137c03c6d8cd9fab1d47ab1628f size: 1569
administrator@k8s-ubuntu-2:~/example-linux$
```

Image has been pushed to a repository on Docker Hub.

Figure 5.28 Execution results of the docker tag and docker push commands.

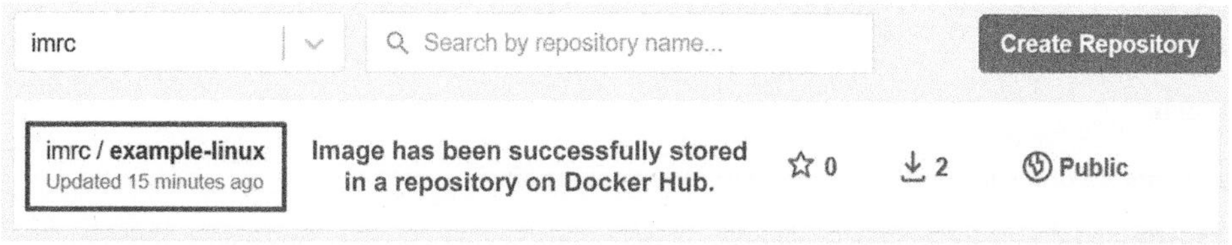

Figure 5.29 Screenshot showing that the "imrc/example-linux" image has been stored on Docker Hub.

```
administrator@k8s-slave-ubuntu-1:~$ docker pull imrc/example-linux
Using default tag: latest
latest: Pulling from imrc/example-linux
Digest: sha256:ec982dd01326b49819508c42f5c6d19d251fb137c03c6d8cd9fab1d47ab1628f
Status: Image is up to date for imrc/example-linux:latest
docker.io/imrc/example-linux:latest
administrator@k8s-slave-ubuntu-1:~$
```

Image has been pulled from a repository on Docker Hub to another Docker host.

Figure 5.30 Execution result of the docker pull command.

```
administrator@k8s-slave-ubuntu-1:~$ docker run -p 8080:80 imrc/example-linux
/docker-entrypoint.sh: /docker-entrypoint.d/ is not empty, will attempt to perform configuration
/docker-entrypoint.sh: Looking for shell scripts in /docker-entrypoint.d/
/docker-entrypoint.sh: Launching /docker-entrypoint.d/10-listen-on-ipv6-by-default.sh
10-listen-on-ipv6-by-default.sh: Getting the checksum of /etc/nginx/conf.d/default.conf
10-listen-on-ipv6-by-default.sh: Enabled listen on IPv6 in /etc/nginx/conf.d/default.conf
/docker-entrypoint.sh: Launching /docker-entrypoint.d/20-envsubst-on-templates.sh
/docker-entrypoint.sh: Configuration complete; ready for start up
```

Run a container from the pulled image.

Figure 5.31 Execution result of the docker run command.

After that, we can pull the image from the registry to the local cache on another Linux Docker host by using the following docker pull command:

docker pull imrc/example-linux

Figure 5.30 shows the docker pull command's execution result in a terminal on the Linux Docker host, named k8s-slave-ubuntu-1. The messages show that the image is pulled from the imrc/example-linux repository on Docker Hub (i.e. docker.io).

Now, we can run a container from the pulled image on the Docker host k8s-slave-ubuntu-1 by using the following docker run command:

docker run –p 8080 : 80 *imrc/example-linux*

where the option "-p 8080:80" instructs Docker to publish and map the host port 8080 to the port 80 on the container. Figure 5.31 shows the docker run command's execution result in a terminal on the Linux Docker host k8s-slave-ubuntu-1.

Finally, we can type the URL "localhost:8080" in a browser. Figure 5.32 shows the screenshot displaying the home page of the running containerized Linux web application.

5.2.3.3 Deployment of a Docker Container Running a Windows Application

In this illustrative example, we first prepare a Dockerfile shown in Figure 5.33 for building a Windows web application image. The Dockerfile instructs Docker to use the aspnet:3.1 image as the base image, then copy the web application's files into the image, and finally set "dotnet aspnetapp.dll" as the default command for the container created from the built image.

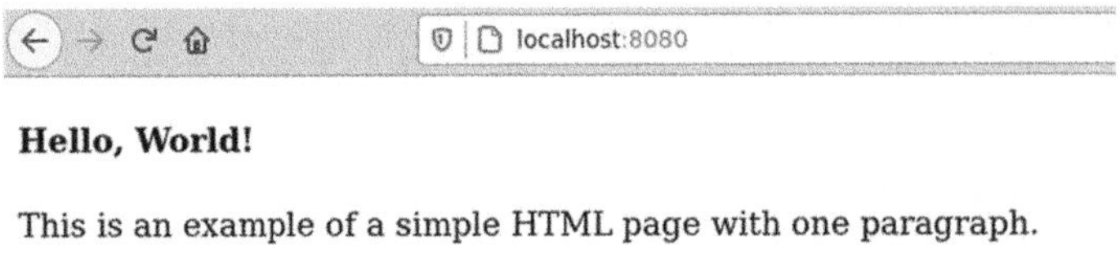
localhost:8080
Hello, World!
This is an example of a simple HTML page with one paragraph.

Figure 5.32 Screenshot displaying the home page of the running containerized Linux web application.

```
# use ASP.NET 3.1 as the base OS image
FROM mcr.microsoft.com/dotnet/core/aspnet:3.1

# define /app as the current working directory
WORKDIR /app

# copy the web app from the host into the container
COPY /app ./

# set starting command for the container
ENTRYPOINT ["dotnet", "aspnetapp.dll"]
```

Figure 5.33 An example Dockerfile for building a Windows web application image.

Then, we build the Windows web application image by using the following docker build command:

docker build. – t imrc/example-windows

where the "." indicates that building the image according to the Dockerfile in the host's current directory, and the option "-t" is used to tag the name "imrc/example-windows" for the image. Figure 5.34 shows the building steps of the Windows web application image. It is noted that the Dockerfile has four instructions, so there are four building steps.

```
C:\example-windows>docker build . -t imrc/example-windows
Sending build context to Docker daemon  5.154MB
Step 1/4 : FROM mcr.microsoft.com/dotnet/core/aspnet:3.1
 ---> 8689f698f6dc
Step 2/4 : WORKDIR /app
 ---> Using cache
 ---> f05ea1ee73a6
Step 3/4 : COPY /app ./
 ---> Using cache
 ---> 0680f4725864
Step 4/4 : ENTRYPOINT ["dotnet", "aspnetapp.dll"]
 ---> Using cache
 ---> 784112b991e0
Successfully built 784112b991e0
Successfully tagged imrc/example-windows:latest
```

Steps of building image according to Dockerfile

Image has been successfully built and tagged.

Figure 5.34 Building steps of the Windows web application image.

```
C:\example-windows>docker push imrc/example-windows
The push refers to repository [docker.io/imrc/example-windows]
bfeadbeca9c4: Pushed
4d58f1f94d05: Pushed
274ce4ac059f: Pushed
30f8b9916e17: Pushed
6f091ca5a20d: Pushed
784b3dcf1e52: Pushed
1f363421b3a5: Pushed
aebacb8b0709: Pushed
b7a95b530b4e: Pushed
604daa52528f: Skipped foreign layer
latest: digest: sha256:7e470abbd5da50611f79cc41612b1b3c9c5ef6b368996fb1a2febb67388013eb
 size: 2587
```

Image has been pushed to a repository on Docker Hub.

Figure 5.35 Execution result of the docker push command.

Next, we can push the image from the local cache to the image registry by using the following docker push command:

docker push imrc/example-windows

where the "imrc" is a Docker hub account name, and the "example-windows" is the repository name used to store this image. That is to say, this image will be uploaded and stored in the "example-windows" repository owned by the "imrc" account on Docker Hub.

Figure 5.35 shows the above docker push command's execution result on a Windows Docker host. The messages show that the push refers to the repository "imrc/example-windows" on Docker Hub (i.e. docker.io). Figure 5.36 is the screenshot showing that the "imrc/example-windows" image has been stored on Docker Hub.

After that, we can pull the image from the registry to the local cache on another Windows Docker host by using the following docker pull command:

docker pull imrc/example-windows

Figure 5.37 shows the docker pull command's execution result on another Windows Docker host. The messages show that the image is pulled from the imrc/example-windows repository on Docker Hub (i.e. docker.io).

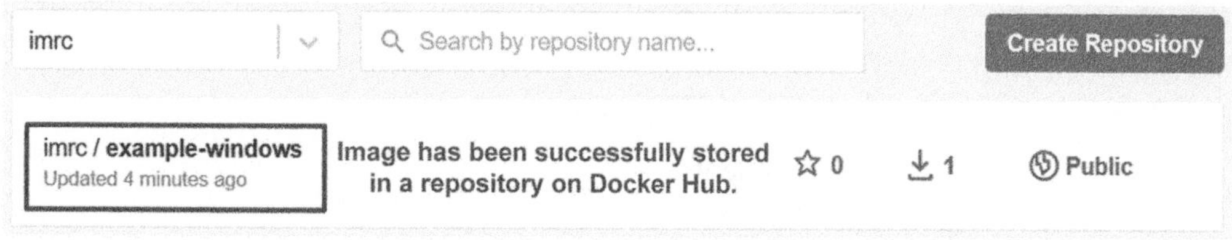

Figure 5.36 Screenshot showing that the "imrc/example-windows" image has been stored on Docker Hub.

```
C:\>docker pull imrc/example-windows
Using default tag: latest
latest: Pulling from imrc/example-windows
Digest: sha256:7e470abbd5da50611f79cc41612b1b3c9c5ef6b368996fb1a2febb67388013eb
Status: Image is up to date for imrc/example-windows:latest
docker.io/imrc/example-windows:latest
```

Image has been pulled from a repository on Docker Hub to another Docker Host.

Figure 5.37 Docker pull command's execution result on a Windows Docker host.

Now, we can run a container from the pulled image on the Windows Docker host by using the following docker run command:

$$docker\ run - p8080:80 imrc/example\text{-}windows$$

where the option "-p 8080:80" instructs Docker to publish and map the host port 8080 to the port 80 on the container. Figure 5.38 shows the docker run command's execution result on the Windows Docker host.

Finally, we can type the URL "localhost:8080" in a browser. Figure 5.39 shows the screenshot displaying the home page of the running containerized Windows web application.

5.2.4 Summary

In this section, we introduce the fundamentals of Docker. Many essential concepts related to Docker from the perspective of architecture are explained, including Docker Engine, a high-level Docker architecture, the architecture of a Linux Docker host, and a Windows Docker architecture

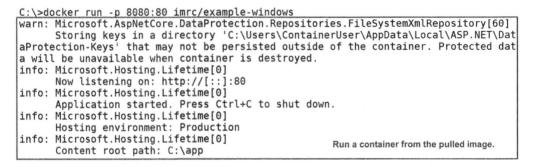

```
C:\>docker run -p 8080:80 imrc/example-windows
warn: Microsoft.AspNetCore.DataProtection.Repositories.FileSystemXmlRepository[60]
      Storing keys in a directory 'C:\Users\ContainerUser\AppData\Local\ASP.NET\Dat
aProtection-Keys' that may not be persisted outside of the container. Protected dat
a will be unavailable when container is destroyed.
info: Microsoft.Hosting.Lifetime[0]
      Now listening on: http://[::]:80
info: Microsoft.Hosting.Lifetime[0]
      Application started. Press Ctrl+C to shut down.
info: Microsoft.Hosting.Lifetime[0]
      Hosting environment: Production
info: Microsoft.Hosting.Lifetime[0]
      Content root path: C:\app
```

Run a container from the pulled image.

Figure 5.38 Execution result of the docker run command.

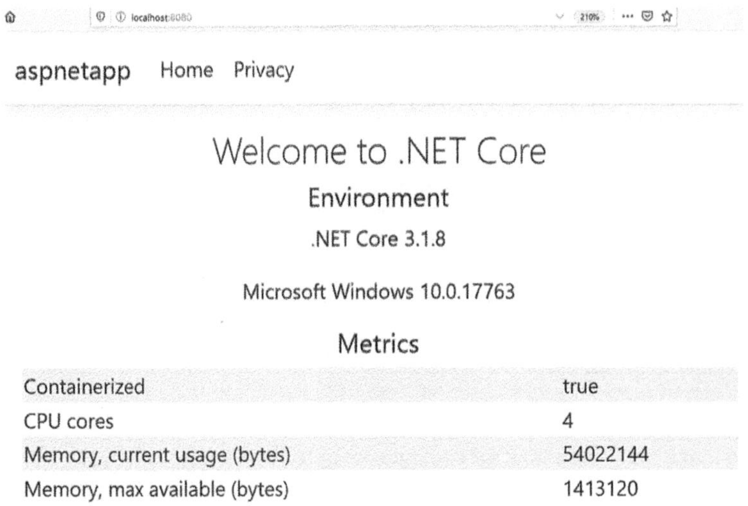

Figure 5.39 Screenshot displaying the home page of the running containerized Windows web application.

host, the architecture of Windows Server containers, and the architecture of Hyper-V containers. We also depict Docker's operational principles, including Docker Image, Dockerfile, Docker Container, Container Network Model, and Docker Networking. Finally, we present some illustrative Docker applications. This section provides a foundation to facilitate the readers to comprehend the Docker-related content in this book. More concepts and materials of Docker worth exploring are available on the Docker website and online documents.

5.3 Fundamentals of Kubernetes

In this section, we introduce the fundamentals of Kubernetes [7], which is a portable open-source system proposed by Google in 2014 and now maintained by the Cloud Native Computing Foundation (CNCF) [9]. Kubernetes aims to orchestrate the deployment, scaling, and management of containerized workloads and services. It provides a framework for deploying and executing distributed containerized applications easily and efficiently. In the meantime, its capabilities include service discovery, load balancing, storage orchestration, automated rollouts and rollbacks, auto-scaling, self-healing, and secret and configuration management. Kubernetes supports various workloads, such as stateless, stateful, and data-processing applications.

This section describes the Kubernetes architecture first, and then presents the operational principles of Kubernetes. After that, we illustrate an application of Kubernetes to show its usability.

5.3.1 Kubernetes Architecture

Figure 5.40 shows the system architecture of Kubernetes, which consists of a Kubernetes control plane node (or master) and a cluster of Kubernetes worker nodes. Kubernetes could deploy containerized workloads and services to the worker nodes in the form of Pods and manage these worker nodes and Pods by the control plane. The workloads and services running in Kubernetes need to be containerized; that is, a workload or service is first packaged into a container image and then run in a container built from that container image. Each Pod contains one or more containers. A worker node could run several Pods.

5.3.1.1 Kubernetes Control Plane Node

The control plane node runs the components of the Kubernetes control plane. These components are responsible for monitoring, scheduling, deploying, and managing containers and Pods in each worker node. The Kubernetes control plane's essential components include kube-controller manager, kube-api-server, etcd, kube-scheduler, and cloud-controller manager, each of which is described as follows.

- **Kube-Controller Manager**
  The kube-controller manager handles various controller processes, including node controller, replication controller, endpoints controller, and service account & token controllers. These built-in controllers provide core capabilities. For example, the node controller monitors the status of nodes. The replication controller maintains the replicas of controller objects. The endpoints controller populates the endpoints object. The service account & token controllers handle access tokens of accounts and APIs for new namespaces. Users could also write a new particular controller to extend the ability of Kubernetes.

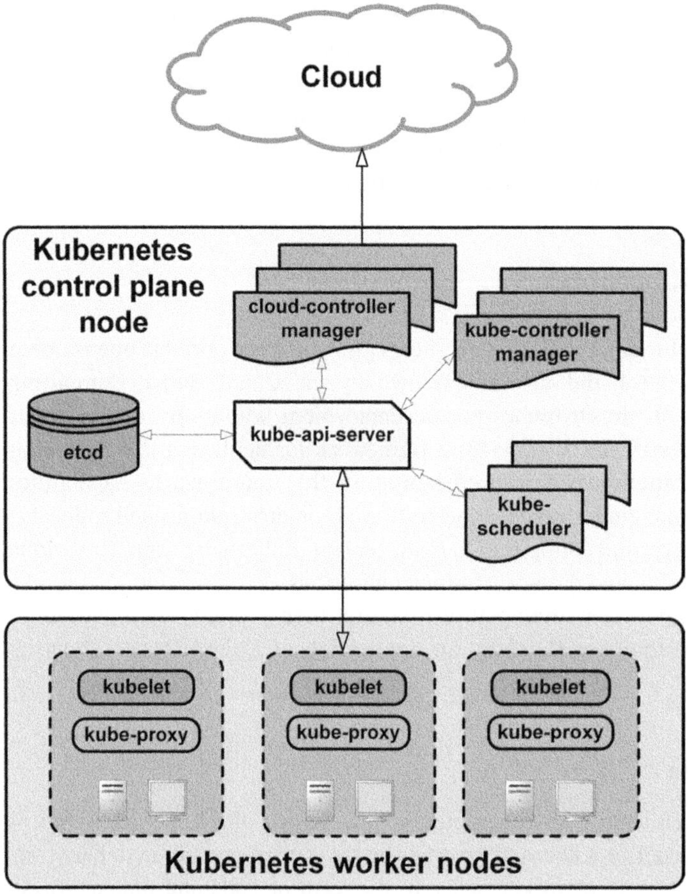

Figure 5.40 Architecture of Kubernetes. *Source:* Reprinted with permission from Ref. [7]; © 2020 The Kubernetes authors, used under CC BY 4.0.

- **Kube-API-Server**
  The kube-api-server provides abundant APIs to support various operations among containers, Pods, nodes, and the control plane. It could scale horizontally by deploying multiple instances of kube-api-server and balancing traffics among them. The kube-api-server is responsible for providing HTTP APIs to send messages among end-users, clusters, and external components and manage the state of objects within Kubernetes APIs. Kubernetes APIs are extensible. The users could add new API resources and new resource fields to meet their requirements. The kube-api-server applies the OpenAPI [32] specification to provide its extensibility.
- **etcd**
  The name "etcd" is a portmanteau word, which is a combination of "/etc" and "d"istributed, to imply that etcd could store configuration information for distributed systems. The etcd [33] is a distributed and reliable key-value store for managing the cluster data. It provides a simple interface to read and write values using standard HTTP tools. The stored data could be represented in hierarchically organized directories. Kubernetes adopts etcd to store the state of the cluster.
- **Kube-Scheduler**
  The kube-scheduler is responsible for scheduling the newly-created Pods to a worker node for execution. The scheduling decision is made according to many factors, including various

resource requirements, hardware constraints, policy constraints, affinity and anti-affinity requirements, data locality, inter-workload interference, and deadlines.

- **Cloud-Controller Manager**
  The cloud-controller manager handles the connection between the Kubernetes cluster and the cloud. It provides cloud-provider-specific controllers. For example, the node controller monitors and manages the nodes in the cloud. The route controller handles the routes in the cloud infrastructure, and the service controller manages the creation, update, and deletion of cloud services. By using these controllers, the Kubernetes cluster could interact with the cloud by decoupling the interoperability logic between Kubernetes and the underlying cloud infrastructure. In general, the cloud-controller manager adopts the plugin approach to provide the connection to different cloud providers.

Kubernetes supports the capability of deploying clusters on specific cloud providers, including Amazon AWS [34], Microsoft Azure [35], CloudStack [36], Google GCE [37], HUAWEI Cloud [38], OpenStack [39], IBM Cloud [40], Baidu Cloud [41], Alibaba Cloud [42], and so on.

5.3.1.2 Kubernetes Worker Nodes

There are three components for maintaining running pods and providing the runtime environment in each worker node: kubelet, kube-proxy, and container runtime. The kubelet is responsible for managing containers created by Kubernetes. It governs containers by the Pod specifications and ensures the health of running containers. The kube-proxy is a network proxy to maintain network rules on a worker node for the communication between Pods or for the outside to access Pods. Kubernetes uses Container Runtime Interface (CRI) to execute different container runtimes. Kubernetes could support any implementation of container runtimes that meet the specification of Kubernetes CRI.

When a worker node is added to a cluster, the kubelet on this worker node could register itself to the kube-api-server in the control plane. The users could also manually add a node object to the cluster. After a node is added to the cluster, the control plane on the cluster will check whether the new node object is valid or not. When all necessary services on this newly added node are ready, this node is valid to run Pods. When the worker node is invalid, Kubernetes keeps checking the status of this node. A worker node's status includes hostname, external/internal IP, resource pressure, health, available resources, and other general information. Users could stop the status checking by deleting the node object.

Initially, Kubernetes creates an object of the worker node and then checks the registering status of the Kubelet in this worker node with matching metadata.name for the kube-api-server. When all necessary services in this worker node are ready, the Pod will be launched. In general, the kubelet will attempt to register itself to the kube-api-server by default. However, the users could manually manage the objects of worker nodes by using kubectl commands. For example, we can place a Pod to a specific worker node by setting the labels on that worker node and the node selectors on the Pod. We could also set a node to be unschedulable to prevent it from accepting new Pods, especially when a node is being rebooted or under maintenance. Kubectl could view each worker node's status, including addresses, conditions of running nodes, capacity and allocable information, etc. Figure 5.41 shows the creation process of a Pod, and the creation steps are listed as follows:

Step 1: The user submits a creation request through the RESTful API of the kube-api-server or the kubectl command-line tool.
Step 2: The kube-api-server processes the user's request and stores Pod data to etcd.

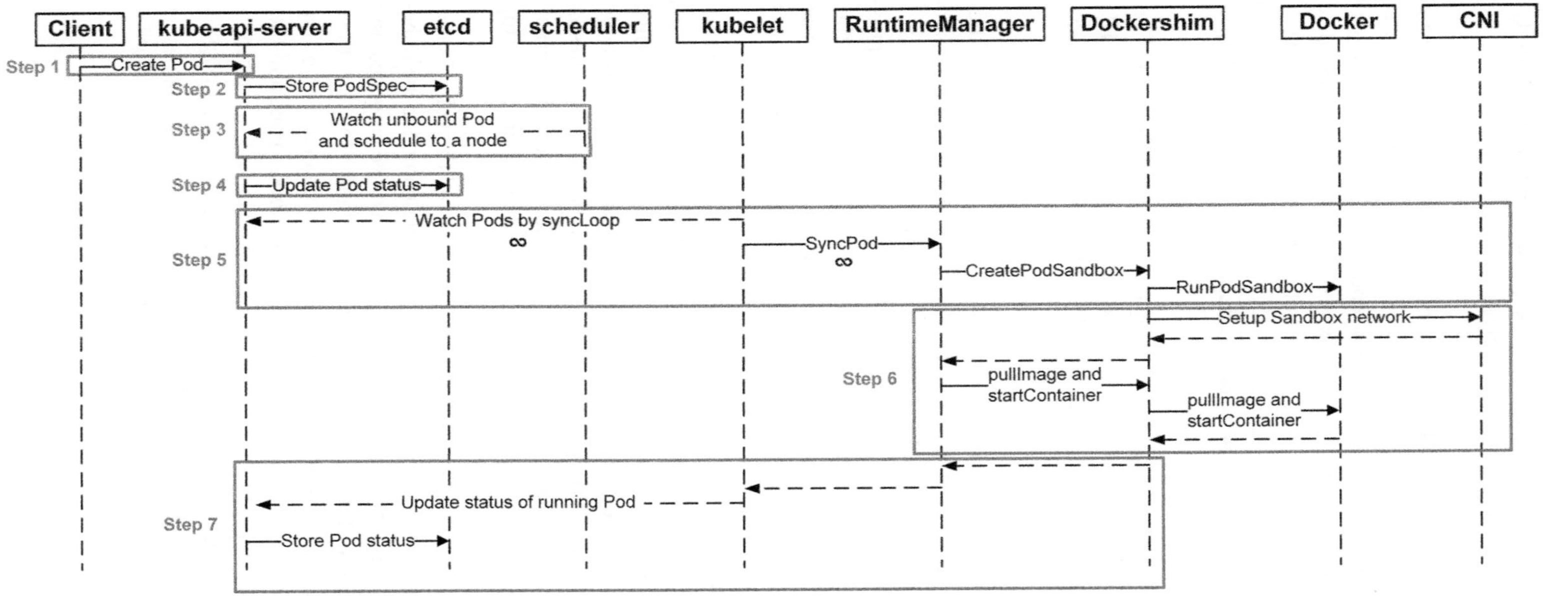

Figure 5.41 **Creation Process of a Pod.** *Source:* Reprinted with permission from Ref. [7]; © 2020 The Kubernetes authors, used under CC BY 4.0.

Step 3: The scheduler watches the unbound Pod through the kube-api-server and tries to assign a node to the Pod.

Step 4: After the scheduler finds the node with the maximal priority, the Pod is bound to this node. Then, the Pod's status is updated in etcd.

Step 5: The kubelet performs the Pod creation process according to the scheduling decision. After the binding process is successful, the scheduler calls the API of the kube-api-server to create a bound Pod object. The kubelet on each worker node regularly synchronizes the bound Pod's information with etcd by calling the API of the kube-api-server.

Step 6: The kubelet instructs RuntimeManager to maintain the different runtime implementations. After the Pod's sandbox is created, and the network is set up, the container is up and running.

Step 7: The running Pod's status is forwarded to Kubelet and is stored to etcd by calling the API of the kube-api-server.

5.3.1.3 Kubernetes Objects

Some essential Kubernetes objects, such as Pod, Label and Label Selector, Service, and Job, are introduced as follows.

- **Pod**
  A Pod object is an atomic unit of deployment in Kubernetes and can provide shared resources for running containers. All the containers within the same Pod share the local storage and the same namespace with the same IP address and port space. The communication between containers within a Pod could use localhost or standard inter-process communications.
- **Label and Label Selector**
  A label object is a key-value pair for grouping objects. Each object may have multiple labels. Each label on an object has to be unique, and each label may be applied to different objects. In the meantime, a label selector selects objects based on their labels.
- **Service**
  A service is an abstract object for publishing the functions of a containerized application running on a set of Pods to the users or other services. In general, native services provided by Kubernetes are exposed by endpoints. Kubernetes could provide the service discovery mechanism and load balance for services by default. Users could treat services as Representational State Transfer (REST) objects and access them by DNS or environment variables.
- **Job**
  In Kubernetes, a job is a resource that runs a Pod or several Pods to execute a task. When the job controller finds the arrival of a new task, the kubelet starts working on a group of Pods to complete this workload. At the same time, the job controller sends messages to the kube-api-server to ask the Kubernetes control plane to create or delete Pods instead of doing these by itself.

Kubernetes contains a mechanism (namely, container runtime) to handle the execution of containers. A container runtime is a piece of software for running containers. There are several container runtimes supported by Kubernetes Container Runtime Interface (CRI), such as Docker [1], containerd [13], CRI-O [12], rktlet [43], and singularity-cri [44]. Kubernetes CRI consists of the specifications and requirements, protobuf [45] API, and runtime libraries to integrate with the kubelet on a worker node. Kubernetes also provides a file system with one or more storage volumes in an image. After version 1.7, Kubernetes adopts the CRI-based implementation for supporting various CRI-compliant container runtimes. Kubernetes applies the gRPC [46] framework to cooperate with kubelet to manage container runtimes on a node.

Although containers are responsible for executing workloads, a Pod is the basic unit for deploying and executing containerized applications. A Pod is composed by one or multiple containers, storage resources, network identity, and other configurations. When an application needs multiple co-working containers for execution, a Pod could encapsulate multiple containers with shared resources to work together. When the application would like to be scaled horizontally, users could run multiple Pods to increase the application's replicas.

After a Pod is created, it would be assigned a unique ID (UID) and be scheduled to one node. Each Pod has a unique IP address. Containers within a Pod share the network namespace and port space and could reach each other via localhost or standard inter-process communications. Containers in different Pods have distinct IP addresses, and they could reach each other via their Pod IP addresses. The containerized application within a Pod could access the shared storage volumes in the Pod. Applications within a Pod could also access the volumes mounted to the Pod via the filesystem.

5.3.2 Kubernetes Operational Principles

This section introduces seven essential functionalities of Kubernetes, including deployment, high availability, ingress, self-healing, replication, scheduler, and autoscaling for handling large scale clusters, each of which is described as follows.

5.3.2.1 Deployment

After a Kubernetes cluster is created, users could deploy their containerized applications on this cluster. Before deploying the containerized applications, users have to create a Kubernetes deployment configuration to guide Kubernetes to build application instances. After creating the application instances, the Kubernetes control plane schedules them to worker nodes in the cluster. In the meantime, a Kubernetes deployment controller starts to monitor the status of these application instances. When a node that hosts an application instance malfunctions, the deployment controller will launch a new application instance on another node to substitute the un-working one. This mechanism provides the capability of self-healing for node failure.

Kubectl provides a command-line interface to manage the deployment configuration. When a new Pod is deployed to a Kubernetes cluster, the user has to input the corresponding command to create a Pod through Kubectl. This command will be passed to the kube-api-server in the control plane after checking the authentication to confirm the user's identity. Then, the kube-api-server backups the command to etcd and then sends a message to the kube-controller manager for creating a new Pod. After receiving the message, the kube-controller manager checks the resource status and creates a new Pod when the resource is enough. When visiting the API server regularly, the scheduler asks the kube-controller manager whether there is a new Pod. If a newly-created Pod is found, the scheduler will deliver this Pod to a suitable node.

5.3.2.2 High Availability and Self-Healing

Kubernetes applies the replication approach to avoid the single point of failure and to achieve high system reliability. In general, a single-control plane cluster is prone to fail due to the control plane node's malfunctions. Consequently, production or production-like environments should use a multiple-control plane Kubernetes cluster for improving the cluster's stability and reliability. Because the Kubernetes control plane adopts the Raft consensus approach to work together, the

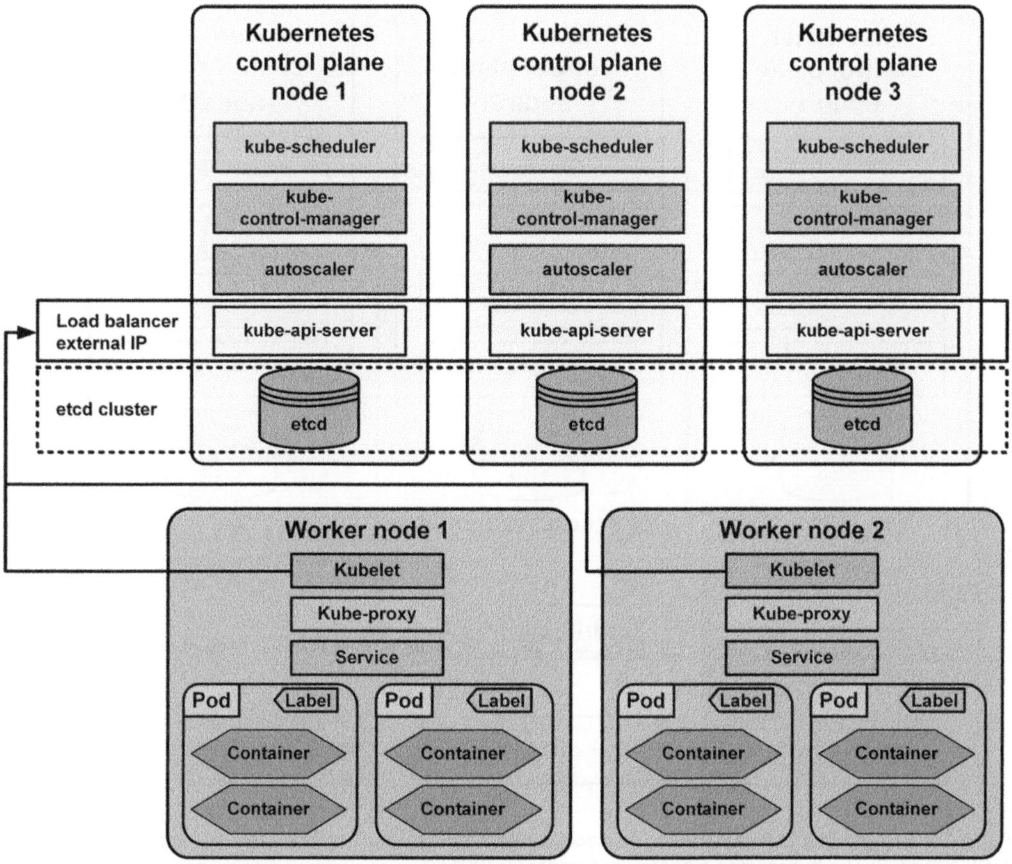

Figure 5.42 System architecture with three Control Plane Nodes. *Source:* Reprinted with permission from Ref. [7]; © 2020 The Kubernetes authors, used under CC BY 4.0.

number of control plane nodes should be odd, usually a minimum of three. When the leader control plane fails, another control plane will take over the leader's role to ensure the operations of the cluster. Figure 5.42 shows the system architecture with three replicas of the control plane. When component failures occur in a dynamic system, we usually want the system to be able to heal itself. Kubernetes replication controllers and replica sets are great examples of self-healing systems. Self-healing starts with automated detection of problems followed by automated resolution. This means that they could be deployed as a cluster (typically across three or five nodes) by the replication controller.

Furthermore, etcd [33] has two topological architectures to implement a Kubernetes cluster's high availability: the stacked etcd topology and the external etcd topology.

- **Stacked etcd Topology**

  Figure 5.43 shows the system architecture with the stacked etcd topology. The stacked etcd topology puts etcd on the same node together with the Kubernetes control plane. Each Control Plane has a local etcd member. This topology simplifies the setup process and replica management. This architecture requires a minimum of three Control Plane nodes to achieve high availability.

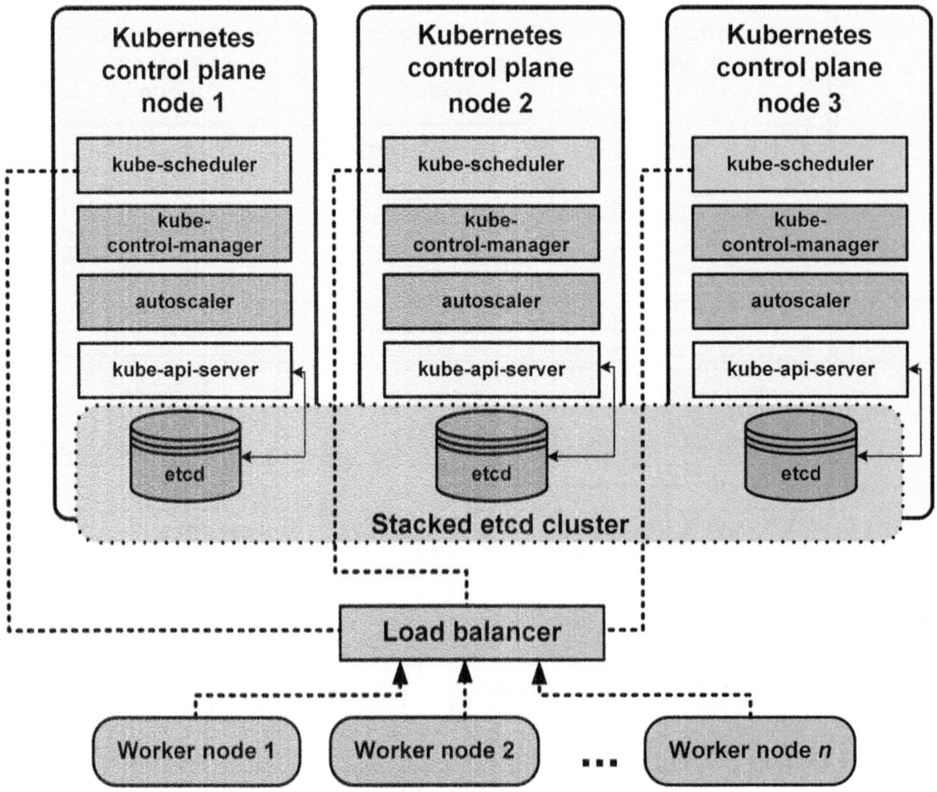

Figure 5.43 System architecture with the stacked etcd topology. *Source:* Reprinted with permission from Ref. [7]; © 2020 The Kubernetes authors, used under CC BY 4.0.

- **External etcd Topology**

  Figure 5.44 shows the system architecture with the external etcd topology. The external etcd topology accesses the external etcd cluster via kube-api-server. In such an architecture, the distributed data storage cluster is separated from the control plane. Each etcd is on a separate host and communicates with the control plane via the kube-api-server. Thus, the system reliability of the external etcd topology is better than that of the stacked etcd topology. However, this architecture needs more machines to support this topology, which needs a minimum of three hosts for the control plane and three hosts for the etcd.

5.3.2.3 Ingress

In general, the services on Pods are accessed via port-mapping when multiple services are running on a node at the same time, as shown in Figure 5.45. Each machine in either public or private clouds equips with its own firewall. This means whenever service objects are created or deleted, the firewall settings have to be adjusted additionally. Therefore, Kubernetes provides ingress to solve this problem. By applying ingress, only an external port number is required for accessing services. Ingress could assign different paths in the configuration to forward the user's requests to different service objects, as shown in Figure 5.46. Ingress controls the routing paths of HTTP or HTTPS by pre-defined rules and forwards the external requests to services within the cluster. The ingress controller is in charge of handling the ingress through the load balancer. Without the ingress controller, an ingress resource is useless.

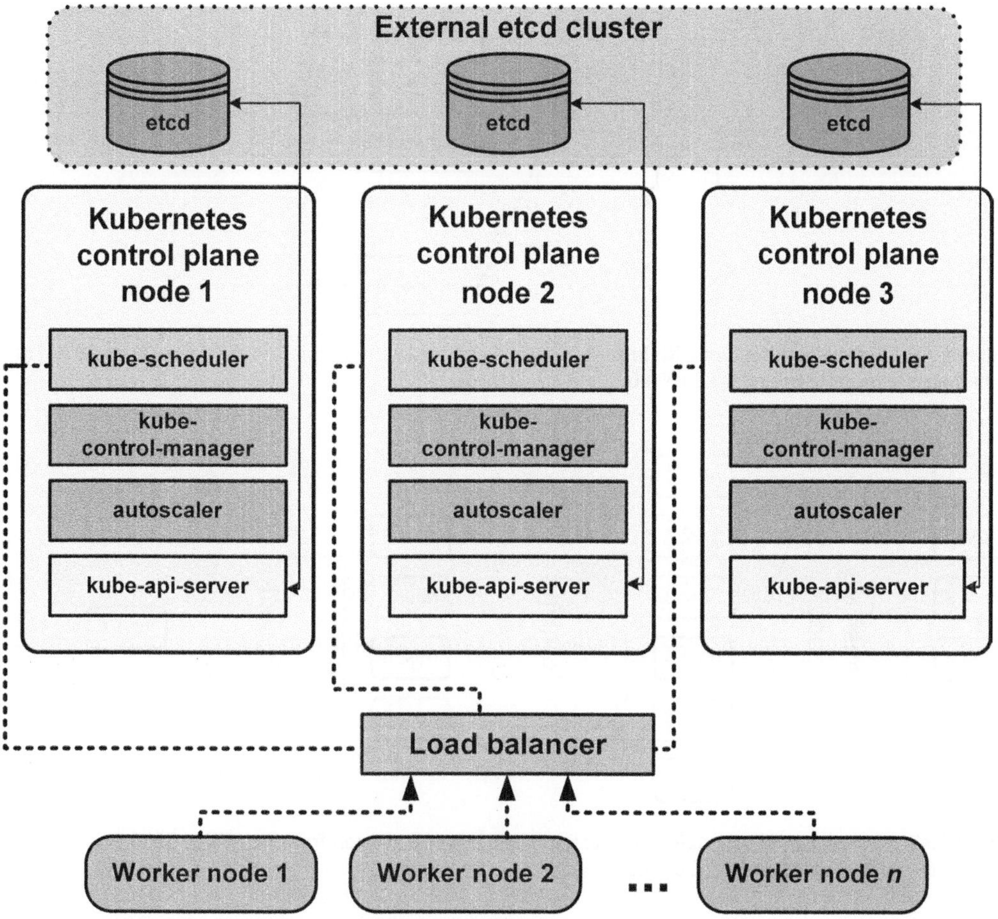

Figure 5.44 System architecture with the external etcd topology. *Source:* Reprinted with permission from Ref. [7]; © 2020 The Kubernetes authors, used under CC BY 4.0.

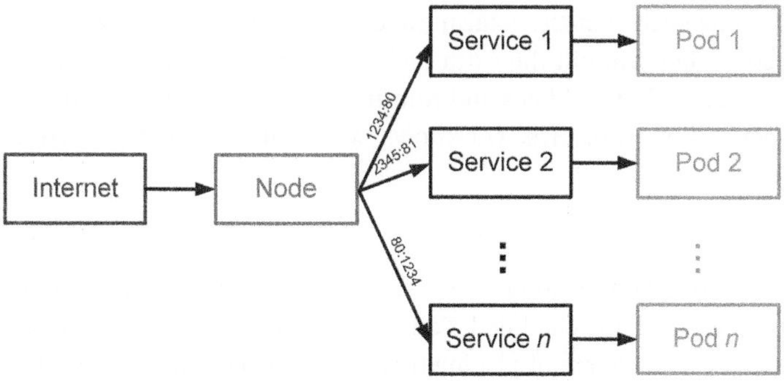

Figure 5.45 System architecture without ingress. *Source:* Reprinted with permission from Ref. [7]; © 2020 The Kubernetes authors, used under CC BY 4.0.

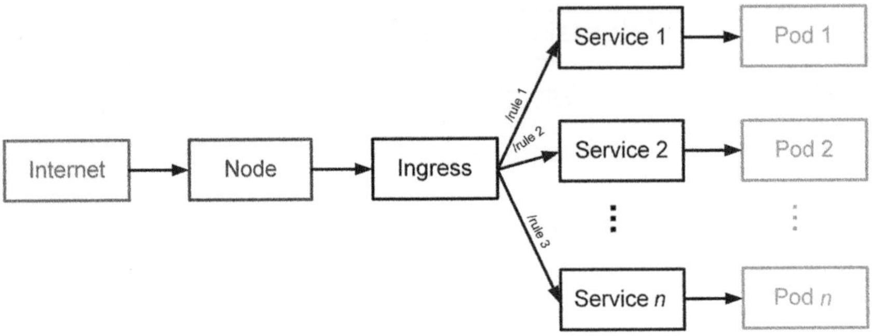

Figure 5.46 System architecture with ingress. *Source:* Reprinted with permission from Ref. [7]; © 2020 The Kubernetes authors, used under CC BY 4.0.

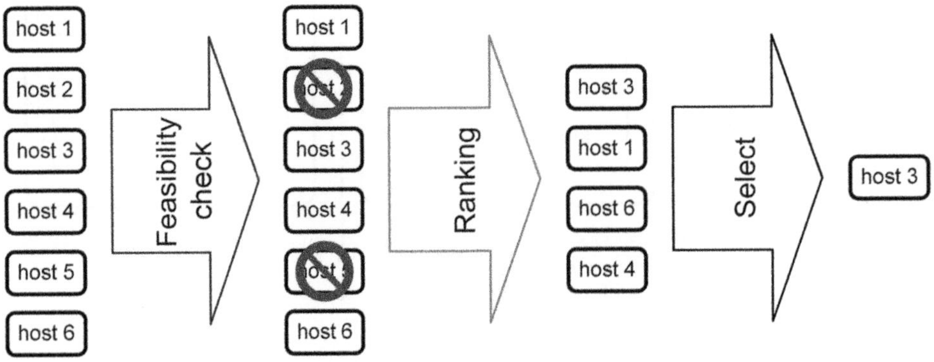

Figure 5.47 Two phases of scheduler. *Source:* Reprinted with permission from Ref. [7]; © 2020 The Kubernetes authors, used under CC BY 4.0.

5.3.2.4 Replication

Kubernetes guarantees that you always have the desired number of running Pods, which is specified in a replication controller or a replica set. Whenever the number of running Pods of an application drops because of a problem with the hosting node or the Pod itself such that the desired state of the application changes, Kubernetes will create new Pods to keep up with the desired state. Similarly, if you manually create Pods such that the total number of running Pods exceeds the desired number, the replication controller will kill the extra Pods. Replication controllers play a crucial role in many workflows, such as rolling updates and running one-off jobs. As Kubernetes evolved, it introduced direct support for many of these workflows, with dedicated objects such as Deployment, Job, and DaemonSet.

5.3.2.5 Scheduler

The scheduler is responsible for allocating Pods to suitable worker nodes. The scheduler regularly monitors whether there are newly-created unscheduled Pods or not. When an unscheduled (unbound) Pod is found, the scheduler is in charge of placing this Pod to the best-suitable worker node for execution.

Kubernetes's default scheduler is kube-scheduler, which can be replaced with the user's desired scheduler. Figure 5.47 shows the two phases of a scheduler. In the first phase, the existing nodes without enough resources are filtered out. In this phase, the remaining nodes are called feasible nodes for a newly created Pod or an unscheduled (unbound) Pod. If there is no feasible node, this Pod keeps unscheduled until a node with enough resources is found.

In the second phase, the feasible nodes are scored by a set of functions, and the feasible node with the highest score would be the candidate node to run this Pod. After the candidate node is selected, the scheduler informs the kube-api-server of the scheduling decision. These scoring functions' parameters include individual and collective resource requirements, hardware/software constraints, policy constraints, affinity and anti-affinity specifications, data locality, inter-workload interference, and so on.

5.3.2.6 Autoscaling

Kubernetes provides the capability of horizontal autoscaling of Pods. To avoid interfering with the replication controller or deployment, the autoscaling controller doesn't create or destroy Pods directly. Kubernetes monitors the status of Pods. When the CPU utilization or some metric crosses a pre-defined threshold, the autoscaling controller starts to adjust the number of replicas of Pods via the replication controller or deployment resources.

CPU utilization is the most common metric for the threshold. When too many requests are forwarded to a Pod, the number of Pod's replicas has to increase. On the contrary, when the CPU utilization of a Pod is low, the number of Pod's replicas could be decreased. Other factors for defining the threshold may include memory usage, the depth of the internal on-disk queue, the average latency per request, the average number of service timeouts, and so on.

5.3.3 Illustrative Applications of Kubernetes

This subsection introduces illustrative applications of Kubernetes to demonstrate the automatic management of Pods on the cluster with multiple worker nodes. Users could specify the resource requirements and configurations to easily start a Kubernetes cluster and run the services on this containerized cluster.

In general, there are two kinds of applications provided by Kubernetes: stateless applications and stateful applications. When an application doesn't handle states in the Kubernetes cluster, the application is called a stateless one. On the contrary, when an application manages its state in the Kubernetes cluster, the application is a stateful one. A stateless application records all of its states on the outside of the cluster. A distributed application usually contains multiple processes executed on multiple machines in parallel. When the distributed application serves requests with massive data volumes over the network, keeping the state of the distributed application within the Kubernetes cluster is not a good idea. Because applying the stateless approach could avoid binding the client to the server, the stateless application could be scaled horizontally by deploying it on multiple hosts.

Before introducing the illustrative application of Kubernetes, we need to build a Kubernetes cluster by the kubectl command-line tool. In general, users have to set up the control plane and worker nodes. In the control plane node, the Docker and Kubernetes are installed first. Then, the Kubernetes cluster is started up, and the configuration of the network is set up. The dashboard could be installed if necessary. After the control plane node is ready, we can follow the command given below to check the control plane node's status.

kubectl get pods--all-namespaces

This command could fetch all pods in all namespaces, and show the status of the control plane node, including namespace, name, ready status, running status, and age. As shown in Figure 5.48,

```
aaa@ubuntu:~$ sudo kubectl get pods --all-namespaces
NAMESPACE              NAME                                         READY   STATUS    RESTARTS   AGE
kube-system            coredns-6dcc67dcbc-nxbfj                     1/1     Running   0          9m51s
kube-system            coredns-6dcc67dcbc-w2j6x                     1/1     Running   0          9m51s
kube-system            etcd-ubuntu                                  1/1     Running   0          8m54s
kube-system            kube-apiserver-ubuntu                        1/1     Running   0          8m43s
kube-system            kube-controller-manager-ubuntu               1/1     Running   0          9m6s
kube-system            kube-flannel-ds-amd64-zn5bl                  1/1     Running   0          9m52s
kube-system            kube-proxy-rpk47                             1/1     Running   0          9m52s
kube-system            kube-scheduler-ubuntu                        1/1     Running   0          8m46s
kubernetes-dashboard   dashboard-metrics-scraper-7f669d7fbb-fwxwj   1/1     Running   0          4m19s
kubernetes-dashboard   kubernetes-dashboard-5845bf99b5-wwhz9        1/1     Running   0          4m19s
```

Figure 5.48 Ready status of the control plane node.

there are two core DNS, one etcd, one kube-api-server, one kube-controller-manager, one kube-flannel, one kube-proxy, one kube-scheduler, one dashboard-metrics-scraper, and one Kubernetes-dashboard. In the meantime, Figure 5.49 shows the dashboard depicting that the control plane node is ready.

In general, we need to establish the bidirectional trust between a node joining the cluster and a control plane node. We can follow the instruction given below to generate the join token for the control plane,

kubeadm token create--print-join-command

where the option "--print-join-command" is to print the full 'kubeadm join' flag needed to join the cluster using the token, as shown in Figure 5.50.

After generating the join token of the control plane node, the worker node could join the cluster using the join token issued from the control plane node, by copying and pasting the command wrapped in the red rectangle in Figure 5.50, as shown in Figure 5.51. Users could check worker nodes' status by the kubectl command-line tool, as shown in Figures 5.52. In Figure 5.52, the status of worker1 is "ready" when worker1 successfully joins the cluster.

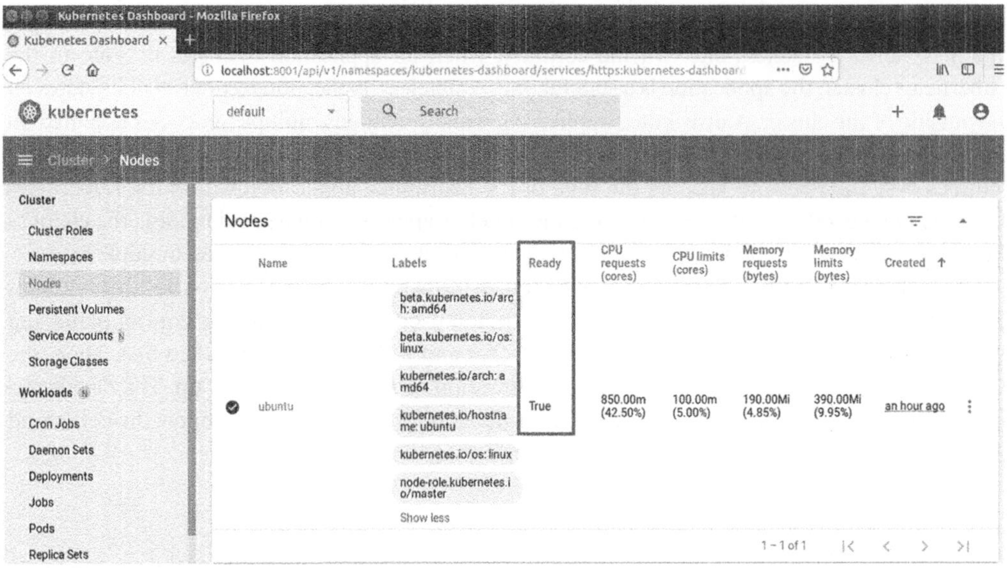

Figure 5.49 Dashboard while the control plane node is ready.

```
aaa@ubuntu:~$ sudo kubeadm token create --print-join-command
[sudo] password for aaa:
kubeadm join 192.168.132.128:6443 --token 1gd1jy.j4j4p73gaxa8vhq1     --discover
y-token-ca-cert-hash sha256:6a924f6c462580a94c710c097b97be129bd7f7f0e85f3bf23334
77b441b35d7b
aaa@ubuntu:~$
```

Figure 5.50 Generating the join token of the control plane.

```
aaa@worker1:~/Desktop$ sudo kubeadm join 192.168.132.128:6443 --token 1gd1jy.j4j
4p73gaxa8vhq1     --discovery-token-ca-cert-hash sha256:6a924f6c462580a94c710c09
7b97be129bd7f7f0e85f3bf2333477b441b35d7b
sudo: unable to resolve host worker1
[preflight] Running pre-flight checks
        [WARNING IsDockerSystemdCheck]: detected "cgroupfs" as the Docker cgroup
 driver. The recommended driver is "systemd". Please follow the guide at https:/
/kubernetes.io/docs/setup/cri/
        [WARNING SystemVerification]: this Docker version is not on the list of
validated versions: 19.03.12. Latest validated version: 18.09
        [WARNING Hostname]: hostname "worker1" could not be reached
        [WARNING Hostname]: hostname "worker1": lookup worker1 on 8.8.8.8:53: no
 such host
[preflight] Reading configuration from the cluster...
[preflight] FYI: You can look at this config file with 'kubectl -n kube-system g
et cm kubeadm-config -oyaml'
```

Figure 5.51 Worker node joining the cluster by the join token.

```
aaa@ubuntu:~$ sudo kubectl get nodes
NAME      STATUS   ROLES    AGE   VERSION
ubuntu    Ready    master   95m   v1.14.10
worker1   Ready    <none>   10m   v1.14.10
worker2   Ready    <none>   50s   v1.14.10
aaa@ubuntu:~$
```

Figure 5.52 Status of the cluster shown by kubectl.

Users also could observe that worker1 successfully joins the cluster from the dashboard, as shown in Figure 5.53.

The service deployment of Apache HTTP Server [47] is demonstrated by the following yet another markup language (YAML) [48] file, as shown in Figure 5.54. YAML is a human-readable data-serialization language commonly used for configuration files. Kubernetes uses the .yaml file to specify the object you want to create. Generally, the following fields have to set values: apiVersion (the API version for creating this object), kind (the kind of object), metadata (data for identifying the object, including a name string, UID, and optional namespace), and spec (the object specification).

Users could deploy a service by using the following command:

kubectl apply -f example.yaml

where the "-f example.yaml" indicates the configuration file name. After executing this command, the deployment object, named "httpd-deployment," and the service object, named "httpd," are created, as shown in Figure 5.55. In the meantime, the dashboard shows the status of Pods and their

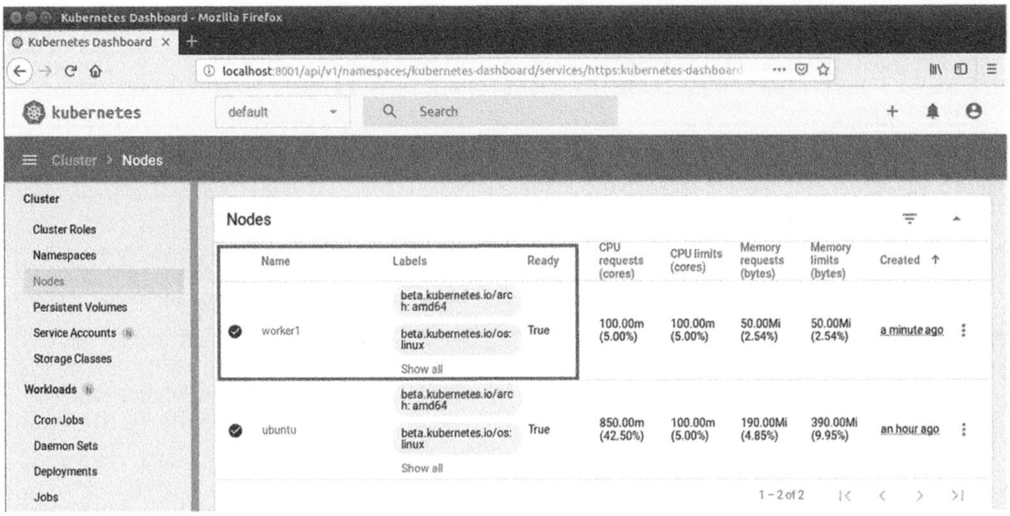

Figure 5.53 Status of the worker1 shown in dashboard.

```
apiVersion: apps/v1
kind: Deployment                  Type of Kubernetes component (Deployment)
metadata:
  name: httpd-deployment          Deployment name
  labels:
    app: httpd
spec:
  replicas: 3                     Number of replica
  selector:
    matchLabels:
      app: httpd
  template:
    metadata:
      labels:                     Labels attached to this component
        app: httpd
    spec:
      containers:
      - name: httpd               Image used for this deployment
        image: httpd
        ports:
        - containerPort: 80       Port used by this deployment
---
apiVersion: v1
kind: Service
metadata:
  name: httpd
  labels:
    app: httpd
spec:
  ports:
  - port: 80
    targetPort: 80
  selector:
    app: httpd
```

Figure 5.54 Example of a YAML file

```
aaa@ubuntu:~$ cd Desktop/
aaa@ubuntu:~/Desktop$ sudo kubectl apply -f example.yaml
deployment.apps/httpd-deployment created
service/httpd created
aaa@ubuntu:~/Desktop$ 
```

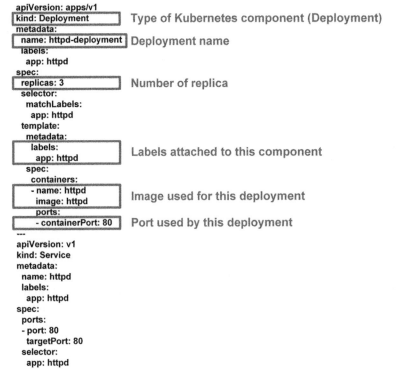

Figure 5.55 Deploying the httpd service by applying example.yaml.

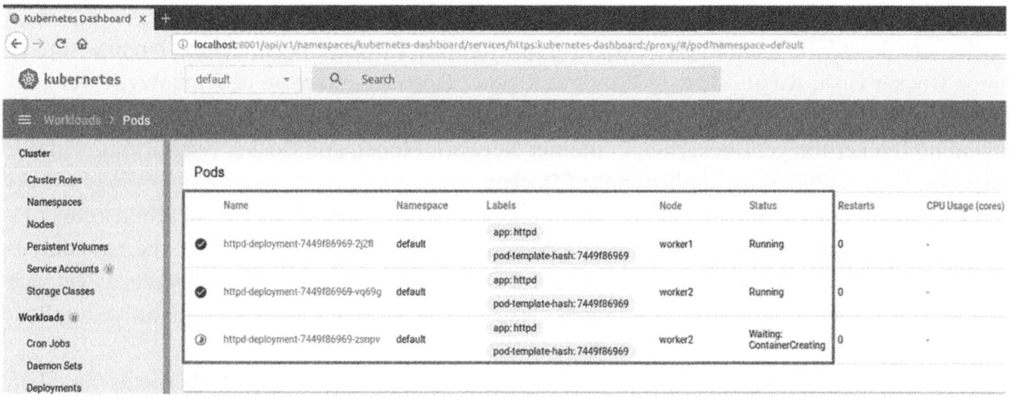

Figure 5.56 Status of Pods shown in dashboard.

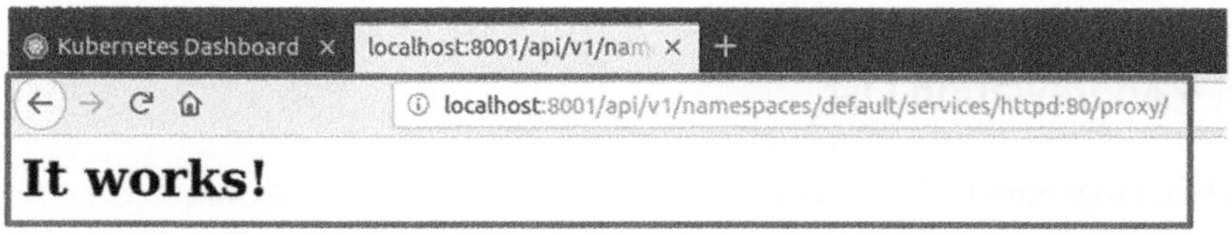

Figure 5.57 Screenshot displaying a workable httpd service.

services, as shown in Figure 5.56. Users could access the httpd service by visiting http://localhost:8001/api/v1/namespaces/default/services/httpd:80/proxy/. Figure 5.57 shows that the service is workable.

5.3.4 Summary

Kubernetes provides the capability of automatic deployment, scaling, and management of containerized workloads and services. It could efficiently orchestrate the cluster resources to satisfy the resource requirement of Pods and services. This section has introduced the fundamentals and essential operational principles of Kubernetes. There are more concepts and materials of Kubernetes on the Kubernetes website and online documents, which are worth to explore.

5.4 Conclusion

This chapter has introduced the fundamentals of Docker and Kubernetes to serve as part of the foundations for Chapters 6, 7, and 9, where Docker, Kubernetes, and various containerized applications are mentioned. In Section 5.1, we describe the importance and benefits of Docker containers, as well as the problems that Docker can solve in the software life cycle. We also explain why we should use Docker in the development and operations of our software applications. Furthermore, we portray the container orchestrator and what tasks it can automatically complete in managing containers. Finally, we briefly compare the top two container orchestrators, Kubernetes and Docker Swarm.

In Section 5.2, we introduce the fundamentals of Docker. Many essential concepts related to Docker are described, including Docker Engine, High-level Docker Architecture, Architecture of Linux Docker Host, Architecture of Windows Docker Host, Architecture of Windows Server containers, and Architecture of Hyper-V containers. We also depict Docker's operational principles, including Docker Image, Dockerfile, Container Network Model, and Docker Networking. Finally, we present some illustrative applications of Docker.

The fundamentals of Kubernetes is elaborated in Section 5.3. We depict a Kubernetes cluster's architecture, which consists of a control plane and many worker nodes. Further, the key components of the Kubernetes control plane and a Kubernetes worker node are described. We also present several essential working principles of Kubernetes. Finally, some illustrative applications of Kubernetes are presented to help readers to better understand.

This chapter has introduced the fundamentals and benefits of Docker and Kubernetes. Based on this chapter, readers should be able to comprehend the contents related to these two technologies in this book. Hopefully, this chapter can motivate the readers to further explore Docker and Kubernetes before adopting them in the development and operation of software applications.

Appendix 5.A – Abbreviation List

API	Application Programming Interface
APPs	Applications
Bins	Binary Codes
CLI	Command-Line Interface
CNCF	Cloud Native Computing Foundation
CNM	Container Network Model
CRI	Container Runtime Interface
DevOps	Development and Operations
DNS	Domain Name System
ECR	Elastic Container Registry
GCR	Google Container Registry
HNS	Host Network Service
HTTP	Hypertext Transfer Protocol
IaaS	Infrastructure as a Service
IPAM	IP Address Management
Libs	Libraries
MCR	Microsoft Container Registry
NAT	Network Address Translation
OS	Operating System
REST	Representational State Transfer
UID	Unique ID

UFS	Union File System
vNIC	Virtual Network Adapter
vSwitch	Virtual Switch
VXLAN	Virtual Extensible LAN
WSCs	Windows Server Containers
YAML	Yet Another Markup Language

References

1 Docker Inc (2020). Docker docs. docs.docker.com (accessed 6 October 2020).
2 Schenker, G.N. (2020). *Learn Docker – Fundamentals of Docker 19.x*, Seconde. UK: Packt Publishing.
3 Morabito, R. (2017). Virtualization on Internet of Things edge devices with Container technologies: a performance evaluation. *IEEE Access* 5: 8835–8850. https://doi.org/10.1109/ACCESS.2017.2704444.
4 Avi. (2019). Docker vs Virtual Machine – Understanding the differences. https://geekflare.com/docker-vs-virtual-machine/ (accessed 13 October 2020).
5 Syrewicze, A. (2019). What is a Container Orchestrator? https://www.altaro.com/msp-dojo/container-orchestrator/ (accessed 13 October 2020).
6 Bernstein, D. (2014). Containers and cloud: From LXC to Docker to Kubernetes. *IEEE Cloud Computing* 1 (3): 81–84. https://doi.org/10.1109/MCC.2014.51.
7 The Kubernetes authors (2020). Kubernetes documentation. https://kubernetes.io/docs/home/ (accessed 6 October 2020).
8 Docker Inc (2020). Swarm mode overview. https://docs.docker.com/engine/swarm/ (accessed 13 October 2020).
9 The Linux Foundation (2020). Cloud Native Computing Foundation. www.cncf.io (accessed 6 October 2020).
10 Binani, H. (2018). Kubernetes vs Docker Swarm—A comprehensive comparison. https://hackernoon.com/kubernetes-vs-docker-swarm-a-comprehensive-comparison-73058543771e/ (accessed 13 October 2020).
11 Mangat, M. (2019). Kubernetes vs Docker Swarm: What are the differences? https://phoenixnap.com/blog/kubernetes-vs-docker-swarm/ (accessed 13 October 2020).
12 Cloud Native Computing Foundation (2020). CRI-O. cri-o.io (accessed 6 October 2020).
13 Containerd authors (2020). Containerd. containerd.io (accessed 6 October 2020).
14 Docker Inc (2020). Docker Overview. https://docs.docker.com/get-started/overview/ (accessed 13 October 2020).
15 Docker Inc (2020). Docker Hub. hub.docker.com (accessed 13 October 2020).
16 Microsoft (2020). Microsoft Container registry. https://azure.microsoft.com/en-us/services/container-registry/ (accessed 13 October 2020).
17 Google (2020). Google Container registry. https://cloud.google.com/container-registry/ (accessed 13 October 2020).
18 Amazon Web Services, Inc (2020).Amazon Elastic Container registry. https://aws.amazon.com/tw/ecr/ (accessed 13 October 2020).

19 Raina, A. (2016). A comparative study of Docker engine on Windows server Vs Linux platform. https://collabnix.com/a-comparative-study-of-docker-engine-on-windows-server-vs-linux-platform/ (accessed 13 October 2020).

20 McCabe, J. and Friis, M. (2017). *Introduction to Windows Containers*. Redmond, WA: Microsoft Press https://docs.microsoft.com/zh-tw/archive/blogs/microsoft_press/free-ebook-introduction-to-windows-containers (accessed 13 October 2020).

21 Docker Inc (2020). Docker images. https://docs.docker.com/engine/reference/commandline/images/ (accessed 13 October 2020).

22 Kisller, E. (2020). A beginner's guide to understanding and building Docker images. https://jfrog.com/knowledge-base/a-beginners-guide-to-understanding-and-building-docker-images/ ().

23 Docker Inc (2020). Docker storage drivers. https://docs.docker.com/storage/storagedriver/select-storage-driver/ (accessed 13 October 2020).

24 Docker Inc (2020). Dockerfile reference. https://docs.docker.com/engine/reference/builder/ (accessed 13 October 2020).

25 Docker Inc (2020). About storage drivers. https://docs.docker.com/storage/storagedriver/ (accessed 13 October 2020).

26 GitHub, Inc (2020). The moby/libnetwork/design.md on Github. https://github.com/moby/libnetwork/blob/master/docs/design.md (accessed 13 October 2020).

27 Calcote, L. (2016). The Container networking landscape: CNI from CoreOS and CNM from Docker. https://thenewstack.io/container-networking-landscape-cni-coreos-cnm-docker/ (accessed 13 October 2020).

28 Kumar, A. (2020). Docker Networking Introduction to Docker Network. https://k21academy.com/docker-kubernetes/docker-networking-different-types-of-networking-overview-for-beginners/ (accessed 13 October 2020).

29 Docker Inc (2020). Networking overview. https://docs.docker.com/network/ (accessed 13 October 2020).

30 Docker Inc (2020). Network containers. https://docs.docker.com/engine/tutorials/networkingcontainers/ (accessed 13 October 2020).

31 Messer. J. (2019). Windows Container networking. https://docs.microsoft.com/en-us/virtualization/community/team-blog/2016/20160505-windows-container-networking/ (accessed 13 October 2020).

32 SmartBear Software (2020). OpenAPI Specification. https://swagger.io/specification/ (accessed 6 October 2020).

33 Etcd authors (2020). Etcd. etcd.io (accessed 6 October 2020).

34 Amazon Web Services, Inc (2020). Amazon AWS. aws.amazon.com (accessed 6 October 2020).

35 Microsoft (2020). Microsoft Azure. azure.microsoft.com (accessed 6 October 2020).

36 The Apache Software Foundation (2020). Apache CloudStack™. cloudstack.apache.org (accessed 6 October 2020).

37 Google (2020). Google GCE. cloud.google.com (accessed 6 October 2020).

38 Aspiegel (2020). HUAWEI Cloud. cloud.huawei.com (accessed 6 October 2020).

39 VEXXHOST (2020). OpenStack. www.openstack.org (accessed 6 October 2020).

40 IBM Corp (2020). IBM Cloud. cloud.ibm.com (accessed 6 October 2020).

41 Baidu (2020). Baidu Cloud. cloud.baidu.com (accessed 6 October 2020).

42 Alibaba Cloud (2020). Alibaba Cloud. www.alibabacloud.com (accessed 6 October 2020).

43 Red Hat, Inc (2020). Rkt. https://coreos.com/rkt/ (accessed 6 October 2020).

44 Sylabs.io (2020).Singularity-cri. sylabs.io (accessed 6 October 2020).

45 Google (2020). Protocol Buffers. https://developers.google.com/protocol-buffers/ (accessed 6 October 2020).

46 GRPC Authors (2020). gRPC. https:grpc.io (accessed 6 October 2020).

47 Docker Inc (2020). Apache HTTP Server Project. https://hub.docker.com/_/httpd (accessed 6 October 2020).

48 YAML. yaml.org (accessed 6 October 2020).

6

Intelligent Factory Automation (iFA) System Platform

Fan-Tien Cheng

Director/Chair Professor, Intelligent Manufacturing Research Center/Institute of Manufacturing Information and Systems, National Cheng Kung University, Tainan, Taiwan, ROC

6.1 Introduction

In the Industry 4.0 era, a smart factory can adopt IoT as its data collection and communication platform. Also, CPSs with predictive algorithms could be constructed on the platform to cooperate and interact seamlessly with real and virtual worlds to provide intelligent capabilities, such as virtual metrology (VM), equipment prognosis, and predictive maintenance. In addition, cloud computing can be leveraged to encapsulate and virtualize distributed manufacturing resources into various cloud manufacturing (CMfg) services that can then be used on demand over networks to support manufacturing activities ranging from product design, simulation, manufacturing, testing, management, and all other tasks in the product life cycle [1, 2]. Thus, how to build an Intelligent Manufacturing platform that can facilitate realizing smart factories is essential and desirable for current manufacturing industries.

A five-stage approach as shown in Figure 1.13 for enhancing production yield and assuring nearly Zero Defects (ZD), with a semiconductor bumping process being the illustrative example, is proposed in Section 1.3.1. To realize the proposed five-stage approach of yield enhancement and ZD assurance, an Intelligent Factory Automation (iFA) system platform based on the so-called Advanced Manufacturing Cloud of Things (AMCoT) framework as presented below is designed.

6.2 Architecture Design of the Advanced Manufacturing Cloud of Things (AMCoT) Framework

In order to accomplish the proposed five-stage approach of yield enhancement and ZD assurance in a systematic and united manner, an AMCoT framework is developed. The architecture design of the AMCoT framework consists of two parts: factory side and cloud side, as shown in Figure 6.1. On the factory side, the so-called Cyber-Physical Agent (CPA) is developed to serve as an agent which makes cyber-physical interactions of AMCoT possible through various protocols/interfaces.

Industry 4.1: Intelligent Manufacturing with Zero Defects, First Edition. Edited by Fan-Tien Cheng.

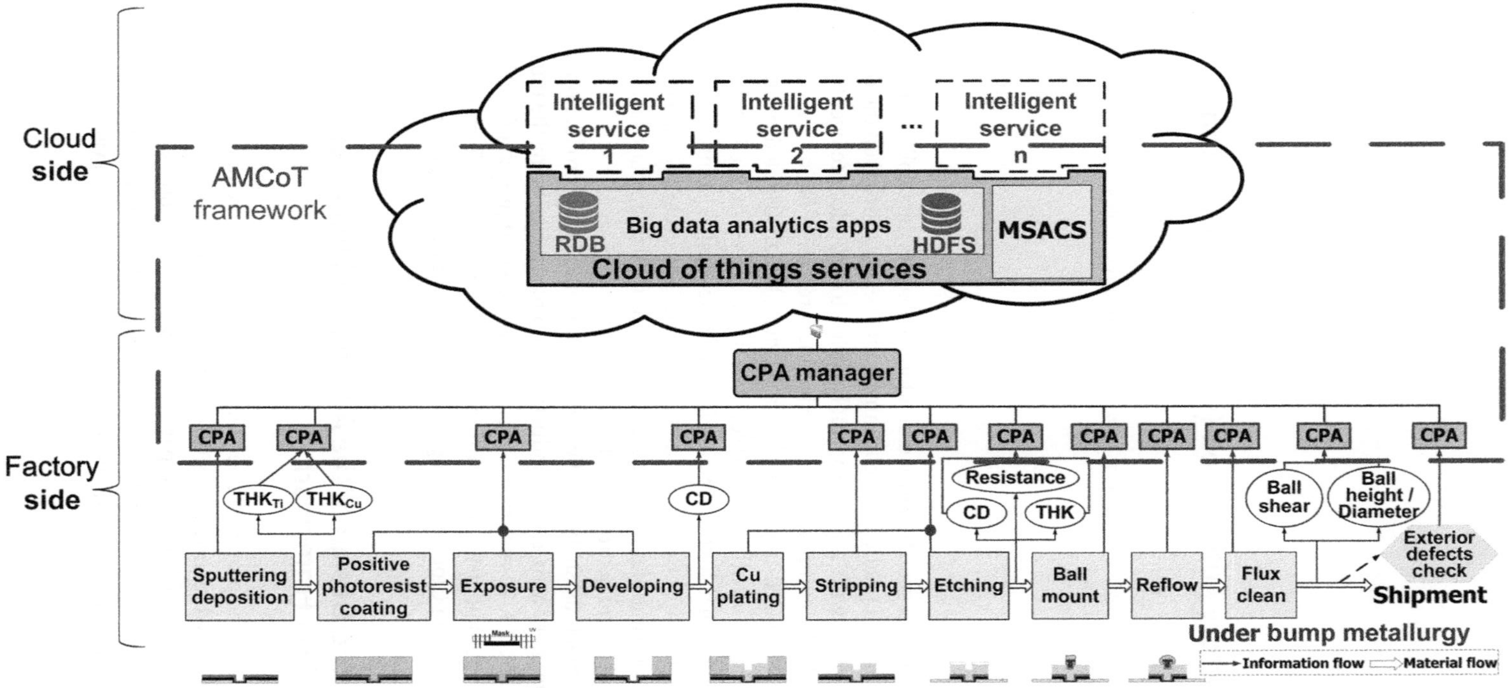

Figure 6.1 Architecture design of the AMCoT framework.

Specifically, the CPA should be able to communicate with a wide range of physical objects (e.g. machine tool, device, and sensor) for the purpose of data collection. In addition, the CPA should be capable of communicating with other CPSs to facilitate human-machine cooperation over the network. Besides, by leveraging some kinds of technologies (e.g. RESTful web services), the CPA-based IoT can be constructed to achieve the horizontal integration among CPAs in the factory side and the vertical integration between CPAs and the cloud side of AMCoT. The detailed design and implementation of CPA are depicted in Chapter 7.

On the cloud side, the big data analytics application modules as well as the interface specifications for pluggable intelligent cloud-based services are developed in the form of web services. Also, a novel automated construction scheme for developing CMfg services denoted Manufacturing Services Automated Construction Scheme (MSACS) [3] is built.

These intelligent cloud-based services are called "cloud of things services" in this study, aiming at enhancing yield of workpieces in the research-and-development (RD) phase and assuring yield as well as achieving nearly ZD of workpieces in the mass-production (MP) phase. Several typical pluggable intelligent services, such as Automatic Virtual Metrology (AVM), Intelligent Predictive Maintenance (IPM), and Intelligent Yield Management (IYM) are briefed as follows.

The AVM cloud service is developed for conducting on-line and real-time fab-wide VM on workpieces, which can be applied to realize the functions specified in Stage 3 of the MP phase. The IPM cloud service is developed for on-line gauging the health status of production tools and predicting the remaining useful lives (RULs) of the sick tools, which can be utilized to realize the missions of Stage 2 in the RD phase and Stage 4 in the MP phase. The IYM cloud service is developed to deal with the high dimensional regression problem and search for the root causes of yield losses, which can be applied to realize the goals of Stage 1 in the RD phase and Stage 5 in the MP phase.

The functions of AVM, IPM, and IYM will be briefed in Sections 6.3–6.5, respectively. The detailed design and implementation of the AMCoT framework, AVM, IPM, and IYM will then be presented in Chapters 7–10, respectively.

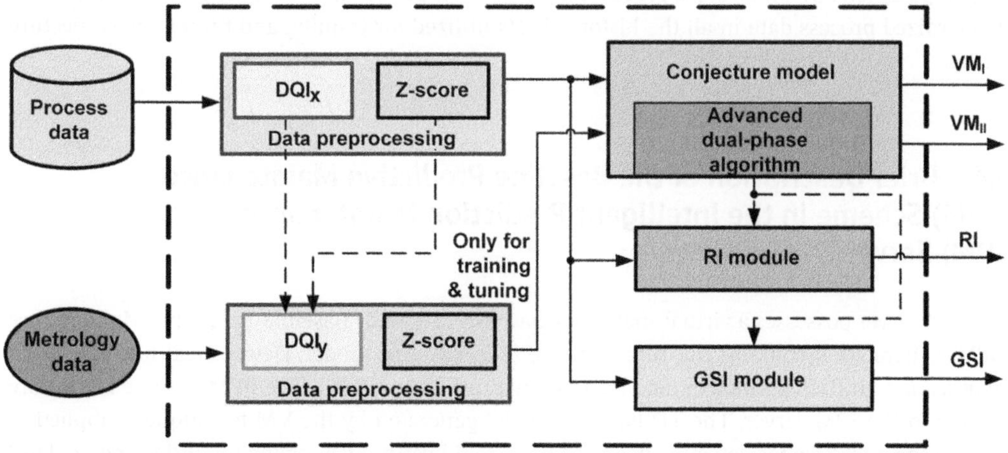

Figure 6.2 Functional block diagram of the AVM server. *Source:* Reprinted with permission from Ref. [4]; © 2012 IEEE.

6.3 Brief Description of the Automatic Virtual Metrology (AVM) Server

Observing Figure 6.2, two data preprocessing modules, a conjecture model with advanced dual-phase VM algorithm, a Reliance Index (RI) module, and a Global Similarity Index (GSI) module comprise the AVM server [4]. While process data of the current workpieces of a manufacturing process and the metrology data of the sampled workpieces being the inputs of the AVM sever, Phase-I VM (VM_I), Phase-II VM (VM_{II}), RI, and GSI values serve as its outputs. The advanced dual-phase VM algorithm, which is the core of the AVM server, will be explained below.

The advanced dual-phase VM algorithm emphasizes both promptness and accuracy at the same time. Phase I calculates and outputs the VM_I value of a workpiece as soon as all workpiece process data are completely gathered to guarantee instantaneousness. In Phase II, on the other hand, right after collecting an actual metrology value of a sampled workpiece in the virtual lot (for model refreshing), VM_{II} values of all workpieces in a virtual lot are re-calculated and outputted to enhance the VM accuracy.

By adopting two different VM prediction models, such as the back-propagation neural networks (BPNN) and partial least square (PLS), the AVM server also generates the accompanying RI and GSI of each VM_I and VM_{II}. RI value is defined between 0 and 1 to represent the percentage of intersection between the statistical normal distribution of BPNN and PLS normalized prediction values. When the two prediction values are exactly the same, the corresponding RI value is 1; and when the two values are far separated, the corresponding RI value would be 0. In other words, users can determine if the VM result is reliable by checking the RI value.

GSI stands for the similarity degree between the current-inputted set of process data and all the historical sets of process data used in the conjecture model for training and tuning purposes. The smaller the GSI value is, the better the similarity degree would be. The purpose of the GSI is to help the RI gauge the reliance level. Generally speaking, the threshold of a process in statistical control is set as triple its standard deviation. Since each standardized Z-score has mean zero and standard deviation one, the control threshold of Z is three. Therefore, the GSI threshold value (GSI_T) is set to be $3^2 = 9$. When the value of GSI exceeds GSI_T, the Individual-Similarity-Index (ISI) analysis [5] should be utilized to distinguish the key process parameters that cause major deviation. The ISI of an individual process parameter is used to indicate the similarity degree between this individual process parameter's standardized process datum of the input set and the same process parameter's standardized process data in all the historical sets utilized for training and tuning the conjecture model.

6.4 Brief Description of the Baseline Predictive Maintenance (BPM) Scheme in the Intelligent Prediction Maintenance (IPM) Server

The IPM server possesses a virtual-metrology-based (VM-based) Baseline-Predictive-Maintenance (BPM) scheme that contains the Target Device (TD) baseline model, Device Health Index (DHI) module, and an RUL predictive module as shown in Figure 6.3 [6]. The BPM scheme is the core module in the IPM server. The TD baseline model generated by the VM technique is applied to serve as the reference for detecting degradation of tool health. By applying the BPM scheme, fault diagnosis and prognosis can be accomplished [7].

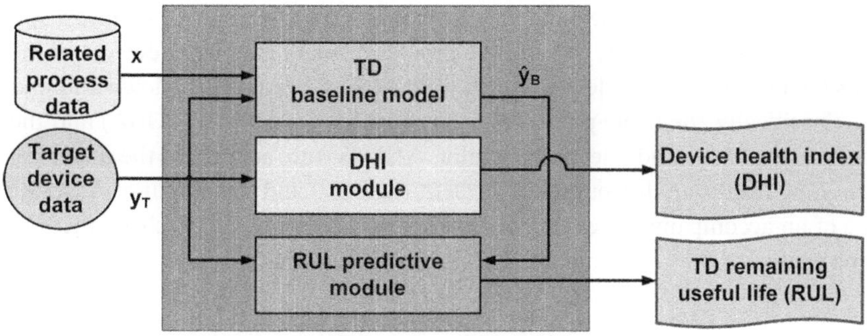

Figure 6.3 Functional block diagram of the BPM scheme in the IPM server. *Source:* Reprinted with permission from Ref. [6]; © 2017 Taylor&Francis Group.

Observing Figure 6.3, BPM generates the baseline of TD ($\hat{y}_B$) according to the important samples of process data (X) and TD data (y_T), and then, DHI is created to judge the current health state of TD. Eventually, RUL is estimated once the TD enters sick state.

A TD includes five states: initial, active, inactive, sick, and dead. Normally, TD is in active state when it is healthy. However, TD enters the sick state once y_T exceeds the threshold. TD will enter dead state from sick state if TD gets worse such that the available resources of the TD are exhausted. As a result, the equipment is down.

6.5 Brief Description of the Key-variable Search Algorithm (KSA) Scheme in the Intelligent Yield Management (IYM) Server

The IYM server utilizes the KSA (Key-variable Search Algorithm) scheme [8], as shown in Figure 6.4, to find out the root causes of yield losses in the condition of '$p>>n$.' The key principle of the KSA scheme is summarized below. The KSA contains two phases: Phase I: yield test results (**Y**) as well as total-inspection inline data (**y**) and/or production routes ($\mathbf{X_R}$) are fed into the KSA

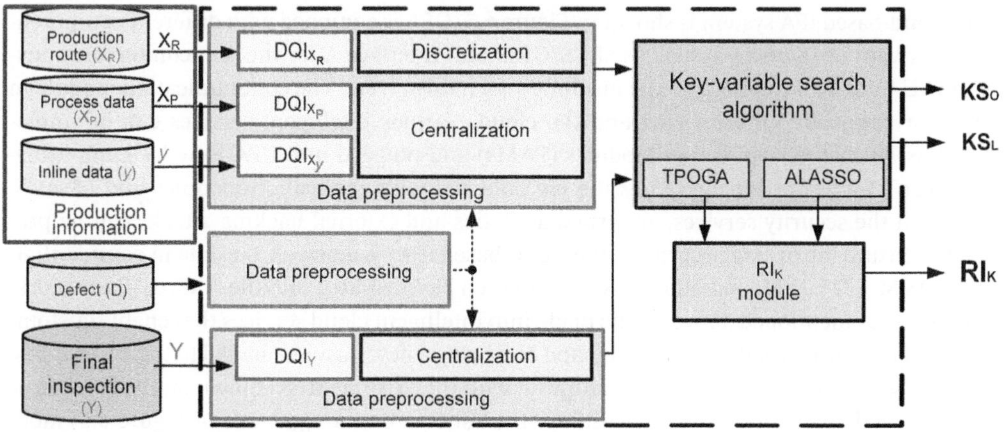

Figure 6.4 Functional block diagram of the KSA scheme in the IYM server. *Source:* Reprinted with permission from Ref. [8]; © 2017 IEEE.

scheme to find top N devices that cause the yield losses; Phase II: after obtaining the most suspicious device from Phase I, yield test results (**Y**) and the process data ($\mathbf{X_P}$) of all the devices of the stage which contains the most suspicious device are fed to the KSA scheme to drill down and find out the root causes. Specifically, the triple phase orthogonal greedy algorithm (TPOGA) and the automated least absolute shrinkage and selection operator (ALASSO) are adopted as the searching algorithms. The searching results are the output of TPOGA (i.e. KS_O) and the output of ALASSO (i.e. KS_L). The value of an accompanying reliance index (RI_K) is also computed to check the reliance of the searching results.

6.6 The iFA System Platform

The novel iFA system platform is realized by integrating technologies such as AMCoT framework with CPAs, AVM, IPM, and IYM, etc., so as to provide the total-solution package for implementing intelligent manufacturing. The iFA system serves as the game changer for both high-tech (semiconductor, flat panel display, solar cell, etc.) and traditional (machine tool, aerospace, carbon fiber, etc.) industries to upgrade Industry 4.0 to Industry 4.1 in pursuit of increasing productivity as well as realizing the goal of ZD of all products.

There are two options of the iFA system platform for the customers to choose from: "cloud-based version" and "server-based version" are provided according to different customer requirements. The contents of the two versions will be illustrated in the following sections, respectively.

6.6.1 Cloud-based iFA System Platform

The on-demand service subscription is a more acceptable solution for many small and medium-sized enterprises whose capital is less able to compete with large consortia to economically adopt the iFA system. As long as the companies install their own internet environment, they can save the high cost of building a data center on their own by just subscribing to the iFA cloud-based services with a relatively low cost of monthly fee. They can access the various intelligent manufacturing services of iFA established on one of the commercialized data centers, such as the Chunghwa Telecom Cloud Data Center.

The cloud-based iFA system is shown in Figure 6.5. CPA is equipped with different equipment drivers for various industries, such as SECS/GEM and Interface A for the semiconductor industry, OPC-UA and MT-Connect for the machine-tool industry, etc. The communication module in CPA is for connecting the internet and the cloud. Various intelligent services can be implemented as Pluggable Application Modules (PAMs) and plugged into CPA for edge computing. The AMCoT framework incorporates the big data analytics application modules and MSACS. Along with the security services, abnormal activities and external hacking attacks can be prevented to ensure the overall security of the cloud-based iFA system with the data in confidentiality. The AVM, IPM, IYM, and other services can then be integrated into the AMCoT framework. All the services mentioned above can be made into intelligent cloud services that enable customers around the world to subscribe on-demand and bring a new business model to the enterprises. The iFA system platform can also communicate with the existing MES (shown on the left side of Figure 6.5) and enterprise resource planning (ERP, shown on the right side of Figure 6.5) modules via CPAs.

The cloud-based iFA system can provide the intelligence capabilities of improving manufacturing productivity as well as enhancing product quality, and preventing unscheduled down of the

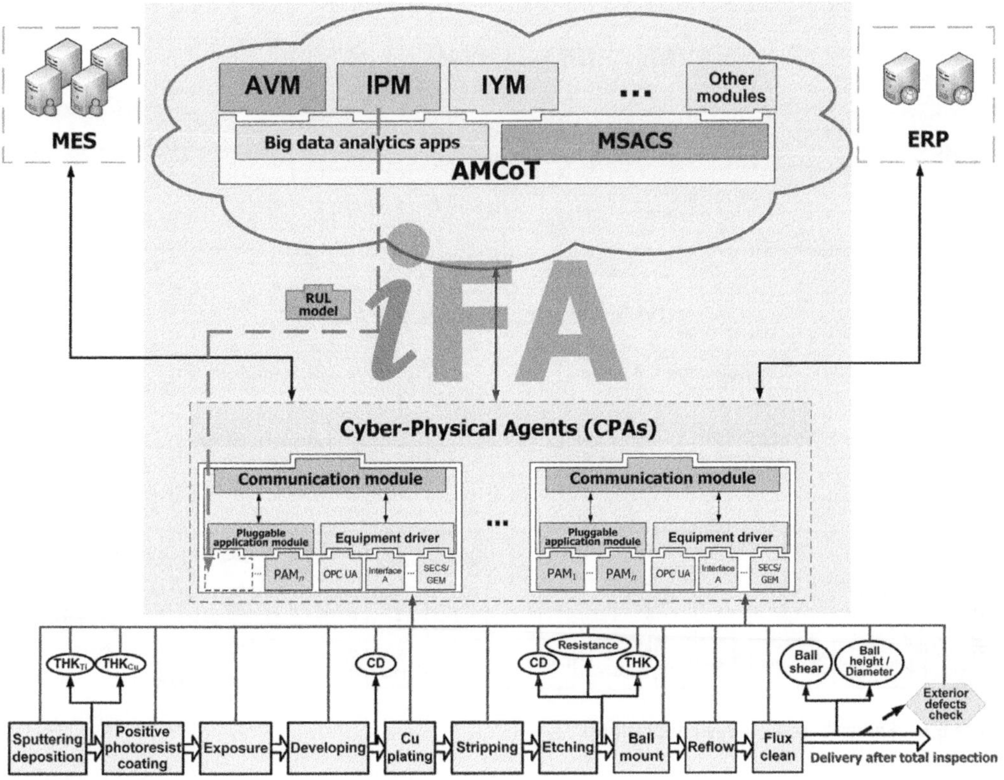

Figure 6.5 Cloud-based iFA system platform.

production machines as well as reducing the maintenance cost of the machines. As such, the goal of nearly ZD of all produced products can be achieved efficiently and economically. This is the state of the so-called Industry 4.1.

6.6.2 Server-based iFA System Platform

On the other hand, for companies with strong capital or low demand of the cloud, a buyout server-based version of the iFA platform as shown in Figure 6.6 is recommended. The IoT technology is utilized to develop a CPA at the machine side. CPA is equipped with various equipment drivers for different industries, such as SECS/GEM and Interface A for the semiconductor industry, OPC-UA and MT-Connect for the machine-tool industry, etc. The communication module in CPA is for connecting internet and intelligent application services. Various intelligent services can be implemented as PAMs and plugged into CPA for edge computing. Then, the big-data-analytics-application framework is adopted to include various intelligent manufacturing services such as AVM, IPM, IYM, etc. AVM can convert offline sampling inspection with metrology delay into online and real-time Total Inspection of all the workpieces; IPM can monitor the health status of a key device and predict its RUL of the device; and IYM is used to quickly find out the root causes that affect the yield. Similarly, the iFA system platform can also communicate with the existing MES, (shown on the left side of Figure 6.6) and ERP, (shown on the right side of Figure 6.6) modules via CPAs.

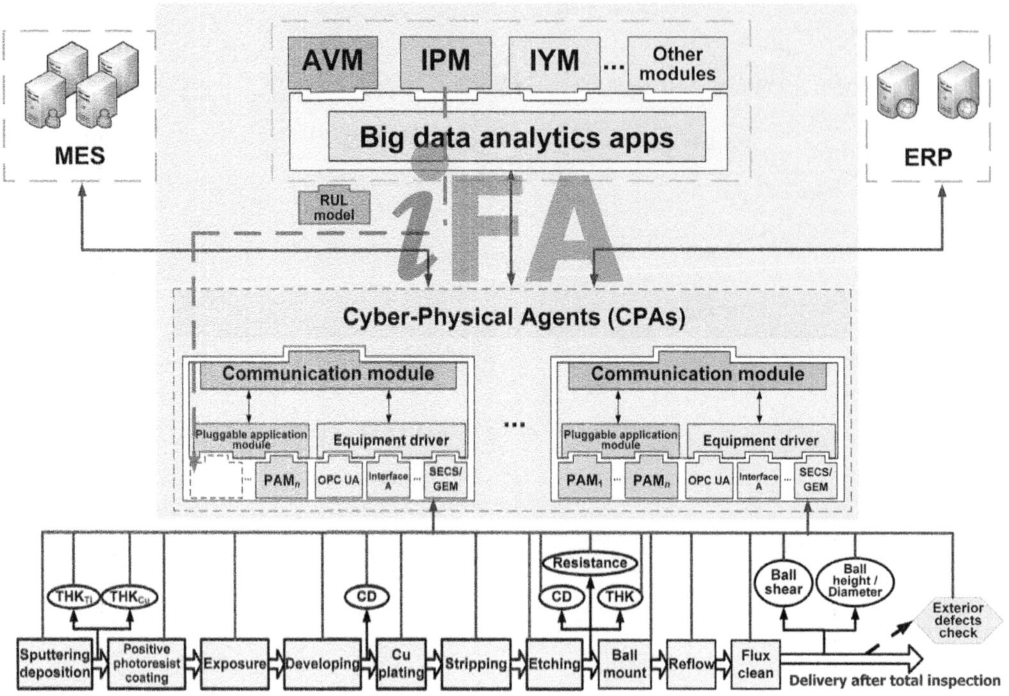

Figure 6.6 Server-based iFA system platform.

6.7 Conclusion

The current Industry 4.0 related technologies emphasize on productivity improvement but not on quality enhancement; in other words, they can only keep the faith of achieving nearly ZD state without realizing this goal. The key reason for this inability is the lack of an affordable online and real-time Total-Inspection technology. By adopting the developed AVM, nearly ZD can be achieved as AVM can provide the Total-Inspection data of all products online and in real time. Further, the developed IYM can be utilized to find out the root causes of the defects for continuous improvement on those defective products. As such, nearly ZD of all products can be achieved.

Appendix 6.A – Abbreviation List

ALASSO	Automated Least Absolute Shrinkage and Selection Operator
AMCoT	Advanced Manufacturing Cloud of Things
AVM	Automatic Virtual Metrology
BPM	Baseline Predictive Maintenance
BPNN	Back-Propagation Neural Networks
CD	Critical Dimensional
CMfg	Cloud Manufacturing

CPA	Cyber-Physical Agent
CPSs	Cyber Physical Systems
DHI	Device Health Index
DQI_x	Data Quality Index of the Process Data
DQI_y	Data Quality Index of the Metrology Data
ERP	Enterprise Resource Planning
GEM	Generic Equipment Model
GSI	Global Similarity Index
GSI_T	GSI Threshold Value
iFA	Intelligent Factory Automation
IoT	Internet of Things
IPM	Intelligent Predictive Maintenance
ISI	Individual Similarity Index
IYM	Intelligent Yield Management
KSA	Key-variable Search Algorithm
MES	Manufacturing Execution System
MP	Mass Production
MSACS	Manufacturing Services Automated Construction Scheme
MT-Connect	A Manufacturing Technical Standard to Retrieve Process Information from Numerically Controlled Machine Tools
OPC-UA	Unified Architecture by Open Platform Communication Foundation
PAMs	Pluggable Application Modules
PLS	Partial Least Square
RD	Research & Development
RI	Reliance Index
RUL	Residual Useful Life
SECS	SEMI Equipment Communications Standard
TD	Target Device
THK	Thickness
TPOGA	Triple Phase Orthogonal Greedy Algorithm
UTM	Unified Threat Management
VM	Virtual Metrology
VM_I	Phase-I VM
VM_{II}	Phase-II VM
ZD	Zero Defects

Appendix 6.B – List of Symbols

$\hat{y}_B$	baseline of TD
X	process data for AVM
y_T	TD data
Y	yield test results
y	total-inspection inline data
$\mathbf{X_R}$	production routes
$\mathbf{X_P}$	process data for KSA
KS_O	output of TPOGA
KS_L	output of ALASSO
RI_K	value of an accompanying reliance index

References

1 Xu, X. (2012). From cloud computing to cloud manufacturing. *Robotics and Computer-Integrated Manufacturing* 28 (1): 75–86. https://doi.org/10.1016/j.rcim.2011.07.002.

2 Hung, M.-H., Li, Y.-Y., Lin, Y.-C. et al. (2017). Development of a novel cloud-based multi-tenant model creation service for automatic virtual metrology. *Robotics and Computer-Integrated Manufacturing* 44: 174–189. https://doi.org/10.1016/j.rcim.2016.09.003.

3 Chen, C.-C., Hung, M.-H., Li, P.-Y. et al. (2018). A novel automated construction scheme for efficiently developing cloud manufacturing services. *IEEE Robotics and Automation Letters* 3 (3): 1378–1385. https://doi.org/10.1109/LRA.2018.2799420.

4 Cheng, F.-T., Huang, H.-C., and Kao, C.-A. (2012). Developing an automatic virtual metrology system. *IEEE Transactions on Automation Science and Engineering* 9 (1): 181–188. https://doi.org/10.1109/TASE.2011.2169405.

5 Cheng, F.-T., Chen, Y.-T., Su, Y.-C. et al. (2007). Evaluating reliance level of a virtual metrology system. *IEEE Transactions on Semiconductor Manufacturing* 21 (1): 92–103. https://doi.org/10.1109/ROBOT.2007.363551.

6 Chiu, Y.-C., Cheng, F.-T., and Huang, H.-C. (2017). Developing a factory-wide intelligent predictive maintenance system based on Industry 4.0. *Journal of the Chinese Institute of Engineers* 40 (7): 562–571. https://doi.org/10.1080/02533839.2017.1362357.

7 Hsieh, Y.-S., Cheng, F.-T., Huang, H.-C. et al. (2013). VM-based baseline predictive maintenance scheme. *IEEE Transactions on Semiconductor Manufacturing* 26 (1): 132–144. https://doi.org/10.1109/TSM.2012.2218837.

8 Cheng, F.-T., Hsieh, Y.-S., Zheng, J.-W. et al. (2017). A scheme of high-dimensional key-variable search algorithms for yield improvement. *IEEE Robotics and Automation Letters* 2 (1): 179–186. https://doi.org/10.1109/LRA.2016.2584143.

7

Advanced Manufacturing Cloud of Things (AMCoT) Framework

Min-Hsiung Hung[1], *Chao-Chun Chen*[2], *and Yu-Chuan Lin*[3]

[1] *Professor, Department of Computer Science and Information Engineering, Chinese Culture University, Taipei, Taiwan, ROC*
[2] *Professor, Institute of Manufacturing Information and Systems, National Cheng Kung University, Tainan, Taiwan, ROC*
[3] *Secretary General, Intelligent Manufacturing Research Center, National Cheng Kung University, Tainan, Taiwan, ROC*

7.1 Introduction

A core vision of Industry 4.0 is to create smart factories that can achieve intelligent manufacturing for providing more flexible and efficient processes to manufacture higher-quality goods at reduced costs by leveraging advanced information and communication technologies (ICTs) [1, 2]. The Advanced Manufacturing Cloud of Things (AMCoT) [3] framework depicted in Chapter 6 is developed by the Intelligent Manufacturing Research Center (iMRC), National Cheng Kung University, Taiwan, ROC, for facilitating building intelligent manufacturing platforms. Several intelligent manufacturing platforms based on the AMCoT framework have been successfully applied to various manufacturing industries, including the semiconductor industry, automotive industry, stretch blow molding machine industry, and aerospace industry. This chapter describes the key components of the AMCoT framework and depicts how to design and implement them by utilizing advanced ICTs, such as cloud computing [4], edge computing [5], IoT [6, 7], big data analytics (BDA) [8, 9], and container technologies [10–14].

The AMCoT framework consists of two parts: the cloud part and the factory part. The cloud part provides various cloud manufacturing (CMfg) [15–19] services, which exploit abundant resources of cloud computing so that they can support factory-wide or even cross-factory manufacturing activities. On the other hand, the factory part provides a variety of edge manufacturing services, which are deployed and run on edge computing devices (briefed as edge devices) for offering equipment-dedicated or time-critical functionalities to equipment. In this chapter, a manufacturing service denotes a software function in the form of Representational State Transfer (REST) [20] web services, which can be called on-demand through networks for supporting manufacturing activities. CMfg services run on virtual machines in the cloud, whereas edge manufacturing services run in edge devices in the factory. A virtual machine functions as a virtual computer, created on a physical computer, with its own CPU, memory, network interface, and storage. An edge device can be a general computer, an embedded system, or an industrial computer.

Compared with cloud computing, edge computing has fewer and limited resources. However, it has less communication latency and faster response between an edge device and the equipment connected with it. Thus, the factory part of the AMCoT framework is designed to comprise many

Industry 4.1: Intelligent Manufacturing with Zero Defects, First Edition. Edited by Fan-Tien Cheng.

edge devices, each of which is called a Cyber-Physical Agent (CPA) [2, 3]. CPA can offer various in-situ edge manufacturing services to the equipment, such as data collection, data preprocessing, data analysis, data storage, communication, and other advanced functionalities (e.g. fault detection and classification, FDC; predictive maintenance, PdM).

A manufacturing factory typically possesses a large amount of equipment. Each piece of equipment may contain several key components needed to be monitored. Therefore, a CPA is required to run many edge manufacturing services simultaneously. Consequently, implementing factory-wide CPAs so that the edge manufacturing services run in each CPA can be pluggable, plug-and-play, and manageable through networks is a complicated and challenging task. For addressing this issue effectively, the AMCoT framework contains a rapid construction scheme of CPAs, namely RCS_{CPA}, based on Docker [10] and Kubernetes [12]. In this chapter, a CPA run with containerized manufacturing services is symbolized as a CPA_C.

As described in Chapter 5, the new-generation virtualization technology – Docker Container has many advantages over the traditional virtual machines. For instance, compared with a virtual machine, a Docker container has a faster booting time, consumes fewer resources, provides a self-contained environment for running isolated applications, and achieves the merit of building a containerized application once and then running it in any environment that has Docker installed. Thus, RCS_{CPA} uses Docker to tackle the development and operations (DevOps) [21] issues of CMfg services running in edge devices, such as resolving the inconsistency between the development and operations environments and providing a uniform and automated approach to deploying containerized CMfg services. Further, RCS_{CPA} utilizes Kubernetes to orchestrate and manage the containerized CMfg services running in edge devices.

The cloud part of the AMCoT framework provides the infrastructure required to host and run CMfg services. These prediction-related and BDA-related CMfg services are implemented in the form of RESTful web services and can thus be used on-demand over networks by the manufacturing industry. For example, the Automatic Virtual Metrology (AVM) [22–24] service can timely perform on-line virtual metrology on workpieces to achieve total inspection of production quality. The Intelligent Predictive Maintenance (IPM) service can gauge the health status of equipment and predict the remaining useful life (RUL) of sick equipment. The Intelligent Yield Management (IYM) service can find the root cause of yield loss for yield enhancement and assurance. The BDA service can store and process factory-wide production-related big data. Due to the abundant resources of the cloud infrastructure, these CMfg services can support factory-wide or even cross-factory intelligent manufacturing activities.

The workflow of creating a CMfg service mainly includes the following steps: (1) Open the integrated development environment (IDE). (2) Create a web service (WS) project. (3) Import the application libraries required to implement the functionalities of the CMfg service. (4) Write the codes of the web methods (or called web APIs, application programming interfaces). (5) Build the WS project to generate the CMfg service package. (6) Test the CMfg service. (7) Debug and revise the codes of the CMfg service if necessary. (8) Deploy the created CMfg service to the target running environment. Thus, manually constructing a CMfg service is cumbersome and error-prone. To solve this issue, the cloud part of the AMCoT framework includes a manufacturing services automated construction scheme (namely MSACS) [25], which allows the user to build and deploy CMfg services efficiently via just operating web GUIs.

Because running applications in containers has become a trend and can bring benefits to the software supply chain for the manufacturing industry, the cloud part of the AMCoT framework also contains an automated construction scheme of containerized manufacturing services (namely $MSACS_C$). $MSACS_C$ is an extended version MSACS by adding the mechanisms of building container images for the CMfg service and deploying the containerized CMfg service to run on the target Kubernetes cluster. The $MSACS_C$ allows the user to build and deploy containerized CMfg services efficiently via just operating web GUIs.

The remaining contents of this chapter are organized as follows: Section 7.2 introduces the key components of the AMCoT framework, including the cloud part and the factory part. Section 7.3 describes the framework designs of the CPA and the CPA_C. Section 7.4 presents the design and application of RCS_{CPA}. The BDA application platform built on the cloud is depicted in Section 7.5. Section 7.6 describes the development of the MSACS, while Section 7.7 introduces the $MSACS_C$. Section 7.8 concludes the chapter.

7.2 Key Components of AMCoT Framework

This section elaborates on the key components of the AMCoT [3] framework. For the sake of easy explanation, the architecture of the AMCoT framework designed in Chapter 6 is shown again in Figure 7.1. The AMCoT framework consists of two parts: the cloud part and the factory part. The cloud part provides the cloud infrastructure with abundant computing resources to build, host, and run various CMfg services and containerized CMfg services that can be used on-demand over networks by the manufacturing industry to support factory-wide or even cross-factory intelligent manufacturing activities. The factory part comprises many CPAs (or CPA_C's) to leverage edge computing for providing more responsive and dedicated manufacturing services to equipment. The key components of the cloud part are described in Section 7.2.1, while those of the factory part are depicted in Section 7.2.2. An example intelligent manufacturing platform based on the AMCoT framework is illustrated in Section 7.2.3.

7.2.1 Key Components of Cloud Part

The cloud part of the AMCoT framework contains the following five key components:

1) Web servers for hosting and running various intelligent CMfg services:
   Because CMfg services may be implemented in different languages, different types of web servers are required. For example, Internet Information Services (IIS) is used to host CMfg services implemented in C#, and Apache Tomcat® is used to host CMfg services implemented in Java. The CMfg services are independent software modules built in the form of RESTful web services. Thus, we can add various CMfg services to the cloud in a plug-and-play manner according to the manufacturing company's requirements, as the intelligent service i, $i = 1, \ldots, n$, shown in Figure 7.1.
2) MSACS [25] for efficiently building intelligent CMfg services and deploying them onto web servers:
   Manually building CMfg services involves a lot of tedious tasks (e.g. creating WS projects, importing dependent libraries, coding web services, debugging, and so on), which are cumbersome and error-prone. Also, the developers of CMfg services require enough knowledge and skills for programming different types of web services. Thus, MSACS is developed and deployed in the cloud to allow engineers to construct CMfg services in a very efficient manner.
3) Docker and Kubernetes Environment for hosting and managing containerized CMfg services:
   The new-generation virtualization technology–Docker Container [10] allows containerized CMfg services to be built once on the development site and then be deployed to and run on any

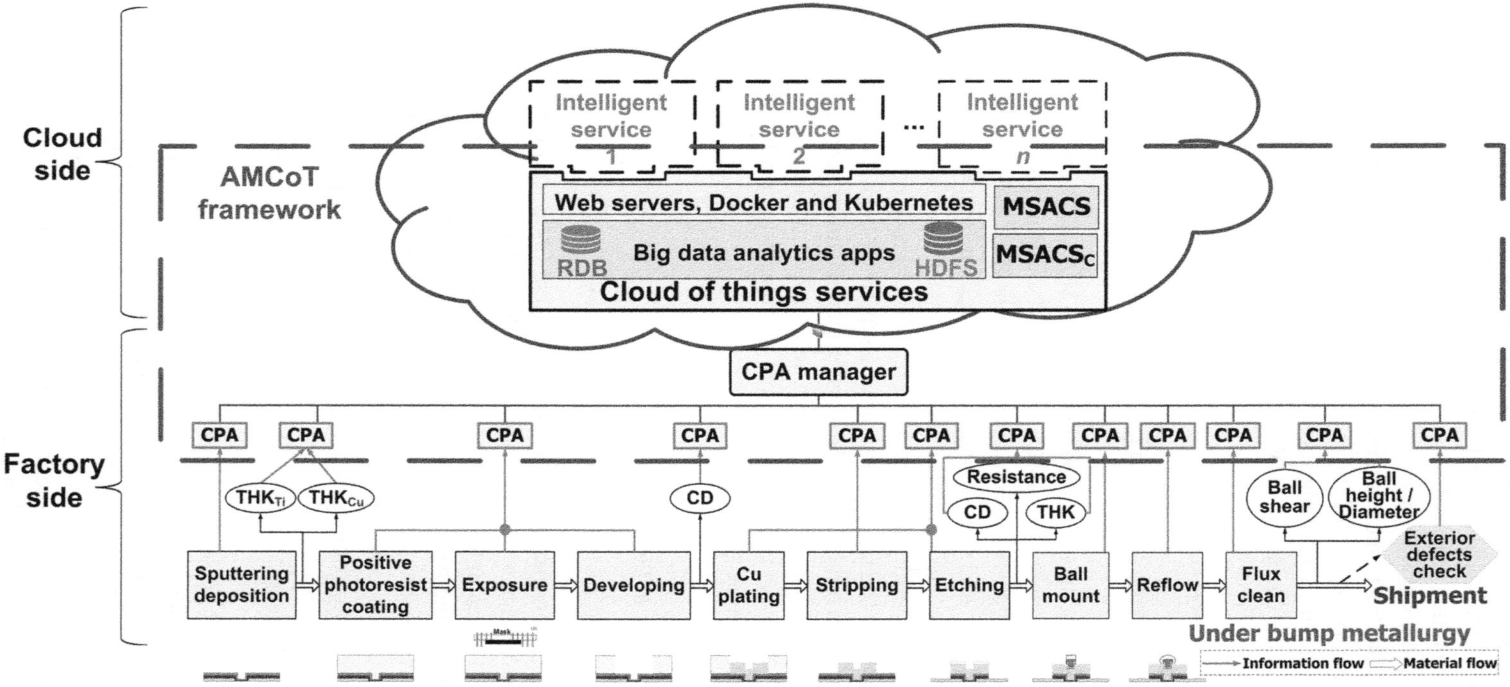

Figure 7.1 Architecture design of the AMCoT framework.

environment with Docker installed. Also, the advantages of Docker containers, such as providing a self-contained environment for running isolated applications, less resource consumption, and faster booting time, can be used to improve the software supply chain of CMfg services, such as resolving the DevOps inconsistency and providing continuous integration/continuous delivery (CI/CD) pipelines. Thus, for taking the advantages of Docker containers to enhance the AMCoT framework, the cloud part contains many Docker hosts to serve as the environment for hosting and running containerized CMfg services, where a Docker host refers to a virtual machine installed with the Docker software. Furthermore, for automatically managing and orchestrating containerized CMfg services, Kubernetes [12] is used to create a hybrid Kubernetes cluster consisting of all Docker hosts, whose operation system (OS) could be Linux (Ubuntu) or Windows (Windows Server 2019).

4) $MSACS_C$ for facilitating building and deploying containerized CMfg services:
   Manually constructing containerized CMfg services involves more tasks than manually building CMfg services. For allowing the users to build and deploy containerized CMfg services efficiently, the cloud part of the AMCoT framework also includes the $MSACS_C$. $MSACS_C$ is an extended version of MSACS. The $MSACS_C$ can package the CMfg services created by the MSACS into Docker images. Then, it can create the Yet Another Markup Language (YAML) files for the desired containerized CMfg service. Finally, the $MSACS_C$ can deploy the generated YAML files to the target Kubernetes cluster, which can then create, run, and orchestrate the desired containerized CMfg service according to the deployed YAML files.
5) BDA Application Platform for storing and processing production-related big data:
   The BDA application platform consists of a Hadoop distributed file system (HDFS), a relational database (RDB), and several BDA-related applications. The HDFS is for storing and replicating historical production-related big data whose size could be up to a scale of terabyte (TB) or petabyte (PB), or even a larger size. The RDB is used to save and provide more recent data with a smaller size for time-critical applications, such as the FDC system.

7.2.2 Key Components of Factory Part

The key components of the factory part comprise many CPAs (or CPA_C's) and a CPA manager. The following descriptions about CPA also apply to CPA_C. Each CPA is implemented and deployed in an edge device. CPA is able to communicate with a wide range of physical objects (e.g. machine tool, equipment, device, and sensor) for the purpose of data collection through various protocols/interfaces. In addition, CPA can communicate with other cyber systems (e.g. CMfg services in the cloud) and other CPAs. Also, CPA can provide manufacturing services in the form of a pluggable application module (PAM) [26] to the equipment connected with it. For managing factory-wide CPAs, a CPA manager is developed. Every CPA deployed in the factory needs to register itself to the CPA manager. The CPA manager can then monitor and manage all the CPAs in the factory through the network.

7.2.3 An Example Intelligent Manufacturing Platform Based on AMCoT Framework

Figure 7.2 shows an example of intelligent manufacturing platform, which is built based on the AMCoT framework and applied to a semiconductor bumping process. The cloud side provides three CMfg services (i.e. the AVM service, IPM service, and IYM service) for supporting intelligent manufacturing activities in the bumping process. Since we can add different CMfg services to the

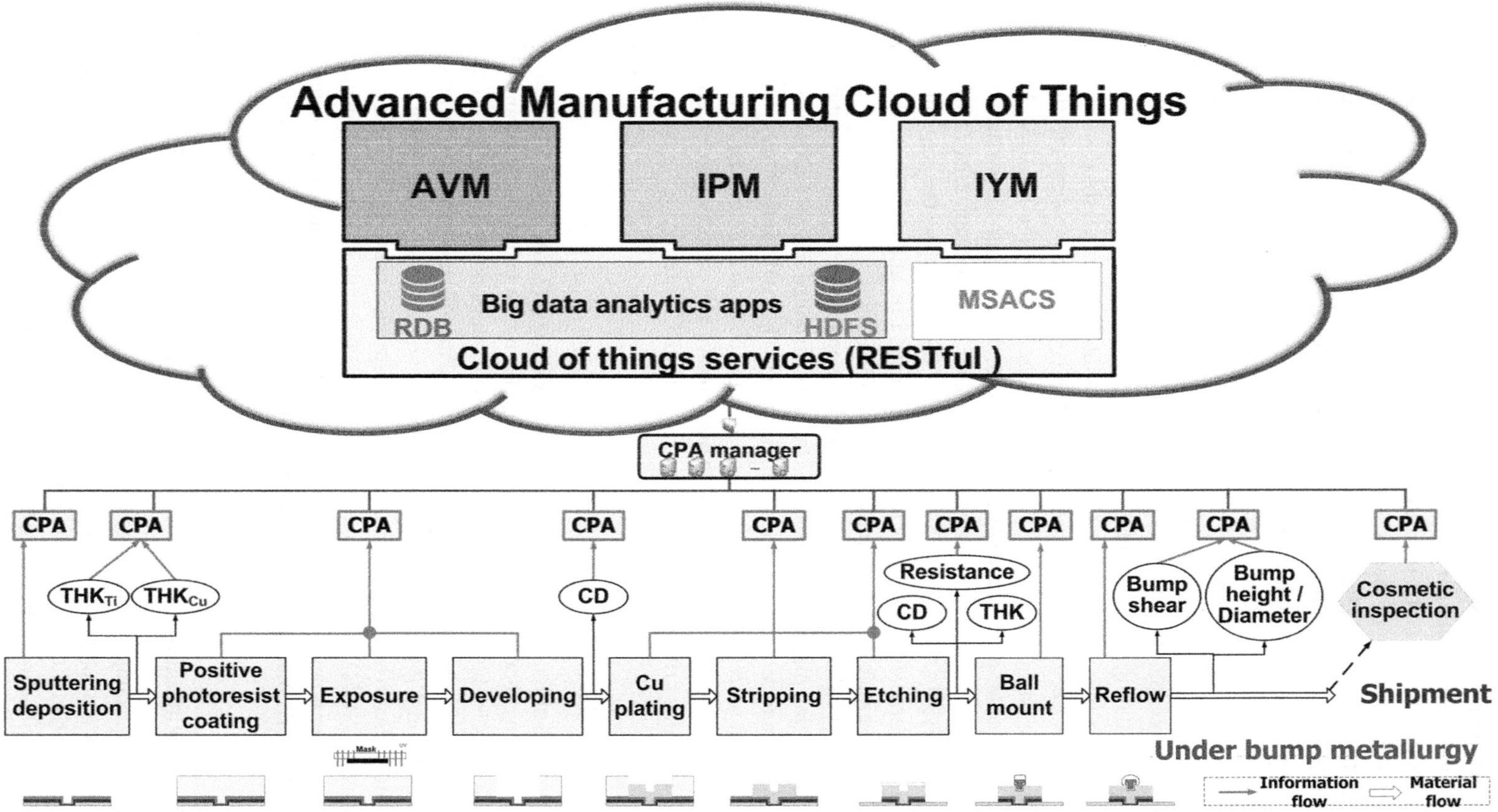

Figure 7.2 An example of intelligent manufacturing platform based on the AMCoT framework applied to a semiconductor bumping process. *Source:* Reprinted with permission from Ref. [3]; © 2017 IEEE.

cloud in a plug-and-play manner according to individual manufacturing company's demand, this intelligent manufacturing platform has also been successfully applied to several other manufacturing industries, including the automotive industry, stretch blow molding machine industry, aerospace industry, etc.

7.2.4 Summary

The AMCoT framework can be used to build intelligent manufacturing platforms for the manufacturing industry. It consists of the cloud part and the factory part. The cloud part provides the cloud infrastructure with abundant computing resources and tools (i.e. MSACS and $MSACS_C$) to build, host, and run various CMfg services and containerized CMfg services, which can be used on-demand over networks by the manufacturing industry to support factory-wide or even cross-factory intelligent manufacturing activities. The factory part comprises many CPAs (or CPA_C's) for leveraging edge computing to provide more responsive and dedicated manufacturing services to equipment. This section describes the key components of the cloud part and the factory part sequentially. Finally, an example intelligent manufacturing platform, which is built based on the AMCoT framework and applied to a semiconductor bumping process, is illustrated to demonstrate the usage of the AMCoT framework.

7.3 Framework Design of Cyber-Physical Agent (CPA)

This section presents the framework designs of CPA [2, 3] and CPA_C. Section 7.3.1 describes the framework of CPA, while Section 7.3.2 depicts the framework of CPA_C.

7.3.1 Framework of CPA

Figure 7.3 shows the framework of CPA, which consists of Control Kernel (CK), Data Collection Manager (DCM), Data Collection Plan (DCP), Data Collection Report (DCR), Equipment Driver Interfaces (EDIs), including OPC Unified Architecture (OPC-UA), EtherCAT, MTConnect, SEMI

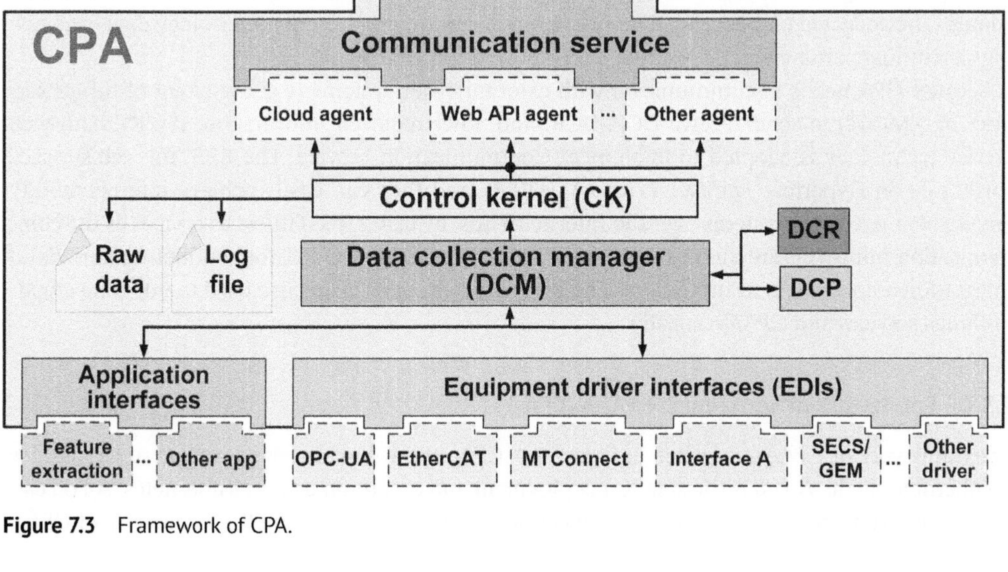

Figure 7.3 Framework of CPA.

Equipment Communications Standard (SECS)/Generic Equipment Model (GEM), Interface A, and other drivers, Application Interfaces (including feature extraction and other applications), and Communication Service (including Web API Agent, Cloud Agent, and other agents).

EDIs are developed to enable CPA to communicate with various kinds of sensors, devices, and machines for the purpose of data collection so that CPA can collect data from a wide range of things in smart factories. Firstly, the OPC-UA driver is designed to allow CPA to access various industrial equipment through its machine-to-machine communication protocol. Secondly, the EtherCAT driver allows CPA to communicate and collect data through Ethernet for real-time automation requirements. Thirdly, the MTConnect driver is included to provide a data exchange protocol to retrieve information from machine tools. Fourthly, the SECS/GEM driver and the Interface A driver are used to establish the connections between CPA and production/metrology tools in the semiconductor industry. Because the EDIs are designed to be pluggable, other types of drivers for data acquisition can be easily added in EDIs to expand the data-gathering capability of CPA. By such design, CPA can serve as an agent that makes cyber-physical interactions possible through all protocols/interfaces mentioned above.

Application Interfaces are designed for CPA to add various pluggable application modules (PAMs) [26] in a plug-and-play manner. Each PAM in CPA is designed to be pluggable for different requirements, allowing users to replace any PAM they need. For instance, data preprocessing (data cleaning, data normalization, data quality check, etc.) and feature extraction are two common PAMs for prediction-related applications.

DCM is responsible for handling and managing data collection through EDIs. DCM may receive a DCP (an Extensible Markup Language (XML) or JavaScript Object Notation (JSON) file specifying the data collection plan) sent from cloud-of-things services of AMCoT through Communication Service. According to the DCP, DCM activates an EDI to start data collection and then packs the collected process data and corresponding metrology data to generate a DCR (an XML or JSON file specifying the data collection report). Afterward, DCM sends the DCR back to the cloud service through Communication Service.

CK is the core of communications among all modules of CPA. The raw data and log file serve as storage in CPA, and they emphasize promptness when compared with a cloud database. During a manufacturing process, DCM continuously collects related engineering data of the workpiece under processing via an EDI and stores them into a raw data file or a log file via the CK. In such a context, the collected engineering data may contain sensor/process data, metrology data, processing/machining parameters, CPA status, user information, and so on.

Because CPA needs to communicate with external cyber systems (e.g. the cloud-of-things services of AMCoT) to send/receive DCP, DCR, and other required information, the RESTful web service technology is adopted to implement Communication Service. The RESTful web services solely rely on Hypertext Transfer Protocol (HTTP), and they can easily achieve interoperability among heterogeneous systems over the Internet. Thus, by using RESTful web services as the communication infrastructure among CPAs, we can form a CPA-based IoT for conducting horizontal integration among CPAs in the factory. The RESTful web services are also used to integrate cloud-of-things services and CPAs vertically.

7.3.2 Framework of Containerized CPA (CPA_C)

The framework of CPA_C is depicted in Figure 7.4. CPA_C is the containerized version of CPA and is a Kubernetes node. A Pod is the smallest deployment and execution unit in Kubernetes. A Pod can contain and run one or more than one application container. In designing the framework of CPA_C,

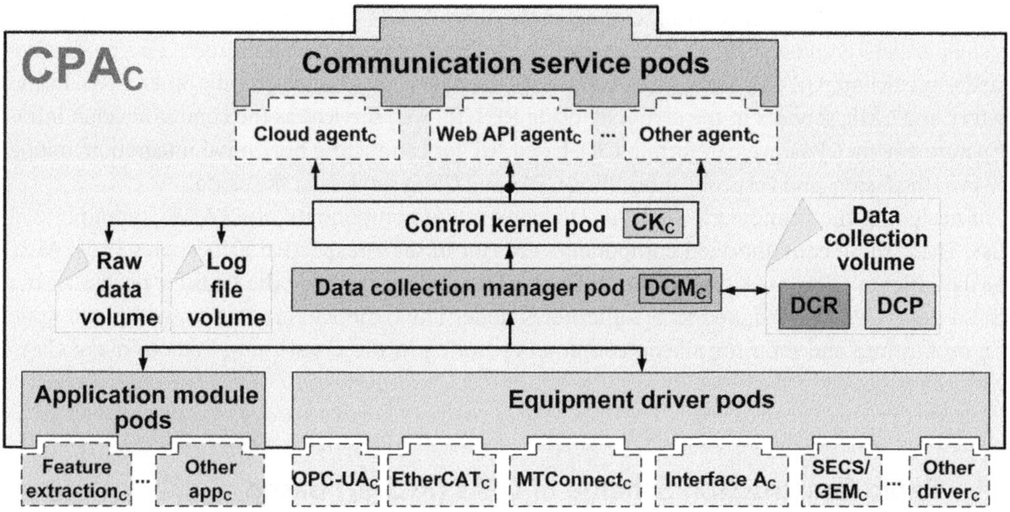

Figure 7.4 Framework of CPA_C.

the constituent components of CPA are containerized first, including containerized control kernel (CK_C), containerized data collection manager (DCM_C), containerized equipment drivers (e.g. $OPC\text{-}UA_C$, $EtherCAT_C$, $MTConnect_C$, $SECS/GEM_C$, Interface A_C, and other containerized drivers), containerized pluggable application modules (e.g. Feature $Extraction_C$ and other App_C), and containerized communication agents (e.g. Cloud $Agent_C$, Web API $Agent_C$, and other $Agent_C$).

The above mentioned containerized components are then run in their respective Kubernetes Pods. CK_C runs in Control Kernel Pod. DCM_C runs in Data Collection Manager Pod. Each containerized equipment driver runs in an Equipment Driver Pod. Each containerized application module runs in an Application Module Pod and each containerized communication agent runs in a Communication Service Pod.

In a Kubernetes node, it is normal that after being created, a Pod may die and is then replaced by a newly-created Pod. If we store data in a Pod, the stored data will be lost after the Pod dies. By contrast, if we store data in a persistent volume, which is outside the Pod, after the Pod dies, the replacing Pod can still access the stored data in the persistent volume. Thus, for persisting the stored data, the data files of CPA_C are stored in various persistent volumes outside the Pods. DCPs and DCRs are stored in Data Collection Volume. The raw data files are stored in RAW Data Volume. And the log files are stored in Log File Volume.

By such design, the Kubernetes control plane node can automatically create the constituent Pods and persistent volumes in a CPA_C according to the YAML files designed for the CPA_C. Then, the Kubernetes control plane node begins to orchestrate and monitor all the containers running in the constituent Pods to make CPA_C work robustly.

7.3.3 Summary

This section describes the framework designs of CPA and CPA_C. CPA is a designed edge device and is the enabler to allow the AMCoT framework to have cyber-physical interactions. CPA_C is the containerized version of CPA.

The main purpose of CPA is data collection. CPA can collect data via various equipment drivers, such as OPC-UA, EtherCAT, MTConnect, Interface A, and SECS/GEM. Also, CPA can simultaneously host and run various manufacturing services in the form of PAM to serve equipment connected with it, such as data preprocessing, feature extraction, fault detection and classification, and production quality prediction. CPA also possesses communication services to interact with other CPAs in the factory and CMfg services in the cloud. By using RESTful web services as the communication infrastructure among CPAs, we can form a CPA-based IoT for conducting horizontal integration among CPAs in the factory and vertically integrating CPAs and CMfg services in the cloud.

In designing the framework of CPA_C, the constituent components of CPA are containerized first. Then, these containerized components are run in their respective Kubernetes Pods. Also, the data files of CPA_C are stored in various persistent volumes outside the Pods for persisting the stored data. CPA_C is designed as a Kubernetes node. Thus, the Kubernetes control plane node can orchestrate and monitor all the containers running in the constituent Pods to make CPA_C work robustly.

7.4 Rapid Construction Scheme of CPAs (RCS_{CPA}) Based on Docker and Kubernetes

This section presents the design of RCS_{CPA} based on Docker and Kubernetes (K8s). Section 7.4.1 describes the background and motivation for developing RCS_{CPA}. Section 7.4.2 explains the system architecture of RCS_{CPA}. Section 7.4.3 depicts the core functional mechanisms of RCS_{CPA}. Section 7.4.4 shows an industrial case study to demonstrate the effectiveness and benefits of RCS_{CPA}.

7.4.1 Background and Motivation

As mentioned in Section 7.1, the new-generation virtualization technology–Docker Container [10] has many advantages compared to the traditional virtualization technology–Virtual Machine. Once a containerized application is built on the development side, it can be shipped and deployed to run in any environment that has Docker installed. Also, Kubernetes is currently the leading orchestrator of containers. Thus, RCS_{CPA} is developed based on Docker and Kubernetes [12] for rapidly constructing CPA_C's (i.e. containerized version of CPA). More specifically, RCS_{CPA} leverages the advantages of Docker containers to resolve the DevOps issue that requires wasting time in adjusting the inconsistency between the development and operation environments. In addition, RCS_{CPA} utilizes Docker to provide a uniform approach to building and deploying PAM_C and other containerized constituent components to CPA_C. Furthermore, RCS_{CPA} employs Kubernetes to host and orchestrate the PAM_C's and other containerized constituent components of CPA_C and to offer mechanisms to make sure they can run robustly in CPA_C.

The merits of RCS_{CPA} are highlighted as follows. First, RCS_{CPA} uses Docker containers to package applications so that each application can run with its dependencies in an environment isolated from other applications in CPA_C. Second, RCS_{CPA} provides an automated creation mechanism of container images for automating the process of building and publishing container images. Third, RCS_{CPA} contains an automated deployment mechanism of PAM_C's for automating the process of deploying and updating PAM_C to CPA_C. Fourthly, RCS_{CPA} possesses a computing-resource management mechanism to allow the users to set up the consumed computing resources limitations for PAM_C in CPA_C. Fifthly, RCS_{CPA} provides the load-balance, horizontal-scaling, failover, and health-inspection mechanisms of PAM_C's to make CPA_C function robustly. Sixthly, RCS_{CPA} has a cloud-based management platform to facilitate the management of CPA_C's and the containerized applications inside each CPA_C.

7.4.2 System Architecture of RCS_{CPA}

Figure 7.5 shows the system architecture of the RCS_{CPA}, which comprises the development side, the cloud side, and the factory side.

On the development side, each containerized application (CA) consists of the following four parts:

1) CA_{CBI} (container base image of the CA): CA_{CBI} consists of the fundamental OS image, the language run-time environment, and the dependencies (Bin/Libs). It serves as the base image for building the container image of the CA.
2) CA_{ASC} (application source codes of the CA): CA_{ASC} could be either plain-text source codes (e.g. application codes written in Python or R) or compiled application codes that are executable in their respective language runtime (e.g. compiled application codes written in Java or C#). By using the Docker build command with a designed Dockerfile, CA_{ASC} is packaged on top of CA_{CBI} to generate the CA's container image, which serves as a template for creating the CA.
3) CA_{KCF} (Kubernetes configuration files of the CA): CA_{KCF} is composed of the YAML files for creating the Kubernetes objects of the CA. By using a Kubectl command, CA_{KCF} can be sent to the Kubernetes control plane node, which then creates the Kubernetes objects of the CA according to the specifications in CA_{KCF} and runs the CA in the Kubernetes nodes.
4) CA_{SS} (shell scripts for automating various tasks related to the CA): To perform tasks related to CA, we need to type in associated commands in the command-line console, such as the CMD or Powershell console in a Windows system and the terminal console in a Linux system. For

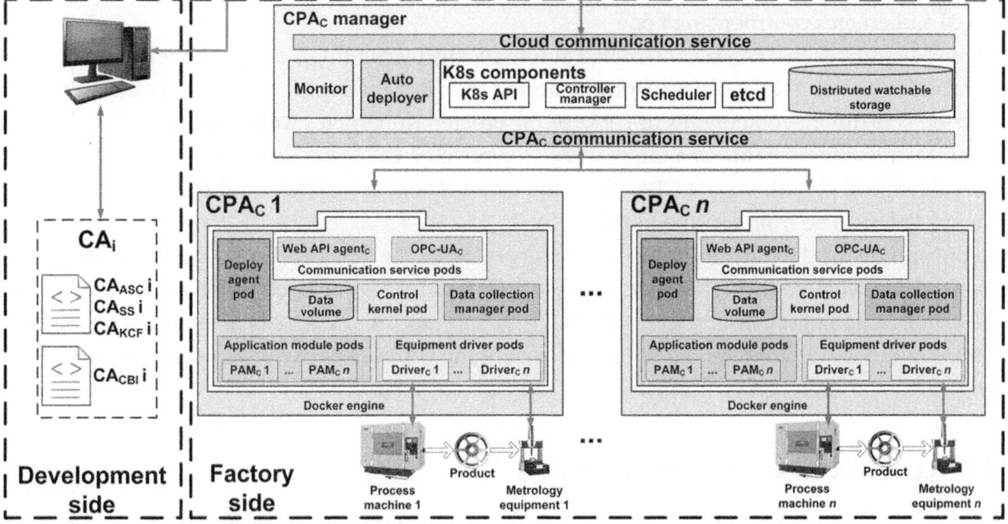

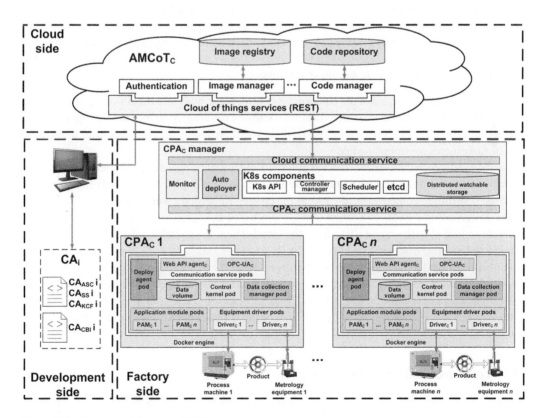

Figure 7.5 System architecture of RCS_{CPA}.

example, we can use Docker commands to build the CA's container images and then push them to the image registry. We can also use Docker commands to pull a container image from the image registry and then run the CA from the container image just downloaded. We need to use Kubectl commands to interact with the Kubernetes control plane node for deploying, updating, and removing CA running in the Kubernetes nodes. Thus, a CA_{SS} file is a collection of Docker and Kubectl commands and other commands list for automating a task of the CA. CA_{SS} consists of many such shell script files.

On the cloud side, an image registry (e.g. Docker Hub) is used to store various CA_{CBI}'s uploaded from the development side, and a code repository (e.g. GitLab) is used to store the developed CA_{ASC}'s, CA_{KCF}'s, and CA_{SS}'s. The web GUIs of Image Manager can be used to manage the Image Registry, while the web GUIs of Code Manager are used to manage the Code Repository. The Authentication web service is used to authenticate users' identities for accessing the Image Manager and Code Manager.

On the factory side, a CPA_C Manager is developed as the Kubernetes control plane node. In addition to Kubernetes control plane node components (i.e. K8s API, Controller Manager, Scheduler, and etcd), the CPA_C Manager also contains Auto Deployer, which is designed to automate the deployment tasks of PAM_C's and create and manage Pods in CPA_C's using shell scripts to call Kubernetes APIs. Many CPA_C's are developed as Kubernetes worker nodes, each of which is connected with a piece of production equipment. As shown in Figure 7.4, a CPA_C is a Kubernetes worker node, thereby having Docker Engine installed. A CPA_C comprises the containerized constituent components (such as Web API $Agent_C$, CK_C, DCM_C, and Equipment $Driver_C$) and PAM_C's, running in their own Pods. In each CPA_C, the Deploy Agent is designed to automate the deployment and update tasks of PAM_C's by using shell scripts to collaborate with the Auto Deployer of the CPA_C Manager. The containerized Deploy Agent (i.e. Deploy $Agent_C$) is run in Deploy Agent Pod.

The workflow of the RCS_{CPA} is described as follows:

1) The CPA_C Manager consists of the Kubernetes control plane node and the Auto Deployer. Thus, the CPA_C Manager can be constructed automatically by running a designed shell script to create and initialize the Kubernetes control plane node and then deploy the Auto Deployer to the created Kubernetes control plane node.
2) Subsequently, each CPA_C is constructed automatically by running a designed shell script to create and initialize a Kubernetes worker node and then deploy Deploy $Agent_C$ and other containerized constituent components of CPA_C to the created Kubernetes worker node.
3) For downloading a PAM_C to a target CPA_C, the selected CA_{CBI} and associated CA_{SS} and CA_{ASC} are downloaded to the target CPA_C from the cloud. Then, the container of the deployed PAM_C is started in a Pod from the downloaded CA_{CBI}.
4) For running the downloaded PAM_C in the target CPA_C, the downloaded CA_{ASC} is transferred to a persistent volume first. Then, the CA_{ASC} is executed by the deployed PAM_C running in a Pod. By such design of separating CA_{ASC} from CA_{CBI} and executing CA_{ASC} in a persistent volume by the container, we can save a lot of update time of the CA because we need not download CA_{CBI} again, and, usually, CA_{CBI} is much bigger than CA_{ASC} in size.

7.4.3 Core Functional Mechanisms of RCS_{CPA}

This section presents some of the core functional mechanisms provided by RCS_{CPA}.

7.4.3.1 Horizontal Auto-Scaling Mechanism

Figure 7.6 shows the horizontal auto-scaling mechanism of PAM_C's provided by RCS_{CPA}, whose workflow is described as follows.

Assume that the user would like to change the number of replicas of a PAM_C (e.g. the containerized baseline predictive maintenance module, BPM_C). According to the demand, the user updates the CA_{KCF} of the BPM_C to increase or decrease the number of replicas of the BPM_C Pod.

Step 1: The updated CA_{KCF} is sent to Auto Deployer and, in turn, to the K8s API in CPA_C Manager.

Step 2: The Kubernetes control plane node monitors the current number of replicas of the BPM_C Pod and knows the difference from the desired number of replicas according to the updated CA_{KCF}.

Step 3: The Kubernetes control plane node automatically increases or decreases the total number of the BPM_C Pods in the Kubernetes worker nodes to complete the horizontal auto-scaling of the BPM_C.

It is worth to note that all the created BPM_C Pods are supervised by the Kubernetes service object of the BPM_C (namely, S_{BPM}), and they can run in different Kubernetes work nodes. The Kubernetes control plane node can decide which BPM_C Pod runs in which Kubernetes worker node. We can also enforce all the underlying BPM_C Pods to run in the same Kubernetes worker node (e.g. CPA_C n).

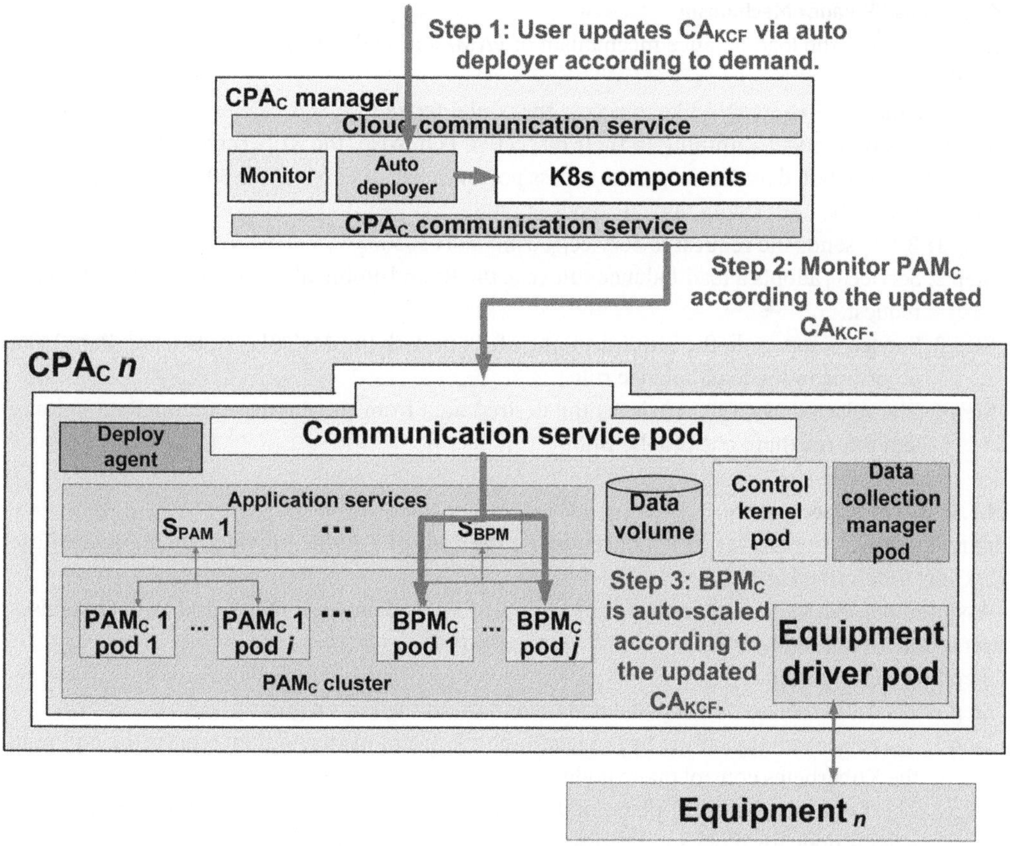

Figure 7.6 Horizontal auto-scaling mechanism of PAM_C's in RCS_{CPA}.

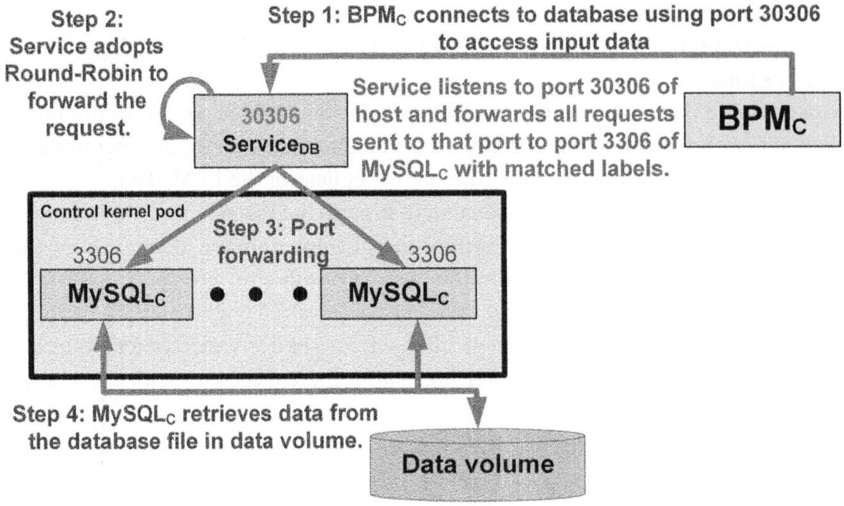

Figure 7.7 Load balance mechanism of PAM_C's in RCS_{CPA}.

7.4.3.2 Load Balance Mechanism

Figure 7.7 shows the load balance mechanism of PAM_C's provided by RCS_{CPA}, whose workflow is described as follows.

Assume that a BPM_C would like to access the containerized MySQL Server (i.e. $MySQL_C$), and $MySQL_C$ has two replicas running in their respective Pod. Also, the Kubernetes service object of the $MySQL_C$ is denoted as $Service_{DB}$, the access port number of $Service_{DB}$ is 30306, and the access port number of the underlying $MySQL_C$ Pods is 3306.

Step 1: BPM_C sends the request to $Service_{DB}$ using Port 30306.

Step 2: $Service_{DB}$ adopts a load-balance rule (e.g. the Round-Robin algorithm) for forwarding the request.

Step 3: $Service_{DB}$ forwards the request to one of its underlying $MySQL_C$ Pods using Port 3306, according to the load-balance rule.

Step 4: The selected $MySQL_C$ retrieves the desired data from the database file in Data Volume and returns them back to the BPM_C.

7.4.3.3 Failover Mechanism

Figure 7.8 shows the failover mechanism among the Pods of a PAM_C provided by RCS_{CPA}, whose workflow is described as follows.

Assume that the Kubernetes service object of a BPM_C is denoted as S_{BPM}, and S_{BPM} supervises several underlying BPM_C Pods.

Step 1: A BPM_C Pod dies.

Step 2: The Kubernetes control plane node monitors all the Pods regularly.

Step 3: S_{BPM} discovers the failure by the health-check mechanism and then reports the failure to the Kubernetes control plane node .

Step 4: The Kubernetes control plane node creates a new BMP_C Pod in the target Kubernetes worker node for replacing the dead one.

Step 5: The newly-created BMP_C Pod joins to the supervision of S_{BPM} to complete the failover of the BPM_C.

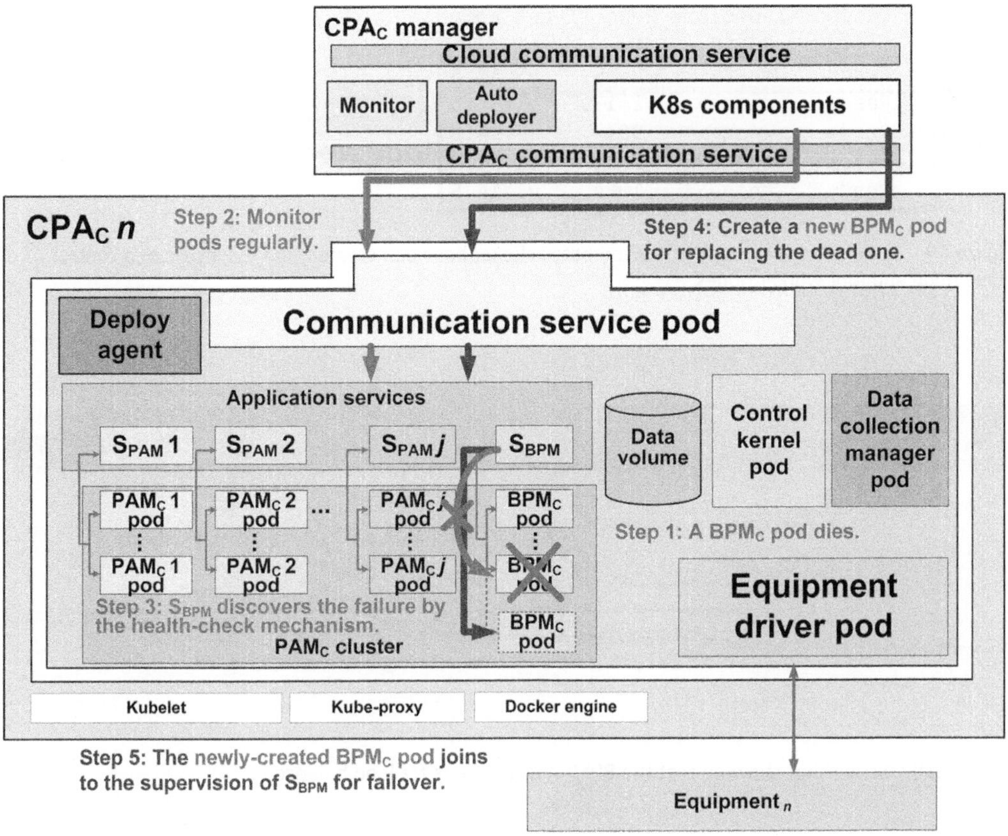

Figure 7.8 Failover mechanism among the Pods of a PAM_C in RCS_{CPA}.

7.4.4 Industrial Case Study of RCS_{CPA}

7.4.4.1 Experimental Setup

UP Board[2] [27] is used as the illustrative edge device for implementing the CPA_C Manager and the CPA_C's. The created Kubernetes cluster contains a control plane node and four worker nodes. The specifications of the UP Board[2] are as follows: OS: Ubuntu 16.04 TLS x64; CPU: 4 Core 2.5 GHz (Intel Pentium); RAM: 8GB DRAM (LPDDR4); Docker: 18.03.1-CE build 9ee9f40; Kubernetes: v1.10.3.

The cloud environment is as follows. The Image Registry and Image Manager are created in Docker Hub [28]. The Code Repository and Code Manager are built in our private cloud. The Linux OS is Ubuntu 16.04.

7.4.4.2 Testing Results

1) Results of Kubernetes Cluster Creation:
   Figure 7.9 is a screenshot showing that a Kubernetes cluster has been created, and it consists of a control plane node (i.e. upb-01) and four worker nodes (i.e. upb-02, upb-03, upb-04, and upb-05).
2) Results of Load Balance of Four BPM_C Pods:
   The Kubernetes web-based dashboard is used to observe the testing results of the load balance mechanism. Figure 7.10 is a screenshot showing that the deployed BPM_C (denoted by

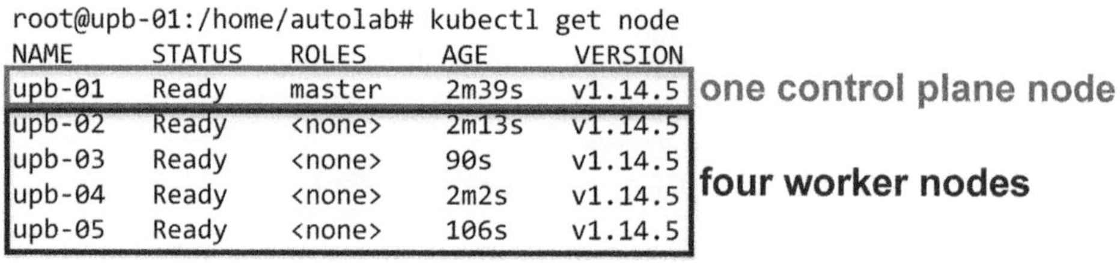
root@upb-01:/home/autolab# kubectl get node
NAME STATUS ROLES AGE VERSION
upb-01 Ready master 2m39s v1.14.5
upb-02 Ready <none> 2m13s v1.14.5
upb-03 Ready <none> 90s v1.14.5
upb-04 Ready <none> 2m2s v1.14.5
upb-05 Ready <none> 106s v1.14.5
one control plane node
four worker nodes

Figure 7.9 Screenshot showing that a Kubernetes cluster has been created, and it consists of a control plane node and four worker nodes.

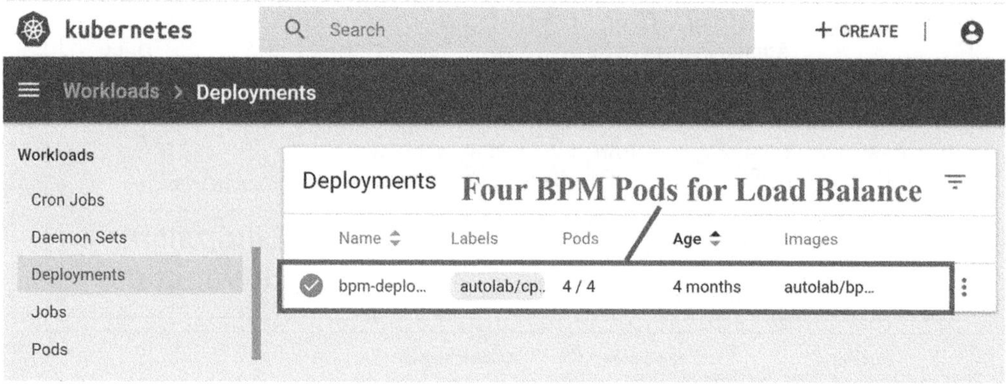

Figure 7.10 Screenshot showing that the BPM_C has four Pods for load balance.

bpm-deployment) has four Pods for load balance, and all of the four Pods are healthy (indicated by 4/4).

Figure 7.11 is a screenshot showing that the four Pods of the BPM_C are distributed to different nodes (i.e. upb-02, upb-03, upb-04, and upb-05) for the purpose of failover and high availability of the application.

3) Results of Running a BPM_C Application:
An industrial case study performing predictive maintenance (PdM) on plasma-enhanced chemical vapor deposition (PECVD) equipment of a solar cell manufacturing factory using the Baseline Predictive Maintenance (BPM) algorithm in [25] is conducted. Figure 7.12 shows the health-gauging web GUI of a running BPM_C. It is noted that the running BPM_C contains four Pods working as a whole to provide its service, and all the four Pods under the BPM service are allocated on different nodes. Thus, by leveraging the load-balance mechanism and the failover mechanism mentioned in Section 7.4.3, the BPM_C can run more robustly, compared with the native BPM running in a noncontainer environment.
4) Analysis of Deployment Time of a PAM_C:
Table 7.1 shows the time analysis of deploying two PAM_C's (i.e. a BPM_C and a $Forging_C$) into a CPA_C. All deployment tasks can be completed automatically without human intervention. The testing results in Table 7.1 show that the download of the container base image CA_{CBI} dominates the total deployment time. Downloading the CA_{CBI} (about 2GB) of the $Forging_C$ takes about 14 minutes and 18 seconds, while downloading the CA_{CBI} (about 344MB) of the BPM_C takes about 2 minutes and 47 seconds. It is worth noting that the download of the CA_{CBI} is needed only at the first time of the deployment of the PAM_C's. In the subsequent deployment

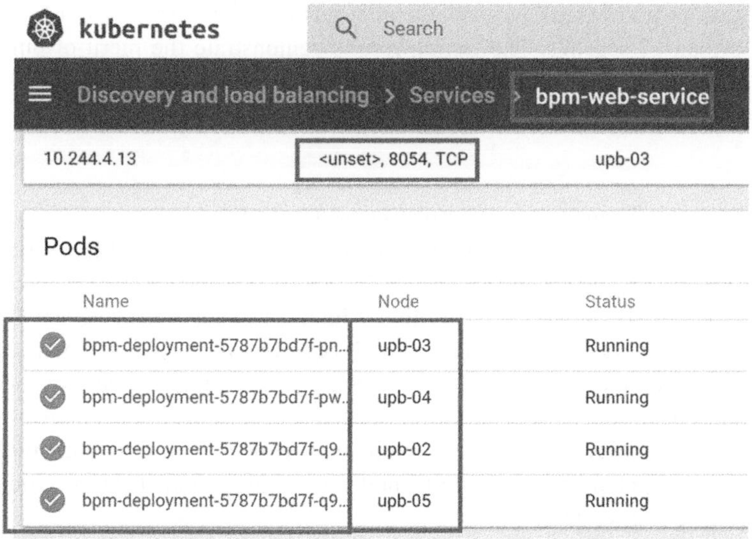

Figure 7.11 Screenshot showing that the four Pods of the BPM_C are distributed to different nodes.

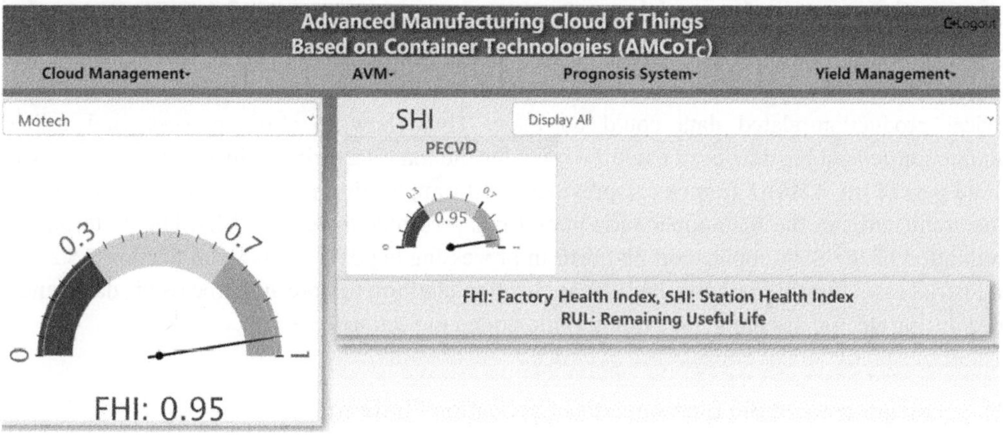

Figure 7.12 Health-gauging web GUI of the BPM_C, which contains four Pods working as a whole. All Pods under the BPM service are allocated on different nodes.

Table 7.1 Time analysis of deploying two PAM_C's into a CPA_C.

#	Task	Spent time
1	Initialize the Kubernetes Control Plane Node(CPA_C Manager Server)	42 s
2	Initialize a Kubernetes Node (CPA_C)	14 s
3	Establish the CPA_C framework for PAM_C's deployment	21 s
4	Download CA_{CBI} of Forging PAM_C's to a CPA_C the first time	14 min 18 s
5	Download CA_{CBI} of BPM_C PAM_C's to a CPA_C the first time	2 min 47 s
6	Deploy/Update PAM_C's of Forging	7 s
7	Deploy/Update PAM_C's of BPM	8 s

or update of the same PAM_C's with revised CA_{ASC}, we do not need to download the CA_{CBI} again, thereby taking only less than 10 seconds. The testing results demonstrate the merit of our design of PAM_C's by separating CA_{ACS} from CA_{CBI}.

7.4.5 Summary

This section presents the RCS_{CPA} based on Docker and Kubernetes. The RCS_{CPA} makes good use of the advantages of the Docker container to have capabilities of automated deployment, horizontal auto-scaling, load balance, self-healing, and failover of PAM_C's. The PAM_C's can be deployed on factory-wide edge devices as manufacturing services to support intelligent manufacturing activities. Finally, an industrial case study performing PdM on PECVD equipment of a solar cell manufacturing factory is conducted to validate the effectiveness of RCS_{CPA}. Testing results show that RCS_{CPA} can facilitate the rapid construction of CPAs, while effectively tackles the related DevOps issues, such as resolving the inconsistency between the development and operations environments and providing robust-operation mechanisms (e.g. horizontal auto-scaling, load balance, and failover) for containerized edge CMfg services.

7.5 Big Data Analytics Application Platform

During production, various kinds of data with different generating rates (e.g. a set of data per millisecond, per second, per minute, per hour, or per day) will be generated. These accumulated historical production-related data could be huge. Thus, how to store, process, and utilize production-related big data is an essential issue for the manufacturing industry. Accordingly, the cloud part of the AMCoT framework provides a BDA application platform to address this issue. The architecture of the BDA application platform is presented in Section 7.5.1. The performance evaluation of the BDA application platform in processing big data is shown in Section 7.5.2. An industrial case study by applying the BDA application platform to store and process big data generated during electric discharge machining is illustrated in Section 7.5.3.

7.5.1 Architecture of Big Data Analytics Application Platform

Figure 7.13 [3] shows the framework of the BDA application platform, which is created in the cloud by using Cloudera Distributed for Hadoop [29]. The BDA application platform consists of the following constituent components:

1) HDFS: a distributed file system, which can support processing and replicating big data with a size of TB or PB, or even with a larger size. The data in HDFS can be stored in four types of format: default format, Parquet, Partition, and "Parquet+Partition." Parquet is a columnar binary storage format. The Partition format uses a clustering technique to partition data into different sub-clusters according to some attributes, such as date.
2) Impala: an SQL query engine, which can bring scalable parallel database technology to Hadoop and allows the users to issue low-latency SQL queries to the data stored in HDFS.
3) Spark: an open-source cluster computing framework, which can perform in-memory cluster computing on data and can thus do computation much faster than Map-Reduce.
4) Hive: a data warehouse infrastructure built on top of Hadoop Map-Reduce, which can give an SQL-like interface for querying the data stored in various databases and file systems that can be

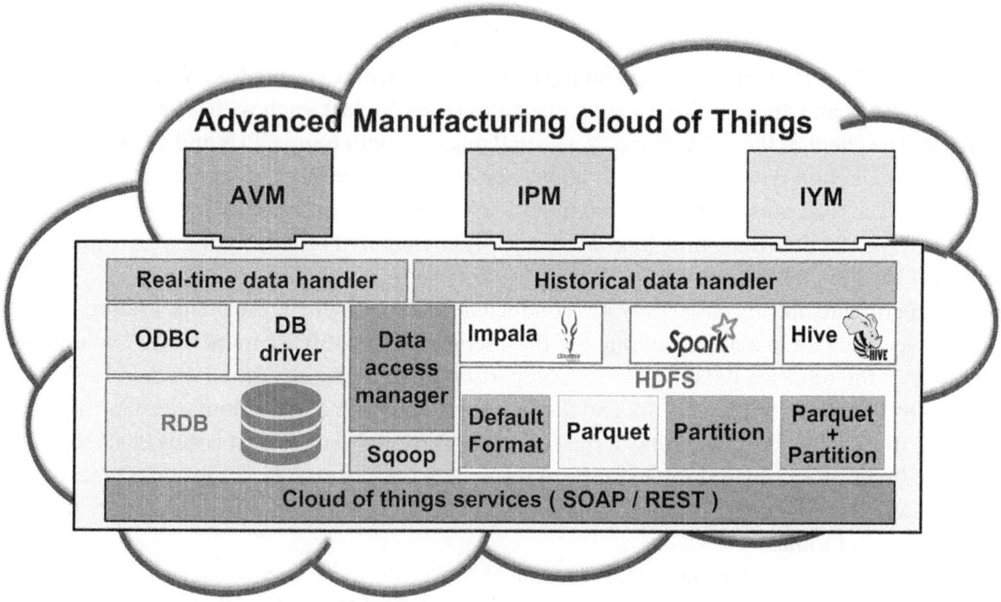

Figure 7.13 Framework of the big-data-analytics application platform. *Source:* Reprinted with permission from Ref. [3]; © 2017 IEEE.

integrated with Hadoop, such as HDFS. If the size of the production data is too large for Spark to process them, Hive is used instead.

5) RDB: a relational database, which is used to store data related to time-critical applications, such as the resulting data of the FDC system.
6) Sqoop: a command-line interface application, which can transfer historical data from RDB to HDFS for keeping the RDB in a good performance.
7) Historical Data Handler: an application module that is developed to automatically select Impala, Hive, or Spark to access historical data in HDFS according to data computation time and data size.
8) Real-Time Data Handler: an application module that is developed to access RDB through ODBC or other DB drivers.

7.5.2 Performance Evaluation of Processing Big Data

Traditionally, an FDC system adopts RDB to store the production-related data. However, the query performance of the RDB becomes poor when the number of records exceeds a couple of millions. For instance, the FDC system of a fab needs to take more than three seconds to query and obtain the historical production data of a specified lot, about 187 records, from 300 million records of data, which cannot meet the desired time constraint. Also, the RDB cannot accommodate the increasingly growing fab-wide production data. For example, a fab has 50 pieces of equipment, each equipment is equipped with 30 sensors, and the sampling rate of sensor data is 1 Hz. As such, the fab will increase more than 4 million records per day and 100 million records per month. For keeping its RDB in a good performance, the extra cost is required to move older historical data into other databases.

To evaluate the performance of the BDA service of the AMCoT framework, we apply the created BDA application platform to process the big production data of a fab. Specifically, we utilize Cloudera to build a Hadoop ecosystem on top of a virtual machine (Name Node) with four vCPUs and 32 GB RAM and another three virtual machines (Data Nodes), each with four vCPUs and 16 GB RAM. The client of the BDA service is a virtual machine with four vCPUs and 4 GB RAM. The historical production data of the fab in about one month come from several pieces of etching equipment. The total number of records is about 100 million, and each record has 32 columns. We store these production data into four data types (i.e. Default Format, Parquet, Partition, and "Parquet+Partition") in HDFS.

Two experiments in this case study are conducted. In experiment 1, we query the production data of a specified lot in a day by calling the BDA service of the AMCoT framework. The number of records of the returned data is 187. The comparison of query times of the four data formats in experiment 1 is shown in Figure 7.14, and the respective query times are summarized in Table 7.2. As shown in the table, the query time of the "Parquet+Partition" data format is only 0.086 seconds, about 510 times shorter than the query time of the default data format. Also, the query times of the Parquet data format and the Partition data format are also much shorter than the query time of the default data format.

In experiment 2, we query the production data of three columns (i.e. contexid, process_timetag, and exh1_flowrate) in a day by calling the BDA service of the AMCoT framework. The number of records of returned data is 40424. The comparison of query times of the four data formats in experiment 2 is shown in Figure 7.15, and the respective query times are summarized in Table 7.2.

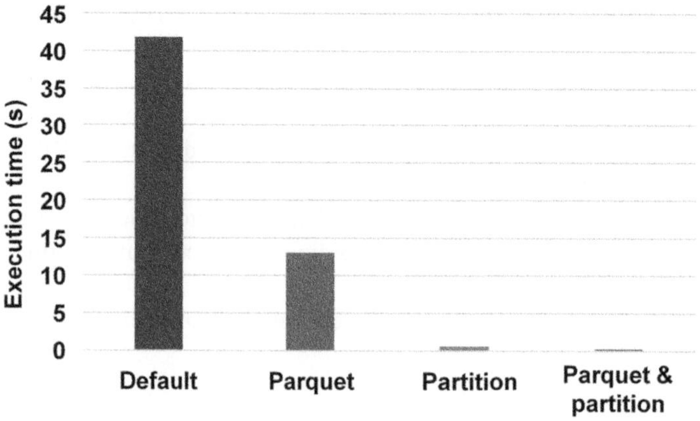

Figure 7.14 Comparison of query times using four types of data table in experiment 1. *Source:* Reprinted with permission from Ref. [3]; © 2017 IEEE.

Table 7.2 Query times using four types of data format in experiments 1 and 2.

Experiment	Default	Parquet	Partition	Parquet & Partition
Experiment 1 (s)	41.718	13.010	0.561	0.086
Experiment 2 (s)	43.332	13.786	2.377	2.252

Source: Reprinted with permission from Ref.[3]; © 2017 IEEE.

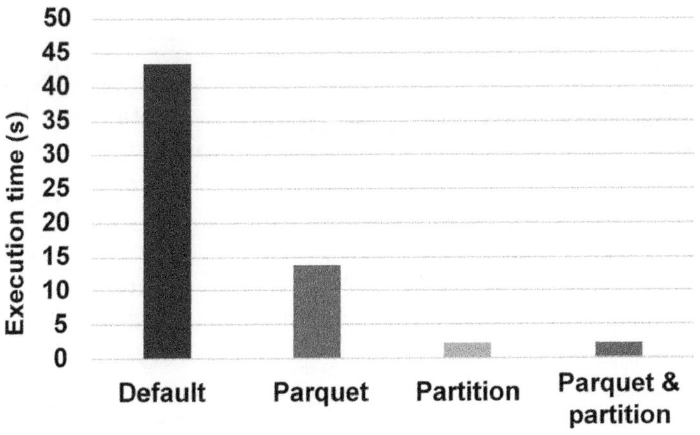

Figure 7.15 Comparison of query times using four types of data format in experiment 2. *Source:* Reprinted with permission from Ref. [3]; © 2017 IEEE.

As shown in the table, the query times of the Partition data format and the "Parquet+Partition" data format are 2.377 seconds and 2.252 seconds, respectively, about 20 times shorter than the query time of the default data format. Also, the query time from the Parquet data table is three times quicker than the query time of the default data format.

The above testing results demonstrate that the BDA service of the AMCoT framework can efficiently store and process big production data. Also, to achieve a good performance in querying data, the raw data should be stored in the Parquet format (column-oriented binary coding) or the Partition format (partitioned by clustering). In this illustrative example, storing the production data in the "Parquet+Partition" format can achieve the best performance for querying historical data.

7.5.3 Big Data Analytics Application in Manufacturing – Electrical Discharge Machining

Electrical discharge machining (EDM) is a manufacturing process where a workpiece is transformed into the desired shape by removing its materials using a series of rapidly recurring current discharges between the tool-electrode and the workpiece-to-workpiece (W2W) control, which are immersed in a dielectric liquid. EDM is one of the most extensively used nonconventional material removal technology, which can be used to machine hard metals or conductive workpieces that are difficult to machine using traditional machining techniques. The common applications of EDM are in the manufacture of mold, die, automotive, aerospace, and surgical parts.

For improving the machining efficiency, very-high-frequency electric pulses are applied to the electrodes of EDM. For example, the normal pulse-on time and pulse-off time of the voltage signal applied on the electrodes are 80 ns and 120 ns, respectively, in micro-EDM machining. Thus, probes with a very high sampling rate are needed to acquire machining raw data from such high-frequency electric pulse signals. For instance, a 50-MHz oscilloscope was used to acquire signals from a voltage sensor and a current sensor installed on the micro-EDM machine, leading to a high data-generation rate, e.g. generating up to 13 GB data in drilling a hole with a depth of 100 μm and a diameter of 3 mm in about 12 minutes. This raises a big data processing issue in feature extraction from machining raw data for conducting online virtual metrology on EDM. In addition, the voltage and current signals in EDM change rapidly and may involve different machining statuses,

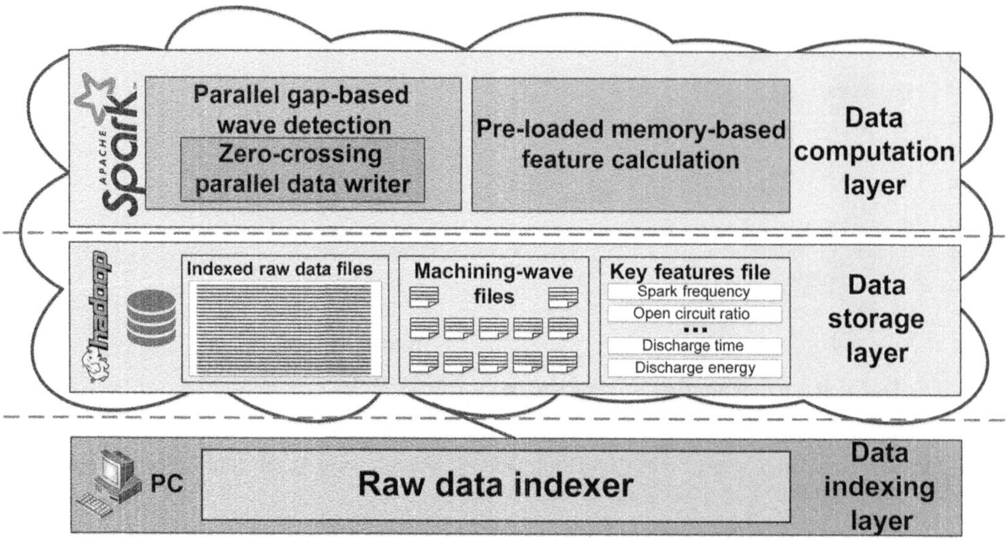

Figure 7.16 Architecture of the proposed BEDPS. *Source:* Reprinted with permission from Ref. [30]; © 2019 IEEE.

including open circuit, short circuit, normal spark discharge, arc, and so on. Thus, how to detect effective discharge waves from big machining raw data of EDM effectively and efficiently is a challenging task.

We develop a novel efficient big data processing scheme for feature extraction in EDM, called Efficient Big Data Processing Scheme (BEDPS) [30], based on Spark and HDFS. Figure 7.16 shows the architecture design of the proposed BEDPS, which consists of three layers. The first layer is the Data Indexing Layer for adding indexes to the machining raw data files. The second layer is the Data Storage Layer built using Hadoop HDFS for storing the indexed raw data files, machining-wave files, and the key-feature file. The third layer is the Data Computation Layer built using Spark for performing the computations of wave detection and feature calculation.

The workflow of the BEDPS is described as follows [30]. First, the data of current and voltage signals, as well as the tool-tip position signal, are acquired from the EDM machine using an oscilloscope and laser rangefinder, respectively, and the acquired sensor data are stored in many raw data files on a PC. Then, the Raw Data Indexer adds indexes to the raw data files and stores the indexed raw data files in the Data Storage Layer. Next, the Parallel Gap-based Wave Detection (PGWD) mechanism in the Data Computation Layer processes the indexed raw data files to detect effective machining waves, produce machining-wave files (one file for each machining wave), and store produced machining-wave files back to the Data Storage Layer in HDFS using the Zero-Crossing Parallel Data Writer method. After that, the Pre-loaded Memory-based Feature Calculation mechanism loads machining-wave files from HDFS, calculates the key features in Spark, and finally stores the key features in a key-feature file in HDFS.

Both the existing system and the proposed BEDPS are tested using real data from a micro-EDM process of a CNC company in Taiwan, ROC. The data used for testing is from the EDM process in drilling a hole with a depth of 100 μm and a diameter of 3 mm in die steel (NAK80) using 2A as the discharge current condition. The generated raw data are stored in about 700 files, with a total size of 13 GB. The original raw data are then replicated to create data of bigger sizes for testing scenarios

Table 7.3 Comparisons of execution times in processing different sizes of data using the existing method and the proposed BEDPS.

	Execution Time (in Minutes)					
	Machining Wave Detection		Feature Calculation		Total Execution Time	
Data Size (GB)	C#	Spark	C#	Spark	C#	Spark
1	1.45	0.53	0.37	0.03	1.82	0.56
5	7.07	1.23	1.95	0.15	9.02	1.38
10	15.27	2.72	4.07	0.28	19.34	3.00
13	18.63	3.33	5.10	0.35	23.73	3.68
25	20.65	5.87	7.17	0.50	27.82	6.37
50	—	13.37	—	0.62	—	13.99
75	—	26.12	—	0.83	—	26.95
100	—	45.52	—	1.27	—	46.79

using synthesized data of up to 100 GB. In addition, some of the original data are also picked to create data of smaller sizes for testing scenarios using real data.

To evaluate the efficiency and effectiveness of the proposed BEDPS, we implement the BEDPS in Spark with six nodes (one control plane node and five worker nodes). Each node is equipped with a CPU of Intel(R) Xeon(R) E3-1220 v3 @ 3.10 GHz and 32 GB RAM. In addition, for the sake of comparisons, the original feature extraction algorithm is implemented in C# in a PC with a CPU of Intel(R) Core (R) i7-6700 @ 3.40 GHz and 32 GB RAM.

Table 7.3 shows the comparisons of execution times in processing different sizes of data using the existing method (labeled as C#) and the proposed BEDPS (denoted as Spark). As shown in the table, the proposed BEDPS is more efficient than the existing method in both detecting machining waves and calculating features in all of the testing scenarios. For example, in processing the data of 13 GB, the BEDPS is 5.6 times faster than the existing method in machining wave detection (3.33 minutes vs. 18.63 minutes), 14.6 times faster in feature calculation (0.35 minutes vs. 5.10 minutes), and 6.5 times faster in the total execution time (3.68 minutes vs. 23.73 minutes). The results also show the existing method can only process data with a size up to 25 GB, while the BEDPS can process the data with a size of up to 100 GB, meaning that the proposed BEDPS is more scalable than the existing method. The EDM machine takes about 12 minutes to make a hole. Thus, according to the testing results, the proposed BEDPS is promising to process big data generated from several EDM machines simultaneously and still keep up with the online Automatic Virtual Metrology (AVM), while the existing method fails to be used in online AVM.

7.5.4 Summary

For storing and processing production-related big data, the cloud part of the AMCoT framework contains a BDA application platform. The BDA application platform consists of an HDFS for storing and accessing big historical data, a Relational Database Management System (RDMS) for storing and accessing recent production-related data, and several BDA applications. This section has described the design, shows the performance, and illustrates the application of the BDA application platform built in the cloud of the AMCoT framework for the manufacturing industry.

7.6 Manufacturing Services Automated Construction Scheme (MSACS)

This section describes the design, and implementation, and application of the MSACS. The background and motivation of developing the MSACS are explained in Section 7.6.1. The three-phase workflow of MSACS is presented in Section 7.6.2. The architecture of MSACS is designed in Section 7.6.3. The design of the core components of MSCAS is depicted in Section 7.6.4. Finally, an industrial case study of applying MSACS to build CMfg services for the manufacturing industry is illustrated in Section 7.6.5 to demonstrate the effectiveness and benefits of MSACS.

7.6.1 Background and Motivation

By leveraging and extending the characteristics of cloud computing (CC), CMfg [15–19] has emerged as a next-generation manufacturing paradigm and has become a hot research topic in both academia and industry. CMfg is considered as a computing and service-oriented manufacturing model and is the manufacturing version of cloud computing. CMfg encapsulates and virtualizes distributed manufacturing resources (both hardware and software) into cloud services that can then be used on-demand over networks to support manufacturing activities, ranging from the design, simulation, manufacturing, testing, management of products to all other tasks in the product life cycle.

CMfg is different from traditional web- and agent-based manufacturing paradigms in terms of computing architecture, data storage, operational process, and business model. CMfg has the potential to satisfy the growing demands of manufacturing for producing customized products more flexibly, supporting global cooperation and collaboration, facilitating knowledge-intensive innovation, and increasing the agility of responding to the changing market. Hence, CMfg is a new paradigm that will revolutionize the manufacturing industry.

In recent years, many CMfg-related research works have proposed a number of techniques and approaches to virtualize and encapsulate manufacturing resources and capabilities as cloud services and to develop CMfg architectures and CMfg-related prototype systems. Tao et al. [18] investigated the applications of IoT and CC in manufacturing, proposed a CC- and IoT-based CMfg service system, and analyzed the relationship among CMfg, IoT, and CC. Chen et al. [31] proposed a CMfg framework with auto-scaling capability for the machine tool industry. Huang et al. [32] developed a cloud-based AVM system for the semiconductor industry. Hung et al. [33] developed a cloud-based multi-tenant model creation service of AVM for the wheel machining industry. Lin et al. [3] developed the AMCoT to serve as a smart manufacturing platform that has been successfully applied in several manufacturing industries, such as the semiconductor industry, machine tool industry, and aerospace industry.

In further promotion of CMfg, how to automatically and efficiently encapsulate and virtualize manufacturing resources into cloud services (i.e. how to build CMfg services in an automatic and efficient manner) is an essential and challenging subject. In the field of services computing [34], a large number of methods have been proposed to address the problem of service composition whose central concept is to integrate multiple existing web services into workflows to create new value-added services. Tao et al. [35] addressed the important supply-demand matching (SDM) problem of manufacturing services and proposed a manufacturing service SDM simulator to serve as a uniformed research platform for related researchers. Jatoth et al. [36] provided a systematic literature review on computational intelligence-based quality of service (QoS) – aware web service composition. The QoS parameters included availability, reliability, response time, cost, reputation, throughput, and security. Paik et al. [37] proposed a scalable architecture for automatic service composition to create new value-added services from existing services in cloud computing environments. Wang et al. [34] proposed a Graphplan-based approach for automatic web service composition with consideration of uncertainty execution effects. However, these previous service-composition-related works assumed that there were some existing services to work on and did not address how to build them automatically. The works [32, 33] did propose systematic approaches for constructing CMfg services using Standalone Software Library Packages (SSLP) that can support manufacturing activities, but they needed developers to construct CMfg services manually.

As described in previous works [32, 33], manually constructing CMfg services using SSLP involves a lot of efforts, as briefed below. First, the developer opens an Integrated Development Environment (IDE) to create a web service project (WSP) by selecting the corresponding WSP template. Next, the SSLP is added to the WSP through the IDE. Then, the developer proceeds to write and complete the source codes of every method (usually called web method) in the web service (WS) by calling desired methods of the SSLP and doing dependency configuration (i.e. managing other required libraries). After that, the developer uses IDE to build (compile and link) the WSP to generate a WS package, which is then deployed onto a cloud web server to become a CMfg service. If a CMfg application needs to implement the functions of many SSLPs, and each SSLP has a great number of methods to be transformed into cloud services, then manually constructing such a CMfg service would become complicated and time-consuming. Also, the developer needs to learn various WS development skills and may commit assorted programming errors in the process of building CMfg services. These development efforts may become a burden for the engineers of the manufacturing industry. The above factors as a whole could hinder the development of CMfg.

This section presents the Manufacturing Services Automated Construction Scheme (namely, MSACS), which can facilitate the rapid construction of CMfg services. The development of MSACS adopts a text-based-template and automated-code-generation approach and leverages technologies of JSON, RESTful web service, and Command Scripting. First, a three-phase workflow of MSACS is developed to clearly address the issues of how to construct CMfg services automatically using SSLP. Next, a system architecture of MSACS is designed to delineate how to implement MSACS. Then, the designs of MSACS's core components are depicted in detail. Finally, MSACS is applied to build the AVM CMfg service and the IYM CMfg service for conducting industrial case studies. Testing results demonstrate that MSACS can automatically construct the target CMfg services in a very efficient manner after uploading the required SSLPs, about 30 times quicker than a skilled professional, and more than 1600 times quicker than a novice.

7.6.2 Design of Three-Phase Workflow of MSACS

JSON is a lightweight (compared to XML) plain-text data format that is built on two structures: a collection of "key:value" pairs and an ordered list (array) of values. A value in JSON can be another JSON object or array, thereby making it suitable to express complex hierarchical data. Another benefit of JSON is the ability to be read by both humans and computers. In MSACS, we adopt JSON to store hierarchical information of SSLPs and service interfaces.

Web service is a technology developed for achieving interoperability among distributed systems over the Internet. As the technology advances, RESTful web services become dominant, compared to traditional SOAP-based web service, because of its several characteristics: relying only on HTTP to exchange messages without extra protocols, needing no service description language, accessing data resources by just sending URLs, supporting both JSON and XML messages, and so on. In MSACS, we construct CMfg services in the form of RESTful web services.

Due to the diversity of software libraries, it is impossible to develop a one-size-fits-all method to extract the key information of all SSLPs. Thus, an SSLP used in MSACS is restricted to be a library archive file, which aggregates many class files and associated metadata and resources into one file for distribution and code reuse. Java Jar files of Java and DLL files of C# and MATLAB are two examples of this kind of SSLPs. According to recent surveys, Java and C# are two mainstream programming languages in terms of popularity and usage. Hence, the restriction of SSLP being a library archive file would not significantly affect the usefulness of MSACS. For instance, the work [25] employed a Jar SSLP to develop a CMfg service that can infer proper machine tools and cutting tools for machining tasks of workpieces. The work [32] utilized a C# DLL SSLP of AVM to develop a CMfg service for conducting fab-wide online virtual metrology of wafers.

The three-phase workflow of MSACS is shown in Figure 7.17. The design philosophy of the workflow for addressing the issues of how to construct CMfg services automatically using SSLP is briefly described below, while the designs of its core components will be depicted in detail in Section 7.6.4.

1) Key components for the library analysis phase:
   We adopt JSON to design a generic library information template (Lib. Info. Template), which is used to specify the data structure of how the hierarchical key information (KI) of SSLP is stored. The KI includes library name, package names, class names, method names, and each method's input parameters' types and output's type. In addition, we design the KI Extractor, which is used to extract the KI of the input SSLP. The extracted KI is then stored into a Lib. Info. file, according to the format specified in Lib. Info. Template.
2) Key components for the service project generation phase:
   The Lib. Info. file is library-centric, which means it contains all KI of an SSLP in a file. Because a web service consists of web methods, each of which is called a service interface (SI), the SI

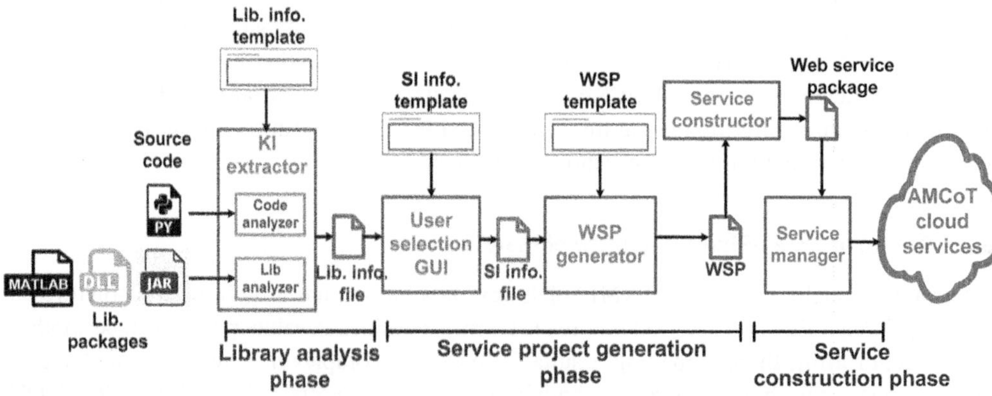

Figure 7.17 Three-phase workflow of MSACS. *Source:* Reprinted with permission from Ref. [25]; © 2018 IEEE.

information should be method-centric, which means it contains only the information of the SIs selected by the user. Thus, we adopt JSON to design a generic SI information template (SI Info. Template), which is used to specify the data structure of how the SI information is stored. In addition, we develop GUI to allow the user to select the target methods to be transformed into web methods and then to generate the service's SI Info. File using the SI Info. Template.

Because the Java WSP and C# WSP have different structures, we design two WSP templates, one for generating Java WSPs and the other for generating C# WSPs. However, both WSP templates adopt the same design principle. The WSP template consists of an annotated source code template file, required library files, and dependency configuration files, which are stored in their respective directories. The annotated source code template file serves as a wrapper for transforming the selected methods of the input SSLP to web methods, and it contains several annotation keywords to annotate the necessary SI information for building the web methods. Then, we design the WSP Generator to automatically search and correlate a lookup table and the input SI Info. File using annotation keywords as keys and further replace the annotation keywords in the annotated source code template file with the corresponding SI information extracted from the SI Info. File. By going through the above process, the WSP Generator can convert the WSP template to generate a WSP.

3) Key components for the service construction phase:
By using the command scripting technology, we design Service Constructor to build the WSP automatically to generate a WS package. Then, we design the Service Manager to decompress and deploy the WS package onto the target cloud web server automatically to become a CMfg service.

7.6.3 Architecture Design of MSACS

Figure 7.18 shows the system architecture of the MSACS, which can delineate how to implement MSACS. MSACS is a cloud-based system and consists of three parts: Manufacturing Service Constructor, Manufacturing Service Manager, and Cloud Web Server.

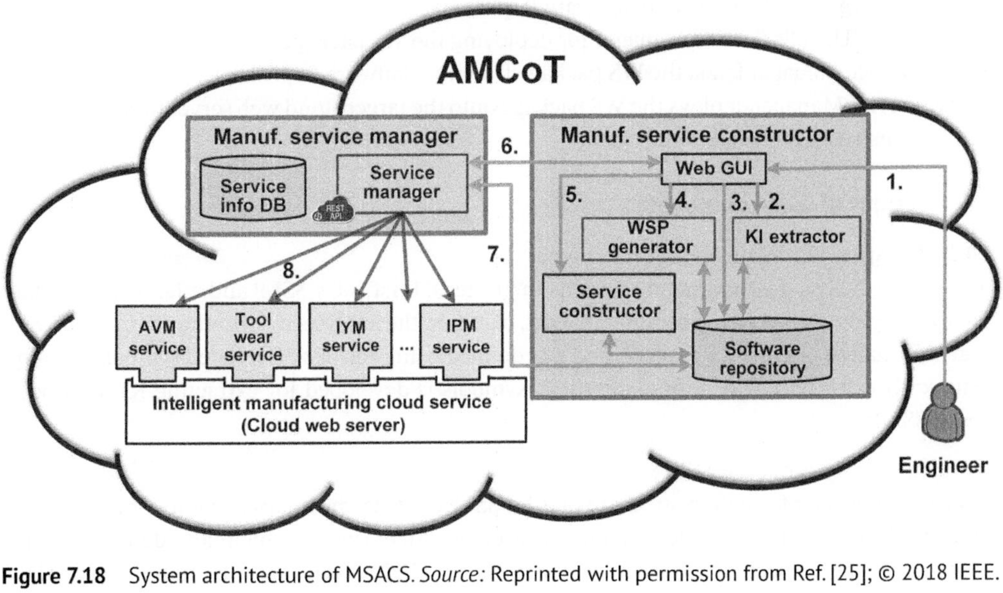

Figure 7.18 System architecture of MSACS. *Source:* Reprinted with permission from Ref. [25]; © 2018 IEEE.

The Manufacturing Service Constructor consists of WSP Generator, KI Extractor, Service Constructor, Software Repository, and Web GUI. The functions of the first three core components have been introduced in Section 7.6.2. The Software Repository is a cloud file system that is used to store assorted files and directories in MSACS, including SSLP files, Lib. Info. Template, Lib. Info. Files, SI Info. Template, SI Info. Files, WSP templates, WSP, and WS packages. The Web GUI is developed using HyperText Markup Language 5 (HTML5) for cross-platform usage. The user can use the Web GUI to upload and parse SSLPs, select target methods of SSLPs, and then automatically construct and deploy WS packages to become CMfg services. The Web GUI also allows the user to manage (e.g. remove and query) the deployed CMfg services.

The Manufacturing Service Manager consists of Service Manager and Service Info DB. The former offers functions of deploying services, removing services, and querying the statuses of services. The latter is a cloud RDMS used to store and manage the information of the deployed CMfg services and the information of the cloud web servers hosting the services.

The Cloud Web Server consists of cloud web servers used to host various CMfg services, such as the AVM service, IYM service, and IPM service. Each cloud web server is installed in a virtual machine.

The operational flow of MSACS for automatically constructing a CMfg service is as follows:

Step 1: The user applies Web GUI to upload an SSLP to the cloud and store it into the Software Repository.

Step 2: Web GUI calls KI Extractor to parse the SSLP, extract the KI, and store the KI into a Lib. Info. File. After that, the Web GUI displays the methods corresponding to the KI on the Web GUI.

Step 3: The user selects some methods on Web GUI. After the user clicks the "Select Done" button, the Web GUI proceeds to generate a SI Info. file of the selected methods and stores it into Software Repository.

Step 4: After the user clicks the "Construct Service" button, Web GUI calls WSP Generator to load the WSP template and the SI Info. File to generate a WSP, which is then stored into Software Repository.

Step 5: Web GUI calls Service Constructor to automatically build the WSP to generate a WS package, which is then saved to Software Repository.

Step 6: Web GUI calls Service Manager for deploying the WS package.

Step 7: Service Manager loads the WS package from the Software Repository.

Step 8: Service Manager deploys the WS package onto the target cloud web server to complete the construction of a CMfg service.

7.6.4 Designs of Core Components

7.6.4.1 Design of Key Information (KI) Extractor

The design principle of KI Extractor contains three steps. First, a C# SSLP and a Jar SSLP are analyzed by using their respective decompiler. Next, based on the analysis results, the common hierarchical information of the SSLP is identified and defined using various designated symbols. Finally, a generic KI extraction algorithm using those symbols is developed for extracting KI from both types of SSLPs.

1) Structure analysis of a Jar SSLP:

   The hierarchical information structure of a Jar SSLP analyzed by a Java decompiler is shown in Figure 7.19. The library information consists of three layers, which are defined as the JarInputStream Object (JISO) layer, JarEntry Object (JEO) layer, and Function Content (FC)

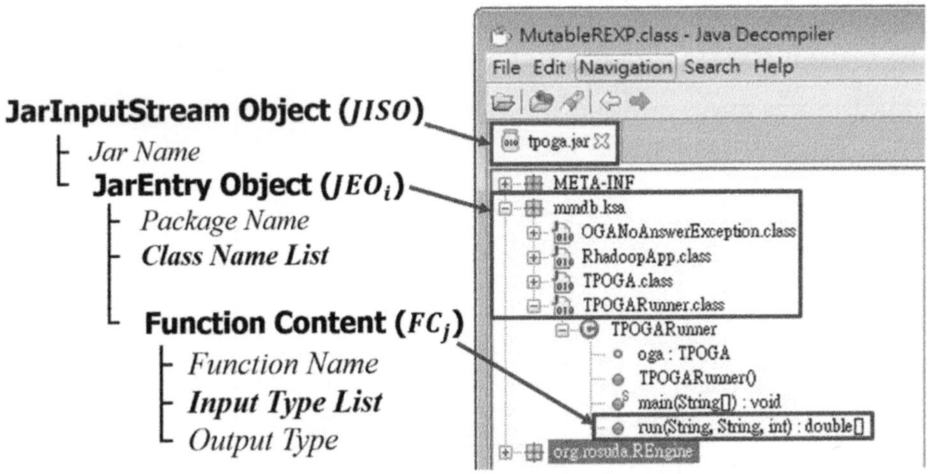

Figure 7.19 Hierarchical information of a Jar SSLP in a Java decompiler. *Source:* Reprinted with permission from Ref. [25]; © 2018 IEEE.

layer, respectively. JISO is the most top layer, whose key information is the Jar file's name, such as tpoga.jar. JEO is the middle layer that may consist of several JEOs. Each JEO (JEO_i) contains the key information of a package, i.e. the package name and the underlying class names. FC is the lowest layer that may consist of several FCs. Each FC (FC_j) contains the key information of a class's function, including the function name, the types of input parameters, and the output type.

2) Structure analysis of a C# DLL SSLP:
   The hierarchical information structure of a C# DLL SSLP analyzed by a C# decompiler is shown in Figure 7.20. The library information consists of three layers, which are defined as the Assembly Object (AO) layer, Type Object (TO) layer, and MethodInfo Object (MO) layer, respectively. AO is the most top layer, whose key information is the DLL file's name, such as BPNN_CSharp. TO is the middle layer that may consist of several Type Objects. Each TO (TO_i) contains the key information of a package, i.e. the namespace name and the underlying class names. MO is the lowest layer that may consist of several MOs. Each MO (MO_j) contains the key information

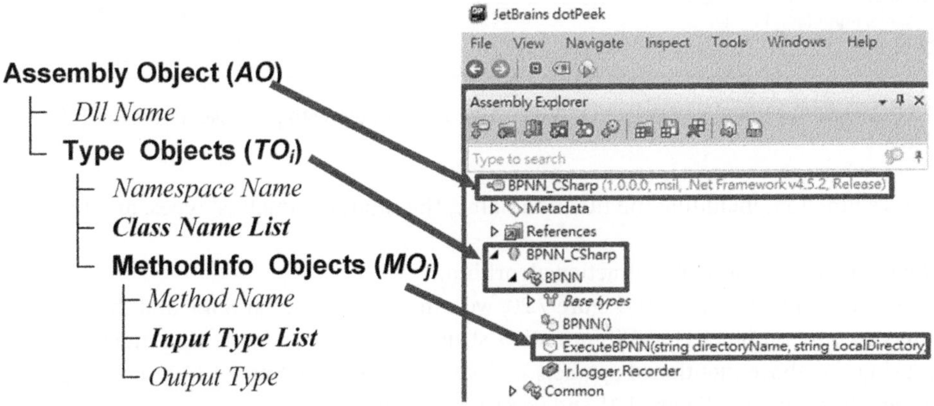

Figure 7.20 Hierarchical information of a C# DLL SSLP in a C# decompiler. *Source:* Reprinted with permission from Ref. [25]; © 2018 IEEE.

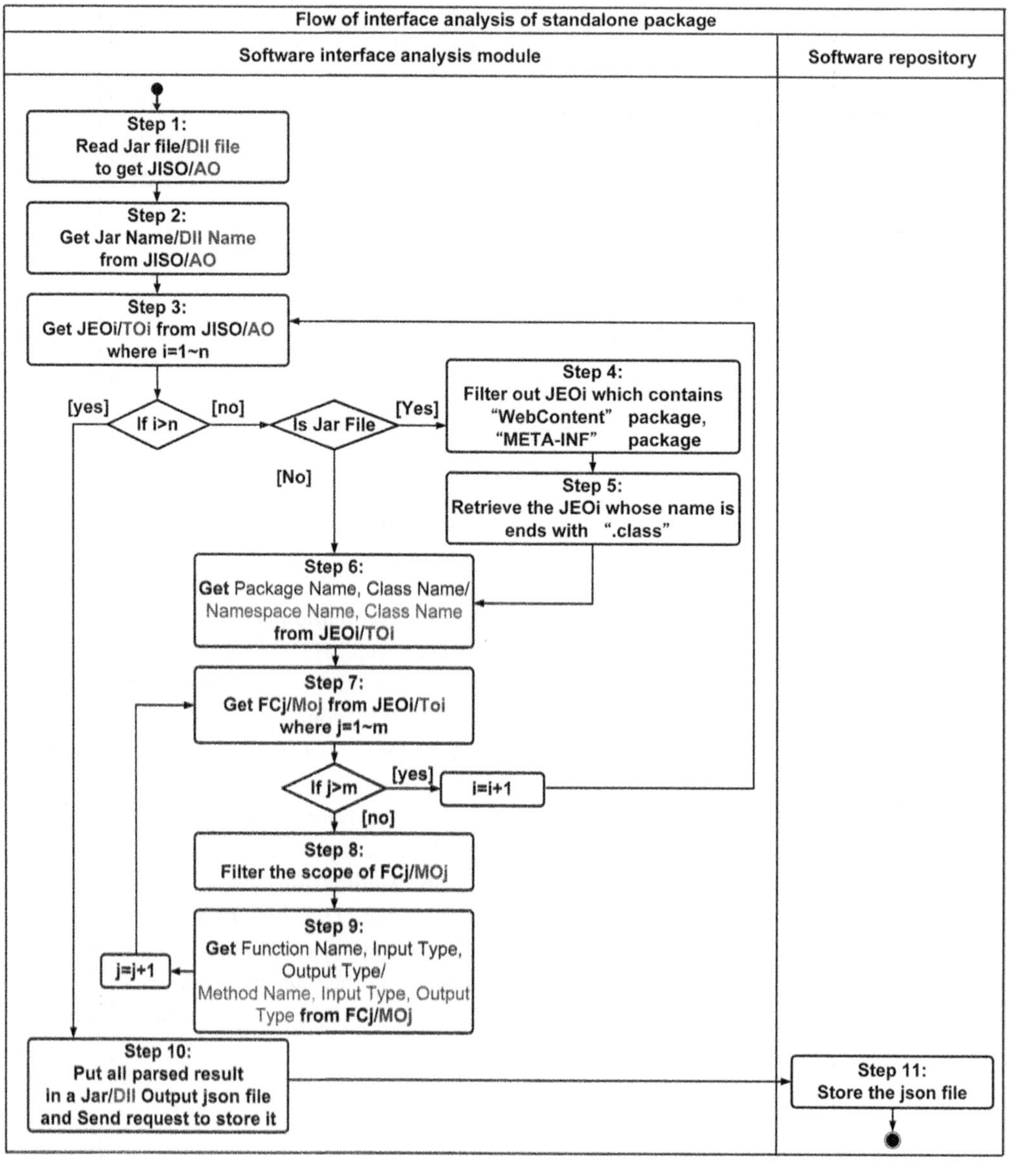

Figure 7.21 Generic KI extraction algorithm of SSLPs. *Source:* Reprinted with permission from Ref. [25]; © 2018 IEEE.

of a class's function, including the function name, the input parameters' types, and the output type.

3) Development of the generic KI extraction algorithm:
   After comparing Figure 7.19 with Figure 7.20, we can find that the Jar SSLP and the C# DLL SSLP has a common hierarchical information structure. Thus, a generic KI extraction algorithm using the above mentioned symbols is developed for extracting KI from both types of SSLPs, which is shown in Figure 7.21 and described below.
   Step 1: Read the Jar file/DLL file to get JISO/AO.
   Step 2: Get the Jar name/DLL name.

Step 3: Get JEO_i/TO_i from JISO/AO, where $i = 1, \ldots,$ n. If i is greater than n, end the parsing process and go to Step 10. If i is equal to or less than n, and the SSLP is a Jar file, then go to Step 4. Otherwise, go to Step 6.

Step 4: Find the JEO_i that contains the WebContent and META-INF directories, and then go to Step 5.

Step 5: Retrieve the JEO_i whose name ends with ".class".

Step 6: Get the package name/namespace name and the underlying class names from JEO_i/TO_i.

Step 7: Get FC_j/MO_j from JEO_i/TO_i, where $j = 1, \ldots, m$. If j is greater than m, increase i by 1, and then go to Step 3. Otherwise, go to Step 8.

Step 8: Filter the scope of FC_j/MO_j.

Step 9: Get the function name and the types of input parameters and the output type from FC_j/MO_j. Increase j by 1, and then go to Step 7.

Step 10: Store all the extracted library information in a JSON file (a Lib. Info. file).

Step 11: Store the JSON file into the software repository.

7.6.4.2 Design of Library Information (Lib. Info.) Template

According to the analysis results of Section 7.6.4.1, Jar SSLP and C# DLL SSLP have a common hierarchical structure. An SSLP file may consist of many packages, each package may contain many classes, each class may contain many methods, and each method may have many input parameters of a certain type and an output type. For storing all of extracted KI from SSLP, a generic Lib. Info. Template is designed using JSON, as shown in Figure 7.22. As the figure shows, both AVMService.dll and tpoga.jar can store their key information using this Lib. Info. Template. Also, the value of InputType of a method is a string of input parameters' types, separated by commas.

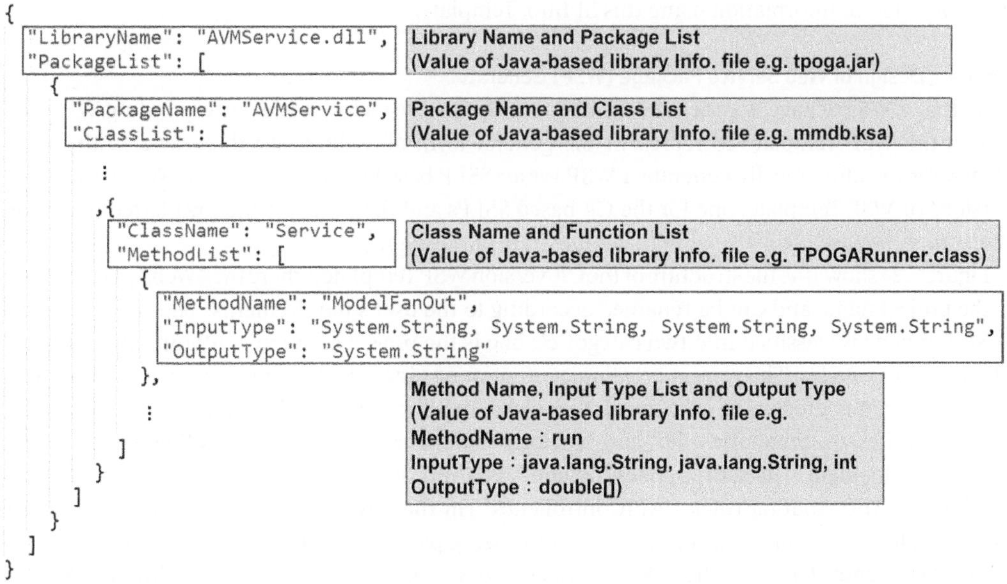
{
"LibraryName": "AVMService.dll",
"PackageList": [
Library Name and Package List
(Value of Java-based library Info. file e.g. tpoga.jar)
"PackageName": "AVMService",
"ClassList": [
Package Name and Class List
(Value of Java-based library Info. file e.g. mmdb.ksa)
"ClassName": "Service",
"MethodList": [
Class Name and Function List
(Value of Java-based library Info. file e.g. TPOGARunner.class)
"MethodName": "ModelFanOut",
"InputType": "System.String, System.String, System.String, System.String",
"OutputType": "System.String"
Method Name, Input Type List and Output Type
(Value of Java-based library Info. file e.g.
MethodName : run
InputType : java.lang.String, java.lang.String, int
OutputType : double[])

Figure 7.22 Illustration of the Lib. Info. Template in JSON. *Source:* Reprinted with permission from Ref. [25]; © 2018 IEEE.

```
[
 {
  "LibraryName": "AVMService.dll",
  "APIType": "POST",
    "PackageList": [
     {
       "PackageName": "AVMService",
         "ClassList": [
          {
            "ClassName": "Service",
              "MethodInfo": [
               {
                  "MethodName": "ModelFanOut",
                  "InputType": "System.String, System.String, System.String, System.String",
                  "OutputType": "System.String"
               },
                :
              ]
          }
         ]
     }
    ]
 }
]
```


Figure 7.23 Illustration of the SI Info. Template in JSON. *Source:* Reprinted with permission from Ref. [25]; © 2018 IEEE.

7.6.4.3 Design of Service Interface Information (SI Info.) Template

As mentioned previously, the SI information should be method-centric, which contains only the information of selected SIs in a file. Thus, we adopt JSON to design a generic SI Info. Template, as shown in Figure 7.23. The selected methods, such as the ModelFanOut method of AVMService.dll, can store their SI information using this SI Info. Template.

7.6.4.4 Design of Web Service Package (WSP) Generator

Note that a WS package is generated by building the source codes in the corresponding WSP. We design the WSP Template to represent the general form of a WSP and then develop the WSP Generator to automatically generate a WSP for an SSLP based on WSP Template. We provide two versions of WSP Template: one for the C#-based SSLPs and the other for the Java-based ones. For illustration, we use the C# version to demonstrate our design.

Figure 7.24 shows the file structure of the C#-version WSP Template, where the "WebApiTemplate" is the project name and can be renamed according to the user's requirements. The content of the WSP Template is classified into two categories: one is the project creation configurations, and the other is source codes. The project creation configurations include: "WebApiTemplate.csproj" and "compile.bat" for processing all compilation tasks, the directory "bin" for placing the generated executable files, the directories "lib" and "packages" for placing required library files and packages during development. The source codes include "WebApiConfig.cs" (in the directory "App_Start") for defining URL routing rules, "APIController.cs" (in the directory "Controller") for providing Web APIs for SSLPs, and "Parameter.cs" for processing parameter variables used in a WSP.

The "APIController.cs" in the WSP Template is designed to be rewritten to fit different SSLPs. The basic idea is to act as a wrapper of the selected methods of an SSLP so as to transform those methods to be web APIs (or called web methods) in a RESTful cloud service. Figure 7.25 shows an

Figure 7.24 C# WSP template. *Source:* Reprinted with permission from Ref. [25]; © 2018 IEEE.

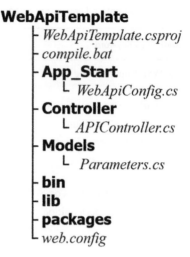

```
WebApiTemplate
├ WebApiTemplate.csproj
├ compile.bat
├ App_Start
│   └ WebApiConfig.cs
├ Controller
│   └ APIController.cs
├ Models
│   └ Parameters.cs
├ bin
├ lib
├ packages
└ web.config
```

Figure 7.25 Example of "APIController.cs" for C# WSP template. *Source:* Reprinted with permission from Ref. [25]; © 2018 IEEE.

```
9  using Newtonsoft.Json.Linq;
10 //@WSP:ImportDllFile
11
12 namespace //@WSP:NamespaceDefinition
13 {
14     public class //@WSP:ClassDefinition
15     {
16         //@WSP:HttpType
17         public string //@WSP:FunctionName
18         {
19             try{
20                 //@WSP:ClassDefinitionForSSLP
21                 //@WSP:InputTypeDefinition
22                 //@WSP:OutputTypeDefinition
23                 //@WSP:ExecutionMethodForSSLP
24                 //@WSP:SuccessReturn
25             }
26             catch(Exception e){
27                 return e.ToString();
28             }
29         }
```

example of "APIController.cs" for the WSP Template, which consists of three parts. The first part is the required packages (Lines 9-11), such as importing "Newtonsoft.Json.Linq" for processing JSON data. The second part contains the definition of the package name (e.g. Line 12: "namespace //@WSP:NamespaceDefinition") and the definition of the class (e.g. Line 14: "public class //@WSP:ClassDefinition"). The third part is the definition of a web method (Lines 16-29). Note that some SSLP attributes and the KI of the web method are stored in the SI Info. file, and they are retrieved at runtime. Hence, the annotated keywords, whose prefix is "//@WSP," are invented for the WSP Template to represent properties of SSLPs so that the WSP Generator can identify them during code generation. In this example, ten annotated keywords are used to indicate the properties of SSLPs from "//@WSP:ImportDllFile" (Line 10) to "//@WSP:SuccessReturn" (Line 24).

For successfully interpreting annotated keywords, the Annotated Keywords-SI Tags Lookup Table (ASLT), as shown in Table 7.4, is invented to present the mapping between annotated

Table 7.4 Annotated keywords-SI tags lookup table.

Annotated Keyword	Tag in SI. Info. file	Code Generation Format
//@WSP:ImportDllFile	PackageName	using {PackageName}_WebApi.Model; using {PackageName};
//@WSP:NamespaceDefinition	PackageName	{PackageName}_WebApi.Controllers
//@WSP:ClassDefinition	ClassName	{ClassName}Controller: ApiController
//@WSP:HttpType	APIType	[Http{APIType}]
//@WSP:FunctionName	MethodName	{MethodName}([FromBody]Parameters parameters)
//@WSP:ClassDefinitionForSSLP	ClassName	{ClassName} obj = new {ClassName}();
//@WSP:InputTypeDefinition	InputType	String[] Data = JSONDecoders; DecoderJsonArray(paremeters.postData); {InputType} input1 = Data[0]; {InputType} input2 = Data[1];
//@WSP:OutputTypeDefinition	OutputType	{OutputType} result;
//@WSP:ExcutionMethodForSSLP	MethodName	rsult=obj.{MethodName}(input1,input2);
//@WSP:SuccessReturn	—	Return result.ToString();

Source: Reprinted with permission from Ref. [25]; © 2018 IEEE.

keywords and the associated tags in the SI Info. File. There are three fields in the ASLT: Annotated Keyword, Tag in the SI Info. file, and Code Generation Format. In the following, the first row is used to explain the usage of ASLT. The first two fields indicate that the annotated keyword "//@WSP:ImportDllFile" corresponds to the tag "PackageName" in the SI Info. File. The third field describes that "//@WSP:ImportDllFile" can be replaced by the format in the third field with substituting values of the tag "PackageName" in the SI. Info. File into "{PackageName}" in the rule. Other rows in ASLT can be interpreted in the same manner.

Figure 7.26 shows the flowchart of generating automated source code: "APIController.cs" for a given SI. Info. File. The first step is to copy the template file "APIController.cs" as the new one for the given SSLP. The second step is to read source codes of a line (say, i) from the new file having n lines inside. If i>n, meaning that all lines are processed, and the process ends. Otherwise, the source code of this line is processed in the following steps. The third step is to check whether the Line i includes the annotated prefix (i.e. "//@WSP"). If no annotated prefix exists in Line i, go to Step 2 for processing the next line. Otherwise (i.e. an annotated prefix exists), continue to the next step. The fourth step is to find the values of "Tag in SI Info. File" and "Code Generation Format" from ASLT for the corresponding annotated keyword in Line i. The fifth step is to retrieve the values of the tag obtained in the last step from the SI Info. File. The sixth step is to generate codes for the annotated keyword of Line i by using the corresponding code generation format retrieved from ASLT in Step 4. Finally, the annotated keyword of Line i in the new "APIController.cs" file is replaced with the generated codes in the seventh step.

Figure 7.27 shows the result of the automated code generation by using the WSP template in Figure 7.25. The process of automatically generating the codes in Figure 7.27 is explained as follows. The generated code in Line 18 of Figure 7.27 indicates that the name of the web API to be built is "ModelFanOut." By referring to the SI Info. Template in Figure 7.23, the KI of this web API stored in the generated SI Info. file includes the following "key":"value" pairs: "MethodName":"ModelFanOut", "ClassName":"Service", "PackageName":"AVMService", "InputType":"System.String, System.String,

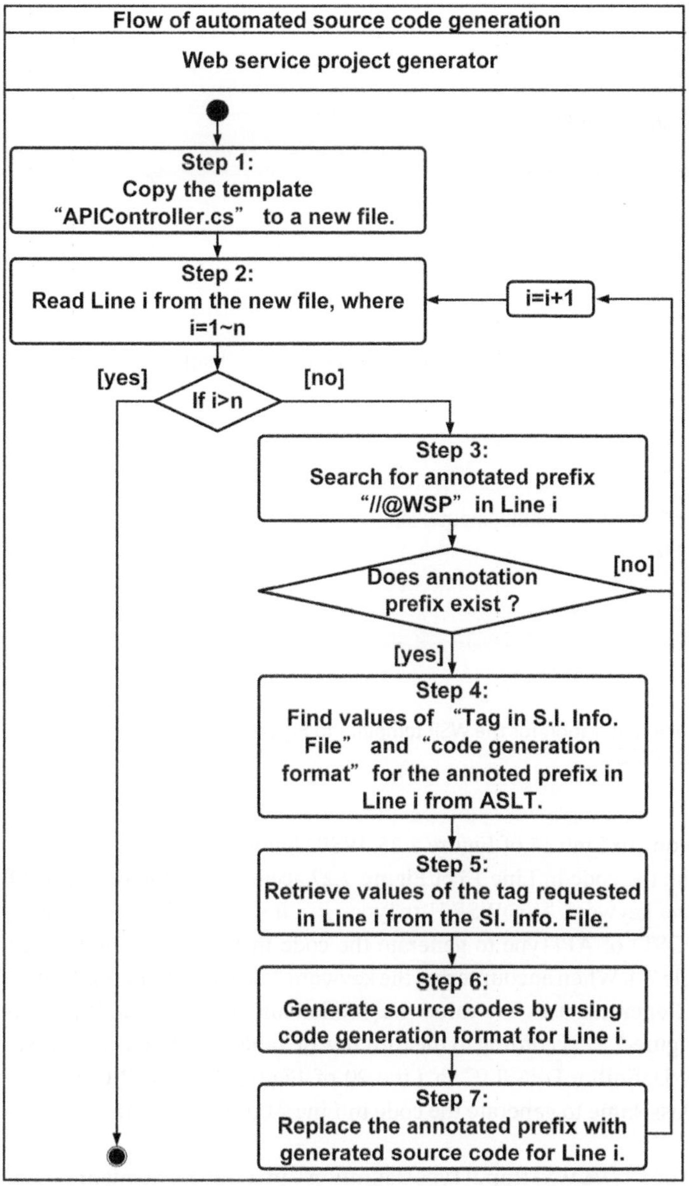

Figure 7.26 Flowchart of the automated source code generation. *Source:* Reprinted with permission from Ref. [25]; © 2018 IEEE.

System.String, System.String", and "OutputType":"System.String". When reading through the WSP template, WSP Generator encounters the first annotated keyword "//@WSP:ImportDllFile" in Line 10 of Figure 7.25, and it then uses the value (i.e. AVMService) of PackageName to generate the codes in Lines 10 and 11 of Figure 7.27 according to the code-generation format defined in the first row of the Annotated Keywords-SI Tags Lookup Table (ASLT) in Table 7.4.

In the same manner, when encountering the keyword "//@WSP:NamespaceDefinition" in Line 12 of Figure 7.25, WSP Generator uses the value (i.e. AVMService) of PackageName to generate the code in Line 13 of Figure 7.27 according to the second row of Table 7.4. When encountering the

```
9  using Newtonsoft.Json.Linq;
10 using AVMService_WebApi.Model;
11 using AVMService;
12
13 namespace AVMService_WebApi.Controllers
14 {
15    public class ServiceController : ApiController
16    {
17       [HttpPost]
18       public string ModelFanOut([FromBody]Parameters parameters)
19       {
20          try{
21             Service obj = new Service();
22             string[] Data = JSONDecoders.DecodeJSONArray(parameters.postData);
23             string input1 = Data[0];
24             string input2 = Data[1];
25             string input3 = Data[2];
26             string input4 = Data[3];
27             string result;
28             result = obj.ModelFanOut(input1, input2, input3, input4);
29             return result.ToString();
30          }
31          catch(Exception e) {
32             return e.ToString();
33          }
34       }
```

Figure 7.27 Result of automated code generation for the WSP template in Figure 7.25. *Source:* Reprinted with permission from Ref. [25]; © 2018 IEEE.

keyword "//@WSP:ClassDefinition" in Line 14 of Figure 7.25, WSP Generator uses the value (i.e. Service) of ClassName to generate the code in Line 15 of Figure 7.27 according to the third row of Table 7.4. When encountering the keyword "//@WSP:HttpType" in Line 16 of Figure 7.25, WSP Generator uses the value (i.e. POST) of APIType to generate the code in Line 17 of Figure 7.27 according to the fourth row of Table 7.4. When encountering the keyword "//@WSP:FunctionName" in Line 17 of Figure 7.25, WSP Generator uses the value (i.e. ModelFanOut) of MethodName to generate the code in Line 18 of Figure 7.27 according to the fifth row of Table 7.4. When encountering the keyword "//@WSP:ClassDefinitionForSSLP" in Line 20 of Figure 7.25, WSP Generator uses the value (i.e. Service) of ClassName to generate the code in Line 21 of Figure 7.27 according to the sixth row of Table 7.4.

When encountering the keyword "//@WSP:InputTypeDefinition" in Line 21 of Figure 7.25, WSP Generator uses the value (i.e. System.String, System.String, System.String, System.String) of InputType to generate the codes in Lines 22–26 of Figure 7.27 according to the seventh row of Table 7.4. When encountering the keyword "//@WSP:OutputTypeDefinition" in Line 22 of Figure 7.25, WSP Generator uses the value (i.e. System.String) of OutputType to generate the code in Line 27 of Figure 7.27 according to the eighth row of Table 7.4. When encountering the keyword "//@WSP:ExcutionMethodForSSLP" in Line 23 of Figure 7.25, WSP Generator uses the value (i.e. ModelFanOut) of MethodName to generate the code in Line 28 of Figure 7.27 according to the ninth row of Table 7.4. When encountering the keyword "//@WSP:SuccessReturn" in Line 24 of Figure 7.25, WSP Generator generates the code in Line 29 of Figure 7.27 according to the tenth row of Table 7.4.

Other lines in the WSP template in Figure 7.25 without annotated keywords are not changed.

7.6.4.5 Design of Service Constructor

Figure 7.28 shows the automated service construction mechanism designed in Service Constructor. Steps 1-3 are used to retrieve the SI Info file to get its key information whose contents are shown in this figure. Steps 4 and 5 are used to duplicate the corresponding annotated text-based WSP

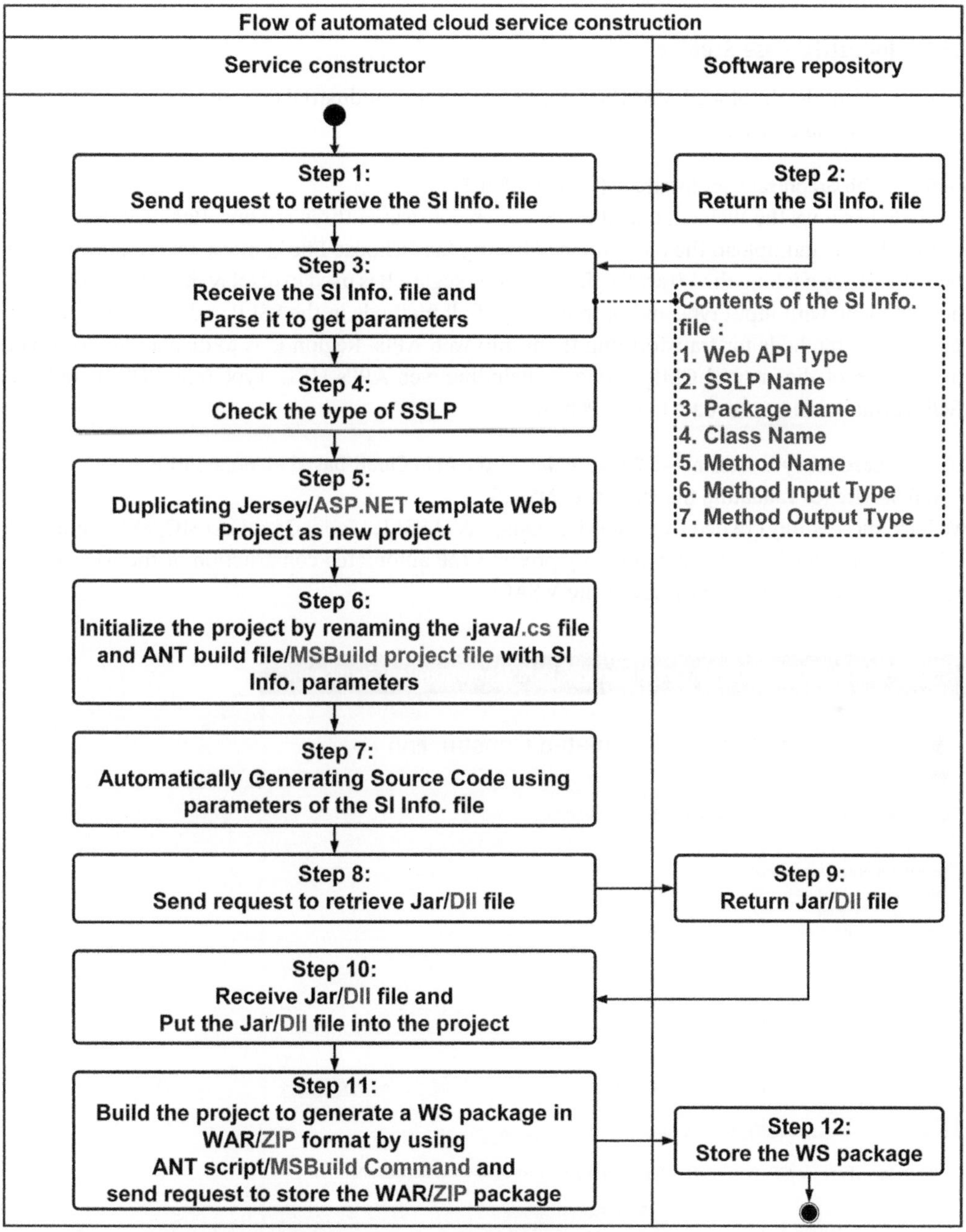

Figure 7.28 Automated service construction mechanism designed in Server Constructor. *Source:* Reprinted with permission from Ref. [25]; © 2018 IEEE.

template as a new WSP according to the type of SSLP, Jar, or DLL. Step 6 is to initialize the new WSP by building it using command scripting (i.e. command-line instructions). Step 7 is used to automatically generate the source code by replacing the annotated texts with the corresponding SI Info. Parameters. Steps 8-10 are used to include the SSLP into the WSP to complete the automated generation of WSP. Step 11 is used to automatically build the WSP using command scripting to generate the WS package, which is then stored into Software Repository in Step 12.

7.6.5 Industrial Case Studies

In this section, the developed MSACS is applied to conduct industrial case studies to demonstrate its feasibility and efficiency.

7.6.5.1 Web Graphical User Interface (GUI) of MSACS

Figure 7.29 shows the Web GUI of MSACS, which consists of three regions. Region A allows the user to choose and upload the SSLP file by clicking the "Choose File" button and to start parsing the SSLP file by clicking the "Start Parsing" button. Region B is used to display the information (i.e. method name with input type and output type) of all methods in the SSLP and allows the user to select target methods for transforming them into web APIs. Region C is to display the web API information of the created CMfg services, including web API's URL, Type (e.g. POST), and the JSON format used to post data to the web API.

7.6.5.2 Case Study 1: Automated Construction of the AVM Cloud-based Manufacturing (CMfg) Service for Validating the Efficacy of MSACS

We have successfully created CMfg services using AVM SSLP, Model Creation (MC) SSLP, and IYM SSLP via the MSACS. The following only presents the automated construction of the AVM CMfg service to demonstrate the efficacy of the MSACS.

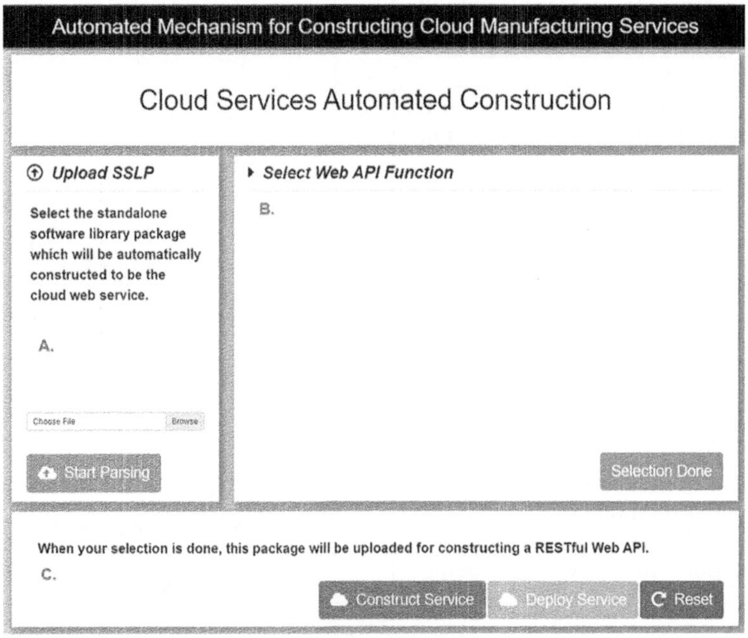

Figure 7.29 Web GUI of MSACS. *Source:* Reprinted with permission from Ref. [25]; © 2018 IEEE.

In this case study, we apply the MSACS to create an AVM CMfg service for conducting virtual metrology for a piece of Ultra Low-k materials Chemical Vapor Deposition (ULKCVD) equipment in a semiconductor fab. This type of equipment contains three process chambers, each of which has two sides. However, to assure the production stability and efficiency, only two chambers are operating in normal production. Because each side of a chamber can process a wafer at a time, a combination is used to refer to a side of a process chamber in this case study. Thus, a combination will request the AVM service to perform a virtual-metrology task that refers to predicting the production quality of a wafer.

In the cloud-based AVM system as manually developed in [32], the virtual machine administrator (VMA) server plays a key role in the system's operations. The VMA server was implemented in the "AVMService.dll" SSLP. It contained the following five methods: (1) the SetCombination method used to assign the combination to the VM server, (2) the ModelFanOut method used to download VM models to the target VM server, (3) the ActivateAVM method used to activate the VM server to perform VM tasks, (4) the TransferCommandToVMM method used to transfer commands to the VM manager, and (5) the RefreshAVM method used to refresh the VM models in the VM server using new measurement data.

The automated construction of the AVM CMfg service using the "AVMService.dll" SSLP via the MSACS is illustrated as follows:

Step 1: By using the web GUI depicted in Figure 7.29, the user logs in the MSACS.

Step 2: The user uploads the "AVMService.dll" SSLP by clicking the "Choose File" button in the Web GUI's Region A shown in Figure 7.29.

Step 3: The user clicks the "Start Parsing" button in region A. Then, the MSACS starts to parse the "AVMService.dll" and display the information (i.e. method name with input type and output type) of all methods in the SSLP on the Web GUI's Region B, which allows the user to select target methods for transforming them into web APIs, as shown in Figure 7.30.

Step 4: The user selects the ModelFanOut, TransferCommandToVMM, RefreshAVM, SetCombination, ModelFanOut, and ActivateAVM methods and clicks "Selection Done" button on the web GUI's Region B to generate the SI Info. File.

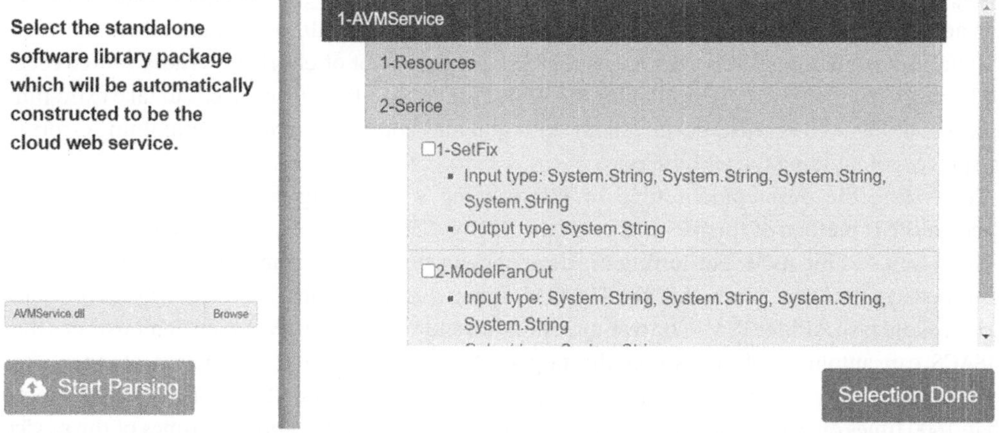

Figure 7.30 GUI showing the method information in the AVMService.dll. *Source:* Reprinted with permission from Ref. [25]; © 2018 IEEE.

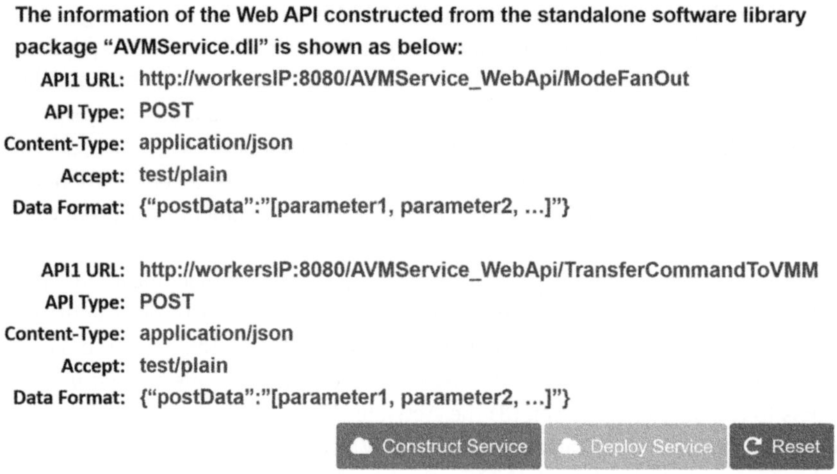
The information of the Web API constructed from the standalone software library package "AVMService.dll" is shown as below:
API1 URL: http://workersIP:8080/AVMService_WebApi/ModeFanOut
API Type: POST
Content-Type: application/json
Accept: test/plain
Data Format: {"postData":"[parameter1, parameter2, ...]"}
API1 URL: http://workersIP:8080/AVMService_WebApi/TransferCommandToVMM
API Type: POST
Content-Type: application/json
Accept: test/plain
Data Format: {"postData":"[parameter1, parameter2, ...]"}
Construct Service
Deploy Service
Reset

Figure 7.31 GUI showing the web API information of the AVM service. *Source:* Reprinted with permission from Ref. [25]; © 2018 IEEE.

Step 5: The user clicks the "Construct Service" button, and then the information of the web APIs of the created AVM CMfg service is displayed on the web GUI's Region C, as shown in Figure 7.31, indicating that the AVM CMfg service has been automatically constructed.

After the AVM CMfg service is created, we further develop a GUI to operate the AVM CMfg service to conduct virtual metrology. As shown in Figure 7.32, the AVM CMfg service can perform virtual metrology well, the predicted values closely following the actual values (star points).

7.6.5.3 Case Study 2: Performance Evaluation of MSACS

To evaluate the performance of MSACS, in this case study we compare the development times of creating a CMfg web service (a web API) using C# SSLP and Java SSLP, respectively, by the MSACS, a skilled programming professional (professional for short), and a novice. Here, the professional refers to a person who has great skills and experiences in both C# and Java web-services programming, while the novice refers to a person who knows how to do programming in C# and Java, but has no experience of building web services. As mentioned in the fifth paragraph of Section 7.6.1, manually constructing a CMfg service using SSLP involves a lot of efforts in writing source codes and doing dependency configuration for the WSP. Particularly, the novice needs to take extra time to learn various web-services development skills and may commit assorted programming errors in the process of building the CMfg service.

For testing the development time of constructing a C# CMfg service, we transform the ExecuteBPNN method of the BPNN class of the BPNN_CSharp package in BPNN_CSharp.dll into a cloud web API for AVM. For testing the development time of constructing a Java CMfg service, we transform the "run" method of the TPOGARunner class of the mmdb.ksa package in tpoga.jar to be a cloud web API for IYM. The testing results are shown in Table 7.5, which demonstrates that MSACS can automatically construct the target CMfg service in a very efficient manner after uploading the required SSLPs, about 30 times on average quicker than the professional and more than 1600 times on average quicker than the novice. Notably, the development times of the novice in Table 7.5 include 22 hours in the C# case and 29 hours in the Java case to learn web-services programming skills and tools. This implies that building CMfg services could become a burden for

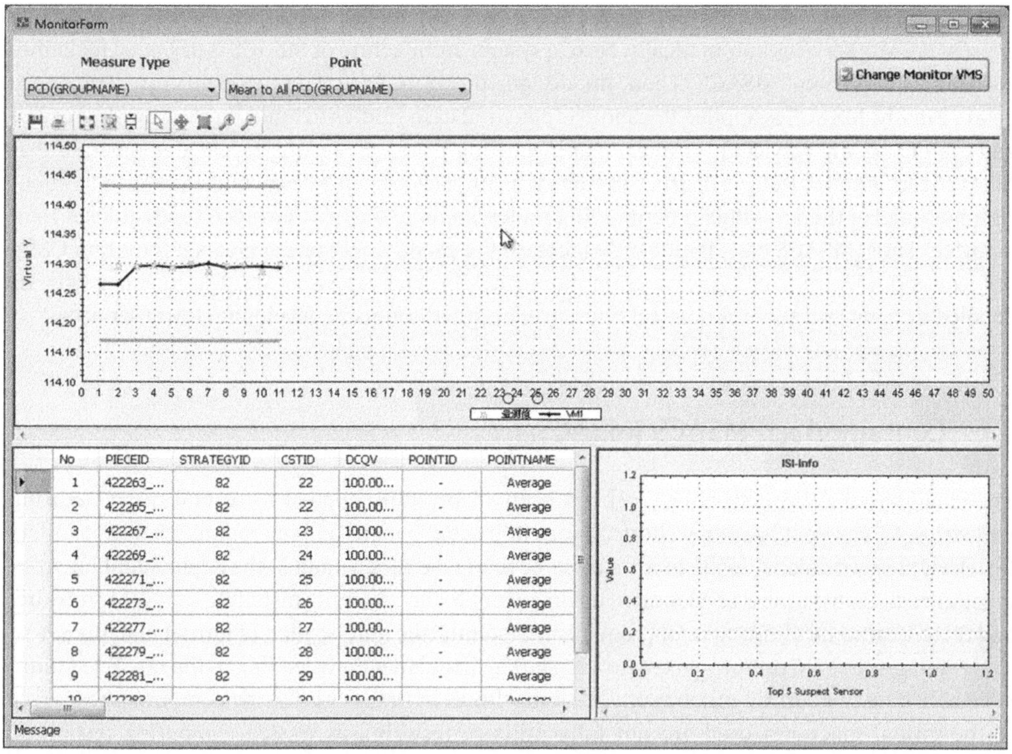

Figure 7.32 GUI for conducting virtual metrology using the created AVM CMfg service. *Source:* Reprinted with permission from Ref. [25]; © 2018 IEEE.

Table 7.5 Comparisons of development times of Novice, Skilled Professional, and MSACS.

Experiments	Novice	Skilled programing professional	MSACS
Construction of an AVM Web API using a C# SSLP	23 h 2 min	10.88 min	0.4 min
Construction of an IYM Web API using a Java SSLP	30 h 30 min	33.18 min	1.12 min

Source: Reprinted with permission from Ref. [25]; © 2018 IEEE.

novice engineers in the factory. Overall, the testing results show that MSACS can significantly increase productivity and reduce the burden of engineers in terms of building CMfg services, particularly when the number of web APIs to be built is large.

7.6.6 Summary

In recent years, CMfg has emerged as a new paradigm that can revolutionize the manufacturing industry. How to build CMfg services in an automated and efficient manner is an essential and challenging subject for further promotion of CMfg. Aiming at facilitating the rapid construction of CMfg services, this section presents the MSACS using SSLPs of two mainstream languages, Java and C#. Specifically, MSACS uses a three-phase workflow to clearly address the issues of how to construct CMfg services automatically. Next, a system architecture of MSACS is designed to delineate how to implement MSACS. Then, the designs of MSACS's core components are depicted in detail. Finally, MSACS is applied to conduct industrial case studies to build the AVM CMfg service and the IYM CMfg service for a smart manufacturing platform. Testing results demonstrate that MSACS can automatically construct the target CMfg services in a very efficient manner, about 30 times quicker than a skilled programming professional, and more than 1600 times quicker than a novice. Thus, MSACS can significantly increase the productivity of engineers in building CMfg services and, in turn, can help the promotion of CMfg. The proposed MSACS, together with its detailed designs, can serve as a useful reference for practitioners in developing CMfg services.

7.7 Containerized MSACS (MSACS$_C$)

The cloud-based AVM system [32, 33] has been successfully applied to several manufacturing industries. However, it has some shortcomings in its deployment and operations. Firstly, it suffers the DevOps inconsistency issue that we often need to take time to adjust the deployment environment to be the same as the development environment for successfully installing the AVM software, and these additional efforts are not productive. Secondly, the auto-scaling of virtual machines is so slow, taking about 5-6 minutes to create and start a virtual machine, leading to the possible failure of establishing the desired number of virtual machines in time. Thirdly, the computing resources of the virtual machines used are not fully utilized, resulting in wasted computing resources. Fourthly, the cloud-based AVM system lacks high-availability designs, such as the load balance, health check, and failover for the robust operations of AVM Servers.

As mentioned in Section 7.1, the newest virtualization technology-Docker Container [10] has many advantages over the traditional virtualization technology-Virtual Machine. Docker containers allow the containerized CMfg services to be built once and then be deployed to and run on any environment with Docker Engine installed. Also, Docker containers can be used to resolve the DevOps inconsistency problem and providing CI/CD pipelines. For leveraging the benefits of container technologies to enhance the CMfg services and resolving the above mentioned shortcomings encountered by the cloud-based AVM system, the cloud part of the AMCoT framework also provides an infrastructure to host and run the containerized CMfg services. Specifically, Docker containers are utilized to encapsulate and run the CMfg services, and Kubernetes (K8s) is used to manage and orchestrate the containerized CMfg services. However, manually constructing containerized CMfg services involves more tasks than manually building CMfg services. Thus, an automated construction scheme of the containerized CMfg services (namely MSACS$_C$) is developed on the cloud part of the AMCoT framework for building and deploying the containerized CMfg services more efficiently.

MSACS$_C$ is an extended version of MSACS. Figure 7.33 shows the four-phase workflow of MSACS$_C$, which consists of the Library Analysis Phase, Web API Project Construction Phase, Manufacturing Service Image Construction Phase, and Containerized Manufacturing Service Deployment Phase. They are briefed below.

1) Library Analysis Phase:
   In this phase, we adopt JSON to design a generic Lib. Info. Template, which is used to specify the data structure of how the hierarchical information of the library is stored. The KI of the library includes library name, class name, method name, and each method's input parameters

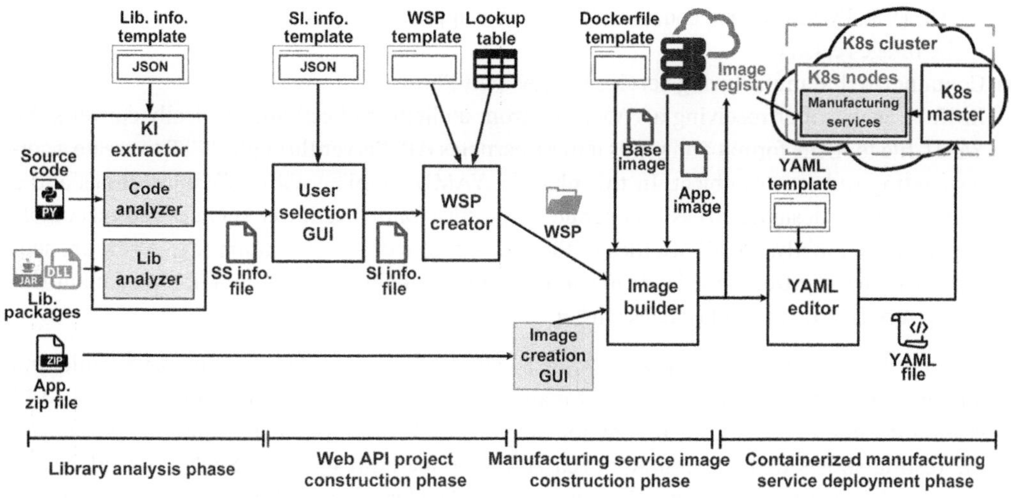

Figure 7.33 Four-phase workflow of MSACS$_C$.

and output parameter. In addition, we design KI Extractor to extract the KI of the input library. The extracted KI is then stored in a Lib. Info. File according to the JSON format specified in Lib. Info. Template. In addition to Java Jar, and C# DLL, MATLAB DLL libraries, the KI extractor of the MSACS$_C$ can extract the KI information of Python source codes as well.

2) Web API Project Construction Phase:
We adopt JSON to design a generic SI Info. Template to specify the data structure of how the SI information is stored. In addition, we develop GUI to allow the user to select the desired methods for transforming them into web methods. The SI information of these selected methods is then stored in the SI info. File according to the JSON format specified in the SI Info. Template.

Because the Java WSP, C# WSP, and Python web API have different structures, we design three WSP templates, one for generating Java WSPs, another for generating C# WSPs, and the other for generating Python web services. However, three WSP templates adopt the same design principle. The WSP template consists of an annotated source code template file, required library files, and dependency configuration files, which are stored in their respective directories. The annotated source code template file serves as a wrapper for transforming the selected methods of the input library to web methods, and it contains several annotation keywords to annotate the necessary SI information for building the web methods. Then, we design the WSP Generator to automatically search and correlate a lookup table and the input SI Info. File using annotation keywords as keys and further replace the annotation keywords in the annotated source code template file with the corresponding SI information extracted from the SI Info. File. By going through the above process, the WSP Generator can convert the WSP template to generate a WSP.

3) Manufacturing Service Image Construction Phase:
A Dockerfile is a script file for building a Docker image. In this phase, Dockerfile templates and corresponding shell script files are utilized to automatically build the images of the web services constructed in the previous phase. We design different Dockerfile templates for different types of web service projects and executable files (i.e. .exe files). The image builder mainly consists of several shell scripts, which can automatically generate the images files of web services of Java

Jar, C# DLL, Python source codes, and executable applications and can then store these images to the image registry.

4) Containerized Manufacturing Service Deployment Phase:
   In Kubernetes, after receiving a YAML file from a client, Kubectl automatically converts the YAML file to JSON format and sends it to Kubernetes API Server through HTTP to create a corresponding Kubernetes object. In this phase, a YAML template with different settings and a YAML editor is designed to easily generate YAML files for different web services and executable applications. Finally, the YAML files of the web services of an application (e.g. AVM) are sent to Kubernetes Control Plane Node to complete the deployment of the containerized manufacturing service onto the Kubernetes cluster.

By using the four-phase workflow of $MSACS_C$ shown in Figure 7.33, we can easily build and deploy the containerized CMfg services. We apply the $MSACS_C$ to build the containerized AVM service for conducting virtual metrology for the wheel machining industry in an industrial case study. Testing shows that the containerized AVM service built by the $MSACS_C$ can effectively resolve the shortcomings of the existing cloud-based AVM system, such as suffering the DevOps inconsistency issue, slow in auto-scaling, computing resources not been fully utilized, and lacking high-availability designs. The $MSACS_C$ consists of system architecture, a generic workflow of building app images of the containerized AVM service, and several core functional mechanisms, such as automatic dispatching of a virtual metrology task, horizontal auto-scaling, load balance, and failover for the robust operations of the containerized AVM service. The $MSACS_C$ is a promising way to bring advantages of container technologies into the virtual metrology and other CMfg applications for manufacturing industries.

7.8 Conclusion

Intelligent manufacturing platforms are essential to realizing smart factories that can provide more flexible and efficient production processes to manufacture higher-quality goods at reduced costs. The AMCoT framework is developed by iMRC, National Cheng Kung University, Taiwan, ROC, which can facilitate building intelligent manufacturing platforms. Several intelligent manufacturing platforms based on the AMCoT framework have been successfully applied to various manufacturing industries. According to the architecture of the AMCoT Framework designed in Chapter 6, this chapter further describes the key components of the AMCoT framework and how to design and implement them by leveraging advanced ICTs, such as CC, Edge Computing, IoT, BDA, and Docker Container.

This chapter begins with an introduction in Section 7.1 to bring in the essentials of each section. Then, the key components of the AMCoT framework are described in Section 7.2. The AMCoT framework consists of two parts: the factory part and the cloud part. The factory part of the AMCoT framework comprises many edge devices, each of which is called a CPA. The CPA can offer equipment various in-situ edge manufacturing services, such as data collection, data preprocessing, communication, and other application modules. As such, the CPA can enable the AMCoT framework to have cyber-physical interactions. Section 7.3 describes the framework designs of CPA and CPA_C. Then, the rapid construction scheme of CPAs, RCS_{CPA}, based on Docker and Kubernetes, is presented in Section 7.4.

The cloud part of the AMCoT framework provides the infrastructure required to host and run the CMfg services. These prediction-related and BDA-related CMfg services are implemented in the form of RESTful web service and can thus be used on-demand over networks by the manufacturing industry. However, manually constructing CMfg services is cumbersome and error-prone. Thus, Section 7.6 depicts the MSACS for building CMfg services efficiently. Furthermore, the $MSACS_C$ based on Docker and Kubernetes is introduced in Section 7.7 for building and deploying the containerized CMfg services efficiently. Also, a BDA application platform built on the cloud is illustrated in Section 7.5 for storing and processing factory-wide big data related to production.

The contents of this chapter can enhance manufacturing industries with the knowledge of how to implement and utilize intelligent manufacturing platforms. Also, the practitioners in manufacturing industries can refer to the contents for building their desired intelligent manufacturing platforms.

Appendix 7.A – Abbreviation List

AMCoT	Advanced Manufacturing Cloud of Things
AO	Assembly Object
ASLT	Annotated Keywords-SI Tags Lookup Table
AVM	Automatic Virtual Metrology
BDA	Big Data Analytics
BEDPS	Efficient Big Data Processing Scheme
BPNN	Back Propagation Neural Network
BPM	Baseline Predictive Maintenance
BPM_C	Containerized Baseline Predictive Maintenance Module
CA	Containerized Application
CA_{ASC}	Application Source Codes of CA
CA_{CBI}	Container Base Image of CA
CA_{KCF}	Kubernetes Configuration Files of CA
CA_{SS}	Shell Scripts related to CA
CC	Cloud Computing
CI/CD	Continuous Integration/Continuous Delivery
CK	Control Kernel
CK_C	Containerized Control Kernel
CMfg	Cloud Manufacturing
CPA	Cyber-Physical Agent
CPA_C	Containerized Cyber-Physical Agent
DCM	Data Collection Manager
DCP	Data Collection Plan
DCR	Data Collection Report
DevOps	Development and Operations
EDI	Equipment Driver Interface
EDM	Electrical Discharge Machining
FC	Function Content
FDC	Fault Detection and Classification
GEM	Generic Equipment Model
HDFS	Hadoop Distributed File System
HTML	Hypertext Markup Language
HTTP	Hypertext Transfer Protocol
ICT	Information and Communication Technology
IDE	Integrated Development Environment

IIS	Internet Information Services
iMRC	Intelligent Manufacturing Research Center
IoT	Internet of Things
IPM	Intelligent Predictive Maintenance
IYM	Intelligent Yield Management
JEO	JarEntry Object
JISO	JarInputStream Objec
JSON	JavaScript Object Notation
K8s	Kubernetes
KI	Key Information
MC	Model Creation
MO	MethodInfo Object
MSACS	Manufacturing Services Automated Construction Scheme
$MSACS_C$	Containerized Manufacturing Services Automated Construction Scheme
$MySQL_C$	Containerized MySQL
OS	Operation System
PAM	Pluggable Application Module
PAM_C	Containerized-PAM
PdM	Predictive Maintenance
PECVD	Plasma-Enhanced Chemical Vapor Deposition
PGWD	Parallel Gap-based Wave Detection
QoS	Quality of Service
RCS_{CPA}	Rapid Construction Scheme of CPA
RDB	Relational Database
RDMS	Relational Database Management System
REST	Representational State Transfer
RI	Reliance Index
RUL	Remaining Useful Life
SCOS	Service Composition Optimal-Selection
SDM	Supply-Demand Matching
SECS	SEMI Equipment Communications Standard
SI	Service Interface
SSLP	Standalone Software Library Package
TO	Type Object
ULKCVD	Ultra Low-k materials Chemical Vapor Deposition
WS	Web Service
WSP	Web Service Project
XML	Extensible Markup Language
YAML	Yet Another Markup Language

Appendix 7.B – Patents (AMCoT + CPA)

USA Patents

1) Cheng, F.T., Huang, G.W., Chen, C.H., et al. (2004). Generic embedded device and mechanism thereof for various intelligent-maintenance applications. US Patent 7,162,394, filed 13 May 2004 and issued 9 January 2007.

2) Chen, C.C., Hung, M.H., Li, P.Y., et al. (2018). Automated constructing method of cloud manufacturing service and cloud manufacturing system. US Patent 10,618,137, filed 22 March 2018 and issued 14 April 2020.

Taiwan, ROC Patents

1) Cheng, F.T., Huang, G.W., Chen, C.H., et al. (2003). Generic embedded device and mechanism thereof for various intelligent-maintenance applications. ROC Patent I225606, filed 13 May 2003 and issued 21 December 2004.
2) Cheng, F.T., Su, Y.C., Huang, H.C., et al. (2005). Predictive maintenance system with generic embedded devices. ROC Patent I265437, filed 18 January 2005 and issued 1 November 2006.
3) Cheng, F.T., Su, Y.C., and Lin, R.C. (2005). Quality prognostics system and method for manufacturing processes with generic embedded devices. ROC Patent I269990, filed 4 February 2005 and issued 1 January 2007.
4) Chen, C.C., Hung, M.H., Li, P.Y., et al. (2018). Automated constructing method of cloud manufacturing service and cloud manufacturing system. ROC Patent I670672, filed 22 March 2018 and issued 1 September 2019.

Japan Patent

1) Cheng, F.T., Huang, G.W., Chen, C.H., et al. (2004). Generic embedded device and mechanism thereof for various intelligent-maintenance applications. JP Patent 4303640, filed 12 May 2004 and issued 1 May 2009.

China Patent

1) Chen, C.C., Hung, M.H., Li, P.Y., et al. (2018). Automated constructing method of cloud manufacturing service and cloud manufacturing system. CN Patent Pending 201810239672.6, filed 22 March 2018.

References

1 Kagermann, H., Wahlster, W., Helbig, J. (2013). *Recommendations for implementing the strategic initiative INDUSTRIE 4.0*. Germany: Forschungsunion.

2 Cheng, F.-T., Tieng, H., Yang, H.-C. et al. (2016). Industry 4.1 for wheel machining automation. *IEEE Robotics and Automation Letters (RA-L)* 1 (1): 332–339. https://doi.org/10.1109/LRA.2016.2517208.

3 Lin, Y.-C., Hung, M.-H., Huang, H.-C. et al. (2017). Development of advanced manufacturing cloud of things (AMCoT) - a smart manufacturing platform. *IEEE Robotics and Automation Letters (RA-L)* 2 (3): 1809–1816. https://doi.org/10.1109/LRA.2017.2706859.

4 Wang, C., Bi, Z., and Xu, L.D. (2014). IoT and cloud computing in automation of assembly modeling systems. *IEEE Transactions on Industrial Informatics* 10 (2): 1426–1434. https://doi.org/10.1109/TII.2014.2300346.

5 Zhou, Z., Chen, X., and Li, E. (2019). Edge intelligence: paving the last mile of artificial intelligence with edge computing. *Proceedings of the IEEE* 107 (8): 1738–1762. https://doi.org/10.1109/JPROC.2019.2918951.

6 Zhou, H. (2013). *The Internet of Things in the Cloud-A Middleware Perspective*. Boca Raton, FL, USA: CRC Press.

7 Bi, Z., Xu, L.D., and Wang, C. (2014). Internet of things for enterprise systems of modern manufacturing. *IEEE Transactions on Industrial Informatics* 10 (2): 1537–1546. https://doi.org/10.1109/TII.2014.2300338.

8 Chen, X.-W. and Lin, X. (2014). Big data deep learning: challenges and perspectives. *IEEE Access* 2: 514–525. https://doi.org/10.1109/ACCESS.2014.2325029.

9 Hu, H., Wen, Y., and Li, X. (2014). Toward scalable systems for big data analytics: a technology tutorial. *IEEE Access* 2: 652–687. https://doi.org/10.1109/ACCESS.2014.2332453.

10 Docker Inc (2020). Docker. www.docker.com ().

11 Morabito, R. (2017). Virtualization on internet of things edge devices with container technologies: a performance evaluation. *IEEE Access* 5: 8835–8850. https://doi.org/10.1109/ACCESS.2017.2704444.

12 The Linux Foundation (2020). Kubernetes Website. kubernetes.io (accessed 27 August 2020).

13 Bernstein, D. (2014). Containers and cloud: From LXC to Docker to Kubernetes. *IEEE Cloud Computing* 1 (3): 81–84. https://doi.org/10.1109/MCC.2014.51.

14 Khan, A. (2017). Key characteristics of a container orchestration platform to enable a modern application. *IEEE Cloud Computing* 4 (5): 42–48. https://doi.org/10.1109/MCC.2017.4250933.

15 Xu, X. (2012). From cloud computing to cloud manufacturing. *Robotics and Computer-Integrated Manufacturing* 28: 75–86. https://doi.org/10.1109/MCC.2017.4250933.

16 Wu, D., Greer, M.J., Rosen, D.W. et al. (2013). Cloud manufacturing: strategic vision and state-of-the-art. *Journal of Manufacturing Systems* 32 (4): 564–579. https://doi.org/10.1016/j.jmsy.2013.04.008.

17 Ren, L., Zhang, L., Wang, L. et al. (2014). Cloud manufacturing: key characteristics and applications. In: *International Journal of Computer Integrated Manufacturing*. https://doi.org/10.1080/0951192X.2014.902105.

18 Tao, F., Cheng, Y., Xu, L.D. et al. (2014). CCIoT-CMfg: cloud computing and internet of things-based cloud manufacturing service system. *IEEE Transactions on Industrial Informatics* 10 (2): 1435–1442. https://doi.org/10.1109/TII.2014.2306383.

19 Wu, D., Rosen, D.W., Wang, L. et al. (2015). Cloud-based design and manufacturing: a new paradigm in digital manufacturing and design innovation. *Computer-Aided Design* 59: 1–14. https://doi.org/10.1016/j.cad.2014.07.006.

20 Fielding, R.T. (2000). *Architectural styles and the design of network-based software architectures-chapter 5: representational state transfer (REST)*. Irvine: University of California.

21 Zhu, L., Weber, I., and Bass, L. (2015). *DevOps: A Software Architect's Perspective*. Boston: Addison-Wesley Professional.

22 Cheng, F.-T., Huang, H.-C., and Kao, C.-A. (2012). Developing an automatic virtual metrology system. *IEEE Transactions on Automation Science and Engineering* 9 (1): 181–188. https://doi.org/10.1109/TASE.2011.2169405.

23 Hung, M.-H., Chen, C.-F., Huang, H.-C. et al. (2012). Development of an AVM system implementation framework. *IEEE Transactions on Semiconductor Manufacturing* 25 (4): 598–613. https://doi.org/10.1109/TSM.2012.2206061.

24 Hsieh, Y.-M., Lin, C.-Y., Yang, Y.-R. et al. (2019). Automatic virtual metrology for carbon fiber manufacturing. *IEEE Robotics and Automation Letters (RA-L)* 4 (3): 2730–2737. https://doi.org/10.1109/LRA.2019.2917384.

25 Chen, C.-C., Hung, M.-H., Li, P.-Y. et al. (2018). A novel automated construction scheme for efficiently developing cloud manufacturing services. *IEEE Robotics and Automation Letters (RA-L)* 3 (3): 1378–1385. https://doi.org/10.1109/LRA.2018.2799420.

26 Liu, Y.-Y., Hung, M.-H., Lin, Y.-C., et al. (2018). A cloud-based pluggable manufacturing service scheme for smart factory. *2018 IEEE International Conference on Automation Science and Engineering (CASE 2018)*. https://doi.org/10.1109/COASE.2018.8560401.

27 Aaeon Europe B.V. (2020). UP Board. up-board.org ().

28 Docker Inc (2020). Docker Hub. hub.docker.com ().

29 Cloudera, Inc. (2020). Cloudera Apache Hadoop. https://www.cloudera.com/products/open-source/apache-hadoop.html (accessed 27 August 2020).

30 Chen, C.-C., Hung, M.-H., Suryajaya, B. et al. (2019). A novel efficient big data processing scheme for feature extraction in electrical discharge machining. *IEEE Robotics and Automation Letters (RA-L)* 4 (2): 910–917. https://doi.org/10.1109/LRA.2019.2891498.

31 Chen, C.-C., Lin, Y.-C., Hung, M.-H. et al. (2016). A novel cloud manufacturing framework with auto-scaling capability for machine tool industry. *International Journal of Computer-Integrated Manufacturing* 29 (7): 786–804. https://doi.org/10.1080/0951192X.2015.1125766.

32 Huang, H.-C., Lin, Y.-C., Hung, M.-H. et al. (2015). Development of cloud-based automatic virtual metrology system for semiconductor industry. *Robotics and Computer-Integrated Manufacturing* 34: 30–43. https://doi.org/10.1016/j.rcim.2015.01.005.

33 Hung, M.-H., Li, Y.-Y., Lin, Y.-C. et al. (2017). Development of a novel cloud-based multi-tenant model creation service for automatic virtual metrology. *Robotics and Computer-Integrated Manufacturing* 44: 174–189. https://doi.org/10.1016/j.rcim.2016.09.003.

34 Wang, P.W., Ding, Z.J., Jiang, C.J. et al. (2016). Automatic web service composition based on uncertainty execution effects. *IEEE Transactions on Services Computing* 9 (4): 551–565. https://doi.org/10.1109/TSC.2015.2412943.

35 Tao, F., Cheng, J., Cheng, Y. et al. (2017). SDMSim: a manufacturing service supply-demand matching simulator under cloud environment. *Robotics and Computer Integrated Manufacturing* 45 (5): 34–46. https://doi.org/10.1016/j.rcim.2016.07.001.

36 Jatoth, C., Gangadharan, G.R., and Buyya, R. (2017). Computational intelligence based QoS-aware web service composition: a systematic literature review. *IEEE Transactions on Services Computing* 10 (3): 475–492. https://doi.org/10.1109/TSC.2015.2473840.

37 Paik, I., Chen, W., and Huhns, M.N. (2014). A scalable architecture for automatic service composition. *IEEE Transactions on Services Computing* 7 (1): 82–95. https://doi.org/10.1109/TSC.2012.33.

8

Automatic Virtual Metrology (AVM)

Fan-Tien Cheng

Director/Chair Professor, Intelligent Manufacturing Research Center/Institute of Manufacturing Information and Systems, National Cheng Kung University, Tainan, Taiwan, ROC

8.1 Introduction

Virtual metrology (VM) can convert sampling inspection with metrology delay into real-time and online total inspection. Taking the thin film transistor-liquid crystal display (TFT-LCD) manufacturing process as an example, Figure 8.1 [2] depicts the comparison between actual metrology and VM. The upper part of the figure is the common approach adopted by most TFT-LCD factories for glass quality monitoring. In TFT-LCD production processes, each cassette contains 30 pieces of glass. For monitoring the abnormality of a production tool and assuring the quality of production glass, real measurement is performed only on the sampled production glass within each cassette. As shown in Figure 8.1, t_i and t_j are the completion time of processing the sampled glasspieces, and real measurement data of monitored glasspieces are obtained at time $t_i + \Delta T$ and $t_j + \Delta T$, respectively. That is, after completing processing sampled glasspieces, a waiting time (ΔT) of about two to four hours is needed to obtain the quality measurement data of those glasspieces. This approach assumes that the process quality of a production tool is unlikely to drop suddenly, and the measurement results of the sampled glasspieces can fully represent the product quality. However, in the real situation, only the quality of the sampled glasspieces is grasped. The quality of other production glasspieces beyond the measuring ones is unknown.

By contrast, through a VM system (VMS) as shown in the lower part of Figure 8.1, the VM values for not only the sampled glasspieces but also all the production glasspieces can be obtained by using on-line process data from the production tool. As depicted in the lower left portion of Figure 8.1, the VM value of the sample at t_{i+3} is out-of-control (OOC), then a warning message will be sent to ask for further quality inspection.

Weber [3] listed ten aspects one should be aware of about VM in January 2007. In fact, VM has been a critical issue and widely discussed for more than a decade. It becomes an even more challenging task as the manufacturing technology nodes advance and metrology measuring becomes costlier and more time-consuming. Therefore, International SEMATECH Manufacturing Initiative (ISMI) added VM into its next-generation factory realization road map of the semiconductor industry [4]. International Technology Road Map for Semiconductors (ITRS) also designated VM as one of the focus areas on Factory Information and Control Systems (FICS) and advanced process

Industry 4.1: Intelligent Manufacturing with Zero Defects, First Edition. Edited by Fan-Tien Cheng.

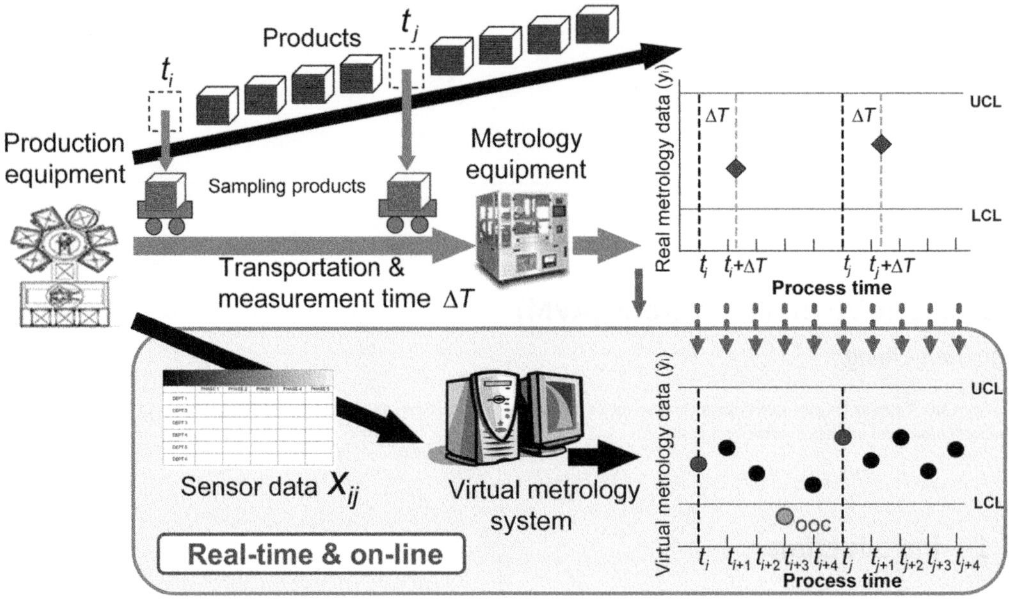

Figure 8.1 Comparison between actual metrology and virtual metrology. *Source:* Reprinted with permission from Ref. [1]; © 2015 IEEE.

control (APC) [5]. Besides, International Business Machines Corporation (IBM) Systems and Technology Group claimed that VM can create an opportunity for manufacturing improvement [6]. Evaluation of economic effects as the basis for assessing VM investment was performed by Koitzsch and Honold [7]. This study indicated that the benefit per fab per year is greater than 30 million US dollars. The calculations were performed for a 200 mm wafer fab of the 0.13 μm technology [7]. To sum up, it is beneficial to develop VM techniques.

8.1.1 Survey of Virtual Metrology (VM)-Related Literature

VM has been a mainstream topic at all three regional (U.S., Europe, and Asia) Advanced Equipment Control/Advanced Process Control (AEC/APC) symposiums. However, these AEC/APC symposium proceedings are not archived. For those archived VM-related articles, the paper presented by Su et al. [8], the first one searched via IEEE Xplore with "virtual metrology" as the key words, proposed a VM scheme based on back-propagation neural networks (BPNN) in 2004. Then, Monahan [9] expressed the need for the transition from actual metrology to VM for enabling design for manufacturability (DFM) and advanced process control (APC) at the 32 nm technology node in 2005. Moreover, Chen et al. [10] suggested applying VM to enable wafer-to-wafer (W2W) control without additional real metrology in the same year (2005). Su et al. [11, 12], then, proposed a processing quality prognostics scheme for plasma sputtering in TFT-LCD manufacturing in 2006. Chang et al. [13] also utilized a piecewise linear neural network and a fuzzy neural network to design a VM scheme in 2006. Hung et al. [14] applied radial basis function networks (RBFN) to develop a VM scheme for predicting chemical-vapor-deposition (CVD) thickness in semiconductor manufacturing in June 2007. Khan et al. published twin papers in November 2007 [15] and 2008 [16] to develop a distributed VM architecture for fab-wide VM and feedback control of semiconductor manufacturing processes by using recursive partial least squares (PLS). Wu et al. [17] studied the performance of run-to-run (R2R) control subject to metrology delay and concluded that applying VM to remedy the metrology-delay problem is justified if the error of the VM method is less than the error caused by stochastic process noise in August 2008. Zeng and Spanos [18] proposed a VM model for plasma etch operations in November 2009. Imai and Kitabata [19] applied VM for prevention of copper interconnection failure in system on chip also in November 2009. Ringwood et al. [20] utilized VM to substitute the direct measurement of key etch variables for estimation and control in semiconductor etch in February 2010. Pan et al. [21] applied a VM system for predicting end-of-line electrical properties in May 2011. Lynn et al. [22] proposed global and local VM models for a plasma etch process in February 2012.

8.1.2 Necessity of Applying VM

This section describes the needs for physical metrology operation in semiconductor manufacturing and the necessity of reducing physical metrology with VM for decreasing manufacturing cost. Figure 8.2 describes the current physical metrology operating scenarios for supporting quality assurance of semiconductor manufacturing. It includes two monitoring operations – off-line (tool) monitoring and on-line (process) monitoring.

Observing the upper portion of Figure 8.2, a continuous process of the normal wafer fabrication is requested to pause for off-line tool quality monitoring; and the fabrication process cannot be resumed until the feedback of metrology data and the quality data of the process tool being within

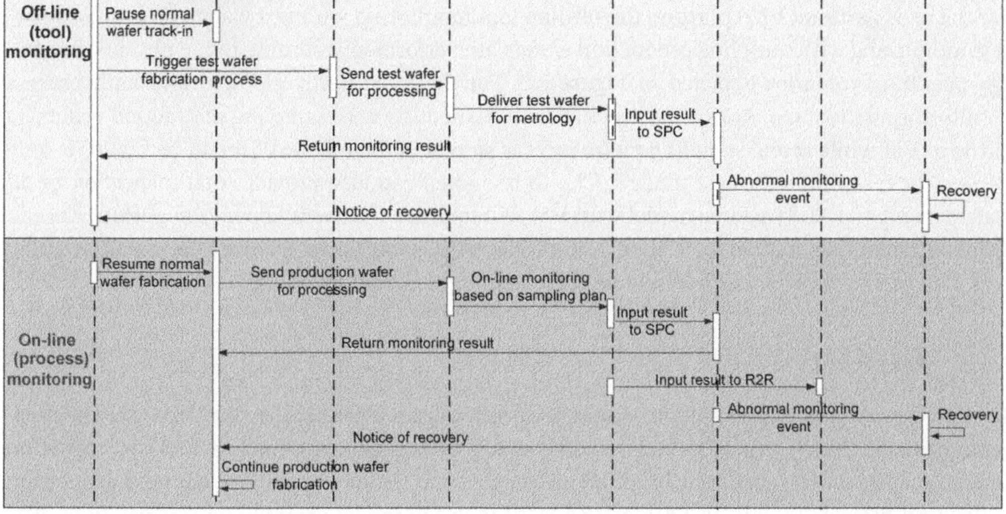

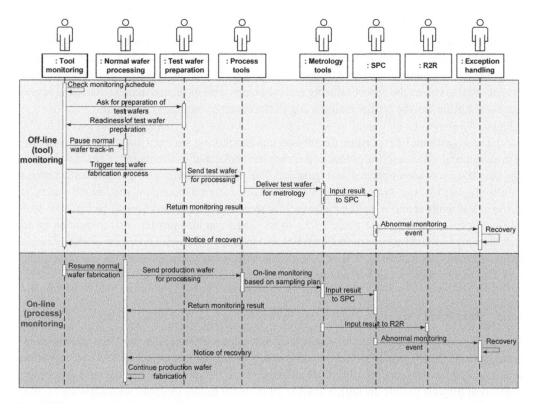

Figure 8.2 Current physical metrology operating scenarios. *Source:* Reprinted with permission from Ref. [23]; © 2011 IEEE.

the control specification. This off-line tool monitoring design is for the operation steps whose performance could not be measured in production wafers, e.g. etching rate. The interruption of production for off-line tool monitoring will cause the productivity loss of the tool. Also, this off-line monitoring requires test wafers (raw materials) and their corresponding preparation that need extra cost. Furthermore, the off-line tool monitoring will prolong the cycle time of wafer fabrication.

The on-line production wafer quality monitoring is depicted in the lower portion of Figure 8.2. In the daily operation of semiconductor manufacturing, the sampling plan will define the sampling rule to choose production wafers for metrology operation to do quality inspection; meanwhile, normal production operation will continue in process tools. The metrology data (such as critical dimension, CD) of production wafers will be fed into statistical process control (SPC) for quality inspection and the data will be sent to the R2R control system for process tuning. It is noted that the cycle time for production will be increased and some other quality issues in the manufacturing process might be incurred because extra manipulation is needed for this physical metrology operation. In the meantime, on-line fault detection monitoring of tools may also be applied by a fault detection and classification (FDC) system [24].

As mentioned above, metrology operation is essential for quality inspection and control of semiconductor manufacturing. Based on actual transaction volume in a manufacturing execution system (MES) of a 300 mm fab, 40–45% of daily manufacturing operation is for metrology operation [25]. Maintaining all the essential metrology operation requires investment of metrology tools, extra operation in process tools and metrology tools, extra test-wafer cost, etc.

VM is a method to conjecture manufacturing quality of a process tool based on data sensed from the process tool and without physical metrology operation. VM can provide users with total metrology information of all wafers in a front opening unified pod (FOUP) by merely measuring a single wafer in the same FOUP [26]. Moreover, VM can convert sampling inspection with metrology delay into real-time and on-line total inspection [3, 5, 27, 28]. Therefore, the promising approach is to apply VM: to reduce the cost of utilizing real metrology (and its related material handling operation) while achieving the goal of maintaining all the essential metrology operation.

Figure 8.3 shows the tool-monitoring and process-monitoring scenarios without and with VM. For the as-is (without VM) portion, the off-line tool monitoring with test wafers and on-line process monitoring with sampling production wafers are performed according to the physical metrology operating scenarios depicted in Figure 8.2. For the to-be (with VM) portion, both process monitoring and tool monitoring are executed simultaneously with sampling production wafers by virtue of VM while actual sampling metrology (as shown in the middle of the to-be figure) is only needed for re-training the VM models. The to-be model can also provide total inspection of all wafers owing to VM. The to-be model with VM will improve the overall operation performance in terms of quality assurance, cycle time, and productivity so as to reduce the total manufacturing cost. The benefit for applying VM will be evaluated in the following section.

8.1.3 Benefits of VM

The driving force for semiconductor-manufacturing improvement is the cost. ISMI set the next-generation-fab (NGF) target for 2012 centered at 50% cycle time reduction and 30% cost reduction so as to keep up with Moore's law [4]. ISMI also designated VM as one of the seven next-generation realization projects [4, 29]. In [29], a DRAM 12-inch fab model example with capacity of 30 K wafer starts plus 4.5 K non production wafers (i.e. test wafers) per month is utilized. This model is applied with ISMI's 45 nm generic logic process flow that has 658 process steps, 36 litho masks, and 268 metrology steps. The metrology steps are mapped as easy/medium/difficult for VM

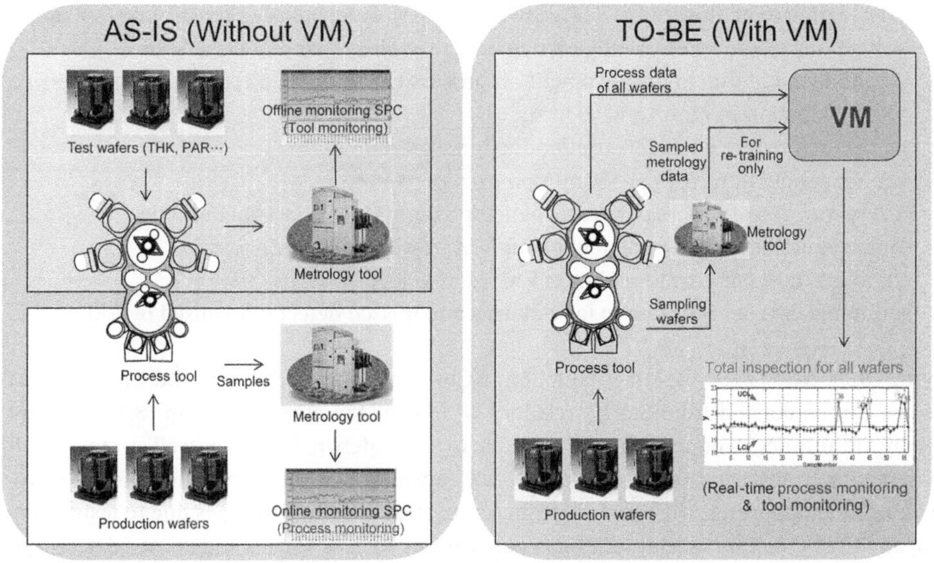

Figure 8.3 Tool and process monitoring without and with VM. *Source:* Reprinted with permission from Ref. [23]; © 2011 IEEE.

implementation and optimistic/realistic/pessimistic scenarios are designed respectively for VM penetration. As a result, the VM replacement rates were proposed in [29] as follows:

- **[Easy]** Film thickness measurements: 50% replacement rate because VM can be easily conjectured from the current tool's process data.
- **[Medium]** CD measurements: 20% replacement rate because VM needs the current tool's process data and pre-process metrology data to be accurately predicted.
- **[Difficult]** Defect inspections and scans: 2% replacement rate because it is difficult to predict the defects induced by the current tool and from the process integration by applying the VM technology.

The metrology-tool costs of film thickness, CD, and defect inspection were proposed to be US$ 1 million, 2 million, and 1 million, respectively [29]. As for the costs to set up and maintain VM per year were US$ 150 K for [Easy], 200 K for [Medium], and 300 K for [Difficult] [29]. It also proposed that each work-in-process (WIP) move requires five minutes [29].

Based upon the assumptions listed above and realistic scenario analysis done by Stark [29], this model yields 8.55% cycle time reduction and US$ 21 million capex reduction, with additional cost of US$ 2.7 million per year to maintain VM [29]. To evaluate the overall benefit of VM, the VM benefit equation of semiconductor manufacturing operation is proposed in (8.1):

$$\begin{aligned}\text{Benefit} = {} & W_O \times \left[\frac{1}{1-(\Delta CT_P + \Delta CT_M)} - 1\right] \times (1+\Delta Y) * (P - C) \\ & + \Delta Cost_M + \Delta Cost_T - Cost_V - Cost_Q \end{aligned} \tag{8.1}$$

where

W_O number of wafer output per year;

ΔCT_P % cycle time reduction due to VM allowing production wafers to skip metrology sampling steps in the on-line process monitoring flow;

ΔCT_M % cycle time reduction due to VM allowing less test wafers used in the off-line tool monitoring process and more intelligent dynamic metrology and process schemes;
ΔY % enhancement due to improvement on process capability (Cpk) (from VM supporting APC), reduction in scrap, and so forth;
P average selling price per 300 mm production wafer;
C average production cost per 300 mm production wafer;
$\Delta Cost_M$ cost saving of test wafers per year when applying VM in all production lines;
$\Delta Cost_T$ capex reduction per year when applying VM in all production lines;
$Cost_V$ additional cost per year to maintain VM;
$Cost_Q$ additional cost per year due to false alarms and missed detections caused by VM.

Observing (8.1), ΔCT_P for cycle-time reduction, $\Delta Cost_T$ for capex reduction, and $Cost_V$ for VM maintenance can be derived from Stark's results [29]. As for the other factors in (8.1), they are developed by the authors. Among them, ΔCT_M is due to test-wafers reduction as well as more intelligent dynamic metrology and process schemes; ΔY considers % enhancement due to improvement on Cpk (VM supporting APC), reduction in scrap, and so forth; $\Delta Cost_M$ is due to test-wafers reduction; and $Cost_Q$ represents additional cost per year due to false alarms and missed detections, which are resulted from the fact that those VM values are not as accurate as their corresponding actual metrology values.

Utilizing the data of ISMI's fab model [29], we have $W_O = 30\,K \times 12 = 360\,K$ and $\Delta CT_P = 8.55\%$. With five-year apportionment, $\Delta Cost_T = 21$ million/5 = US$ 4.2 million. And, $Cost_V$ = US$ 2.7 million. The current average market prices are P = US$ 4,000 and C = US$ 2,400.

In this fab model, the test wafers to production wafers' ratio is 4.5 K/30 K = 6.7%. Therefore, ΔCT_M can have about 1% cycle time reduction. In this 1% reduction, 0.67% is due to VM allowing less test wafers used in the off-line tool monitoring processes and off-line monitoring requires less operating steps than those of on-line monitoring; and the other 0.33% is due to VM enabling more intelligent dynamic metrology and process schemes to further reduce cycle time. As for $\Delta Cost_M$, the average cost of test wafers is about US$ 40 per wafer because test wafers may be recycled; moreover, a half of all the test wafers are assumed to be saved thanks to VM, then $\Delta Cost_M = 40 \times 4.5\,K \times 12 \times 0.5$ = US$ 1,080 K.

Moyne pointed out in [5] that the VM solution must incorporate VM data quality prediction when the W2W control consumes the VM data. Khan et al. studied this VM quality issue in [15, 16]. Cheng et al. [30] also proposed a Reliance Index (RI) to gauge the reliability level of VM. Furthermore, Kao et al. presented an APC system and method utilizing VM with RI in [31]. As demonstrated in [31], RI can be utilized to tune the R2R controller gain, α, such that R2R control errors due to VM variability can be compensated. In fact, α is an exponentially weighted moving average (EWMA) coefficient ranged between 0 and 1 [32]. Assume that $\alpha 1$ represents the EWMA coefficient when actual metrology value is adopted as the feedback of the R2R controller and $\alpha 2$ stands for the EWMA coefficient if VM value is applied. As stated in [32], $\alpha 2$ should depend on the quality or reliability of VM and $\alpha 2 < \alpha 1$. Because the value of RI is a good VM reliability evaluation index and $0 < RI < 1$, higher RI means better VM reliability [30], we can then naturally set: $\alpha 2 = RI \times \alpha 1$ [31]. The details of R2R control utilizing VM with RI will be elaborated in Section 8.4.

ΔY percentage enhancement due to improvement on Cpk (from VM supporting APC), reduction in scrap, and so forth is estimated below. Assume that the equivalent overall Cpk improves from 1.000 to 1.333, its corresponding defect rate can then be reduced from 0.27% to 0.0063% [33]. Therefore, the improvement (ΔY) yields 0.2637%. Note that the improvement (ΔY) mentioned here merely counts for defect-rate reduction. In fact, improvement on Cpk can also increase the

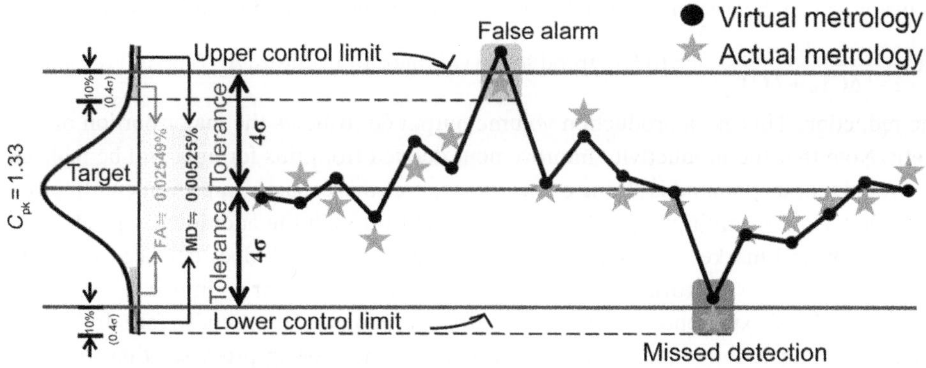

Figure 8.4 Illustration of a false alarm and a missed detection. *Source:* Reprinted with permission from Ref. [23]; © 2011 IEEE.

yield of high-quality product. Therefore, beyond the defect-rate reduction, the benefit of process-capability improvement (from VM supporting APC) would be much higher when considering the price difference for different quality categories due to high-quality yield improvement. This portion will not be addressed here because it is mostly treated as the benefit of applying APC.

Additional cost per year due to handling VM false alarms and missed detections ($Cost_Q$) is considered as follows. $Cost_Q$ is mainly caused by those VM values which are not as accurate as their corresponding actual metrology values. This accuracy issue may cause false alarms and missed detections as shown in Figure 8.4. Before determining the false alarm rate, the required VM solution accuracy should be defined first. In general, the acceptable-measurement-error bound of actual metrology is 10% of the tolerance spec. Therefore, in this study, the ideal 10% error bound is selected as the accuracy requirement when VM is applied. Assume that the Cpk is 1.333, the tolerance spec is then 4σ [30]. As shown in Figure 8.4, if the value of the actual metrology is within the ($0.4\sigma = 4\sigma - 3.6\sigma$) 10% error bound, then its corresponding VM value may cause a false alarm. This ($3.6\sigma \sim 4\sigma$) 10% error bound is equivalent to 0.02549%. The average annual salary of an engineer is about US$ 40 K, then the hourly pay is $40{,}000/(12\times22\times8) = 18.94$. It generally requires an hour for an engineer to handle an out-of-control action plan (OCAP) from a VM alarm report. On average, there are totally 30 VM operations for a generic logic process flow. Therefore, the false-alarm portion of $Cost_Q = 360{,}000 \times 0.02549\% \times 18.94 \times 30 =$ US$ 52,140.

On the other hand, if the value of the actual metrology is within the other ($0.4\sigma = 4.4\sigma - 4\sigma$) 10% error bound, then its corresponding VM value may cause a missed detection. This ($4\sigma \sim 4.4\sigma$) 10% error bound is equivalent to 0.00525%. As a result, the missed-detection portion of $Cost_Q = 360{,}000 \times 0.00525\% \times 2{,}400 =$ US$ 45,360.

In fact, VM solution accuracy depends on the prediction algorithm utilized [34], on-line model refreshing method applied [28], process and metrology data quality evaluation approaches adopted [35], and so on. It may not be easy to meet the ideal 10% VM accuracy requirement bound. If the actual VM accuracy is larger than 10% (such as 15% or 20%) of the tolerance spec, (the false-alarm and miss-detection portions of) $Cost_Q$ may be recalculated by the same approach as stated above.

Substituting all the parameter values mentioned above into (8.1), the benefit of VM yields US$ 63,458,792 per year for this fab model with capacity of 30 K wafer starts plus 4.5 K test wafers per month.

In conclusion, given the assumptions noted above, VM fab-wide implementation is expected to gain extra $\left[\frac{1}{1-(\Delta CT_P+\Delta CT_M)}\right]$ (=10.56%) production volume output due to ($\Delta CT_P + \Delta CT_M = 9.55\%$) cycle-time reduction. This extra production volume output contributes the major portion of VM total benefit. Note that the productivity improvement derived from this formula will be realized when the overall market demand is strong enough to support this capacity improvement. In other words, the first term (W_O multiplier) only provides benefit through the additional capacity realized. While the overall market demand is not strong such that the fab does not need the extra capacity provided with the improved cycle time, then this term has less or no impact.

Based on the authors' experience of 300 mm foundry operations [25], generally, foundry's high-product-mix operation requires four to five times more of monitoring processes than those of DRAM's low-product-mix operation. From the actual manufacturing management experience, the total test-wafer cost of a monthly 30 K foundry could be easily over US$ 9 million per year. Therefore, the benefit of deploying VM will be even better in the advanced foundry service facility. The cost saving mentioned here is purely the test-wafer-cost reduction. Besides test-wafer-usage reduction, if we also consider the corresponding FOUP-usage reduction, stocker-bin-storage reduction, and AMHS-investment-cost saving, the benefit will be higher.

Cheng et al. [36] have developed the so-called Automatic Virtual Metrology (AVM) system to implement and deploy the VM operations and functions in a fab automatically. Evolution of VM and invention of the AVM system is presented next.

8.2 Evolution of VM and Invention of AVM

Traditional VM schemes surveyed in Section 8.1.1, have four deficiencies listed below and require improvements for better accuracy and shorter deployment time:

1) Promptness and accuracy of traditional VM may not be achieved simultaneously. When promptness is emphasized, accuracy is poor; while accuracy is emphasized, promptness cannot be achieved.
2) Traditional VM values are provided without their accompanied quality indicators (such as reliance indexes (RIs)) thus users do not know whether VM values are reliable or not. This phenomenon is attributed to the so-called applicability/manufacturability problem of VM.
3) Physical characteristics of different tools of the same type are not identical. To maintain VM accuracy for a semiconductor fab tool consisting of multiple chambers, VM models of a specific chamber must be created based on the historical process-and-metrology data acquired from that chamber. For fab-wide implementation, the number of VM models increases rapidly with the growing number of tools. In this case, taking the traditional method of creating models one by one with a lot of historical data will result in huge labor expenses and long model-creation time.
4) Traditional VM schemes are not able to perform on-line and real-time quality evaluations of process-and-metrology data collected. As such, abnormalities in process data or metrology data cannot be excluded and will be added to the model tuning or re-training processes, resulting in deteriorated VM accuracy.

To overcome the drawbacks mentioned above, an effective and refined VM system should be developed.

The fundamental capability of a VM system is to convert off-line sampling inspection with metrology delay into on-line and real-time total inspection. However, the four deficiencies mentioned above are frequently encountered when applying VM technology to TFT-LCD, semiconductor, and/or other manufacturing processes. As such, an effective and refined VM system should be proposed and developed with the capabilities of solving those four deficiencies. This refined VM system is the AVM system [36].

8.2.1 Invention of AVM

As proposed in [1, 36] and shown in Figure 8.5, the AVM system consists of a model-creation (MC) server, a VM manager, several VM clients, and many AVM servers. The core of the AVM system is the AVM server whose features are summarized below.

The AVM server, which contains data preprocessing [35], conjecture model with advanced dual-phase VM algorithm [36], Reliance Index (RI) module [30], and Global Similarity Index (GSI) module [30], is shown in Figure 8.6. Besides the fundamental functions of VM on-line conjecturing and VM quality evaluation, the AVM server also possesses the functions of automatic data quality evaluation and automatic model refreshing [36]. The inputs of an AVM server include process data of the current tool and, if needed, the previous tools as well as the metrology data of

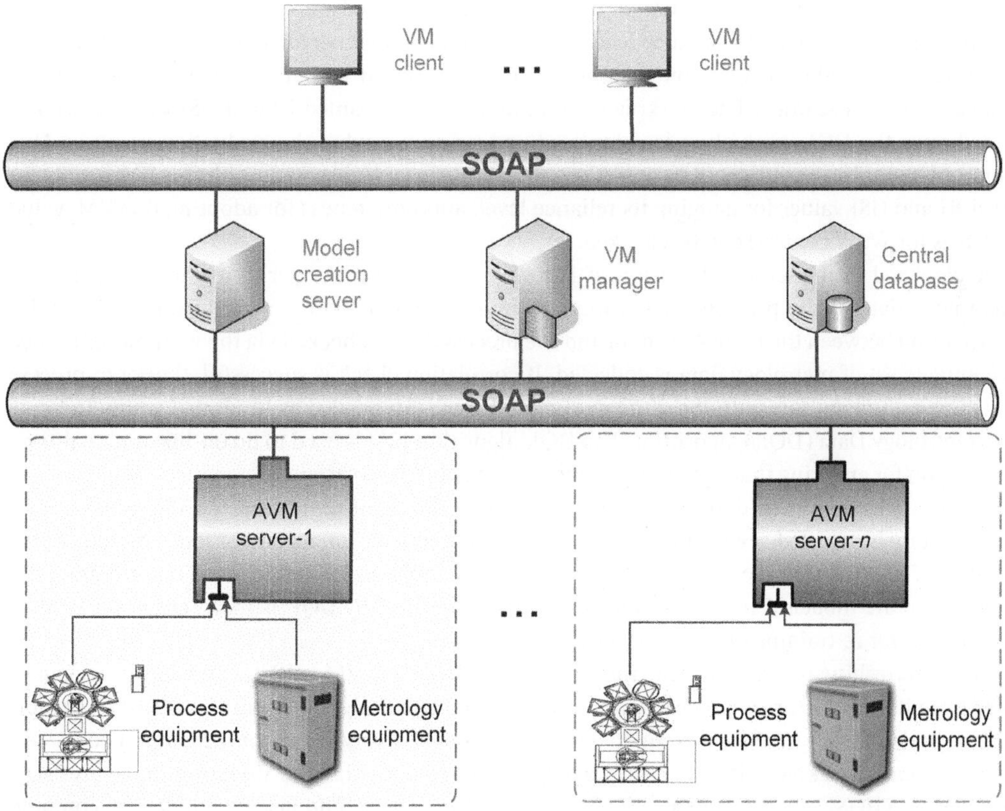

Figure 8.5 Automatic Virtual Metrology (AVM) system. *Source:* Reprinted with permission from Ref. [1]; © 2015 IEEE.

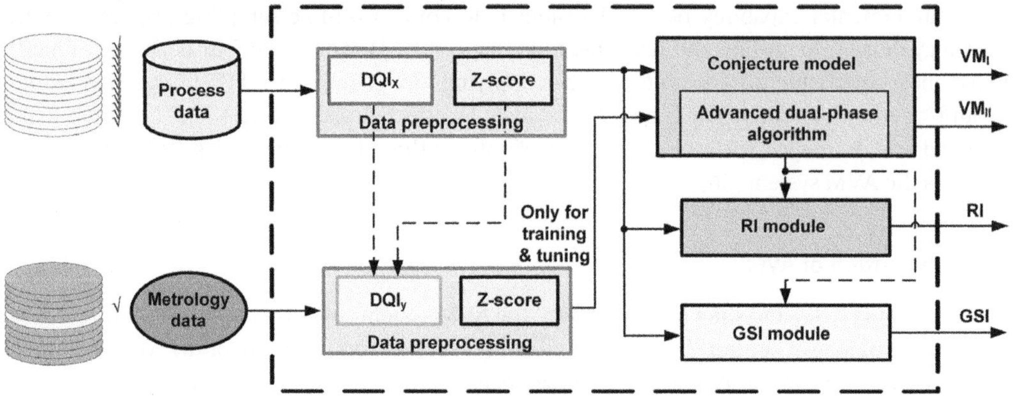

Figure 8.6 AVM server. *Source:* Reprinted with permission from Ref. [1]; © 2015 IEEE.

the current tool. The outputs are Phase-I VM (VM_I), Phase-II VM (VM_{II}), RI, and GSI values. The kernel of the AVM server is the advanced dual-phase VM algorithm [36] shown in Figure 8.7.

Observing the right-hand portion of Figure 8.7, the Phase-I algorithm starts to collect the process data of each processing workpiece after the conjecture model is built. The Data Quality Index of the Process Data (DQI_X) algorithm [35] will be applied to evaluate the quality of the collected process data once process data collection of the workpiece mentioned above is completed. If an abnormality is detected, a warning signal will be sent to the process engineer for analysis and confirmation to make sure whether it is an abnormality or a clean outlier. If it is an abnormality, this process data should be discarded. The workpiece's VM_I and its accompanied RI and GSI values are computed once the DQI_X evaluation is completed and no matter what the evaluation result is. This computation takes several seconds only; therefore, promptness is assured. With this VM_I accompanied RI and GSI values for gauging its reliance level, appropriateness for adopting this VM_I value (such as for W2W control) can be checked.

As depicted in the left-hand portion of Figure 8.7, the Phase-II algorithm starts to collect the metrology data of the pre-specified workpiece in a cassette after the conjecture model is built. Correlation between the metrology data and the process data is checked via the workpiece ID once a complete set of metrology data is collected. If correlation check is successful, the set of process data and metrology data with the same workpiece ID will be checked by the Data Quality Index of the Metrology Data (DQI_y) algorithm. The DQI_y algorithm [35] is used to do on-line and real-time evaluation for ensuring that both the metrology datum and its corresponding process data are normal before this metrology datum can be considered for tuning or re-training the VM models. If an abnormality is detected, a warning signal will be sent to process engineers for analysis and confirmation. If abnormality is confirmed, the metrology data will be deleted to avoid deteriorating the VM conjecture models. The normal metrology datum verified by DQI_y is then sent to the conjecture model for re-training or tuning usage.

Model refreshing is essential if the first set of DQI_X/DQI_y and VM & RI/GSI models is not generated from its own historical process and metrology data. The DQI_X/DQI_y and VM & RI/GSI models will be updated once they are re-trained or tuned. Subsequently, VM_{II} and its accompanied RI/GSI values of each workpiece in the entire cassette are re-computed. After updating the DQI_X/DQI_y and VM & RI/GSI models in Phase II, these updated models should also be adopted to compute

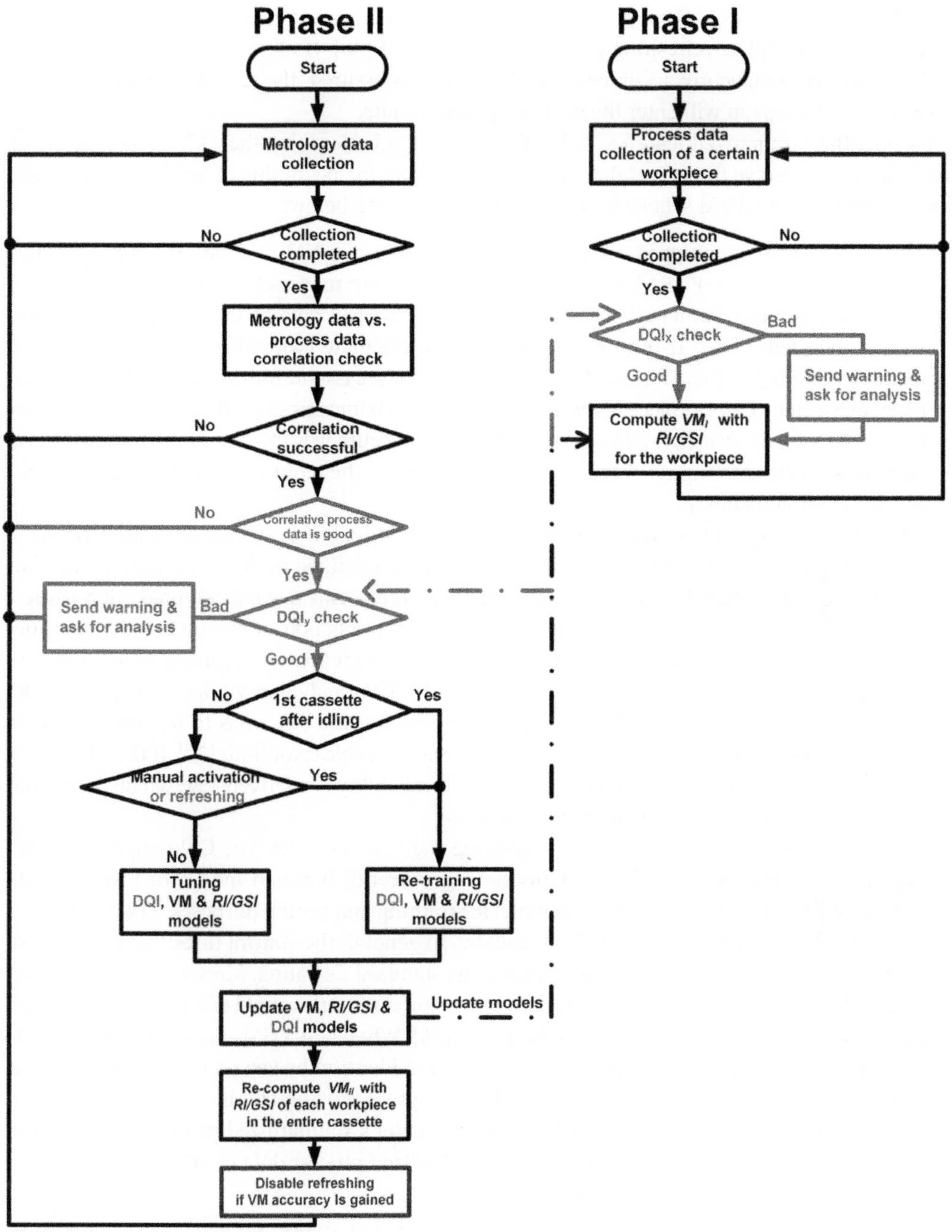

Figure 8.7 Advanced dual-phase VM algorithm. *Source:* Reprinted with permission from Ref. [1]; © 2015 IEEE.

the subsequent DQI_X/DQI_y, VM_I and its accompanied RI/GSI values. Finally, the conditions for a successful and complete refreshing procedure are checked [36]. If all of the conditions are met, which means the conjecture accuracy of the VM server is assured, the refreshing request will be disabled and the system will enter the normal operation state.

Viewing the summarized features of the AVM system, it is clear that the AVM system is one of the solutions to resolve those four deficiencies (as stated in the beginning of Section 8.2) encountered in the traditional VM schemes. The details are presented below:

1) The AVM system is equipped with the advanced dual-phase VM method [36] that generates dual-phase VM values. Phase I emphasizes promptness to immediately calculate and output the VM_I of a workpiece (wafer or glass) once the entire process data of the workpiece are completely collected. Phase II improves accuracy to re-calculate (with the newly refreshed VM models) and output the VM_{II} of all the workpieces in a cassette immediately after an actual metrology value of a workpiece in the cassette is collected (for refreshing the models). If promptness is the major concern, then VM_I is selected; and if accuracy is the first priority, then VM_{II} is chosen. As such, the outputs of the AVM system can meet the requirements of promptness and accuracy simultaneously.
2) The AVM system generates the accompanied RI and GSI [30] of each VM_I and VM_{II}. The AVM system has two different VM prediction models, and RI is defined as the percentage of intersection between the statistical normal distributions of those two normalized prediction values. Therefore, the RI value is between 0 and 1. If those two prediction values are exactly the same, the corresponding RI value is 1, and if those two values are far apart, the corresponding RI value is 0. Therefore, users can check the reliability of the VM prediction via its corresponding RI value. To distinguish how good the RI is, RI threshold value (RI_T) has to be defined. RI_T is defined as the RI value corresponding to the maximal tolerable error limit [30]. If the RI exceeds the RI_T, then the VM value is reliable; otherwise, its reliability is relatively low, which means the VM conjecture result requires further verification.

   The GSI is utilized to assist the RI in gauging the reliance level. The GSI is defined as the degree of similarity between the set of process data currently inputted and all the historical sets of process data used in the conjecture model for training and tuning purposes. The smaller the GSI is, the better the degree of similarity will be. In general, the control threshold of a process in statistical control is set to be three times of its standard deviation. Since each standardized Z-score has mean zero and standard deviation one, the control threshold of Z is three. Therefore, the GSI threshold value (GSI_T) is set to be $3^2 = 9$ [36]. When $GSI > GSI_T$ occurs, the individual-similarity-index (ISI) analysis should be performed to identify the key process parameters that cause major deviation [30]. The ISI of an individual process parameter is defined as the degree of similarity between this individual process parameter's standardized process datum of the input set and the same process parameter's standardized process data in all the historical sets utilized for training and tuning the conjecture model.
3) The AVM system possesses automatic fanning out and model refreshing capabilities to enable VM models of a specific chamber to be propagated and refreshed to other chambers of the same type of tools for maintaining VM accuracy of all chambers with lower labor expenses and less model-creation time [36]. Hence, the AVM system facilitates fab-wide VM implementation.
4) The AVM system is equipped with DQI_X algorithm and DQI_y algorithm [35]. Upon receiving a set of process data, the AVM system uses the DQI_X algorithm to perform on-line and real-time detection of a possible process-data abnormality. Whenever an actual metrology datum of a

workpiece is obtained, the DQI_y algorithm is then used to perform on-line and real-time evaluation of a possible metrology-datum abnormality before this metrology datum can be considered for tuning or re-training the VM models. As a result, deterioration of VM accuracy due to model refreshing with bad-quality data is waived [35].

Up to now, parts of the key functional modules of AVM systems have been tested and/or implemented in an etch process of Taiwan Semiconductor Manufacturing Company, Ltd. (tsmc)'s Fab 14 in the semiconductor industry in 2007, portions of the verification results were published in [30, 34, 37]; the CVD and photo processes of Chi Mei Optoelectronics Corporation, Ltd. (CMO)'s 5th and 6th fabs in the TFT-LCD industry during 2006 to 2010, portions of the verification results were published in [34–36, 38, 39], and [1]; and the plasma-enhanced-chemical-vapor-deposition (PECVD) process of Motech Inc. in the solar-cell industry during 2010 to 2011, portions of the verification results were published in [40, 41]. Moreover, not only etch, CVD, and photo, but also other processes exist in the complete manufacturing processes of a semiconductor or TFT-LCD fab. After 2012, AVM has also been gradually applied to other industries; such as wheel machining automation of Far East Machinery Industry Corporation, Ltd. (FEMCO) in the automotive industry during 2014 to 2016, portions of the verification results were published in [42]; flange-hole processing of airplane engine-case manufacturing of Aerospace Industrial Development Corporation, Ltd. (AIDC) in the aerospace industry during 2017 to 2018, portions of the verification results were published in [43]; and Carbon-Fiber manufacturing of Formosa Plastics Corporation, Ltd. (FPC) in the chemical industry during 2017 to 2019, portions of the verification results were published in [44].

In Chapter 11, the complete TFT-LCD production tools and manufacturing processes are introduced. Then, the generic deployment schemes of the VM technology for the TFT-LCD tools using the AVM system for fab-wide implementation systematically are illustrated. Besides TFT-LCD, the actual AVM implementation cases in the other six industries, including solar cell, semiconductor, automotive, aerospace, carbon fiber, and blow molding, are also presented in Chapter 11.

8.3 Integrating AVM Functions into the Manufacturing Execution System (MES)

The Manufacturing Execution System (MES) is a shop floor control system which includes manual and/or automatic labor, production reporting, on-line inquiries, and links to tasks that take place on the production floor. The MES provides links to work orders, receipt of goods, shipping, quality control, maintenance, scheduling, and other related tasks [45]. The MES is one of the core components of the advanced e-Manufacturing model depicted in Figure 1.2. The mission of MES is to increase productivity and yield.

ISMT developed a SEMATECH computer-integrated manufacturing (CIM) framework [46] to specify the common MES infrastructure and the software functions of MES applications and incorporate those MES applications into a coherent system. By specifying the standard interfaces and functions of the common MES components, manufacturers can collect system components from multiple suppliers. Thus, manufacturers can develop systems by extending the common components and substituting old components with improved ones of the same interfaces and functions. The authors have also proposed a MES framework [47] and a holonic manufacturing execution system (HMES) framework [48] by applying distributed and object-oriented technologies and the concepts of holon and holarchy.

Most of the current commercial MESs (such as IBM's SiView) contain statistical process control (SPC) functional module to handle on-line quality monitoring. However, SPC can merely perform quality monitoring on those sampling products that are measured by real metrology tools. The AVM system presented in Section 8.2 has the capability to convert sampling inspection with metrology delay into real-time and on-line total inspection of all products. Therefore, to enable the MES to have the capability of real-time and on-line total inspection, the AVM system should be integrated into the MES.

Based upon the frameworks of [46–48], the AVM system can also be developed as a pluggable component of the MES framework shown in Figure 8.8 The MES components include scheduler, SPC, equipment manager, alarm manager, material manager, etc.

The capabilities of AVM include (i) converting sampling inspection with metrology delay into real-time and on-line total inspection; (ii) control/monitoring workpieces reduction/elimination; and (iii) supporting W2W control. To bring the AVM's capabilities into full play in a novel manufacturing system, the relationships among AVM, MES components and R2R controllers are depicted in Figure 8.9. The AVM module is plugged into the MES. The AVM module collects process data of process equipment and metrology data of metrology tool via the equipment manager. The AVM outputs (including VM_I, VM_{II}, and their accompanying RI/GSI values) are sent respectively to the SPC module for W2W total quality inspection, to the alarm manager when RI and/or GSI values exceed their thresholds, and to the scheduler module for golden-route consideration, which dispatches important products to stable equipment so as to improve yield stability. The VM_I value is delivered to the R2R controller of the current equipment for feedback R2R control [28].

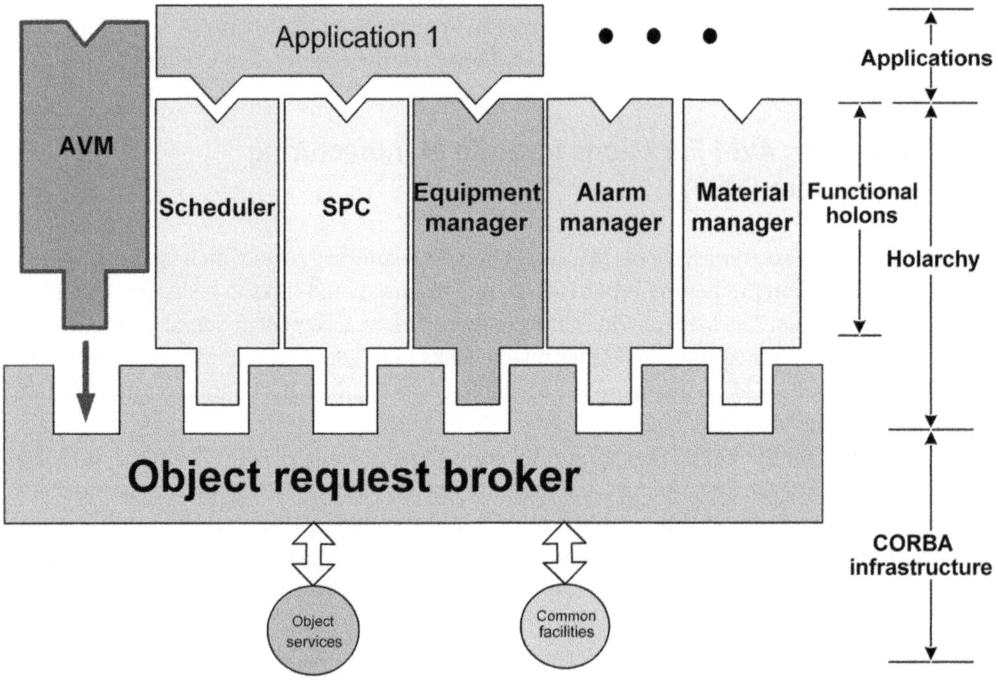

Figure 8.8 Plugging AVM into the MES framework. *Source:* Reprinted with permission from Ref. [23]; © 2011 IEEE.

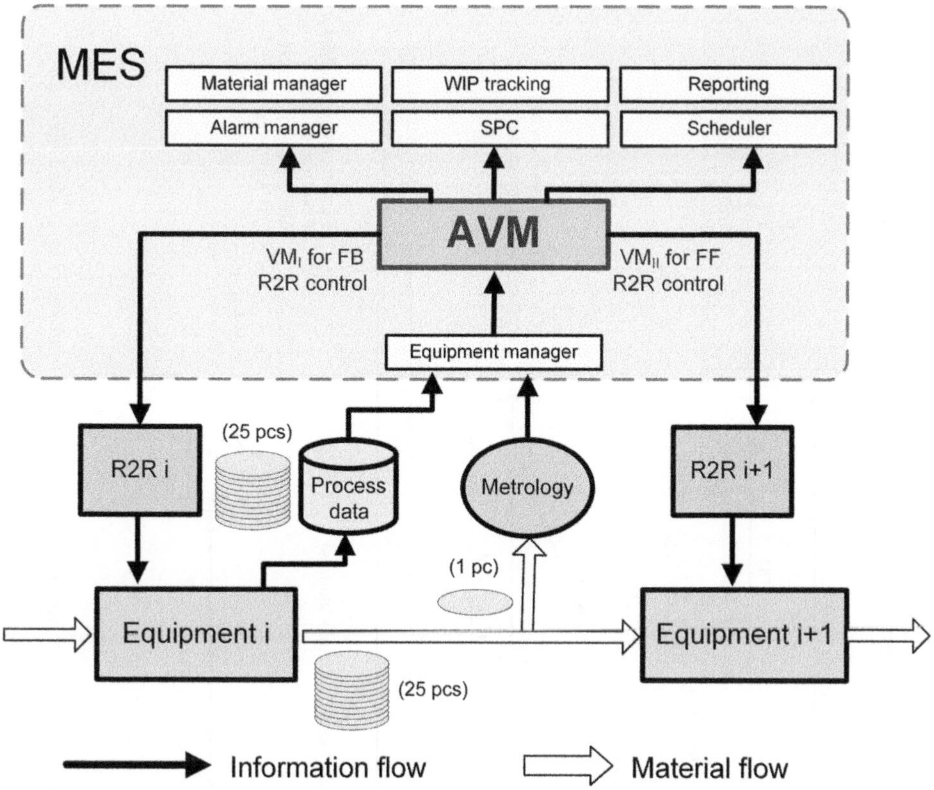

Figure 8.9 Relationships among AVM, MES components, and R2R controllers. *Source:* Reprinted with permission from Ref. [23]; © 2011 IEEE.

Finally, the VM_{II} value is dispatched to the R2R controller of the subsequent equipment for feedforward R2R control [28]. The detailed operating scenarios of a novel manufacturing system with AVM capabilities are presented next.

8.3.1 Operating Scenarios among AVM, MES Components, and Run-to-Run (R2R) Controllers

A novel manufacturing system with AVM capabilities shall at least consist of a MES with an embedded AVM module and R2R controllers. The detailed operating scenarios among AVM, MES components, and R2R controllers are shown in Figure 8.10 and are described below. Note that, the operating scenarios should align with the advanced dual-phase VM algorithm shown in Figure 8.7.

1. and 2. Scheduler dispatches a lot via Equipment Manager to Process Equipment I (PE i).
3. and 4. PE i delivers process data (PD) via Equipment Manager to AVM.
5. AVM calculates VM_I and its accompanying RI, GSI, and DQI_X; then checks if any alarm(s) occurs.
6. If any alarm(s) occurs, then AVM sends alarm(s) to Alarm Manager.
7. AVM reports VM_I (with accompanying RI, GSI, and DQI_X) to SPC for W2W total inspection.
8. AVM reports VM_I (with accompanying RI) to R2R i for supporting feedback R2R control.
9. SPC triggers an out-of-control action plan (OCAP) to Alarm Manager if an alarm is detected.

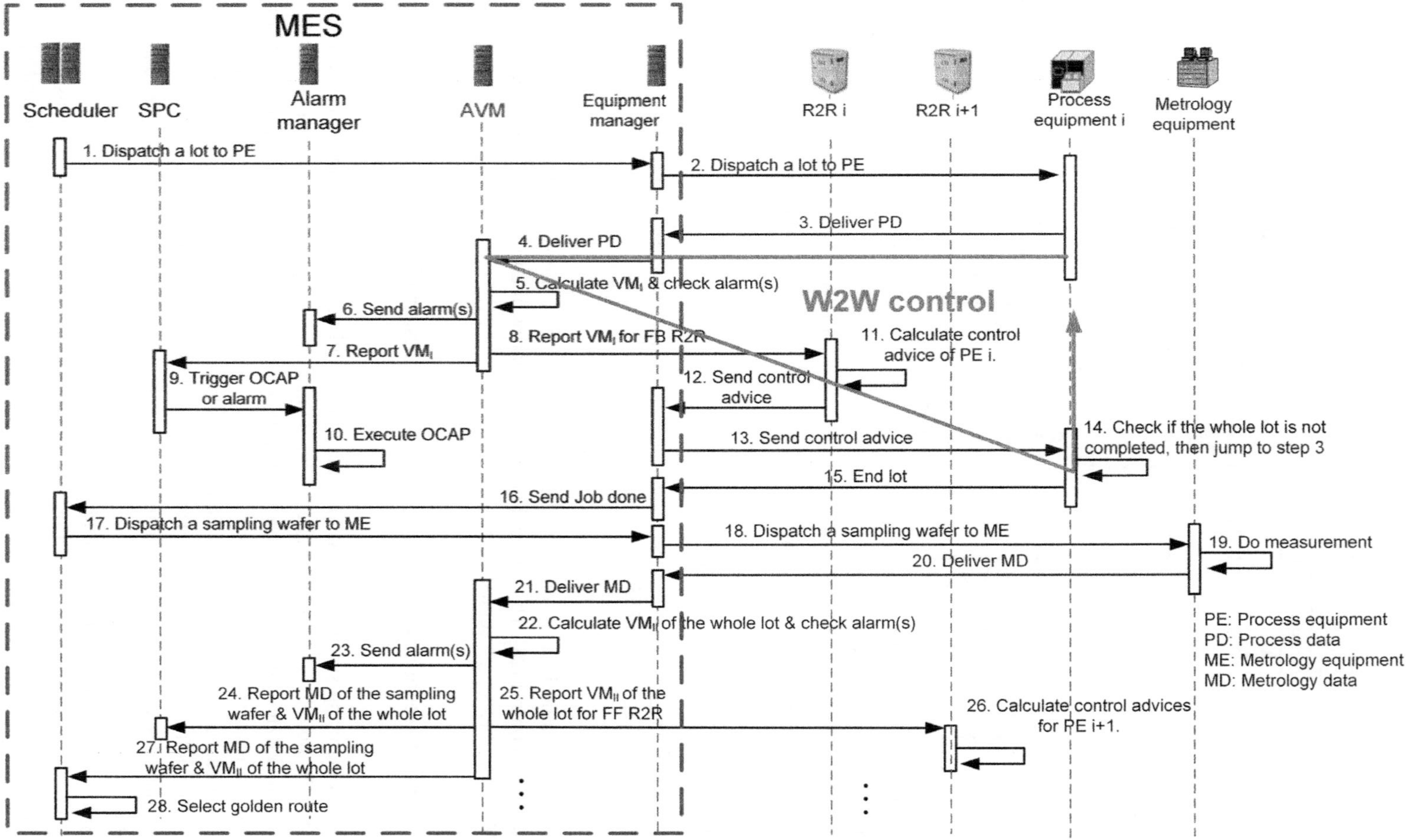

Figure 8.10 Operating scenarios among AVM, MES components, and R2R controllers. *Source:* Reprinted with permission from Ref. [23]; © 2011 IEEE.

10. Alarm Manager executes OCAP.
11. R2R i calculates control advice of PE i.
12. and 13. R2R i sends control advice via Equipment Manager to PE i.
14. PE i checks if the whole lot is not completed then the procedure jumps to Step 3 for processing the subsequent wafer in the lot.
15. and 16. PE i checks if the whole lot is completed then the job-done message is sent via Equipment Manager to Scheduler.
17. and 18. Scheduler dispatches a sampling wafer via Equipment Manager to Metrology Equipment (ME).
19. ME performs measurement on the sampling wafer.
20. and 21. ME delivers metrology data (MD) via Equipment Manager to AVM.
22. AVM calculates VM_{II} and its accompanying RI, GSI, and DQIy; then checks if any alarm(s) occurs.
23. If any alarm(s) occurs, then AVM sends alarm(s) to Alarm Manager.
24. AVM reports MD of the sampling wafer and the VM_{II} (with accompanying RI, GSI, and DQIy) of the whole lot to SPC for W2W total inspection.
25. AVM reports VM_{II} (with accompanying RI) to R2R i+1 for supporting feedforward R2R control of the subsequent PE i+1.
26. R2R i+1 calculates control advice of PE i+1.
27. AVM reports MD of the sampling wafer and the VM_{II} (with accompanying RI, GSI, and DQI_y) of the whole lot to Scheduler for golden-route consideration.
28. Scheduler selects a golden route based on MD and VM_{II} results.

    The major differences between the flow diagrams by adopting AVM versus high speed (e.g. integrated) metrology are

a) Besides VM_I and VM_{II}, AVM can also generate and send RI, GSI, DQI_X, and DQI_y to the SPC, alarm manager, scheduler, and so forth for gauging the VM reliability level, monitoring the equipment health status, and evaluating the process and metrology data qualities, respectively such that more applications (e.g. intelligent dynamic metrology and process schemes leading to further reduction in cycle time) may be developed; while integrated metrology can only provide actual metrology data.
b) AVM needs sampled metrology data to re-train or tune the VM models; while integrated metrology requires calibration wafers to re-calibrate itself.

Observing the red triangle shown in Figure 8.10, it explicitly shows the operating scenario of AVM supporting the W2W control. The corresponding collaboration diagram of the operating scenarios shown in Figure 8.10 is depicted in Figure 8.11. The framework messages shown in Figures 8.10 and 8.11 are aligned (with the same sequence numbers). Both the operating-scenario diagram and the collaboration diagram are the essential developing guides for designing the framework messages [47, 48] among AVM, MES components, and R2R controllers. With these framework messages designed, the AVM module can be configured as a pluggable component of the MES and the AVM module can send VM_I and VM_{II} to the R2R controllers for supporting the W2W control. Observing Sequence 8 in Figure 8.10: "AVM reports VM_I (with accompanying RI) to R2R i for supporting feedback R2R control," the purpose of reporting VM_I with accompanying RI to R2R i is to tune $\alpha 2$ with $\alpha 2 = \text{RI} \times \alpha 1$ when W2W control with VM_I is applied, which will be elaborated in Section 8.4.

In conclusion, a novel manufacturing system with AVM capabilities is presented in this section. This novel manufacturing system is composed of at least a MES with a pluggable AVM module and

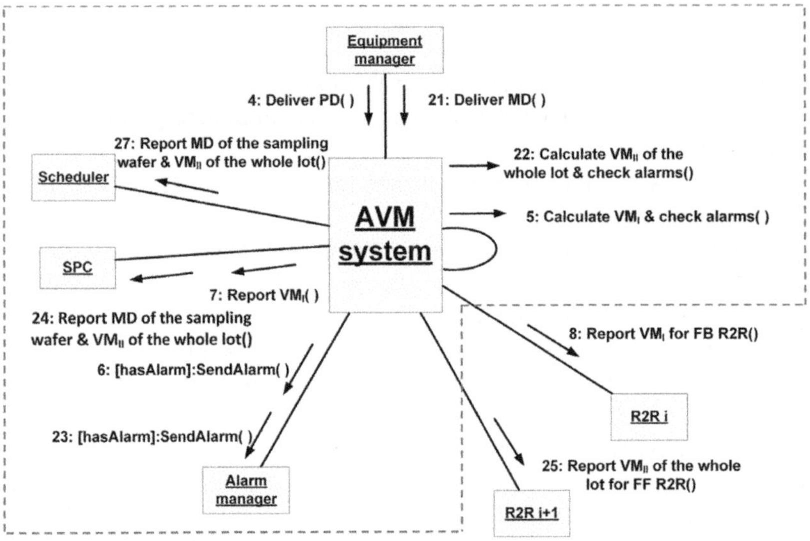

Figure 8.11 Collaboration diagram for integrating AVM into MES. *Source:* Reprinted with permission from Ref. [23]; © 2011 IEEE.

R2R controllers. The operating scenarios of the novel manufacturing system and the collaboration diagram of the AVM pluggable module are elaborated in this section. Furthermore, a benefit equation is established to evaluate the profitability of applying VM fab-wide in Section 8.1.3. The evaluation result shows that VM fab-wide implementation is expected to gain roughly 10% extra production volume output due to cycle-time reduction. Therefore, applying the manufacturing system with AVM capabilities is one of the effective approaches to accomplish the requirements of both cycle-time and cost reductions of the next generation fab.

8.4 Applying AVM for Workpiece-to-Workpiece (W2W) Control

Run-to-Run (R2R) control is the technique of modifying recipe parameters or the selection of control parameters between runs to improve process performance [49]. In the semiconductor industry, R2R control is widely applied to improve Cpk. A run can be a batch, a lot, or an individual workpiece. A workpiece may represent a wafer in the semiconductor industry or a glass in the TFT-LCD industry. Lot-to-lot (L2L) advanced process control (APC) is now widely implemented for dealing with advanced technologies [10]. When L2L control is applied, only a single workpiece in the lot is required to be measured for feedback and feedforward control purposes. However, as the device dimension shrinks further, tighter process control is needed. In this case, L2L may not be accurate enough and therefore workpiece-to-workpiece (W2W) becomes essential for critical stages. As a result, each workpiece in the lot should be measured. To fulfill this requirement, large amounts of metrology tools will often be required and production cycle time will also be increased significantly. Furthermore, metrology delays, which are inevitable as real measurements are performed, will not only cause complicated control problems but also degrade the APC performance [17].

To resolve the problem mentioned above, virtual metrology (VM) was proposed. VM is a technology to predict metrology variables using information about the state of the process for every workpiece [3]. If the VM conjecture model is fresh and accurate enough, it can generate a VM value within seconds after collecting the complete tool process data of a workpiece. Therefore, this VM value can be applied for real-time W2W control [28].

Not much literature concerning R2R control utilizing VM was found. Among them, Besnard and Toprac [50] concluded that VM is a key enabler for W2W data granularity and can dramatically improve Cpk in conjunction with classical R2R control. However, the quality issue of VM was not addressed in [50]. Khan et al. [15, 16, 32] proposed a W2W control scheme to utilize VM and to consider the data quality of the VM value into the R2R control scheme. The highlights of [15, 16, 32] will be described in Section 8.4.1. Wu et al. [17] studied the performance of R2R control subject to metrology delay and concluded that applying VM to remedy the metrology-delay problem is justified if the error of the VM method is less than the error caused by stochastic process noise [32]. In other words, the performance of R2R control with VM may be worse than that of R2R control with actual metrology subject to metrology delay if the error of the VM method is more than the error caused by stochastic process noise [32].

A preliminary study of R2R control utilizing VM with RI was proposed in [51]. In the preliminary study, only RI is considered for gauging the reliance level and the simulation results of a single round is shown. In this section, not only RI but also GSI [30] are applied to tune the R2R controller gain. Moreover, simulation results of five random-generated rounds are performed. These random-generated rounds simulate the situation as if five modifications are performed on the process or the equipment due to predictive maintenances (PMs). As such, the issue of the R2R controller-gain management in real-production environment may be addressed whenever a modification is performed on the process or the equipment.

The chemical mechanical polishing (CMP) process is chosen for illustration because it is a common tool in the semiconductor industry, and APC is commonly adopted to achieve improved CMP quality control for remedying the problem that the nominal removal rate of the CMP process varies with the parts usage count.

The remainder of this section is organized as follows. The background materials are presented in Section 8.4.1. Section 8.4.2 briefly introduces the AVM system, RI, and GSI. Section 8.4.3 proposes the R2R control scheme that utilizes VM as the measurement feedback and VM's corresponding RI and GSI to tune α_2. Then, Section 8.4.4 presents an illustrative example by applying Case 1: R2R with in-situ metrology, Case 2: R2R+VM without RI, Case 3: R2R+VM with RI, Case 4: R2R+VM with (1-RI), and Case 5: R2R+VM with RI & (1-RI) switching [also denoted RI.S.(1-RI)] to a simulated CMP tool for comparison and evaluation. Finally, a summary and conclusions are made in Section 8.4.5.

8.4.1 Background Materials

A schematic block diagram showing a conventional model of exponentially weighted-moving-average (EWMA) R2R control is depicted in Figure 8.12. Let us consider a process with linear input and output relationship:

$$y_k = \beta_0 + \beta_1 u_k + \eta_k \tag{8.2}$$

where y_k is the plant output, u_k the control action taken for run k, β_0 the initial bias of process, β_1 the process gain, and η_k the disturbance model input [17].

Given a process predictive model Au_k where A is a gain parameter (e.g. removal rate for CMP) estimated for the system; its initial values can be obtained based on the actual tool/recipe performance.

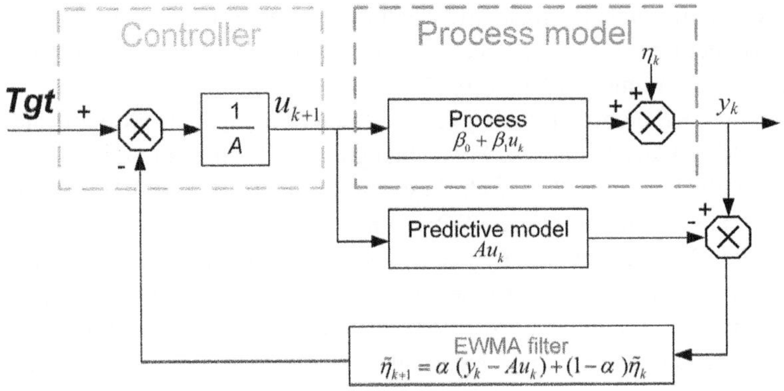

Figure 8.12 Model of EWMA R2R control. *Source:* Reprinted with permission from Ref. [31]; © 2013 IEEE.

Using an EWMA filter, the model offset or disturbance of the $(k+1)^{th}$run is estimated to be

$$\tilde{\eta}_{k+1} = \alpha\left(y_k - Au_k\right) + \left(1-\alpha\right)\tilde{\eta}_k \tag{8.3}$$

where α is an EWMA coefficient ranged between 0 and 1.

The $(k+1)^{th}$ run control action is derived by

$$u_{k+1} = \frac{Tgt - \tilde{\eta}_{k+1}}{A} \tag{8.4}$$

where ***Tgt*** represents the target value.

Khan et al. [15, 16, 32] proposed a W2W control scheme utilizing VM as shown in Figure 8.13, where y_Z is the Zth actual metrology data of the sampled product (workpiece), $\hat{y}_k$ is the kth run VM data, and X_k is the kth run process data of the equipment.

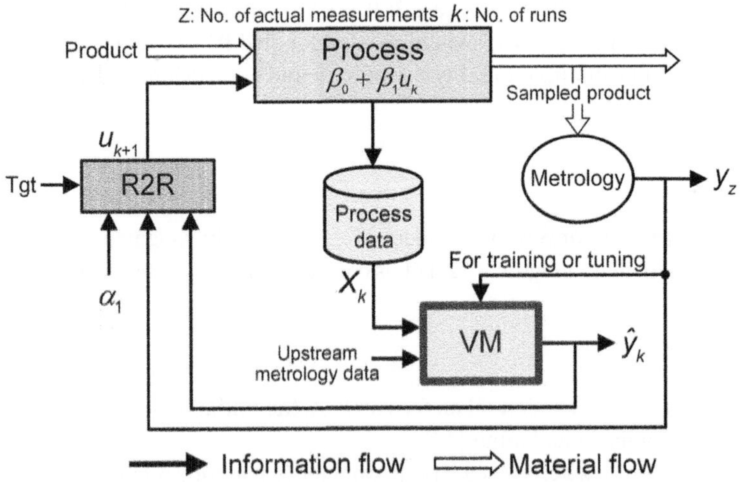

Figure 8.13 W2W control scheme utilizing VM [15, 16, 32]. *Source:* Reprinted with permission from Ref. [31]; © 2013 IEEE.

Khan et al. also proposed to consider the data quality of the VM value into the R2R control scheme. Based on [15, 16, 32], (8.3) was modified as follows:

- When y_k is measured by an actual metrology tool, it becomes y_z, an EWMA coefficient α_1 is used in

$$\tilde{\eta}_{k+1} = \alpha_1\left(y_z - Au_k\right) + \left(1 - \alpha_1\right)\tilde{\eta}_k \tag{8.5}$$

- When y_k is conjectured or predicted by a VM system, it becomes $\hat{y}_k$, an EWMA coefficient α_2 is used in

$$\tilde{\eta}_{k+1} = \alpha_2\left(\hat{y}_k - Au_k\right) + \left(1 - \alpha_2\right)\tilde{\eta}_k \tag{8.6}$$

Khan et al. [15, 16, 32] pointed out that $\alpha_1 > \alpha_2$ (usually, depending on the relative quality of VM data). Now, the controller-gain problem of applying VM is focused on how to set α_2, wherein the rule of thumb is that α_2 should depend on the quality or reliability of VM and $\alpha_2 < \alpha_1$. Khan et al. [15, 16, 32] proposed two VM quality metrics to consider incorporating VM quality into the R2R controller gain:

1) Prediction error at metrology runs:

$$\text{Error} = y - \hat{y} \tag{8.7}$$

2) If y and $\hat{y}$ are zero-mean Gaussian deviations from targets, then Min mean-square-error (MSE) estimator of y based on $\hat{y}$ is

$$y_{\text{mmse}} = \rho\frac{\sigma_y}{\sigma_{\hat{y}}} \tag{8.8}$$

where the correlation coefficient

$$\rho = \frac{cov\left[y, \hat{y}\right]}{\sigma_y \sigma_{\hat{y}}} \tag{8.9}$$

and σ_y and $\sigma_{\hat{y}}$ are standard deviations of y and $\hat{y}$, respectively.

Nevertheless, both metrics proposed above have the following disadvantages:

1) Both (8.7) and (8.8) need actual metrology data "y"; however, if actual metrology data (y) are available, then VM data ($\hat{y}$) are not needed at all.
2) There is no proven optimal calculation or simulation that translates the value of these data quality metrics to a value of α_2 with α_1 given.

As a result, it may not be easy to combine the data quality metrics as in (8.7) and (8.8) into the R2R model. Hence, there is a need to develop a R2R control scheme utilizing VM with RI and GSI [30] to effectively consider the data quality of VM into the R2R model.

8.4.2 Fundamentals of Applying AVM for W2W Control

The AVM system described in Section 8.2 consists of a model-creation (MC) server, a VM manager, several VM clients, and many AVM servers. The MC server will generate the first set of data quality evaluation models, VM conjecture models, and VM reliability evaluation models of a certain tool type. Under fab-wide VM deployment, the VM manager can fan out the first set of models generated to all of the AVM servers of the same tool type. Also, the AVM server of each individual

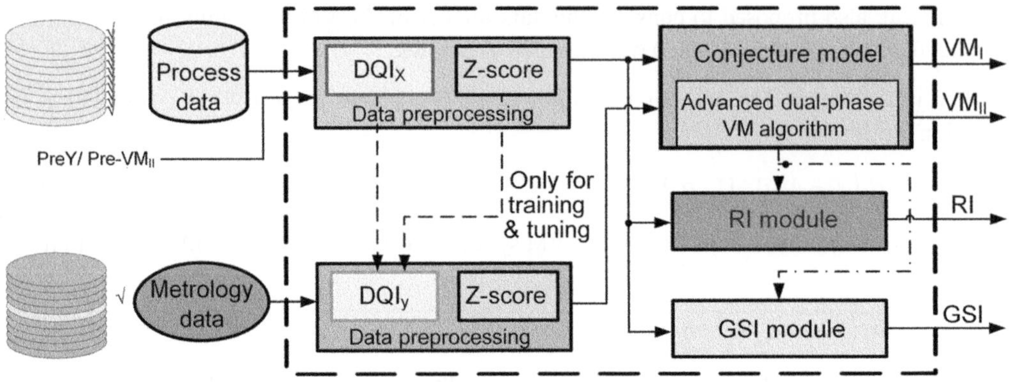

Figure 8.14 AVM server with PreY input. *Source:* Reprinted with permission from Ref. [31]; © 2013 IEEE.

fan-out-accepter can perform an automatic model refreshing process to promptly refresh its own model set. Consequently, the VM accuracy of each AVM server can be maintained and the AVM server is then ready to serve various VM applications [36].

The AVM server with pre-metrology (PreY) input, which contains data preprocessing, conjecture model with advanced dual-phase VM algorithm, RI module, and GSI module, is shown in Figure 8.14. Besides the fundamental functions of VM on-line conjecturing and VM quality evaluation, the AVM server also possesses the functions of automatic data quality evaluation and automatic model refreshing. The inputs of the AVM server include process data of the current tool and PreY data of the previous tool as well as the metrology data of the current tool. The process data and PreY data are also called Type-1 data and Type-2 data, respectively, in [15, 16]. The PreY data may be either actual or VM data. The outputs are Phase-I VM (VM_I), Phase-II VM (VM_{II}), RI, and GSI values. The functions of RI and GSI will be briefly introduced next.

The RI module generates the VM accompanying RI value to estimate the reliance level of the VM value [30]. Observing Figure 8.15, the RI is defined as the intersection-area value (overlap area A) between the statistical distribution $Z_{\hat{y}_{N_i}}$ of the VM value from the conjecture model (built by such as the BPNN algorithm) and the statistical distribution $Z_{\hat{y}_{r_i}}$ of the reference prediction value from the reference model (built by such as the multi-regression (MR) algorithm). As a result, the RI equation is listed below [30]:

$$\mathrm{RI} = 2\int_{\frac{Z_{\hat{y}_{N_i}}+Z_{\hat{y}_{r_i}}}{2}}^{\infty} \frac{1}{\sqrt{2\pi}\sigma} e^{-\frac{1}{2}\left(\frac{x-\mu}{\sigma}\right)^2} \mathrm{d}x$$

$$\text{with } \mu = Z_{\hat{y}_{N_i}} \text{ if } Z_{\hat{y}_{N_i}} < Z_{\hat{y}_{r_i}} \qquad (8.10)$$

$$\mu = Z_{\hat{y}_{r_i}} \text{ if } Z_{\hat{y}_{r_i}} < Z_{\hat{y}_{N_i}}$$

and σ is set to be 1

where $Z_{\hat{y}_{N_i}}$ is the statistical distribution of the VM value built by NN and $Z_{\hat{y}_{r_i}}$ is the statistical distribution of the reference prediction value built by MR.

The RI increases with increasing overlap area A. This phenomenon indicates that the result obtained using the conjecture model is closer to that obtained from the reference model, and thus

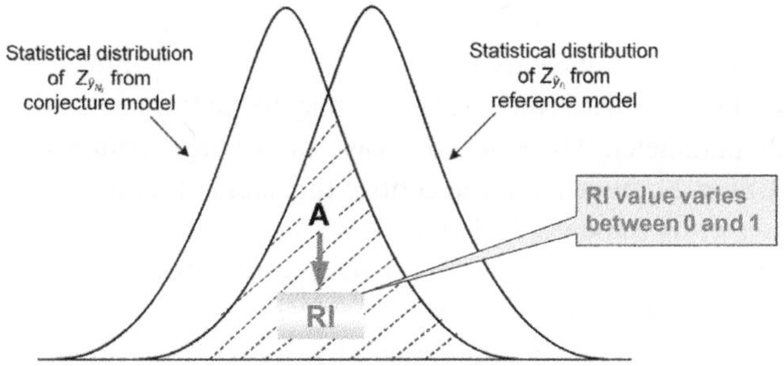

Figure 8.15 Schematic Diagram of Defining RI for Normal Distribution. *Source:* Reprinted with permission from Ref. [31]; © 2013 IEEE.

the corresponding VM value is more reliable. Otherwise, the reliability of the corresponding measurement value reduces with decreasing RI. When the distribution $Z_{\hat{y}_{N_i}}$ estimated from $\mathbf{S}_p^c$ is fully overlapped with the distribution $Z_{\hat{y}_{r_i}}$ estimated from $\mathbf{S}_p^c$, according to the distribution theory of statistics, the RI value equals 1; on the contrary, when those two distributions are almost separate, the RI value approaches 0. Therefore, the RI value is defined as lying between 0 and 1.

To distinguish how good the RI is, RI threshold (RI_T) value has to be defined. RI_T is defined as the RI value corresponding to the maximal tolerable error limit (E_L), which has been clearly explained in [30]. If the RI exceeds the RI_T, then the VM value is reliable; otherwise, its reliability is relatively low, which means the VM conjecture result is not reliable or requires further verification [30].

When VM is applied, no actual measurement value is available to verify the accuracy of the virtual measurement value. Therefore, instead of the standardized actual measurement value $\hat{x}$, the standardized MR prediction value $Z_{\hat{y}_{r_i}}$ is adopted to calculate the RI. This substitution may cause inevitable gauging errors in the RI. For example, when deviations occur in the process data such that the similarity between this new set of process data and the model set of all historical process data is bad, then the accuracy of the VM value conjectured by NN and the reference prediction value predicted by MR would not be good. However, it is still possible that the conjectured VM value and the predicted reference value happen to be nearly the same such that the RI value is higher than RI_T. To compensate for this unavoidable substitution, the GSI is proposed to help the RI gauge the reliance level of VM.

The GSI assesses the degree of similarity between any set of process data and the model set of process data. This model set is derived from all of the historical sets of process data used for building the conjecture model [30].

This work uses the calculated GSI (via Mahalanobis distance) size to determine whether the newly entered set of process data resembles the model set of process data. If the calculated GSI is small, the newly input set is relatively similar to the model set. That is, the virtual measurement value of the newly input (high-similarity) set is relatively accurate. On the contrary, if the calculated GSI is too large, the newly input set is somewhat different from the model set. Consequently, the virtual measurement value estimated in accordance with the newly input (low-similarity) set has low reliance level in terms of accuracy [30].

The formula of GSI for the standardized λ_{th} set process data ($\boldsymbol{Z}_\lambda$) is listed as follows [30]:

$$\text{GSI} = D_\lambda^2 / p \tag{8.11}$$

where $D_\lambda^2 = Z_\lambda^T R^{-1} Z_\lambda$ stands for the Mahalanobis distance of the standardized λ_{th} set process data (Z_λ) and R^{-1} is the inverse matrix of correlation coefficients among the standardized parameters. p stands for the number of process parameters. Observing (8.11), GSI may be interpreted as the mean Mahalanobis distance for each parameter. Then, for the case of a single parameter, $\boldsymbol{R} = \boldsymbol{R}^{-1} = \boldsymbol{I}$ and p=1, consequently, GSI = $\boldsymbol{Z}^2$. In general, the control threshold of a process in statistical control is set to be three times of its standard deviation (3σ). Since each standardized Z-score has mean zero and standard deviation one, namely $Z \sim N(0,1)$, the control threshold of Z is three. Therefore, the threshold of GSI (GSI_T) is set to be $3^2 = 9$ [36].

The calculation of RI mentioned above is based on an assumption of a normal distribution of data. When the data are not normally distributed, such as a Weibull distribution [52], the specific RI formula for Weibull (RI_W) can also be derived according to Figure 8.16 as follows:

$$RI_w \cong \int_{-\infty}^{\frac{\hat{y}_r+\hat{y}_N}{2}} \left(\frac{\beta_N}{\alpha_N}\right)\cdot\left(\frac{x}{\alpha_N}\right)^{\beta_N-1}\cdot e^{-\left(\frac{x}{\alpha_N}\right)^{\beta_N}} dx + \int_{\frac{\hat{y}_r+\hat{y}_N}{2}}^{\infty} \left(\frac{\beta_r}{\alpha_r}\right)\cdot\left(\frac{x}{\alpha_r}\right)^{\beta_r-1}\cdot e^{-\left(\frac{x}{\alpha_r}\right)^{\beta_r}} dx \tag{8.12}$$

where

- $\hat{y}_N$ NN VM value;
- $\hat{y}_r$ MR VM value;
- α_N scale parameter of the Weibull distribution of NN conjecturing model;
- β_N shape parameter of the Weibull distribution of NN conjecturing model;
- α_r scale parameter of the Weibull distribution of MR predictive model;
- β_r shape parameter of the Weibull distribution of MR predictive model.

Note that Weibull distribution is a skewed distribution. Therefore, the intersection of Weibull distributions $W(\alpha_N, \beta_N)$ and $W(\alpha_r, \beta_r)$, the intersected point in the middle of Figure 8.16, is hard to express. However, we may adopt $\frac{(\hat{y}_r + \hat{y}_N)}{2}$ instead for approximation.

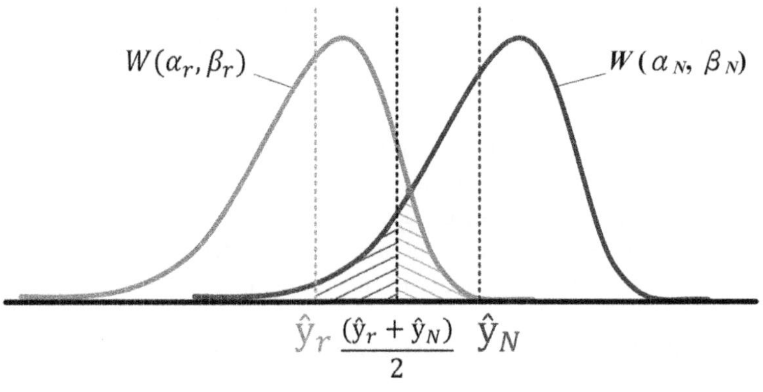

Figure 8.16 Schematic diagram of defining RI_W for Weibull distribution. *Source:* Reprinted with permission from Ref. [31]; © 2013 IEEE.

8.4.3 R2R Control Utilizing VM with Reliance Index (RI) and Global Similarity Index (GSI)

As mentioned in Section 8.4.1, the controller-gain problem of applying VM in R2R control is focused on how to set the α_2 in (8.6). The rule of thumb is that α_2 should depend on the quality or reliability of VM and $\alpha_2 < \alpha_1$. Because the value of RI is a good VM reliability evaluation index and $0 < RI < 1$, higher RI means better VM reliability, we can then naturally set:

$$\alpha_2 = RI \times \alpha_1 \tag{8.13}$$

Equation (8.13) will be applied when the R2R controller needs relatively high gain. The situations that need high controller gain are y_k is apart from the target value or the production process is relatively unstable. On the contrary, if y_k is near the target or the production process is relatively stable, then the controller gain should be small. For generating a small controller gain, we may also set:

$$\alpha_2 = (1 - RI) \times \alpha_1 \tag{8.14}$$

Equations (8.13) and (8.14) are valid only when RI is good enough; in other words, RI should be greater than RI_T. If $RI < RI_T$, this VM value cannot be adopted for tuning the R2R controller gain. Further, due to the fact that the GSI is designed to help the RI gauge the reliance level of VM, when $GSI > GSI_T$, its corresponding VM value cannot be adopted, either. In conclusion, if $RI < RI_T$ or $GSI > GSI_T$, then α_2 is set to be zero (0).

The issue of the R2R controller-gain management in real-production environment whenever a modification is performed on the process or the equipment is considered as follows. In general, the production process of the first lot (just after a modification is performed) is relatively unstable; therefore, the controller gain should be relatively high. After finishing the production of the first lot, the production process will become comparatively stable. In other words, the rest of the lots should have small controller gains.

In summary, α_2 can be set as:

$$\alpha_2 = f(RI, GSI) \times \alpha_1 \tag{8.15}$$

$$f(RI, GSI) = \begin{cases} 0, \; if\; RI < RI_T \;\; or \;\; GSI > GSI_T \\ RI, \; if\; RI \geq RI_T \;\; and\;\; GSI \leq GSI_T \;\; and\, for\; k \leq C \\ 1 - RI, \; if\; RI \geq RI_T \;\; and\;\; GSI \leq GSI_T \;\; and\, for\; k > C \end{cases} \tag{8.16}$$

C in (8.16) is a constant to be set. Equation (8.15) is also called the RI.S.(1-RI) scheme. While (8.16) will be shown to be effective in providing W2W control in an example, it is not guaranteed to be optimal. Determining an optimal set of equations for VM control with a quality metric is a topic of future research.

A modification of the R2R control is needed once a discontinuity in a production process occurs. The events of discontinuity mentioned above are (i) PM, (ii) product/recipe change, (iii) target value change, (iv) tool adjustment, etc. Whenever an event of discontinuity occurs, the RI.S.(1-RI) scheme should be activated to initiate the R2R control. In general, a convenient way for setting the breaking point to switch from RI to 1-RI is just after finishing the process of the first lot. As such, $C = 25$ since 1 lot = 25 wafers.

The detailed diagram of the W2W control scheme utilizing AVM with RI and GSI is shown in Figure 8.17, in which VM_I and VM_{II} represent phase-I VM value and phase-II VM value of the dual-phase VM scheme, respectively. The EWMA filters with α_1 and α_2 shown in Figure 8.17 are governed by (8.5) and (8.6), respectively. The method of calculating α_1 can be found in [53, 54]. The formula of RI is expressed in (8.10), and the $(k+1)^{th}$ run R2R control action is governed by (8.4) and Figure 8.17b.

8.4.4 Illustrative Examples

Due to the fact that a polishing pad of a CMP tool should be changed after polishing 600 wafers, a scheduled periodic maintenance (PM) of the CMP tool is defined by the maximum usage count of the pad. Therefore, the W2W control of a CMP tool with a scheduled PM cycle being 600 pieces of

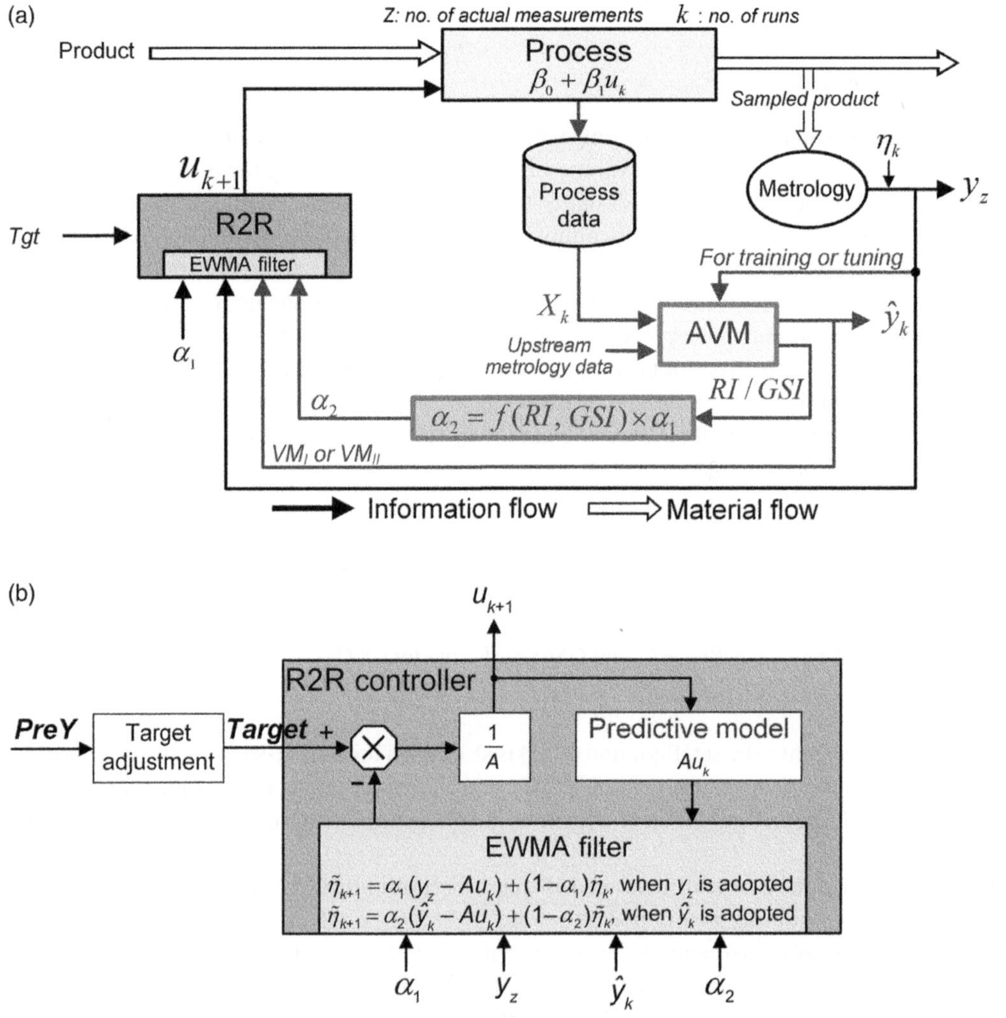

Figure 8.17 W2W control scheme utilizing AVM with RI and GSI. (a) Complete W2W control scheme. (b) R2R controller. *Source:* Reprinted with permission from Ref. [31]; © 2013 IEEE.

wafers (pcs) is selected as the illustrative example for evaluation and comparisons. The simulation conditions and scenarios are listed as follows:

1) y_k is the actual removal amount measured from the metrology tool and $PostY_k$ is the actual post CMP thickness of run k. The specification of $PostY_k$ is 2800±150 Angstrom (Å) with 2800 being the target value denoted by Tgt_{PostY}. Therefore, we have

$$\text{Post}Y_K = \text{Pre}Y_K - y_K \tag{8.17}$$

with

$$y_K = ARR_k \times u_k \tag{8.18}$$

where ARR_k is the actual removal rate of run k and u_k represents the polish time in this example.

The well-known Preston equation, empirically found from the experiment of the glass polishing in 1927, has been proposed to predict the material removal rate of CMP [55, 56]. According to the Preston equation, the material removal rate is affected by the contact pressure (also denoted as tool stress) distribution at contact point, magnitude of the relative velocity (also denoted as tool rotation speed) at contact point between wafer and polishing pad, and constant representing the effect of the other remaining parameters including the slurry fluid speed, pad property, and so on. Therefore, ARR_k is simulated by:

$$ARR_k = \left(A_k \times \left(\frac{Stress1 + Stress2}{1000} \right) \times \left(\frac{Rotspd1 + Rotspd2}{100} \right) \times \left(\frac{Sfuspd1 + Sfuspd2}{100} \right) \right) + \left(PM1 + PM2 \right) + Error \tag{8.19}$$

The meanings of *Stress1, Stress2, Rotspd1, Rotspd2, Sfuspd1, Sfuspd2, PM1, PM2,* and *Error* are tabulated in Table 8.1. The A_k in (8.19) is the nominal removal rate, which is empirically simulated by a polynomial curve fitting of parts usage count between PMs (denoted by PU varying from 1 to 600):

$$A_k = \left(4 \times 10^{-6}\right) \times \left(PU - 1\right)^3 - \left(3.4 \times 10^{-3}\right) \times \left(PU - 1\right)^2 + \left(6.9 \times 10^{-3}\right) \times \left(PU - 1\right) + \left(1.202 \times 10^3\right) \tag{8.20}$$

2) $Post\hat{Y}_k$ represents the predictive value of $PostY_k$, then, from (8.17) and (8.18) we have

$$\hat{y}_k = A\hat{R}R_k \times u_k \tag{8.21}$$

$$Post\hat{Y}_k = PreY_k - \hat{y}_k = PreY_k - A\hat{R}R_k \times u_k \tag{8.22}$$

where

$$A\hat{R}R_k = f\left(Stress, Rotspd, Sfuspd, PU, PU^2, PU^3\right) \tag{8.23}$$

is the VM_I value of ARR_k with *Stress* (= *Stress1*+ *Stress2*), *Rotspd* (=*Rotspd1*+*Rotspd2*), *Sfuspd* (=*Sfuspd1*+*Sfuspd2*), *PU*, PU^2, PU^3 as the process parameters. The reason of adopting *Stress, Rotspd, Sfuspd, PU,* PU^2, and PU^3 as the process parameters is based on the Preston equation, (8.19) and (8.20). The setting values of the simulated process parameters are tabulated in Table 8.1.

Table 8.1 Simulation-parameter definitions and setting values.

Abbreviation	Definition	Setting Values	
		Mean	VAR
Error	Random error represented by white noise	0	300
PM1	Error due to tool-parts' variation that was caused by periodic maintenance (PM)	0	100
PM2	Random disturbance of tool-parts' variation	0	6
Stress1	Tool stress error due to re-assembly during PM	1000	2000
Stress2	Random disturbance of tool stress	0	20
Rotspd1	Tool rotation-speed error due to re-assembly during PM	100	25
Rotspd2	Random disturbance of tool rotation speed	0	1.2
Sfuspd1	Slurry fluid-speed error due to re-assembly during PM	100	25
Sfuspd2	Random disturbance of slurry fluid speed	0	1.2
$PreY_k$	Pre-process (etching depth) value that affects the process result of run k	3800	2500

Source: Reprinted with permission from Ref. [31]; © 2013 IEEE.

3) Based on (8.4), the $(k+1)^{th}$ run control action is derived by

$$Tgt_{k+1} = PreY_{k+1} - Tgt_{PostY} \tag{8.24}$$

$$u_{k+1} = \frac{Tgt_{k+1} - \tilde{\eta}_{k+1}}{A_{k+1}} \tag{8.25}$$

4) When $PostY_k$ is measured by an actual metrology tool, then, from (8.5),

$$\tilde{\eta}_{k+1} = \alpha_1 \left(y_Z - A_k u_k \right) + \left(1 - \alpha_1 \right) \tilde{\eta}_k \tag{8.26}$$

When $PostY_k$ is conjectured or predicted by a VM system, then, from (8.6), (8.15), and (8.16), we have

$$\tilde{\eta}_{k+1} = \alpha_{2,k} \left(y_Z - A_k u_k \right) + \left(1 - \alpha_{2,k} \right) \tilde{\eta}_k \tag{8.27}$$

with

$$\alpha_{2,k} = f\left(RI_k, GSI_k \right) \times \alpha_1 \tag{8.28}$$

where

$$f(RI, GSI) = \begin{cases} 0, \text{ if } RI_k < RI_T \text{ or } GSI_k > GSI_T \\ RI_k, \text{ if } RI_k \geq RI_T \text{ and } GSI_k \leq GSI_T \text{ and for } k \leq C \\ 1 - RI_k, \text{ if } RI_k \geq RI_T \text{ and } GSI_k \leq GSI_T \text{ and for } k > C \end{cases} \tag{8.29}$$

For this example, $C = 25$.

5) 1 Lot = 25 pcs in which the 2nd wafer being the sampling wafer that applies α_1 and (8.26) to calculate $\tilde{\eta}_{k+1}$.

$$Cpk(\text{Process Capability}) = \min\left\{\frac{UCL - mean(PostY)}{3 \times std(PostY)}, \frac{mean(PostY) - LCL}{3 \times std(PostY)}\right\} \tag{8.30}$$

is utilized to evaluate the performance of R2R control with $UCL = 2950$ and $LCL = 2650$.

6) The process mean-absolute-percentage error ($MAPE_P$) with respect to the target value:

$$MAPE_P = \frac{\sum_{i=1}^{k}\left|\left(PostY_i - Tgt_{PostY}\right)/Tgt_{PostY}\right|}{k} \times 100\% \tag{8.31}$$

is also used to evaluate the performance of R2R control.

7) Extra random disturbances caused by *Sfuspd2* with mean = 0 and variance = 0.36 are also added at Samples 50, 111, 179, 251, 349, and 503. In other words, the combined variances of *Sfuspd2* at Samples 50, 111, 179, 251, 349, and 503 are 1.2 + 0.36 = 1.56. With these extra random disturbances, the RI and/or GSI values may exceed their thresholds.

Five rounds with different random seeds are performed to evaluate and compare the performance. Based on the authors' experience, the value of controller-gain, α_1, usually varies between 0.2 and 0.5 in the semiconductor R2R control. Therefore, the α_1 value is set to be 0.35 in this paper. For each round, the simulation results of $PreY_k$, Tgt_k, A_k, and ARR_k for $k = 1\sim600$ should be generated firstly based on the setting values shown in Table 8.1, (8.24), (8.20), and (8.19), respectively. Then, let $\alpha_1 = 0.35$ and $\tilde{\eta}_1 = 0$ to calculate u_1 as well as apply (8.18), (8.26), (8.25), and (8.17) to calculate y_k, $\tilde{\eta}_{k+1}$, u_{k+1}, and $PostY_k$, respectively for $k = 1$ and 2 for all of the five cases. As mentioned in the simulation condition 5), all the sampling wafers apply α_1 and (8.26) to calculate $\tilde{\eta}_{k+1}$. For all the nonsampling wafers and for $k = 3\sim600$, control schemes for those five cases are different and are described below:

Case 1: R2R with in-situ metrology

Let $\alpha_1 = 0.35$. Apply (8.18), (8.26), (8.25), and (8.17) to calculate y_k, $\tilde{\eta}_{k+1}$, μ_{k+1}, and $PostY_k$, respectively for $k = 3\sim600$.

Case 2: R2R+VM without RI

Let $\alpha_2 = \alpha_1 = 0.35$. Apply (8.21), (8.27), (8.25), (8.22), and (8.17) to calculate $\hat{y}_k$, $\tilde{\eta}_{k+1}$, μ_{k+1}, $Post\hat{Y}_k$, and $PostY_k$, respectively for $k = 3\sim600$.

Case 3: R2R+VM with RI

Let $\alpha_1 = 0.35$. If RI < RI_T or GSI > GSI_T, then let $\alpha_2 = 0$; otherwise, let $\alpha_{2,k} = RI_k \times \alpha_1$; as well as apply (8.21), (8.27), (8.25), (8.22), and (8.17) to calculate $\hat{y}_k$, $\tilde{\eta}_{k+1}$, μ_{k+1}, $Post\hat{Y}_k$, and $PostY_k$, respectively for $k = 3\sim600$.

Case 4: R2R+VM with (1-RI)

Let $\alpha_1 = 0.35$. If RI < RI_T or GSI > GSI_T, then let $\alpha_2 = 0$; otherwise, let $\alpha_{2,k} = (1 - RI_k) \times \alpha_1$; as well as apply (8.21), (8.27), (8.25), (8.22), and (8.17) to calculate $\hat{y}_k$, $\tilde{\eta}_{k+1}$, μ_{k+1}, $Post\hat{Y}_k$, and $PostY_k$, respectively for $k = 3\sim600$.

Case 5: R2R+VM with RI.S.(1-RI)

Let $\alpha_1 = 0.35$. Apply the RI.S.(1-RI) scheme as shown in (8.28) and (8.29) to set α_2; as well as apply (8.21), (8.27), (8.25), (8.22), and (8.17) to calculate $\hat{y}_k$, $\tilde{\eta}_{k+1}$, μ_{k+1}, $Post\hat{Y}_k$, and $PostY_k$, respectively for $k = 3\sim600$.

Both Cpk and $MAPE_P$ [as expressed in (8.30) and (8.31), respectively] are applied to evaluate and compare the performance of those five cases. The Cpk is a measure of process capability [57] and the $MAPE_P$ is the process mean-absolute-percentage error with respect to the target value. The Cpk and $MAPE_P$ values of those five cases are tabulated in Tables 8.2 and 8.3, respectively.

Observing Tables 8.2 and 8.3 and treating Case 1 as the baseline, it is obvious that the performance of Case 2, which does not consider RI/GSI, is the worst. Case 3, which filters out the bad-quality $Post\hat{Y}_k$ (VM) values and lets $\alpha_2 = RI \times \alpha_1$, is the most natural approach and has acceptable performance. The performance of Case 4, which filters out the bad-quality $Post\hat{Y}_k$(VM) values and lets $\alpha_2 = (1-RI) \times \alpha_1$, is better than that of Case 3 on average except for Round 1. Case 5, which filters out the bad-quality $Post\hat{Y}_k$(VM) values and applies the RI.S.(1-RI) scheme shown in (8.29), fixes the problem of Case 4 in Round 1; and Case 5's performance is compatible with that of Case 1. The reasons support the observations and conclusions mentioned above are presented in the following paragraphs.

Simulation Results of Round 1 for those five cases are shown in Figure 8.18. Due to the fact that extra random disturbances caused by *Sfuspd2* with mean = 0 and variance = 0.36 are added at Samples 50, 111, 179, 251, 349, and 503, bad $Post\hat{Y}_k$(VM) values are generated and displayed in Figure 8.18. According to the authors' experiences of testing the AVM system in the Taiwan Semiconductor Manufacturing Company (tsmc), Ltd., Taiwan, ROC [30] and Chi Mei Optoelectronics (CMO), Ltd., Taiwan, ROC [35], similar bad VM values (as shown in Figure 8.18) are common to occur and can be detected by RI and/or GSI. Please refer to Figure 6 of [30] and 8 of [35] for details.

In this example, the maximal tolerable E_L is defined to be 32 Å or 1.14% (=32/2800). The RI value corresponding to the maximal tolerable E_L is about 0.7, therefore the RI_T is set at 0.7. The method of converting E_L into RI_T is clearly explained in [30]. Also, the GSI_T value is set at 9 as explained in Section 8.4.2. The cases that $RI < RI_T$ and $GSI > GSI_T$ at Sample 50 of Round 1 as well as $GSI > GSI_T$ at Sample 349 of Round 1 are enlarged and depicted in Figures 8.19 and 8.20, respectively.

As shown in Figure 8.19, the Sample 50's $Post\hat{Y}_{50}$ (VM) values of various cases are deviated with extra variance 0.36 adding to *Sfuspd2*. Owing to the fact that $RI < RI_T$ and $GSI > GSI_T$, those $Post\hat{Y}_{50}$ values of Cases 3, 4, and 5 are filtered out by setting $\alpha_2 = 0$; while the $Post\hat{Y}_{50}$ value of Case 2 is still adopted to adjust the R2R controller gain with $\alpha_2 = \alpha_1 = 0.35$. The effect of filtering out the bad-quality $Post\hat{Y}_{50}$ value is displayed in Sample 51, which shows that the $PostY_{51}$ value of Case 2 is pulled down by the R2R controller since the $Post\hat{Y}_{50}$ value at Sample 50 is too high. As for the other cases, $PostY_{50}$ and $PostY_{51}$ have no much difference.

Observing Figure 8.20, the Sample 349's $Post\hat{Y}_{349}$ values of various cases are deviated with extra variance 0.36 adding to *Sfuspd2*, again. In this case, only GSI exceeds its threshold. By the same token, these bad $Post\hat{Y}_{349}$ values are discarded in Cases 3, 4, and 5 yet not in Case 2. As such, improper R2R control result of Case 2, which generates a surge $PostY_{350}$, is shown in Figure 8.20. The evidences displayed in Figures 8.19 and 8.20 reveal that the result of adopting an unreliable VM value is worse than if no VM at all is utilized.

As mentioned in Section 8.4.3, we can naturally set $\alpha_2 = RI \times \alpha_1$ when $PostY_k$ is apart from the target value or production process is relatively unstable. On the contrary, if $PostY_k$ is near the target or production process is relatively stable, we may then set $\alpha_2 = (1-RI) \times \alpha_1$.

Rounds 1 and 4 simulation results of Cases 3–5 for the first 200 samples are displayed in Figure 8.21. Observing Figure 8.21a, the initial $PostY_1$ value is 2924, which is somewhat apart from the target value (2800). Therefore, the performance of Case 3 is comparatively better than that of Case 4, as shown in Tables 8.2 and 8.3. After applying the W2W control to the first lot (25 wafers) in Case 3, the $PostY_{25}$ value (2821) is near the target. In this situation, we may then switch α_2 to $\alpha_2 = (1-RI) \times \alpha_1$ for the subsequent W2W control so as to improve the performance. In fact, by doing so, this would become the Case 5. Therefore, as far as Round 1 is concerned, the performance of Case 5 is the best, then Case 3, and Case 4 is the worst.

Table 8.2 Cpk values of 5-cases APC methods (α_1 =0.35).

	Case 1: Insitu			Case 2: VM			Case 3: VM+RI			Case 4: VM+(1-RI)			Case 5: VM+RI.S.(1-RI)		
Round	**1~25**	**1~200**	**1~600**	**1~25**	**1~200**	**1~600**	**1~25**	**1~200**	**1~600**	**1~25**	**1~200**	**1~600**	**1~25**	**1~200**	**1~600**
1	1.09	1.58	1.62	1.14	1.42	1.31	1.14	1.54	1.49	1.12	1.29	1.38	1.14	1.57	1.55
2	1.73	1.89	1.86	1.51	1.64	1.72	1.51	1.73	1.77	1.89	2.00	2.04	1.51	1.71	1.74
3	1.60	1.74	1.77	1.72	1.64	1.77	1.72	1.72	1.80	1.76	1.79	1.87	1.72	1.85	1.90
4	1.43	1.95	1.87	1.45	1.74	1.72	1.45	1.89	1.76	1.51	1.95	1.87	1.45	1.94	1.87
5	1.32	1.85	1.81	1.41	1.78	1.71	1.41	1.83	1.79	1.33	1.77	1.81	1.41	1.89	1.86
Mean	**1.43**	**1.80**	**1.80**	**1.45**	**1.64**	**1.65**	**1.45**	**1.74**	**1.72**	**1.52**	**1.76**	**1.80**	**1.45**	**1.79**	**1.78**

Source: Reprinted with permission from Ref. [31]; © 2013 IEEE.

Table 8.3 $MAPE_P$ values of 5-cases APC methods ($\alpha_1 = 0.35$, Unit: %).

	Case 1: Insitu			Case 2: VM			Case 3: VM+RI			Case 4: VM+(1-RI)			Case 5: VM+RI.S.(1-RI)		
Round	**1~25**	**1~200**	**1~600**	**1~25**	**1~200**	**1~600**	**1~25**	**1~200**	**1~600**	**1~25**	**1~200**	**1~600**	**1~25**	**1~200**	**1~600**
1	1.13	0.86	0.86	1.35	1.00	1.07	1.35	0.94	1.00	2.52	1.45	1.15	1.35	0.98	0.97
2	0.85	0.75	0.76	0.98	0.87	0.81	0.97	0.84	0.79	0.80	0.72	0.69	0.97	1.05	0.82
3	0.84	0.85	0.82	0.94	0.86	0.81	0.94	0.83	0.79	1.03	0.80	0.76	0.94	0.77	0.75
4	0.93	0.73	0.76	1.11	0.83	0.83	1.11	0.78	0.82	1.07	0.84	0.79	1.11	0.84	0.79
5	0.99	0.75	0.78	1.05	0.78	0.82	1.05	0.77	0.80	1.41	0.83	0.78	1.05	0.77	0.76
Mean	**0.95**	**0.79**	**0.80**	**1.08**	**0.87**	**0.87**	**1.09**	**0.83**	**0.84**	**1.37**	**0.93**	**0.83**	**1.09**	**0.88**	**0.82**

Source: Reprinted with permission from Ref. [31];

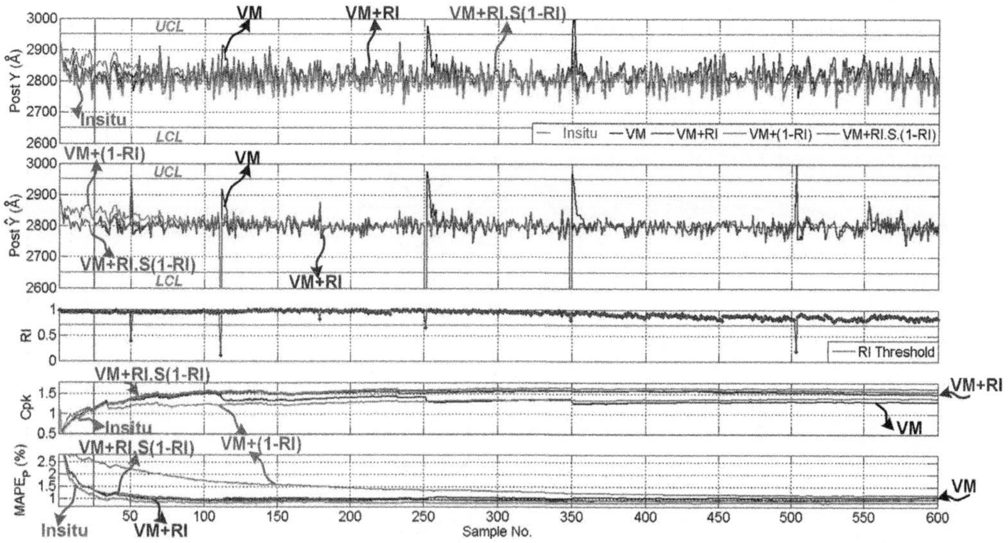

Figure 8.18 Simulation results of 5-cases APC methods of Round 1. *Source:* Reprinted with permission from Ref. [31]; © 2013 IEEE.

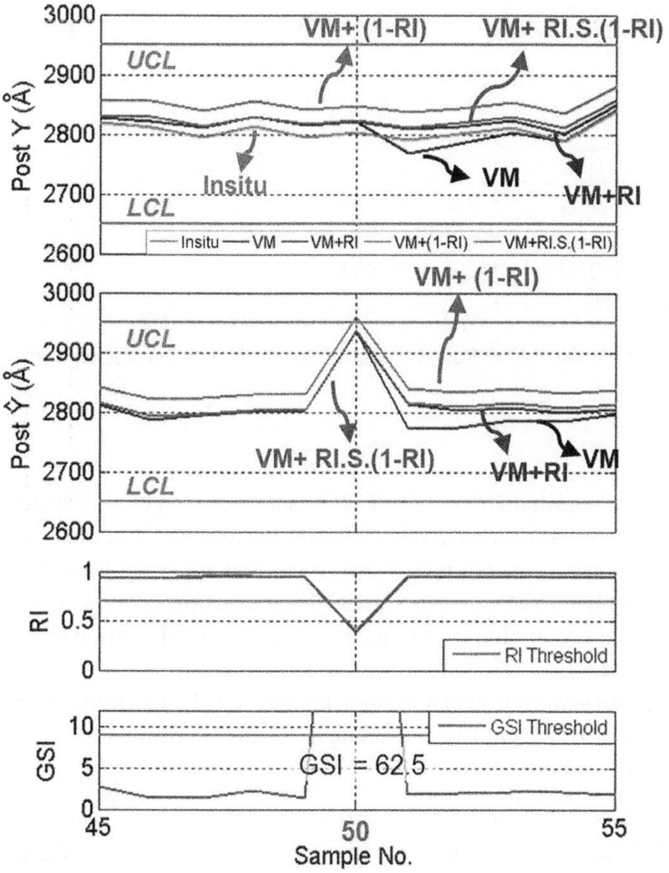

Figure 8.19 RI and GSI exceed their thresholds at Sample 50 of Round 1. *Source:* Reprinted with permission from Ref. [31]; © 2013 IEEE.

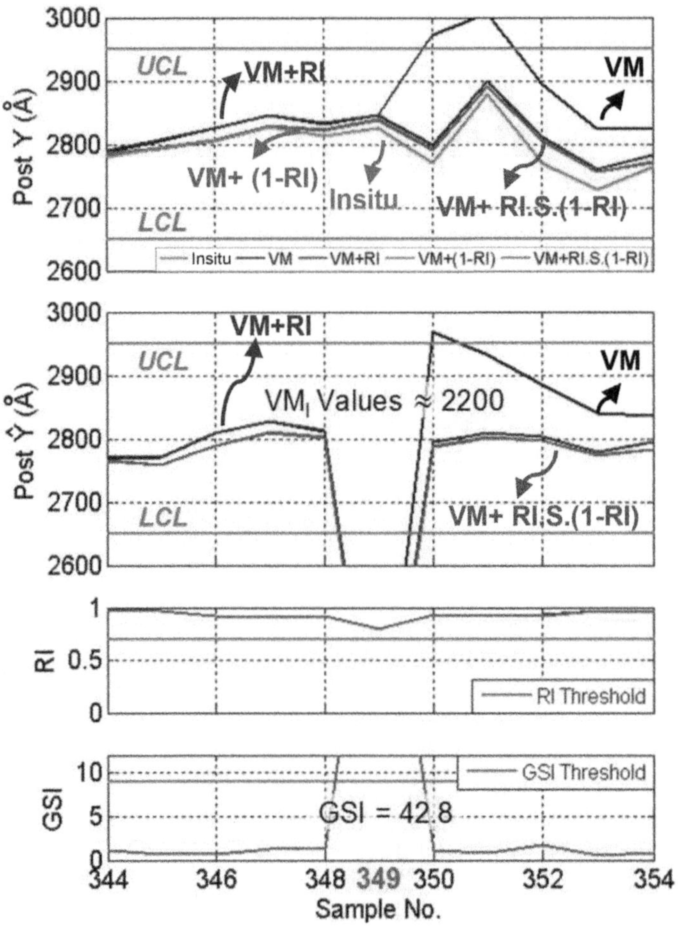

Figure 8.20 GSI exceeds its threshold at Sample 349 of Round 1. *Source:* Reprinted with permission from Ref. [31]; © 2013 IEEE.

Examining Round 4 in Figure 8.21b, the initial $PostY_1$ value (2812) is near the target. As a result, the performance of Case 4 is better than that of Case 3 (check Tables 8.2 and 8.3). On the other hand, the performance of Case 5 (with the α_2 switching scheme) is as good as that of Case 4. In conclusion, the R2R control scheme applying VM with the RI.S.(1-RI) scheme may take care of the R2R controller-gain management in real-production environment whenever a modification is performed on the process or the equipment.

The above simulations are based on an assumption of a normal distribution of data. Simulations of the nonnormal Weibull distribution with RI and RI_W for tuning α_1 are also performed. The formula of Weibull distribution [52] is shown below:

$$f(x)=\left(\frac{\beta}{\alpha_W}\right)\cdot\left(\frac{x}{\alpha_W}\right)^{\beta-1}\cdot e^{-\left(\frac{x}{\alpha_W}\right)^{\beta}} \quad \alpha_W,\beta,x>0 \tag{8.32}$$

where α_W is the scale parameter and β is the shape parameter.

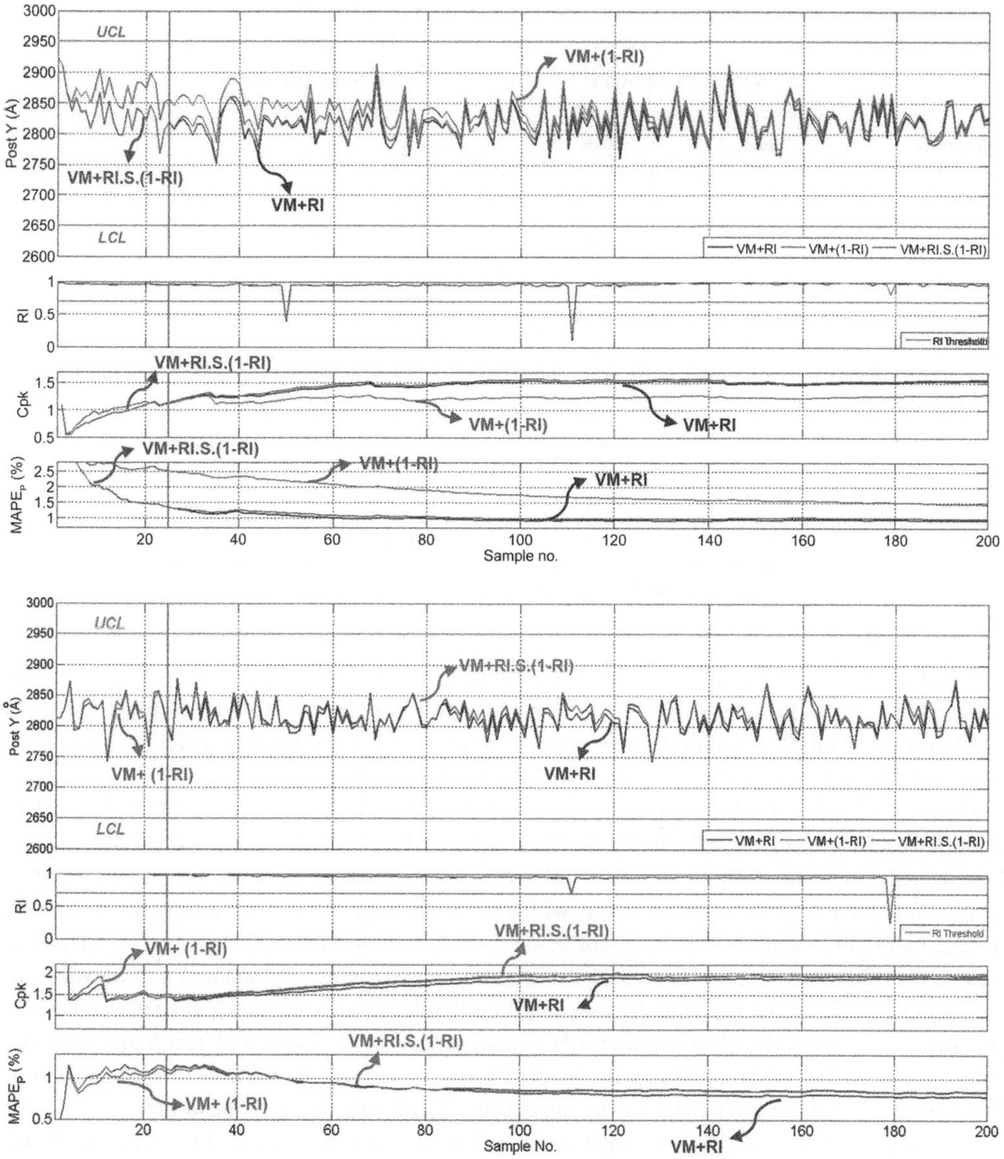

Figure 8.21 Simulation results of Cases 3-5 for the first 200 samples. (a) Round 1. (b) Round 4. *Source:* Reprinted with permission from Ref. [31]; © 2013 IEEE.

To begin with, the α_W and β of Weibull distribution of each simulation parameter shown in Table 8.1 are identified and assigned. Take *Sfuspd1* and *Sfuspd2* as the examples. Assign mean = 100 and variance = 25 for *Sfuspd1* as well as mean = 0 and variance ≐ 1.2 for *Sfuspd2*, then generate all the random data of *Sfuspd1* and *Sfuspd2* by normal distribution. And then sum the random data of *Sfuspd1* and *Sfuspd2* together to become the random data *Sfuspd*. According to (8.18), *Sfuspd/100* will be utilized to identify the α_W and β values for *Sfuspd's* Weibull distribution, which are 0.917 and 250, respectively and are shown in Table 8.4. By the same token, the α_W and β values for all the simulation parameters are identified and displayed in Table 8.4. However, due to the fact

Table 8.4 Simulation parameters and setting values (Normal and Weibull distributions).

	Normal distribution		Weibull distribution	
Parameter	**Mean**	**VAR**	α_W	β
Error	0	300	108.5	7.2
PM1	0	100	2800.1	850.1
PM2	0	6		
Stress1	1000	2000	0.951	150
Stress2	0	20		
Rotspd1	100	25	0.986	100
Rotspd2	0	1.2		
Sfuspd1	100	25	0.917	250
Sfuspd2	0	1.2		
$PreY_k$	3800	2500	3826.3	77.83

Source: Reprinted with permission from Ref. [31]; © 2013 IEEE.

that Weibull distribution cannot generate negative values, biases of 100 and 2800 are added to the random data produced by normal distribution of *Error* and *PM1*, respectively, to identify the α_w and β values of Weibull distribution; then these biases of 100 and 2800 will be deducted from every *Error* and *PM1* value created by Weibull distribution respectively.

With α_1 =0.35, the $MAPE_P$ values of 5-Cases APC methods for Test A: normal with RI; Test B: Weibull with RI_W; and Test C: Weibull with RI are tabulated in Table 8.5. The $MAPE_P$ values of Test A are exactly the same as those shown in Table 8.3. The purpose of showing Test A here again is simply for comparison. Comparing Test A with Test B, the major difference occurs in the first lot (1~25); the $MAPE_P$ values of the first lot of Test B are larger than those of Test A. The reason is that Weibull distribution is a kind of skewed distributions that will pull the initial *PostY* value away from the target. However, after applying the various APC methods the *PostY* value will be driven back to the target. As a result, the $MAPE_P$ values of (1 ~ 600) of Cases 1, 2, 3, and 5 of Test B are about the same as those of Test A. The reason why the performance of Case 4 in Test B is worse than that in Test A is due to (1-RI) is too small to bring the deviated *PostY* value back to the target.

Test B applies RI_W as in (8.12) to calculate the reliance index. As indicated in (8.12), to obtain RI_W the values of α_N, β_N, α_r, and β_r should be available. However, those scale and shape parameters may not be easy to get on-line. Of cause, we may obtain the default scale and shape parameters from the historical model creation data and design an on-line refreshing scheme to update those α_N, β_N, α_r, and β_r such that RI_W can be calculated on-line. Nevertheless, this way is tedious and time-consuming. To remedy the above problem, we may apply RI as in (8.10) instead for approximation, which is easy and computationally efficient. That is the reason why Test C is performed. Comparing Tests B with C, the results show that the $MAPE_P$ values of Cases 1, 2, 3, and 5 of those two tests are almost the same; while Case 4 has slight difference. Therefore, it is concluded that the RI formula as in (8.10) can be applied efficiently and accurately to calculate the reliance index of VM for normal distribution and nonnormal (Weibull) distribution data.

The illustrative examples mentioned above for all 5-cases APC methods select α_1= 0.35 only. To determine the optimal α_1 value to the noise level of the above illustrative examples, further

Table 8.5 $MAPE_P$ values of 5-Cases APC methods for A. Normal with RI; B. Weibull with RI_W; and C. Weibull with RI (α_1 = 0.35, Unit: %).

		Case 1: Insitu			Case 2: VM			Case 3: VM+RI			Case 4: VM+(1-RI)			Case 5: VM+RI.S.(1-RI)		
	Round	**1~25**	**1~200**	**1~600**	**1~25**	**1~200**	**1~600**	**1~25**	**1~200**	**1~600**	**1~25**	**1~200**	**1~600**	**1~25**	**1~200**	**1~600**
A. Normal with RI	1	1.13	0.86	0.86	1.35	1	1.07	1.35	0.94	1.00	2.52	1.45	1.15	1.35	0.98	0.97
	2	0.85	0.75	0.76	0.98	0.87	0.81	0.97	0.84	0.79	0.80	0.72	0.69	0.97	1.05	0.82
	3	0.84	0.85	0.82	0.94	0.86	0.81	0.94	0.83	0.79	1.03	0.80	0.76	0.94	0.77	0.75
	4	0.93	0.73	0.76	1.11	0.83	0.83	1.11	0.78	0.82	1.07	0.84	0.79	1.11	0.84	0.79
	5	0.99	0.75	0.78	1.05	0.78	0.82	1.05	0.77	0.80	1.41	0.83	0.78	1.05	0.77	0.76
	Mean	**0.95**	**0.79**	**0.8**	**1.08**	**0.87**	**0.87**	**1.09**	**0.83**	**0.84**	**1.37**	**0.93**	**0.83**	**1.09**	**0.88**	**0.82**
B. Weibull with RI_W	1	1.39	0.84	0.82	1.45	0.86	0.9	1.44	0.85	0.86	2.99	1.47	1.06	1.44	0.83	0.84
	2	1.26	0.81	0.82	1.27	0.83	0.87	1.27	0.79	0.83	2.44	1.28	1	1.27	0.84	0.84
	3	1.26	0.8	0.79	1.27	0.84	0.84	1.26	0.8	0.81	2.42	1.26	0.96	1.26	0.83	0.81
	4	1.37	0.81	0.82	1.42	0.86	0.88	1.41	0.83	0.86	2.55	1.25	0.98	1.41	0.89	0.86
	5	1.06	0.76	0.79	1.29	0.81	0.84	1.28	0.78	0.82	2.71	1.26	0.97	1.28	0.79	0.81
	Mean	**1.27**	**0.8**	**0.81**	**1.34**	**0.84**	**0.87**	**1.33**	**0.81**	**0.84**	**2.62**	**1.3**	**0.99**	**1.33**	**0.84**	**0.83**
C. Weibull with RI	1	1.39	0.84	0.82	1.45	0.86	0.9	1.43	0.85	0.86	3.12	1.61	1.16	1.43	0.84	0.84
	2	1.26	0.81	0.82	1.27	0.83	0.87	1.27	0.79	0.83	2.55	1.41	1.17	1.27	0.87	0.86
	3	1.26	0.8	0.79	1.27	0.84	0.84	1.25	0.8	0.81	2.53	1.39	1.12	1.25	0.84	0.81
	4	1.37	0.81	0.82	1.42	0.86	0.88	1.41	0.83	0.86	2.66	1.36	1.1	1.41	0.93	0.87
	5	1.06	0.76	0.79	1.29	0.81	0.84	1.28	0.78	0.82	2.83	1.44	1.12	1.28	0.82	0.82
	Mean	**1.27**	**0.8**	**0.81**	**1.34**	**0.84**	**0.87**	**1.33**	**0.81**	**0.84**	**2.74**	**1.44**	**1.13**	**1.33**	**0.86**	**0.84**

Source: Reprinted with permission from Ref. [31]; © 2013 IEEE.

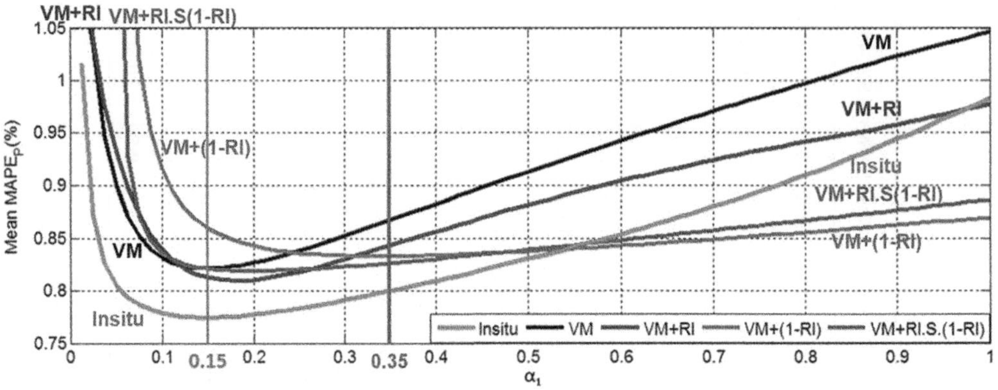

Figure 8.22 Mean MAPE$_P$ curves as functions of α_1 of 5-cases APC methods. *Source:* Reprinted with permission from Ref. [31]; © 2013 IEEE.

simulations setting α_1 between 0 and 1 with the interval of 0.0125 are performed. In this way, mean MAPE$_P$ curves serve as functions of α_1 of 5-cases APC methods are shown in Figure 8.22. As indicated in Table 8.3, with α_1= 0.35, the mean MAPE$_P$ values of all five cases are 0.80%, 0.87%, 0.84%, 0.83%, and 0.82%, respectively. Those values are correctly shown in Figure 8.22. Observations of all the five curves are described below.

Case 1: R2R with in-situ metrology

The shape of the MAPE$_P$ curve of Case 1 is similar to that of the mean-squared-error (MSE) curve in Figure 4 of [58]. Observing Case 1's MAPE$_P$ curve, the optimal α_1 value appears to be 0.15 for this example.

Case 2: R2R+VM without RI

The trend of the MAPE$_P$ curve of Case 2 is similar to that of Case 1. However, due to the fact that process data contain noise which affects the conjecturing accuracy of VM, the MAPE$_P$ curve of Case 2 is higher than that of Case 1.

Case 3: R2R+VM with RI

Observing the MAPE$_P$ curve of Case 2, when $0.15<\alpha_1<1$, the larger the α_1 is, the more serious the impact of noise is, this means larger MAPE$_P$ value. As for Case 3, yet $\alpha_2 = \text{RI}\times\alpha_1$ is rather smaller than α_1, it means that Case 3 with VM+RI has the effect of releasing the effect of noise. Thus, as α_1 increasing, the MAPE$_P$ value of Case 3 gets smaller and smaller in comparison with that of Case 2.

On the contrary, viewing the MAPE$_P$ curve of Case 2, when $0<\alpha_1<0.15$, the smaller α_1 is, the larger MAPE$_P$ gets. Thus, observing the MAPE$_P$ curve of Case 3 for $0<\alpha_1<0.15$, due to the fact that α_2 is smaller than α_1, MAPE$_P$ of Case 3 is larger than that of Case 2. Note that the mean of RI of these simulations is about 0.89.

Case 4: R2R+VM with (1-RI)

Since the corresponding $\alpha_2[=(1-RI)\times\alpha_1]$ will be quite small, it has moderate effect on MAPE$_P$; therefore, the MAPE$_P$ values of Case 4 within the range of $0.15<\alpha_1<1$ do not have much variation. Note that, as mentioned in the simulation condition 5), all the sampling wafers with actual metrology still apply α_1 and (8.26) to calculate $\tilde{\eta}_{k+1}$. Besides the first wafer after PM, there are 24 sampling wafers in (600/25) lots. As such, those sampling wafers with α_1 as the controller gain will help the nonsampling wafers with α_2 to drive the *PostY* value back to the target if deviations occur.

For $0 < \alpha_1 < 0.15$, α_2 will be much smaller in Case 4. Thus, once the initial *PostY₁* value is somewhat apart from the target value, it is hard to drive the *PostY* value back to the target. As a result, the $MAPE_P$ curve of Case 4 is the highest among all the five cases during the range of $0 < \alpha_1 < 0.15$.

Case 5: R2R+VM with RI.S.(1-RI)
The trend of the $MAPE_P$ curve of Case 5 should be similar to that of Case 4. However, since α_2 of the first lot is controlled by $RI \times \alpha_1$ in Case 5, it should be easier to drive the *PostY* value back to the target if the initial $PostY_1$ value deviates from the target. Therefore, when $0 < \alpha_1 < 0.35$, the $MAPE_P$ curve of Case 5 is obviously lower than that of Case 4. Also, when α_1 gets closer to the optimal value (0.15), the $MAPE_P$ curve of Case 5 is close to those of Cases 2 and 3.

8.4.5 Summary

This work proposes a novel scheme of R2R control that utilizes VM with RI and GSI in the feedback loop. The controller-gain problem of applying VM in R2R control is focused on how to set the EWMA coefficient α_2 when VM value is adopted as the feedback. And, a simulated CMP tool with five random-generated rounds is adopted as the illustrative example for comparison and evaluation; each round contains five cases. Simulation results show that the RI and GSI of the AVM system can be effectively utilized to filter out the bad-quality VM values such that the performance of R2R control utilizing VM can be assured. Moreover, with $\alpha_1 = 0.35$, the proposed RI.S.(1-RI) scheme may handle the R2R controller-gain management in real production environment whenever a modification is performed on the process or the equipment. Normal distribution is assumed for the calculation of RI. The RI_W formula for the nonnormal Weibull distribution is also derived. From the simulation results, it is concluded that the RI formula can be applied efficiently to calculate the reliance index of VM for normal distribution and nonnormal (Weibull) distribution data. Finally, besides $\alpha_1 = 0.35$, further simulations setting α_1 between 0 and 1 with the interval of 0.0125 are performed. As such, mean $MAPE_P$ curves serve as functions of α_1 of 5-cases APC methods are drawn; and observations of these five curves are made. According to the observations, the optimal value of α_1 of the illustrative example is found (0.15) and the effects of those 5-cases APC methods: Insitu, VM, VM+RI, VM+(1-RI), and VM+RI.S.(1-RI) on the $MAPE_P$ values become clear. In conclusion, this technique has been shown to be effective in providing W2W control in an example, and the authors believe it would be effective in a majority of W2W control cases.

8.5 AVM System Deployment

After elaborating the theoretical illustrations of the AVM system, this section introduces the deployment considerations and requirements of the AVM system, which include automation levels of virtual metrology systems and the AVM system deployment management.

8.5.1 Automation Levels of VM Systems

An AVM system should be able to provide all kinds of VM applications and to realize automatic and fab-wide VM deployment. To accomplish the mission of AVM, not only the VM kernel functions should be maturely developed but also the enabling information-technology efforts should be adequately applied. The VM kernel functions include data preprocessing, process/metrology data quality evaluations, VM conjecturing, VM reliance level evaluation, model creation, on-line learning, and model refreshing, etc. The enabling information-technology efforts contain data

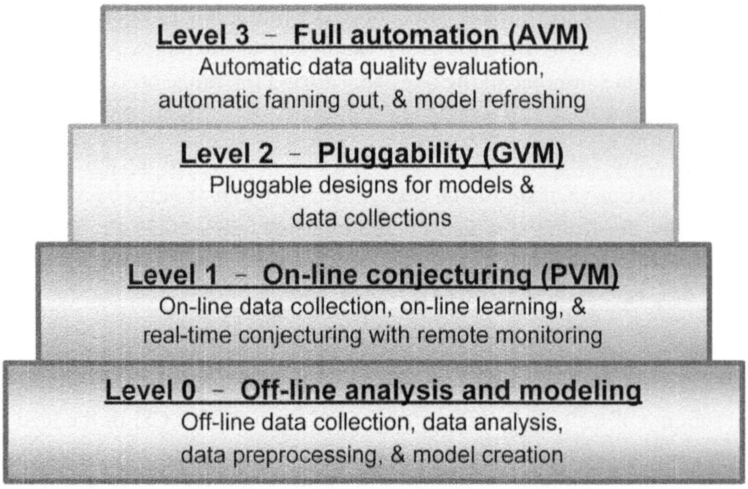

Figure 8.23 Automation levels of virtual metrology systems. *Source:* Reprinted with permission from Ref. [36]; © 2012 IEEE.

collection, graphical user interface (GUI) design and remote monitoring, pluggable design for generalization, and various automation programming technologies to enable/automate those VM kernel functions.

It is difficult to accomplish AVM thoroughly in a single bound; instead, AVM shall be achieved step by step. Referring to the methodology used for classifying the e-diagnostics capabilities [59] and according to the VM deployment and automation degree, the VM systems can be classified into four levels as shown in Figure 8.23 and depicted in the following paragraphs.

Level 0 – Off-line Analysis and Modeling

Level 0 is the most fundamental level in terms of VM automation degree. The Level 0 VM system is capable of executing off-line collection and analysis of the historical process and metrology data. It can also create the first DQI_X and DQI_y models, VM conjecture models, RI model, and GSI model. The accuracy of these models should also be evaluated in Level 0.

Level 1 – On-line Conjecturing (PVM)

Being the preliminary VM (PVM) system, the Level 1 VM system focuses on on-line conjecturing. The functions of Level 0 VM system for constructing the first set of DQI_X/DQI_y and VM & RI/GSI models must be established prior to building the PVM. The PVM system should possess capabilities of on-line data collection and on-line learning algorithm (such as the dual-phase VM scheme). As a result, Phase-I VM value (denoted VM_I) and Phase-II VM value (denoted VM_{II}) along with their accompanying RI/GSI can be generated in real time. Furthermore, the PVM system has the capability of empowering its remote clients to monitor the system conjecture results at any time and any place.

Level 2 – Pluggability (GVM)

With further improvement on Level 1, the Level 2 generic VM (GVM) system aims at module pluggability. The GVM framework not only inherits the on-line conjecturing capability from PVM, but also possesses the pluggability which enables easy exchanges of its data collection driver, VM & RI/GSI models, and communication agent. For example, a GVM system for a CVD tool can also be applied to an etching tool with only a few replacements (i.e. data collection module and VM & RI/GSI models) in the pluggable modules.

Level 3 – Full Automation (AVM)
Level 3, denoted automatic VM (AVM), is the highest level (full automation) in VM deployment and automation. Besides the functions of PVM and GVM, an AVM system is also equipped with the capabilities of automatic data quality evaluation, automatic fanning out, and automatic model refreshing.

Physical characteristics of different tools are not the same. For maintaining VM conjecture accuracy, DQI_X/DQI_y and VM & RI/GSI models must be created based on the historical process and metrology data acquired from each chamber (a semiconductor manufacturing tool is usually composed of 1 to 6 chambers). Therefore, when considering fab-wide VM implementation, we must be aware that the number of DQI_X/DQI_y and VM & RI/GSI models will increase rapidly with the growing number of tools. Under such condition, if we still take the traditional method to create those models one by one with a lot of historical data, huge labor expenses and model-creation time will make fab-wide VM implementation impossible. To solve this problem, the capabilities of automatic fanning out and automatic model refreshing must be implemented in the AVM system to enable those models to be automatically spread and refreshed to the same type of tools for maintaining the VM conjecture accuracy and saving tremendous labor expenses and model-creation time.

8.5.2 Deployment of the AVM System

Six steps are required to deploy the AVM system into manufacturing processes. They are (step 1) operation analysis, (step 2) AVM system setup, (step 3) data collection, (step 4) AVM modeling, (step 5) AVM function and integration tests, and finally (step 6) system release. These steps are elaborated as follows.

Step 1: Operation Analysis
Before deploying the AVM system into a manufacturing process, the first step is to analyze the operation flow and interfaces of the manufacturing process. Both AS-IS and TO-BE operation flows should be analyzed. AS-IS operation flow means the current standard operation procedure (SOP). All the steps in the AS-IS operation flow need to be presented in the flow diagram, and the operation rules including standard specification, customer/engineering instruction, and related materials should be documented. AS-IS operation flow is the basic reference such that the project team can figure out the desired/expected TO-BE operation flow when taking the AVM system into consideration. Comparing to AS-IS operation flow, the project team should consider the new man-machine-interfaces (MMI) in the physical environment when including the AVM system into the existing manufacturing systems (such as manufacturing execution system (MES), statistical process control (SPC), fault detection and classification (FDC), etc.). The new MMI should be designed and included in TO-BE operation flow.

Moreover, the system integration task among the AVM system and existing manufacturing systems needs to be performed. The data flow should be designed by referring to TO-BE operation flow. The data flow described the data items of input, process, and output of each systems. As such, the requirements of data exchange among various systems can be defined. The data exchange requirements should include data items, formats, frequencies, and communication protocols.

There are several communication protocols for connecting equipment, such as SECS/GEM, Interface A, OPC-UA, MTConnect, etc., and for integrating systems, such as web services, web application programming interface (API), windows communication foundation (WCF), etc. The project team should review the TO-BE data exchange requirements and system's capabilities of integration, then choose proper communication protocols to integrate the AVM system.

Step 2: AVM System Setup

As mentioned in Chapter 6, the novel Intelligent Factory Automation (iFA) system platform is developed by integrating technologies such as AMCoT framework with Cyber-Physical Agents (CPAs), AVM, IPM, and IYM, etc., so as to provide the total-solution package for implementing intelligent manufacturing. Therefore, the AVM system is deployed via the iFA system platform. There are two options of the iFA system platform for the users to choose from. "cloud-based version" shown in Figure 6.5 and "server-based version" depicted in Figure 6.6 are provided according to different requirements.

In the step of operation analysis, the operation flow and data exchange requirements are defined, then the gap analysis of the AVM system standard version will be performed by the project team. As a result, the scope and schedule of the AVM system's customization will be addressed. After obtaining the agreement among users and the project team, the installation of the iFA system platform will start. At the same time, the hardware of servers and communication networks need to be prepared. Referring to Figure 8.5, the AVM module of the iFA system platform consists of a model-creation (MC) server, a VM manager, several VM clients, and many AVM servers. The detailed specifications of servers need to be confirmed by taking data exchange requirements into consideration.

Furthermore, the IT infrastructure expert will arrange the physical network layout topology to enable the servers of the iFA system platform to collect manufacturing-tool process data as well as measurement-equipment metrology data and communicate with existing manufacturing systems. Also, the WIP tracking scheme for assuring workpiece traceability should be ready. The integration system tests will then be performed accordingly when the software coding tasks and hardware configuration are completed.

Step 3: Data Collection

As shown in Figures 6.5 and 6.6, CPAs are adopted by the equipment side to perform data collection. CPA offers several equipment drivers for various industries including SECS/GEM and Interface A for semiconductor and TFT-LCD industry, as well as OPC-UA and MTConnect for machine-tool industry, etc. Moreover, various intelligent services can be implemented as pluggable application modules (PAMs) and plugged into CPA for edge computing. For example, the tasks of data cleaning and feature extraction after raw data collection via equipment drivers can be implemented as PAMs and plugged into CPA.

Step 4: AVM Modeling

AVM modeling can be proceeded after data collection. The paired samples including process data and their corresponding metrology data correlated by the WIP tracking mechanism are required for AVM modeling. The rule of thumb for preparing the number of samples required for modeling is about three times of that of the key process parameters.

The trained power user should have the ability to conduct the AVM modeling. As the design of the AVM system, the model creation (MC) server shown in Figure 8.5 can connect to various data sources including MES, SPC, FDC, Electronic Design Automation, etc., and select the needed data items via VM client GUI.

During the period of AVM modeling, about 70% of modeling samples are adopted to build conjecture models and 30% are applied to validate the accuracy of VM models. The acceptance criteria such as mean absolute percentage error (MAPE) and maximum error are adopted to judge the accuracy of VM models. Note that, the data quality of modeling samples will impact the accuracy of model. The power user needs to double-check the correctness of the data quality so as to assure the accuracy of the AVM models.

The architecture of the AVM System is shown in Figure 8.5. Each tool is equipped with an AVM server that contains a VM conjecture model for conjecturing the process quality of the equipment on-line. Also, the AVM server possesses a mechanism to automatically tune or re-train the VM conjecture model so that the VM accuracy of each AVM server can be maintained, and the AVM server is then ready to serve various VM applications. In addition, the AVM server is able to compute the quality of process data and metrology data to select good-quality data for on-line VM conjecture or refreshing the VM models.

When factory-wide VM deployment is considered, a VM manager is needed for managing all the AVM servers. The VM manager shall also be designed to supervise the processes of new-model creations, fanning-out, and refreshing. The MC server is responsible for generating the first set of data quality evaluation models (DQI_X/DQI_y), VM conjecture models, and VM reliance evaluation models (RI/GSI) of a certain tool type. The VM manager can fan out the first set of models generated to all the AVM servers of the same tool type and can monitor the real-time statuses of the AVM servers. The VM client is in charge of defining configuration of the AVM servers, assisting the MC server in model creation, on-line monitoring the VM results, and searching the historical VM results.

A real example for deploying the AVM system into a fifth-generation TFT-LCD factory is illustrated as follows. There are totally 18 CVD tools of the same type and each tool contains six (6) chambers. To achieve the goal of full automation, the AVM system operates as follows to automatically deploy and refresh the first set of DQI_X/DQI_y and VM & RI/GSI models to all of the $18 \times 6 = 108$ chambers of CVD tools. Please refer to Figure 8.5 for the following explanations.

Step 4.1: Download AVM Configuration
The VM client checks all the configurations of the AVM servers. Then, the VM client will make necessary changes and download the new configuration tables to the AVM servers for setting up the CVD VM environment.

Step 4.2: Create the First Set of Models
The VM client assigns AVM Server-1 to connect to the Chamber A of CVD Tool-1. This Chamber A will serve as the first chamber to create the first set of DQI_X/DQI_y and VM & RI/GSI models. Then, the AVM Server-1 starts to collect enough process and metrology data from Chamber A. Subsequently, those process and metrology data are sent to the MC server for creating and verifying the first set of models. The VM client then cooperates with the MC server to accomplish the model creation/verification task. After successful verification, the MC server notifies the VM client. Finally, the VM client requests the MC server to store this first set of models to the central database via the VM manager.

Step 4.3: Deploy the First Set of Models
The VM client requests the VM manager to find the first set of models for the CVD tools from the central database and then to download this first set of models into all the AVM servers that handle the 108 chambers of the CVD tools. Basically, each AVM server can handle six (6) chambers that belong to the same CVD tool. Therefore, 18 VM servers are required for this deployment.

Step 4.4: Refresh Models
After receiving the first set of models, each AVM server will initiate its advanced dual-phase VM algorithm to refresh the models in each chamber. After the refreshing process is accomplished, the AVM server will report its readiness to the VM client.

Step 4.5: Upload VM Results
The VM client prepares the data collection plan (DCP) for defining the desired VM applications, downloads the DCP to the AVM server, and then activates the AVM server to commence the VM applications. Finally, the AVM server uploads the VM results to the VM manager according to the DCP in real time. The VM manager then displays the real-time VM results on the VM client and/or stores those results into the central database for various VM applications, such as workpiece-to-workpiece (W2W) quality monitoring and/or supporting W2W control, and so on.

Step 5: AVM Function and Integration Tests
Prior to releasing the AVM system for production run, the AVM function and integration tests should be performed based on the TO-BE operation flow defined in Step 1. The AVM system is developed by applying the object-oriented technology. The system designer will analyze the gap between the existing use cases of the standard AVM version and those of the version in TO-BE operation flow so as to define the extra use cases that are required in this project. Then, the new class diagrams and sequence diagrams including the changes of the existing classes and new required classes are generated according to the new version of use-case diagrams.

The function and integration tests will then be conducted according to the new class diagrams and sequence diagrams. First, the test plan and test scenarios need to be confirmed to fulfill the requirements of users. Then, the project team starts the AVM function and integration test according to the test plan and test scenarios. Besides the changes due to adding the AVM System, the existing manufacturing systems could also need to be modified based on TO-BE operation flow. Therefore, the existing manufacturing systems may also be required to test the interfaces for new data exchanges.

An example of a new sequence diagram of TO-BE operation flow is illustrated below. Referring to Figures 8.9 and 8.10 and taking the integration Steps 1-10 between MES and AVM for example, the scheduler of MES dispatches a lot to Process Equipment through Equipment Manager. Then, Equipment Manager sends the process data to the AVM system via Equipment Manager. After that, AVM calculates VM_I result and sends an alarm to Alarm Manager of MES if an abnormality occurs when the VM_I result is generated. Also, AVM sends VM_I to SPC of MES and triggers an out-of-control action plan (OCAP). Finally, the action of OCAP will be executed by Alarm Manager of MES. Accordingly, the project team needs to do the AVM function and integration tests based on the new sequence diagram illustrated above.

Step 6: System Release
After all the functions and interfaces have been validated, the project team will move on to the next step: System Release. The reliability of the AVM system should be able to run 24 hours a day and 365 days a year. Therefore, the acceptance criteria of the AVM system's stability and reliability, such as mean time between failures (MTBF) and mean time to repair (MTTR), should be added into the acceptance check list. The project team and user should inspect all the items in the checklist until all issues closed.

Moreover, the MAPE and maximum error are the VM accuracy acceptance criteria, and the online AVM models will receive the real-time actual process and metrology data from production lines. In this step, the accuracy of all the VM results needs to be verified with the actual metrology results.

Besides the AVM system's function and accuracy validations, the related staffs of factory also need to practice the TO-BE operation flow. For example, what action should be taken and verified once process engineers receive alarms of OOC from the AVM system.

8.6 Conclusion

This chapter introduces the evolution, necessity, as well as benefits of VM. Then, the invention of AVM is explained. After that, integrating AVM functions into MES and applying AVM for W2W control are also elaborated. Finally, the procedure of AVM system deployment is described. As far as the AVM application cases of various industries, they will be presented in Chapter 11.

Appendix 8.A – Abbreviation List

AEC	Advanced Equipment Control
APC	Advanced Process Control
API	Application Programming Interface
AVM	Automatic Virtual Metrology
BPNN	Back-Propagation Neural Networks
CD	Critical Dimension
CIM	Computer-Integrated Manufacturing
CMO	Chi Mei Optoelectronics
CMP	Chemical Mechanical Polishing
CPA	Cyber-Physical Agent
Cpk	Process Capability
CVD	Chemical-Vapor-Deposition
DCP	Data Collection Plan
DFM	Design for Manufacturability
DQI_X	Data Quality Index of the Process Data
DQI_y	Data Quality Index of the Metrology Data
E_L	Error Limit
EWMA	Exponentially Weighted Moving Average
FDC	Fault Detection and Classification
FICS	Factory Information and Control Systems
FOUP	Front Opening Unified Pod
GSI	Global Similarity Index
GSI_T	Global Similarity Index Threshold Value
GUI	Graphical User Interface
GVM	Generic VM
HMES	Holonic Manufacturing Execution System
IBM	International Business Machines Corporation
iFA	Intelligent Factory Automation
IPM	Intelligent Predictive Maintenance
ISI	Individual Similarity Index
ISMI	International SEMATECH Manufacturing Initiative
ITRS	International Technology Road Map for Semiconductors
IYM	Intelligent Yield Management
L2L	Lot-to-Lot

MAPE	Mean Absolute Percentage Error
$MAPE_P$	Process Mean-Absolute-Percentage Error
MC	Model Creation
MD	Metrology Data
ME	Metrology Equipment
MES	Manufacturing Execution System
MMI	Man-Machine-Interface
MR	Multi Regression
MSE	Mean Square Error
MTBF	Mean Time Between Failures
MTTR	Mean Time to Repair
NGF	Next-Generation-Fab
NN	Neural Networks
OCAP	Out-of-Control Action Plan
OOC	Out-of-Control
PAMs	Pluggable Application Modules
PD	Process Data
PE i	Process Equipment I
PECVD	Plasma Enhanced Chemical Vapor Deposition
PLS	Partial Least Squares
PM	Periodic Maintenance
PM	Predictive Maintenance
PreY	Pre-Metrology
PVM	Preliminary VM
R2R	Run-to-Run
RBFN	Radial Basis Function Networks
RI	Reliance Index
RI_T	RI Threshold Value
RI_W	RI Formula for Weibull
SOP	Standard Operation Procedure
SPC	Statistical Process Control
TFT-LCD	Thin Film Transistor-Liquid Crystal Display
tsmc	Taiwan Semiconductor Manufacturing Company
VM	Virtual Metrology
VM_I	Phase-I VM Value

VM_{II}	Phase-II VM Value
VMS	VM System
W2W	Wafer-to-Wafer
W2W	Workpiece-to-Workpiece
WCF	Windows Communication Foundation
WIP	Work-in-Process

Appendix 8.B – List of Symbols in Equations

t_i, t_j	completion time of processing the sampled glasspieces
ΔT	waiting time before real measurement
W_O	number of wafer output per year
ΔCT_P	% cycle time reduction due to VM allowing production wafers to skip metrology sampling steps in the on-line process monitoring flow
ΔCT_M	% cycle time reduction due to VM allowing less test wafers used in the off-line tool monitoring process and more intelligent dynamic metrology and process schemes
ΔY	% enhancement due to improvement on process capability (from VM supporting APC), reduction in scrap, and so forth
P	average selling price per 300 mm production wafer
C	average production cost per 300 mm production wafer
$\Delta Cost_M$	cost saving of test wafers per year when applying VM in all production lines
$\Delta Cost_T$	capex reduction per year when applying VM in all production lines
$Cost_V$	additional cost per year to maintain VM
$Cost_Q$	additional cost per year due to false alarms and missed detections caused by VM
α	EWMA coefficient ranging between 0 and 1
α_1	EWMA coefficient when actual metrology value is adopted as the feedback of the R2R controller
α_2	EWMA coefficient if VM value is applied
y_k	plant output of run *K*
u_k	control action taken for run *K*
β_0	initial bias of process
β_1	process gain
η_k	disturbance model input
Au_k	process predictive model
A	gain parameter
$\boldsymbol{Tgt}$	target value

$\mathbf{y_z}$	*zth* actual metrology data of the sampled product (workpiece)
$\hat{y}_k$	*kth* run VM data
$\mathbf{X_k}$	*kth* run process data of the equipment
$\boldsymbol{y}$	actual metrology data
$\hat{y}$	virtual metrology data
σ	standard deviations
σ_y	standard deviations of y
$\sigma_{\hat{y}}$	standard deviations of $\hat{y}$
PreY	pre-metrology data
$Z_{\hat{y}_{N_i}}$	statistical distribution of the VM value built by NN
$Z_{\hat{y}_{r_i}}$	statistical distribution of the reference prediction value built by MR
E_L	error limit
$\boldsymbol{Z}_\lambda$	standardized λ_{th} set process data
R^{-1}	inverse matrix of correlation coefficients among the standardized parameters
p	number of process parameters
$\hat{y}_N$	NN VM value
$\hat{y}_r$	MR VM value
α_N	scale parameter of the Weibull distribution of NN conjecturing model
β_N	shape parameter of the Weibull distribution of NN conjecturing model
α_r	scale parameter of the Weibull distribution of MR predictive model
β_r	shape parameter of the Weibull distribution of MR predictive model
C	a constant to be set
RI.S.	1-RI
y_k	actual removal amount measured from the metrology tool
$PostY_k$	actual post CMP thickness of run k
Å	angstrom
ARR_k	actual removal rate of run k
u_k	polish time of run k
A_k	nominal removal rate
$Post\hat{Y}_k$	predictive value of $PostY_k$
Error	random error represented by white noise
PM1	error due to tool-parts' variation that was caused by periodic maintenance (PM)
PM2	random disturbance of tool-parts' variation
Stress1	tool stress error due to re-assembly during PM
Stress2	random disturbance of tool stress

Rotspd1	tool rotation-speed error due to re-assembly during PM
Rotspd2	random disturbance of tool rotation speed
Sfuspd1	slurry fluid-speed error due to re-assembly during PM
Sfuspd2	random disturbance of slurry fluid speed
$PreY_k$	pre-process (etching depth) value that affects the process result of run k
α_w	scale parameter
β	shape parameter

Appendix 8.C – Patents (AVM)

USA Patents

1) Cheng, F.T., Su, Y.C., Huang, G.W., et al. (2005). Quality prognostics system and method for manufacturing process. US Patent 7,493,185 B2, filed 2 June 2005 and issued 17 February 2009.
2) Cheng, F.T., Chen, Y.T., and Su, Y.C. (2006). Method for evaluating reliance level of a virtual metrology system in product manufacturing. US Patent 7,593,912 B2, filed 29 December 2006 and issued 22 September 2009.
3) Cheng, F.T., Huang, H.C., and Kao, C.A. (2007). Dual-phase virtual metrology method. US Patent 7,603,328 B2, filed 18 July 2007 and issued 13 October 2009.
4) Cheng, F.T., Huang, H.C., Huang, Y.T., et al. (2008). System and method for automatic virtual metrology. US Patent 8,095,484 B2, filed 10 September 2008 and issued 10 January 2012.
5) Cheng, F.T., Kao, C.A., Huang, H.C., et al. (2010). Manufacturing execution system with virtual-metrology capabilities and manufacturing system including the same. US Patent 8,983,644 B2, filed 20 May 2010 and issued 17 March 2015.
6) Cheng, F.T., Kao, C.A., and Wu, W.M. (2011). Advanced process control system and method utilizing virtual metrology with reliance index. US Patent 8,688,256 B2, filed 29 July 2011 and issued 1 April 2014.
7) Cheng, F.T. and Wu, W.M. (2012). Method for screening samples for building prediction model and computer program product thereof. US Patent 8,862,525 B2, filed 2 November 2012 and issued 14 October 2014.
8) Yang, H.C., Tieng, H., Hung, M.H., et al. (2013). Method for predicting machine quality of machine tool. US Patent 9,508,042 B2, filed 1 November 2013 and issued 29 November 2016.
9) Cheng, F.T., Chen, C.F., Huang, H.H., et al. (2015). Metrology sampling method and computer program product thereof. US Patent 9,829,415 B2, filed 24 March 2015 and issued 28 November 2017.
10) Cheng, F.T., Chen, C.F., Lyu, J.R., et al. (2016). Metrology sampling method with sampling rate decision scheme and computer program product thereof. US Patent 10,269,660, filed 19 May 2016 and issued 23 April 2019.
11) Chen, C.F., Tieng, H., Cheng, F.T., et al. (2017). Product quality prediction method for mass customization. US Patent 10,345,794, filed 26 October 2017 and issued 9 July 2019.
12) Yang, H.C., Lo, Y.L., Hsiao, H.C., et al. (2019). Addictive manufacturing system and method and feature extraction method. US Patent Pending 16/591,613, filed 2 October 2019. US Patent CIP Pending 16/854,927, filed 22 April 2020.

13) Yang, H.C., Huang, C.H., and Cheng, F.T. (2019). An intelligent metrology architecture with AVM for metal additive manufacturing. US Patent Provisional Pending 62/808,865, filed 22 February 2019.

Taiwan, ROC Patents

1) Cheng, F.T., Su, Y.C., Huang, G.W., et al. (2004). Quality prognostics system and method for manufacturing process. ROC Patent I267012, filed 3 June 2004 and issued 21 November 2006.
2) Cheng, F.T., Chen, Y.T., and Su, Y.C. (2006). Method for evaluating reliance level of a virtual metrology system in product manufacturing. ROC Patent I315054, filed 10 May 2006 and issued 21 September 2009.
3) Cheng, F.T., Huang, H.C., and Kao, C.A. (2007). Dual-phase virtual metrology method. ROC Patent I338916, filed 8 June 2007 and issued 11 March 2011.
4) Cheng, F.T., Huang, H.C., Huang, Y.T., et al. (2008). System and method for automatic virtual metrology. ROC Patent I349867, filed 20 May 2008 and issued 1 October 2011.
5) Cheng, F.T., Kao, C.A., Huang, H.C., et al. (2010). Manufacturing execution system with virtual-metrology capabilities and manufacturing system including the same. ROC Patent I412906, filed 13 April 2010 and issued 21 October 2013.
6) Cheng, F.T., Kao, C.A., and Wu, W.M. (2011). Advanced process control system and method utilizing virtual metrology with reliance index. ROC Patent I427722, filed 26 July 2011 and issued 21 February 2014.
7) Cheng, F.T. and Wu, W.M. (2011). Method for screening samples for building prediction model and computer program product thereof. ROC Patent I451336, filed 20 December 2011 and issued 1 September 2014.
8) Yang, H.C., Tieng, H., Hung, M.H., et al. (2013). Method for predicting machine quality of machine tool. ROC Patent I481978, filed 1 November 2013 and issued 21 April 2015.
9) Cheng, F.T., Chen, C.F., Huang, H.H., et al. (2014). Metrology sampling method and computer program product thereof. ROC Patent I521360, filed 26 March 2014 and issued 11 February 2016.
10) Cheng, F.T., Chen, C.F., Lyu, J.R., et al. (2015). Metrology sampling method with sampling rate decision scheme and computer program product thereof. ROC Patent I539298, filed 27 May 2015 and issued 21 June 2016.
11) Chen, C.F., Tieng, H., Cheng, F.T., et al. (2016). Product quality prediction method for mass customization. ROC Patent I614699, filed -30 December 2016 and issued 11 February 2018.
12) Yang, H.C., Lo, Y.L., Hsiao, H.C., et al. (2019). Addictive manufacturing system and method and feature extraction method. ROC Patent Pending 108134930, filed 26 September 2019.
13) Cheng, F.T., Hsieh, Y.M., and Lu, J.W. (2020). Method for predicting occurrence of tool processing event and virtual metrology application and computer program product thereof. ROC Patent Pending 109118836, filed 4 June 2020.

Japan Patents

1) Cheng, F.T., Su, Y.C., Huang, G.W., et al. (2005). Quality prognostics system and method for manufacturing process. JP Patent 4601492, filed 2 June 2005 and issued 8 October 2010.
2) Cheng, F.T., Huang, H.C., and Kao, C.A. (2007). Dual-phase virtual metrology method. JP Patent 4584295, filed 15 October 2007 and issued 10 September 2010.
3) Cheng, F.T., Huang, H.C., Huang, Y.T., et al. (2009). System and method for automatic virtual metrology. JP Patent 4914457, filed 27 January 2009 and issued 27 January 2012.

4) Cheng, F.T., Kao, C.A., and Wu, W.M. (2011). Advanced process control system and method utilizing virtual metrology with reliance index. JP Patent 5292602, filed 1 August 2011 and issued 21 June 2013.
5) Cheng, F.T. and Wu, W.M. (2012). Method for screening samples for building prediction model and computer program product thereof. JP Patent 5515125, filed 4 December 2012 and issued 11 April 2014.
6) Cheng, F.T., Chen, C.F., Lyu, J.R., et al. (2016). Metrology sampling method with sampling rate decision scheme and computer program product thereof. JP Patent 6285494, filed 26 May 2016 and issued 9 February 2018.

German Patent

Cheng, F.T., Chen, C.F., Lyu, J.R., et al. (2016). Metrology sampling method with sampling rate decision scheme and computer program product thereof. DE Patent 102016109232, filed 19 May 2016 and issued 29 March 2018.

Austria Patent

Cheng, F.T., Chen, C.F., Lyu, J.R., et al. (2016). Metrology sampling method with sampling rate decision scheme and computer program product thereof. AT Patent Pending A50480.2016, filed 27 May 2016.

China Patents

1) Cheng, F.T., Huang, H.C., and Kao, C.A. (2007). Dual-phase virtual metrology method. CN Patent 823284, filed 8 June 2007 and issued 10 August 2011.
2) Cheng, F.T., Huang, H.C., Huang, Y.T., et al. (2008). System and method for automatic virtual metrology. CN Patent 843932, filed 5 June 2008 and issued 21 September 2011.
3) Cheng, F.T., Kao, C.A., Huang, H.C., et al. (2010). Manufacturing execution system with virtual-metrology capabilities and manufacturing system including the same. CN Patent 1464514, filed 19 May 2010 and issued 20 August 2014.
4) Cheng, F.T., Kao, C.A., and Wu, W.M. (2011). Advanced process control system and method utilizing virtual metrology with reliance index. CN Patent 1205265, filed 1 August 2011 and issued 5 June 2013.
5) Cheng, F.T. and Wu, W.M. (2012). Method for screening samples for building prediction model and computer program product thereof. CN Patent 2117690, filed 13 November 2012 and issued 22 June 2016.
6) Yang, H.C., Tieng, H., Hung, M.H., et al. (2013). Method for predicting machine quality of machine tool. CN Patent 2625066, filed 21 November 2013 and issued 29 September 2017.
7) Cheng, F.T., Chen, C.F., Huang, H.H., et al. (2014). Metrology sampling method and computer program product thereof. CN Patent 2822926, filed 9 May 2014 and issued 16 February 2018.
8) Cheng, F.T., Chen, C.F., Lyu, J.R., et al. (2015). Metrology sampling method with sampling rate decision scheme and computer program product thereof. CN Patent 3092894, filed 12 October 2015 and issued 28 September 2018.
9) Chen, C.F., Tieng, H., Cheng, F.T., et al. (2016). Product quality prediction method for mass customization. CN Patent 4602042, filed 30 December 2016 and issued 6 August 2021.
10) Yang, H.C., Lo, Y.L., Hsiao, H.C., et al. (2019). Addictive manufacturing system and method and feature extraction method. CN Patent Pending 201910942668.0, filed 30 September 2019.

Korea Patents

1) Cheng, F.T., Huang, H.C., and Kao, C.A. (2007). Dual-phase virtual metrology method. KR Patent 10-0915339, filed 13 September 2007 and issued 27 August 2009.
2) Cheng, F.T., Huang, H.C., Huang, Y.T., et al. (2008). System and method for automatic virtual metrology. KR Patent 10-1098037, filed 30 December 2008 and issued 16 December 2011.
3) Cheng, F.T., Kao, C.A., and Wu, W.M. (2011). Advanced process control system and method utilizing virtual metrology with reliance index. KR Patent 10-1335896, filed 1 August 2011 and issued 26 November 2013.
4) Cheng, F.T. and Wu, W.M. (2012). Method for screening samples for building prediction model and computer program product thereof. KR Patent 10-1440304, filed 11 December 2012 and issued 4 September 2014.
5) Cheng, F.T., Chen, C.F., Lyu, J.R., et al. (2016). Metrology sampling method with sampling rate decision scheme and computer program product thereof. KR Patent 10-1930420, filed 26 May 2016 and issued 12 December 2018.

References

1 Cheng, F.T., Kao, C.A., Chen, C.F. et al. (2015). Tutorial on applying the VM technology for TFT-LCD manufacturing. *IEEE Transactions on Semiconductor Manufacturing* 28 (1): 55–69. https://doi.org/10.1109/TSM.2014.2380433.

2 Lin, T.H., Hung, M.H., Lin, R.C. et al. (2006). A virtual metrology scheme for predicting CVD thickness in semiconductor manufacturing. *Proceedings of the 2006 IEEE International Conference on Robotics and Automation*, Orlando, Florida, USA (15-19 May 2006). USA: IEEE.

3 Weber, A. (2007). Virtual metrology and your technology watch list: ten things you should know about this emerging technology. *Future Fab International* 22 (4): 52–54.

4 Rothe, O. (2009). ISMI next-generation factory program industry briefing. *Proceedings of the ISMI Next-Generation Factory Program Industry Briefing*, San Francisco, CA, USA (15 July 2009). USA: SEMATECH.

5 Moyne, J. (2009). International technology road Map for semiconductors (ITRS) perspective on AEC/APC. *Proceedings of the ISMI AEC/APC Symposium XXI*, Ann Arbor, MI, USA (27-30 September 2009). USA: SEMATECH.

6 Passow, M. (2009). Solar APC semiconductor APC Déjà-vu. *Proceedings of the ISMI AEC/APC Symposium XXI*, Ann Arbor, MI, USA (27-30 September 2009). USA: SEMATECH.

7 Koitzsch, M. and Honold, A. (2011). Evaluation of economic effects as the basis for assessing virtual metrology investment. *Future Fab International* 37 (7): 89–92.

8 Su, Y.C., Cheng, F.T., Huang, G.W. et al. (2004). A quality prognostics scheme for semiconductor and TFT-LCD manufacturing processes. *Proceedings of the 30th Annual Conference of the IEEE Industrial Electronics Society (IECON 2004)*, Busan, Korea (2-6 November 2004). USA: IEEE.

9 Monahan, K.M. (2005). Enabling DFM and APC strategies at the 32 nm technology node. *Proceedings of the 2005 IEEE International Symposium on Semiconductor Manufacturing (ISSM 2005)*, San Jose, CA, USA (13-15 Sept. 2005). USA: IEEE.

10 Chen, P.H., Wu, S., Lin, J.S. et al. (2005). Virtual metrology: a solution for wafer to wafer advanced process control. *Proceedings of the IEEE International Symposium on Semiconductor Manufacturing (ISSM 2005)*, San Jose, CA, USA (13-15 Sept. 2005). USA: IEEE.

11 Su, Y.C., Hung, M.H., Cheng, F.T. et al. (2006). A processing quality prognostics scheme for plasma sputtering in TFT-LCD manufacturing. *IEEE Transactions on Semiconductor Manufacturing* 19 (2): 183–194. https://doi.org/10.1109/TSM.2006.873514.

12 Su, Y.C., Cheng, F.T., Hung, M.H. et al. (2006). Intelligent prognostics system design and implementation. *IEEE Transactions on Semiconductor Manufacturing* 19 (2): 195–207. https://doi.org/10.1109/TSM.2006.873512.

13 Chang, Y.J., Kang, Y., Hsu, C.L. et al. (2006) Virtual metrology technique for semiconductor manufacturing. *Proceedings of the 2006 International Joint Conference on Neural Networks (IJCNN'06)*, Vancouver, BC, Canada (16-21 July 2006). USA: IEEE.

14 Hung, M.H., Lin, T.H., Cheng, F.T. et al. (2007). A novel virtual metrology scheme for predicting CVD thickness in semiconductor manufacturing. *IEEE/ASME Transactions on. Mechatronics* 12 (3): 308–316. https://doi.org/10.1109/TMECH.2007.897275.

15 Khan, A.A., Moyne, J.R., and Tilbury, D.M. (2007). An approach for factory-wide control utilizing virtual metrology. *IEEE Transactions on Semiconductor Manufacturing* 20 (4): 364–375. https://doi.org/10.1109/TSM.2007.907609.

16 Khan, A.A., Moyne, J.R., and Tilbury, D.M. (2008). Virtual metrology and feedback control for semiconductor manufacturing process using recursive partial least squares. *Journal of Process Control* 18 (10): 961–974. https://doi.org/10.1016/j.jprocont.2008.04.014.

17 Wu, M.F., Lin, C.H., Wong, D.S.H. et al. (2008). Performance analysis of EWMA controllers subject to metrology delay. *IEEE Transactions on Semiconductor Manufacturing* 21 (3): 413–425. https://doi.org/10.1109/TSM.2008.2001218.

18 Zeng, D. and Spanos, C.J. (2009). Virtual metrology modeling for plasma etch operations. *IEEE Transactions on Semiconductor Manufacturing* 22 (4): 419–431. https://doi.org/10.1109/TSM.2009.2031750.

19 Imai, S.I. and Kitabata, M. (2009). Prevention of copper interconnection failure in system on chip using virtual metrology. *IEEE Transactions on Semiconductor Manufacturing* 22 (4): 432–437. https://doi.org/10.1109/TSM.2009.2031757.

20 Ringwood, J.V., Lynn, S., Bacelli, G. et al. (2010). Estimation and control in semiconductor etch: practice and possibilities. *IEEE Transactions on Semiconductor Manufacturing* 23 (1): 87–98. https://doi.org/10.1109/TSM.2009.2039250.

21 Pan, T.H., Sheng, B.Q., Wong, S.H. et al. (2011). A virtual metrology system for predicting end-of-line electrical properties using a MANCOVA model with tools clustering. *IEEE Transactions on Industrial Informatics* 7 (2): 187–195. https://doi.org/10.1109/TII.2010.2098416.

22 Lynn, S., Ringwood, J., and MacGearailt, N. (2012). Global and local virtual metrology models for a plasma etch process. *IEEE Transactions on Semiconductor Manufacturing* 25 (1): 94–103. https://doi.org/10.1109/TSM.2011.2176759.

23 Cheng, F.T., Chang, J.Y.C., Huang, H.C. et al. (2011). Benefit model of virtual metrology and integrating AVM into MES. *IEEE Transactions on Semiconductor Manufacturing* 24 (2): 261–272. https://doi.org/10.1109/TSM.2011.2104372.

24 Imai, S., Sato, N., Kitabata, M. et al. (2006). Fab-wide equipment monitoring and FDC system. *Proceedings of the 2006 IEEE International Symposium on Semiconductor Manufacturing*, Tokyo, Japan (25-27 September 2006). USA:IEEE.

25 Chang, J. and Cheng, F.T. (2005). Application development of virtual metrology in semiconductor industry. *Proceedings of the 31st Annual Conference of the IEEE Industrial Electronics Society (IECON 2005)*, Raleigh, NC, USA (6-10 November 2005). USA: IEEE.

26 Patel, N.S., Miller, G.A., and Jenkins, S.T. (2002). In situ estimation of blanket polish rates and wafer-to-wafer variation. *IEEE Transactions on Semiconductor Manufacturing* 15 (4): 513–522. https://doi.org/10.1109/TSM.2002.804900.

27 Huang, Y.T., Huang, H.C., Cheng, F.T. et al. (2008). Automatic virtual metrology system design and implementation. *Proceedings of the 2008 IEEE International Conference on Automation Science and Engineering*, Arlington, VA, USA (23-26 Aug. 2008). USA:IEEE.

28 Cheng, F.T., Huang, H.C., and Kao, C.A. (2007). Dual-phase virtual metrology scheme. *IEEE Transactions on Semiconductor Manufacturing* 20 (4): 566–571. https://doi.org/10.1109/TSM.2007.907633.

29 Stark, D. (2008). Data usage. e-Manufacturing Workshop, SEMICON West, San Francisco.

30 Cheng, F.T., Chen, Y.T., Su, Y.C. et al. (2008). Evaluating reliance level of a virtual metrology system. *IEEE Transactions on Semiconductor Manufacturing* 21 (1): 92–103. https://doi.org/10.1109/TSM.2007.914373.

31 Kao, C.A., Cheng, F.T., Wu, W.M. et al. (2013). Run-to-run control utilizing virtual metrology with reliance index. *IEEE Transactions on Semiconductor Manufacturing* 26 (1): 69–81. https://doi.org/10.1109/TSM.2012.2228243.

32 A. A. Khan, J. R. Moyne, and D. M. Tilbury. (2007). On the quality of virtual metrology data for use in R2R process control. *Proceedings of the ISMI AEC/APC Symposium XIX*, Indian Wells, CA. USA (September 15-20, 2007). USA: SEMATECH.

33 Groover, M.P. (2001). *Automation, Production Systems, and Computer-Integrated Manufacturing*, 2nde. London: Prentice Hall.

34 Su, Y.C., Lin, T.H., Cheng, F.T. et al. (2008). Accuracy and real-time considerations for implementing various virtual metrology algorithms. *IEEE Transactions on Semiconductor Manufacturing* 21 (3): 426–434. https://doi.org/10.1109/TSM.2008.2001219.

35 Huang, Y.T. and Cheng, F.T. (2011). Automatic data quality evaluation for the AVM system. *IEEE Transactions on Semiconductor Manufacturing* 24 (3): 445–454. https://doi.org/10.1109/TSM.2011.2154910.

36 Cheng, F.T., Huang, H.C., and Kao, C.A. (2012). Developing an automatic virtual metrology system. *IEEE Transactions on Automation Science and Engineering* 9 (1): 181–188. https://doi.org/10.1109/TASE.2011.2169405.

37 Lin, T.H., Cheng, F.T., Wu, W.M. et al. (2009). NN-based key-variable selection method for enhancing virtual metrology accuracy. *IEEE Transactions on Semiconductor Manufacturing* 22 (1): 204–211. https://doi.org/10.1109/TSM.2008.2011185.

38 Wu, W.M., Cheng, F.T., Lin, T.H. et al. (2011). Selection schemes of dual virtual-metrology outputs for enhancing prediction accuracy. *IEEE Transactions on Automation Science and Engineering* 8 (2): 311–318. https://doi.org/10.1109/TASE.2010.2089451.

39 Wu, W.M., Cheng, F.T., and Kong, F.W. (2012). Dynamic-moving-window scheme for virtual-metrology model refreshing. *IEEE Transactions on Semiconductor Manufacturing* 25 (2): 238–246. https://doi.org/10.1109/TSM.2012.2183398.

40 Cheng, F.T. and Chiu, Y.C. (2013). Applying the automatic virtual metrology system to obtain tube-to-tube control in a PECVD tool. *IIE Transactions* 45 (6): 671–682. https://doi.org/10.1080/0740817X.2012.725507.

41 Lin, L.R., Chiu, Y.C., Mo, W.C. et al. (2011). Run-to-run control utilizing the AVM system in the solar industry. *Proceedings of the 2011 e-Manufacturing & Design Collaboration Symposium & International Symposium on Semiconductor Manufacturing (ISSM)*, Hsinchu, Taiwan (5-6 September. 2011). USA:IEEE.

42 Yang, H.C., Tieng, H., and Cheng, F.T. (2016). Automatic virtual metrology for wheel machining automation. *International Journal of Production Research* 54 (21): 6367–6377. https://doi.org/10.1080/00207543.2015.1109724.

43 Tieng, H., Tsai, T.H., Chen, C.F. et al. (2018). Automatic virtual metrology and deformation fusion scheme for engine-case manufacturing. *IEEE Robotics and Automation Letters* 3 (2): 934–941. https://doi.org/10.1109/LRA.2018.2792690.

44 Hsieh, Y.M., Lin, C.Y., Yang, Y.R. et al. (2019). Automatic Virtual Metrology for Carbon Fiber Manufacturing. *IEEE Robotics and Automation Letters* 4 (3): 2730–2737. https://doi.org/10.1109/LRA.2019.2917384.

45 TechTarget (2007). Manufacturing execution systems. http://www.bitpipe.com/rlist/term/Manufacturing-Execution-Systems.html (accessed 17 Aug 2020).

46 SEMATECH (1998). Computer integrated manufacturing (CIM) framework specification version 2.0. *SEMATECH Technology Transfer # # 93061697J-ENG*. https://bit.ly/3hDCvJH (accessed 27 Aug 2020).

47 Cheng, F.T., Shen, E., and Deng, J.-Y. (1999). Development of a system framework for the computer-integrated manufacturing execution system: a distributed object-oriented approach. *International Journal of Computer Integrated Manufacturing* 12 (5): 384–402. https://doi.org/10.1080/095119299130137.

48 Cheng, F.T., Chang, C.F., and Wu, S.L. (2004). Development of holonic manufacturing execution systems. *Journal of Intelligent Manufacturing* 15 (2): 253–267. https://doi.org/10.1023/B:JIMS.0000018037.63935.a1.

49 Moyne, J.E., Castillo, d., and Hurwitz, A.M. (eds.) (2001). *Run-to-Run Control in Semiconductor Manufacturing*. Boca Raton, FL: CRC Press.

50 Besnard, J. and Toprac, A. (2006). Wafer-to-wafer virtual metrology applied to run-to-run control. *Proceedings of the AEC/APC Symposium XVIII*, Westminster, CO, USA (30 September - 4 October 2006). USA: SEMATECH.

51 Kao, C.A., Cheng, F.T. and Wu, W.M. (2011). Preliminary study of run-to-run control utilizing virtual metrology with reliance index. *Proceedings of the 2011 IEEE International Conference on Automation Science and Engineering*, Trieste, Italy (24-27 August 2011). USA: IEEE.

52 Weibull, W. (1951). A statistical distribution function of wide applicability. *Journal of Applied Mechanics* 18: 293–297.

53 Patel, N. and Anderson, M. (2003). Considerations in deploying wafer-level control across the fab. AEC/APC Symposium XV, Colorado Springs, CO, USA.

54 Castillo, E.D. and Rajagopal, R. (2002). A multivariate double EWMA process adjustment scheme for drifting processes. *IIE Transactions* 34 (12): 1055–1068. https://doi.org/10.1080/07408170208928934.

55 Preston, F.W. (1927). The theory and design of plate glass polishing machines. *Journal of the Society of Glass Technology* 11: 214–256.

56 Feng, T. (2007). Nonuniformity of wafer and pad in CMP: kinematic aspects of view. *IEEE Transactions on Semiconductor Manufacturing* 20 (4): 451–463. https://doi.org/10.1109/TSM.2007.907625.

57 Montgomery, D.C. (2009). *Introduction to Statistical Quality Control*, 6the. New York: Wiley.

58 Smith, T.H. and Boning, D.S. (1997). A self-tuning EWMA controller utilizing artificial neural network function approximation techniques. *IEEE Transactions on Components, Packaging, and Manufacturing Technology: Part C* 20 (2): 121–132. https://doi.org/10.1109/3476.622882.

59 Wohlwend, H. (2005). e-diagnostics guidebook, version 2.1. *ISMI Technology Transfer #01084153D-ENG*.

9

Intelligent Predictive Maintenance (IPM)

Yu-Chen Chiu[1], Yu-Ming Hsieh[2], Chin-Yi Lin[3], and Fan-Tien Cheng[4]

[1] *Specialist, Intelligent Manufacturing Research Center, National Cheng Kung University, Tainan, Taiwan, ROC*
[2] *Associate Research Fellow, Intelligent Manufacturing Research Center, National Cheng Kung University, Tainan, Taiwan, ROC*
[3] *Postdoctoral Research Fellow, Intelligent Manufacturing Research Center, National Cheng Kung University, Tainan, Taiwan, ROC*
[4] *Director/Chair Professor, Intelligent Manufacturing Research Center/Institute of Manufacturing Information and Systems, National Cheng Kung University, Tainan, Taiwan, ROC*

9.1 Introduction

Production equipment is the essential part for any manufacturing factory. Failure of a component, module, or device (e.g. heater, pressure module, and throttle valve, etc.) in the production equipment may cause production abnormalities which lead to poor product quality and/or low production capacity and thus cause significant losses.

In general, the most-commonly utilized approach for remedying the problems mentioned above is scheduled preventive maintenance (PM). That is, to execute maintenance-related operations in a predetermined time interval. This time interval is basically decided according to the mean-time-between-failure (MTBF) of the Target Device (TD) [1]. As such, how to schedule a proper PM is usually a key issue for the factories. An improper scheduled PM may increase the maintenance cost or lower the production capacity [2].

To improve equipment maintenance programs for increasing fab performance, the International SEMATECH Manufacturing Initiative (ISMI) proposed an initiative of predictive and preventive maintenance (PPM) [1–5]. As defined by ISMI, PPM comprises PM, condition-based maintenance (CbM), predictive maintenance (PdM), and breakdown maintenance (BDM). Among them, ISMI claimed that CbM and PdM capabilities should be developed and available as individual or incremental modules so that the end user can choose to implement one, some, or all of the capabilities. CbM is the approach to activate the maintenance task after one or more indicators show that equipment is going to fail or that equipment performance is deteriorating [1]. And the technique of fault-detection-and-classification (FDC) is an approach related to CbM. PdM, on the other hand, is the technique of applying a prediction model to relate facility-state information to maintenance information for forecasting the remaining useful life (RUL) and the demand for maintenance events used to alleviate unscheduled downtime.

Industry 4.0, the rise of new digital industrial technology, intends to make the factory be able to enable faster, more flexible, and more efficient processes to manufacture higher-quality goods at reduced costs [6]. PdM has been featured as a key theme "Predictive Maintenance 4.0" of Industry

Industry 4.1: Intelligent Manufacturing with Zero Defects, First Edition. Edited by Fan-Tien Cheng.

4.0. PdM monitors the equipment health and indicates when a maintenance event will be required in the future, which has been raised to the top priority as the key enabler for maximizing tool availability [7].

As stated in Stage 4 of the five-stage strategy of yield enhancement and zero-defects assurance in Chapter 1, an Intelligent Predictive Maintenance (IPM) system should be constructed to perform tool health monitoring and execute tool RUL prediction. The purpose of this chapter is to describe the details of developing the IPM system. In the following, the necessity of the Baseline Predictive Maintenance (BPM) scheme in the IPM system is first presented. Then, the prediction algorithms of RUL are introduced. After that, the requirements of the factory-wide IPM system is summarized.

9.1.1 Necessity of Baseline Predictive Maintenance (BPM)

As mentioned in [8], intelligent fault diagnostic and prognostic methods can be roughly categorized into two major classes: model-based and data-driven methods. Model-based approaches require an accurate mathematical model to be developed and use residuals as features, where residuals are the outcomes of consistency checks between the sensed measurements of a real system and the outputs of a mathematical model. However, it is usually difficult to develop accurate fault-growth models in most real-world applications [8].

In many applications, measured input/output data are the major source of knowledge to understand the system degradation behavior. The data-driven approaches rely on the assumption that the statistical characteristics of data are relatively consistent unless a malfunctioning event occurs in the system. Without an understanding of the physics of the fault progression, data-driven approaches are one of the feasible options [8].

Most conventional FDC approaches are to find out the TDs required for monitoring and the TDs' related key parameters needed to be monitored and then to detect the faults by applying the statistical-process-control (SPC) approach [9]. As described in [10], several researchers utilized model-based algorithms to define the key-parameters' SPC control limits that detect the faults. The methods proposed in [7, 11] are two examples of the conventional FDC approaches mentioned above.

However, in a practical situation, an abnormal key-parameter value may not be solely caused by its own TD; instead, it may also be affected by other related parameters. Take practical plasma-enhanced-chemical-vapor-deposition (PECVD) throttle-valve angles for example. As shown in Figure 9.1, the center is at 27° and the upper-control limit and lower-control limit (UCL and LCL) are 32° and 22°, respectively as defined by the maintenance engineers. 450 samples in total are monitored. As indicated in Figure 9.1, this conventional SPC method concludes that those samples in circles 1, 2, and 4 are outliers while the sample in circle 3 is within the control limit.

After careful inspections, the fact is that those samples in circles 2 and 4 are indeed abnormal and are caused by the throttle-valve's malfunction. As for the sample in circle 1, this abnormality is not caused by the throttle-valve itself; it is resulted from the deviation of the related-parameter "Ammonia (NH_3)". Also, the deflection shown in the circle 3 is due to the deviation of the related-parameter "Tube Pressure". As such, the conventional SPC method cannot detect and diagnose the faults at the samples in circle 1 and circle 3. Hsieh et al. presented a preliminary study of a virtual-metrology-based (VM-based) FDC scheme to remedy the problem mentioned above in [13].

Liu et al. [14] used the similarity-based method for the spindle load prediction and diagnosis. They created plural match matrices utilizing a lot of historical information. The central concept is to have a library of degradation patterns from previous run-to-failure data sets and estimate the

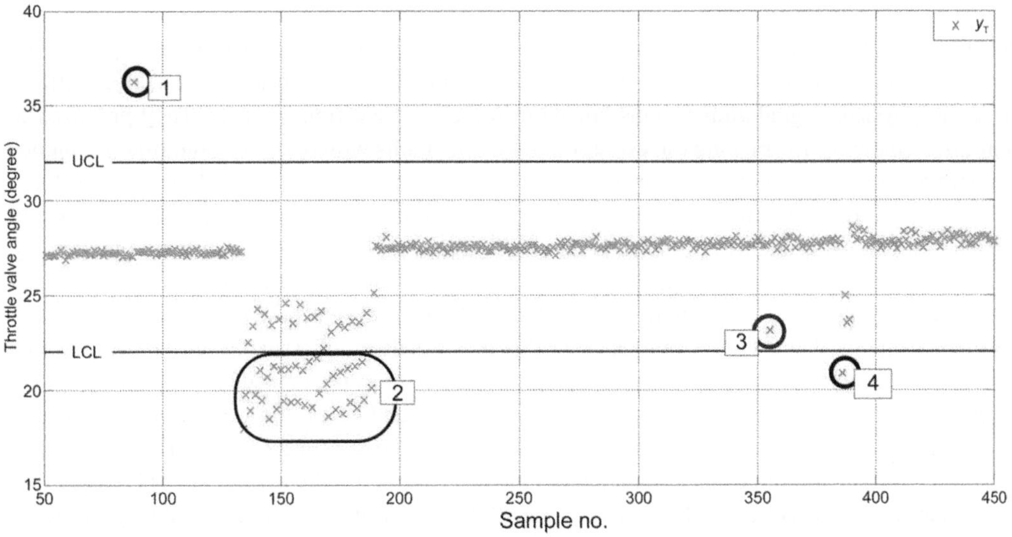

Figure 9.1 SPC control chart of throttle-valve angles in a PECVD tool. *Source:* Reprinted with permission from Ref. [12]; © 2013 IEEE.

current remaining life by matching the current degradation pattern with the most similar ones in the degradation library. Kim et al. [15] utilized the support-vector-machine classifier as a tool for estimating health-state probability of machine degradation process. Through prior analysis of historical failure data, discrete failure-degradation stages were employed to estimate the health-state probability for accurate long-term machine prognosis. Susto et al. [16] utilized machine learning with multi-class classification to generate the health and cost indexes for establishing the relationship between operating costs and fault risks, so as to minimize downtime and related costs.

The RUL prediction methods utilized in [14–16] require a library of degradation patterns from previous run-to-failure data sets. Without those massive historical failure data, these two RUL prediction methods may not function properly.

To remedy the problems mentioned above, a VM-based BPM scheme is required and proposed. The BPM scheme possesses the capabilities of not only FDC but also PdM. The BPM scheme is composed of the TD baseline model, FDC logic, and an RUL prediction module. The TD baseline model is generated by referring to the technique of Automatic Virtual Metrology (AVM) presented in Chapter 8 to serve as the reference for detecting the fault. By applying this BPM scheme, fault diagnosis and prognosis can be accomplished, the problem of the conventional SPC method mentioned above can be resolved, and the requirement of massive historical failure data can also be released. The details of the BPM scheme are elaborated in Section 9.2.

9.1.2 Prediction Algorithms of Remaining Useful Life (RUL)

Nowadays, many scholars found out that most TD's aging features appear to degrade along with time. Among all the prediction models of the degrading trend, the exponential-curve-fitting (ECF) model is one of the most popular methods. Gebraeel et al. [17] constructed the ECF model to predict the RUL and adopted the Bayesian method to combine the metrology data. Zhang et al. [18] proposed to integrate the ECF model and the particle filter to predict the cycle time of lithium-ion batteries. In addition to the ECF model, scholars also developed some other RUL prediction methods. Xia et al. [19] proposed a prognostic real-time rolling grey forecasting method to generate machine health prediction, as well as to analyze the effects of influencing factors such as operating load. Singleton et al. [20] suggested that it is the main challenge for bearing prediction without accurate physical degradation models but with limited labeled training data. They proposed to solve this challenge by tracking the evolution of bearing faults with both time and time-frequency domain features. Once features are extracted, extended Kalman filter is utilized to predict RUL.

The original BPM [12] also applied the ECF model to construct the RUL prediction module and adopted only a single algorithm to extract the aging features. However, as mentioned previously, there are several problems for applying the ECF model to predict RUL. First, when TD's aging features rise or drop instantly, the RUL predicted by the ECF model could greatly drop or rise due to the sensitivity of the TD's aging features; this may result in excessive deviation between the previous and the following runs. Second, being affected by the smooth or nearly flat trend of TD's aging features, the ECF model could generate falsely prolonged RUL. Third, solely using indicators as the aging features can cause too much noise that leads to inaccurate prediction. To tackle with these three problems of inaccurate RUL prediction, the Time Series Prediction (TSP) [21] algorithm is proposed, which will be embedded into the RUL prediction module of the newly proposed Advanced Baseline Predictive Maintenance (ABPM). TSP will conduct predictor selection according to the information criterion so as to adapt to the complicated future trends as well as to solve the issue of TD unscheduled down. The details of the TSP algorithm are explained in Section 9.3.

9.1.3 Introducing the Factory-wide IPM System

The requirements of the factory-wide IPM system are (i) powerful infrastructure and rapid data communication-and-processing platform; (ii) effective fault diagnostic and prognostic engine; (iii) manageable hierarchy of health indexes from a factory-wide view.

Industry 4.0 creates a feasible environment to fit today's PdM requirements. Internet of things (IoT) acts as a data communication platform; cyber-physical systems (CPS) with prediction algorithms are constructed on the platform and cooperate seamlessly with real and virtual spaces; then, cloud computing provides the capabilities of various essential services. However, the traditional PdM system was only designed for a single tool; as a result, the resource allocation will become extremely complicated when hundreds of tools work together in a factory. A manageable hierarchy and various health indexes are required for factory-wide equipment maintenance.

To solve the problem mentioned above, a factory-wide IPM system by applying the so-called Advanced Manufacturing Cloud of Things (AMCoT) framework in the Intelligent Factory Automation (iFA) system platform to fulfill the requirements of Industry 4.1 [22] is proposed and will be elaborated in Section 9.4.

9.2 BPM

As shown in Figure 9.2, the BPM possesses two portions: FDC and PdM. The FDC portion consists of two parts, the left-hand part generates TD and ISI_B (Baseline-Individual-Similarity-Index) baseline models as well as DHI (Device-Health-Index) and BEI (Baseline-Error-Index) modules, which are utilized to generate the baseline of TD ($\hat{y}_B$), the indexes of ISI_B, DHI, and BEI, respectively. The right-hand part defines FDC logic to convert DHI, BEI, and ISI_B values into the healthy status of TD.

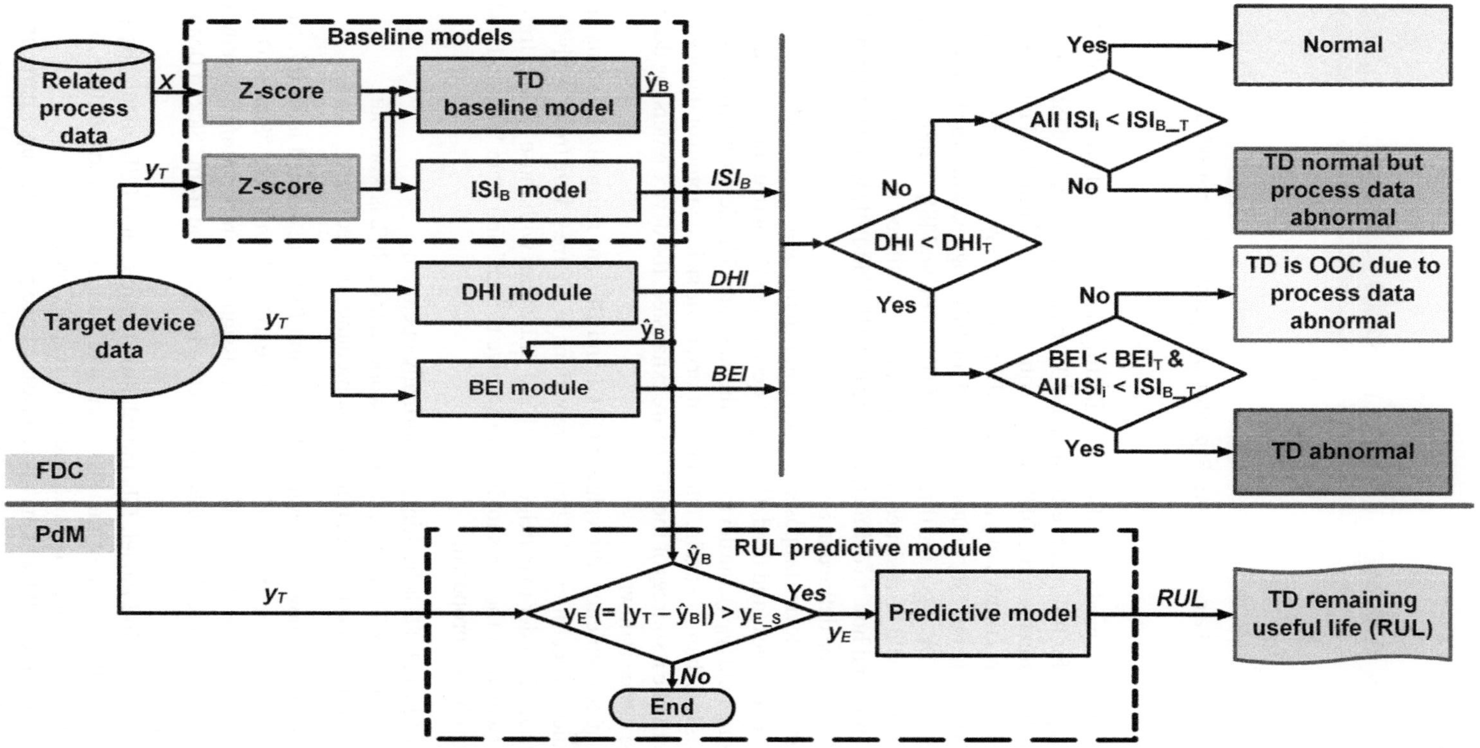

Figure 9.2 BPM scheme. *Source:* Reprinted with permission from Ref. [12]; © 2013 IEEE.

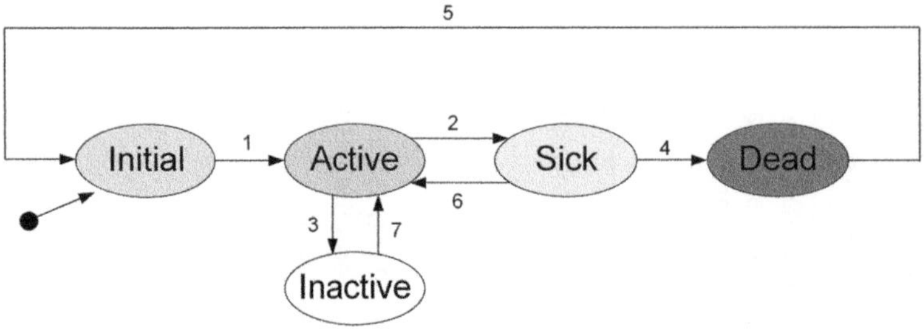

Figure 9.3 State diagram of a device. *Source:* Reprinted with permission from Ref. [12]; © 2013 IEEE.

The TD baseline model predicts $\hat{y}_B$ through collecting the TD-related process data (**X**); while the TD data (y_T) is applied for tuning or re-training the TD baseline model. The ISI_B model transforms **X** into the ISI_B value. Moreover, the DHI module converts $|y_T - \bar{y}_T|$ into DHI index where $\bar{y}_T$ represents the mean of fresh samples of y_T; and the BEI module converts $y_E\left(=|y_T - \hat{y}_B|\right)$ into the BEI index. At last, DHI, BEI, and ISI_B values are sent to the FDC logic to infer the healthy status of TD.

Before applying the FDC logic, the thresholds of DHI, BEI, and ISI_B (denoted as DHI_T, BEI_T, and ISI_{B_T}, respectively) should be assigned. DHI_T and BEI_T are utilized to judge whether the TD is in the sick state or not, and how to define the values of DHI_T and BEI_T will be explained as follows. The value of ISI_{B_T} will be defined in Section 9.2.2.

The state diagram of a device proposed in [12] is shown in Figure 9.3, which includes five states: initial, active, inactive, sick, and dead. Usually, a device is in the active state. However, when y_E is larger than its threshold (y_{E_S}), the device enters the sick state. The device returns to the active state when y_E becomes less than y_{E_S} again. On the contrary, if the sick symptom of the device gets worse such that the available resources of the device are exhausted, the device enters the dead state or, in other words, the device is down.

For the sake of convenience, the values of DHI and BEI are normalized to vary from 1 to 0; higher value indicates healthier TD. The key transitions in Figure 9.3 are from "Active to Sick" and from "Sick to Dead." In order to match the DHI and BEI values with those two key transitions, the following rules (which set the thresholds from "Active to Sick" and from "Sick to Dead" to be 0.7 and 0.3, respectively) are proposed:

When $1 > \text{DHI/BEI} > 0.7$, TD is healthy and normally-operated;
When $0.7 > \text{DHI/BEI} > 0.3$, TD is sick, it cannot work after its RUL is worn out;
When $0.3 > \text{DHI/BEI} > 0$, TD is dead, it needs maintenance immediately.

According to the rules above, the values of DHI_T and BEI_T are defined to be 0.7. Note that it is just an example to assign the thresholds from "Active to Sick" and from "Sick to Dead" to be 0.7 and 0.3, respectively in this chapter. Users may re-assign any proper numbers between 0 and 1 (such as 0.8 and 0.2) to be the thresholds from "Active to Sick" and from "Sick to Dead" according to their requirements.

The details of TD baseline model, ISI_B model, DHI module, and BEI module are presented in Sections 9.2.1–9.2.4. Illustration of the FDC logic and flow chart of the baseline FDC execution procedure are described in Sections 9.2.5 and 9.2.6, respectively.

The PdM portion of the BPM scheme also consists of two parts. The upper part contains the TD baseline model while the lower part includes a RUL prediction module, which will be delineated in Section 9.2.7.

9.2.1 Important Samples Needed for Creating Target-Device Baseline Model

The purpose of the TD baseline model is to generate the healthy baseline of the TD ($\hat{y}_B$). To begin with, the procedure shown in Figure 9.4 for collecting the necessary important samples to create the TD baseline model is performed. The TD baseline model-creation procedure consists of two stages; Stage 1: off-line operations by performing the Keep-Important-Sample (KIS) scheme to select the concise and healthy (C&H) historical samples; and Stage 2: on-line operations to collect the fresh samples just after maintenance, with each sample including y_T and its corresponding X.

The off-line KIS scheme includes two steps: (i) (Step 1 in Figure 9.4) ensure the appropriateness of every historical sample collected, which means the sample was generated under healthy status and its data quality is good; (ii) (Step 2 in Figure 9.4) utilize the Dynamic-Moving-Window (DMW) scheme [12] for picking out the C&H samples from all of the healthy historical samples selected in Step 1. The DMW scheme adds a new sample into the model and applies a clustering technology to do similarity clustering. Next, the number of elements in each cluster is checked. If the largest number of the elements is greater than the predefined threshold, then the oldest sample in the cluster of the largest population is deleted.

The on-line operations are executed just after maintenance and contain three steps: (i) (Step 3 in Figure 9.4) collect the fresh samples that just finished the maintenance process; (ii) (Step 4 in Figure 9.4) check whether the number of fresh and healthy samples collected is enough for modeling or not; (iii) (Step 5 in Figure 9.4) if the samples collected are enough, then add those fresh samples to the C&H historical samples produced in Step 5 of Figure 9.4.

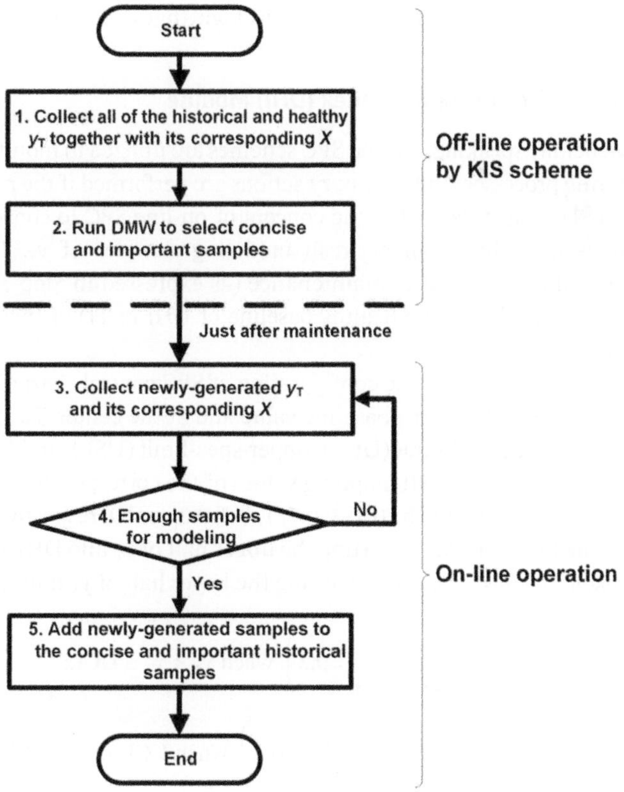

Figure 9.4 Procedure of collecting the important samples needed for creating TD baseline model. *Source:* Reprinted with permission from Ref. [12]; © 2013 IEEE.

The rule of thumb for determining the number of samples needed for creating the baseline model is about 10 times the number of the related process data; and the ratio between the C&H historical samples and the fresh samples are 3 to 1. All of the samples created in Step 5 of Figure 9.4 are utilized to construct the TD baseline model. Therefore, this baseline model not only contains the C&H samples but also possesses fresh data just after maintenance such that the healthy baseline of the TD can be generated. With the necessary important samples being completely collected, the TD baseline model can be built and it can function properly. The execution procedure of TD baseline model will be explained in Section 9.2.6.

9.2.2 Samples Needed for Creating Baseline Individual Similarity Index (ISI_B) Model

As shown in Figure 9.2, except for the TD baseline model, the ISI_B model should also be created in the box of baseline models. The authors invented the concept of the Individual Similarity Index (ISI) in Section 8.2 to identify the key-process parameters that cause major deviation when the Global Similarity Index (GSI) of the input set of process parameters exceed its threshold (GSI_T) in the AVM system.

As stated in Section 8.2, ISI of an individual process parameter is defined as the degree of similarity between this individual process-parameter's standardized process datum of the input set and the same process-parameter's standardized process data in all the historical sets that are used for training and tuning the ISI model. In this chapter, the fresh process data (X) collected just after maintenance (as expressed in Step 3 of Figure 9.4) are adopted to train the ISI model such that the healthy baseline of ISI in TD of this particular maintenance cycle can be generated; hence baseline-ISI (ISI_B) is named here. As indicated in Figure 9.2, the ISI_B model transforms X into the ISI_B value, which is then considered in the FDC logic that will be detailed in Section 9.2.5.

9.2.3 Device-Health-Index (DHI) Module

Generally speaking, on-line SPC schemes are utilized to monitor the quality status during manufacturing processes, and necessary actions are performed if the process is out of statistical control [23]. The DHI module applies the concept of on-line SPC to convert $|y_T - \bar{y}_T|$ into DHI index where $\bar{y}_T$ represents the mean of fresh modeling samples of y_T. The fresh y_T samples for modeling are collected just after maintenance (as expressed in Step 3 of Figure 9.4) to construct the DHI module such that the healthy baseline of DHI in TD of this particular maintenance cycle can be generated.

Figure 9.5a shows the configuration of SPC control chart of y_T with the mean of the y_T modeling samples ($\bar{y}_T$) being the baseline value and at the center. The Min y_T, lower-spec-limit (LSL), LCL, $\bar{y}_T$, upper-control-limit (UCL), upper-spec-limit (USL), and Max y_T are shown on the control chart of y_T; further, the DHI mapping values of the corresponding y_T ones mentioned above are normalized and assigned to be 0, 0.3, 0.7, 1, 0.7, 0.3, and 0, respectively.

The formulas for converting the upper half of y_T into DHI are expressed in Eq. (9.1). By the same token, the formulas for converting the lower half of y_T into DHI can also be derived.

$$\mathrm{DHI} = 1 - \left(\frac{y_T - \bar{y}_T}{\mathrm{UCL} - \bar{y}_T} \times 0.3 \right), \text{ when } \bar{y}_T < y_T \le \mathrm{UCL};$$

$$\mathrm{DHI} = 0.7 - \left(\frac{y_T - \mathrm{UCL}}{\mathrm{USL} - \mathrm{UCL}} \times 0.4 \right), \text{ when } \mathrm{UCL} < y_T \le \mathrm{USL};$$

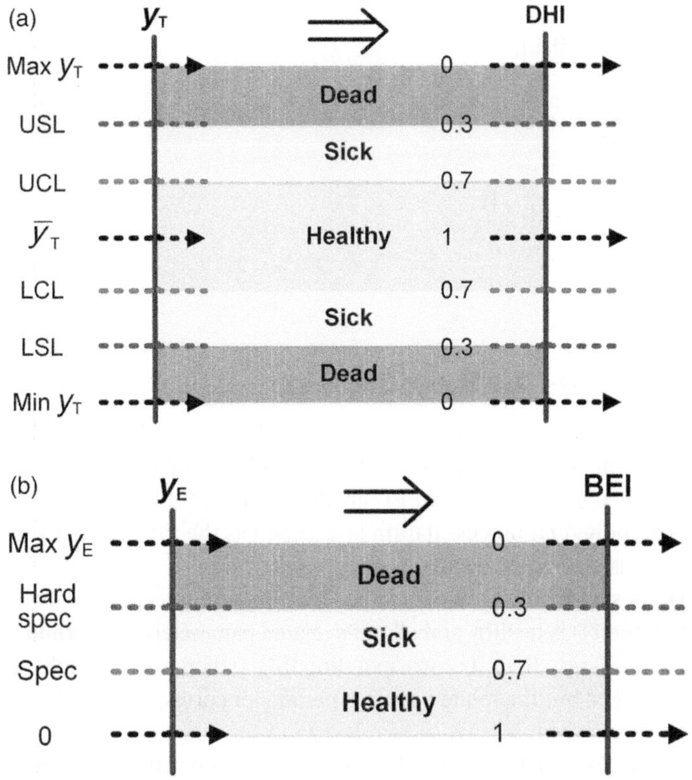

Figure 9.5 Configurations of SPC control charts of DHI and BEI. (a) Converting y_T into DHI. (b) Converting y_E into BEI. *Source:* Reprinted with permission from Ref. [12]; © 2013 IEEE.

$$\text{DHI} = 0.3 - \left(\frac{y_T - \text{UCL}}{\text{Max } y_T - \text{USL}} \times 0.3 \right), \text{ when USL} < y_T. \tag{9.1}$$

In this chapter, the UCL/LCL and USL/LSL are corresponding to the process Spec (the border of sickness) and HardSpec (the border of death) of the TD, and the associated DHI values are 0.7 and 0.3, respectively.

9.2.4 Baseline-Error-Index (BEI) Module

The purpose of the BEI module is to transform the difference between y_T and $\hat{y}_B$, i.e. $y_E \left(= \left| y_T - \hat{y}_B \right| \right)$, into the BEI index. Figure 9.5b depicts the configuration of SPC control chart of y_E with zero (0) being the baseline value and at the bottom. The Spec, HardSpec $- \bar{y}_T$, and Max y_E are shown on the control chart of y_E; further, the BEI mapping values of the corresponding y_E ones mentioned above are normalized and assigned to be 1, 0.7, 0.3, and 0, respectively. The formulas for converting y_E into BEI are expressed in Eq. (9.2).

$$\text{BEI} = 1 - \left(\frac{y_E}{\text{Spec}} \times 0.3 \right), \text{ when } 0 < y_E \leq \text{Spec};$$

$$\text{BEI} = 0.7 - \left(\frac{y_E - \text{Spec}}{\left(\text{HardSpec} - \bar{y}_T\right) - \text{Spec}} \times 0.4 \right),$$

$$\text{when Spec} < y_E \leq \left(\text{HardSpec} - \bar{y}_T\right);$$

$$\text{BEI} = 0.3 - \left(\frac{y_E - \left(\text{HardSpec} - \bar{y}_T\right)}{\text{Max}\, y_E - \left(\text{HardSpec} - \bar{y}_T\right)} \times 0.3 \right),$$

$$\text{when} \left(\text{HardSpec} - \bar{y}_T\right) < y_E. \tag{9.2}$$

9.2.5 Illustration of Fault-Detection-and-Classification (FDC) Logic

The right-hand part of the baseline FDC scheme shown in Figure 9.2 defines the FDC logic to convert DHI, BEI, and ISI_B values into the healthy status of TD. Before applying the FDC logic, the DHI_T, BEI_T, and $\text{ISI}_{\text{B\_T}}$ should be assigned. Both DHI_T and BEI_T are defined to be 0.7 previously. Six times of standard deviations of each individual process datum is assigned as the threshold of the ISI_B; hence, $\text{ISI}_{\text{B\_T}} = 6$. The FDC logic is elaborated as follows.

If $\text{DHI} > \text{DHI}_\text{T}$ and all ISI_B values are smaller than their corresponding $\text{ISI}_{\text{B\_T}}$, then a green light is shown. The green light indicates that the TD is healthy and all of its related parameters are normal.

If $\text{DHI} > \text{DHI}_\text{T}$ and at least one ISI_B is larger than its corresponding $\text{ISI}_{\text{B\_T}}$, then a purple light is shown. This implies that the TD is normal but the related process parameter corresponding to the largest ISI_B is abnormal and should be checked.

If $\text{DHI} < \text{DHI}_\text{T}$ and $\text{BEI} > \text{BEI}_\text{T}$; or if $\text{DHI} < \text{DHI}_\text{T}$ and at least one ISI_B is larger than its corresponding $\text{ISI}_{\text{B\_T}}$, then a yellow light is shown. This yellow light means that the TD itself is healthy; while the reason for its out-of-control (OOC) is due to abnormality of the related process parameter corresponding to the largest ISI_B which should be checked then.

If $\text{DHI} < \text{DHI}_\text{T}$ and $\text{BEI} < \text{BEI}_\text{T}$ and all ISI_B values are smaller than their corresponding $\text{ISI}_{\text{B\_T}}$, then a red light is shown. The red light reveals that the TD is abnormal and caused by itself. This TD needs to be maintained.

9.2.6 Flow Chart of Baseline FDC Execution Procedure

The baseline FDC execution procedure is shown in Figure 9.6 and is explained below.

Step 1: Collect a y_T and its corresponding X of a new workpiece until the collection is successful.

Step 2: Add the y_T and its corresponding X of the new workpiece into the TD baseline modeling set and re-train the TD baseline model; while the ISI_B model does not need re-creation.

Step 3.1: Calculate DHI of this new workpiece.

Step 3.2: Calculate $\hat{y}_B$, ISI_B, and BEI of this new workpiece.

Step 4: Use the DHI, ISI_B, and BEI to infer the healthy status of TD via the FDC logic.

Step 5: Delete y_T and its corresponding X of the new workpiece from TD baseline modeling set.

9.2.7 Exponential-Curve-Fitting (ECF) RUL Prediction Module

The ECF RUL prediction module includes a detection scheme and a prediction model. The detection scheme checks whether $y_\text{E} > y_{\text{E\_S}}$ or not; where $y_\text{E} = \left| y_\text{T} - \hat{y}_\text{B} \right|$ and $y_{\text{E\_S}}$ represent the threshold

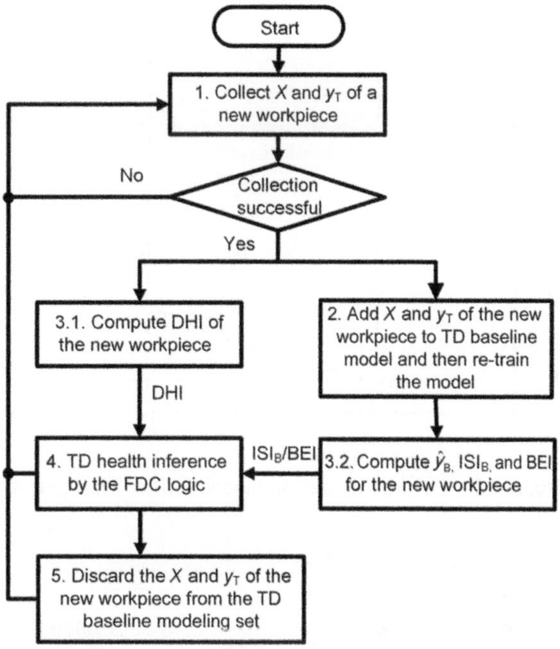

Figure 9.6 Flow chart of baseline FDC execution procedure. *Source:* Reprinted with permission from Ref. [12]; © 2013 IEEE.

of y_E for detecting the sick state of the TD. To avoid a false alarm, two consecutive detections of $y_E > y_{E\_S}$ are required to make sure that the device is entering the sick state. The ECF prediction model will be activated to predict the RUL once the device is in the sick state.

The conventional data-driven methods for predicting RUL include simple projection models, such as exponential smoothing, autoregressive model, and regression-based methods. One major advantage of these techniques is the simplicity of their calculations [24]. Considering limited amount of run-to-failure data sets, a regression-based prediction method is a suitable choice.

Failures of a large population of statistically identical and independent items often exhibit a typical bathtub curve with the following three phases: (i) early failures, (ii) failures with a constant (or nearly so) failure rate, and (iii) wear-out failures [25]. Generally, the sick state of a device occurs in the wear-out phase. A failure in the wear-out phase is resulted from aging, wear-out, or fatigue, etc., whose failure rate increases exponentially with time [26]. Therefore, a regression-based ECF is adopted in this chapter to implement the RUL prediction model.

Define $y_{E\_B} = \bar{y} + 3\sigma_{y_E}$, where $\bar{y}_E$ and σ_{y_E} represent the mean and standard deviation of y_E of the healthy baseline samples collected just after maintenance (as expressed in Step 3 of Figure 9.4). Also, let y_{E_S} and y_{E_D} stand for the values of y_E when the device enters the sick and dead states, respectively. Referring to Figure 9.5b, we have $y_{E\_S} = \text{Spec}$ and $y_{E\_D} = \text{HardSpec} - \bar{y}_T$.

The ECF RUL prediction model is shown in Figure 9.7 and the flowchart for calculating RUL is depicted in Figure 9.8, which is explained as follows.

Phase I (Off-line):

Step 1: Calculate y_{E_B} and find k_B, which is the sample number corresponding to y_{E_B}. Then, define y_{E_D} and y_{E_S}.

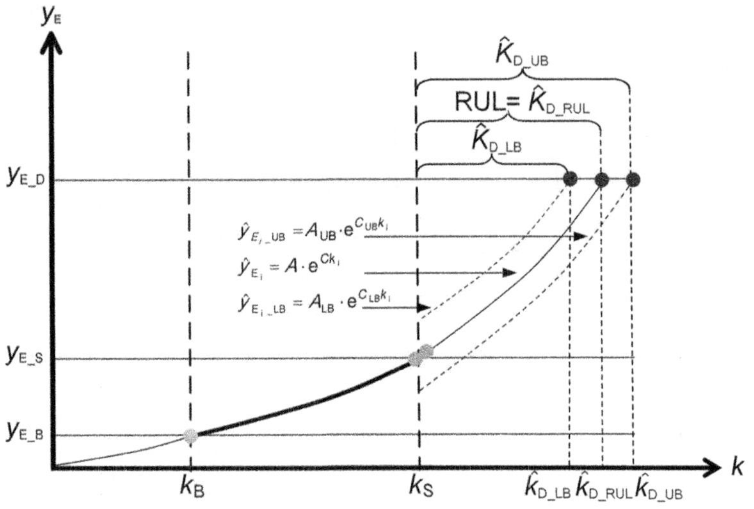

Figure 9.7 ECF RUL prediction model. *Source:* Reprinted with permission from Ref. [12]; © 2013 IEEE.

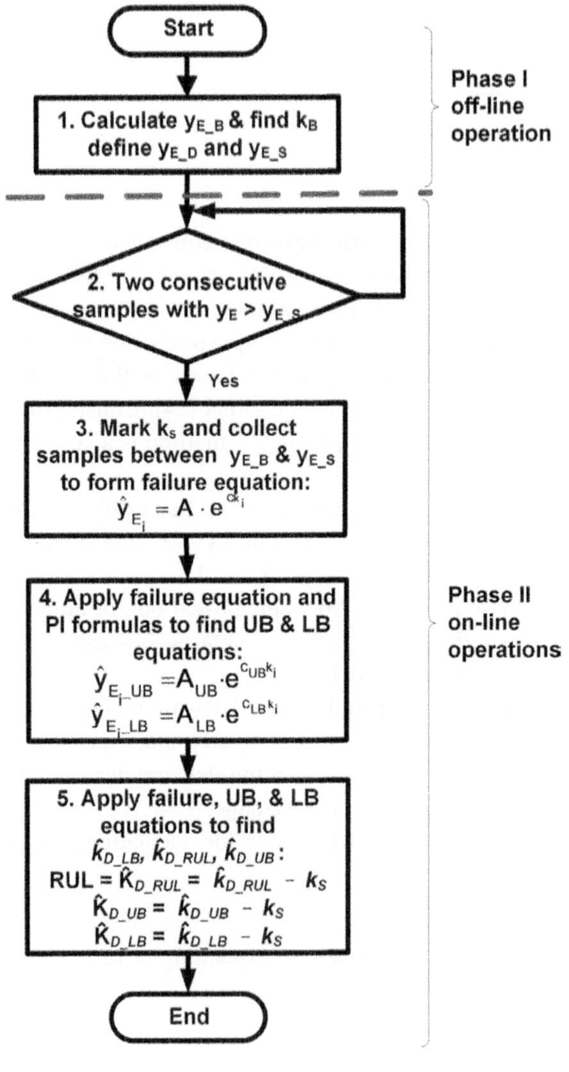

Figure 9.8 Flowchart for calculating ECF RUL. *Source:* Reprinted with permission from Ref. [12]; © 2013 IEEE.

Phase II (On-line):

Step 2: If the condition of two consecutive detections of $y_E > y_{E\_S}$ is confirmed, then the TD is entering the sick state and the flow jumps to Step 3.

Step 3: Collect all of the samples between y_{E_B} and y_{E_S} and apply the ECF formula [26] to obtain the wear-out failure equation:

$$\hat{y}_{E_i} = A \cdot e^{Ck_i}, i = B, B+1, \ldots, S \tag{9.3}$$

with

$$C = \frac{\sum_{i=B}^{S} (k_i - \bar{k}) \left[\ln(y_{E_i}) - \overline{\ln(y_E)} \right]}{\sum_{i=B}^{S} (k_i - \bar{k})^2} \tag{9.4}$$

$$A = e^{\left(\overline{\ln(y_E)} - C\bar{k}\right)} \tag{9.5}$$

where

$$\bar{k} = \frac{\sum_{i=B}^{S} k_i}{k_S - k_B + 1}$$

$$\overline{\ln(y_E)} = \frac{\sum_{i=B}^{S} \ln(y_{E_i})}{k_S - k_B + 1}$$

And

k_i ith sample number;
$\hat{y}_{E_i}$ ith y_E predictive value corresponding to k_i;
y_{E_i} ith y_E actual value corresponding to k_i;
k_B Bth sample number corresponding to y_{E_B};
k_S Sth sample number corresponding to y_{E_S}.

After obtaining the wear-out failure equation $\hat{y}_{E_i} = A \cdot e^{Ck_i}$, the upper-bound (UB) equation, $\hat{y}_{E_i\_UB} = A_{UB} \cdot e^{C_{UB}k_i}$, and the lower-bound (LB) equation, $\hat{y}_{E_i\_LB} = A_{LB} \cdot e^{C_{LB}k_i}$, are also derived as follows.

Step 4: Apply the failure equation $\hat{y}_{E_i} = A \cdot e^{Ck_i}$ to predict the $\hat{y}_{E_i}$ values started from $k_i = k_{S+1}$ until the $\hat{y}_{E_i}$ value is equal to or greater than the y_{E_D} value; its corresponding sample number is denoted as $\hat{k}_{D\_RUL}$. Then, utilize the two-side 95% prediction-interval (PI) formula [27] to calculate all the 95% UB and LB values from $k_i = k_{S+1}$ till $\hat{k}_{D\_RUL}$ with

$$\hat{y}_{E_i\_UB_{S+j}} = \hat{y}_{E_{S+j}} - \Delta_{S+j}, \quad j = \left[1, 2, \ldots, (D\_RUL - S)\right] \tag{9.6}$$

$$\hat{y}_{E_i\_LB_{S+j}} = \hat{y}_{E_{S+j}} + \Delta_{S+j}, \quad j = \left[1, 2, \ldots, (D\_RUL - S)\right] \tag{9.7}$$

$$\Delta_{S+j} = t_{\frac{\alpha}{2}}(k_S - k_B + 1 - p)\sqrt{s_{y_E}^2 \times \left[1 + (k_{S+j})^2 (K^T K)^{-1}\right]} \tag{9.8}$$

where

$$s_{y_E} = \frac{\sum_{i=B}^{S}\left(\hat{y}_{E_i} - \bar{y}_E\right)}{k_S - k_B + 1}$$

$$\bar{y}_E = \frac{\sum_{i=B}^{S} y_{E_i}}{k_S - k_B + 1}$$

$$\boldsymbol{K} = \left[k_B, k_{B+1}, \ldots, k_S\right]^T$$

and

$\hat{y}_{E_i\_UB_{S+j}}$ predictive $S+j$th UB value;
$\hat{y}_{E_i\_LB_{S+j}}$ predictive $S+j$th LB value;
$\hat{k}_{D\_RUL}$ predictive sample number corresponding to y_{E_D};
Δ_{S+j} 95% PI value corresponding to the $S+j$th sample;
$t_{\frac{\alpha}{2}}\left(k_S - k_B + 1 - p\right)$ t-distribution with $k_S - k_B + 1 - p$ degrees of freedom with p being the number of parameters and $a = 0.05$.

After obtaining all the 95% UB and LB values from $k_i = k_{S+1}$ till $\hat{k}_{D\_RUL}$, the UB equation

$$\hat{y}_{E_i\_UB} = A_{UB} \cdot e^{C_{UB} k_i} \tag{9.9}$$

and the LB equation

$$\hat{y}_{E_i\_LB} = A_{LB} \cdot e^{C_{LB} k_i} \tag{9.10}$$

can then be derived by the same ECF formula [26] and approaches that were applied to obtain the failure equation: $\hat{y}_{E_i} = A \cdot e^{Ck_i}$.

Step 5: Apply the failure, UB, and LB equations (Eqs. (9.3), (9.9), and (9.10)) to find RUL, $\hat{K}_{D\_UB}$, and $\hat{K}_{D\_LB}$ that are corresponding to y_{E_D} with

$$\text{RUL} = \hat{K}_{D\_RUL} = \hat{k}_{D\_RUL} - k_S \tag{9.11}$$

$$\hat{K}_{D\_UB} = \hat{k}_{D\_UB} - k_S \tag{9.12}$$

$$\hat{K}_{D\_LB} = \hat{k}_{D\_LB} - k_S \tag{9.13}$$

where

$\hat{k}_{D\_RUL}$ D_RUL-th sample number corresponding to y_{E_D};
$\hat{k}_{D\_UB}$ D_UB-th sample number corresponding to y_{E_D};
$\hat{k}_{D\_LB}$ D_LB-th sample number corresponding to y_{E_D}.

After presenting BPM, the TSP algorithm for predicting the RUL of TD is elaborated as follows.

9.3 Time-Series-Prediction (TSP) Algorithm for Calculating RUL

Among the PdM technologies proposed by many scholars, the ECF model was commonly applied to predict the RUL of TD. However, due to the algorithm limitations, when TD is about to die, whether the TD's aging feature suddenly rises or becomes smooth, the ECF model may not be able to keep up with the real-time prediction or even falsely predicts long RUL. To solve the problem of inaccurate RUL prediction, the authors propose the TSP algorithm. TSP applies the time series analysis model built by information criterion to adapt to the complicated future trend of solving TD fault prediction. Also, the Pre-Alarm Module (PreAM) to make an alert of immediate maintenance when a TD is likely to shut down shortly as well as the Death Correlation Index (DCI) to reveal the possibility of entering the dead state are proposed. How to select the most effective predictors and adjust the predictor weights to construct high-performance prediction model will also be illustrated in Sections 11.3.3 and 11.4.2.

9.3.1 ABPM Scheme

Initially, the function of BPM needs to be enhanced. The core algorithm of BPM is the baseline predictive maintenance algorithm based on virtual metrology. Yet a single algorithm is not sufficient for diversified fab-wide requirements. ABPM is then proposed to this objective. As shown in Figure 9.9, ABPM executes data quality check on all the samples of TD data; after filtering out the abnormalities, the data are turned into indicators (e.g. root mean square (RMS), mean, etc.)

One of the differences between ABPM and BPM is that ABPM is designed to possess the algorithm pool so that users can select the most suitable algorithm in the pool to extract the appropriate aging feature (Y_T) for DHI to determine the TD current health status in response to different equipment requirements. Then, the RUL prediction module will start to predict RUL when TD enters the sick state. The algorithm pool contains several algorithms, in which baseline global similarity index (GSI_B) is used to extract the aging features with multivariate; moving window (MW) can weaken the interference when there is too much indicator noise; and SPC is adopted to deal with the control of indicator upper and lower limits, etc.

When an abnormality occurs among hundreds of TDs, the threshold values of various DHIs would be too complicated to monitor. Therefore, a method of converting different DHI values to normalized indexes ~0~1 is proposed. Larger DHI value indicates better TD heath. Also, the sick

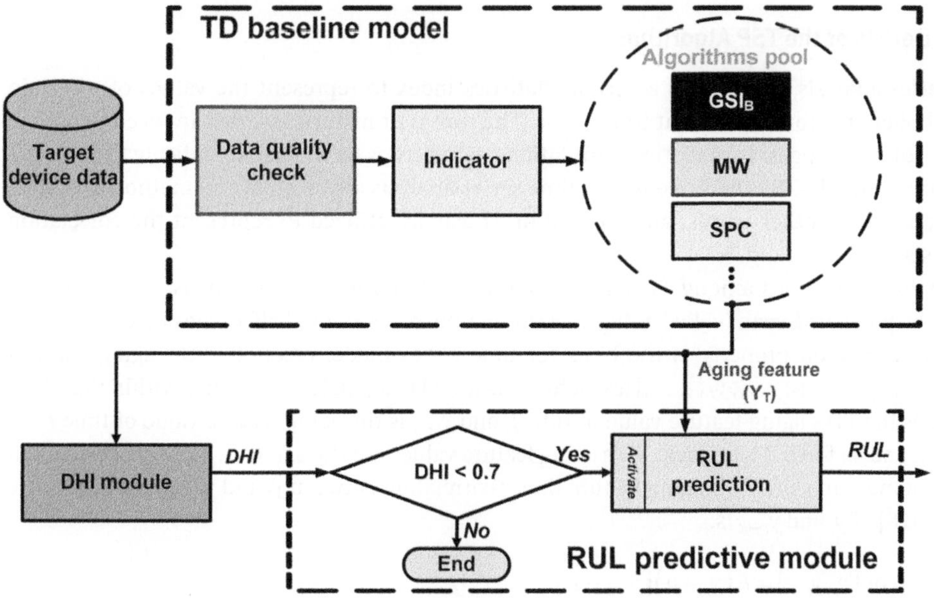

Figure 9.9 Advanced BPM (ABPM) scheme. *Source:* Reprinted with permission from Ref. [21]; © 2019 IEEE.

threshold is set as 0.7 and the dead one as 0.3. A TD has five states: initial, active, inactive, sick, and dead. Under normal circumstances, TD in healthy state is defined by DHI > 0.7 as in the active state. Once 0.7 > DHI > 0.3, it becomes the sick state. If the sick TD is not improved and continues to get worse, TD is getting into the dead state with DHI < 0.3, which means its resource has run out and equipment is down. After the RUL prediction module makes sure that DHI < 0.7 and the TD is in the sick state, RUL prediction is activated. The problems for applying the ECF model to predict RUL will be discussed as follows.

9.3.2 Problems Encountered with the ECF Model

The ECF model cannot correctly predict RUL under two different aging feature trends, which means it may not effectively predict the future RUL according to historical data and that will lead to inaccurate RUL prediction.

Case A: TD's Aging Feature Goes Too Smooth

Figure 9.10 displays the TD's aging trend, and the ECF model is applied to predict TD RUL. Take the curves in the t_1 square for example, when TD's aging trend goes too smooth, the prediction error between RUL prediction value $\hat{y}_{\text{RUL1}}$ and RUL actual value y_{RUL1} grows larger as the smooth time prolongs. Observing Figure 9.10, as the smooth time goes on, RUL predicted by the ECF model would grow gradually and make the predicted RUL longer than the actual one, which results in inaccurate RUL prediction values.

Case B: TD's Aging Feature Rises or Drops Drastically

Take the curves in the t_2 square of Figure 9.10 for instance, when the aging trend rises up drastically, the ECF model prediction will not be able to catch up with the aging trend, and this leads to the increase of prediction inaccuracy between RUL prediction value $\hat{y}_{\text{RUL2}}$ and RUL actual value y_{RUL2}. To solve these problems, the TSP algorithm is proposed.

9.3.3 Details of the TSP Algorithm

Time series analysis [28] adopts a certain statistics index to represent the values of a certain physical phenomenon at different time spans. The trends of historical series are used to predict possible future changes. In statistics, correlation coefficient is often used to indicate the correlation of two variables. By the same token, time series analysis also adopts this method to denote time sequence; in other words, autocorrelation is usually utilized to represent the correlation of time series.

There are correlations among TD's aging features (Y_{T}) in different time spans; that is, the TD state of the previous k span (called as lag operator in time series) would affect the current TD state. This autocorrelation property is the basis for time series model creation. TD's aging features $Y_{\text{T}} = \{\ldots, y_{t-2}, y_{t-1}, y_t, y_{t+1}, y_{t+2}, \ldots\}$ are defined as the TD's aging feature values within time 1~T, where y_t is the TD's aging feature value of time t, and y_{t-1} is the aging feature value of time $t-1$, and so on and so forth. The typical TD's aging feature values are RMS, mean, etc. Autocorrelation coefficient measures the TD's aging feature linear correlation between y_t and y_{t-k}. The covariance function [28] of y_t and y_{t-k} is:

$$\gamma_k = \text{cov}\left(y_t, y_{t-k}\right) = E\left(y_t - \mu\right)\left(y_{t-k} - \mu\right) \tag{9.14}$$

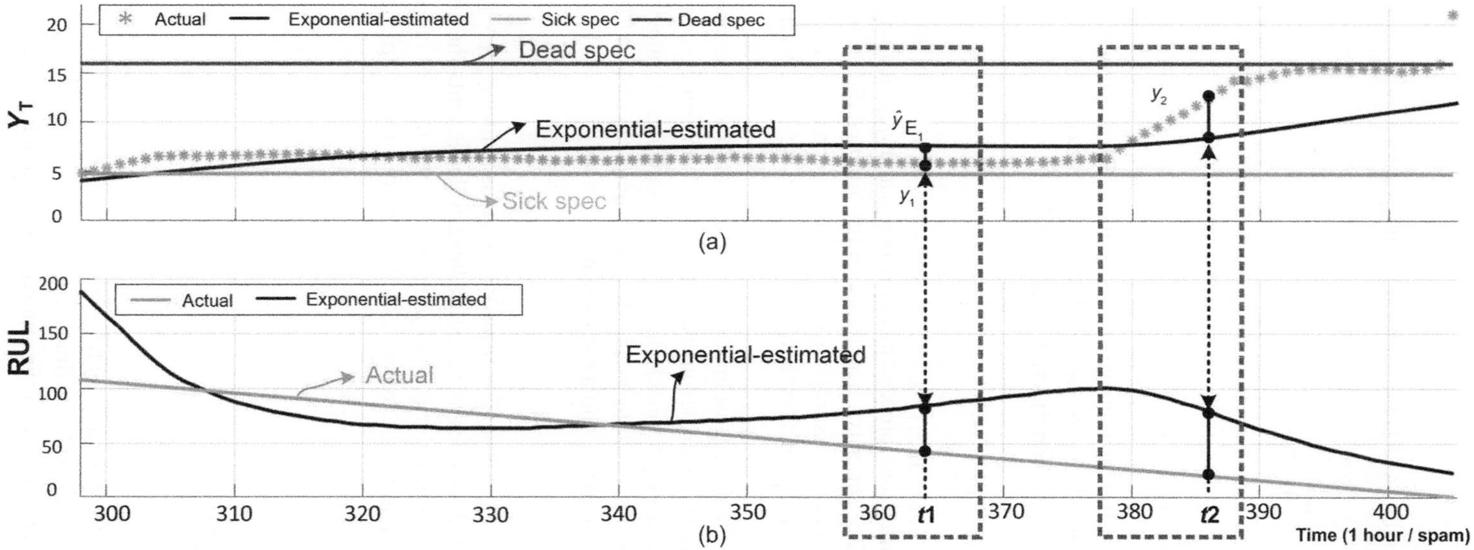

Figure 9.10 Prediction results of the ECF model. (a) Aging feature prediction. (b) RUL prediction. *Source:* Reprinted with permission from Ref. [21]; © 2019 IEEE.

with

$E[\cdot]$ expected value;

μ mean of y_t.

The autocorrelation function (ACF) [28] between y_t and y_{t-k} is:

$$\rho_k = \frac{\text{cov}(y_t, y_{t-k})}{\sqrt{\text{Var}(y_t)}\sqrt{\text{Var}(y_{t-k})}} = \frac{\gamma_k}{\gamma_0} \tag{9.15}$$

where

$$\text{Var}(y_t) = \text{Var}(y_{t-k}) = E(y_t - \mu)^2 = \gamma_0.$$

In addition to the autocorrelation between y_t and y_{t-k}, TD manufacturing effect between y_{t-1}~y_{t-k+1} could also be excluded to derive the interference-free correlation between y_t and y_{t-k}, which is called partial autocorrelation function (PACF) [28]:

$$\rho\rho_k = \text{Corr}(y_t, y_{t-k} \mid y_{t-1}, y_{t-2}, \ldots, y_{t-k+1}). \tag{9.16}$$

However, if $Y_T = \{\ldots, y_{t-2}, y_{t-1}, y_t, y_{t+1}, y_{t+2}, \ldots\}$ is assumed to be a total irrelevant random variable, then Y_T becomes a white noise process, meaning that y_t and y_{t-k} are totally irrelevant.

The basic definition of time series is a hypothesis on the immutability of time shift. Therefore, time series fulfill two conditional assumptions: (i) mean functions are constants that will not change over time; (ii) covariance function ($\text{cov}(y_t, y_{t-k}) = \gamma_k$) is time-irrelevant, which is called as weak stationarity.

After the introduction of the three most important concepts of time series including autocorrelation, white noise, and weak stationarity, four basic models of time series, namely autoregressive model (AR) [29], moving average model (MA) [30], autoregressive moving average model (ARMA) [31], and autoregressive integrated moving average model (ARIMA) [32] will be illustrated below with AR and MA formula clearly stated.

9.3.3.1 AR Model

The autoregressive model adopts Y_T and utilizes y_t's previous values ($y_{t-1} \sim y_{t-p}$) to predict the performance of time t (y_t) and assumes that they are linearly dependent. The notation AR(p) indicates an autoregressive model of order p, with p being the total number of lagged observations required to initialize the autoregressive component of the model. The AR(p) model is defined as

$$\hat{y}_t = \sum_{i=1}^{p} \phi_i y_{t-i} \tag{9.17}$$

$$\varepsilon_t = y_t - \hat{y}_t = y_t - \left(\sum_{i=1}^{p} \phi_i y_{t-i} \right) \tag{9.18}$$

with

y_t actual TD's aging feature value at time t;

$\hat{y}_t$ predicted TD's aging feature value at time t;

φ_i least square estimated coefficient of the autoregressive model, $i = 1, 2, \ldots, p$;

y_{t-i} actual aging feature value at time $t - i$, $i = 1, 2, \ldots, p$;

ε_t white noise error terms at time t.

9.3.3.2 MA Model

$\hat{y}_t$ expressed in Eq. (9.19) can be explained with different weights given by historical stochastic error, and this is the so-called MA model. The notation MA(q) refers to the moving average model of order q, with q being the total number of lagged innovations necessary to initialize the moving average component of the model.

$$\hat{y}_t = \sum_{j=1}^{q} \theta_j \, \varepsilon_{t-j} \tag{9.19}$$

with

θ_j least square estimated coefficient of the MA model, $j = 1, 2, \ldots, q$;
ε_{t-j} white noise error terms at time $t-j, j = 1, 2, \ldots, q$.

9.3.3.3 ARMA and ARIMA Models

The ARMA model constitutes of the autoregressive and moving average models. The ARMA (p, q) model contains p autoregressive operators and q moving average operators, and the enhanced ARMA model becomes ARIMA (p, d, q) [32]. ARIMA is different from ARMA as ARIMA only conducts the ARMA model creation after executing dth difference on data till it is back to the stationary. The ARIMA (p, d, q) model can then be expressed as:

$$\hat{y}_t = \sum_{i=1}^{p} \phi_i \, y_{t-i} + \sum_{j=1}^{q} \theta_j \, \varepsilon_{t-j}. \tag{9.20}$$

9.3.3.4 TSP Algorithm

Though ARIMA model can predict the future trend based on historical data, the combination that AR(p) and MA(q) require for model creation are unclear. On top of that, in PdM, most TD's aging features are degrading and not stationary. However, ARIMA can only conduct the prediction under the condition of weak stationarity and cannot decide the best model via determining the number of predictors. The proposed TSP algorithm deals with the issues of predictor selection and overfitting of ARIMA through Bayesian information criterion (BIC) and a series of hypothesis tests. Moreover, TSP only needs to do the model creation for once when DHI < 0.7 is first encountered, then the new TSP-generated model can be applied for good. The TSP algorithm contains seventeen steps as shown in Figure 9.11.

Step 1: Input TD's aging features ($Y_{\mathrm{T}} = \{\ldots, y_{t-2}, y_{t-1}, y_t, y_{t+1}, y_{t+2}, \ldots\}$) into the DHI module and RUL prediction module.

Step 2: If DHI < 0.7, go to Step 3; else go back to Step 1.

Step 3: Determine the sampling span M for model creation and define the model matrix Y_{M}, $Y_{\mathrm{M}} \subseteq Y_{\mathrm{T}}$. Suppose $M = 30$ (sample amount of more than 30 usually satisfies normal distribution hypothesis [33]), and input Y_{T} spans $t-1 \sim t-30$ to the model, then $Y_{\mathrm{M}} = \{y_{t-30}, y_{t-29}, \ldots, y_{t-2}, y_{t-1}\}$.

Step 4: If Y_{T} trend grows as $y_t = (1+\alpha) \times y_{t-1}$, where α is the increase rate, then go to Step 5; else jump to Step 6. This indicates that Y_{T} increases over time and Var(y_t) grows along with α, which is against the time series assumption: Var(y_t) = Var(y_{t-k}) = γ_0.

Step 5: If variance becomes larger with time, then the assumption of time series stationary is not satisfied. Thus, let Y_{T} go through Log transformation, and $Y_{\mathrm{T}} = \log(Y_{\mathrm{T}})$ is applied to force the increase rate distribution of data to possess certain regularity [32].

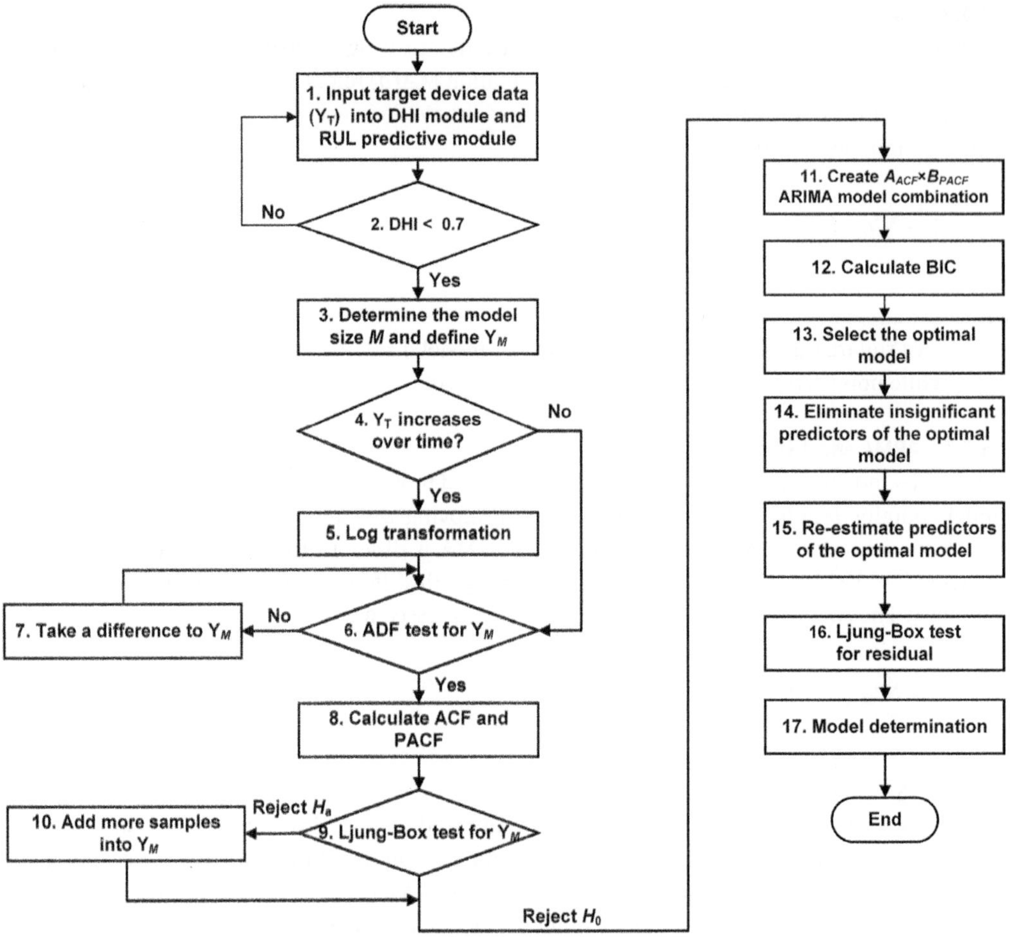

Figure 9.11 Flow chart of the TSP algorithm. *Source:* Reprinted with permission from Ref. [21]; © 2019 IEEE.

Step 6: Perform the augmented Dickey–Fuller (ADF) test [34] on the modeling samples. ADF test is utilized to test if an autoregressive model is stationary or not. When H_0 is not rejected, Y_M is not stationary. If Y_M is not stationary, go to Step 7; else jump to Step 8. Null hypothesis and alternative hypothesis are calculated as:

$$H_0 : \phi_l = 1 \tag{9.21}$$

$$H_a : \phi_l < 1. \tag{9.22}$$

Step 7: Execute dth difference on Y_M, in which d is a nonnegative integer indicating the degree of differencing operator polynomial in the time series, while $\nabla^d y_{t-i}$ means the difference operator, and $i = 1, 2, \ldots, M$. Non-stationary series can be converted into stationary ones [28] through the process of difference, which is formulated as:

$$\nabla^d y_{t-i} = y_{t-i} - y_{t-i-1}. \tag{9.23}$$

Step 8: Calculate ACF as in Eq. (9.15) and PACF as in Eq. (9.16), and mark the most y_{t-1}-related ACF time as A_{ACF} and PACF time as B_{PACF}.

$$A_{\text{ACF}} = \arg\max\left(\rho_{\text{k}}\right) \tag{9.24}$$

$$B_{\text{PACF}} = \arg\max\left(\rho\rho_{\text{k}}\right) \tag{9.25}$$

Step 9: Execute the Ljung–Box test [35] on Y_{M}. Ljung–Box test is applied to check whether any set of auto correlation is 0 in time series. When H_0 is not rejected, Y_{M} is white noise. If Y_{M} is white noise, go to Step 10; else jump to Step 11. Null hypothesis and alternative hypothesis are calculated as:

$$H_0 : \rho_1 = \rho_2 = \ldots = \rho_M = 0 \tag{9.26}$$

$$H_{\text{a}} : \rho_i \neq 0, \text{where}\, i = 1, 2, \ldots, M. \tag{9.27}$$

Step 10: Add more samples into Y_{M}.

Step 11: According to the ARMA formula Eq. (9.20), several ARIMA (p, d, q) model combinations are created on Y_{M}, $p = 1, 2, \ldots, B_{\text{PACF}}$, $q = 1, 2, \ldots, A_{\text{ACF}}$, $d = d$th. As such, there are $A_{\text{ACF}} \times B_{\text{PACF}}$ model combinations: ARIMA(1, d, 1), ARIMA(1, d, 2), . . ., ARIMA(B, d, A). In terms of time series, the autoregressive model usually observes PACF, thus the biggest autoregressive lag of PACF is chosen as the biggest lag combination; on the other hand, the moving average model usually observes ACF, therefore, the biggest moving average lag of ACF is selected as the biggest lag combination [36].

Step 12: Calculate BIC [37] of each ARIMA (p, d, q) model combination. In statistics, BIC is a criterion for model selection among a finite set of models; thus, the model with the lowest BIC is preferred.

$$\text{BIC}\left(p+q\right) = \log\left(\frac{\text{SSE}\left(p+q\right)}{M}\right) + \left(p+q+1\right)\frac{\log\left(M\right)}{M} \tag{9.28}$$

where

$$\text{SSE}\left(p+q\right) = \sum_{i=1}^{M}\left(y_i - \hat{y}_i\right)^2$$

and

SSE sum of squared errors;

M sample size of Y_{M}.

Step 13: Pick the ARIMA model combination with the lowest BIC as the best model.

Step 14: Eliminate the insignificant predictors of the optimal model. Check if there's insignificant predictor. When the estimated coefficient is over 95% confidence interval, it is an insignificant predictor. With assumption of normal distribution, 95% confidence interval equals 1.96, which means this predictor can't be utilized to explain the future spans, and thus will be deleted [36]:

$$\left|\phi_i\right| > 1.96 \times \text{s.e.}\left(\phi_i\right) \tag{9.29}$$

$$\left|\theta_j\right| > 1.96 \times \text{s.e.}\left(\theta_j\right) \tag{9.30}$$

where $i = 1, 2, \ldots, p$, $j = 1, 2, \ldots, q$

with

s. e. (φ_i) standard error of the autoregressive coefficient;

s. e. (θ_j) standard error of the MA coefficient.

Step 15: Re-estimate the model after removing the insignificant predictors. For example, the preliminary estimation model ARIMA (3, d, 2) is

$$\hat{y}_t = \varepsilon_t + \phi_1 y_{t-1} + \phi_2 y_{t-2} + \phi_3 y_{t-3} + \theta_1 \varepsilon_{t-1} + \theta_2 \varepsilon_{t-2}$$

but the estimation coefficient φ_2 is insignificant, so delete y_{t-2} and re-calculate the estimation coefficient.

Step 16: Execute Ljung–Box test on the model residual. When the model predictors are explanatory, the residual should be the white noise.

Step 17: Confirm the model and complete model creation.

After applying TSP to create the ARIMA model as shown in Eq. (9.20), the RUL can then be obtained as explained below. To predict $t+n$ span at time t when only the actual aging feature value of $t-k$ is known, the predictive aging feature value should take the place of the actual one for the prediction as below:

$$\hat{y}_{t+n} = \sum_{i=1}^{p} \phi_i y_{t-i} + \sum_{i=t+1}^{t+n-1} \phi_i \hat{y}_i + \sum_{j=1}^{q} \theta_j \varepsilon_{t-j} + \sum_{j=t+1}^{t+n-1} \theta_j \hat{y}_j. \tag{9.31}$$

Let the aging feature value of tool death be y_{death}. When $\hat{y}_{t+n} \geq y_{\text{death}}$, the TD is most likely to be dead. Hence, define the predictive death value of aging feature at time $t+n$, $\hat{y}_{t+n}$, to be the predictive y_{death} ($\hat{y}_{\text{death}}$), then RUL equals n.

9.3.3.5 Pre-Alarm Module

The major advantage of TSP is that it excels short-term predictions. Yet it has the following two flaws. (i) When TD goes up steadily in a short term, TSP predicts that the future trend would be the same; but if any drastic change occurs in the future, TSP might not adapt accordingly due to its algorithm limitation. (ii) When TD's aging feature oscillates along the dead state, TSP would generate oscillating RUL predictions near the dead state according to the aging-feature's past trend. As such, users cannot know when to execute maintenance. To solve these flaws, a PreAM is proposed to make the alert of immediate maintenance when a TD is likely to shut down soon.

The flow of PreAM after acquiring RUL is shown in Figure 9.12. The decision node of Level I is that TD could shut down any time, and the alert would be sent when the current RUL declines for 30% compared to the previous RUL (RUL_{t-1}), as it indicates that TD is in an unstable status. The decision of Level II is $\text{RUL}_t < \text{BT}$, when the currently estimated RUL is shorter than the buffer time (BT), TD needs maintenance ahead of the BT span. Four situations are disclosed below.

1) Both $\dfrac{(\text{RUL}_{t-1} - \text{RUL}_t)}{\text{RUL}_{t-1}} \geq 0.3$ and $\text{RUL}_t < \text{BT}$ are false – green light: TD is in the sick state but it does not deteriorate or enter the maintenance BT. Therefore, TD is not in danger of dead status.
2) $\dfrac{(\text{RUL}_{t-1} - \text{RUL}_t)}{\text{RUL}_{t-1}} \geq 0.3$ is false and $\text{RUL}_t < \text{BT}$ is true – blue light: RUL does not deteriorate drastically, yet it is shorter than the BT span, and TD is entering BT. Once a blue light shows up again in the next span, TD maintenance should be scheduled immediately.

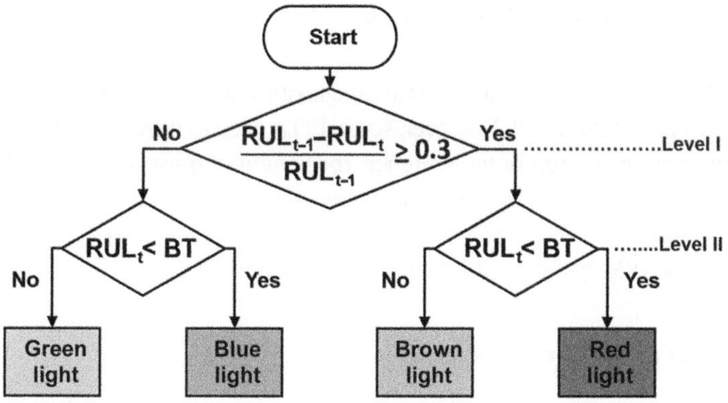

Figure 9.12 Flow chart of the pre-alarm module (PreAM). *Source:* Reprinted with permission from Ref. [21]; © 2019 IEEE.

3) $\frac{(\mathrm{RUL}_{t-1} - \mathrm{RUL}_t)}{\mathrm{RUL}_{t-1}} \geq 0.3$ is true and $\mathrm{RUL}_t < \mathrm{BT}$ is false – brown light: TD deteriorates in a sudden, even RUL does not enter the maintenance BT. If 30% deterioration occurs for three spans in consecution, TD will enter the dead state. For the next span, if the brown light shows again, TD should be checked.
4) Both $\frac{(\mathrm{RUL}_{t-1} - \mathrm{RUL}_t)}{\mathrm{RUL}_{t-1}} \geq 0.3$ and $\mathrm{RUL}_t < \mathrm{BT}$ are true – red light: TD deteriorates drastically and enters BT directly (TD could die soon), it requires immediate maintenance.

9.3.3.6 Death Correlation Index

To avoid the situation that PreAM shows blue light at present but TD suddenly shuts down in the next span, the DCI is proposed to estimate the correlation between y_{t-1} and $\hat{y}_{\mathrm{death}}$. DCI is defined to be within 0 to 1; and higher value means sicker TD. When the value of DCI is higher than its threshold ($\mathrm{DCI_T}$), under 95% confidence level, there is a positive correlation between the current TD status and predicted down time, which means that TD's aging feature is closer to the dead state, and it requires immediate maintenance. On the contrary, when the calculated DCI is closer to 0, it means the current TD status is irrelevant with the predictive TD death value; in other words, even TD is sick, it is not in danger of entering the dead state. According to ACF as in Eq. (9.15), DCI can be calculated as:

$$\mathrm{DCI} = \frac{\mathrm{cov}\left(\hat{y}_{\mathrm{death}}, y_{t-1}\right)}{\mathrm{Var}\left(\hat{y}_{\mathrm{death}}\right)}. \tag{9.32}$$

The DCI threshold ($\mathrm{DCI_T}$) can be defined as

$$\mathrm{DCI_T} = 1.96 \times \mathrm{s.e.}(\mathrm{DCI}) \tag{9.33}$$

where s. e. (DCI) is the standard error of DCI.

After introducing the TSP algorithm for predicting RUL, the factory-wide IPM management framework is elaborated as follows.

9.4 Factory-Wide IPM Management Framework

Currently, most conventional PdM systems can only execute fault diagnostic and/or prognostic algorithms on each individual device in a production tool. However, several TDs need to be monitored for a single tool, and a factory contains a large amount of tools. Hence, the benefit of most conventional PdM systems has been limited as they focus merely on monitoring the equipment level, rather than the entire factory-wide metrics. To monitor all the health statuses from each individual TD up to the whole factory, an effective factory-wide IPM system architecture should be developed.

9.4.1 Management View and Equipment View of a Factory

The management and equipment views of a solar-cell manufacturing factory are depicted in Figure 9.13. They are described below.

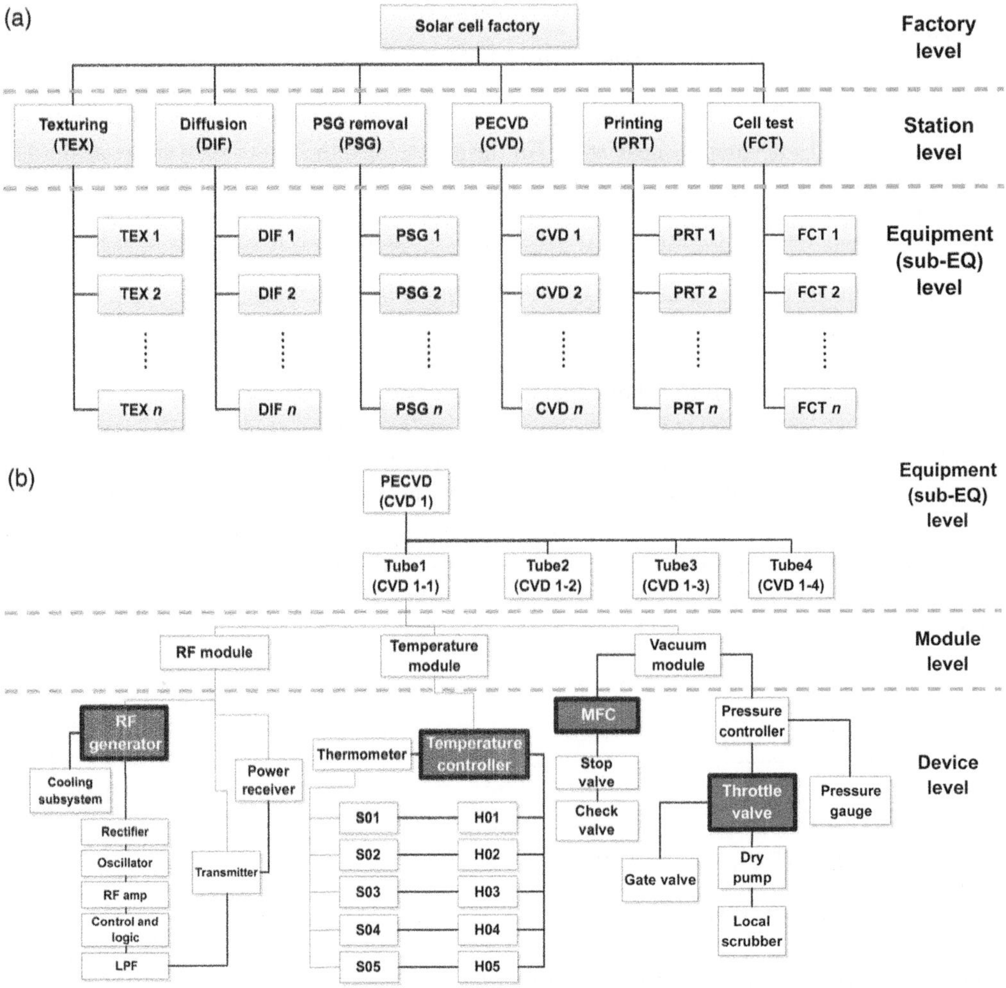

Figure 9.13 Management and equipment views of a solar-cell manufacturing factory. (a) Management view. (b) Equipment view. *Source:* Reprinted with permission from Ref. [38]; © 2017 JCIE.

Management View

The management view is composed of factory, station, and equipment levels by referring to the e-Manufacturing hierarchy depicted in Figure 1.1. As shown in Figure 9.13a, the management view of the solar-cell factory consists of six types of process stations: texturing, diffusion, phosphorous silicate glass (PSG) removal, PECVD, printing, and cell test. Each individual station contains the same type of production tools.

Equipment View

Equipment view describes the physical structure of each specific equipment by referring to the specification of common equipment model (CEM) [39]. CEM is composed of various classes organized in a logical hierarchy. An example of the equipment view, as shown in Figure 9.13b, indicates that a PECVD tool comprises equipment, module, and device levels.

9.4.2 Health Index Hierarchy (HIH)

Overall equipment effectiveness (OEE), as a comprehensive index for individual equipment performance, has been proposed to evaluate the effectiveness of a tool's manufacturing operation [40]. However, since a large number of tools operate jointly on production lines, a single OEE index is not sufficient to evaluate the performance of the entire production line. Therefore, more indexes concerning all the tools factory-wide are required to be defined to complement the insufficiency of OEE in real production environment. Several approaches to deal with this issue are surveyed and described below.

Overall tool group efficiency (OGE) was proposed to monitor equipment performances from groups to single tool via longitudinal analysis, to observe the equipment performance at the tool group level [41]. Overall equipment effectiveness of a manufacturing line (OEEML) was developed based on OEE, to evaluate the effectiveness of the whole production line [42]. Moreover, SEMI E124 defined overall factory effectiveness (OFE) to provide factory-level productivity metrics [43]. These indexes can be used for tracking factory performance in a way that rewards good operational decisions. They can be used in a process of ongoing improvements that are visible to all levels of a manufacturing organization.

To fulfill the requirements of factory-wide equipment maintenance, the Health Index Hierarchy (HIH) is proposed for a factory-wide IPM system. HIH is defined to cover all levels of management and equipment views depicted in Figure 9.13. Observing the detailed structure of HIH as shown in Figure 9.14, management view contains Factory Health Index (FHI), Station Health Index (SHI), and Equipment Health Index (EHI) for the factory level, station level, and equipment level, respectively. Equipment view consists of EHI, Module Health Index (MHI), and DHI for the equipment level (defined in Section 9.2 for BPM), module level, and device level, respectively.

An example shown in Figure 9.13 is taken to explain the relationships among all the indexes of different levels in HIH. Observing Figure 9.13b, all the devices shown in the device level have their individual DHIs. Similarly, all the modules, equipment, stations, and factories in the corresponding module level, equipment level, station level, and factory level own their individual MHIs, EHIs, SHIs, and FHI, respectively. As mentioned above, DHI has been defined in Section 9.2 for BPM; while the relationships among DHI, MHI, EHI, SHI, and FHI are shown in Eqs. (9.34)–(9.37) with all the index values being normalized to vary between 0 and 1. In fact, Eqs. (9.34)–(9.37) imply that the health-index value of upper level is equal to the minimum among the health-index values of bottom level.

$$\text{MHI} = \min\left(\text{DHI}_i\right) \tag{9.34}$$

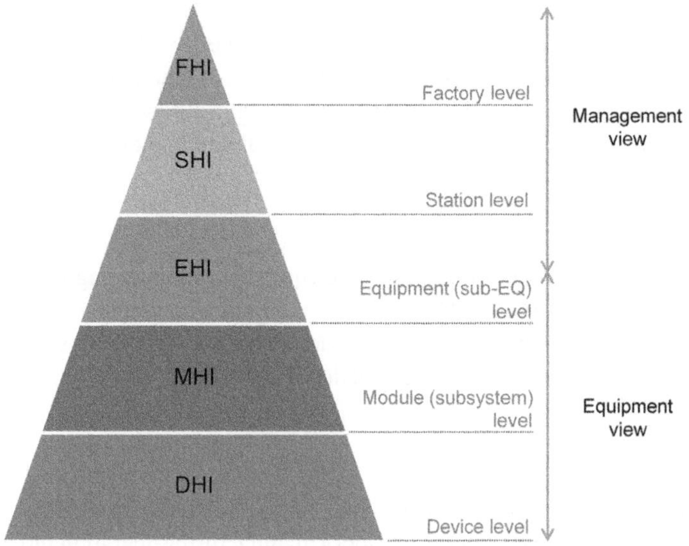

Figure 9.14 Health index hierarchy. *Source:* Reprinted with permission from Ref. [38]; © 2017 JCIE.

$$\mathrm{EHI} = \min\left(\mathrm{MHI}_i\right) \tag{9.35}$$

$$\mathrm{SHI} = \min\left(\mathrm{EHI}_i\right) \tag{9.36}$$

$$\mathrm{FHI} = \min\left(\mathrm{SHI}_i\right) \tag{9.37}$$

where $i = 1, 2, 3, \ldots, n$.

An illustrative example of HIH is shown in Table 9.1. The domain expert defines the throttle valve, mass flow controller (MFC), temperature controller, and radio frequency (RF) generator (as shown in gray background of Figure 9.13b) to be the TDs of PECVD through failure mode and effects analysis (FMEA). In the equipment view of chemical vapor deposition (CVD) 1-1, the DHIs of those four devices before throttle valve repairing are 0.21, 0.87, 0.84, and 0.93, respectively, as shown in Table 9.1a. The values of DHIs of the other devices are approaching 1. Then, according to Eq. (9.34), the MHIs of vacuum, temperature, and RF modules are 0.21, 0.84, and 0.93, respectively. Based on Eq. (9.35), the EHI of CVD 1-1 becomes 0.21. In the management view shown in Table 9.1a, according to Eq. (9.36), the SHIs of those six stations are 0.86, 0.75, 0.91, 0.21, 0.77, and 0.89, respectively. Finally, based on Eq. (9.37), FHI becomes 0.21 in red.

Because the DHI value (0.21) of throttle valve is the lowest and in red, it should be maintained immediately. After repairing, the DHI value of throttle valve becomes 0.96 as depicted in Table 9.1b. Consequently, the new DHIs, MHIs, and EHIs are displayed in Table 9.1b with FHI being 0.75 because Station Diffusion (DIF) has the lowest SHI value.

9.4.3 Factory-wide IPM System Architecture

The system architecture of factory-wide IPM system is shown in Figure 9.15. The system is composed of five major parts: (i) concise-and-healthy creation server (CCS); (ii) IPM server; (iii) IPM manager; (iv) IPM client; and (v) central database. As red curves (clouds) framing in Figure 9.15,

Table 9.1 Scenario of HIH. (a) Before repair of throttle valve. (b) After repair of throttle valve.

(a)												
FHI	0.21											
SHI	TEX	DIF	PSG	CVD 1							PRT	FCT
	0.86	0.75	0.91	0.21							0.77	0.89
EHI	—	—	—	CVD 1-1				CVD 1-2	CVD 1-3	CVD 1-4	—	—
				0.21				0.77	0.94	0.88		
MHI				Vacuum		Temp.	RF	—	—	—		
				0.21		0.84	0.93					
DHI				Throttle valve	MFC	Temp. controller	RF generator					
				0.21	0.87	0.84	0.93					
(b)												
FHI	0.75											
SHI	TEX	DIF	PSG	CVD 1							PRT	FCT
	0.86	0.75	0.91	0.77							0.77	0.89
EHI	—	—	—	CVD 1-1				CVD 1-2	CVD 1-3	CVD 1-4	—	—
				0.84				0.77	0.94	0.88		
MHI				Vacuum		Temp.	RF	—	—	—		
				0.96		0.84	0.93					
DHI				Throttle valve	MFC	Temp. controller	RF generator					
				0.96	0.87	0.84	0.93					

Source: Reprinted with permission from Ref. [38]; © 2017 JCIE.

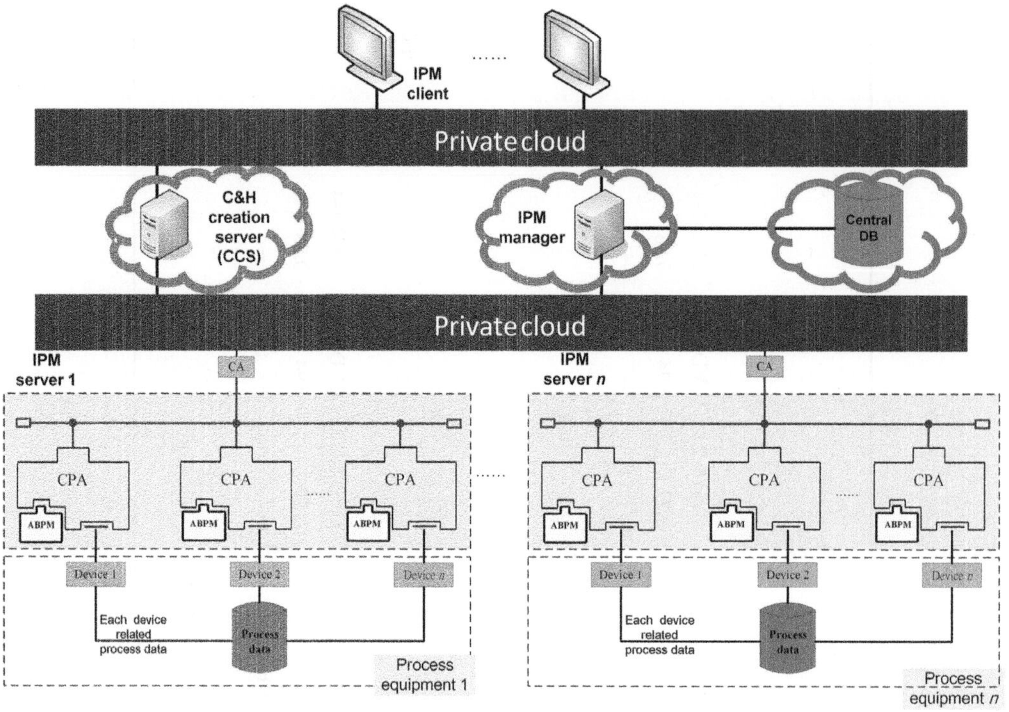

Figure 9.15 Intelligent predictive maintenance (IPM). *Source:* Reprinted with permission from Ref. [38]; © 2017 JCIE.

the CCS, IPM Manager, and central database of the factory-wide IPM system can also be implemented in a private cloud-computing environment such as AMCoT to become the IPM intelligent service of the iFA system shown in Figure 6.5. These five major parts are described below.

(i) Concise-and-healthy Creation Server (CCS)

The CCS is responsible for collecting C&H historical samples to establish the baseline models. CCS performs data processing function to define and collect indicators from equipment sensor data; then, CCS downloads the data collection plan to the IPM server. The so-called automatic baseline sample selection (ABSS) scheme which consists of healthy samples selection (HSS) and DMW for automatic C&H samples selection [44] is executed in CCS.

(ii) IPM Server

As indicated in Figure 9.15, an IPM server contains several Cyber-Physical Agents (CPAs). Each TD selected by the domain expert is connected to a CPA, and sensor data of the TD would be continually collected to the CPA. Each CPA is embedded into a specific equipment to collect its equipment sensor data using various communication standards, such as SEMI equipment communication standard/generic equipment model (SECS/GEM) and Interface A for semiconductor and thin film transistor-liquid crystal display (TFT-LCD) industry, as well as open platform communications unified architecture (OPC-UA) and manufacturing technical (MT)-Connect for machine-tool industry, etc. CPA also possesses a pluggable application module (PAM), which can be configured with customized intelligent applications for stand-alone predictive-maintenance or quality-prognostic tasks. The ABPM module acts as the PAM for judging DHIs and conjecturing RUL of the equipment online.

(iii) IPM Manager

The IPM manager has four major functions. (i) IPM server status monitoring: Since factory-wide IPM system deployment is considered, an IPM manager is essential for managing all IPM servers. All the IPM servers are registered to the IPM manager and then update their health statuses online and in real time. With the ABPM scheme being applied online, the IPM server stores the refreshed ABPM model in the central database via the IPM manager. (ii) Alarm reporting and management: Once the IPM server detects an abnormal status of the equipment, an alarm will be sent from the IPM server to the IPM manager; then, the IPM manager will store the alarm records in the central database, and deliver messages to related task owners by the function of auto-mail. (iii) Client access and security controlling: Authority is required in the IPM system; users should log in with account ID and password. The administrator could assign/modify the access rights of each account. (iv) Health index handling: For the purpose of factory-wide equipment management, the IPM manager collects DHIs from all of the IPM servers, and then generates MHI, EHI, SHI, and FHI according to Eqs. (9.34)–(9.37).

(iv) IPM Client

The IPM client provides the graphical user interfaces (GUI) for users to operate the IPM system. Users can configure data collection plan, modify the configurations, build the ABPM models, and monitor the statuses of the IPM servers via the IPM client. The IPM client also contains an IPM dashboard, which displays management view and equipment view of HIH including real-time FHI, SHI, EHI, MHI, and DHI.

(v) Central Database

The central database is required by the factory-wide IPM system; the configurations, alarm records, and all the historical health indexes are stored in the central database.

9.5 IPM System Implementation Architecture

As described in Chapter 5, Docker container technology allows applications to be written once and then run in any environment that has Docker installed. Also, the container orchestrator Kubernetes (aka K8s) can run and manage containerized applications in an automatic manner and provide them with robustly-operational functionalities, such as load balance, health check, failover, and resource limitation. This section describes how to implement the IPM system by leveraging the container technology. More specifically, an implementation architecture of the containerized IPM system (i.e. IPM_C) based on Docker and Kubernetes is presented in Section 9.5.1. Then, the construction and implementation of the IPM_C are depicted in Section 9.5.2.

9.5.1 Implementation Architecture of IPM_C based on Docker and Kubernetes

Figure 9.16 shows an implementation architecture of the IPM_C, called IPM_C-IA, based on Docker and Kubernetes. The IPM_C-IA consists of a Kubernetes cluster and the IPM_C running in the Kubernetes cluster and can be divided into the cloud side and the factory side.

The cloud side of the IPM_C-IA comprises the Kubernetes Control Plane Node and several Kubernetes Worker Nodes. The former is responsible for managing the cluster and all the applications running in the cluster. The latter is in charge of running key components of the IPM_C in the cloud, including the containerized CCS (i.e. CCS_C), the containerized IPM Manager (i.e. IPM

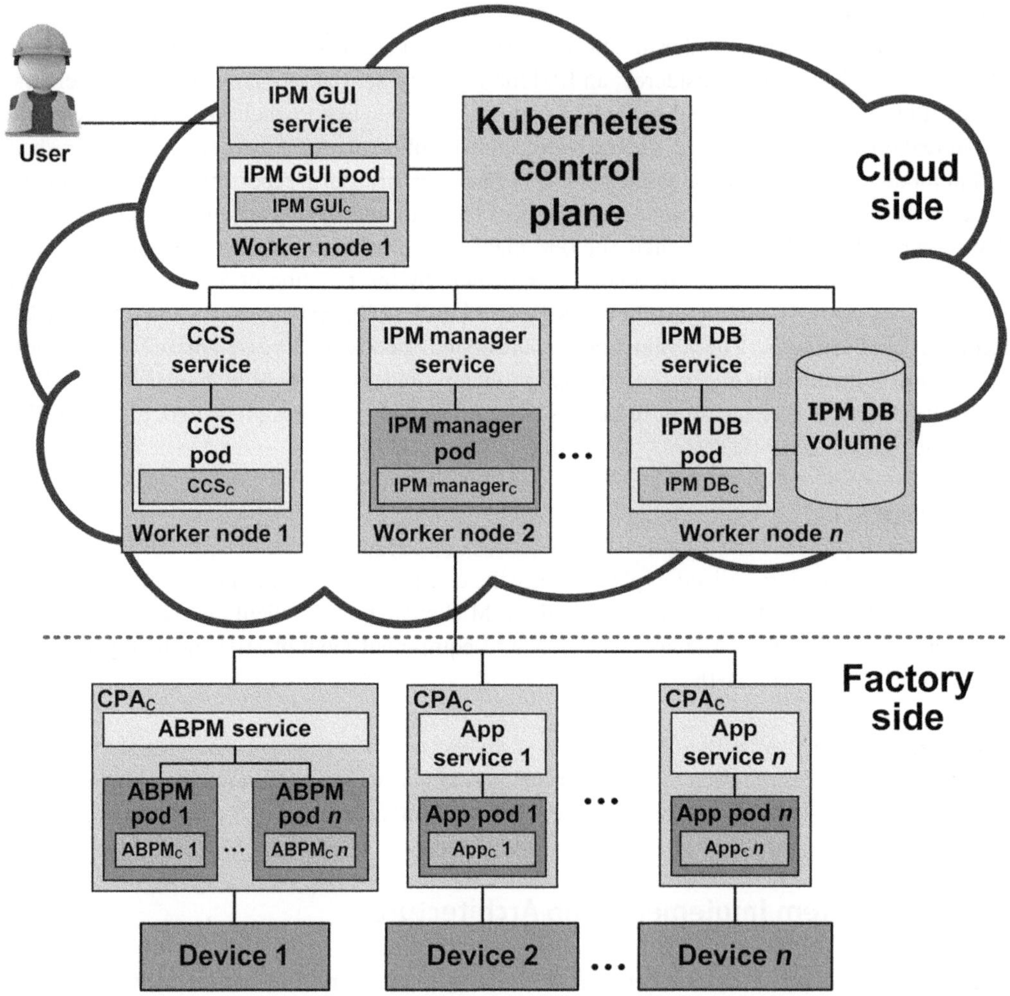

Figure 9.16 Implementation architecture of the IPM_C (i.e. IPM_C-IA) based on Docker and Kubernetes.

$Manager_C$), the containerized IPM GUI (i.e. IPM GUI_C), and the containerized IPM database (i.e. IPM DB_C), together with a data volume (i.e. IPM DB Volume) used to store the IPM database files for data persistency.

The CCS_C is responsible for using the C&H historical data to create the ABPM models, which can then be downloaded from the cloud to CPA_C in the factory for monitoring the tools' health status and predicting their RUL. The IPM $Manager_C$ is in charge of handling messages, commands, and communications among functional components of the IPM_C. The IPM_C, which can be applied to factory-wide equipment, requires an IPM DB_C to store and manage a variety of data, such as configurations, alarm records, historical process data, metrology data, and health indexes of the key devices. Due to containers' short lifetimes (having an average lifespan of about a couple of hours to several days), we cannot preserve data in containers. Thus, we store the data files of IPM DB_C in a persistent volume (i.e. IPM DB Volume) and mount it to the IPM DB_C. By doing so, if the current IPM DB_C dies, the newly-created IPM DB_C that replaces the dead one can still access the data stored in the database, thereby achieving data persistency. The IPM GUI_C provides the GUIs for users to operate the IPM_C. By using the IPM GUI_C, users can configure data collection plans, modify configurations, and build the ABPM models. The IPM GUI_C also contains a dashboard that can display the management view and the equipment view of HIH, including real-time FHI, SHI, EHI, MHI, and DHI as shown in Figure 11.38.

The Kubernetes cluster contains hybrid nodes, where the control plane node must run Linux, and the worker nodes can run either Windows or Linux depending on the workload needs. Every node in the Kubernetes cluster is a Docker host that has Docker installed for running containers. Each Docker host can be created on a virtual machine or a physical computer (a.k.a bare metal). Because the IPM is a Windows-based application, on the cloud side of the IPM_C-IA, the Kubernetes Control Plane is created on a virtual machine that has Ubuntu installed and each Kubernetes worker is a Windows server 2019 node created on a virtual machine.

Because the atomic unit of deployment for Kubernetes is a pod, each container of the IPM_C runs in a pod. The CCS_C runs in the CCS Pod, the IPM $Manager_C$ runs in the IPM Manager Pod, the IPM DB_C runs in the IPM DB Pod, and the IPM GUI_C runs in the IPM GUI Pod. In addition, for each containerized key component of the IPM_C, a Kubernetes service is created to provide both an abstraction that defines a logical set of pods using labels and a policy for accessing the pods underlying the service. In other words, when a client wants to use the functionality of a containerized component in a pod, the client just needs to send the request to the service that supervises the desired pod. Then, the service will forward the request to the target pod or one of the replicas of pods supervised by service using a pre-defined load-balancing rule. As Figure 9.16 shows, the CCS service, the IPM Manager service, the IPM DB service, and the IPM GUI service are created for the clients to access the CCS pod, the IPM Manager pod, the IPM DB pod, and the IPM GUI pod, respectively.

The factory side of the IPM_C-IA comprises many containerized Cyber-Physical Agents (i.e. CPA_C's). As introduced in Section 7.3, each CPA_C is implemented and deployed in an edge device. CPA_C is able to communicate with a wide range of physical objects (e.g. machine tool, equipment, device, and sensor) for the purpose of data collection through various protocols/interfaces. In addition, CPA_C can communicate with other cyber systems (e.g. cloud-based manufacturing (CMfg) services in the cloud) and other CPA_C's. Also, CPA_C can provide manufacturing services in the form of a PAM to the equipment connected with it. The detailed framework and usage of CPA_C can refer to Section 7.3. Here, each CPA_C is both a Docker host and a Kubernetes worker node.

The ABPM module of the IPM system has functionalities of FDC and RUL prediction for key devices of equipment. It is more proper to deploy it in an edge processor connected with the monitored TD, instead of the cloud, to provide responsive designed-functionalities to the monitored key devices. As shown at the bottom of Figure 9.16, each containerized ABPM ($ABPM_C$) runs in an ABPM pod, and we may run several replicas of the ABPM pod under the ABPM service in avoidance of redundancy. If we have at least two replicas of the ABPM pod running under the ABPM service, when an ABPM pod dies, the client can still access the functionalities of the $ABPM_C$. Also, Kubernetes will promptly create a new ABPM pod to replace the dead one to achieve failover. Other applications can run in CPA_C in a similar manner. Each containerized application runs in a pod, and we may run several replicas of the application pod under the application service.

9.5.2 Construction and Implementation of the IPM_C

We can implement and run a containerized application in a Kubernetes cluster by sending the associated yet another markup language (YAML) files of the application to the Kubernetes Control Plane. After receiving the YAML files, the Kubernetes Control Plane will create the corresponding Kubernetes objects that constitute the containerized application, such as pods, deployment, persistent volume, and service, on the Kubernetes Worker Nodes according to those YAML files. After that, Kubernetes orchestrates the containerized application in a declarative manner, instead of imperatively. In other words, Kubernetes continuously monitors the current states of the running containerized application and checks whether they meet the desired states declared in YAML files. Kubernetes will automatically recover the states of the containerized application back to the desired states once the current states differ from the desired ones. By contrast, if a container orchestrator maintains the states of the applications running in it in an imperative manner, we need to provide the container orchestrator with a step-by-step procedure with commands to specifically tell it how to restore the deviated states to the desired states, which is tedious and difficult.

Figure 9.17 shows the workflow for constructing and deploying the IPM_C in a Kubernetes cluster, which is explained as follows.

Step 1: Prepare the YAML files for IPM_C deployment, volume, and service:
The IPM_C comprises several constituent components, including CCS, IPM Manager, IPM DB, IPM GUI, and ABPM. We usually prepare a deployment YAML file and a service YAML file for each of those components. The deployment YAML file is used by the Kubernetes Control Plane to create the Kubernetes pod object as well as deployment object and maintain their desired states, while the service YAML file is for the Kubernetes Control Plane to create the Kubernetes service object that allows the client to access its underlying pods. The volume YAML file is used by the Kubernetes Control Plane to create Kubernetes persistent volume object, which is for storing persistent data files and can be mounted to containers that need to access the stored data.

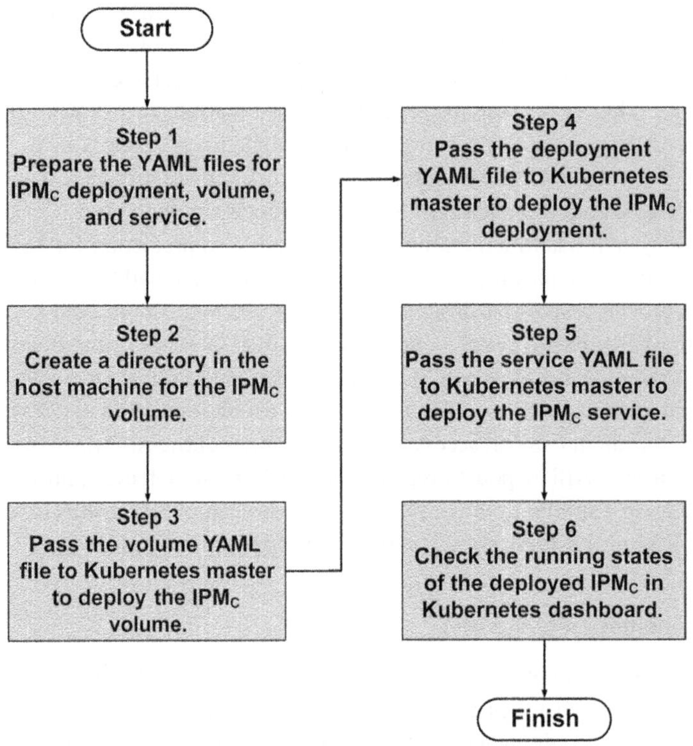

Figure 9.17 Workflow for constructing and deploying the IPM_C in a Kubernetes cluster.

Step 2: Create a directory in the host machine for the IPM_C volume:
As mentioned in the previous section, the IPM_C volume can store the database files of the IPM DB_C for data persistency. The IPM_C volume needs to be located outside the pod due to short lifetimes of pods. We can create a directory in the host machine to serve as the IPM_C volume.

Step 3: Pass the Volume YAML file to Kubernetes Control Plane to deploy the IPM_C Volume:
To create the IPM_C volume, we need to pass the IPM_C volume YAML file to the Kubernetes Control Plane. After receiving the IPM_C volume YAML file, the Kubernetes Control Plane will create the IPM_C volume at the location specified in the YAML file. After that, the IPM_C volume is ready to be mounted to containers that need to access data in it.

Step 4: Pass the Deployment YAML file to Kubernetes Control Plane to deploy the IPM_C Deployment:
To create the deployment object of each constituent component of the IPM_C, we need to pass the deployment YAML of that component to the Kubernetes Control Plane. For example, we can send the deployment YAML file of the IPM manager to the Kubernetes Control Plane for creating the deployment object and associated pods of the IPM manager. After receiving the deployment YAML file, the Kubernetes Control Plane will create the deployment object and associated pods on the suitable worker nodes.

Step 5: Pass the Service YAML file to Kubernetes Control Plane to deploy the IPM_C Service:
To create the service object for each deployment object created in Step 4, we need to pass the service YAML for that deployment object to the Kubernetes Control Plane. For example, after receiving the service YAML file for the IPM manager deployment object, the Kubernetes Control Plane will create the IPM Manager service object that allows the client to access the functionalities of the IPM $Manager_C$ running in the IPM Manager pod.

Step 6: Check the running states of the deployed IPM_C in Kubernetes Dashboard:
After all the Kubernetes objects of the IPM_C have been created, the IPM_C starts to run in the Kubernetes Worker Nodes, and the Kubernetes Control Plane continuously orchestrates and monitors the IPM_C. Then, we can check the states of the IPM_C running in the Kubernetes cluster using the Kubernetes dashboard web GUI.

Figure 9.18 shows an example IPM_C Volume YAML file, which contains two parts separated by three dashes "---." The part above three dashes declares that the kind is a persistent volume with a name IPM-ABPM, the capacity of the volume is 1 GB, the path of the volume is at "C:\path_in_host," and the name of its referring persistent volume claim is IPM-ABPM. The part below three dashes in Figure 9.18 declares that the kind is a persistent volume claim with a name IPM-ABPM, and the requested storage space is 1 GB.

Figure 9.19 shows an example IPM_C Deployment YAML file. It declares that the kind is a deployment with a name IPM-ABPM, the deployment has a label of "app.kubernetes.io/name: IPM," the number of

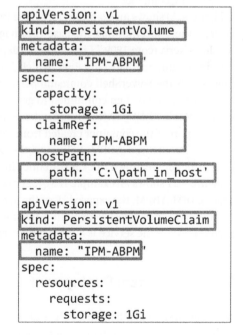

```
apiVersion: v1
kind: PersistentVolume
metadata:
  name: "IPM-ABPM"
spec:
  capacity:
    storage: 1Gi
  claimRef:
    name: IPM-ABPM
  hostPath:
    path: 'C:\path_in_host'
---
apiVersion: v1
kind: PersistentVolumeClaim
metadata:
  name: "IPM-ABPM"
spec:
  resources:
    requests:
      storage: 1Gi
```

Figure 9.18 Example IPM_C volume YAML file.

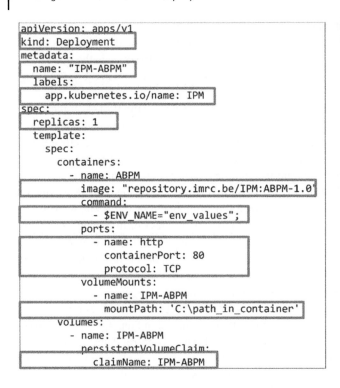

```
apiVersion: apps/v1
kind: Deployment
metadata:
  name: "IPM-ABPM"
  labels:
    app.kubernetes.io/name: IPM
spec:
  replicas: 1
  template:
    spec:
      containers:
        - name: ABPM
          image: "repository.imrc.be/IPM:ABPM-1.0"
          command:
            - $ENV_NAME="env_values";
          ports:
            - name: http
              containerPort: 80
              protocol: TCP
          volumeMounts:
            - name: IPM-ABPM
              mountPath: 'C:\path_in_container'
      volumes:
        - name: IPM-ABPM
          persistentVolumeClaim:
            claimName: IPM-ABPM
```

Figure 9.19 Example IPM_C deployment YAML file.

replicas of the pod that runs the ABPM container is one, the ABPM container is created from the image "respository.imrc.be/IPM:ABPM-1.0," and the exposed port of the container is 80. Also, the name of the volume mounted to the container is IPM-ABPM, and the volume is located at "C:\path_in_container."

Figure 9.20 shows an example Dockerfile for creating the ABPM image that consists of multiple layers. Line 1 specifies that the base OS image is the first layer and is from "mcr.microsoft.com/windows/servercore:1809." Line 2 copies "dependency/setup.exe" from the local host to "C:/setup.exe" in the intermediate container, which is then committed to generate the second layer image. Line 3 runs the PowerShell command Start-Process to execute "C:\setup.exe" in the intermediate container, which is then committed to generate the third layer image. Line 4 defines an environment variable "ENV_VARIABLES" whose default value is env_values. Line 5 specifies that the exposed port of the container created from this image is 80. Line 6 specifies that the starting command is "dotnet.exe ABPM.dll" when the container runs. More details about how to write a Dockerfile for creating a Docker container image can refer to Section 5.2.

Figure 9.21 shows an example IPM_C Service YAML file. It declares that the kind is a service with a name IPM-ABPM, the exposed IP of the service is 10.110.20.21 of the type ClusterIP, the selector of the service can select Pods with a matched label of "app.kubernetes.io/name: IPM," and the exposed port of the service is 8080, which is mapped to the port 80 of the underlying pod.

9.6 IPM System Deployment

After elaborating the theoretical illustrations of the IPM system, this section introduces the deployment considerations and requirements of the IPM system. Six steps are required to deploy the IPM system into manufacturing processes. Step 1: TD selection and operation analysis, Step 2: IPM

```
FROM mcr.microsoft.com/windows/servercore:1809 as final

COPY dependency/setup.exe C:/setup.exe

RUN Start-Process C:/setup.exe -ArgumentList '-mode silent'

ENV ENV_VARIABLES=env_values

EXPOSE 80

ENTRYPOINT ["dotnet.exe", "ABPM.dll"]
```

Figure 9.20 Example Dockerfile for creating the ABPM image.

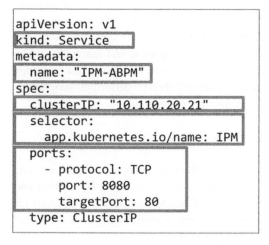

```
apiVersion: v1
kind: Service
metadata:
  name: "IPM-ABPM"
spec:
  clusterIP: "10.110.20.21"
  selector:
    app.kubernetes.io/name: IPM
  ports:
    - protocol: TCP
      port: 8080
      targetPort: 80
  type: ClusterIP
```

Figure 9.21 Example IPM_C service YAML file.

system setup, Step 3: data collection, Step 4: IPM modeling, Step 5: IPM function and integration tests, and Step 6: system release.

Step 1: TD Selection and Operation Analysis

The purpose of IPM is to monitor the health statuses of TDs so that it can predict their RULs to prevent unscheduled down. From the viewpoint of cost-effectiveness, only the key devices, which are TDs, of production equipment need to be monitored by IPM. Therefore, the first step of IPM deployment is TD selection.

Referring to the history of equipment maintenance, the domain expert can prioritize the availability of production equipment based on MTBF and meantime to repair (MTTR). MTBF indicates the frequency of a certain failure that may occur; while MTTR reflects the loss of production capacity.

Besides TD selection, the operation analysis of applying the IPM system for maintenance operations needs to be defined. The AS-IS operation flows of equipment maintenance without IPM are documented first; then, the TO-BE operation flows with IPM should be planned. For example, the IPM system can be integrated with manufacturing execution system (MES) such that the health status of all the TDs can be shown on the IPM equipment status dashboard in MES as depicted in Figure 11.39.

Step 2: IPM System Setup

As mentioned in Chapter 6, the novel iFA system platform is developed by integrating technologies such as the AMCoT framework with CPAs, AVM, IPM, and Intelligent Yield Management (IYM), etc., so as to provide the total-solution package for implementing Intelligent Manufacturing. Therefore, the IPM system is deployed via the iFA system platform. There are two options of the iFA system platform for the users to choose from. "Cloud-based version" shown in Figure 6.5 and "Server-based version" depicted in Figure 6.6 are provided to fulfill different requirements.

In the step of TD selection and operation analysis, all the TDs as well as the operation flows and data exchange requirements are defined, then the gap analysis of the IPM system standard version will be performed by the project team. As a result, the scope and schedule of the IPM system's customization will be addressed. After obtaining the agreement among users and the project team, the installation of the iFA system platform will start. At the same time, the hardware of servers and communication networks need to be prepared. Referring to Figure 9.15, the IPM module of the iFA system platform consists of a CCS, an IPM manager, several IPM clients, and many CPAs. The detailed specifications of data-collection servers need to be confirmed by taking data exchange requirements into consideration. The integration system tests will then be performed accordingly when the software coding tasks and hardware configuration are completed.

Step 3: Data Collection

As shown in Figures 6.5, 6.6, and 9.15, CPAs are adopted by the equipment side to perform data collection. CPA offers several Equipment Drivers for various industries including SECS/GEM and Interface A for semiconductor and TFT-LCD industry, as well as OPC-UA and MT-Connect for machine-tool industry, etc. Moreover, each ABPM module is implemented as a PAM and plugged into each specific CPA for edge computing as shown in Figure 9.15.

Step 4: IPM Modeling

IPM modeling can be proceeded after data collection. Conventionally, the entire process data covering the whole degradation pattern from brand-new till break-down of a specific TD are required for IPM modeling. However, with BPM as explained in Section 9.2.1, only the C&H historical samples are required for collecting the necessary important samples to create the TD baseline model.

As depicted in Figure 9.15, all of the TD process data of the historical samples are collected via CPAs and sent to the CCS server. The trained power user, then, extracts the aging features, which will be converted to DHIs, and sets the sick and dead thresholds of RUL prediction for conducting the IPM modeling by manipulating one of the IPM clients. After finishing modeling, as shown in Figures 6.5 and 6.6, the IPM models are packed as a PAM for edge computing and downloaded to the designated CPA connected to the TD.

Step 5: IPM Function and Integration Tests

Prior to releasing the IPM system for production run, the IPM function and integration tests should be performed based on the TO-BE operation flow defined in Step 1. The IPM system is designed to be integrated with MES. A real example of deploying the IPM System into a solar cell manufacturing factory is illustrated in Section 11.3.3. On the layout of the equipment status dashboard in MES, the health status icons of TDs are designed next to the equipment status icons. When the IPM system is running in the production lines, the MES equipment status dashboard shows all the health status icons in green, yellow, or red to indicate the healthy, sick, or dead statuses via the IPM dashboard. Once the IPM dashboard shows that a certain TD becomes sick, the color of its icon will change from green to yellow. Thus, MES will be notified to lock the production equipment that contains the TD, and a warning message will be sent to maintenance engineers for checking this sick TD.

Step 6: System Release

After all the functions and interfaces have been validated, the project team will move on to the next step: system release. The reliability of the IPM system should guarantee to run 24 hours a day and 365 days a year. To verify the stability and reliability of the IPM system, its MTBF and MTTR are adopted as the acceptance criteria. Moreover, the accuracy of RUL may be the other acceptance criteria as well. RUL is used to predict the remaining time period before the TD breakdown so as to avoid unscheduled down. However, the accuracy of RUL is not easy to verify. As such, the saving of maintenance/repairing time is proposed to be the acceptance criteria rather than the accuracy of RUL.

Besides the IPM system's function and accuracy validations, the related staff of a factory also needs to practice the TO-BE operation flow. They need to know what actions should be taken and what items should be verified once maintenance engineers receive warnings or alarms from the IPM system.

9.7 Conclusion

This chapter introduces the concept of PdM and the necessity of BPM. Then, the VM-based BPM scheme and TSP algorithm for calculating RUL are elaborated. After that, the factory-wide IPM management framework is illustrated. And then, the IPM system implementation architecture is described. Finally, the procedure of IPM system deployment is explained. Additionally, the IPM application cases of various industries will be presented in Chapter 11.

Appendix 9.A – Abbreviation List

ABPM	Advanced Baseline Predictive Maintenance
$ABPM_C$	Containerized ABPM
ACF	Autocorrelation Function
ADF test	Augmented Dickey–Fuller Test
AMCoT	Advanced Manufacturing Cloud of Things
AR	Autoregressive Model
ARIMA	Autoregressive Integrated Moving Average Model
ARMA	Autoregressive Moving Average Model
BDM	Breakdown Maintenance

BEI	Baseline Error Index
BEI_T	Threshold of BEI
BIC	Bayesian Information Criterion
BPM	Baseline Predictive Maintenance
BT	Buffer Time
C&H	Concise and Healthy
CbM	Condition-Based Maintenance
CCS	Concise-and-Healthy Creation Server
CCS_C	Containerized CCS_C
CEM	Common Equipment Model
CMfg	Cloud Manufacturing
CPA_C	Containerized Cyber-Physical Agent
CPS	Cyber-Physical Systems
DCI	Death Correlation Index
DCI_T	Threshold of DCI
DHI	Device Health Index
DHI_T	Threshold of DHI
DMW	Dynamic Moving Window
ECF	Exponential Curve Fitting
EHI	Equipment Health Index
FDC	Fault Detection and Classification
FHI	Factory Health Index
FMEA	Failure Mode and Effects Analysis
GSI	Global Similarity Index
GSI_B	Baseline Global Similarity Index
GSI_T	Threshold of GSI
GUI	Graphical User Interfaces
HIH	Health Index Hierarchy
HSS	Healthy Samples Selection
IoT	Internet of Things
IPM	Intelligent Predictive Maintenance
IPM DB_C	Containerized IPM Database
IPM GUI_C	Containerized IPM GUI
IPM $Manager_C$	Containerized IPM Manager
IPM_C	Containerized IPM
IPM_C-IA	Implementation Architecture of IPM_C
ISI	Individual Similarity Index

ISI_B	Baseline Individual Similarity Index
ISI_{B_T}	Threshold of ISI_B
ISMI	International SEMATECH Manufacturing Initiative
K8s	Kubernetes
KIS	Keep Important Sample
LB	Lower-Bound
LCL	Lower Control Limit
LSL	Lower Spec Limit
MA	Moving Average Model
MHI	Module Health Index
MTBF	Mean-Time-Between-Failure
MTTR	Meantime to Repair
MW	Moving Window
OEE	Overall Equipment Effectiveness
OEEML	Overall Equipment Effectiveness of a Manufacturing Line
OFE	Overall Factory Effectiveness
OGE	Overall Tool Group Efficiency
OOC	Out of Control
PACF	Partial Autocorrelation Function
PAM	Pluggable Application Module
PdM	Predictive Maintenance
PECVD	Plasma Enhanced Chemical Vapor Deposition
PI	Prediction Interval
PM	Preventive Maintenance
PPM	Predictive and Preventive Maintenance
PreAM	Pre-Alarm Module
PSG	Phosphorous Silicate Glass
RUL	Remaining Useful Life
SHI	Station Health Index
SPC	Statistical Process Control
TD	Target Device
TFT-LCD	Thin Film Transistor-Liquid Crystal Display
TSP	Time Series Prediction
UB	Upper-Bound
UCL	Upper Control Limit
USL	Upper Spec Limit
VM-based	Virtual Metrology-Based
YAML	Yet Another Markup Language

Appendix 9.B – List of Symbols in Equations

y_{T}	actual value of TD
$\overline{y}_{\mathrm{T}}$	mean of y_{T}
$\hat{y}_{\mathrm{B}}$	baseline predictive value of TD
y_{E}	absolute value of $\hat{y}_{\mathrm{B}}$ minus y_{T}
$y_{\mathrm{E\_S}}$	sick value of y_{E}
$\overline{y}_{E}$	mean of y_{E}
σ_{y_E}	standard deviation of y_{E}
$y_{\mathrm{E\_B}}$	begin value of y_{E}
$y_{\mathrm{E\_D}}$	dead value of y_{E}
k_i	ith sample number
$\hat{y}_{\mathrm{E}_i}$	ith y_{E} predictive value corresponding to k_i
y_{Ei}	ith y_{E} actual value corresponding to k_i
k_{B}	Bth sample number corresponding to $y_{\mathrm{E\_B}}$
k_{S}	Sth sample number corresponding to $y_{\mathrm{E\_S}}$
A	interception value of the ECF formula
C	slope value of the ECF formula
$\hat{y}_{\mathrm{E}_i\mathrm{\_UB}_{S+j}}$	predictive $S+j$th UB value
$\hat{y}_{\mathrm{E}_i\mathrm{\_LB}_{S+j}}$	predictive $S+j$th LB value
$\hat{k}_{\mathrm{D\_RUL}}$	D_RULth sample number corresponding to $y_{\mathrm{E\_D}}$
Δ_{S+j}	95% PI value corresponding to the $S+j$th sample
$t_{\frac{\alpha}{2}}(k_{\mathrm{S}}-k_{\mathrm{B}}+1-p)$	a t-distribution with $k_{\mathrm{S}}-k_{\mathrm{B}}+1-p$ degrees of freedom with p being the number of parameters and $a = 0.05$
$\hat{k}_{\mathrm{D\_UB}}$	D_UBth sample number corresponding to $y_{\mathrm{E\_D}}$
$\hat{k}_{\mathrm{D\_LB}}$	D_LBth sample number corresponding to $y_{\mathrm{E\_D}}$
k_i	ith sample number
$\hat{y}_{\mathrm{E}_i}$	ith y_{E} predictive value corresponding to k_i
γ_k	Covariance
$E[\cdot]$	expected value
μ	mean of y_t
p	total number of lagged observations required to initialize the autoregressive component of the model
q	total number of lagged innovations necessary to initialize the moving average component of the model
ρ_k	value of autocorrelation
$\mathrm{Var}(\cdot)$	Variance
$\rho\rho_k$	value of partial autocorrelation

Y_T	TD's aging feature
y_t	TD's aging feature value at time t
y_{t-i}	TD's aging feature value at time $t-i$, $i = 1, 2, \ldots, p$
y_{t-1}	TD's aging feature value at time $t-1$
y_{t-k}	TD's aging feature value at time $t-k$
$\hat{y}_t$	predicted TD's aging feature value at time t
φ_i	least square estimated coefficient of the autoregressive model, $i = 1, 2, \ldots, p$
θ_j	least square estimated coefficient of the moving average model, $j = 1, 2, \ldots, q$
ε_{t-j}	white noise error terms at time $t-j, j = 1, 2, \ldots, q$
H_0	null hypothesis
H_a	alternative hypothesis
dth	nonnegative integer indicating the degree of differencing operator polynomial in the time series
Y_M	model matrix of TD's aging feature
$\nabla^d y_{t-i}$	difference operator
A_{ACF}	most y_{t-1}-related ACF time
B_{PACF}	most y_{t-1}-related PACF time
BIC(·)	Bayesian information criterion function
SSE	sum of squared errors
M	sample size of Y_M
s. e. (φ_i)	standard error of the AR coefficient
s. e. (θ_j)	standard error of the MA Coefficient
y_{death}	death value of aging feature
$\hat{y}_{t+n}$	predicted TD's aging feature value at time $t+n$
$\hat{y}_{death}$	predictive death value of aging feature

Appendix 9.C – Patents (IPM)

USA Patents

1) Cheng, F.T., Hsieh, Y.S., Wang, C.R., et al. (2013). Baseline predictive maintenance method for target device and computer program product thereof. US Patent 10,242,319 B2, filed 18 March 2013 and issued 26 March 2019.
2) Yang, H.C. and Cheng, F.T. (2018). Tool wear monitoring and predicting method. US Patent 10,695,884 B2, filed 23 March 2018 and issued 30 June 2020.
3) Kao, Y.C., Yang, H.C., and Cheng, F.T. (2018). System and method for machine tool maintenance and repair. US Patent Pending 15/933,379, filed 23 March 2018.
4) Lin, C.Y., Hsieh, Y.M., Cheng, F.T., et al. (2020). Predictive maintenance method for component of production tool and computer program product thereof. US Patent Pending 16/857,178, filed 24 April 2020.

Taiwan, ROC Patents

1) Cheng, F.T., Hsieh, Y.S., Wang, C.R., et al. (2012). Baseline predictive maintenance method for target device and computer program product thereof. ROC Patent I463334, filed 20 July 2012 and issued 1 December 2014.
2) Yang, H.C. and Cheng, F.T. (2018). Tool wear monitoring and predicting method. ROC Patent I640390, filed 23 March 2018 and issued 11 November 2018.
3) Kao, Y.C., Yang, H.C., and Cheng, F.T. (2018). System and method for machine tool maintenance and repair. ROC Patent I662440, filed 23 March 2018 and issued 11 June 2019.
4) Lin, C.Y., Hsieh, Y.M., Cheng, F.T., et al. (2019). Predictive maintenance method for component of production tool and computer program product thereof. ROC Patent I708197, filed 26 April 2019 and issued 21 October 2020.

Japan Patent

1) 9.8. Cheng, F.T., Hsieh, Y.S., Wang, C.R., et al. (2013). Baseline predictive maintenance method for target device and computer program product thereof. JP Patent 5643387, filed 14 June 2013 and issued 7 November 2014.

European Patent

1) 9.9. Cheng, F.T., Hsieh, Y.S., Wang, C.R., et al. (2013). Baseline predictive maintenance method for target device and computer program product thereof. EP Patent 2687935, filed 16 May 2013 and issued 25 March 2020. (Officially effective in UK, DE, and FR)

China Patents

1) Cheng, F.T., Hsieh, Y.S., Wang, C.R., et al. (2013). Baseline predictive maintenance method for target device and computer program product thereof. CN Patent 2608456, filed 16 May 2013 and issued 15 September 2017.
2) Yang, H.C. and Cheng, F.T. (2018). Tool wear monitoring and predicting method. CN Patent 3810797, filed 23 March 2018 and issued 22 May 2020.
3) Kao, Y.C., Yang, H.C., and Cheng, F.T. (2018). System and method for machine tool maintenance and repair. CN Patent 3840559, filed 23 March 2018 and issued 16 June 2020.
4) Lin, C.Y., Hsieh, Y.M., Cheng, F.T., et al. (2020). Predictive maintenance method for component of production tool and computer program product thereof. CN Patent Pending 202010313837.7, filed 20 April 2020.

Korea Patent

1) Cheng, F.T., Hsieh, Y.S., Wang, C.R., et al. (2013). Baseline predictive maintenance method for target device and computer program product thereof. KR Patent 10-1518448, filed 29 May 2013 and issued 30 April 2015.

References

1 International SEMATECH Manufacturing Initiative (2010). *ISMI Predictive and Preventive Maintenance Equipment Implementation Guidelines. Technology Transfer #10105119A-TR*. California, USA: International SEMATECH Manufacturing Initiative (ISMI).

2 Stark, D. (2009). *Predictive Preventive Maintenance Impact on Factory Cost and Cycle Time. International SEMATECH Manufacturing Initiative*. California, USA: International SEMATECH Manufacturing Initiative (ISMI).

3 International SEMATECH Manufacturing Initiative (2007). *ISMI Consensus Preventive and Predictive Maintenance Vision Guideline: Version 1.1. Technology Transfer #06114819C-ENG*. California, USA: International SEMATECH Manufacturing Initiative (ISMI).

4 Hollister, J. and McGuire, P. (2009). Research on the Current Status of Predictive Maintenance (PdM) Algorithms and Applications. ISMI Project Report #336623. *ISMI Predictive and Preventive Maintenance Initiative*.

5 International SEMATECH Manufacturing Initiative (2010). *ISMI Predictive Preventive Maintenance (PPM) Data Requirements. Technology Transfer #10105120A-TR*. California, USA: International SEMATECH Manufacturing Initiative (ISMI).

6 Communication Promoters Group of Industry-Science Research Alliance (2013). Recommendations for Implementing the Strategic Initiative INDUSTRIE 4.0 Final Report of the Industrie 4.0 Working Group. https://bit.ly/2XPsSj5 (accessed 12 August 2020).

7 Sun, B., Luh, P.B. O'Neill, Z. and Song, F. (2011). Building energy doctors: SPC and Kalman filter-based fault detection. *Proceedings of the IEEE Conference on Automation Science and Engineering*, Trieste, Italy (24–27 August 2011). USA: IEEE.

8 Vachtsevanos, G., Lewis, F.L., Roemer, M. et al. (2006). *Intelligent Fault Diagnosis and Prognosis for Engineering Systems*. Hoboken, NJ: Wiley.

9 Shewhart, W.A. (1931). *Economic Control of Quality of Manufactured Product*. New York: Van Nostrand.

10 Inman, D.J., Farrar, C.R., Steffan, V. et al. (2005). *Damage Prognosis: For Aerospace, Civil and Mechanical Systems*. New York: Wiley.

11 Letellier, C, Hoblos, G. and Chafouk, H. (2011). Robust fault detection based on multimodel and interval approach. Application to a throttle valve. *Proceedings of the 2011 19th Mediterranean Conference on Control & Automation (MED)*, Corfu, Greece (20–23 June 2011). USA: IEEE.

12 Hsieh, Y.S., Cheng, F.T., Huang, H.C. et al. (2013). VM-based baseline predictive maintenance scheme. *IEEE Transactions on Semiconductor Manufacturing* 26 (1): 132–144. https://doi.org/10.1109/TSM.2012.2218837.

13 Hsieh, Y.S., Cheng, F.T. and Yang, H.C. (2012). Virtual-metrology-based FDC scheme. *Proceedings of the 2012 IEEE Conference on Automation Science and Engineering*, Seoul, South Korea (20–24 August 2012). USA: IEEE.

14 Liu, J., Djurdjanovic, D., Ni, J. et al. (2007). Similarity based method for manufacturing process performance prediction and diagnosis. *Computers in Industry* 58 (6): 558–566. https://doi.org/10.1016/j.compind.2006.12.004.

15 Kim, H.E., Tan, A.C.C., Mathew, J. et al. (2012). Bearing fault prognosis based on health state probability estimation. *Expert Systems with Applications* 39 (5): 5200–5213. https://doi.org/10.1016/j.eswa.2011.11.019.

16 Susto, G.A., Schirru, A., Pampuri, S. et al. (2015). Machine learning for predictive maintenance: a multiple classifier approach. *IEEE Transactions on Semiconductor Manufacturing* 11 (3): 812–820. https://doi.org/10.1109/TII.2014.2349359.

17 Gebraeel, N.Z., Lawley, M.A., Li, R. et al. (2005). Residual-life distributions from component degradation signals: a bayesian approach. *IIE Transactions* 37 (6): 543–557. https://doi.org/10.1080/07408170590929018.

18 Zhang, L.J., Mu, Z.Q., and Sun, C.Y. (2018). Remaining useful life prediction for lithium-ion batteries based on exponential model and particle filter. *IEEE Access* 6: 17729–17740. https://doi.org/10.1109/ACCESS.2018.2816684.

19 Xia, T.B., Jin, X.N., Xi, L.F. et al. (2015). Operating load based real-time rolling grey forecasting for machine health prognosis in dynamic maintenance schedule. *Journal of Intelligent Manufacturing* 26: 269–280. https://doi.org/10.1007/s10845-013-0780-8.

20 Singleton, R.K., Strangas, E.G., and Aviyente, S. (2015). Extended Kalman filtering for remaining-useful-life estimation of bearings. *IEEE Transactions on Industrial Electronics* 62 (3): 1781–1790. https://doi.org/10.1109/TIE.2014.2336616.

21 Lin, C.Y., Hsieh, Y.M., Cheng, F.T. et al. (2019). Time series prediction algorithm for intelligent predictive maintenance. *IEEE Robotics and Automation Letters* 4 (3): 2807–2814. https://doi.org/10.1109/LRA.2019.2918684.

22 Cheng, F.T., Tieng, H., Yang, H.C. et al. (2016). Industry 4.1 for wheel machining automation. *IEEE Robotics and Automation Letters* 1 (1): 332–339. https://doi.org/10.1109/LRA.2016.2517208.

23 Groover, M.P. (2001). *Automation, Production Systems, and Computer Integrated Manufacturing*, 2nde. Englewood Cliffs, NJ: Prentice-Hall.

24 Si, X.S., Wang, W., Hu, C.H. et al. (2011). Remaining useful life estimation: a review on the statistical data driven approaches. *European Journal of Operational Research* 213 (1): 1–14. https://doi.org/10.1016/j.ejor.2010.11.018.

25 Lewis, E.E. (1996). *Introduction to Reliability Engineering*, 2nde. New York: Wiley.

26 Steven, C.C. and Raymond, C.P. (2002). *Numerical Methods for Engineers: With Software and Programming Applications*, 4the. New York: McGraw-Hill.

27 Meli, R. and Santillo, L. (1999). Function point estimation methods: a comparative overview. *European Software Measurement Conference*, Amsterdam, Netherlands.

28 Box, G.E.P., Jenkins, G.M., Reinsel, G.C. et al. (1994). *Time Series Analysis: Forecasting and Control*. New York: Prentice-Hall.

29 Kelejian, H.H. and Prucha, I.R. (1998). A generalized spatial two-stage least squares procedure for estimating a spatial autoregressive model with autoregressive disturbances. *Journal of Real Estate Finance and Economics* 17 (1): 99–121.

30 Durbin, J. (1959). Efficient estimation of parameters in moving-average models. *Biometrika* 46 (3): 306–316. https://doi.org/10.2307/2333528.

31 Akaike, H. (1973). Maximum likelihood identification of gaussian autoregressive moving average models. *Biometrika* 60 (2): 255–265. https://doi.org/10.2307/2334537.

32 Box, G.E.P. and Pierce, D.A. (1970). Distribution of residual autocorrelations in autoregressive-integrated moving average time series models. *Journal of the American Statistical Association* 65 (332): 1509–1526. https://doi.org/10.2307/2284333.

33 Hogg, R.V. and Tanis, E.A. (1977). *Probability and Statistical Inference*. Iowa: Pearson Educational International.

34 Elliott, G., Rothenberg, T.J., and Stock, J.H. (1996). Efficient tests for an autoregressive unit root. *Econometrica* 64 (4): 813–836. https://doi.org/10.2307/2171846.

35 Ljung, G.M. and Box, G.E.P. (1978). On a measure of lack of fit in time series models. *Biometrika* 65 (2): 297–303. https://doi.org/10.1093/biomet/65.2.297.

36 Shibata, R. (1976). Selection of the order of an autoregressive model by akaike's information criterion. *Biometrika* 63 (1): 117–126. https://doi.org/10.2307/2335091.

37 Akaike, H. (1979). A bayesian extension of the minimum AIC procedure of autoregressive model fitting. *Biometrika* 66 (2): 237–342. https://doi.org/10.2307/2335654.

38 Chiu, Y.C., Cheng, F.T., and Huang, H.C. (2017). Developing a factory-wide intelligent predictive maintenance system based on Industry 4.0. *Journal of the Chinese Institute of Engineers* 40 (7): 562–571. https://doi.org/10.1080/02533839.2017.1362357.

39 SEMI E120-1104 (2004). *SEMI E120-1104: Specification for the Common Equipment Model (CEM).* California, USA: SEMI https://bit.ly/3h97Kfa (accessed 12 August 2020).

40 Ames, V.A., Gililland, J., Konopka, J. et al. (1995). *Semiconductor Manufacturing Productivity: Overall Equipment Effectiveness (OEE) Guidebook. ISMI Technology Transfer # 95032745A-GEN.* California, USA: International SEMATECH Manufacturing Initiative (ISMI).

41 Chien, C.F., Chen, H.K., Wu, J.Z. et al. (2007). Constructing the OGE for promoting tool group productivity in semiconductor manufacturing. *International Journal of Production Research* 45 (3): 509–524. https://doi.org/10.1080/00207540600792515.

42 Braglia, M., Frosolini, M., and Zammori, F. (2008). Overall equipment effectiveness of a manufacturing line (OEEML): an integrated approach to assess systems performance. *Journal of Manufacturing Technology Management* 20 (1): 8–29. https://doi.org/10.1108/17410380910925389.

43 SEMI E124-1107 (2007). *SEMI E124-1107: Guide for Definition and Calculation of Overall Factory Efficiency (OFE) and Other Associated Factory-level Productivity Metrics.* California, USA: SEMI https://bit.ly/3hP0N3v (accessed 12 August 2020).

44 Cheng, F.T., Chen, C.F., Hsieh, Y.S. et al. (2013). Automatic baseline-sample-selection scheme for baseline predictive maintenance. *Proceedings of the IEEE International Conference on Automation Science and Engineering*, Madison, Wisconsin, USA (17–21 August 2013). USA: IEEE.

10

Intelligent Yield Management (IYM)

Yu-Ming Hsieh[1], Chin-Yi Lin[2], and Fan-Tien Cheng[3]

[1]Associate Research Fellow, Intelligent Manufacturing Research Center, National Cheng Kung University, Tainan, Taiwan, ROC
[2]Postdoctoral Research Fellow, Intelligent Manufacturing Research Center, National Cheng Kung University, Tainan, Taiwan, ROC
[3]Director/Chair Professor, Intelligent Manufacturing Research Center/Institute of Manufacturing Information and Systems, National Cheng Kung University, Tainan, Taiwan, ROC

10.1 Introduction

With the advancement of semiconductor, thin film transistor-liquid crystal display (TFT-LCD) and bumping manufacturing technologies, their manufacturing processes are becoming more and more sophisticated. As a result, how to maintain feasible production yield of these sophisticated manufacturing processes becomes an essential issue.

For the sake of easy explanation, the changing curves of yield and cost during the product life cycle in Chapter 1 is shown again in Figure 10.1. The figure depicts that product yield (blue line) will gradually rise up in the ramp-up phase, and then keep steady in the mass-production phase. On the contrary, product cost (red line) will decrease as the phases proceed. Company's competitiveness would be effectively enhanced if the blue/red solid lines could be improved into their corresponding segmented lines. This implies rapidly increasing the yield in the ramp-up phase, and promptly transferring products into the mass-production phase; then enhancing product quality in the mass-production phase as well as finding out and resolving the yield-loss problem encountered.

Most companies use yield management systems to search for the root causes of yield loss and to improve the yield. A yield management system is a big-data-analysis system composed of a number of analytical devices, such as engineering data analysis [1], fault detection and classification (FDC) [2], and big data analytics (BDA) [3], etc. Lee and Smith [4] discussed the typical yield enhancement method of the semiconductor manufacturing processes. Generally, when analyzing the yield loss caused by devices, the typical yield enhancement method utilizes the already-known corresponding conditions of good lots and bad lots to create the database. So that if bad lots occur in the future, the method directly explores the historical data to nail down the target devices, as well as to observe the changes in parameters for better adjustment.

The yield management system aims to enhance product yield. However, the number of workpieces is small at the research-and-development (RD) and ramp-up phases, and that makes it hard to find out

Industry 4.1: Intelligent Manufacturing with Zero Defects, First Edition. Edited by Fan-Tien Cheng.

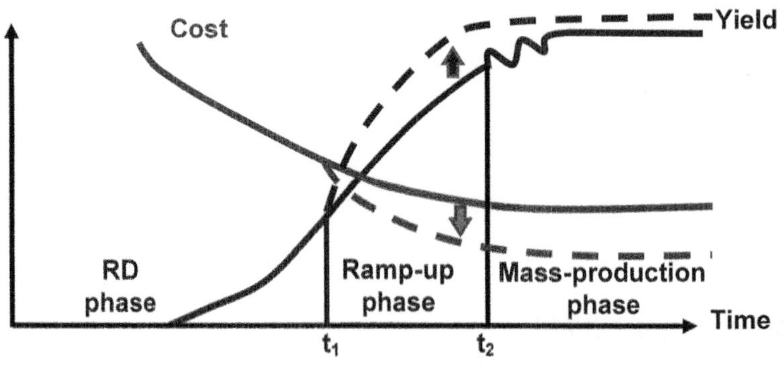

Figure 10.1 Changing curves of yield and cost during the product life cycle. *Source:* Reprinted with permission from Ref. [11]; © 2017 IEEE.

the root causes of defects among all production tools. Thus, how to identify the key variables affecting yield among numerous production tools with limited samples is a widely concerned issue. This challenge is the so-called high-dimensional variable selection problem, which is also denoted as the issue of $p >> n$. "p" in $p >> n$ represents the number of explanatory variables in the models and means all process-related parameters in the semiconductor or TFT-LCD process. "n", on the other hand, is the sampling number that is also the amount of end products in the semiconductor or TFT-LCD process.

Several approaches have been proposed to solve the high-dimensional variable selection problem. Among them, Chen and Hong proposed sample-efficient regression trees (SERT) [5] for semiconductor yield loss analysis, which combined the methodologies of forward selection in regression analysis and regression tree. Compared to the conventional decision trees like CART (classification and regression trees), SERT is able to handle combination effect in the so-called high-dimensional ($p >> n$) problem. However, in order to handle different types of input variables (continuous, ordinal, binary, and categorical), SERT needs to transfer all types of input data into binary form during data preprocessing.

Data-mining approaches were adopted by Chien et al. [6] to infer possible causes of yield losses and manufacturing process variations. The inputs of this work are chip-probing (CP) yield results as well as process data, equipment data, and lot history during wafer fabrication. This work aimed to develop a framework for data mining and knowledge discovery from database that consists of a Kruskal–Wallis test, K-means clustering, and the variance reduction splitting criterion to investigate the huge amount of semiconductor manufacturing data.

Hsu et al. [7] also proposed a data mining framework for finding root causes of defects for TFT-LCD manufacturing. Rough set theory (RST) was applied to generate the candidate rules and then these rules were utilized to locate the root causes after several rounds of cross-validation. Decision tree (DT) was also used to determine the root causes. Different attribute evaluation criteria were applied by these two methods. RST aims to eliminate redundant or surplus attributes in the decision table and DT explains the data based on the maximum information value during each splitting tree process.

Ing and Lai [8] proposed to apply orthogonal greedy algorithm (OGA) together with high-dimensional information criterion (HDIC) and backward trimming (Trim) to solve the high-dimensional variable selection problem. OGA is a fast stepwise regression method which forward selects input variables to enter a p-dimensional linear regression model ($p \gg n$). Comparing to the usual model selection criteria, HDIC + Trim work well in the high-dimensional case, and also resolve the issue of spurious variable selection when the number of variables is much larger than the sample size.

Unlike stepwise regression, a least absolute shrinkage and selection operator (LASSO) regression [9, 10] was also proposed to solve the high-dimensional variable selection problem. LASSO is a shrinkage method which minimizes the residual sum of squares subject to the sum of the absolute value of the coefficients being less than a constant. Because of the nature of this constraint, it tends to produce some coefficients that are exactly zero and hence gives interpretable models. The significant difference between LASSO and the usual stepwise regression is that all the independent variables can be processed at the same time via LASSO but not the usual stepwise regression.

The yield management systems surveyed above emphasized on adopting various data-mining and/or regression algorithms to resolve the problem of figuring out the root causes of yield losses under the high-dimensional environment; while the issues of input data quality and reliability level of the search results were not addressed.

10.1.1 Traditional Root-Cause Search Procedure of a Yield Loss

The traditional root-cause search procedure of a yield loss is depicted in Figure 10.2 and described below.

Step 1: When a yield test failure is encountered, the root-cause search procedure is launched.
Step 2: The defect mode database is searched to see if this defect mode exists in the database or not.
Step 3: If this defect mode exists, jump to Step 5; otherwise, continue to Step 4.
Step 4: Due to the fact that this defect mode is new and its characteristics are not well known yet, the sampling rate related to the in-line inspections of this defect should be increased so as to enhance the defect mode database.
Step 5: Apply a suitable software tool, such as engineering data analysis [1], to find the root cause of defect.
Step 6: If the root cause is found, jump to Step 7; otherwise, move back to Step 4.
Step 7: Issue a notice to the relevant departments for fixing the problem and for continuous improvement.

The major drawbacks of the traditional root-cause search process are (i) the information contained in the defect mode database may not be complete such that associated sampling rates need to be raised, which will increase the production cost; (ii) even the current defect mode is in the database, the search tool (such as engineering data analysis) may not find out the root cause easily.

In fact, this root-cause search issue is also a high-dimensional variable selection problem. To solve this problem economically and efficiently, the Intelligent Yield Management (IYM) system is proposed as follows.

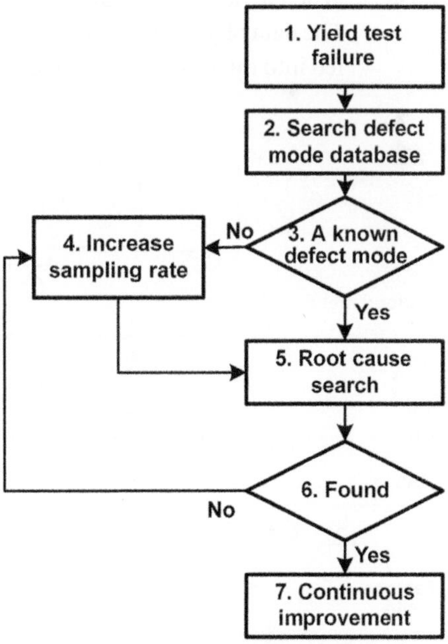

Figure 10.2 Traditional root-cause search process of a yield loss. *Source:* Reproduced with permission from Ref. [11]; © 2017 IEEE.

10.1.2 IYM System

The purpose of the IYM system is to figure out the root causes which affect the yield of the final product effectively. Therefore, the inputs of the IYM system should cover all the big data related to the entire manufacturing process of this product. These big data include workpiece production routes provided by the manufacturing execution system (MES), process data of all the production tools, and in-line measurement data of all the metrology tools as well as the final yield inspection data. To collect all the big data mentioned above, the Intelligent Factory Automation (iFA) system platform proposed in Chapter 6 is adopted. As a result, the IYM system applied to the bumping process of the semiconductor industry is shown in Figure 10.3. The kernel of the IYM system is the Key-variable Search Algorithm (KSA) scheme, which will be elaborated in Section 10.2. The KSA scheme not only contains new root-cause search algorithms for solving the high-dimensional variable selection problem but also includes data preprocessing modules for checking the quality of all input data as well as reliance-index module for evaluating the reliability of the search results. IYM is suitable for a wide range of manufacturing systems, such as semiconductor, TFT-LCD, machine-tool, etc.

A three-phase process is suggested in the next section to apply the KSA scheme for searching the root causes of a yield loss with Phase A for searching the top N key stages; Phase B for selecting the key process parameters within the most suspicious stage found in Phase A; and Phase C for suggesting the FDC thresholds of the key process parameters identified in Phase B.

10.1.3 Procedure for Finding the Root Causes of a Yield Loss by Applying the Key-variable Search Algorithm (KSA) Scheme

The procedure of three-phase process by applying the KSA scheme is shown in Figure 10.4 and explained below.

Phase A: When a yield test failure is encountered, the root-cause search procedure is launched. The data of yield test result (**Y**) as well as inline data (***y***) and production routes ($\mathbf{X}_R$) are fed into the KSA scheme to search for the top N key stages that affect the yield the most.

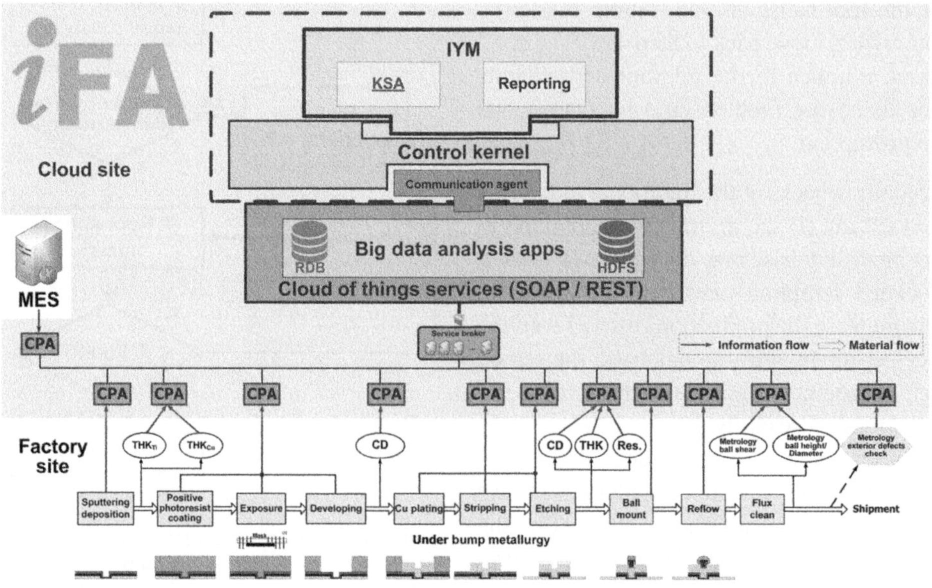

Figure 10.3 Intelligent yield management system.

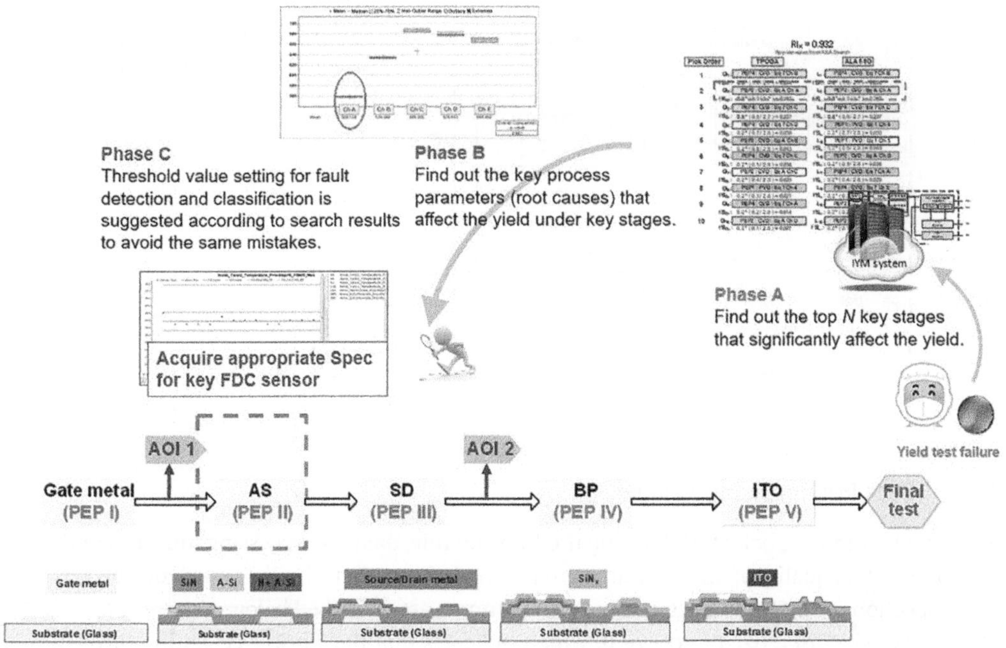

Figure 10.4 Procedure for finding the root causes of a yield loss by applying the KSA scheme.

Phase B: The same data of yield test result (**Y**) and the process data ($\mathbf{X}_P$) of all the devices of the stage to which the most suspicious device belongs are fed into the KSA scheme to search for the key variables that cause the yield loss.

Phase C: Threshold value setting for FDC is suggest according to research results to avoid the same mistakes.

The KSA scheme will be elaborated as follows.

10.2 KSA Scheme

The KSA scheme [11] contains various data preprocessing modules, KSA module, reliance index (RI_K) module, blind-stage search algorithm (BSA), and interaction-effect search algorithm (IESA) as shown in Figure 10.5.

Besides the fundamental functions of KSA and search quality evaluation, the KSA scheme also possesses the functions of automatic data quality evaluation. The inputs of a KSA scheme include production routes ($\mathbf{X}_R$), process data ($\mathbf{X}_P$), and in-line metrology values ($\mathbf{y}$) as well as the final inspection data (**Y**). Note that $\mathbf{y}$ may be provided by actual measurement and/or virtual metrology. Defects (D) could happen in any production stage, thus they are also imported into the KSA scheme for analysis. Then, the processed D may join the production information group or final inspection group depending on the nature of D. The outputs are the search result of Triple Phase Orthogonal Greedy Algorithm (TPOGA): KS_O and that of Automated Least Absolute Shrinkage and Selection Operator, (ALASSO): KS_L, and their accompanying reliance index: RI_K, and Blind-stage index: BS, as well as Interaction-effect index: IE_O, IE_T, and IESA reliance index: RI_I. These functional modules are described below.

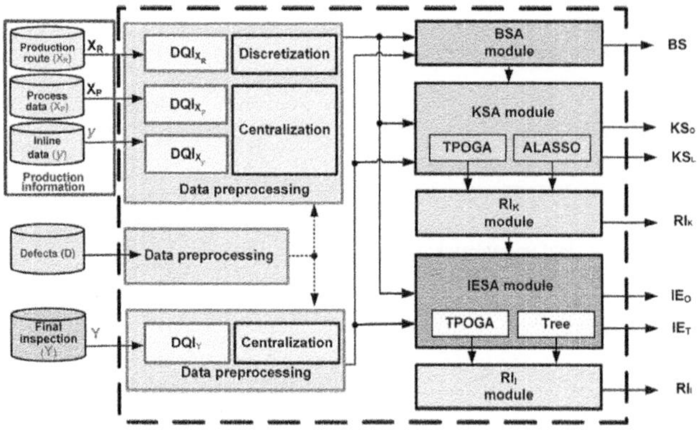

Figure 10.5 The KSA scheme. *Source:* Reproduced with permission from Ref. [18]; © 2018, IEEE.

10.2.1 Data Preprocessing Module

Observing Figure 10.5, before performing the KSA module, data preprocessing must be conducted to assure the data quality of all the inputs: production information ($\mathbf{X_R}$, $\mathbf{X_P}$, and $\boldsymbol{y}$), defects (D), and final inspections ($\mathbf{Y}$). The characteristics of these inputs are described below.

$\mathbf{X_R}$ needs to be discretized into 1 or 0, which indicates that the workpiece getting through this stage or not. $\mathbf{X_P}$ contains tool process data (such as voltage, pressure, temperature, etc.) which need to be centralized. $\boldsymbol{y}$ stands for inline inspection data (such as critical dimension, thickness, etc.) which need to be centralized. As for D, different companies have different definitions of defects, thus discussion with domain experts is required before executing data-preprocessing and quality check. Finally, $\mathbf{Y}$ stands for the yield test results that should be centralized.

The data quality evaluation algorithm of $\mathbf{X_R}$, DQI_{X_R}, evaluates the following four facts: (i) a stage may contain several devices of the same type, while a stage utilizes only one device; if a process should get through three devices, it then has three stages; (ii) if a device is used in different processes, the same device in a different process would be considered as a different stage; (iii) there are only two possibilities for a workpiece passing through the device: get through (1) or not (0); (iv) a workpiece cannot get through any device that doesn't belong to that stage.

Similarly, the data quality evaluation algorithms of $\mathbf{X_P}$ and $\boldsymbol{y}$ are denoted as DQI_{X_P} and DQI_{X_y}, respectively. Both DQI_{X_P}, and DQI_{X_y} adopt the algorithm similar to the process data quality evaluation scheme utilized in Automatic Virtual Metrology (AVM) presented in Chapter 8. Finally, the data quality evaluation algorithm of $\mathbf{Y}$ is denoted as DQI_Y. DQI_Y applies the algorithm similar to the metrology data computation layer quality evaluation scheme used in AVM.

10.2.2 KSA Module

To double-check the reliability of the search results, the KSA module as shown in the upper right corner of Figure 10.5 (the KSA scheme) contains two algorithms: TPOGA and ALASSO. They are described below.

10.2.2.1 Triple Phase Orthogonal Greedy Algorithm (TPOGA)

The greedy algorithm is a stepwise regression method that considers the correlation between all the causing parameters ($\mathbf{X}$) and the results ($\mathbf{Y}$). In this study, $\mathbf{X}$ includes all the related variables of production: $\mathbf{X_R}$, $\mathbf{X_P}$, and $\boldsymbol{y}$; while $\mathbf{Y}$ represents the final inspection.

Pure greedy algorithm (PGA) and orthogonal greedy algorithm (OGA) are commonly used in the literature for solving the high-dimensional regression problem. In general, OGA performs better than PGA in high-dimensional linear regression [8]. The steps of OGA are briefly described below.

Step 1: Define $R_0 = \mathbf{Y} = (Y_1, Y_2, \ldots, Y_n)^T$. Choose the variable defined as $x_{\hat{S}_1}$ that is most correlated with R_0 in $\{\mathbf{X} = x_1, x_2, \ldots, x_p\}$. Then, the corresponding residual will be

$$\hat{R}_1 = R_0 - \hat{\beta}^1_{\hat{S}_1} x_{\hat{S}_1} \tag{10.1}$$

where

p number of parameters;
n sample size;
$x_{\hat{S}_1}$ highest correlation variable with R_0 in $\mathbf{X}$;
$\hat{\beta}^1_{\hat{S}_1}$ regression coefficient of R_0 for $x_{\hat{S}_1}$.

Step 2: Choose another variable, $x_{\hat{S}_2}$, which is most correlated with $\hat{R}_1$ in $\mathbf{X}$. Then, the corresponding residual will be

$$\hat{R}_2 = R_0 - \hat{\beta}^2_{\hat{S}_1} x_{\hat{S}_1} - \hat{\beta}^2_{\hat{S}_2} x_{\hat{S}_2} \tag{10.2}$$

where $\hat{\beta}^2_{\hat{S}_2}$ is the regression coefficient of R_0 for $X_{\hat{S}_2}$.

Step 3: Return to Step 2 and repeat m times so as to stepwise choose $x_{\hat{S}_1}, x_{\hat{S}_2}, \ldots, x_{\hat{S}_m}$ and calculate the corresponding regression coefficients $(\hat{\beta}^m_{\hat{S}_1}, \hat{\beta}^m_{\hat{S}_2}, \ldots, \hat{\beta}^m_{\hat{S}_m})$. Then, the corresponding residual will be

$$\hat{R}_m = R_0 - \hat{\beta}^m_{\hat{S}_1} x_{\hat{S}_1} - \hat{\beta}^m_{\hat{S}_2} x_{\hat{S}_2} - \ldots - \hat{\beta}^m_{\hat{S}_m} x_{\hat{S}_m}. \tag{10.3}$$

However, it is hard to select exact m features to recover the entire original signal. Therefore, Ing and Lai proposed a termination condition, High-Dimensional Information Criterion (HDIC) [8], to choose along the OGA path that has the smallest value of a suitably chosen criterion. Let

$$\text{HDIC}(J) = n \log \hat{\vartheta}^2_J + \#(J) w \, log \, p \tag{10.4}$$

with $\hat{\vartheta}^2_J = \frac{1}{n} \sum_{i=1}^{n} \left(Y_i - \hat{Y}_{i;J}\right)^2$ where

J set of variables selected in the model $(x_{\hat{S}_1}, x_{\hat{S}_2}, \ldots, x_{\hat{S}_m})$;
#(.) number of the variable sets;
$\hat{\vartheta}^2_J$ mean square error of the corresponding model;
w general constant penalties > 0;
Y_i *ith* sample of actual value of final inspection;
$\hat{Y}_{i;J}$ *ith* sample of predictive value of final inspection corresponding to J.

Define $\hat{J}_m$ to be the set of all the variables selected after executing OGA+HDIC via (10.4), where m represents the number of variables selected in the model. However, there are still some irrelevant variables in $\hat{J}_m$. The method of removing irrelevant variables is needed.

Trimming is the method of rechecking all the variables selected in the model so as to remove those irrelevant ones. The trimming steps [8] are described below.

Step 1: Based on the concept of OGA+HDIC, define a subset $\hat{N}$ of $\hat{J}_m$, where $\hat{N}$ is the real relevant variables in the model.

Step 2: $\hat{N} = \left\{\hat{J}_l : HDIC\left(\hat{J}_m - \left\{\hat{J}_l\right\}\right) > HDIC\left(\hat{J}_m\right), 1 \leq l \leq m\right\}$ if $m > 1$. $\hat{N} = \left\{\hat{J}_1\right\}$ if $m = 1$.

Step 3: Repeat Step 2 for $m - 1$ times.

Finally, combining the processes of OGA, HDIC, and Trimming, the so-called TPOGA is formed.

10.2.2.2 Automated Least Absolute Shrinkage and Selection Operator (ALASSO)

In general regression models, coefficients are obtained with a least-squares method. Set the loss function as $||Y\text{-}XB||^2$. The penalty function f (λ, β) is added into the regularization regression, where λ being defined as penalty and β as regression coefficient. When the penalty function equals $\lambda||\beta||^2$, this method is called ridge regression; on the other hand, when the penalty function equals $\lambda||\beta||$, then the method is known as LASSO regression [9, 10]. In LASSO regression, the value of λ may affect the coefficients of the variables. In addition, if the coefficients equal 0, it means that these variables are not suggested of being selected into the model. The LASSO formulation is stated as follows [9, 10]:

$$\hat{\beta}^{lasso} = \underset{\beta}{argmin} \sum_{i=1}^{n} \left(y_i - \beta_0 - \sum_{j=1}^{p} x_{ij} \beta_j \right)^2 \tag{10.5}$$

$$\text{subject to} \sum_{j=1}^{p} \left|\beta_j\right| \leq t \tag{10.6}$$

where

- y_i final inspection result of the ith sample;
- x_{ij} jth parameter of the ith sample;
- n sample size;
- p number of parameters;
- t tuning parameter which restricts the sum of $|\beta_j|$;
- β_j regression coefficient of jth parameter;
- $\hat{\beta}^{lasso}$ optimized coefficients.

Equation (10.6) is the constraint that sets the sum of β absolute values to be less than t. With this constraint, residual sum of squares are minimized so as to make some regression coefficients be equal to 0 for obtaining an explanatory model.

To solve (10.5), combine (10.5) and (10.6) such that the LASSO problem is written in the equivalent Lagrangian form:

$$\hat{\beta}^{lasso} = \underset{\beta}{argmin} \left\{ \frac{1}{2} \sum_{i=1}^{n} \left(y_i - \beta_0 - \sum_{j=1}^{p} x_{ij} \beta_j \right)^2 + \lambda \sum_{j=1}^{p} \left|\beta_j\right| \right\} \tag{10.7}$$

The value of λ should be set before solving (10.7). The larger value of λ is set, the less number of key variables will be generated. If the value of a regression coefficient is zero shown in the solution, then it means that its corresponding variable is deleted.

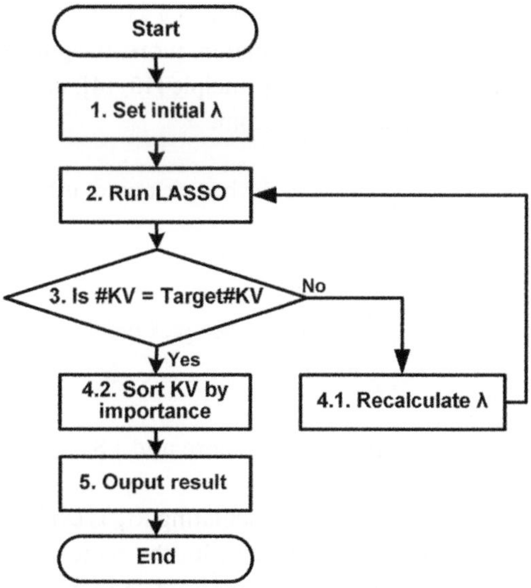

Figure 10.6 Flowchart of ALASSO with automated λ adjusting. λ: penalty; KV: key variables; #KV: number of KV; Target#KV: target number of KV. *Source:* Reproduced with permission from Ref. [11]; © 2017, IEEE.

To reduce the puzzle of setting λ and obtain proper results, the so-called Automated LASSO (ALASSO), which contains the automated λ adjusting method, is proposed. The flowchart of ALASSO is shown in Figure 10.6 and described below.

Step 1: The initial value of λ is set to be 5 by experience.

Step 2: Execute LASSO.

Step 3: Select a proper target number of key variables (Target#KV); for example, 10. Then, check that the number of key variables (#KV) selected by LASSO equals Target#KV or not. If no, go to Step 4.1 to recalculate λ; otherwise go to Step 4.2. The root causes of a yield loss should be focused. Therefore, Target#KV is set to avoid picking up too many unnecessary variables.

Step 4.1: With *Lowerbound* being defined as the lower bound of λ. The initial value of *Lowerbound* is 0 (because $\lambda \geq 0$). If #KV < Target#KV, $\lambda_{new} = (\lambda_{old} + \mathit{Lowerbound})/2$. On the other hand, if #KV > Target#KV, $\lambda_{new} = \lambda_{old} + (\lambda_{old} + \mathit{Lowerbound})/2$ and re-assign *Lowerbound* $= \lambda_{old}$. Finally, return to Step 2.

Step 4.2: Sort the KV by importance in descending order.

Step 5: Output the sorted KV as the result.

10.2.2.3 Reliance Index of KSA (RI_K) Module

RI_K is designed to evaluate the reliability of KSA results with reference to the reliance index (RI) concept of the AVM system [12]. By comparing the results of TPOGA and ALASSO and taking overlapping and weights into considerations, RI_K is re-calculated and set between 0 and1. With $RI_{KT} = 0.7$ being the threshold, if RI_K is greater than RI_{KT}, good search results are obtained; otherwise, the search results need to be re-examined. The calculation of RI_K is described below.

Suppose that up to 10 key variables are selected by both TPOGA and ALASSO. The order of each key variable being picked out is meaningful, thus the original scores of the first to tenth variables descend from 1 to 0.1 sequentially.

Since the sequential order of key variables searched by TPOGA and ALASSO is important as the variable chosen earlier matters more, weights are assigned not only according to the sequential order but also the 80–20 principle [13, 14], to ensure higher score goes to the crucial minority. As a result, the final scores become

$$\mathrm{FS}_l = \begin{cases} \dfrac{\mathrm{OS}_l}{\sum_{l=1}^{3}\mathrm{OS}_l} \times 0.8, \text{when pick order is } 1,2,3. \\ \dfrac{\mathrm{OS}_l}{\sum_{l=4}^{10}\mathrm{OS}_l} \times 0.2, \text{when pick order is } 4,5,\ldots,10. \end{cases} \tag{10.8}$$

where OS_l is the original score and FS_l is the final score with $l = 1, 2, \ldots, 10$ being the pick order.

The rule of thumb of calculating $\mathrm{RI_K}$ is explained below. If a certain variable has been chosen by both TPOGA and ALASSO with the same sequential order, the score of this variable is counted. Then, if a certain variable has been chosen by both TPOGA and ALASSO with different sequential orders, the score of this variable is also counted. Finally, if TPOGA and ALASSO pick out different variables without overlapping, the score of this variable is not counted. Therefore, $\mathrm{RI_K}$ of the search results is calculated by

$$\mathrm{RI_K} = \sum_{i=1}^{10}\sum_{j=1}^{10}\left(\frac{FS_{O_i} + FS_{L_j}}{2}\right) \text{if } O_i = L_j \tag{10.9}$$

where

$\mathrm{FS}_{\mathrm{O}i}$ final score of O_i;
$\mathrm{FS}_{\mathrm{L}j}$ final score of L_j;
O_i ith pick variable of TPOGA, $i = 1, 2, 3, \ldots, 10$;
L_j jth pick variable of ALASSO, $j = 1, 2, 3, \ldots, 10$.

The KSA Module defines data such as production tools and process data as the explanatory variables, and sets the value as 1 if the workpiece gets through and as 0 if it does not. It also adopts TPOGA and ALASSO algorithm to quickly locate the key devices that result in poor yield for yield enhancement.

However, under certain production restrictions, the KSA module may not be able to find the device that may be a key factor due to the characteristics of the algorithm. These potential key factors are defined as blind-stages. Therefore, in the next section, we will introduce the solution when encountering the blind-stages problem.

10.2.3 Blind-stage Search Algorithm (BSA) Module

When there is only one device in a certain production stage, all the explanatory variables become 1. It is discussed in [15], when all explanatory variables equal to 1, these variables are treated as intercept terms. As a result, TPOGA and ALASSO adopted in KSA cannot distinguish these intercept terms. In other words, if the root courses of a yield loss are embedded in these intercept terms, they cannot be identified. Moreover, both TPOGA and ALASSO cannot identify the problematic devices if they have the same defect density. These two cases mentioned above result in the so-called blind-stage problem. To solve the blind-stage problem, this section discusses a Blind-stage Search Algorithm (BSA) [16] to improve the performance of KSA.

10.2.3.1 Blind Cases

While analyzing the key stages that cause yield loss, two blind cases could occur in the KSA searching module, which might lead to the chance of missing out on the actual root causes of affecting yield during the search. These two blind cases are illustrated below.

Blind Case I

The first blind case happens when there is a production stage that contains only one device. In other words, all the workpieces shall pass through this device during a manufacturing process. As a result, neither TPOGA nor ALASSO can discover this device. The explanation is as follows.

Define

$$\mathbf{X_R} = \begin{bmatrix} x_{1R1} & \cdots x_{1Rj} \cdots & x_{1Rp} \\ \vdots & \ddots & \vdots \\ x_{nR1} & \cdots x_{nRj} \cdots & x_{nRp} \end{bmatrix} \tag{10.10}$$

$$\mathbf{Y} = \left[\mathrm{Y}_1 \ \ldots \ \mathrm{Y}_n \right]^{\mathrm{T}}. \tag{10.11}$$

In (10.10), $\mathbf{X_R}$ represents the production routes, p is the number of parameter which means total device number a workpiece may go through, and n is the number of samples. As for (10.11), $\mathbf{Y}$ means the final inspection data and again n is the number of samples.

In practical manufacturing, the situation of only one device being utilized for production in a certain stage of the whole production line may happen. Hence, set device j as the only device in the situation mentioned above, its $\mathbf{X_R}$ will be shown as in (10.12).

$$\mathbf{X_{Rj}} = \left[x_{1Rj} \ x_{2Rj} \cdots x_{nRj} \right]^{\mathrm{T}} = \left[1 \ 1 \ldots 1 \right]^{\mathrm{T}}. \tag{10.12}$$

As mention in Section 10.2.2.1, TPOGA defined $\mathrm{R}_0 = \mathbf{Y} = (\mathrm{Y}_1, \mathrm{Y}_2, \ldots, \mathrm{Y}_n)^{\mathrm{T}}$. Choose the variable defined as $x_{\hat{S}_1}$, which is the most relevant with R_0 in $\{\mathbf{X} = x_1, x_2, \ldots, x_p\}$. Following the algorithm of TPOGA and (10.1), TPOGA will keep searching for the $\mathbf{X_R}$ that has the highest correlation. $\hat{\beta}^1_{\hat{S}_1}$ is the least squares estimation value obtained with regression of $x_{\hat{S}_1}$ on R_0. Assuming the original $x_{\hat{S}_1}$ value as 1, after being centralized in the KSA preprocessing module, $x_{\hat{S}_1}$ becomes 0. As a result, $\hat{\beta}^1_{\hat{S}_1}$ cannot be derived by fitting this explanatory variable to the regression model. Thus this explanatory variable becomes an intercept term.

As for the case of ALASSO, the LASSO regression formulation is as (10.5). Assuming all x_{ij} equal 1 in (10.12), after being centralized in the KSA preprocessing module, all x_{ij} become 0. Similar to the TPOGA case mentioned above, β_j cannot be estimated. If the coefficients equal 0, it means that these variables are not suggested to be included in the model.

Blind Case II

The second blind case occurs when both TPOGA and ALASSO cannot find out the problematic devices if they have the same defect density. Utilization, which means the total sampling counts pass through a device, is calculated as in (10.13); and defect density, which indicates the abnormal product status that go through this device, is defined in (10.14).

$$\text{Utilization} = \sum_{i=1}^{n} x_{iRa} \tag{10.13}$$

$$\text{Defect Density} = \frac{\sum_{i=1}^{n} \left(x_{iRa} \times \mathrm{Y}_i \right)}{\mathit{Utilization}} \tag{10.14}$$

where x_{iRa} represents the ith sample of production route of the ath device.

Assuming there are only two devices (devices k and l) machining in the same process, if a workpiece is machined by device k, it will not go through device l; on the other hand, if device l picks up a workpiece for processing, then this workpiece will not pass through device k. Set the condition of passing through as 1, and as 0 if it does not. The $\mathbf{X_R}$ data are calculated as in (10.15) and (10.16), and the corresponding yield is derived as in (10.17).

$$\mathbf{X_{Rk}} = \left[x_{1Rk}\ x_{2Rk} \ldots x_{nRk}\right]^{\mathrm{T}} \tag{10.15}$$

$$\mathbf{X_{Rl}} = \left[x_{1Rl}\ x_{2Rl} \ldots x_{nRl}\right]^{\mathrm{T}} \tag{10.16}$$

$$\mathbf{Y} = \left[\mathrm{Y}_1 \quad \mathrm{Y}_2 \quad \ldots \quad \mathrm{Y}_n\right]^{\mathrm{T}}. \tag{10.17}$$

If $x_{iRk} = 1$, then $x_{iRl} = 0$, and if $x_{iRk} = 0$, then $x_{iRl} = 1$, $i = 1, 2, \ldots, n$.

where

x_{iRk} ith device of production route of kth device;
x_{iRl} ith device of production route of lth device.

The correlation coefficient, γ, is defined in (10.18).

$$\gamma = \frac{\sum_{i=1}^{n}\left(x_{iRa} - \bar{x}_{Ra}\right)\left(Y_i - \bar{\mathbf{Y}}\right)}{\sqrt{\sum_{i=1}^{n}\left(x_{iRa} - \bar{x}_{Ra}\right)^2}\sqrt{\sum_{i=1}^{n}\left(Y_i - \bar{\mathbf{Y}}\right)^2}} \tag{10.18}$$

where

$\bar{\mathrm{Y}}$ mean of the final inspection;
$\bar{x}_{\mathrm{R}_a}$ mean of production route of the ath device.

For regression analysis, it is possible to accurately observe the changes of the dependent variables by locking the important independent variables, which can be determined by the correlation coefficient values. TPOGA and ALASSO, the core algorithms of the KSA scheme, are similar to the concept of regression analysis. The reasons why TPOGA and ALASSO cannot distinguish the problematic devices if they have the same defect density are explained below.

At first, TPOGA is studied. TPOGA determines the key parameters according to the correlation of the devices on yield. Suppose that there are only two devices, k and l, in the same process stage and their defect densities are identical; also suppose that workpieces 1, 2, . . ., m are processed by device k, and workpieces $m + 1$, $m + 2$, . . ., n are machined by device l, then the numerator of (10.18) for $a = k$ becomes

$$\begin{aligned}
&\sum_{i=1}^{n}\left(x_{iRa} - \bar{x}_{Ra}\right)\left(\mathrm{Y}_i - \bar{\mathbf{Y}}\right) \\
&= \left(x_{1Rk} - \bar{x}_{Rk}\right)\left(\mathrm{Y}_1 - \bar{\mathbf{Y}}\right) + \left(x_{2Rk} - \bar{x}_{Rk}\right)\left(\mathrm{Y}_2 - \bar{\mathbf{Y}}\right) + \ldots + \left(x_{nRk} - \bar{x}_{Rk}\right)\left(\mathrm{Y}_n - \bar{\mathbf{Y}}\right) \\
&= \sum_{i=1}^{m}\left(x_{iRk} - \bar{x}_{Rk}\right)\left(\mathrm{Y}_i - \bar{\mathbf{Y}}\right) + \sum_{i=m+1}^{n}\left(x_{iRk} - \bar{x}_{Rk}\right)\left(\mathrm{Y}_i - \bar{\mathbf{Y}}\right) \\
&= m\left(x_{iRk} - \bar{x}_{Rk}\right)\sum_{i=1}^{m}\left(\mathrm{Y}_i - \bar{\mathbf{Y}}\right) + \left(n - m\right)\left(x_{iRk} - \bar{x}_{Rk}\right)\sum_{i=m+1}^{n}\left(\mathrm{Y}_i - \bar{\mathbf{Y}}\right).
\end{aligned}$$

If the defect densities of these two devices are the same, then

$$\frac{\sum_{i=1}^{n}\left(x_{iRk}\times Y_i\right)}{\text{Utilization}}=\frac{\sum_{i=1}^{n}\left(x_{iRl}\times Y_i\right)}{\text{Utilization}}.$$

Thus, the means of the final inspection of devices k and l are identical as shown below:

$$\frac{\sum_{i=1}^{m}Y_i}{m}=\frac{\sum_{i=m+1}^{n}Y_i}{n-m}=\bar{\mathbf{Y}}.$$

Due to the means of the final inspection of the devices k and l are the same, it can be inferred that the numerator in (10.18) is 0, that is:

$$\sum_{i=1}^{m}\left(Y_i-\bar{\mathbf{Y}}\right)=0,\ \sum_{i=m+1}^{n}\left(Y_i-\bar{\mathbf{Y}}\right)=0\Rightarrow\sum_{i=1}^{n}\left(x_{iRk}-\bar{x}_{Rk}\right)\left(Y_i-\bar{\mathbf{Y}}\right)=0.$$

By the same token, with $a=l$,

$$\sum_{i=1}^{n}\left(x_{iRl}-\bar{x}_{Rl}\right)\left(Y_i-\bar{\mathbf{Y}}\right)=0.$$

Therefore, when the defect densities of all the devices in the same process are the same, then $\gamma=0$ that is the reason why TPOGA cannot identify the problematic devices if they have the same defect density.

On the other hand, ALASSO selects the key parameters according to the β_j of the device on yield. β_j in (10.5) is calculated as in (10.19).

$$\beta_j=\frac{\sum_{i=1}^{n}\left(x_{ij}-\bar{x}_j\right)\left(Y_i-\bar{\mathbf{Y}}\right)}{\sum_{i=1}^{n}\left(x_{ij}-\bar{x}_j\right)^2}. \tag{10.19}$$

When there are only two devices processing in the same stage and their defect densities are the same, (10.19) for device k can be rewritten as below:

$$\beta_{Rk}=\frac{\sum_{i=1}^{n}\left(x_{iRk}-\bar{x}_{Rk}\right)\left(Y_i-\bar{\mathbf{Y}}\right)}{\sum_{i=1}^{n}\left(x_{iRk}-\bar{x}_{Rk}\right)^2}.$$

Then, after knowing that all the defect densities are the same, the numerator is referred to be 0, thus $\beta_{Rk}=0$. This verifies that ALASSO cannot identify the problematic devices if they have the same defect density, either.

In the following, hypothesis test is also applied to analyze the situation of several devices with the same defect density in a certain stage. The partial F-statistic is used for forward regression analysis in [17]. It also proved that two criteria produce the same selection results for the candidate set consisting of only binary attributes [5]. The advantage of the partial F-statistic is that the corresponding F-distribution from the calculated p-value can be used to indicate the significance of the parameter contribution. The partial F-statistic of a candidate $\mathbf{X_R}$ is calculated based on (10.20) and (10.21).

$$H_0: Y_i=\beta_0+\varepsilon_j \tag{10.20}$$

$$H_1 : Y_i = \beta_0 + \beta_j x_{ij} + \varepsilon_j \tag{10.21}$$

where

Y_i final inspection of the ith sample;
x_{ij} jth parameter of the ith sample;
β_0 intercept term;
β_j regression coefficient of jth parameter;
ε_j error term.

In regression analysis, if $F_{value} > 1$ (F_{value} is calculated as in (10.22)), the null hypotheses (10.20) are rejected, which means it is accepted, then it indicates that there is no significant impact on yield of workpieces going through this device. Therefore, the device cannot be identified.

$$F_{value}(x_j) = \frac{MSR(x_j)}{MSE(x_j)} = \frac{SSR(x_j)/1}{SSE(x_j)/n-2} = \frac{\sum_{i=1}^{n}(\hat{Y}_i - \bar{Y})^2/1}{\sum_{i=1}^{n}(Y_i - \hat{Y}_i)^2/n-2} \tag{10.22}$$

where $\hat{Y}_i$ represents the predictive final inspection of the ith sample.

Assuming Devices A and B have the same defect density in the same stage, and the F values by (10.22) are both < 1 after testing, then these two devices have similar yet not significant impact on the yield. Therefore, all the devices with the same defect density in a certain stage cannot be picked out by the KSA module.

As mentioned above, both TPOGA and ALASSO have the problems of Blind Cases I and II. To solve these problems, BSA is proposed and added into the KSA scheme, as shown in the bottom-right corner of Figure 10.5, and will be presented next.

10.2.3.2 Blind-stage Search Algorithm

As described in Section 10.2.3.1, when Blind Case I occurs, it means that there is a production process constitutes of only one device existing in a specific stage. Assuming that this device is the key factor affecting the final yield, the KSA module is not able to identify this device as the root cause due to the mathematical features of the TPOGA and ALASSO algorithms. Moreover, when Blind Case II happens, it indicates that several devices in a certain stage have the same defect density.

The main purpose of the BSA module is to remedy the KSA module's deficiencies by identifying the blind devices, and then adding more production and/or maintenance data of the blind devices to search for the possible yield-loss root causes embedded in these blind devices. The flow chart of the BSA module is shown in Figure 10.7, and the explanation is stated below.

In the beginning, data preprocessing and quality check must be performed to ensure the quality of the input data: production information ($\mathbf{X_R}$, $\mathbf{X_P}$, and $\boldsymbol{y}$), defect (D), and final inspection (**Y**). Then, the BSA process is divided into two phases: find out the blind devices of Cases I and II in Phase I and discover the possible yield-loss root causes embedded in these blind devices in Phase II.

A Phase I: Find Out the Blind Devices

Execute Rule I: The purpose of Rule I is to discover Blind Case I: all the workpieces go through a single device in a manufacturing process. Observing Figure 10.8, there are six steps for Rule I as explained below.

Step 1: Input an $n \times p$ production routes ($\mathbf{X_R}$) matrix, where n is the number of samples, and p is the number of devices, as shown in (10.10).

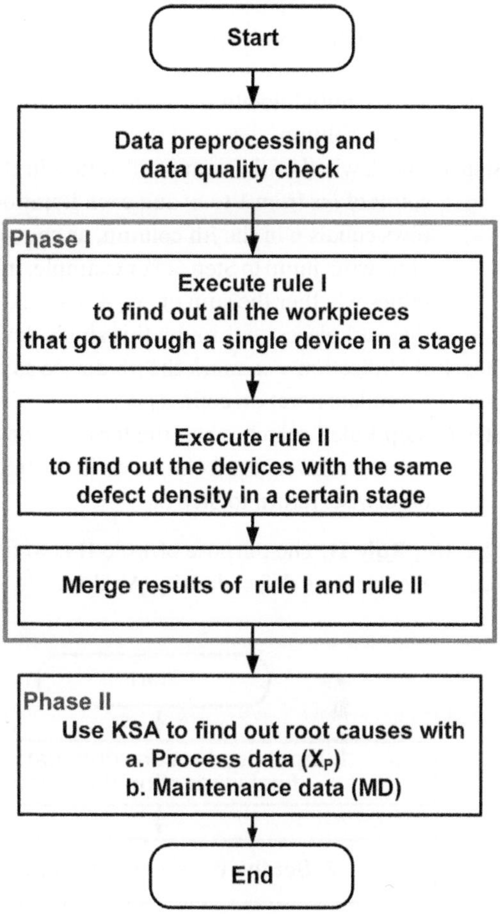

Figure 10.7 Flow chart of the BSA module. *Source:* Reproduced with permission from Ref. [16]; © 2017, IEEE.

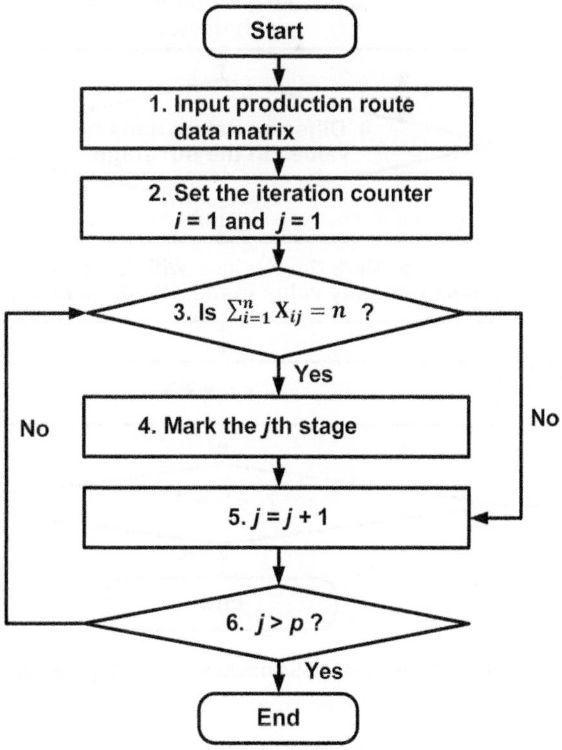

Figure 10.8 Rule I in the BSA module. *Source:* Reproduced with permission of Ref. [16]; © 2017, IEEE.

Step 2: Set the initial iteration counter $i = 1$ and $j = 1$, i is used to record the matrix row, and j for matrix column.

Step 3: Check whether the sum of all values in the jth column is equal to n. The value of iteration count of j is from 1 to n; and each iteration calculates to check if the sum of all values of n rows equals n in the jth column, as shown in (10.13). If Utilization $= n$, then enter Step 4; otherwise, jump to Step 5. For example, in the first column of the $\mathbf{X_R}$ matrix, this step determines whether the sum of $(x_{11} + x_{21} + \ldots + x_{n1})$ is equal to n, which is used to check if all the workpieces go through this single device.

Step 4: If Utilization $= n$, mark the jth device as a blind one, and then go to Step 5.

Step 5: Cumulate iteration counts $j = j + 1$.

Step 6: Stop Rule I search when the termination criterion ($j > p$) is satisfied (which means when the recorded matrix column variable amount is greater than the number of the devices); otherwise, return to Step 3.

Execute Rule II: The purpose of Rule II is to find out Blind Case II: several devices in a certain stage have the same defect density. As shown in Figure 10.9, there are seven steps for Rule II.

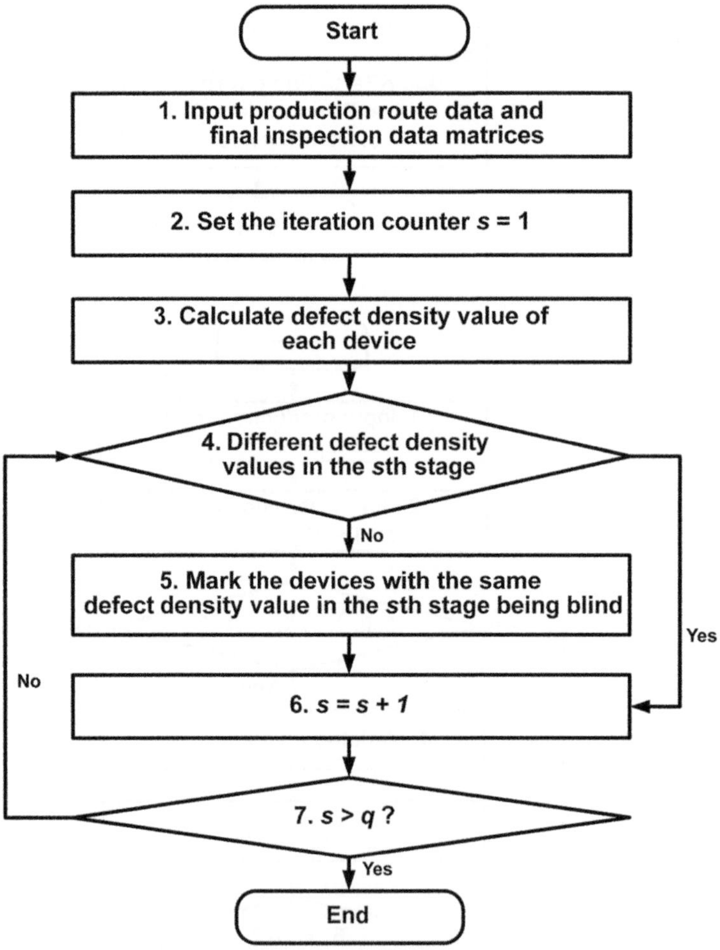

Figure 10.9 Rule II in the BSA module. *Source:* Reproduced with permission from Ref. [16]; © 2017, IEEE.

Step 1: Input an $n \times p$ production routes ($\mathbf{X_R}$) matrix and $n \times 1$ final inspection data ($\mathbf{Y}$) matrix ($\mathbf{Y} = [Y_1, Y_2, \ldots, Y_n]^T$), where n is the number of samples.
Step 2: Set the initial iteration count $s = 1$, s is the number of stages used to indicate which stage has been calculated.
Step 3: Calculate the defect density of each device, as shown in (10.14). Note that there may be more than one device utilized in each production stage.
Step 4: Check if the devices in the sth stage have the same defect density; if they do, then this stage has 0 correlation on yield (please refer to Section 10.2.3.1 for details). If there are devices of different defect densities under the stage, jump to Step 6; otherwise, go to Step 5.
Step 5: Mark the devices under the sth stage with the same defect density as blind.
Step 6: Cumulate iteration counts $s = s + 1$ to $s = q$. The total number of stages (q) is the number of all the required production stages.
Step 7: Stop the Rule II search when s is greater than q; otherwise, return to Step 4.

Merge Results of Rules I and II: Merge the results of Rules I & II. Rules I & II help to determine whether blind device(s) exist. To further check if the blind device(s) possess the root cause(s) affecting the final inspection, the second phase of BSA scheme is necessary.

B *Phase II: Check Whether a Blind Device Possesses the Root Cause of Yield Loss*

Consider the blind device's process data ($\mathbf{X_P}$) or maintenance data (MD) in the KSA scheme to determine whether this device is the key that affects the final inspection yield.

After introducing the BSA module, we will discuss the issue of interaction-effect in yield. When an interaction effect exists between a key device/variable and the other device/variable; also, the impact of this effect is greater than those impacts of the original devices/variables, the KSA module may not correctly ascertain the root causes. To remedy this insufficiency, the next section introduces the "Interaction-Effect Search Algorithm (IESA)" module [18].

10.2.4 Interaction-Effect Search Algorithm (IESA) Module

As manufacturing processes getting more and more complicated, the interactions between various devices are found to have certain impact on the yield [19, 20]. Considering key variables only may not be sufficient to characterize the interaction between the response and explanatory variables since its interaction effect between variables could also cause a yield loss. The interaction effect is elaborated below.

10.2.4.1 Interaction-Effect

When an interaction variable is added into the search, and the overall variation degree of its impact on the final yield decreases, the explanatory level of the final yield would increase. Also, when the variation reduction of the interaction variable is greater than that of the original variables, it means that the interaction effect has more impact on the yield than the original variables do.

As shown in (10.10), $\mathbf{X_R}$ represents the production routes, p is the number of variables (the maximum total device number a workpiece may go through), n is the number of samples in subset τ, and τ is the search space that may contain $\mathbf{X_R}$, $\mathbf{X_P}$, etc. As for (10.11), $\mathbf{Y}$ represents the final inspection data.

Assuming there are two devices (devices k and l) operating in the search space:

$$\mathbf{X_{Rk}} = \begin{bmatrix} x_{1Rk} & x_{2Rk} & \cdots & x_{nRk} \end{bmatrix}^T \tag{10.23}$$

$$\mathbf{X_{Rl}} = \begin{bmatrix} x_{1Rl} & x_{2Rl} & \cdots & x_{nRl} \end{bmatrix}^T. \tag{10.24}$$

Then a new variable $\mathbf{X_{Rkl}}$ can be derived from the interaction of $\mathbf{X_{Rk}}$ and $\mathbf{X_{Rl}}$ as in (10.25):

$$\begin{aligned}\mathbf{X_{Rkl}} &= \begin{bmatrix} x_{1Rk}\cdot x_{1Rl} & x_{2Rk}\cdot x_{2Rl} & \dots & x_{nRk}\cdot x_{nRl}\end{bmatrix}^{T} \\ &= \begin{bmatrix} x_{1Rkl} & x_{2Rkl} & \dots & x_{nRkl}\end{bmatrix}^{T}\end{aligned} \tag{10.25}$$

with x_{iRkl} being the interaction variable generated by devices k & l of the ith sample, where $i = 1, 2, 3, \dots, n$.

Define sample variance, $v(\tau)$, as in (10.26):

$$v(\tau) = \frac{1}{n}\sum_{i\in\tau}\left(Y_i - \bar{Y}(\tau)\right)^2 \tag{10.26}$$

with

v sample variance;
Y_i final inspection of the ith sample;
$\bar{Y}(\tau)$ sample mean of the final inspection in subset τ.

Assuming interaction effect $\mathbf{X_{Rkl}}$ is selected in the search space with its variation reduction $\Delta v(\mathbf{X_{Rkl}})$being:

$$\Delta v\left(\mathbf{X_{Rkl}}\right) = v(\tau) - \frac{n_0}{n}\times v\left(\tau | x_{iRkl} = 0\right) - \frac{n_1}{n}\times v\left(\tau | x_{iRkl} = 1\right) \tag{10.27}$$

with

n_0 number of samples that do not pass through the interaction variable;
n_1 number of samples that pass through the interaction variable.

Then, the explanatory variance of $\mathbf{X_{Rkl}}$ can be calculated as shown below from the perspective of regression analysis:

$$\begin{aligned} SSR\left(\mathbf{X_{Rkl}}\right) &= SST - SSE\left(\mathbf{X_{Rkl}}\right) \\ &= \sum_{i\in\tau}\left(Y_i - \bar{Y}(\tau)\right)^2 - \sum_{i\in\tau}\left(Y_i - \hat{Y}_i\right)^2 \\ &= \sum_{i\in\tau}\left(Y_i - \bar{Y}(\tau)\right)^2 - \left[\left(\sum_{i\in\tau}\left(Y_i - \bar{Y}_0\right)^2 | x_{iRkl} = 0\right) + \left(\sum_{i\in\tau}\left(Y_i - \bar{Y}_1\right)^2 | x_{iRkl} = 1\right)\right] \\ &= n\left[\frac{\sum_{i\in\tau}\left(Y_i - \bar{Y}(\tau)\right)^2}{n} - \frac{n_0\left(\sum_{i\in\tau}\left(Y_i - \bar{Y}_0\right)^2 | x_{iRkl} = 0\right)}{n\cdot n_0} - \frac{n_1\left(\sum_{i\in\tau}\left(Y_i - \bar{Y}_1\right)^2 | x_{iRkl} = 1\right)}{n\cdot n_1}\right] \\ &= n\left[v(\tau) - \frac{n_0}{n}\times v\left(\tau | x_{iRkl} = 0\right) - \frac{n_1}{n}\times v\left(\tau | x_{iRkl} = 1\right)\right] \\ &= n\cdot\Delta v\left(\mathbf{X_{Rkl}}\right)\end{aligned}$$

with

SSR sum of squared residuals;
SST total sum of squares;
SSE sum of squared errors;
$\hat{Y}_i$ least squared estimate of Y_i;

$\bar{Y}_0$ sample mean of the final inspection that does not pass through the interaction variable;
$\bar{Y}_1$ sample mean of the final inspection that passes through the interaction variable.

When the explanatory variance of $\mathbf{X_{Rkl}}$ is greater than those of $\mathbf{X_{Rk}}$ or $\mathbf{X_{Rl}}$ [as expressed in (10.28)] then $\mathbf{X_{Rkl}}$'s corresponding F-value, the impact of interaction variable on the yield, is greater than those of the original devices themselves, $\mathbf{X_{Rk}}$ or $\mathbf{X_{Rl}}$, [as expressed in (10.36)]. This statement is verified below:

Let

$$\Delta v(\mathbf{X_{Rkl}}) > \Delta v(\mathbf{X_{Rk}}) \tag{10.28}$$

Minus $\Delta v(\mathbf{X_{Rk}})$ on both sides of (10.28), to obtain

$$\Delta v(\mathbf{X_{Rkl}}) - \Delta v(\mathbf{X_{Rk}}) > 0. \tag{10.29}$$

With the regression analysis perspective, (10.29) is converted as

$$SSR(\mathbf{X_{Rkl}}) - SSR(\mathbf{X_{Rk}}) > 0. \tag{10.30}$$

Then, $SSR(\mathbf{X_{Rkl}})$ and $SSR(\mathbf{X_{Rk}})$ multiple SST respectively and positive and negative mathematical terms are added as well:

$$SSR(\mathbf{X_{Rkl}})SST - SSR(\mathbf{X_{Rkl}})SSR(\mathbf{X_{Rk}}) - SSR(\mathbf{X_{Rk}})SST + SSR(\mathbf{X_{Rkl}})SSR(\mathbf{X_{Rk}}) > 0. \tag{10.31}$$

According to the distribution law, (10.31) can be express as

$$SSR(\mathbf{X_{Rkl}})\left[SST - SSR(\mathbf{X_{Rk}})\right] - SSR(\mathbf{X_{Rk}})\left[SST - SSR(\mathbf{X_{Rkl}})\right] > 0. \tag{10.32}$$

Divide (10.32) by$[SST - SSR(\mathbf{X_{Rk}})] \cdot [SST - SSR(\mathbf{X_{Rkl}})]$ yielding

$$\frac{SSR(\mathbf{X_{Rkl}})}{SST - SSR(\mathbf{X_{Rkl}})} - \frac{SSR(\mathbf{X_{Rk}})}{SST - SSR(\mathbf{X_{Rk}})} > 0. \tag{10.33}$$

Divide the denominator in (10.33) by $(n-2)$:

$$\frac{SSR(\mathbf{X_{Rkl}})}{\dfrac{SSE(\mathbf{X_{Rkl}})}{(n-2)}} - \frac{SSR(\mathbf{X_{Rk}})}{\dfrac{SSE(\mathbf{X_{Rk}})}{(n-2)}} > 0. \tag{10.34}$$

(10.34) can then be express as follows via the F-value formula:

$$F_{value}(\mathbf{X_{Rkl}}) - F_{value}(\mathbf{X_{Rk}}) > 0. \tag{10.35}$$

Add $F_{value}(\mathbf{X_{Rk}})$ to both sides of (10.35) for derivation of the desired conclusion:

$$F_{value}(\mathbf{X_{Rkl}}) > F_{value}(\mathbf{X_{Rk}}). \tag{10.36}$$

After the interaction effect is discovered, the associated threshold of the key variable that causes the interaction effect should be identified such that the corresponding key devices should be tuned to avoid the same yield loss in the subsequent processes. The methods for discovering an interaction effect and identifying the associated threshold will be included in the proposed IESA module that is elaborated below.

10.2.4.2 Interaction-Effect Search Algorithm

To remedy the KSA module's deficiency of dealing with the interaction effect, the IESA module is added into the original KSA scheme as shown in the bottom-right corner of Figure 10.5. The IESA module mainly contains two phases. Phase I identifies the existence of an interaction effect; and Phase II determines the threshold of the associated key variable that causes the interaction effect. Finally, the key variable is tuned to be within the threshold for avoiding the interaction effect.

A *Phase I: Identifying Interaction Effect*

To double-check the reliability of search result, the IESA module contains two algorithms: Regression Tree and TPOGA. IESA then make a comparison on the results of these two different algorithms and provide a reliance index (RI_I) of the IESA result. They are described below.

- **Regression Tree**

The Regression Tree is a decision tree with continuous variables serving as the response. Usually, a variable is chosen as the decision variable by the root node and the samples are divided by this decision variable. In the decision narration, the samples that pass through the jth node ($x_{\hat{R}_j} = 1$) are assigned to the right sub-node; the others ($x_{\hat{R}_j} = 0$) are then assigned to the left sub-node. The samples are separated into the sub-node on both sides with the same procedure, and then new variables are selected as the decision variables respectively. In this way, the samples will be rapidly consumed with the lower sub-node level. Therefore, to avoid quick consumption of samples leading to the failure of the most relevant variable selection, this work proposes to use the most correlated interpretation variables as the decision variables. The interaction between each key node is then utilized to check if the interaction F_{value} is greater than the parent node's F_{value}. There are six steps for the Regression Tree as shown in Figure 10.10.

Step 1: Define $\mathbf{Y} = (Y_1, Y_2, \ldots, Y_n)^T$. Choose the parent-node variable defined as $x_{\hat{R}_1}$, which is the most correlated with $\mathbf{Y}$ in $\mathbf{X} = \{x1, x2, \ldots, x_p\}$, and then its F_{value} (as expressed in 10.22), the parent node's impact degree on $\mathbf{Y}$, is calculated below:

$$F_{value}\left(x_{\hat{R}_j}\right) = \frac{SSR\left(x_{\hat{R}_j}\right)}{\frac{SSE\left(x_{\hat{R}_j}\right)}{(n-2)}}$$

where

$x_{\hat{R}_j}$ is the jth node of Regression Tree.

Then, take the parent node $x_{\hat{R}_1}$ as the decision variable for sample segmentation to two sub-nodes. The samples go through $x_{\hat{R}_1}$ are denoted as 1 and put into the right sub-node, while the samples do not go through the parent node would be denoted as 0 and go to the left sub-node, as depicted in Step 1 of Figure 10.11.

Step 2: After the sample segmentation is done, select the key variables that are the most correlated with $\mathbf{Y}$ in $\mathbf{X} = \{x_1, x_2, \ldots, x_p\}$ from both sub-nodes as the decision variables, which are denoted as $x_{\hat{R}_2}$ and $x_{\hat{R}_3}$, respectively, as shown in Step 2 of Figure 10.11.

Step 3: Check the interaction between the parent node $x_{\hat{R}_1}$ and the sub-nodes $x_{\hat{R}_2}$ and $x_{\hat{R}_3}$ individually with partial-F test, and match them to the n response variables $(Y_1, Y_2, \ldots, Y_n)^T$ as below. (Note that, since the factors of minor effects usually will not have significant interactions [22], only the interactions with strong-heredity influence between the parent node and sub-nodes of the adjacent layers are checked here.)

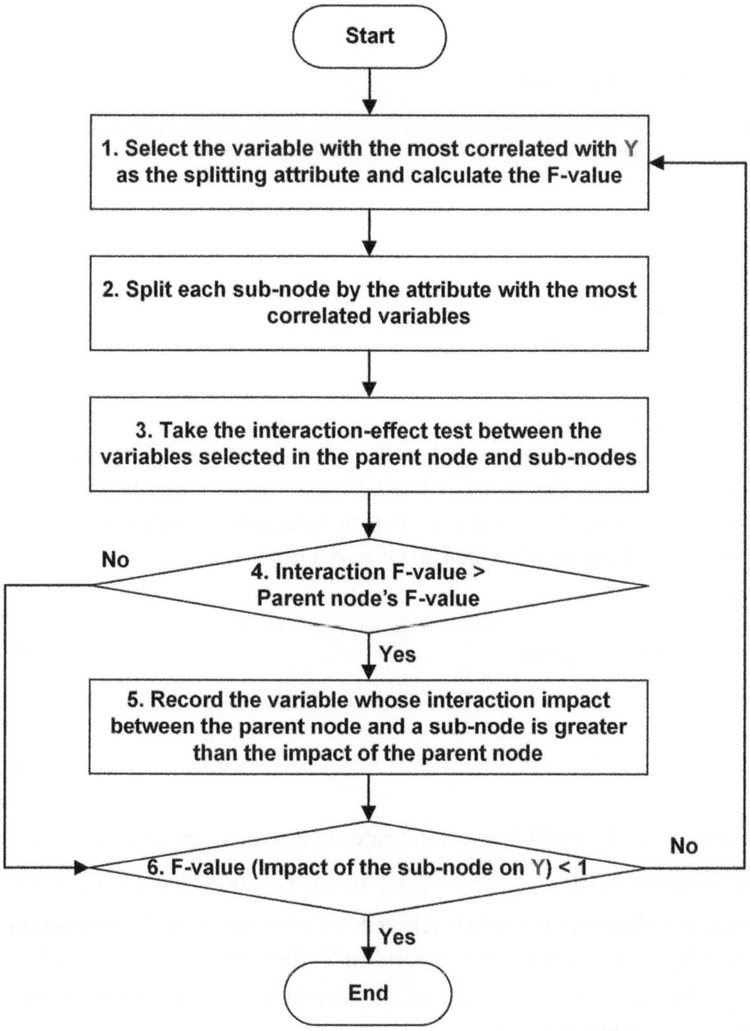

Figure 10.10 Flow Chart of the Regression Tree. *Source:* Reproduced with permission from Ref. [18]; © 2018, IEEE.

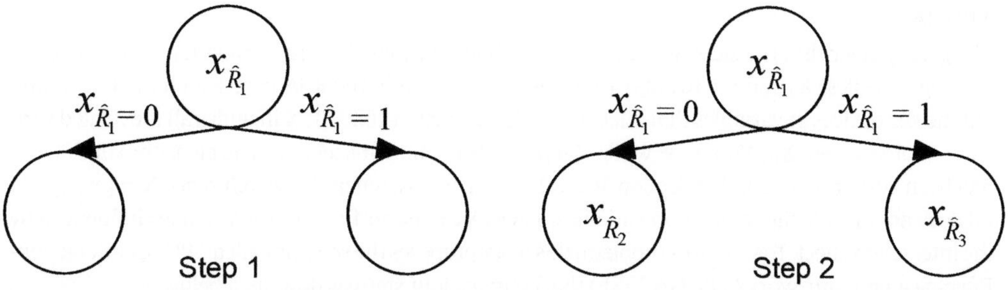

Figure 10.11 Description of Regression Tree Step 1 and Step 2. *Source:* Reproduced with permission from Ref. [21]; © 2018, IEEE.

In the left sub-node:

$$\hat{Y}_i = \beta_0 + \beta_1 x_{\hat{R}_1} + \beta_2 x_{\hat{R}_2} + \beta_{1,2} x_{\hat{R}_1} x_{\hat{R}_2} + \varepsilon \tag{10.37}$$

$$H_0 : \beta_{1,2} = 0 \tag{10.38}$$

$$H_1 : \beta_{1,2} \neq 0 \tag{10.39}$$

with

Y_i final inspection of the ith sample;
β_j regression coefficient of the jth variable;
ε error term.

Referring to regression analysis [22], if $F_{1,2}$-value < 1, do not reject H_0 in (10.38) as it indicates that when $x_{\hat{R}_1} = 1$ or $x_{\hat{R}_1} = 0$, the impact resulted from the interaction between these two variables makes no difference. On the other hand, if $F_{1,2}$-value is larger than or equal to 1, then H_0 in (10.38) is rejected; which means that the interaction between these two variables will have different impacts on **Y** when given $x_{\hat{R}_1} = 1$ or $x_{\hat{R}_1} = 0$.
In the right sub-node:

$$\hat{Y}_i = \beta_0 + \beta_1 x_{\hat{R}_1} + \beta_3 x_{\hat{R}_3} + \beta_{1,3} x_{\hat{R}_1} x_{\hat{R}_3} + \varepsilon \tag{10.40}$$

$$H_0 : \beta_{1,3} = 0 \tag{10.41}$$

$$H_1 : \beta_{1,3} \neq 0. \tag{10.42}$$

If $F_{1,3}$-value <1, do not reject H_0 in (10.41) as it indicates that when $x_{\hat{R}_1} = 1$ or $x_{\hat{R}_1} = 0$, the interaction between these two variables makes no difference. On the other hand, if $F_{1,2}$-value is larger than or equal to 1, then H_0 in (10.41) is rejected; which means that the interaction between these two variables will have different impacts on **Y** when given $x_{\hat{R}_1} = 1$ or $x_{\hat{R}_1} = 0$.

Step 4: If the interaction-F_{value} between the parent node and a sub-node is greater than the parent node, then go to Step 5; else jump to Step 6.
Step 5: Record the variable whose interaction impact between the parent node and a sub-node is greater than the impact of the parent node.
Step 6: If the F_{value}, the impact of the sub-node on **Y**, is smaller than 1 (meaning the impact of the sub-node on **Y** is not significant [21]), then stop searching; else jump to Step 1.

- **TPOGA**
  The greedy algorithm is a stepwise regression method that considers the correlation between all the causing variables (**X**) and the results (**Y**). TPOGA is adopted in the KSA core search module to find out the key devices or variables that affect the yield. In Section 10.2.2.1, **X** includes all the related variables of production: $\mathbf{X_R}$, $\mathbf{X_P}$, and $\boldsymbol{y}$; while **Y** represents the final inspection. The algorithm of TPOGA had been described in detail in Section 10.2.2.1. Nevertheless, when the search space $\mathbf{X} = \{x_1, x_2, \ldots, x_p\}$ does not include the interaction variables, TPOGA can never find out the variables influenced by the interaction effect. To solve this problem, this work proposes the scheme called TPOGA Interaction-Effect Search Framework (TPOGA IESF) that contains four steps as described below.
  Step 1: Activate the TPOGA IESF when a yield loss is discovered.
  Step 2: Apply the original KSA to find top ***N*** variables that cause the yield loss.

Step 3: After finding out the top **N** variables that might affect the yield, add interaction variables with the strong heredity condition into the search space $\mathbf{X} = \{x_1, x_2, \ldots, x_p\}$ and re-run KSA. Then utilize TPOGA to search again the new generated interaction variable $\mathbf{X}_{\hat{s}_m \times \hat{s}_q}$ as follows:

$$\mathbf{X}_{\hat{s}_m \times \hat{s}_q} = \left[x_{1\hat{s}_m} \cdot x_{1\hat{s}_q} \quad x_{2\hat{s}_m} \cdot x_{2\hat{s}_q} \quad \cdots \quad x_{n\hat{s}_m} \cdot x_{n\hat{s}_q} \right]^T \tag{10.43}$$

with $m = 1, 2, \ldots \mathbf{N}$, $q = 1, 2, \ldots, \mathbf{N}$, $m \neq q$.

The strong heredity principle is stated below. Only high-impact variables corresponding to high-impact effect are added into the model, as the variables with less impact tend to have minor interaction effect [22]. Please refer to the illustrative example presented in Section 11.4.3.2 for the demonstration of adding new interaction-effect variables.

Step 4: Check if there is an interaction variable that has greater influence than those of the original KSA results. If this interaction variable exists, then those two devices that cause this interaction effect should be tuned such that the effect of this interaction will not happen again.

- **$\mathrm{RI_I}$ Module**

$\mathrm{RI_I}$ is designed to evaluate the reliability of IESA results with reference to the reliance index ($\mathrm{RI_K}$) concept, described in Section 10.2.2.3, of the KSA scheme as shown in Figure 10.5. By comparing the results of Regression Tree and TPOGA and taking overlapping and weights into considerations, $\mathrm{RI_I}$ is in between 0 and 1.

Referring to the concept of statistical correlation, when correlation coefficient > 0.7, it means that the two variables are highly correlated [23]. By the same token, it is defined that when $\mathrm{RI_I} > 0.7$, the search results of IESA are reliable; otherwise, the results need further verification. The algorithm of $\mathrm{RI_I}$ is described below.

IESA select Top 10 key variables by both Regression Tree and TPOGA, and the sequence of these variables indicates different levels of influence. Unlike KSA $\mathrm{RI_K}$, Regression Tree and TPOGA find out the key variables with **Y** correlation. Thus, this study proposes to sort out the original scores of Top 10 to Top 1 variables according to their weights of **Y** correlation with ascending order as shown below:

$$\mathrm{OS}_l = \frac{C_l}{\sum_{t=1}^{10} C_l} \tag{10.44}$$

where OS_l is the original score with $l = 1, 2, \ldots, 10$ being the pick order. C_l is the lth correlation variable with **Y**.

Since the sequential order of the key variables searched by Regression Tree and TPOGA is important, weights are assigned according to not only the sequential order but also the 80–20 principle, to ensure higher score goes to the crucial minority. As a result, the final scores become

$$FS_l = \begin{cases} \dfrac{\mathrm{OS}_l}{\sum_{l=1}^{3} \mathrm{OS}_l} \times 0.8, \text{ when pick order is } 1, 2, 3 \\ \dfrac{\mathrm{OS}_l}{\sum_{l=4}^{10} \mathrm{OS}_l} \times 0.2, \text{ when pick order is } 4, 5, \ldots, 10 \end{cases} \tag{10.45}$$

where FS_l is the final score with $l = 1, 2, \ldots, 10$ being the pick order.

If Regression Tree and TPOGA pick out the same sequential order, the score of this variable is counted. On the other hand, when Regression Tree and TPOGA choose different variables without overlapping, the score of this variable is not counted. Therefore, RI_I of the search results is calculated as

$$RI_I = \sum_{i=1}^{10}\sum_{j=1}^{10}\left(\frac{FS_{Ri} + FS_{Oj}}{2}\right) \text{if } R_i = O_j \tag{10.46}$$

with

FS_{Ri} final score of R_i;
FS_{Oj} final score of O_j;
R_i ith pick variable of Regression Tree, $i = 1, 2, \ldots, 10$;
O_j jth pick variable of TPOGA, $j = 1, 2, \ldots, 10$.

B *Phase II: Determining the Threshold of the Key Variable*
After discovering the existence of an interaction effect, the threshold of the associated key variable that causes the interaction effect should be determined such that the key variable can be tuned to be within the threshold for avoiding the interaction effect of the subsequent processes. To determine the threshold, the selection method of a continuous variables [21] is proposed below.

1) **Threshold of the Key Variable with Regression Tree**
   When Regression Tree selects a continuous variable as a decision node, the decision node cannot utilize narration of whether to pass through or not for sample segmentation ($x_{\hat{R}_j} = 1$ or 0). This continuous variable should be converted into binary variables for sample segmentation. Assuming there are n sample data and p independent variables $\mathbf{X} = \{x_1, x_2, \ldots, x_p\}$, where the wth independent variable (x_w) has n different continuous values, they can be reorganized into a sequence $\{t_1, t_2, \ldots, t_n\}$. The values greater than or equal to the threshold are denoted as 1, and the ones smaller than the threshold as 0. The variable conversion criteria are

$$x_{w(\geq t_i)} = \begin{cases} 1, & when\, x_w \geq t_i \\ 0, & otherwise \end{cases}, 1 \leq t_i \leq n \tag{10.47}$$

   where t_i is the ith threshold of x_w. A matrix of $n \times n$ will be derived after the variable conversion.
   Regression Tree uses the most correlated $x_{w(\geq t_i)}$ in $\{x_{w(\geq t_1)}, x_{w(\geq t_2)}, \ldots, x_{w(\geq t_n)}\}$ with $\mathbf{Y}$ to search for a decision variable and keep segmenting to find out the sub-nodes. If the threshold value found out by Regression Tree is $x_{w(\geq t_i)}$ which means that with the condition of $x_w \geq t_i$, this wth independent variable will have significant impact on the yield.
2) **Threshold of the Key Variable with TPOGA**
   When TPOGA is applied, TPOGA searches for the most R_0-correlated variable $x_{\hat{S}_1}$ in $\mathbf{X_T} = \left\{x_{w(\geq t_1)}, x_{w(\geq t_2)}, \ldots, x_{w(\geq t_n)}\right\}$, and find out the value that might affect the yield with the most residual after several iterations. The same selection method utilized in the approach of Regression Tree is also adopted here. If the search result is $x_{w(\geq t_i)}$; it means that $x_w \geq t_i$ has significant impact on the yield.

10.3 IYM System Deployment

After elaborating the theoretical illustrations of the IYM system, this section introduces the deployment considerations and requirements of the IYM system. Five steps are required to deploy the IYM system into manufacturing processes. They are (i) operation analysis, (ii) IYM system setup, (iii) data collection, (iv) IYM function and integration tests, and (v) system release.

Step 1: Operation Analysis

The purpose of IYM is to find the root causes of yield loss and to improve the yield. From the viewpoint of root causes analysis, all data generated during the manufacturing process from various sources that may possibly cause a yield loss need to be collected by IYM. Therefore, the first step of the IYM system deployment is to identify all the production routes ($\mathbf{X}_R$), manufacturing process data ($\mathbf{X}_P$), and in-line metrology values ($\boldsymbol{y}$), as well as the final yield inspection data ($\mathbf{Y}$) needed to be collected.

Besides data collection, the operation analysis of applying the IYM system for yield management operations needs to be defined. The AS-IS operation flows of yield management without IYM are documented first; then, the TO-BE operation flows with IYM should be planned as shown in Figure 10.4. For example, the IYM system needs to be integrated with the FDC system such that the FDC threshold of the key process data can be set automatically as depicted in Phase C of Figure 10.4.

Step 2: IYM System Setup

As mentioned in Chapter 6, the novel iFA system platform is developed by integrating technologies such as the AMCoT framework with Cyber-Physical Agents (CPAs), AVM, Intelligent Predictive Maintenance (IPM), and IYM, etc., so as to provide the total-solution package for implementing Intelligent Manufacturing. Therefore, the IYM system is deployed via the iFA system platform. There are two options of the iFA system platform for the users to choose from. "Cloud-based version" shown in Figure 6.5 and "Server-based version" depicted in Figure 6.6 are provided according to different requirements.

In the step of operation analysis, all the data as well as the operation flows and data exchange requirements are defined, then the gap analysis of the IYM system standard version will be performed by the project team. As a result, the scope and schedule of the IYM system's customization will be addressed. After obtaining the agreement among users and the project team, the installation of the iFA system platform will start. At the same time, the hardware of servers and communication networks need to be prepared. Referring to Figure 10.3, the IYM module of the iFA system platform consists of the KSA module and the Reporting module, as well as many CPAs. The detailed specifications of data-collection servers need to be confirmed by taking data exchange requirements into consideration. The integration system tests will then be performed accordingly when the software coding tasks and hardware configuration are completed.

Step 3: Data Collection

As shown in Figures 6.5, 6.6, and 10.3, CPAs are adopted by the factory side to perform data collection. CPA offers several Equipment Drivers for various industries including SECS/GEM and Interface A for semiconductor and TFT-LCD industry, as well as OPC-UA and MT-Connect for machine-tool industry, etc.

Step 4: IYM Function and Integration Tests

Prior to releasing the IYM system for production run, the IYM function and integration tests should be performed based on the TO-BE operation flow defined in Step 1 and Figure 10.4. The IYM system needs to be integrated with the MES and FDC systems as well as those systems that provide manufacturing big data needed for KSA analysis. The integration tests of IYM can just follow the Phases A, B, and C defined in Section 10.1.3 and Figure 10.4. A real example of deploying the IYM system into the bumping Process of semiconductor manufacturing is illustrated in Section 11.4.3.

Step 5: System Release

After all the functions and interfaces have been validated, the project team will move on to the next step: System Release. The reliability of the IYM system should be able to run 24 hours a day and 365 days a year. To verify the stability and reliability of the IYM system, its mean time between failures (MTBF) and mean time to repair (MTTR) are adopted as the acceptance criteria.

Besides the IYM system's function and accuracy validations, the related staffs of factory also need to practice the TO-BE operation flow. For example, what action should be taken and verified once integration engineers receive the results from the IYM system.

10.4 Conclusion

This chapter introduces the traditional root-cause search procedure of a yield loss. Then, a more systematic and efficient tool: the IYM system for finding the root causes of a yield loss is presented. The kernel of the IYM system: the KSA scheme is then elaborated. The KSA scheme possesses the KSA, BSA, and IESA modules. Finally, the procedure of IYM system deployment is explained. In Regards to the IYM application cases of various industries, they will be presented in Chapter 11.

Appendix 10.A – Abbreviation List

ALASSO	Automated Least Absolute Shrinkage and Selection Operator
AMCoT	Advanced Manufacturing Cloud of Things
AVM	Automatic Virtual Metrology
BDA	Big Data Analytics
BSA	Blind-stage Search Algorithm
CART	Classification and Regression trees
CC	Cloud Computing
CP	Chip-Probing
CPA_S	Cyber-Physical Agents
DT	Decision Tree
FDC	Fault Detection and Classification
HDIC	High-Dimensional Information Criterion

IESA	Interaction-Effect Search Algorithm
IPM	Intelligent Predictive Maintenance
IYM	Intelligent Yield Management system
KSA	Key-variable Search Algorithm
LASSO	Least Absolute Shrinkage and Selection Operator
MTBF	Mean Time Between Failures
MTTR	Mean Time to Repair
MES	Manufacturing Execution System
MD	Maintenance Data
OGA	Orthogonal Greedy Algorithm
PGA	Pure Greedy Algorithm
RI	Reliance Index
RST	Rough Set Theory
SERT	Sample-Efficient Regression Trees
TFT-LCD	Thin Film Transistor-Liquid Crystal Display
TPOGA	Triple Phase Orthogonal Greedy Algorithm
TPOGA IESF	TPOGA Interaction-Effect Search Framework

Appendix 10.B – List of Symbols in Equations

$\mathbf{Y}$	final inspection data
$\mathbf{X}$	related variables of production
$\mathbf{X_R}$	production routes
$\mathbf{X_P}$	process data
$\boldsymbol{y}$	total-inspection inline data
$\mathbf{D}$	defect data
KS_O	output of TPOGA in KSA module
KS_L	output of ALASSO in KSA module
RI_K	value of KSA reliance index
IE_O	output of TPOGA in IESA module
IE_T	output of regression tree in IESA module
RI_I	value of IESA reliance index
BS	output of blind stage index
P	number of parameters

n	sample size
$x_{\hat{S}_1}$	highest correlation variable with R_0 in **X**
x_{iRk}	*ith* sample of production route of *kth* device
x_{iRl}	*ith* sample of production route of *lth* device
x_{iRa}	*ith* sample of production route of the *ath* device
x_{iRkl}	being the interaction variable generated by devices k & l of the *ith* sample, where $i = 1, \ldots, n$.
$x_{iR\theta}$	*ith* sample of production route of θth device θ, where $\theta = 1, \ldots, p$ and $i = 1, \ldots, n$
x_{ij}	*jth* parameter of the *ith* sample
$\overline{x}_{Ra}$	mean of production route of the *ath* device
$\mathbf{X}_{\hat{S}_m \times \hat{S}_q}$	interaction variable of TPOGA, where $m = 1, 2, \ldots \mathbf{N}$, $q = 1, 2, \ldots, \mathbf{N}$, $m \neq q$.
$x_{\hat{R}_j}$	*jth* node of regression tree
x_w	*wth* independent variable, where $w = 1, 2, \ldots, p$.
$\hat{\beta}^1_{\hat{S}_1}$	regression coefficient of R_0 for $x_{\hat{S}_1}$
β_j	regression coefficient of *jth* parameter
β_0	intercept term
$\hat{\beta}^{lasso}$	optimized coefficients of ALASSO
y_i	*ith* sample of actual value of final inspection
$\hat{y}_{i;J}$	*ith* sample of predictive value of final inspection corresponding to J
$\overline{\mathbf{Y}}$	mean of the final inspection
Y_i	final inspection of the *ith* sample
$\overline{Y}_0$	sample mean of the final inspection that does not pass through the interaction variable
$\overline{Y}_1$	sample mean of the final inspection that passes through the interaction variable
$\hat{Y}_i$	predictive final inspection of the *ith* sample
$\overline{\mathbf{Y}}(\tau)$	sample mean of the final inspection in subset τ
t	tuning parameter which restricts the sum of $\lvert\beta j\rvert$
t_i	*ith* threshold of x_w
ε_j	error term
J	set of variables selected in the model $(x_{\hat{s}_1}, x_{\hat{s}_2}, \ldots, x_{\hat{s}_m})$
$\hat{\vartheta}^2_J$	mean square error of the corresponding model
w	general constant penalties > 0
FS_{Oi}	final score of O_i in KSA module
FS_{Lj}	final score of L_j in KSA module
O_i	*ith* pick variable of TPOGA in KSA module, $i = 1, 2, 3, \ldots, 10$
L_j	*jth* pick variable of ALASSO in KSA module, $j = 1, 2, 3, \ldots, 10$
ν	sample variance

n_0	number of samples that do not pass through the interaction variable
n_1	number of samples that pass through the interaction variable
SSR	sum of squared residuals
SSE	sum of squared errors
SST	total sum of squares
OS_l	original score with $l = 1, 2, \ldots, 10$ being the pick order in IESA Module
C	*ith* correlation variable with **Y** in IESA module
FS_l	the final score with $l = 1, 2, \ldots, 10$ being the pick order in IESA module
$\mathrm{FS}_{\mathrm{R}i}$	final score of Ri in IESA module
$\mathrm{FS}_{\mathrm{O}j}$	final score of Oj in IESA module
R_i	*ith* pick variable of regression tree in IESA module, $i = 1, 2, \ldots, 10$
O_j	*jth* pick variable of TPOGA in IESA module, $j = 1, 2, \ldots, 10$

Appendix 10.C – Patents (IYM)

USA Patents

1) Cheng, F.T., Hsieh, Y.S., and Zheng, J.W. (2016). System and method for identifying root causes of yield loss. US Patent 10,935,962 B2, filed 9 September 2016 and issued 2 March 2021.
2) Lin, C.Y., Hsieh, Y.M., and Cheng, F.T. (2018). System and method that consider tool interaction effects for identifying root causes of yield loss. US Patent 10,948,903 B2, filed 14 December 2018 and issued 16 March 2021.

Taiwan, ROC Patents

1) Cheng, F.T., Hsieh, Y.S., and Zheng, J.W. (2016). System and method for identifying root causes of yield loss. ROC Patent I623830, filed 2 November 2016 and issued 11 May 2018.
2) Lin, C.Y., Hsieh, Y.M., and Cheng, F.T. (2018). System and method that consider tool interaction effects for identifying root causes of yield loss. ROC Patent I660322, filed 15 November 2018 and issued 21 May 2019.

China Patents

1) Cheng, F.T., Hsieh, Y.S., and Zheng, J.W. (2016). System and method for identifying root causes of yield loss. CN Patent 3405221, filed 9 November 2016 and issued 7 June 2019.
2) Lin, C.Y., Hsieh, Y.M., and Cheng, F.T. (2018). System and method that consider tool interaction effects for identifying root causes of yield loss. CN Patent Pending 201811445872.3, filed 29 November 2018.

Korea Patent

1) Cheng, F.T., Hsieh, Y.S., and Zheng, J.W. (2017). System and method for identifying root causes of yield loss. KR Patent 10-2003961, filed 19 June 2017 and issued 19 July 2019.

References

1 Briand, L.C., Basili, V.R., and Thomas, W.M. (1992). A pattern recognition approach for software engineering data analysis. *IEEE Transactions on Software Engineering* 18 (11): 931–942. https://doi.org/10.1109/32.177363.

2 Hong, S.J., Lim, W.Y., Cheong, T. et al. (2012). Fault detection and classification in plasma etch equipment for semiconductor manufacturing e-Diagnostics. *IEEE Transactions on Semiconductor Manufacturing* 25 (1): 83–93. https://doi.org/10.1109/TSM.2011.2175394.

3 Tsuda, T., Inoue, S., Kayahara, A. et al. (2015). Advanced semiconductor manufacturing using big data. *IEEE Transactions on Semiconductor Manufacturing* 28 (3): 229–235. https://doi.org/10.1109/TSM.2015.2445320.

4 Lee, F. and Smith, S. (1998). Yield analysis and data management using yield manager™. *Proceedings of the IEEE/SEMI Advanced Semiconductor Manufacturing Conference and Workshop*, Boston, MA, USA (19–30 September 1998). USA: IEEE.

5 Chen, A. and Hong, A. (2010). Sample-efficient regression trees (SERT) for semiconductor yield loss analysis. *IEEE Transactions on Semiconductor Manufacturing* 23 (3): 358–369. https://doi.org/10.1109/TSM.2010.2048968.

6 Chien, C.F., Wang, W.C., and Cheng, J.C. (2007). Data mining for yield enhancement in semiconductor manufacturing and an empirical study. *Expert Systems with Applications* 33 (1): 192–198. https://doi.org/10.1016/j.eswa.2006.04.014.

7 Hsu, C.Y., Chien, C.F., Lin, K.Y. et al. (2010). Data mining for yield enhancement in TFT-LCD manufacturing: an empirical study. *Journal of the Chinese Institute of Industrial Engineers* 27 (2): 140–156. https://doi.org/10.1080/10170660903541856.

8 Ing, C.K. and Lai, T.L. (2011). A stepwise regression method and consistent model selection for high-dimensional sparse linear models. *Statistica Sinica* 21 (4): 1473–1513. https://doi.org/10.5705/ss.2010.081.

9 Tibshirani, R. (1996). Regression shrinkage and selection via the LASSO. *Journal of the Royal Statistical Society. Series B* 58 (1): 267–288. http://www.jstor.org/stable/2346178.

10 Hastie, T., Tibshirani, R., and Friedman, J. (2009). *The Elements of Statistical Learning: Data mining, Inference, and Prediction*, 2nde. USA: Springer.

11 Cheng, F.T., Hsieh, Y.S., Zheng, J.W. et al. (2017). A scheme of high-dimensional key-variable search algorithms for yield improvement. *IEEE Robotics and Automation Letters* 2 (1): 179–186. https://doi.org/10.1109/LRA.2016.2584143.

12 Cheng, F.T., Chen, Y.T., Su, Y.C. et al. (2008). Evaluating reliance level of a virtual metrology system. *IEEE Transactions on Semiconductor Manufacturing* 21 (1): 92–103. https://doi.org/10.1109/TSM.2007.914373.

13 Port, D., Kazman, R., Nakao, H. et al. (2007). Practicing what is preached: 80/20 rules for strategic IV & V assessment. *Proceedings of the 2007 IEEE International Conference on Exploring Quantifiable IT Yields*, Amsterdam, Netherlands (19–21 March 2007). USA: IEEE.

14 Singson, M. and Hangsing, P. (2015). Implication of 80/20 rule in electronic journal usage of UGC-Infonet consortia. *The Journal of Academic Librarianship* 41 (2): 207–219. https://doi.org/10.1016/j.acalib.2014.12.002.

15 Draper, N.R. and Smith, H. (1998). *Applied Regression Analysis*, 3rde. New York: Wiley.

16 Cheng, F.T., Lin, C.Y., Chen, C.F. et al. (2017). Blind-stage search algorithm for the key-variable search scheme. *IEEE Robotics and Automation Letters* 2 (4): 1840–1847. https://doi.org/10.1109/LRA.2017.2708132.

17 Sen, A. and Srivastava, M (1997). *Regression Analysis: Theory*. In: *Methods and Applications*. New York: Springer.

18 Lin, C.Y., Hsieh, Y.M., Cheng, F.T. et al. (2018). Interaction-effect search algorithm for the KSA scheme. *IEEE Robotics and Automation Letters* 3 (4): 2778–2785. https://doi.org/10.1109/LRA.2018.2838323.

19 Dumitru, I., Li, F., Wiley, J.B. et al. (2006). Interaction effects analysis of FMR spectra on dense nanowire systems. *IEEE Transactions on Magnetics* 42 (10): 3225–3227. https://doi.org/10.1109/TMAG.2006.880139.

20 Trusca, O.C., Cimpoesu, D., Lim, J.H. et al. (2008). Interaction effects in Ni nanowire arrays. *IEEE Transactions on Magnetics* 44 (11): 2730–2733. https://doi.org/10.1109/TMAG.2008.2001501.

21 Hong, A. and Chen, A. (2012). Piecewise regression model construction with sample efficient regression tree (SERT) and applications to semiconductor yield analysis. *Journal of Process Control* 22 (7): 1307–1317. https://doi.org/10.1016/j.jprocont.2012.05.017.

22 Hao, N. and Zhang, H.H. (2014). Interaction screening for ultrahigh-dimensional data. *Journal of the American Statistical Association* 109 (507): 1285–1301. https://doi.org/10.1080/01621459.2014.881741.

23 Taylor, R. (1990). Interpretation of the correlation coefficient: A basic review. *Journal of Diagnostic Medical Sonography* 6 (1): 35–39. https://doi.org/10.1177%2F875647939000600106.

11

Application Cases of Intelligent Manufacturing

Fan-Tien Cheng[1], Yu-Chen Chiu[2], Yu-Ming Hsieh[3], Hao Tieng[3], Chin-Yi Lin[4], and Hsien-Cheng Huang[5]

[1] *Director/Chair Professor, Intelligent Manufacturing Research Center/Institute of Manufacturing Information and Systems, National Cheng Kung University, Tainan, Taiwan, ROC*
[2] *Specialist, Intelligent Manufacturing Research Center, National Cheng Kung University, Tainan, Taiwan, ROC*
[3] *Associate Research Fellow, Intelligent Manufacturing Research Center, National Cheng Kung University, Tainan, Taiwan, ROC*
[4] *Postdoctoral Research Fellow, Intelligent Manufacturing Research Center, National Cheng Kung University, Tainan, Taiwan, ROC*
[5] *Deputy Director, e-Manufacturing Research Center, National Cheng Kung University, Tainan, Taiwan, ROC*

11.1 Introduction

The actual Intelligent Manufacturing implementation cases adopting all the Industry 4.1 related techniques involving Cyber-Physical Agent (CPA), Advanced Manufacturing Cloud of Things (AMCoT), Automatic Virtual Metrology (AVM), Intelligent Predictive Maintenance (IPM), and Intelligent Yield Maintenance (IYM) in seven industries, including flat panel display, semiconductor, solar cell, automobile, aerospace, carbon fiber, and blow molding, are presented in this chapter.

11.2 Application Case I: Thin Film Transistor Liquid Crystal Display (TFT-LCD) Industry

Both AVM and IYM have been deployed in the thin film transistor-liquid crystal display (TFT-LCD) industry. The deployment of AVM is illustrated first followed by the presentation of the IYM deployment.

11.2.1 Automatic Virtual Metrology (AVM) Deployment Examples in the TFT-LCD Industry

The authors would like to thank Chi Mei Optoelectronics Corporation, Ltd. (CMO) in Taiwan, ROC for providing the raw data used in this AVM application case.

Industry 4.1: Intelligent Manufacturing with Zero Defects, First Edition. Edited by Fan-Tien Cheng.

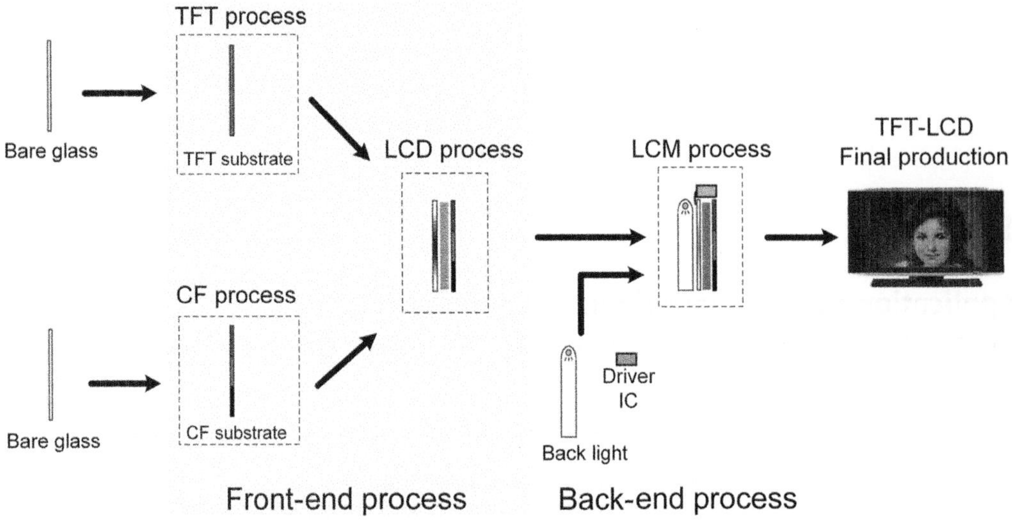

Figure 11.1 Process flow of TFT-LCD manufacturing. *Source:* Reprinted with permission from Ref. [1]; © 2015 IEEE.

11.2.1.1 Introducing the TFT-LCD Production Tools and Manufacturing Processes for AVM Deployment

The TFT-LCD manufacturing flow consists of four processes: thin film transistor (TFT), color filter (CF), liquid crystal display (LCD), and liquid crystal module (LCM), as shown in Figure 11.1. First, in the TFT process, the transistors are fabricated on a piece of glass substrate. Second, color sources including red, green, and blue are produced on a glass in the CF process. Third, a TFT substrate is joined with a CF substrate in the LCD process. The space between these two substrates is filled with liquid crystal. Finally, in the LCM process, additional components, such as driver integrated circuits and backlight units, are connected to comprise the final fabricated product. Among those four processes, the TFT, CF, and LCD processes are aggregated as the front-end process, which requires high-degree automation. On the other hand, the LCM process is also denoted back-end process, which is mostly labor-oriented. This study focuses on the front-end process.

TFT Process

The TFT process is divided into five layers, including gate, semiconductor, data, protection, and indium tin oxide (ITO) layers. Each layer contains the following six processes: film deposition, positive photoresist coating, exposure, developing, etching, and stripping. The processes of coating, exposure, and developing can be combined and denoted as the photo step. The process flow of the semiconductor layer, as shown in Figure 11.2, is the most complex among those five layers.

The first step of the semiconductor process flow is film deposition that is deposited twice by chemical vapor deposition (CVD) tools to enhance the production yield. As shown in Figure 11.3, three layers of different thin films (G, I, and N layers) are deposited on the substrate. The G layer serves as a gate insulator that comprises silane, ammonia, and nitrogen. The I layer, which contains intrinsic amorphous silicon, is formed via the CVD reaction between silane and hydrogen. The N layer, which contains n-plus amorphous silicon, is deposited with silane, hydrogen, and phosphine. The G layer is further decomposed into two layers, G1 and G2, to enhance the production yield. Observing the lower-left portion of Figure 11.2a, the G1 layer is plated in the first

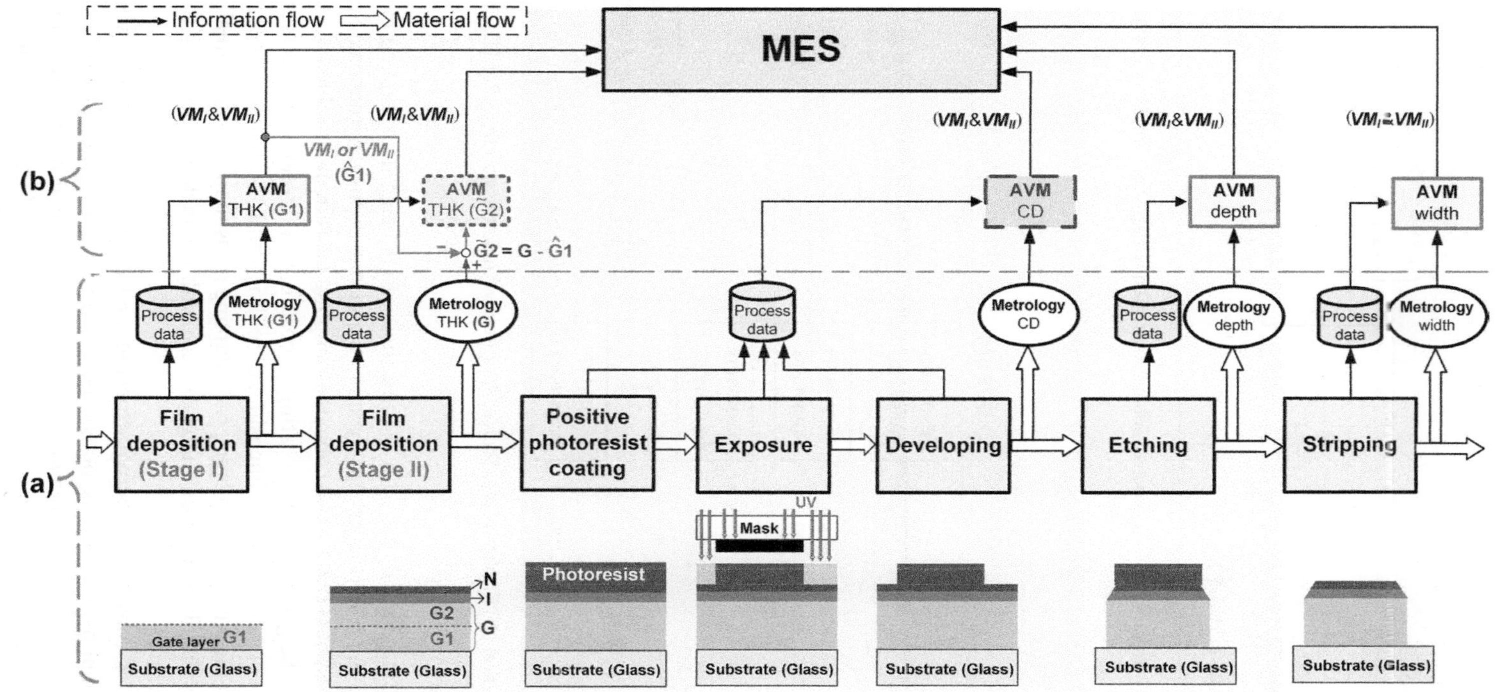

Figure 11.2 Semiconductor layer of the TFT process flow with deployment of AVM servers. (a) Semiconductor layer process flow. (b) Deployment of AVM servers. *Source:* Reprinted with permission from Ref. [1]; © 2015 IEEE.

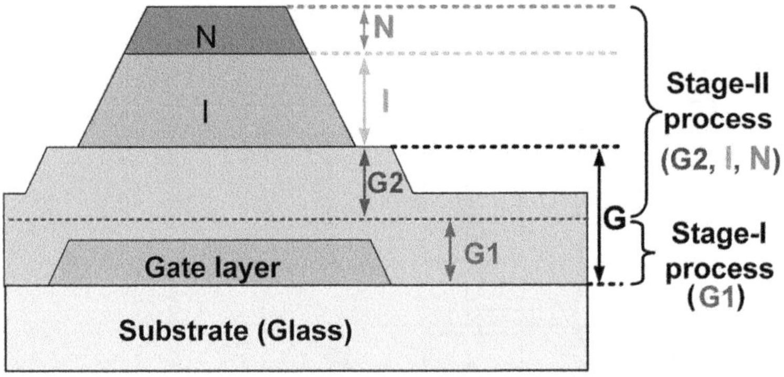

Figure 11.3 Thin-film structure in CVD process. *Source:* Reprinted with permission from Ref. [1]; © 2015 IEEE.

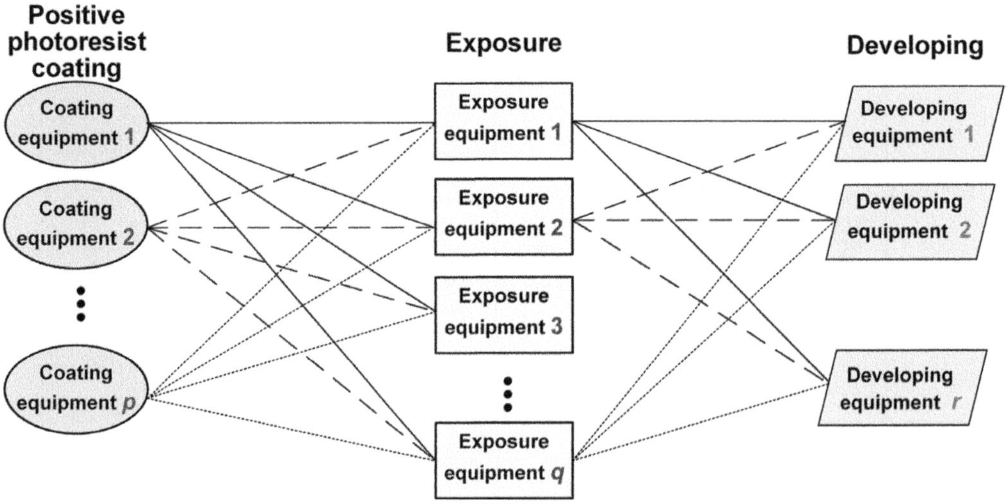

Figure 11.4 Combination of TFT photo step. *Source:* Reprinted with permission from Ref. [1]; © 2015 IEEE.

deposition (denoted Stage-I) process with the thickness of G1 being the sampling measurement value; whereas, the G2, I, and N layers are coated in the second deposition (denoted Stage-II) process with the thickness of G (=G1 + G2) being the sampling measurement value.

In the photo step, the glasspiece runs coating, exposure, and developing processes before the sampling measurement value (critical dimension, CD of line-width) is obtained. Then, the etching process follows with the etching depth being the sampling measurement value. Finally, the stripper process removes the photoresist on the glasspiece with the metal width being the sampling measurement value.

The TFT photo step consists of coating, exposure, and developing processes, which may be accomplished by one of p coating tools, one of q exposure tools, and one of r developing tools as shown in Figure 11.4. Therefore, it requires $p \times q \times r$ sets of virtual metrology (VM) models to maintain the accuracy of VM conjecturing. In this case, if the traditional approach is applied to create VM models one by one with its own historical data, then huge labor expenses and model-creation time will make VM deployment impossible. Therefore, a refined virtual metrology system (VMS) should possess the capabilities of automatic fanning out and model refreshing so as to take care of the problem mentioned above.

CF Process

As shown in Figure 11.5a, the CF manufacturing process includes black matrix (BM), red (R), green (G), blue (B), ITO, and photo space (PS) layers. The main functions of these layers are described below. BM covers the light leak of TFT; R, G, and B generate colors; ITO controls the direction of crystal; and PS is the support between TFT and CF. Except for the ITO layer, the processes of the other ones are the same as those of the TFT photo step. However, the TFT photo step is accomplished by mixed tools of coating, exposure, and developing as shown in Figure 11.5; while the R, G, B, and PS processes of CF adopt inline-type tools of coating, exposure, and developing. Take the PS layer of CF photo process as the example and observe Figure 11.6a, the glasspiece is processed with negative photoresist coating, exposure, and developing with the PS thickness being the sampling measurement value.

LCD Process

Six steps are included in the LCD process as shown in Figure 11.7a: polyimide print, seal dispense, liquid crystal drop, assemble, scribe, and polarizer attach. The physical operations of the six steps are depicted at the bottom of Figure 11.7a. Among these six steps, the sampling measurements of polyimide print, scribe, and polarizer attach are performed immediately after their manufacturing processes are completed, and the measurement values are print thickness, scribe location, and polarizer location, respectively. As for the seal-dispense process, the sampling measurement (seal width) will not be performed until the assemble process is finished. Finally, the sampling measurements of liquid crystal drop and assemble are cell gap and mark shift, and will not be executed until the assemble process is finished.

11.2.1.2 AVM Deployment Types for TFT-LCD Manufacturing

As indicated in Figure 8.6, the inputs of an AVM server are process data and sampled metrology data; where the process data contain the key factors that affect the process performance presented by the corresponding metrology data. Based on the principle mentioned above, the deployments of AVM servers at the TFT, CF, and LCD manufacturing phases are displayed in Figures 11.2b, 11.5b, and 11.7b, respectively. Three deployment types namely single-stage, dual-stage, and cooperative-tools (with the AVM-block in solid squares, dotted square, and segmented squares, respectively) are found in Figures 11.2b, 11.5b, and 11.7b.

Single-Stage AVM Deployment [Solid Squares in Figures 11.2b, 11.5b, and 11.7b]

The single-stage type simply collects the process data from a target tool and the sampled metrology data of the workpieces processed by that target tool. The AVM server shown in Figure 8.6 is adopted to serve as the single-stage VM architecture.

Dual-Stage VM Deployment [Dotted Square in Figure 11.2b]

The dual-stage type is mainly designed for conjecturing the G2 value of the Stage II of film deposition indicated in Figure 11.3. As shown in the left portion of Figure 11.2, the dual-stage deployment scheme involves two process tools. The detailed dual-stage indirect VM architecture is depicted in Figure 11.8.

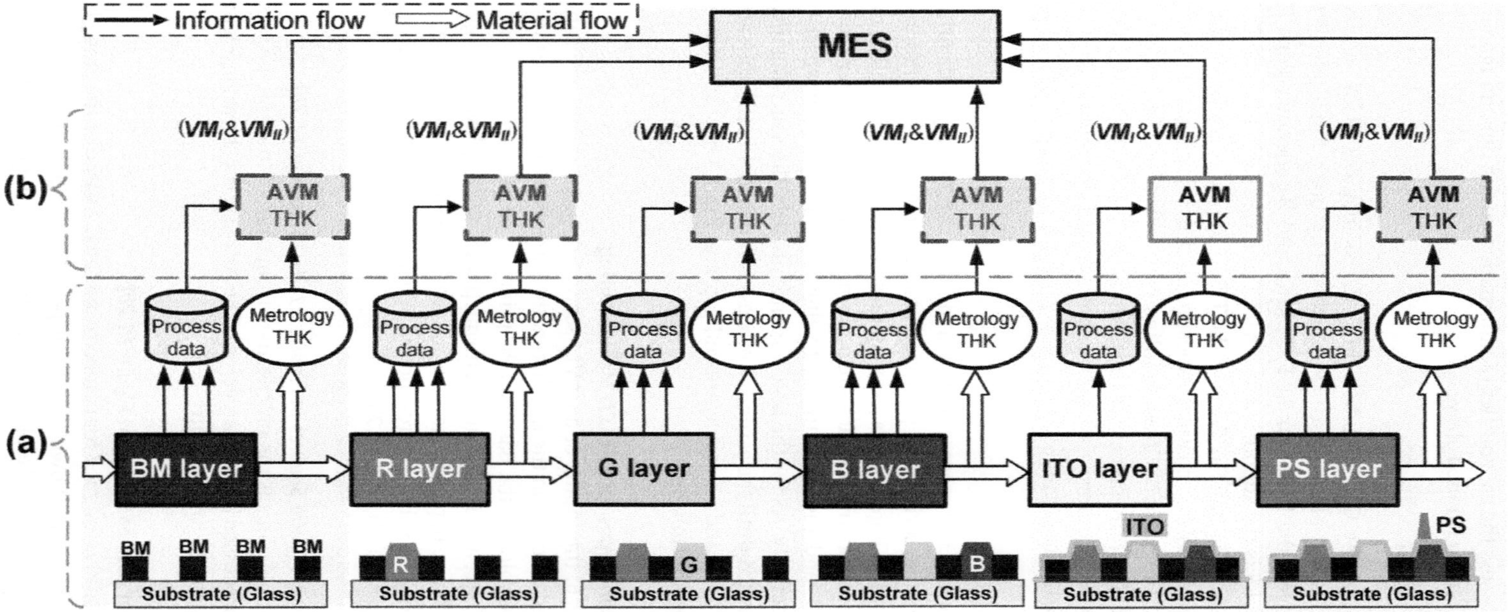

Figure 11.5 CF manufacturing process flow with deployment of AVM servers. (a) CF manufacturing process flow. (b) Deployment of AVM servers. *Source:* Reprinted with permission from Ref. [1]; © 2015 IEEE.

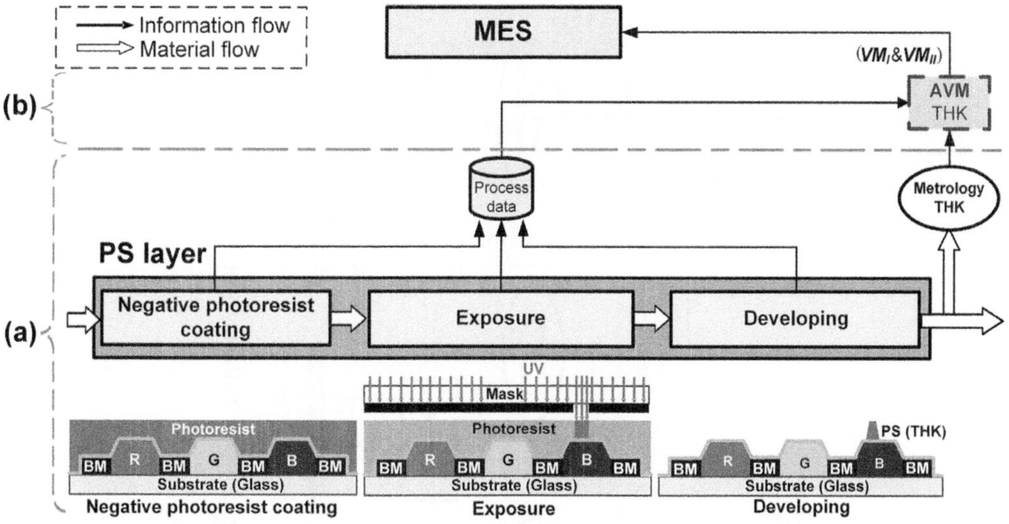

Figure 11.6 PS layer flow of the CF manufacturing process with deployment of AVM servers. (a) PS layer flow. (b) Deployment of AVM servers. *Source:* Reprinted with permission from Ref. [1]; © 2015 IEEE.

In general, one out of 20 cassettes is chosen to select five or six glasspieces (one for each chamber) as the metrology glasspieces for quality monitoring. Hence, for a tool with five/six chambers, one out of 100/120 glasspieces is chosen as a metrology glasspiece for a specific chamber. A typical example shows that Stage-I tool has five chambers while Stage-II tool has six ones. For this example, the chance of a specific glasspiece to have both G1 and G thicknesses been measured is only $(1/100)\times(1/120) = 1/12{,}000$. Since the chance is so slim, it becomes clear about the reason why the dual-stage indirect VM architecture shown in Figure 11.8 is required to accomplish the mission of monitoring the quality of G1 and G2 thicknesses for every individual glasspiece.

In Stage I, the VM scheme of the fore-tool is established by the single-stage type; while the indirect VM scheme of the rear-tool is built in Stage II. The required process data of the Stage-II AVM server are collected from the rear-tool; while the required metrology data ($\tilde{G}2$) are derived from Stage-I VM output ($\hat{G}1$) and the actual metrology (AM) data measured just after finishing the Stage-II process (G) with

$$\tilde{G}2 = G - \hat{G}1. \tag{11.1}$$

Equation (11.1) is implemented and shown in the dotted square of Figure 11.8. Observing Equation (11.1) $\tilde{G}2$ represents the synthesized G2 and $\hat{G}1$ is the G1 VM value (Stage-I output) which is available as long as AVM_{G1} (AVM Server of Stage I for G1 Conjecturing) server functions normally. Here, the quality of $\hat{G}1$ can be checked by evaluating its accompanied Reliance Index (RI)/Global Similarity Index (GSI) values ($RI_{\hat{G}1}/GSI_{\hat{G}1}$). Feeding the process data of the Stage II tool (which deposits the G2 layer) and $\tilde{G}2$ to AVM_{G2} (AVM Server of Stage II for G2 Conjecturing) server, the G2 VM value ($\hat{G}2$) and its accompanied RI/GSI values ($RI_{\hat{G}2}/GSI_{\hat{G}2}$) can then be obtained for evaluating the quality of $\hat{G}2$.

Cooperative-Tools AVM Deployment [Segmented Squares in Figures 11.2b, 11.5b, and 11.7b]

Finally, the cooperative-tools type needs to collect the process data from several cooperative tools one after another and the sampled metrology data measured just after finishing the process

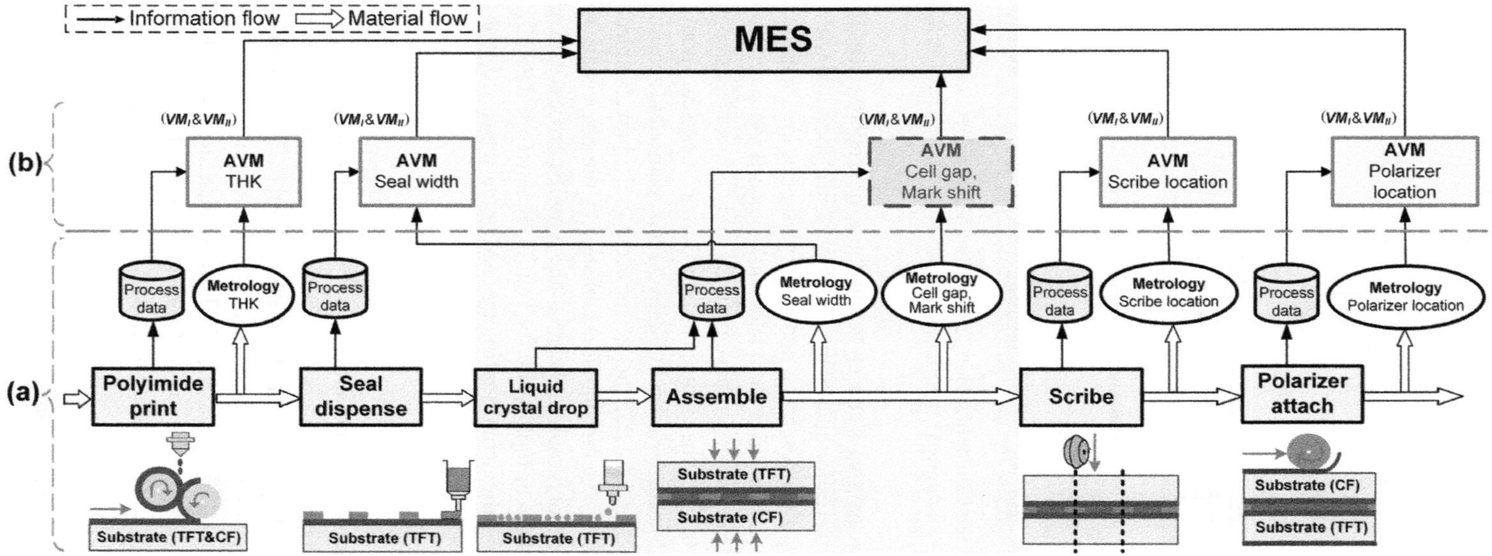

Figure 11.7 LCD manufacturing process flow with deployment of AVM servers. (a) LCD manufacturing process flow. (b) Deployment of AVM servers. *Source:* Reprinted with permission from Ref. [1]; © 2015 IEEE.

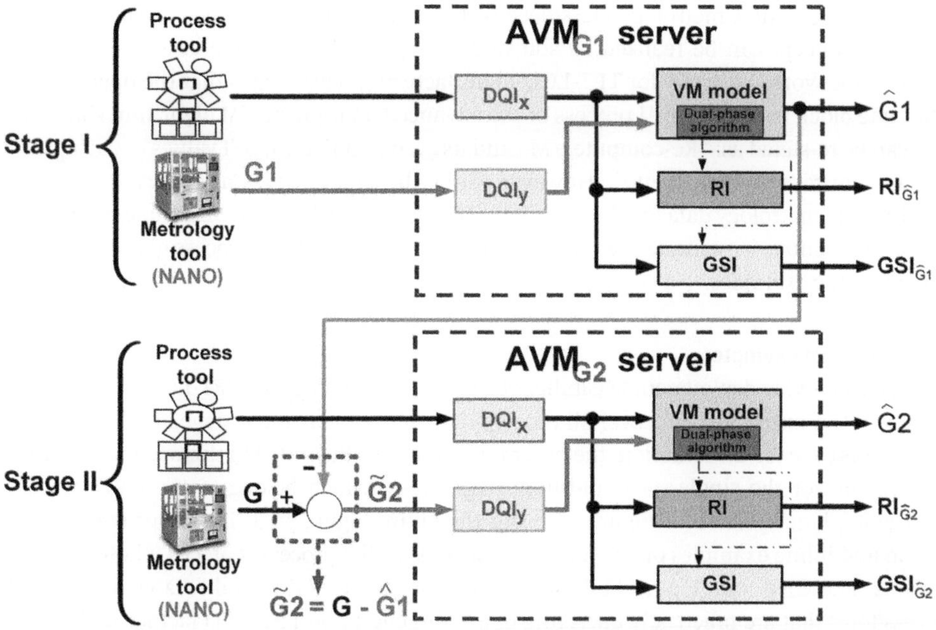

Figure 11.8 Dual-stage indirect VM architecture. *Source:* Reprinted with permission from Ref. [1]; © 2015 IEEE.

performed by the cooperative-tools. For example, as shown in the middle of Figure 11.2, the AVM server (segmented square) collects process data of a specific glasspiece from a positive-photoresist-coating tool, then an exposure tool, and finally a developing tool (so-called cooperative tools); the AVM server also collects the sampled metrology data (CD) just after finishing the process performed by the developing tool. Two varieties of cooperative-tools exist in TFT-LCD manufacturing; one is combination (as shown in Figure 11.4) and the other is inline (as displayed in Figure 11.6).

Concept of a Virtual Cassette

For TFT-LCD manufacturing, each cassette contains 30 pieces of glass (also called glasspieces). Each tool may contain five or six chambers. To reduce cycle time, after a cassette arrives at a tool station, all the glasspieces in the same cassette will be unloaded one by one and sent to available chambers of the same tool randomly. Those after-process glasspieces will be loaded into a new cassette in a first-come-first-serve fashion. Consequently, each cassette contains glasspieces that are processed from various chambers. Generally, five or six glasspieces of sampled cassettes (one for each chamber) are chosen as the metrology glasspieces for quality monitoring. Based upon the fact mentioned above and for obtaining an acceptable conjecturing accuracy, we cannot apply a single VM model (created and dedicated to the same chamber) to conjecture the VM values of all the glasspieces in the same real cassette because most of the glasspieces are not processed by the same chamber that creates the VM model. To resolve this problem, the concept of a virtual cassette is proposed.

A virtual cassette is defined to contain many process glasspieces and one metrology glasspiece; those process and metrology glasspieces are processed by the same chamber. Moreover, a virtual cassette starts collecting process glasspieces that are processed by a specific chamber one by one and ends collecting until a metrology glasspiece of the same chamber has been obtained. This virtual-cassette concept can be realized by software. By applying the virtual cassette, the dual-phase VM scheme works perfectly for TFT-LCD manufacturing. This means that the operation of the last but one block in the Phase-II process of the advanced dual-phase VM algorithm shown in Figure 8.7 can be re-stated as: "Re-compute VM_{II} and its accompanied RI/GSI values of each glasspiece in the entire virtual cassette". However, only the metrology glasspieces have their corresponding real metrology data for checking VM accuracy of the illustrative examples presented in the following section; for concise presentation, the VM values of those process glasspieces will not be shown in the examples.

11.2.1.3 Illustrative Examples

To demonstrate fab-wide deployment feasibility of the AVM technology for TFT-LCD manufacturing, four illustrative examples that cover all the deployment types of single-stage, dual-stage, and cooperative-tools (one for combination, the other for inline) are illustrated in this section. The first two examples are for the single-stage and dual-stage types, which belong to the CVD process shown in the left-hand portion of Figure 11.2 and the entire Figure 11.3. The other two are the combination and inline examples of cooperative-tools type in the processes of TFT photo (middle of Figure 11.2) and CF photo (Figure 11.6), respectively. All the experimental data were collected from process tools that are practically operating in a fifth-generation TFT-LCD factory in Chi Mei Optoelectronics Corporation, Ltd. (CMO) in Taiwan, ROC.

The back-propagation neural networks (BPNN) and partial least squares (PLS) are adopted as the designated algorithms for creating the VM conjecture models. Given the conjecture VM value $\hat{y}_i$, AM value y_i, target value y, and the sample size n, the conjecture accuracy is presented by the mean absolute percentage error (MAPE) and maximum error (Max Error) denoted in (11.2) and (11.3), respectively. The closer the MAPE and Max Error are to zero, the better the conjecture accuracy is achieved.

$$MAPE = \frac{\sum_{i=1}^{n} \left| \left(\hat{y}_i - y_i \right) / y \right|}{n} \times 100\%. \tag{11.2}$$

$$Max\ Error = \max\left\{ \frac{\left| \left(\hat{y}_i - y_i \right) \right|}{y} \times 100\%,\ i = 1, 2, \ldots, n \right\}. \tag{11.3}$$

Single-Stage Example

The Stage-I film deposition process is the first step of the TFT process displayed in Figure 11.2. The thin-film structure in CVD process is shown in Figure 11.3. The G1 layer is plated in the Stage-I process with the thickness of G1 being the sampling measurement value. The AVM_{G1} server for Stage-I VM architecture is depicted in the upper portion of Figure 11.8.

Each CVD tool contains five chambers. To monitor the processing quality, 19 positions are measured on a 26″ product glass. Chamber A and Position 2 are chosen for demonstration in this example. According to the physical properties of a CVD tool and the expert knowledge of equipment engineers, ten significant process parameters including gas flow, temperature, pressure, RF power, and the others are selected as the inputs of the VM model. The example involves 68 sets of corresponding real metrology data and process data. Among those 68 sets of data, the first 44 data

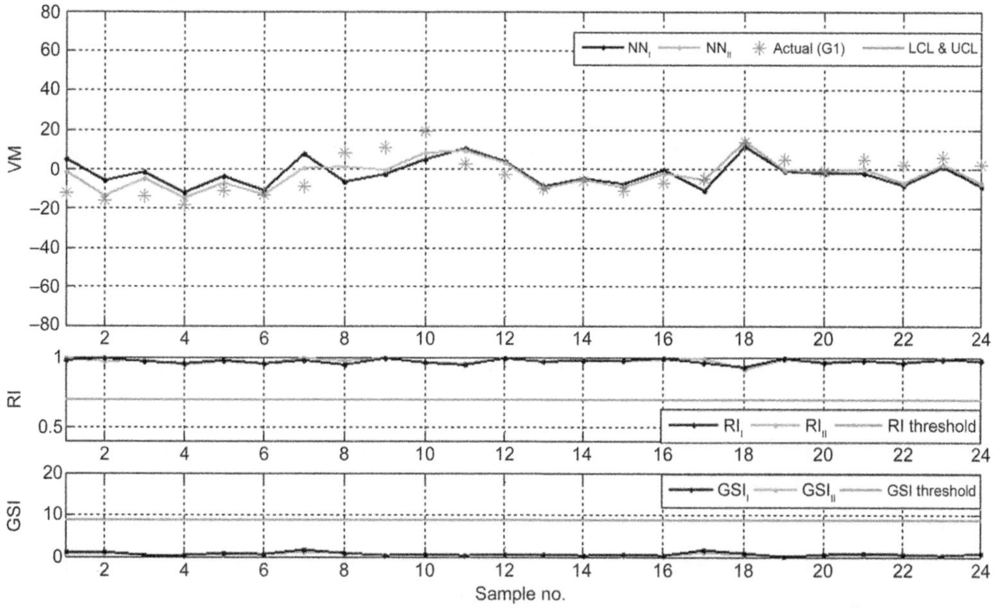

Figure 11.9 Single-stage example: Stage-I VM results at Position 2. LCL, lower control limit; UCL, upper control limit. *Source:* Reprinted with permission from Ref. [1]; © 2015 IEEE.

sets are used to build the AVM_{G1} model, while the other 24 data sets are applied for evaluating G1 VM conjecture accuracy.

The Position 2 VM results are illustrated in Figure 11.9 with the MAPEs of Phase-I NN-based VM value (NN_I) and Phase-II NN-based VM value (NN_{II}) being 0.45 and 0.30%, respectively. In fact, the average NN_I MAPE of all the 19 positions is 0.58%, which is within the accuracy specification (2%). Further, all the Phases-I & II RI (RI_I and RI_{II}) values are greater than the RI threshold value (RI_T) and all the Phases-I & II GSI (global similarity index of Phase I (GSI_I) and global similarity index of Phase II (GSI_{II})) values are smaller than the threshold of GSI (GSI_T) for the samples shown in Figure 11.9, this indicates that all the NN_I and NN_{II} values are reliable and no property drift or shift occurred.

Dual-Stage Example

The Stage-II film deposition process is the second step of the TFT process displayed in Figure 11.2. As shown in Figure 11.3, the G2 layer is deposited in the Stage-II process with the thickness of G (=G1 + G2) being the sampling measurement value. The AVM_{G2} server for Stage-II VM architecture is displayed in the lower portion of Figure 11.8 with $\tilde{G}2 = G - \widehat{G1}$.

Each Stage-II CVD tool contains six chambers. To monitor the processing quality, 19 positions are measured on a 26″ product glass. Again, Chamber A and Position 2 are chosen for demonstration. The ten key process parameters of the CVD tool are selected as the inputs of the VM model. Inheriting the corresponding Stage-I VM conjecture ($\widehat{G1}$) results from the single-stage example to perform the indirect AVM_{G2} scheme for this demonstration. The example employs 47 samples, in which the first 26 samples are adopted to establish the AVM_{G2} model, and the other 21 samples are applied to evaluate the G2 VM conjecture accuracy.

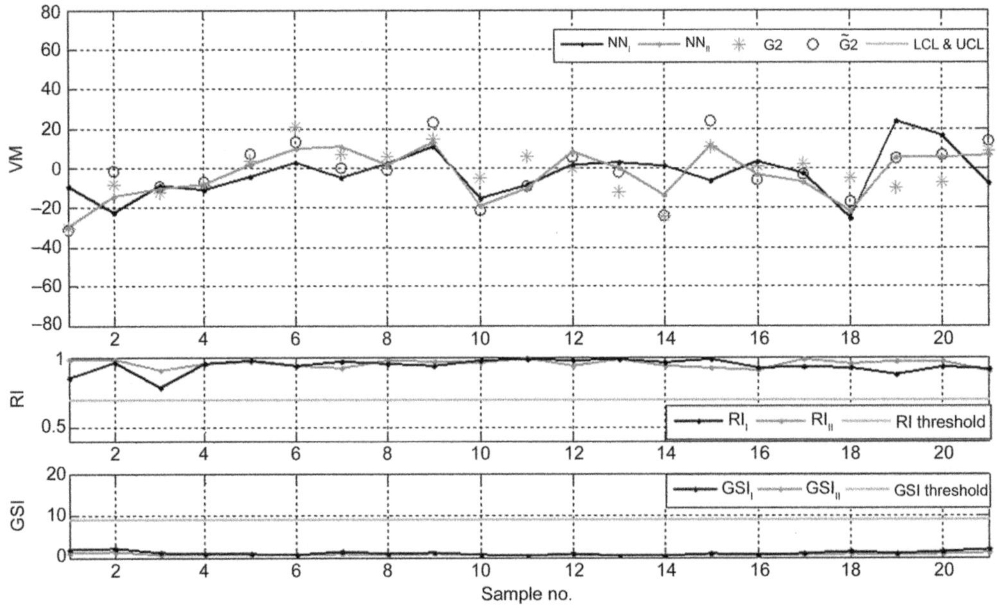

Figure 11.10 Dual-stage example: Stage-II VM results at Position 2. LCL, lower control limit; UCL, upper control limit. *Source:* Reprinted with permission from Ref. [1]; © 2015 IEEE.

The VM results are shown in Figure 11.10. Note that, both G2 and $\tilde{G}2$ are displayed in Figure 11.10. The actual G2 (=G − G1) values are obtained in special arrangements to directly measure the G and G1 values on the same metrology glasspiece and are utilized for calculating the MAPEs. Instead, the $\tilde{G}2(= G - \widehat{G1})$ values are applied for re-training the VM models during Phase-II calculation of the dual-phase VM algorithm. The NN_I and NN_{II} MAPEs of Position 2 are 0.74 and 0.43%, respectively, and the average NN_I MAPE of all the 19 positions is 0.77%, which is within the accuracy specification (2%). Moreover, all the RI_I, RI_{II}, GSI_I, and GSI_{II} values of the testing samples are within their corresponding thresholds; therefore, all the VM results are reliable.

Cooperative-Tools Example [Combination Case]

The TFT photo (coating/exposure/developing) process shown in the middle of Figure 11.2 is adopted as the combination-case example of the cooperative-tools whose combination-explication is depicted in Figure 11.4. The illustrative fifth-generation TFT fab has eight coating, seven exposure, and ten developing tools; hence, 560 combinations exist in this cooperative-tools example. Among those 560 combinations, the 7-7-6 combination is the most frequent rout; therefore, it has the largest population of historical samples and is selected for building the first VM models. The VM models of all the other combinations can then be constructed by the automatic model refreshing scheme described in Chapter 8.

A 14.1″ product glass has 16 measurement positions. The VM results of CD at Position 13 are selected for demonstration in this example. According to the physical properties of the photo process tools, 21 key process parameters, such as coating temperature & glass thickness for coating equipment, lamp illumination & exposure energy for exposure equipment, act-process time, and spray temperature for developing equipment, etc., are selected for building the VM models. 178 sets of corresponding metrology and process data are collected in this example. The first 110 sets are applied to establish the VM models, whereas the other 68 data sets are used to evaluate the conjecture accuracy. The NN_I results at Position 13 are displayed in Figure 11.11.

The AVM system possesses the data quality index of the metrology data (DQI_y) algorithm to avoid abnormal metrology data being used for tuning or training the VM models, such that the VM conjecture accuracy can be maintained. Observing Figure 11.11, the metrology data quality of Samples 36, 43, 44, 54, and 55 are detected abnormal by the DQI_y algorithm; and the abnormalities are caused by erroneous measurement of these samples. The illustration of an erroneous measurement is shown in Figure 11.12. Instead of the correct length from L1 to R1, the erroneous length from L1 to R2 is sensed by the metrology equipment.

Significance of the DQI_y algorithm is also demonstrated in Figure 11.11. As shown in the left portion of Figure 11.11, the NN_I MAPE of the first 35 samples is 0.59% regardless whether the DQI_y algorithm is applied or not because no metrology data are abnormal in this period. However, in the period of Samples 36–56, the NN_I MAPE values are 0.63% with DQI_y and 1.11% without DQI_y. Note that the MAPE values in the period of Samples 36–56 do not count those abnormal-metrology samples (No. 36, 43, 44, 54, and 55) for the sake of fairness. In conclusion, the VM conjecturing has better performance when the DQI_y algorithm is applied.

To demonstrate the performance of the VM predictor when the actual (and predicted) deviates away from the target towards near the lower control limit (LCL) or upper control limit (UCL), 12 experimental sets (Samples 57–68) are adopted. Among these 12 sets, Samples 59–65 are especially chosen to perform a CD spread test with the adjustment of a major parameter (Act-Process Time) on the photo equipment. Observing Sample 59, the value of Act-Process Time deviates from 21 to 31. As a result, the value of GSI_I becomes 150.63, which is greater that GSI_T (=9), and the value of RI_I (0.58) is less that RI_T (=0.7). These two indexes warn that the VM quality is bad, it cannot be adopted for VM applications. Under this circumstance, a real measurement should be performed if a reliable metrology value is required. In fact, the VM_I and real metrology values of Sample 59 are 21.84 and 21.08, respectively; it represents about 3.5% error that may be too rough for applications. By applying the dual-phase algorithm, the process data and corresponding real metrology value of Sample 59 will be adopted to tune the VM, RI, and GSI models such that the predictive values of VM_I, RI_I, and GSI_I at Sample 60 are accurate enough for various VM applications even if the value of the major parameter (Act-Process Time) at Sample 60 remains at 30.6. Then, as shown in Figure 11.11 and started from Sample 61, the values of Act-Process Time are sequentially adjusted to be 25.5, 25.6, 15.8, 15.9, and 11.6; finally back to 20.7, 20.9, and 20.8. Observing the RI_I and GSI_I values of Samples 61–68, only the RI_I value (0.68) at Sample 65 is less than RI_T (0.7). Again, it means that the VM_I value of Sample 65 is not reliable.

Cooperative-Tools Example [Inline Case]

The PS layer flow of the CF manufacturing process, which includes coating/exposure/developing inline process shown in Figure 11.6 is adopted as the inline-case example of the cooperative-tools. According to the physical properties of these PS cooperative-tools, seven significant process parameters including lamp identity, exposure accumulation, total discharge, pump rate, bake time, stage temperature, and mask count are selected as the inputs of VM models. To inspect the processing quality, 36 positions are measured on a 31.5″ product glass. The VM results at Position 1 are chosen for demonstration. The example includes 150 sets of sample data. The first 100 sets are used to establish the VM models and the other 50 sets are adopted to evaluate the conjecture accuracy.

Illustrated in Figure 11.13a, the NN_I and Phase-I partial-least-square-based VM (PLS_I) MAPEs of Position 1 are 0.52 and 0.53%, respectively. Moreover, the RI value at Sample 26 is less than RI_T (0.7). This implies that the VM result is not reliable due to the NN_I and PLS_I VM values being

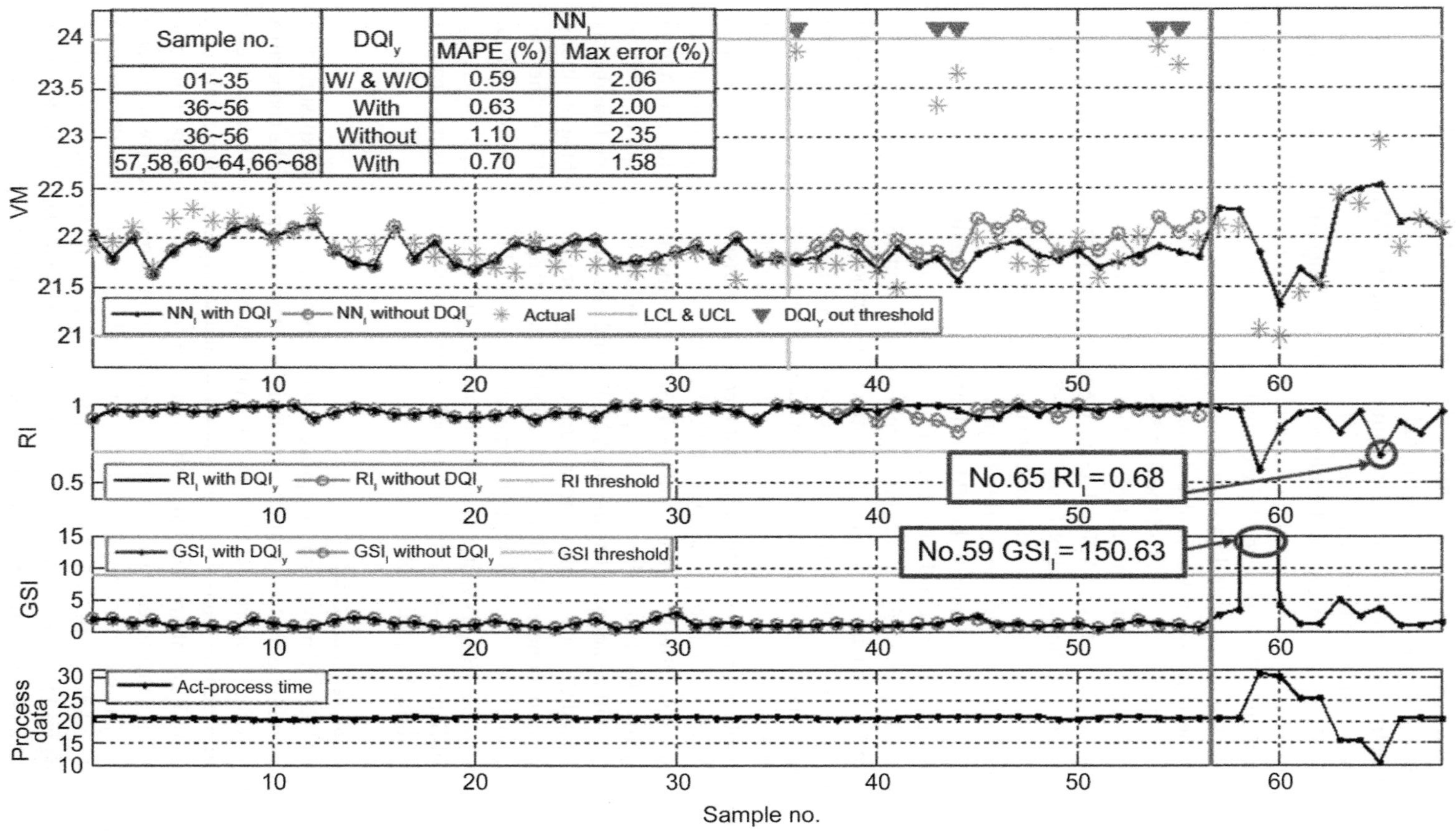

Figure 11.11 Combination example of cooperative-tools: VM$_I$ result at Position 13. *Source:* Reprinted with permission from Ref. [1]; © 2015 IEEE.

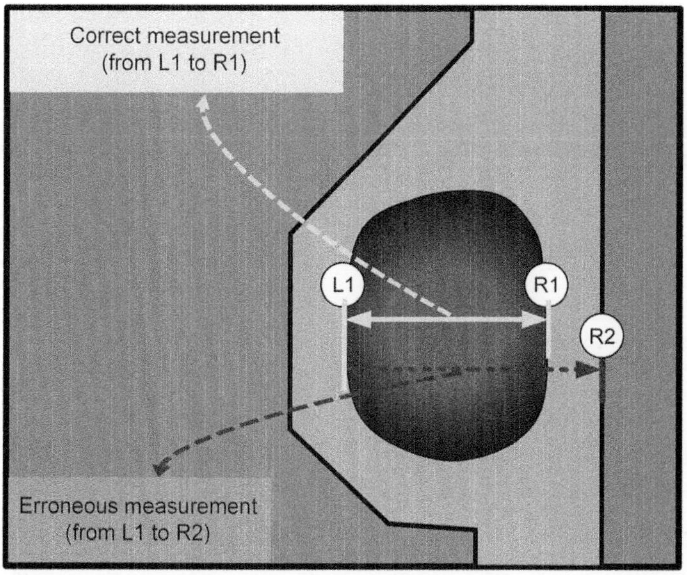

Figure 11.12 Illustration of an erroneous measurement. *Source:* Reprinted with permission from Ref. [1]; © 2015 IEEE.

dissimilar as shown in Figure 11.13a. Further, the GSI values of Samples 13, 14, 26, 27, and 44 are higher than GSI_T (9). Therefore, the Individual-Similarity-Index (ISI) analysis is applied to identify the root causes that generate those deviations.

The ISI analyses of Samples 13 and 14 reveal that the root cause is due to the parameter deviation of process datum: Lamp Identity occurred just before Sample 13 as shown in Figure 11.13b. And, the ISI analyses of Samples 26, 27, and 44 indicate that the main reason is due to the parameter deviations of process datum: Total Discharge happened just before Samples 26 and 44, respectively, as displayed in Figure 11.13b.

To demonstrate how the advanced dual-phase VM algorithm (shown in Figure 8.7) corresponds to the abovementioned anomalies (parameter deviations) for model refreshing/calibration, Figure 11.13c displays both the NN_I and NN_{II} results of those 50 test sets. Note that Figure 11.13c is a graph shown in a VM client of the AVM system depicted in Figure 8.5. The NN_I and NN_{II} results of the first 12 samples shown in Figure 11.13c are almost identical due to the fact that the manufacturing process is stable during this period. Then, at Sample 13, the NN_I prediction error is enlarged due to the value of Lamp Identity being changed from 4.1 to 7.2 as indicated in Figure 11.13b. This parameter deviation also enlarges the GSI_I value to 68.8 (see Figure 11.13a). Observing Figure 11.13c, a blue light is present at Sample 13. This blue light is an indication that warns users about the fact that $GSI_I > GSI_T$ at Sample 13. That means this blue light is an early-warning/benefit that addresses the concern about the VM quality at Sample 13.

Upon receiving the AM value of Sample 13, the Phase II of the advanced dual-phase algorithm performs model refreshing/calibration such that the NN_{II} value of Sample 13 is close to the AM value. Subsequently, the process data of Sample 14 are collected and the newly updated VM and GSI models are applied to calculate the NN_I and GSI_I values at Sample 14. As shown in Figure 11.13a, the GSI_I value of Sample 14 is reduced to 18.9; however, it is still greater than GSI_T such that a blue light is again shown at Sample 14. By the same token, the advanced dual-phase VM algorithm is applied to all the following samples such that their GSI_I values are reduced gradually to be less

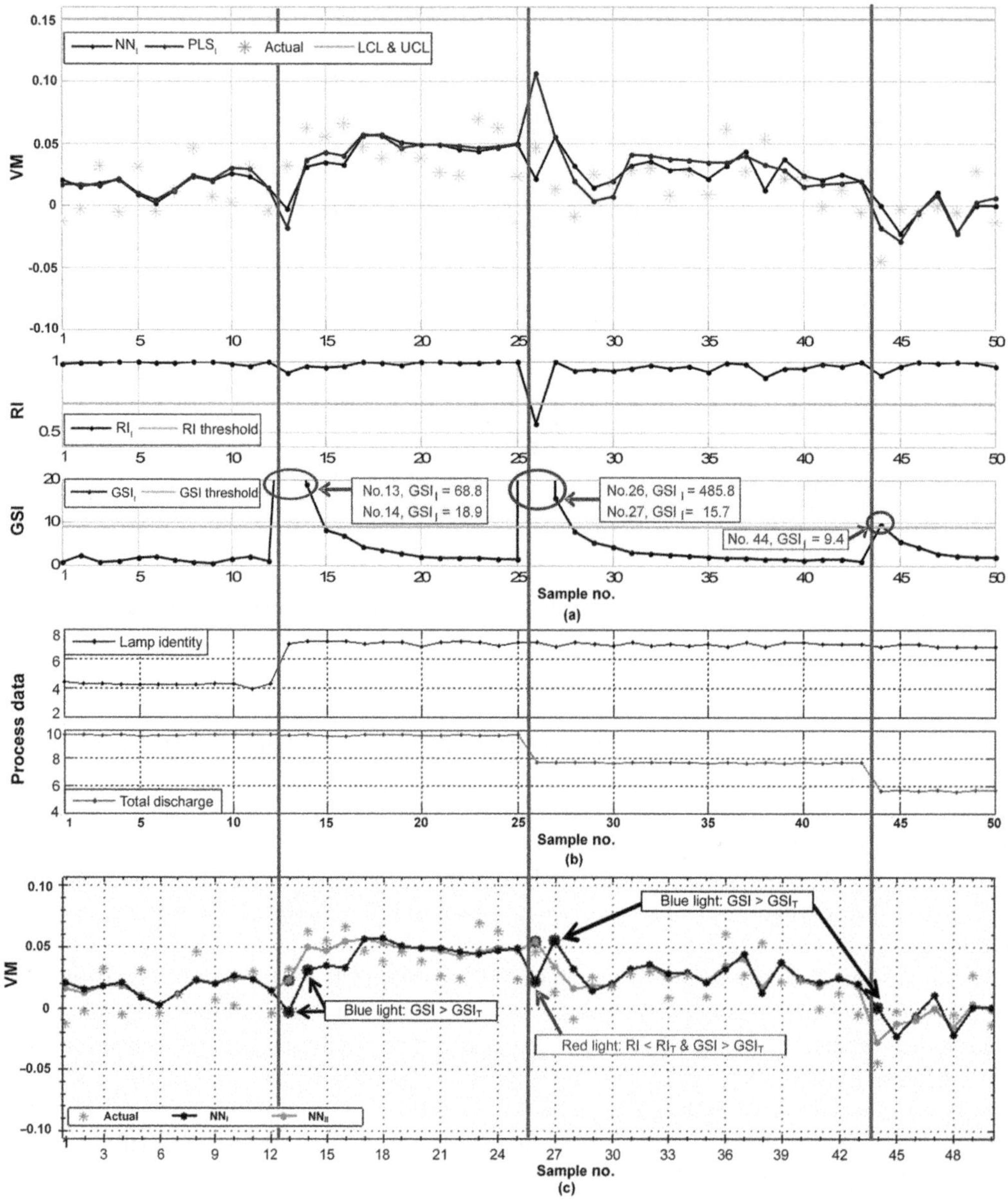

Figure 11.13 Inline example of cooperative-tools at Position 1: (a) NN_I and PLS_I with accompanied RI and GSI; (b) Lamp identity and total discharge; (c) NN_I and NN_{II} with blue light indicating $GSI > GSI_T$ and red light representing $RI < RI_T$ & $GSI > GSI_T$. *Source:* Reprinted with permission from Ref. [1]; © 2015 IEEE.

than the GSI_T after Sample 15. Moreover, the NN_I values are approaching their corresponding NN_{II} values gradually and finally the NN_I and NN_{II} values coincide with each other; also, the NN_I values are close to their corresponding AM values after Sample 17.

In other words, upon encountering an anomaly at Sample 13, the advanced dual-phase algorithm refresh/calibrate VM models automatically such that the VM accuracy is regained after refreshing four samples (#13–#16). Similarly, two other parameter deviations occur at Samples 26 and 44 so the automatic VM models' refreshing/calibrating capability of the advanced dual-phase algorithm recovers the VM accuracy at Samples 29 and 46, respectively. Observing Figure 11.13c, a red light is shown at Sample 26. This red light is an indication that warns users about the fact that $RI_I < RI_T$ & $GSI_I > GSI_T$ at Sample 26. It implies that this red light is an early-warning/benefit that addresses the concern about the VM quality at Sample 26. Similarly, blue lights are present at Samples 27 and 44 due to the concerns of VM quality.

Note that, parameter deviations mentioned above may be caused by tool maintenance operations, recipe/parameter adjustments, or VM models fanning-out. That is, with Tool A and Tool B being of the same type (such as both are CVD tools), by deploying Tool A's VM models to Tool B, then Tool B can apply the automatic VM models' refreshing/calibrating capability of the advanced dual-phase algorithm to retrieve the VM accuracy of Tool B by refreshing three to five samples.

11.2.1.4 Summary

The TFT-LCD manufacturing flow and production tools are introduced in this section. Fab-wide deployments of the AVM system to the TFT, CF, and LCD manufacturing processes are also demonstrated. Three deployment types namely single-stage, dual-stage, and cooperative-tools are defined and illustrative VM application examples of all the three types are presented.

The target value for the liquid-crystal-drop process in LCD manufacturing will mainly depend on the thickness of PS processed in CF manufacturing, while the thicknesses of R, G, and B may also slightly affect the target value. As such, an inter-fab VM application that forwards the PS, R, G, and B VM values of each individual glass generated in the CF fab to the LCD fab for glass-to-glass liquid crystal volume control in the liquid-crystal-drop process can then be proceeded. It is believed that by adopting the methodology proposed in this section, the VMS can be applied to TFT-LCD manufacturing fab-wide, and various VM applications can be accomplished efficiently in the TFT-LCD industry.

11.2.2 Intelligent Yield Management (IYM) Deployment Examples in the TFT-LCD Industry

The IYM deployment examples in the TFT-LCD industry are presented below. As shown in Figure 10.3, the IYM system is composed of the Key-variable Search Algorithm (KSA) module and Reporting module. The key functions of the KSA module are illustrated below.

11.2.2.1 Introducing the TFT-LCD Production Tools and Manufacturing Processes for IYM Deployment

Four processes constitute the TFT-LCD manufacturing flow as displayed in Figure 11.1, which are TFT, CF, LCD, and LCM. Among them, the TFT, CF, and LCD processes require high-degree automation and are integrated into the front-end process; while the LCM process, which is mostly labor-oriented, is denoted as back-end process. The front-end process is adopted in these examples.

As shown in Figure 11.14, the TFT process consists of five layers: gate, semiconductor, data, protection, and ITO layers. Each layer includes the so-called photo engraving processes (PEP) such as film deposition, positive photoresist coating, exposure, developing, etching, and stripping. The coating, exposure, and developing processes can be combined into the photo step. The process flow of the semiconductor layer, as depicted in Figure 11.15, is the most complicated layer among those five ones.

The first step of the semiconductor process flow is to deposit film via CVD tools with the film thickness being the sampling measurement value. In the photo step, before acquiring the sampling

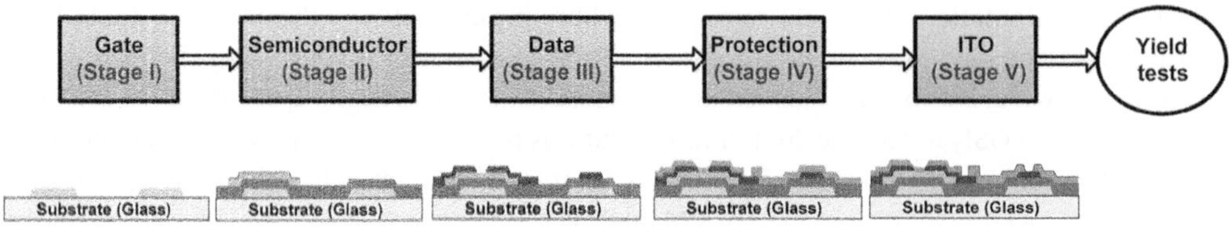

Figure 11.14 TFT manufacturing process. *Source:* Reprinted with permission from Ref. [2]; © 2017 IEEE.

measurement value (CD of line-width), the workpiece executes coating, exposure, and developing processes. After that, the etching process follows where the etching depth is taken as the sampling measurement value. Finally, with the metal width being the sampling measurement value, the stripper process removes the photoresist on the workpiece.

After going through those five stages shown in Figure 11.14, yield tests will be conducted. Yield tests may encounter Type 1–Type 10 yield losses that are resulted from various defects caused by electric-test failures, particles, etc.

To finish the TFT manufacturing process, PEP should be adopted onto each layer; thus there are 5 PEPs in total in the entire TFT process. This work adopts the TFT process for illustration.

11.2.2.2 KSA Deployment Example

The TFT array process data for TFT-LCD manufacturing are adopted as the illustrative example. Lot is the basic tracking unit of TFT production process. There are a total of 28 sheets in a full lot, and each sheet can be cut into six chips. There are five PEP layers in the TFT process and each layer consists of different number of devices. The process route descriptions, including layers, processes, and number of devices are shown in Table 11.1.

Both etching and film deposition machines contain multiple chambers. Each chamber is considered as a different device in this study, and each device in the algorithm is treated as a variable in $\mathbf{X}_R$. By summing up all the numbers of devices shown in Table 11.1, there are a total of 789 variables in this example. Note that all kinds of variables in $\mathbf{X}_R$ are denoted as devices in this work. Therefore, a chamber in a production tool, a metrology station, or a stocker, and so on is treated as a device as shown in Table 11.1.

The count of defects is by chip, along with 10 different yield-test types. Defects will be allocated in percentage into these 10 types according to their causes. For example, a certain chip has defects of Types 1 & 3 with percentages of 66 and 33% respectively, then the yield-test result of this chip will be $[0.66\ \ 0\ \ 0.33\ \ 0\ \ 0\ \ 0\ \ 0\ \ 0\ \ 0\ \ 0]$. If there are no defects detected in this chip, the result is $[0\ \ 0\ \ 0\ \ 0\ \ 0\ \ 0\ \ 0\ \ 0\ \ 0\ \ 0]$. Each sheet contains six chips. Therefore, the counts of defects of a sheet are to add up the yield-test results of all 6 chips. After discussing with domain experts, Type 2 Loss is the most critical defect, thus the analysis of this defect is conducted first within the time period of nine days.

A total of 113 lots were processed during this nine-days period. However, only 104 lots are kept after filtering through data quality index of X_R (DQI_{XR}). The accumulated Type 2 Loss results of these 104 lots are shown in Figure 11.16 with the largest value (6.9) being occurred at Lot 49. Thus, the root causes of Type 2 Loss at Lot 49 are searched as follows.

As indicated in Step 2 of Figure 11.17, besides $\mathbf{Y}$, both $\boldsymbol{y}$ and $\mathbf{X}_R$ are required to be the inputs for searching the Top N devices. To do so, total inspection is required for in-line metrology. However, the inline data, $\boldsymbol{y}$, have merely sampling inspection in this example; therefore, only $\mathbf{X}_R$ is adopted as input in this case.

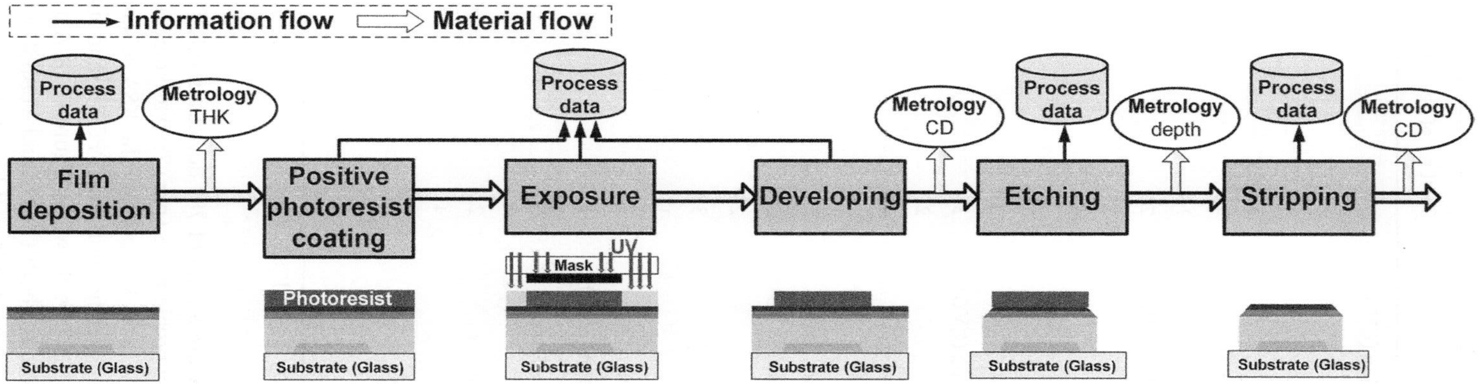

⇨

Figure 11.15 PEP flow of the semiconductor layer. *Source:* Reprinted with permission from Ref. [2]; © 2017 IEEE.

Table 11.1 Total number of devices in the TFT process.

Layer	Process	Total no. of devices
PEP 1: Gate	Film deposition (CVD)	34
	Lithography	25
	Etching	14
	Stripping	11
	Measurement	43
	Others	13
PEP 2: Semiconductor	Film deposition (CVD)	70
	Lithography	25
	Etching	38
	Stripping	11
	Measurement	33
PEP 3: Data	Film deposition (CVD)	34
	Lithography	25
	Etching	45
	Stripping	11
	Measurement	43
PEP 4: Protection	Film deposition (CVD)	70
	Lithography	25
	Etching	38
	Stripping	11
	Measurement	33
PEP 5: ITO	Film deposition (CVD)	17
	Lithography	25
	Etching	4
	Measurement	22
Others	Annealing	4
	Electrical test	13
	Final repair	26
	Stocker	26

Source: Reprinted with permission from Ref. [2]; © 2017 IEEE.

Lot 49 contains 28 sheets while there are a total of 789 devices in this TFT process. Therefore, this high dimensional variable selection problem has (p =) 789 variables of $\mathbf{X}_{\mathrm{R}}$ and (n =) 28 samples. The yield test result (**Y**) is Type 2 Loss. By applying the KSA scheme, the Top 10 devices searched by both triple-phase orthogonal greedy algorithm (TPOGA) and automated least absolute shrinkage and selection operator (ALASSO) are shown in Figure 11.18. A bar of solid outline indicates that a certain variable has been chosen by both TPOGA and ALASSO with the same sequential order, and the score of this variable is counted; a brown bar of segmented outline means that a certain variable has been chosen by both TPOGA and ALASSO with different sequential orders,

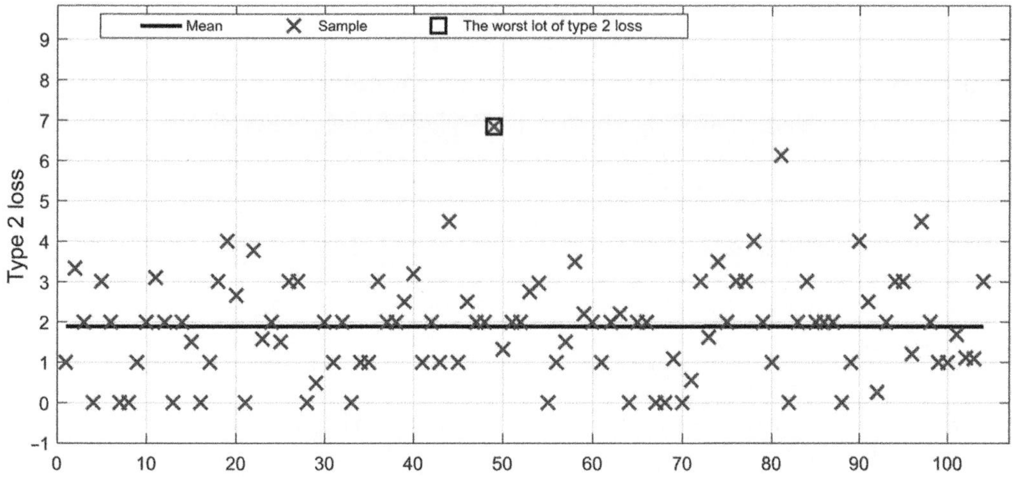

Figure 11.16 Accumulated Type 2 loss results. *Source:* Reprinted with permission from Ref. [2]; © 2017 IEEE.

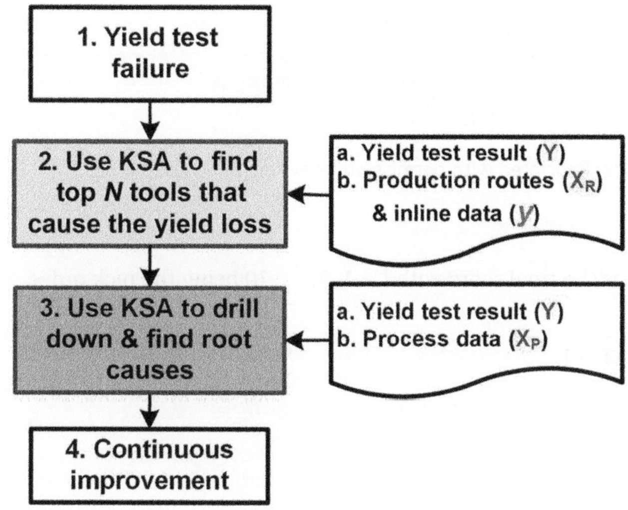

Figure 11.17 Procedure for finding the root causes of a yield loss by applying the KSA scheme. *Source:* Reprinted with permission from Ref. [2]; © 2017 IEEE.

the score of this variable is also counted; and a white bar shows that TPOGA and ALASSO pick out different variables without overlapping, the score of this variable is not counted. As such, RI_K is calculated by (11.4) and (11.5). The result of RI_K is 0.932 that is larger than the threshold (0.7). This implies that the search results of TPOGA and ALASSO are almost the same, and thus the results are reliable. The top two devices are investigated as follows.

$$FS_l = \begin{cases} \dfrac{OS_l}{\sum_{l=1}^{3} OS_l} \times 0.8, \text{when pick order is } 1,2,3. \\ \dfrac{OS_l}{\sum_{l=4}^{10} OS_l} \times 0.2, \text{when pick order is } 4,5,\ldots,10. \end{cases} \quad (11.4)$$

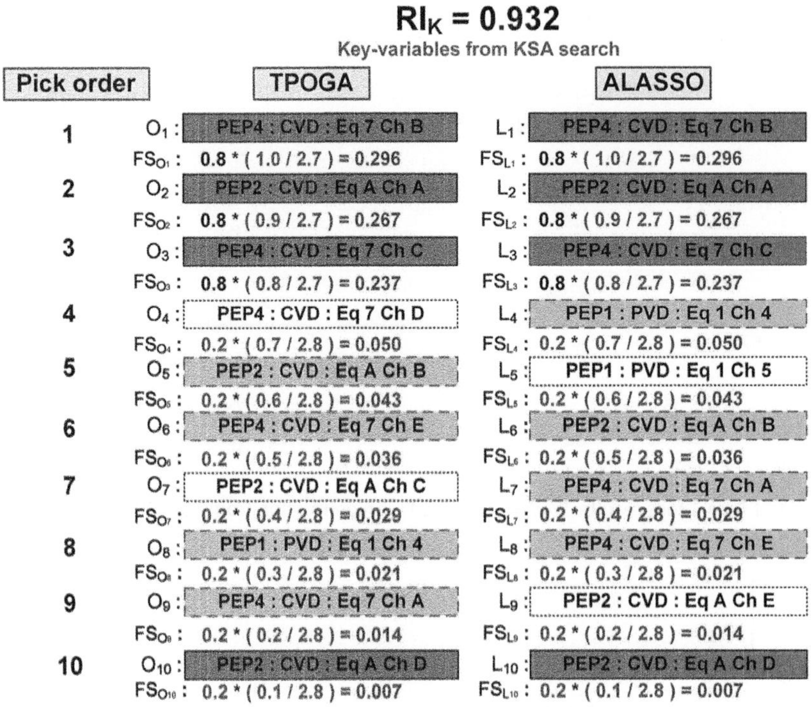
RIK = 0.932
Key-variables from KSA search
Pick order
TPOGA
ALASSO
1
O1 : PEP4 : CVD : Eq 7 Ch B
FSO1 : 0.8 * (1.0 / 2.7) = 0.296
L1 : PEP4 : CVD : Eq 7 Ch B
FSL1 : 0.8 * (1.0 / 2.7) = 0.296
2
O2 : PEP2 : CVD : Eq A Ch A
FSO2 : 0.8 * (0.9 / 2.7) = 0.267
L2 : PEP2 : CVD : Eq A Ch A
FSL2 : 0.8 * (0.9 / 2.7) = 0.267
3
O3 : PEP4 : CVD : Eq 7 Ch C
FSO3 : 0.8 * (0.8 / 2.7) = 0.237
L3 : PEP4 : CVD : Eq 7 Ch C
FSL3 : 0.8 * (0.8 / 2.7) = 0.237
4
O4 : PEP4 : CVD : Eq 7 Ch D
FSO4 : 0.2 * (0.7 / 2.8) = 0.050
L4 : PEP1 : PVD : Eq 1 Ch 4
FSL4 : 0.2 * (0.7 / 2.8) = 0.050
5
O5 : PEP2 : CVD : Eq A Ch B
FSO5 : 0.2 * (0.6 / 2.8) = 0.043
L5 : PEP1 : PVD : Eq 1 Ch 5
FSL5 : 0.2 * (0.6 / 2.8) = 0.043
6
O6 : PEP4 : CVD : Eq 7 Ch E
FSO6 : 0.2 * (0.5 / 2.8) = 0.036
L6 : PEP2 : CVD : Eq A Ch B
FSL6 : 0.2 * (0.5 / 2.8) = 0.036
7
O7 : PEP2 : CVD : Eq A Ch C
FSO7 : 0.2 * (0.4 / 2.8) = 0.029
L7 : PEP4 : CVD : Eq 7 Ch A
FSL7 : 0.2 * (0.4 / 2.8) = 0.029
8
O8 : PEP1 : PVD : Eq 1 Ch 4
FSO8 : 0.2 * (0.3 / 2.8) = 0.021
L8 : PEP4 : CVD : Eq 7 Ch E
FSL8 : 0.2 * (0.3 / 2.8) = 0.021
9
O9 : PEP4 : CVD : Eq 7 Ch A
FSO9 : 0.2 * (0.2 / 2.8) = 0.014
L9 : PEP2 : CVD : Eq A Ch E
FSL9 : 0.2 * (0.2 / 2.8) = 0.014
10
O10 : PEP2 : CVD : Eq A Ch D
FSO10 : 0.2 * (0.1 / 2.8) = 0.007
L10 : PEP2 : CVD : Eq A Ch D
FSL10 : 0.2 * (0.1 / 2.8) = 0.007

Figure 11.18 RI_K result of X_R search. *Source:* Reprinted with permission from Ref. [2]; © 2017 IEEE.

where OS_l is the original score and FS_l is the final score with $l = 1, 2, \ldots, 10$ being the pick order.

$$RI_K = \sum_{i=1}^{10}\sum_{j=1}^{10}\left(\frac{FS_{O_i} + FS_{L_j}}{2}\right) \quad \text{if} \quad O_i = L_j \tag{11.5}$$

where

- FS_{Oi} final score of O_i;
- FS_{Li} final score of L_j;
- O_i ith pick variable of TPOGA, $i = 1, 2, 3, \ldots, 10$;
- L_j jth pick variable of ALASSO, $j = 1, 2, 3, \ldots, 10$.

As shown in Figure 11.19, the Top 1 device is Chamber B of Equipment 7 of Film Deposition process in Protection layer; while Top 2 device is Chamber A of Equipment A of Film Deposition process in Semiconductor layer. Figure 11.19a shows that among those 28 sheets, 3 out of 8 Type 2 Loss samples were processed by the Top 1 device. Also, Figure 11.19b indicates that another 3 out of 8 Type 2 Loss samples were processed by the Top 2 device. To find out the root causes, Step 3 of Figure 11.17 should be performed by inputting **Y** and $\mathbf{X}_P$ into the KSA scheme.

The Top 2 device is selected for illustration. The process data ($\mathbf{X}_P$) of the Top 2 device has 27 variables. After conducting the KSA analysis on all the devices of the same stage to which the Top 2 device belongs, its RI_K value of X_P search is 0.864 (>0.7) as shown in Figure 11.20. Therefore, the search result is reliable with the Top 1 variable being Control Voltage.

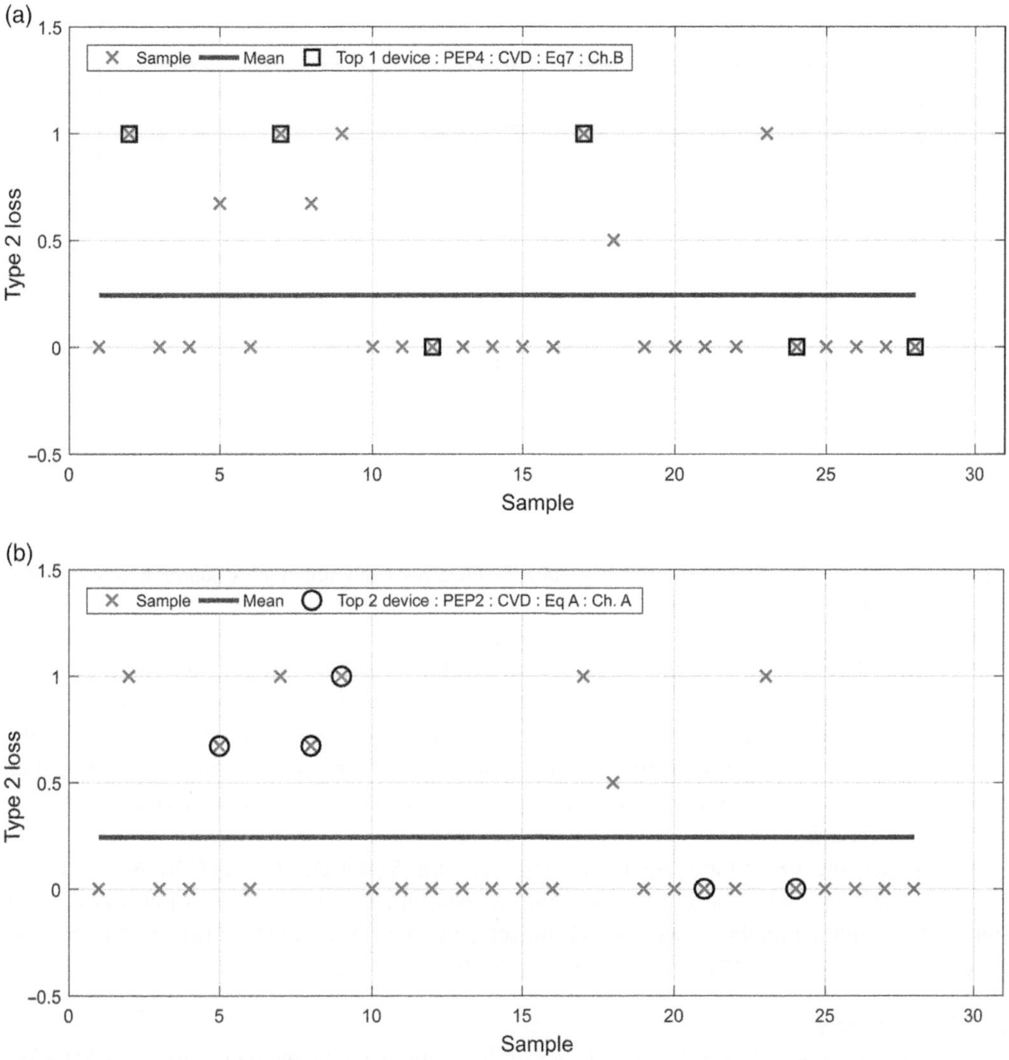

Figure 11.19 KSA search results of Type 2 loss on Lot 49. (a) Top 1 Device: Chamber B of Equipment 7 of film deposition process in protection layer. (b) Top 2 Device: Chamber A of Equipment A of film deposition process in semiconductor layer. *Source:* Reprinted with permission from Ref. [2]; © 2017 IEEE.

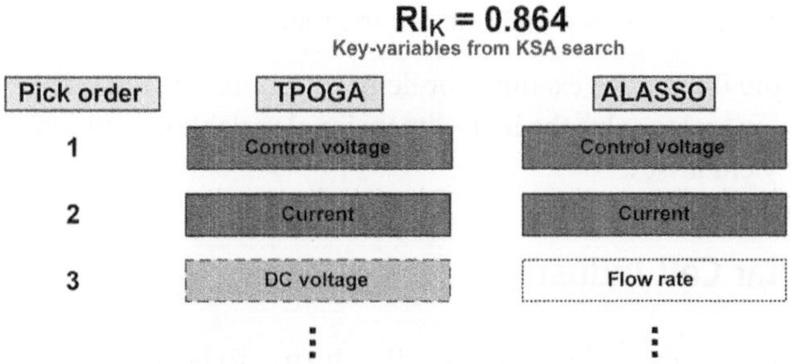

Figure 11.20 RI_K result of X_P search. *Source:* Reprinted with permission from Ref. [2]; © 2017 IEEE.

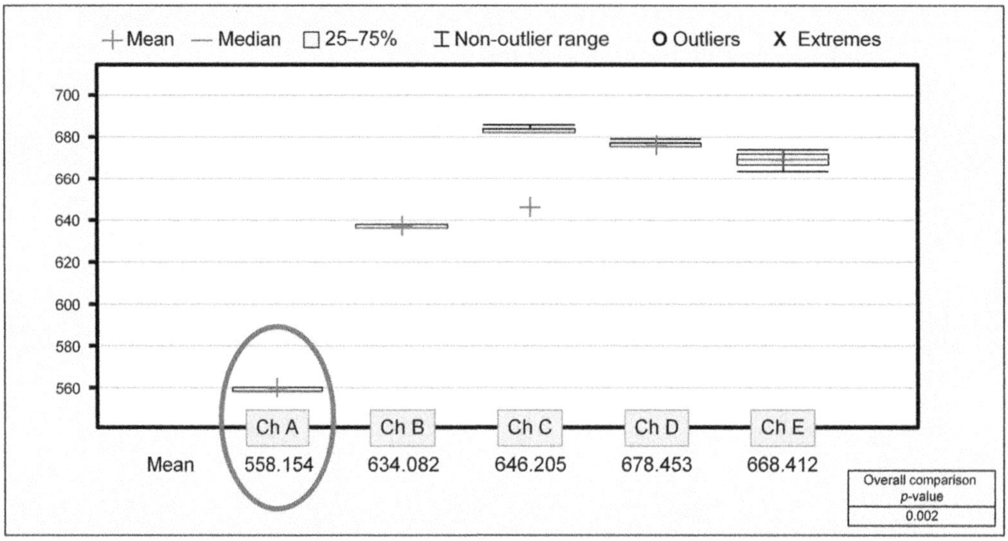

Figure 11.21 Root cause analysis of control voltage on Chamber A of Equipment A. *Source:* Reprinted with permission from Ref. [2]; © 2017 IEEE.

To confirm that "Control Voltage" is the root cause, the "Control Voltage" values of all the five chambers (A–E) of Equipment A are drawn and verified by a box plot chart and the hypothesis test is executed as shown in Figure 11.21. It shows that *p*-value (=0.002) is less than 0.05, which indicates that Chamber A's "Control Voltage" value in the circle is indeed less than those of the other chambers. As such, the root cause of this Type 2 Loss is due to the abnormality found in Chamber A's "Control Voltage".

The KSA scheme was implemented in a computer with i5-3450@3.10 GHz CPU, 6.0 GB RAM, and 1 TB hard disk. With the illustrative example mentioned above, the execution times of the Steps 2 (for searching the key stages) and 3 (for selecting the key variables in the most suspicious stage) shown in Figure 11.17 are 1.22 seconds and 1.36 seconds, respectively.

11.2.2.3 Summary

A two-phase process is suggested to apply the KSA scheme for searching the root causes of yield losses:

- Phase I: Feed yield test results ($\mathbf{Y}$) as well as total-inspection inline data ($\boldsymbol{y}$) and/or production routes ($\mathbf{X}_{\mathrm{R}}$) to the KSA scheme to find Top N devices that cause the yield losses.
- Phase II: After obtaining the most suspicious device from Phase I, input Tube to Tube (T2T) control ($\mathbf{Y}$) as well as the process data ($\mathbf{X}_{\mathrm{P}}$) of all the devices of the stage to which the most suspicious device belongs to the KSA scheme to drill down and find out the root causes.

The TFT-LCD process is adopted as the illustrative example for demonstration. The test results reveal that the KSA scheme is a promising tool to solve the high-dimensional variable search problem so as to find out the root causes of yield losses.

11.3 Application Case II: Solar Cell Industry

T2T control with AVM and IPM has been deployed in the solar cell industry. To begin with, the solar cell manufacturing process and requirement analysis of intelligent manufacturing are introduced. Then, the deployment of T2T control with AVM is illustrated followed by the presentations of the factory-wide IPM deployments. The authors would like to thank Motech Industries, Inc., Taiwan, ROC for providing the raw data used in these application cases.

11.3.1 Introducing the Solar Cell Manufacturing Process and Requirement Analysis of Intelligent Manufacturing

New and clean energy is becoming more important because of the global warming issue, and the solar system is an option of the new and clean energy. Solar cell is the core of the solar system that generates electricity from sunlight. One of the challenges for solar-cell manufacturing is to handle massive production with stable quality and low cost.

The manufacturing processes and control levels in a c-Si solar-cell factory are described as follows [3]. Six main processes are performed on each silicon wafer that contains many solar cells.

A) Texturing: The wafer is treated to remove defects and create minute pyramidal structures on the wafer surface to reduce reflectivity and increase power output.
B) Diffusion: A P-N junction is created on the wafer surface by using high temperature to bind a layer of phosphorous over the silicon surface to create electricity in the wafer.
C) Phosphosilicate Glass (PSG) Removal: The wafer is shaved off using plasma etching to insulate the top and bottom sides and prevent unwanted current flow.
D) Plasma-Enhanced Chemical Vapor Deposition (PECVD): This process is also denoted as an anti-reflection coating. A layer of silicon nitride (SiN) is deposited on the wafer surface to further reduce reflection and improve power output.
E) Printing: Silver electrodes and an aluminum layer are screen-printed on the front and back of the wafer respectively. The wafer is completed when ceramic powder is bonded to the wafer surface through a high-heat process called sintering.
F) Cell Test: Electricity and optical tests are performed to ensure the quality of final products.

Among the manufacturing processes and control levels in c-Si solar-cell, PECVD is a key process to determine the efficiency of the solar cell. The efficiency of solar cell will influence the construction area of solar system. Therefore, the stability of quality should be confident enough to handle the massive production of solar cells. As described in Section 8.4, Run-to-Run (R2R) control is commonly utilized to enhance the process capability. R2R control is the technique of modifying recipe parameters or the selection of control parameters between runs to improve processing performance. The R2R control is also called the T2T control for the PECVD case. In the deposition process of the solar-cell manufacturing, its sampling rate is less than 10%. Therefore, AVM is proposed to provide 100% total inspection on the PECVD process such that the T2T control can be applied to tune the recipe parameters for each tube precisely.

Moreover, a key problem prohibiting effective utilization of VM in T2T control is the inability to take the reliance level in the VM feedback loop of T2T control into consideration. The reason is that the result of adopting an unreliable VM value may be worse than if no VM at all is utilized, please refer to Section 8.4 for details.

The authors presented a preliminary study of R2R control utilizing VM with RI to handle the quality issue of VM in [4]. In the preliminary study, only RI [5] is considered to gauge the reliability of the VM results and tune the R2R controller gain, and the scheme of R2R+VM with RI is only tested by simulation runs with the simulation model of the chemical mechanical polishing (CMP) process.

Besides RI, the GSI is also considered for tuning the T2T controller gain (α) and the scheme of T2T+VM with RI&GSI is implemented and tested in a real PECVD equipment. The T2T controller will utilize the VM value and its accompanying RI&GSI of the current (k) run as well as the information of a batch at the first run to calculate the suggestion value of the deposition time for the following (k+2) run so as to improve the process capability index (Cpk). The T2T with AVM deployment examples will be presented in Section 11.3.2.

Moreover, to improve the overall equipment effectiveness (OEE), the factory-wide IPM system based on the Intelligent Factory Automation (iFA) system platform presented in Chapter 6 is proposed. The so-called CPA and AMCoT are adopted to construct the iFA infrastructure, upon which the IPM system is built. The scheme of Health Index Hierarchy (HIH) that contains several health indexes varying from the fundamental equipment level (device) up to the highest management level (factory) is proposed to monitor the health statuses of all the devices in an entire factory. The solar-cell manufacturing factory is adopted as an illustrative example to demonstrate how this IPM system with the scheme of HIH functions. Finally, the factory-wide IPM deployment examples will be presented in Section 11.3.3.

11.3.2 T2T Control with AVM Deployment Examples

Generally, a T2T controller is designed for adjusting the following (k+1)th run by using the result of the current kth run. However, it may not be the case in real production environment due to reasons such as metrology delay, data-collection constraints, cycle-time-reduction consideration, etc. [6].

A PECVD tool possesses four manufacturing tubes. A buffer (which contains six boats) is deployed in front of a PECVD tool. Each boat carries about 200 wafers. A boat in the buffer will be immediately selected and randomly loaded into any empty tube of the PECVD tool for cycle-time-reduction consideration. As such, adjusting the following (k+1)th run by using the result of the current (kth) run cannot be achieved in this PECVD T2T control because computation delay of a T2T control scheme is inevitable. Therefore, as shown in Figure 11.22, adjusting the (k+2)th run by using the result of the current (kth) run is adopted in the T2T control scenario of the PECVD manufacturing process. Also indicated in Figure 11.22, the T2T controller may utilize AM or VM for controller feedback computation [6].

There is no ID on each wafer in solar-cell manufacturing such that it is quite difficult to trace each wafer. Therefore, the solar-cell manufacturing process is performed based on batch or lot management. One lot contains about 5000 non-ID wafers in solar-cell manufacturing. Since each boat can carry about 200 wafers, those wafers in the same lot are divided and loaded into roughly 25 boats for PECVD process. Each PECVD tool has four tubes; hence, a certain tube may perform 1 to 6 process runs of a single lot on average.

All of the wafers in the same lot share similar physical characteristics. However, physical properties of each individual lot may be different from one another due to different suppliers, quality classification, performance of pre-process, etc. Because the fab has no suitable measurement tools to measure or evaluate those physical properties automatically, it requires operators to evaluate the properties of wafers (in the same lot) manufactured by the previous process step with manual inspections. At the same time, the operators manually select the suitable recipe of the first boat of a certain lot based on the manual-inspections results. After that, operator tunes the recipe parameters of the following (k+2)th runs according to the PECVD process result of the current (kth) run manually until all of the boats of the same lot are processed [6].

Therefore, the control scheme of T2T+VM with RI&GSI will be applied to automate the process of adjusting the recipe parameters of the following (k+2)th runs according to the PECVD process result of the current (kth) run. However, the inspections and recipe selection of the first run of a certain lot still require manual operations.

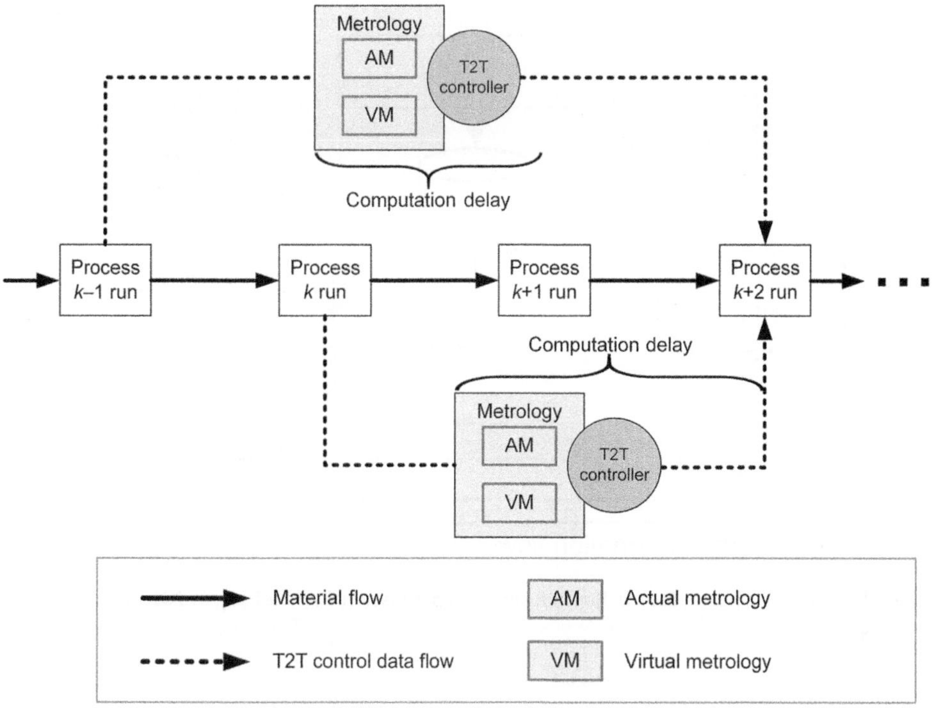

Figure 11.22 T2T control scenario of the PECVD manufacturing process. *Source:* Reproduced with permission from Ref. [6]; © 2013 IIE.

11.3.2.1 T2T+VM Control Scheme with RI&GSI

The control scheme of T2T+VM with RI&GSI to adjust the recipe parameters of the following ($k+2$)th runs according to the actual-metrology (AM)/virtual-metrology (VM) of the current (kth) run is shown in Figure 11.23.

The details of the T2T controller are depicted in Figure 11.24.

In the PECVD process, the thickness is influenced by gas flow, boat run numbers, deposition time, temperature, pressure, and so on. After careful examination and consideration, the thickness (y_k) of the PECVD process outputs is chosen as the metrology data to be regulated; and the disposition time (u_{k+2}) is selected to be the key parameter for adjusting the thickness (y_{k+2}) by T2T control.

As shown in Figure 11.23, either AM (y_k) or VM$_\mathrm{I}$ ($\hat{y}_k$) is adopted by the T2T controller where VM$_\mathrm{I}$ represents phase-I VM value of the dual-phase VM scheme as described in Chapter 8. After finishing the process, the AM provides the measurement value (y_k) to the T2T controller by a sampling plan. If AM is utilized in the T2T controller, then the exponentially weighted moving average (EWMA) coefficient is denoted as α_1. Meanwhile, the AVM system provides VM$_\mathrm{I}$ and its accompanying RI&GSI to the T2T controller. The RI and GSI values are utilized to generate the EWMA coefficient (α_2) when VM$_\mathrm{I}$ is chosen by the T2T controller.

The goal of T2T control is to keep the process output approaching the desired target value. However, the plant noise will impact the process output. The kth run model offset or disturbance (η_k) is the deviation between the target value (Tgt) and the process output (y_k) measured from AM:

$$\eta_k = Tgt - y_k. \tag{11.6}$$

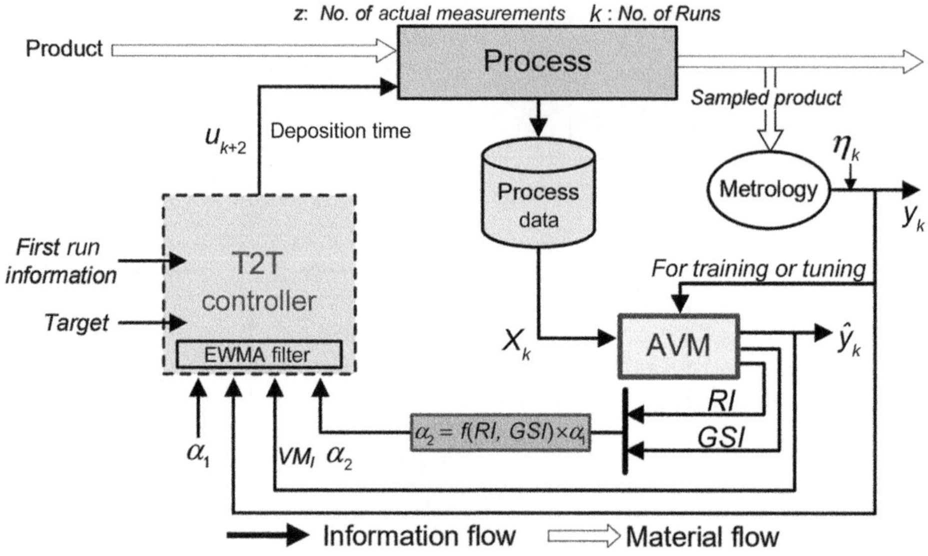

Figure 11.23 T2T control scheme. *Source:* Reproduced with permission from Ref. [6]; © 2013 IIE.

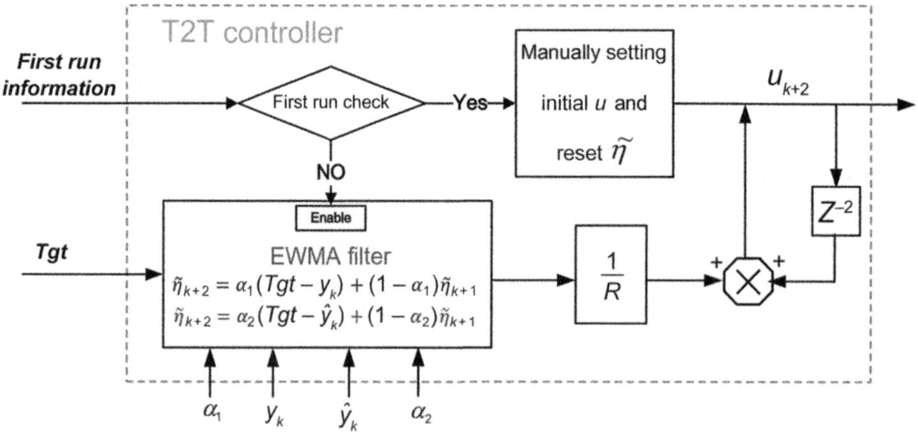

Figure 11.24 T2T controller. *Source:* Reproduced with permission from Ref. [6]; © 2013 IIE.

In the preliminary study [4], the relationship between α_2 and α_1 is expressed as: $\alpha_2 = RI \times \alpha_1$ for considering the reliability of VM. As explained in [5], when VM is applied, no actual measurement value is available to verify the accuracy of the virtual measurement value. Therefore, instead of the standardized actual measurement value Z_{y_i}, the standardized multi-regression (MR) prediction value $Z_{\hat{y}_{r_i}}$ is adopted to calculate the RI. This substitution may cause inevitable gauging errors in the RI. For example, when deviations occur in the process data such that the similarity between this new set of process data and the model set of all historical process data is bad, then the accuracy of the VM value conjectured by NN and the reference prediction value predicted by MR would not be good. However, it is still possible that the conjectured VM value and the predicted reference value happen to be nearly the same such that the *RI* value is higher than RI_T. To compensate for this unavoidable substitution, the *GSI* is proposed to help the *RI* gauge the reliance level of VM. Therefore, the relationship between α_2 and α_1 is modified to be

$$\alpha_2 = f\left(RI, GSI\right) \times \alpha_1 \tag{11.7}$$

where *f*(*RI*, *GSI*) is an adjustment function to consider the reliability of VM as follows:

$$f\left(RI, GSI\right) = 1 \quad \text{if } RI \geq RI_T \text{ and } GSI \geq GSI_T. \tag{11.8}$$

$$f\left(RI, GSI\right) = 0 \quad \text{if } RI < RI_T \text{ or } GSI < GSI_T. \tag{11.9}$$

Equations (11.8) and (11.9) are the conditions that the T2T controller adopts or rejects VM value for EWMA controller gain (α) calculation. When *RI* is greater than RI_T and *GSI* is greater than GSI_T, the VM value is reliable and can be adopted, therefore, α_2 is set to be equal to α_1. On the contrary, if *RI* is smaller than RI_T or *GSI* is smaller than GSI_T, it means the VM value is not reliable and may not be adopted for tuning the T2T controller gain. In this case, α_2 is set to zero (0) such that the current VM value is discarded and the estimated model disturbance of (*k*+*2*)th run ($\tilde{\eta}_{k+2}$) is equal to that of (*k*+*1*)th run ($\tilde{\eta}_{k+1}$).

In conclusion, according to the PECVD manufacturing process mentioned above, the T2T controller is designed for (*k*+*2*)th run. Using an EWMA filter with AM, the model disturbance of the (*k*+*2*)th run is estimated to be:

$$\tilde{\eta}_{k+2} = \alpha_1\left(Tgt - y_k\right) + \left(1 - \alpha_1\right)\tilde{\eta}_{k+1}. \tag{11.10}$$

The method of calculating α_1 can be found in [7].

When y_k is conjectured or predicted by a VMS, it becomes $\hat{y}_k$. By adopting $\hat{y}_k$ in the EWMA filter calculation, the model disturbance of the (*k*+*2*)th run is then estimated to be:

$$\tilde{\eta}_{k+2} = \alpha_2\left(Tgt - \hat{y}_k\right) + \left(1 - \alpha_2\right)\tilde{\eta}_{k+1}. \tag{11.11}$$

Equations (11.10) and (11.11) are included in the down-left corner of Figure 11.24. In order to regulate the deviation between the target value of the process output (thickness) and the output (y_k) of the current (*k*th) run, the T2T controller considers the deposition time of the *k*th run (u_k) as the feedback value. Then, the estimated disturbance of the (*k*+*2*)th run ($\tilde{\eta}_{k+2}$) is converted and added to the deposition time of the (*k*+*2*)th run (u_{k+2}) for bringing the process output to its target value as follows:

$$u_{k+2} = u_k + \frac{1}{R}\tilde{\eta}_{k+2} \tag{11.12}$$

where *R* represents the deposition rate and its value is assigned by the process engineer according to experiences. The implementation of (11.12) is depicted on the right side of Figure 11.24, which shows that u_{k+2} consists of two terms: the first term is u_k ($= u_{k+2} * Z^{-2}$) and the second term is $\frac{1}{R} * \tilde{\eta}_{k+2}$.

11.3.2.2 Illustrative Examples of T2T Control with AVM

The scheme of T2T+VM with RI&GSI as shown in Figures 11.23 and 11.24 has been implemented and tested in a real PECVD equipment. The scenario of applying the AVM system to the PECVD

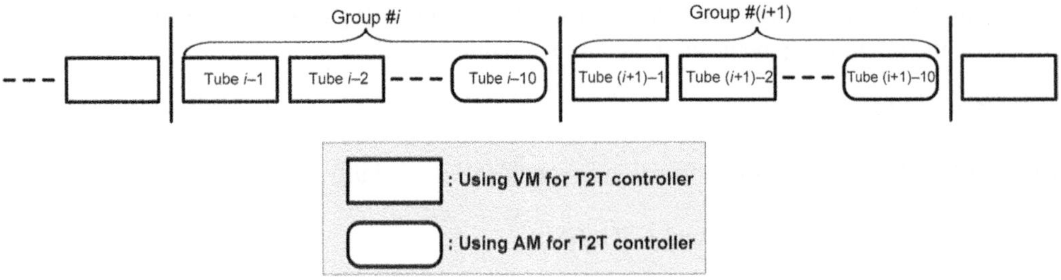

Figure 11.25 Scenario of Applying the AVM system to the PECVD process. *Source:* Reproduced with permission from Ref. [6]; © 2013 IIE.

process is shown in Figure 11.25. All the tubes being processed in the PECVD tool are divided into groups as depicted in Figure 11.25. Each group contains 10 tubes. The thickness of the last tube in a group is measured by AM while those of the other nine tubes are conjectured by VM. The actual thickness measurement is performed manually for this PECVD process. For T2T control, every tube should be measured. Therefore, 90% of manpower for thickness measurement is saved by applying the AVM system.

Based on the PECVD manufacturing process explained above, the VM value and its accompanying RI&GSI of the current (kth) run as well as the first-run information of a lot are utilized to calculate the suggestion value of the deposition time for the following ($k+2$)th run (u_{k+2}) so as to improve the process capability index (Cpk) and the process mean-absolute-percentage error with respect to the target value (MAPE_P). The formulas of Cpk and MAPE_P are listed below:

$$Cpk\left(Process\ Capability\right) = \min\left\{\frac{UCL - mean\left(y_k\right)}{3\times std\left(y_k\right)}, \frac{mean\left(y_k\right) - LCL}{3\times std\left(y_k\right)}\right\} \tag{11.13}$$

$$MAPE_P = \frac{\sum_{i=1}^{k}\left|\left(y_k - Tgt\right)/Tgt\right|}{k}\times 100\% \tag{11.14}$$

where y_k is the process output; Tgt is the target value; UCL and LCL are the upper-control-limit and lower-control-limit of y_k, respectively.

Five cases are included in this section to illustrate the validation procedure of the T2T+VM with RI&GSI scheme. To begin with, Case 0 is performed to verify that the VM accuracy is good enough for supporting the T2T control. Then, Case 1 is executed to verify that the T2T control scheme by applying AM (T2T+AM) can indeed improve Cpk. The purpose of Case 2 is to certify that the T2T control scheme with VM (T2T+VM) is compatible with that of T2T+AM. Subsequently, Case 3 adds two bad-quality VM samples and applying T2T+VM without RI&GSI to demonstrate that adopting an unreliable VM sample in the T2T control calculation may be worse than if no VM at all is utilized. Finally, Case 4 is acted to show that the T2T+VM with RI&GSI scheme can detect and delete the bad-quality VM samples in the T2T control calculation and sustain the process quality. The experimental results of Cases 0–4 are presented as follows.

Case 0: VM Accuracy Verification

Statistical process control (SPC) is a popular method using the statistical theories to ensure the quality of production. However, SPC monitoring data are based on sampling plan. To accomplish this requirement, large amounts of measurement are needed, and therefore production cycle time

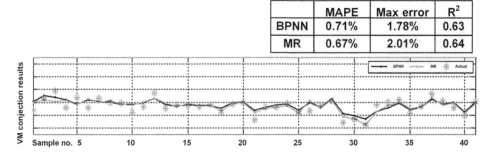

	MAPE	Max error	R^2
BPNN	0.71%	1.78%	0.63
MR	0.67%	2.01%	0.64

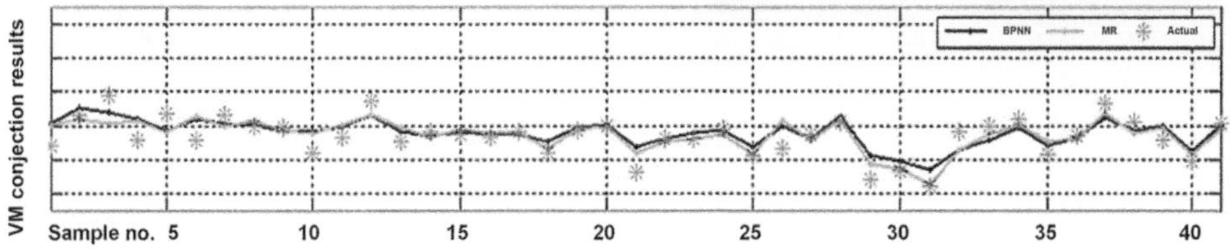

Figure 11.26 VM accuracy verification. *Source:* Reproduced with permission from Ref. [6]; © 2013 IIE.

and cost increases. To resolve the problem mentioned above, VM was proposed. If the VM conjecture model is fresh and accurate enough, it can generate a VM value within seconds after collecting the complete tool process data of a tube. Therefore, T2T control is realized in PECVD tools of a solar cell manufacturing factory. Thus, the accuracy of AVM conjecture needs to be good enough before T2T control applied.

According to the authors' experiences, the R^2 values of VM should be greater than 0.5 so that the VM values may be adopted for R2R-control computation [7]. Forty-one (41) testing samples collected from the real PECVD tool are utilized in Case 0. To ensure that the VM accuracy is good enough, observing Figure 11.26, the R^2 values of VM by applying the BPNN and MR are 0.63 and 0.64, respectively. Therefore, these VM prediction results can be utilized for supporting T2T control. After confirming that the VM accuracy is good enough, Cases 1–4 can then be commenced by setting $\alpha_1 = 0.35$ and $\tilde{\eta}_1 = 0$ initially. Both Cpk and $MAPE_P$ [as expressed in (11.13) and (11.14), respectively] are applied to evaluate and compare the performance of Cases 1–4.

Case 1: T2T with AM

As explained, the T2T control scheme is applied to automate the process of adjusting the recipe parameters of the following ($k+2$)th runs according to the PECVD process result of the current (kth) run. However, the inspections and recipe selection of the first run of a certain lot still require manual operations. Also, a certain tube may perform 1 to 6 process runs of a single lot on average.

In this Case, AM is adopted first and the experimental results are shown in Figure 11.27. Note that, instead of the real thickness, normalized thickness with its target value being zero and its unit being micro-meter is adopted. Figure 11.27a displays all samples; while the first-run and ($k+1$)th samples, which cannot be controlled by the T2T scheme, are shown with the highlighted blocks. The up-right corner of Figure 11.27a indicates that the Cpk improvement is small (from 1.24 to 1.28); it is due to the fact that only 20 out of 41 samples are adjusted by the T2T control scheme. To be exact, only those 20 advanced-process-control (APC) samples controlled by T2T scheme should be compared for Cpk improvement evaluation. As shown in Figure 11.27b, the thickness values after T2T control are much closer to the target value than those without T2T control such that the Cpk is improved from 1.30 to 1.93 (48% up) and the $MAPE_P$ is reduced from 1.92 to 0.97%. The line of AM values shown in the middle of Figure 11.27b is the target value. Therefore, it is confirmed that the T2T+AM control scheme can indeed improve Cpk.

Case 2: T2T with VM

As indicated in Figure 11.28b for the case of APC samples, the Cpk is improved from 1.30 to 1.90 (46% up) and the $MAPE_P$ is reduced from 1.92 to 1.10%. The Cpk and $MAPE_P$ values of Cases 1 and

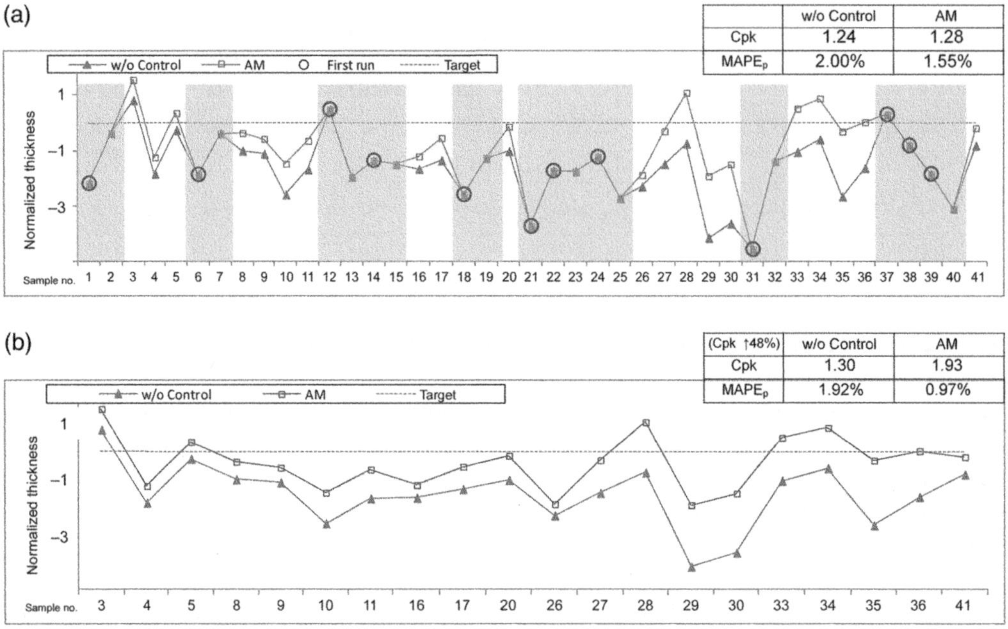

Figure 11.27 T2T with AM. (a) All samples. (b) APC samples. *Source:* Reproduced with permission from Ref. [6]; © 2013 IIE.

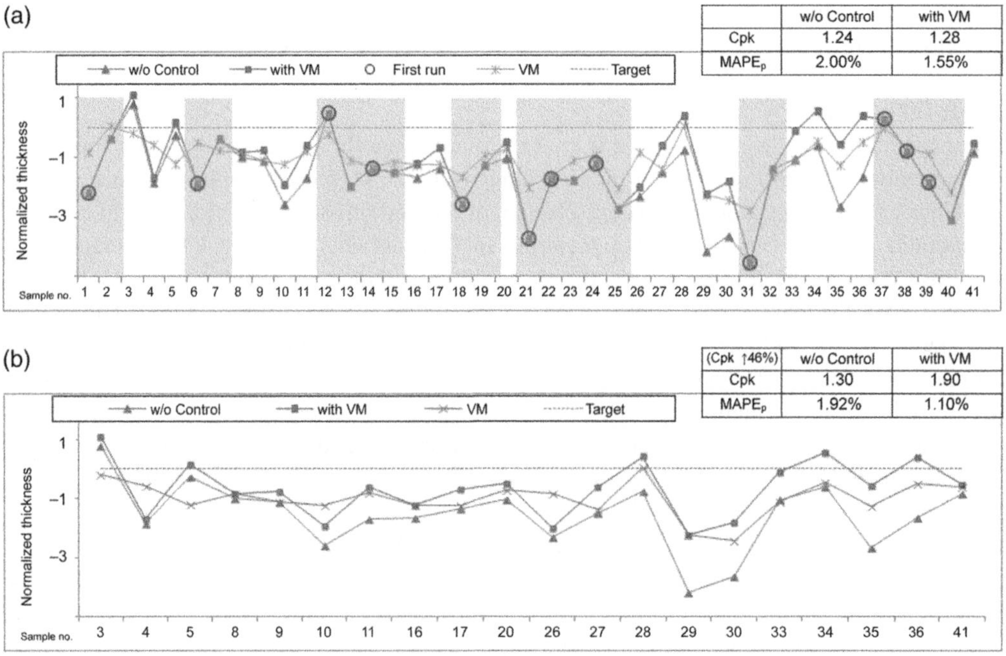

Figure 11.28 T2T with VM. (a) All samples. (b) APC samples. *Source:* Reproduced with permission from Ref. [6]; © 2013 IIE.

Table 11.2 Cpk and MAPE$_P$ values of Cases 1 and 2.

	Without control	CASE 1 T2T+AM (Cpk ↑48%)	CASE 2 T2T+VM (Cpk ↑46%)
Cpk	1.30	1.93	1.90
MAPE$_p$	1.92%	0.97%	1.10%

Source: Reproduced with permission from Ref. [6]; © 2013 IIE.

2 are tabulated in Table 11.2 for easy comparison. Comparing the values in Table 11.2, it is testified that the effect of T2T+VM is compatible with that of T2T+AM.

Case 3: Add Two Bad-Quality VM Samples & Apply T2T+VM w/o RI&GSI

Samples 7 and 25 of the original 41 samples are replaced by two bad-quality VM values and the T2T+VM control scheme is applied without considering RI and GSI. The experimental results of all-samples' case are shown in Figure 11.29a with Cpk reduced from 1.24 to 0.87; and, those of APC-samples' case are depicted in Figure 11.29b with Cpk reduced from 1.30 to 0.67 (48% down). Therefore, it is obvious that adopting an unreliable VM value may be worse than if no VM at all is utilized.

According to the authors' experiences of testing the AVM system in the Taiwan Semiconductor Manufacturing Company (tsmc), Ltd. [5] and Chi Mei Optoelectronics Corporation, Ltd. (CMO) [8], similar bad VM values (as shown in Figure 11.29a) are common to occur and can be detected by RI and/or GSI. Please refer to Figure 6 of [5] and Figure 4 of [8] for details.

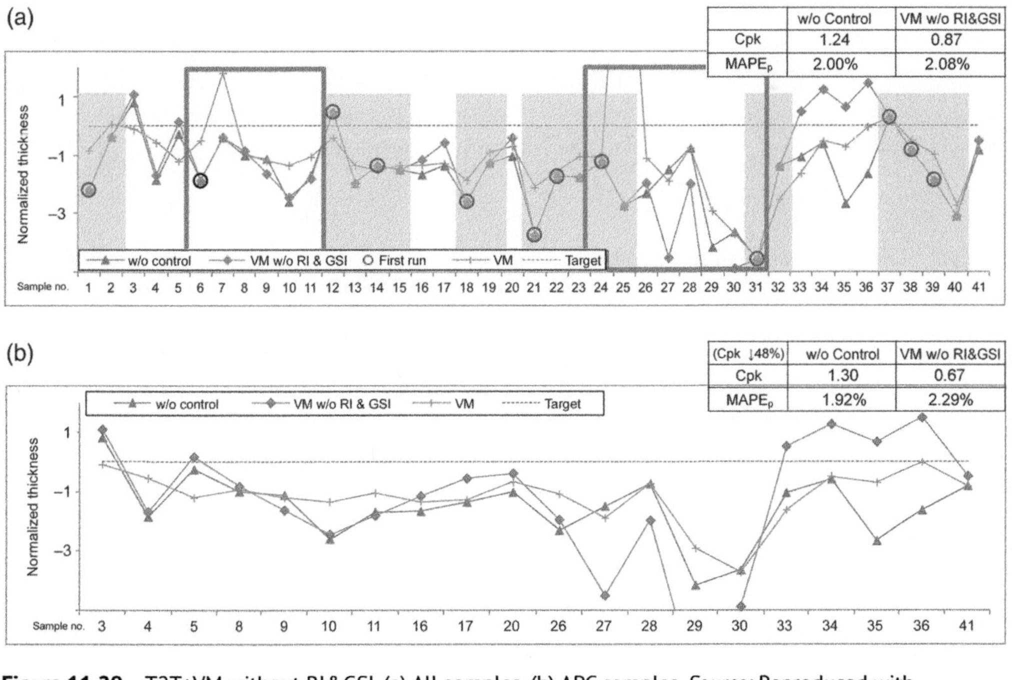

Figure 11.29 T2T+VM without RI&GSI. (a) All samples. (b) APC samples. *Source:* Reproduced with permission from Ref. [6]; © 2013 IIE.

Case 4: Add Two Bad-Quality VM Samples & Apply T2T+VM with RI&GSI

Samples 7 and 25 of the original 41 samples are, again, replaced by two bad-quality VM values in this case; while the T2T+VM control scheme is applied with RI&GSI being considered. The experimental results of all-samples' case are shown in Figure 11.30a with Cpk increased from 1.24 to 1.29; and, those of APC-samples' case are depicted in Figure 11.30b with Cpk increased from 1.30 to 1.63 (25% up). Therefore, it is apparent that adopting T2T+VM with RI&GSI can detect & delete the bad-quality VM samples and sustain the process quality.

To further analyze the differences between the scheme of T2T+VM without RI&GSI and that of T2T+VM with RI&GSI, the experimental results of both schemes at Samples 19 to 31 with Sample 25 being the bad-quality one are plotted in the same Figure 11.31. In this example, both RI_T and GSI_T are set to be 0.7. Owing to the fact that $RI < RI_T$ and $GSI < GSI_T$, the VM value of Case 4 are filtered out by setting $\alpha_2 = 0$; while the VM value of Case 3 is still adopted to adjust the T2T controller gain with $\alpha_2 = \alpha_1 = 0.35$. The effect of filtering out the bad-quality VM value is displayed in Sample 27, which shows that its actual-thickness value of Case 3 is pulled down by the T2T controller since the VM value at Sample 25 is too high. As for Case 4, the actual-thickness values of Samples 25 and 27 show no much difference and are close to the actual-thickness value produced by the T2T+AM scheme.

In addition to Cpk enhancement, the production rate of the PECVD process can also be improved by applying T2T+VM with RI&GSI (also denoted T2T+AVM). As indicated in Figure 11.32, the cycle time without APC is $T_P + T_S + T_C + T_M$ while that with T2T+AVM is $T_P + T_A$. Therefore, the cycle time reduction (ΔT) is

$$\Delta T = T_S + T_C + T_M - T_A \tag{11.15}$$

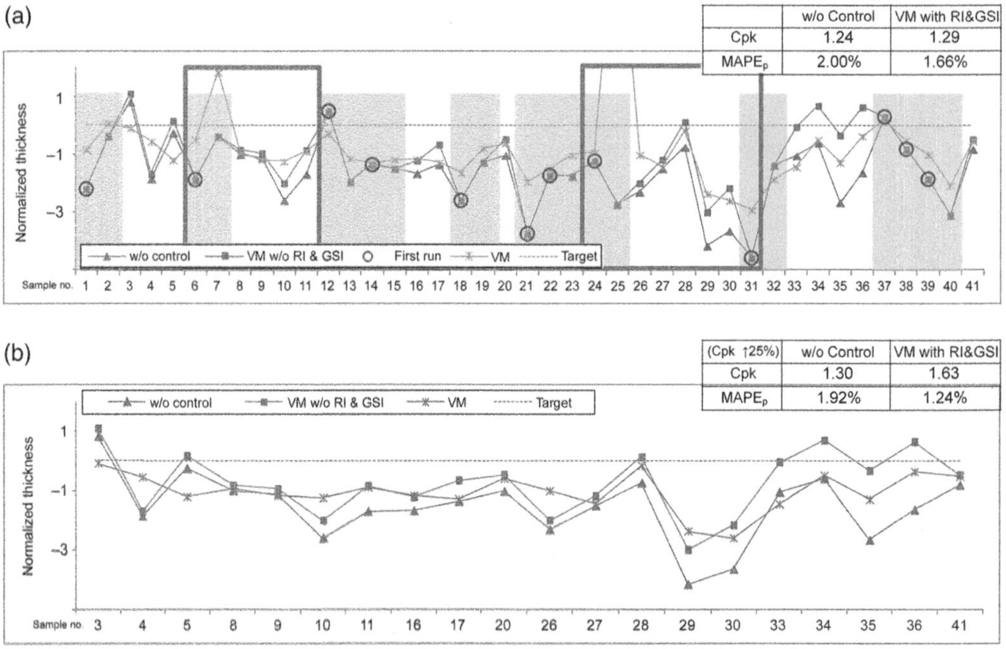

Figure 11.30 T2T+VM with RI&GSI. (a) All samples. (b) APC samples. *Source:* Reproduced with permission from Ref. [6]; © 2013 IIE.

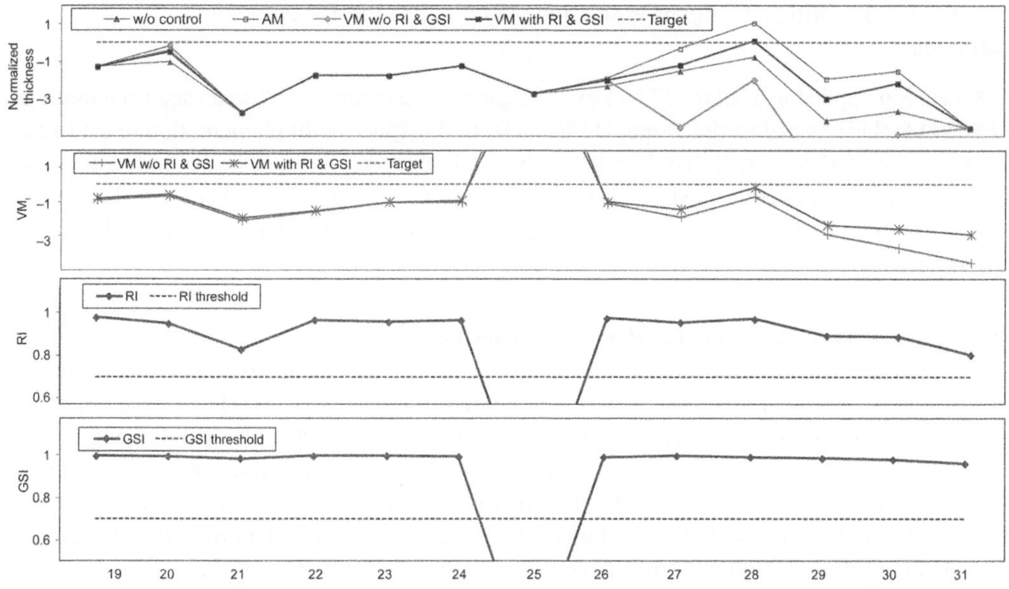

Figure 11.31 RI and GSI are lower than RI_T and GSI_T, respectively at Sample 25. *Source:* Reproduced with permission from Ref. [6]; © 2013 IIE.

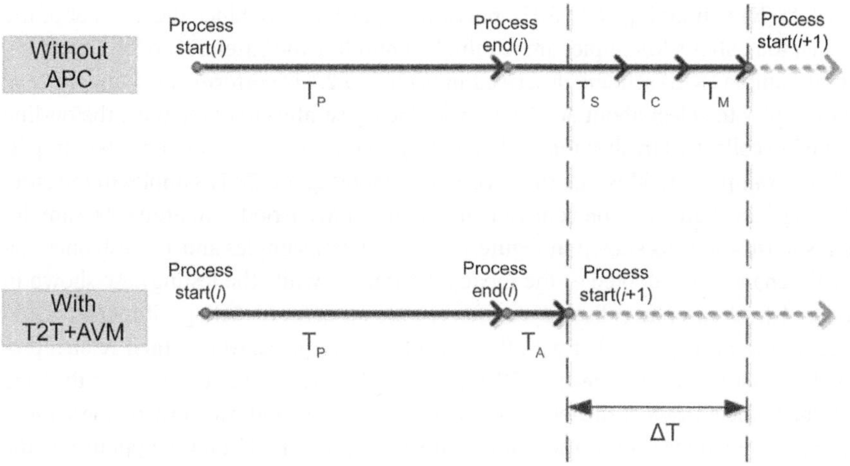

Figure 11.32 Cycle-time improvement by applying T2T+AVM. *Source:* Reproduced with permission from Ref. [6]; © 2013 IIE.

where T_P is the PECVD process time; T_S is the manual recipe-inspection time; T_C is the manual recipe-parameter-calculation time; T_M is the manual recipe-parameter-modification time; and T_A is the T2T+AVM calculation time. Define $\%\Delta T = \Delta T/(T_P + T_A)$, then T2T+AVM implementation is expected to gain extra $[\frac{1}{1-\%\Delta T} - 1]$ production volume output with good Cpk due to ΔT cycle-time reduction. For this example, $T_P + T_A$ and ΔT are about 60 and 5 minutes, respectively. Therefore, the production-rate improvement with good-quality output is about 8.3%.

11.3.3 Factory-Wide Intelligent Predictive Maintenance (IPM) Deployment Examples

IPM has been applied in a solar-cell factory. To begin with, a throttle valve in a vacuum module of a PECVD tool is selected as the Target Device (TD) to illustrate the implementation of a Virtual-Metrology-based (VM-based) Baseline-Predictive-Maintenance (BPM) scheme and remaining-useful-life (RUL) predictive modules described in Sections 9.2 and 9.3. Then, a real solar-cell factory example of the factory-wide IPM management framework presented in Section 9.4 is demonstrated.

11.3.3.1 Illustrative Examples of BPM and RUL Prediction

Three cases of a throttle valve in a vacuum module of a PECVD tool for solar-cell manufacturing are adopted as the illustrative examples. As such, the throttle valve is the TD in these examples. Based on the domain knowledge, the angle of the TD may be affected by the following related parameters: NH_3, SiH_4, pressure, and radio-frequency (RF) power. Three illustrations are included in this section: (i) necessity of including the concise and healthy (C&H) samples in the baseline model, (ii) fault detection and classification (FDC) portion, and (iii) RUL portion including exponential-curve-fitting (ECF) model and Time-Series-Prediction (TSP) Algorithm.

Necessity of Adopting the C&H Samples in the Baseline Model

The procedure shown in Figure 9.4 is followed to collect the important samples needed for creating the TD baseline model. In this example, the angle of the TD (throttle valve) is designated as y_T while its related process data (**X**) containing NH_3, SiH_4, pressure, and RF power. Since the number of the related process data is four, about 40 samples are required for building the baseline model. Therefore, the Keep-Important-Sample (KIS) scheme described in Section 9.2.1 is performed off-line on 2602 healthy historical samples to select about 30 C&H ones. Then, just after maintenance, the on-line operation is activated to collect 10 fresh samples. Each sample contains y_T and its corresponding **X**.

The purpose of this example is to illustrate the necessity of adopting the C&H samples in the baseline model. The testing data of an entire preventive-maintenance (PM) period containing 390 samples are examined in this illustration. As shown in Figure 11.33a, 30 C&H samples and 10 fresh ones collected just after maintenance are adopted as the modeling samples; while the testing case shown in Figure 11.33b adopts 40 fresh samples collected just after maintenance for modeling without any C&H samples. The angles of the TD (y_T) together with the baseline values ($\hat{y}_B$) as well as their related process data (NH_3, SiH_4, pressure, and RF power) of Samples 296–380 are displayed. Note that the TD's angles and their related process data of the testing runs in Figure 11.33a and b are exactly the same.

Comparing the baseline values ($\hat{y}_B$) in the testing runs of Figure 11.33a and b, apparently, the ones in Figure 11.33b are relatively rough, which are incorrect. The reason is due to the fact that Figure 11.33a's modeling samples contain 30 C&H samples while the ones in Figure 11.33b do not. As shown in the left portion of Figure 11.33a and b, the peak-to-peak variations of all the TD's related process data of Figure 11.33a are much larger than those of Figure 11.33b. As a result, the representativeness of the model-operating space of the case shown in Figure 11.33b is rather weak. Hence, the reason that the C&H samples are required for modeling is clear.

Illustration of the FDC Portion of the BPM Scheme

The purpose of this example is to demonstrate the FDC functions of the BPM scheme. The 450 testing samples used in the example illustrated in Section 9.1.1 are re-utilized for easy comparison. Again, the procedure shown in Figure 9.4 is followed to collect the important samples needed for

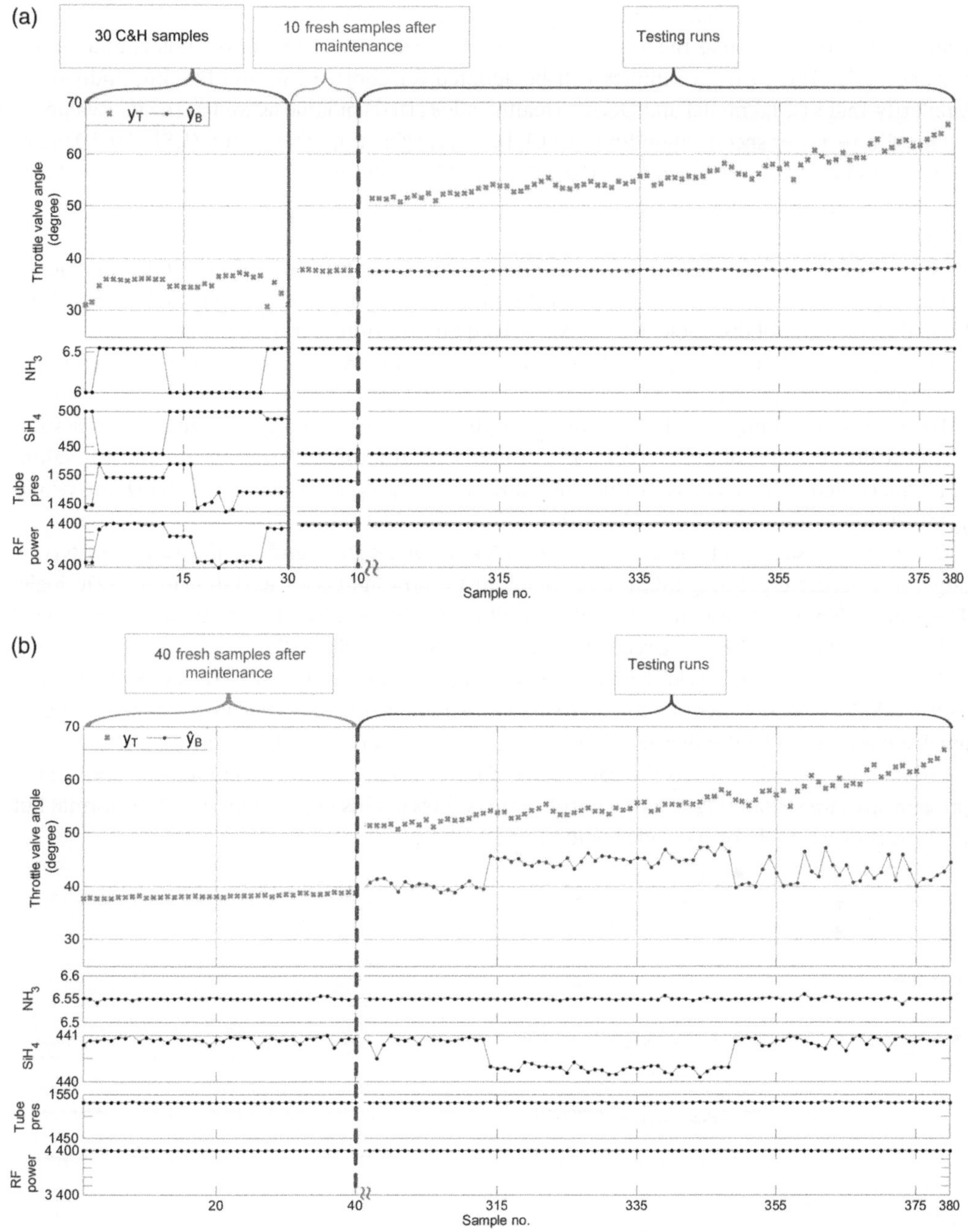

Figure 11.33 Illustration of the necessity of adopting the C&H modeling samples. (a) BPM results with C&H modeling samples. (b) BPM results without C&H modeling samples. *Source:* Reprinted with permission from Ref. [9]; © 2013 IEEE.

creating the TD baseline model. Therefore, 30 C&H samples plus 10 fresh ones collected just after maintenance are adopted as the baseline modeling samples. Each sample contains y_T and its corresponding **X**. The 10 fresh samples will be adopted for constructing the Baseline-Individual-Similarity-Index (ISI_B) model and Device-Health-Index (DHI) module as well as the y_{E_B} value.

The Min y_T, lower specification limit (LSL), LCL, $\bar{y}_T$, UCL, upper spec limit (USL), and Max y_T of this example for converting y_T to DHI are 0, 5, 22, 27, 32, 50, and 90, respectively. Also, the Spec, HardSpec − $\bar{y}_T$, and Max y_E for transforming y_E to Baseline-Error Index (BEI) are 5, 23, and 63, respectively.

The maximal allowable deviation of each related process datum is 10% of each individual nominal value. Therefore, by assigning 10% of each individual nominal value to be equivalent to 6 of the ISI_B value, the value of threshold of ISI_B (ISI_{B_T}) is equal to 6. Moreover, as mentioned in Section 9.2, both threshold of DHI (DHI_T) and threshold of BEI (BEI_T) are assigned to be 0.7, which is the threshold for entering the sick state.

The results of running the FDC portion of the BPM scheme are shown in Figure 11.34. Besides y_T, the healthy baseline of the TD ($\hat{y}_B$) is also displayed. Furthermore, DHI, BEI, and the ISI_B values of all the related parameters: NH_3, SiH_4, pressure, and RF power are also shown below the y_T and $\hat{y}_B$ results.

Observing the sample at circle 1 in Figure 11.34, its DHI < DHI_T and the ISI_B (=120) of NH_3 is larger than its corresponding ISI_{B_T} (=6); therefore a yellow light is on according to the FDC logic. This yellow light means that the TD itself is healthy; while the reason for its out-of-control (OOC) is due to the glitch occurs in the related NH_3, which should be checked.

All the samples within circles 2 and 4 satisfy the conditions that DHI < DHI_T and BEI < BEI_T as well as all ISI_B values of these samples are smaller than their corresponding ISI_{B_T}, hence red lights are displayed. A red light indicates that the TD is abnormal and caused by itself.

The sample in circle 3 meets the conditions of DHI > DHI_T as well as the ISI_B value (=13.5) of pressure are larger than ISI_{B_T}; thus a purple light is shown. This implies that the TD is normal but the tube pressure is abnormal and should be checked [9].

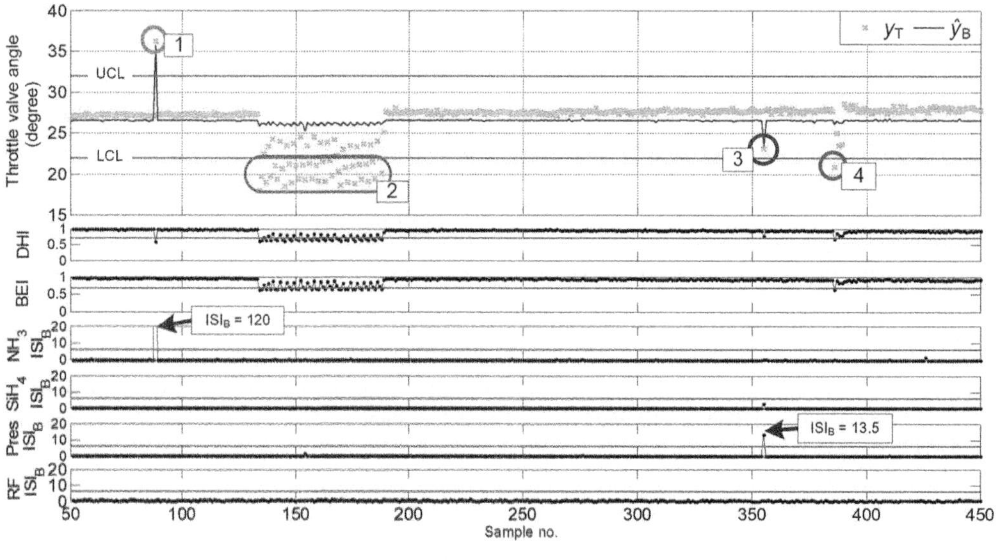

Figure 11.34 Results of the FDC portion of the BPM scheme. *Source:* Reproduced with permission from Ref. [9]; © 2013 IEEE.

Illustration of the RUL Portion of the BPM Scheme

Both the ECF model and the TSP algorithm are applied to perform RUL prediction. They are presented as follows.

Exponential-Curve-Fitting (ECF) Model An entire PM period containing 390 samples utilized in Section 11.3.3.1.A are re-examined in this example. Thirty C&H samples are collected and included in the modeling samples. The first 10 fresh samples are utilized for constructing the ISI_B model and DHI module as well as the y_{E_B} value. The following 380 samples with the BPM-related data and indexes are shown in Figure 11.35. As depicted in Figure 11.35, the angles of the TD (y_T) increase gradually due to the aging effect; while the baseline values ($\hat{y}_B$) are relatively flat due to the fact that their related process data: NH_3, SiH_4, pressure, and RF power are stable with little variations. The conditions of DHI < DHI_T and BEI < BEI_T with all ISI_B values being smaller than their corresponding ISI_{B_T} occur at around Sample 213; therefore, Sample 213 is the entry point of the sick state. This event activates the RUL predictive process. All the ECF RUL predictive results are shown in Figure 11.36 and are explained as follows.

According to the flowchart for calculating ECF RUL in Figure 9.8, we have the following.

Phase I (Off-line)

Step 1: Calculate $y_{E\_B}\left(=\bar{y}_E+3\sigma_{y_E}\right)=0.28$ and find $k_S = 13$. Then, define y_{E_D} (=HardSpec-$\bar{y}_T$) = $50 - 37.74 = 12.26$, and y_{E_S} (=Spec) = 5.

Phase II (On-line)

Step 2: The condition of two consecutive detections of $y_E > y_{E\_S}$ is confirmed at $k_S = 213$.

Step 3: Collect all the samples between $k_B = 13$ and $k_S = 213$ and apply (9.3)–(9.5) in Chapter 9, the failure equation is obtained:

$$\hat{y}_{E_i} = 0.5753 \times e^{0.00112 k_i}. \tag{11.16}$$

Step 4: Apply (11.16) to predict the $\hat{y}_{E_i}$ values started from $k_i = k_{S+1} = 214$ until the $\hat{y}_{E_i}$ value is equal to or greater than the y_{E_D} value (12.26); this sample number is denoted

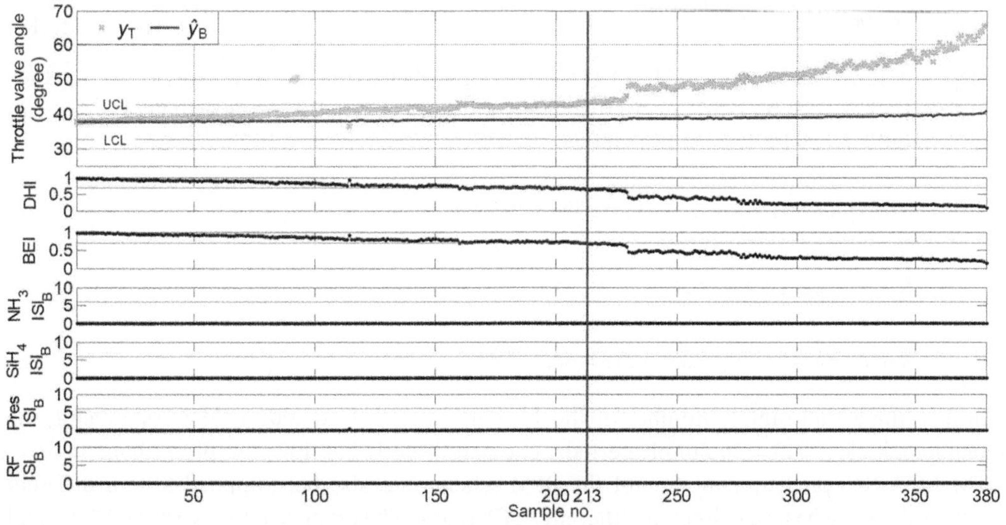

Figure 11.35 BPM-related data and indexes of an entire PM period. *Source:* Reproduced with permission from Ref. [9]; © 2013 IEEE.

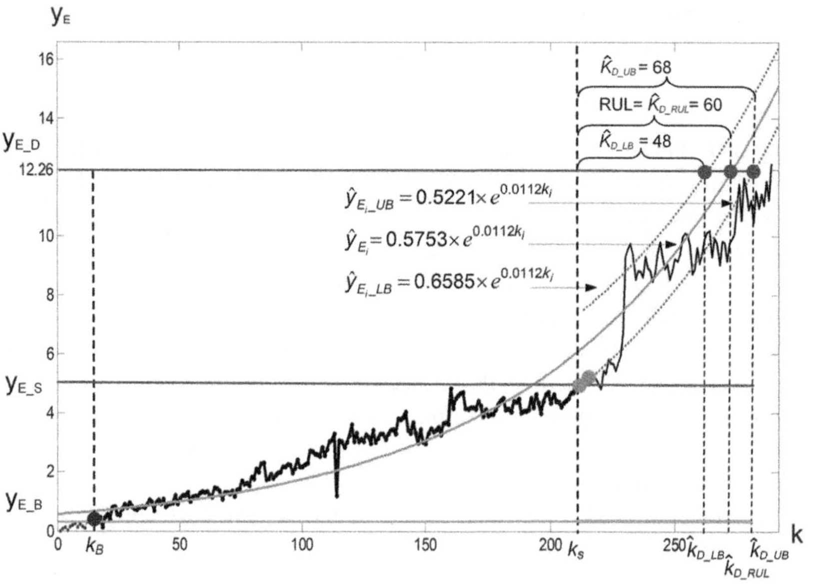

Figure 11.36 ECF RUL predictive results. *Source:* Reproduced with permission from Ref. [9]; © 2013 IEEE.

as $\hat{k}_{D\_RUL}$, which is 273. Then, calculate all the 95% upper bound (UB) and LB values from 214 till 273 with (9.6)–(9.8). Finally, the UB equation is derived as

$$\hat{y}_{E_i\_UB} = 0.5221 \times e^{0.00112k_i}. \tag{11.17}$$

and the LB equation is

$$\hat{y}_{E_i\_LB} = 0.6585 \times e^{0.00112k_i}. \tag{11.18}$$

Step 5: With $y_{E\_D} = 12.26$ and (11.16)–(11.18) being applied to find $\hat{k}_{D\_RUL}$, $\hat{k}_{D\_UB}$, and $\hat{k}_{D\_LB}$; they are 273, 281, and 261, respectively. Then, utilize (9.11)–(9.13) to calculate RUL, $\widehat{K}_{D\_UB}$, and $\widehat{K}_{D\_LB}$ with $k_S = 213$ as follows:

$$RUL = \widehat{K}_{D\_RUL} = \hat{k}_{D\_RUL} - k_S = 273 - 213 = 60,$$

$$\widehat{K}_{D\_UB} = \hat{k}_{D\_UB} - k_S = 281 - 213 = 68,$$

$$\widehat{K}_{D\_LB} = \hat{k}_{D\_LB} - k_S = 261 - 213 = 48.$$

In fact, the actual sample number of the TD entering the dead state is 289. As a result, the actual RUL is 289 − 213 = 76 sample periods. The sample period of this example is about an hour.

The ECF RUL predictive module of the BPM scheme has also been applied to four different throttle valves. Each valve belongs to different PECVD tool. The predictive RULs are 29, 17, 19, and 60; while their corresponding actual RULs are 32, 19, 21, and 68, respectively. The test results of the RUL predictive module show that the predictive accuracy is reasonably fine.

Time-Series-Prediction (TSP) Algorithm The TSP algorithm presented in Section 9.3 is also applied to predict the RUL of the TD (throttle valve) mentioned above. Based on domain knowledge, the angle of the throttle valve is adopted as the TD's aging feature (Y_T). TD data are collected once per hour, and 109 samples are collected. Apply the samples to the TSP procedures in Figure 9.11, then the results are shown below.

Step 1: Input TD's aging feature ($Y_T = \{. . ., y_{295}, y_{296}, . . .\}$) into the DHI module and RUL predictive module.
Step 2: When DHI < 0.7, start TSP model creation.
Step 3: 30 samples ($Y_M = \{y_{267}, . . ., y_{295}, y_{296}, . . .\}$) before the sick state (DHI ≥ 0.7) are adopted for TSP modeling.
Step 4: The variates are found to get larger gradually along with time.
Step 5: Let Y_T go through Log transformation to make variates smaller.
Step 6: Execute augmented Dickey-Fuller test (ADF test) on Y_M to ensure Y_M be in the stationary state.
Step 7: As it is stationary, there is no need to make the difference to Y_M.
Step 8: Pick out the most correlation span from autoregressive model (9.16) and moving average model (9.15), the correlations between AR and MA are the degradations on spans *t*-9 and *t*-10 as shown in Table 11.3, thus by (9.24) $A_{\text{ACF}} = 10$ and by (9.25) $B_{\text{PACF}} = 9$.
Step 9: If the spans of Y_M arc correlated (reject H_0), jump to Step 11.
Step 11: Find out the most correlation span using Step 8 to establish 10×9 kinds of autoregressive integrated moving average model (ARIMA) combinations.
Step 12: Calculate the Bayesian information criterion (BIC) of 10×9 ARIMA model combinations.
Step 13: ARIMA (2, 0, 1) is derived to be the combination with the lowest BIC, thus it is chosen as the best model temporarily.
Step 14: Observing Table 11.4, exclude MA(1) coefficient$|-0.1343| > 1.96 \times 0.0423$ as it is an insignificant predicator.

Table 11.3 ACF and PACF.

	t-1	*t*-2	*t*-3	*t*-4	*t*-5	*t*-6	*t*-7	*t*-8	*t*-9	*t*-10	*t*-11
AR (PACF)	0.536	0.09	0.14	0.202	0.009	0.076	0.159	0.135	0.4	0.259	0.025
MA (ACF)	0.536	0.314	0.219	0.059	0.017	0.032	0.172	0.006	0.236	0.374	0.157

Source: Reproduced with permission from Ref. [10]; © 2019 IEEE.

Table 11.4 Significance of predictors.

Predictor	Coefficient	*s.e.*	Significance
AR(1)	1.226	0.7669	Yes
AR(2)	−0.577	0.3224	Yes
MA(1)	−0.1343	0.0423	No

Source: Reproduced with permission from Ref. [10]; © 2019 IEEE.

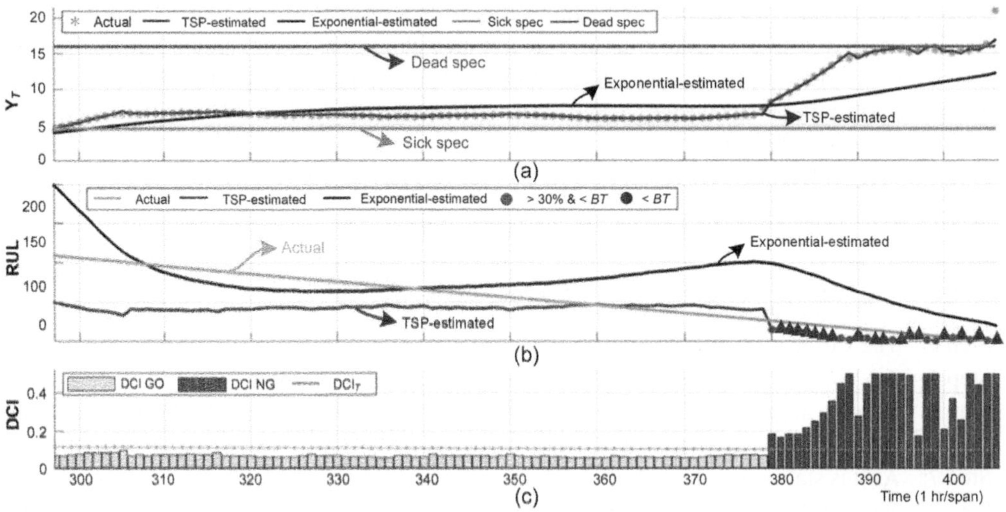

Figure 11.37 Throttle valve RUL predictive results of the TSP algorithm. (a) Aging-feature prediction. (b) RUL prediction. (c) DCI analysis. *Source:* Reproduced with permission from Ref. [10]; © 2019 IEEE.

Step 15: Recalculate ARIMA(2, 0, 0) coefficient after excluding the insignificant predicator and obtain $\hat{y}_t = 1.7783y_{t-1} - 0.7783y_{t-2}$.

Step 16: Check the Y_M residual that the result is the white noise.

Step 17: Make sure ARIMA(2, 0, 0) is the final model and $\hat{y}_t = 1.7783y_{t-1} - 0.7783y_{t-2}$, the model generated by TSP can then be added for prediction.

Since it takes long time span for a throttle valve to change, buffer time (BT) in the pre-alarm module (PreAM) is set to be 30 hours. Figure 11.37a is the aging feature prediction trend of the throttle valve angle, Y_T, with * being the actual aging feature of throttle valve angle, blue curve (TSP-estimated line) being the predicted value generated by ARIMA (2, 0, 0) of TSP, and black curve (exponential-estimated line) being the predicted value acquired from the ECF model. Figure 11.37b is the RUL prediction, with pink curve (actual-value line) being the actual RUL, blue curve (TSP-estimated line) being the RUL estimated by ARIMA (2, 0, 0) of TSP, and black curve (exponential-estimated line) being the RUL predicted by the ECF model. Observing the RUL curves, when Y_T is within 330–379 spans and stays steady, the ECF-estimated RUL curve shows an up-rising trend that departs from the real RUL curve, which is the CASE A (TD's Aging Feature Goes too Smooth) problem of the ECF model mentioned previously. In contrast, during the same 330–379 spans, the TSP-estimated RUL curve sustains at a certain level without showing an up-rising trend. This demonstrates that the TSP algorithm can solve the CASE A problem of the ECF model.

Though the TSP-estimated RUL curve does not follow up the trend instantly when Y_T suddenly rises up at time 379, it reacts at time 380 and sends the alarm of red round dots (TSP predicts RUL shorter than *BT* and the decline rate is over 30% compared to the previous span) to inform users for immediate maintenance. Forward 10 spans from time 380, and suppose that the aging feature does not stay at high points but rise proportionally instead, the final result could be that TD dies at time 390. This means that it is accurate to send an alarm of red round dots at time 380 for immediate maintenance. However, the ECF-estimated RUL curve at time 380 cannot follow up the aging trend in a short time, which could cause underestimation of the sick status as mentioned in the Case B (TD's Aging Feature Rises or Drops Drastically) problem of the ECF model.

Observing the prediction trend in Figure 11.37a, the curve sways along the dead spec line during span 390–404 and then shuts down suddenly at time 405, which shows that TD is highly unstable before it dies. This also verifies the necessity that when the RUL decline rate is over 30% and shorter than BT, the PreAM module will inform users by turning on the red light (0–0.3). Figure 11.37c displays the Death Correlation Index (DCI) values; before time 380, DCI GO ($DCI < DCI_T$), but at and after time 380, DCI no go (NG) ($DCI \geq DCI_T$), which indicates that under 95% confidence level, current TD status is near dead and demands close attention.

11.3.3.2 Illustrative Example of Factory-Wide IPM System

As explained in Section 9.6 for IPM system deployment, the IPM system can be integrated with manufacturing execution system (MES) such that the health status of all the TDs can be shown on the equipment status IPM dashboard in MES. The IPM system with HIH has been deployed on the production line of Motech Industries, Inc., Taiwan, ROC. The production line is composed of two Texturing (TEX), three Diffusion (DIF), two PSG Removal, four PECVD, three Printing (PRT), and three Cell Test (FCT) tools. With this IPM system framework, all health indexes on the production line are updated in real time and supervised according to HIH.

The visualized IPM dashboard, which is the graphic interface on the IPM client, allows users to handle all the factory-wide health indexes. The management view of the IPM dashboard is shown in Figure 11.38a. Factory Health Index (FHI) is displayed on the top-left side of the IPM dashboard; Station Health Indexes (SHIs) of the six stations are shown at the top-right corner. Moreover, Equipment Health Indexes (EHIs) are depicted at the bottom half of Figure 11.38a; users are able to switch the web tab to show all the EHIs of a certain station, which is CVD, as depicted in Figure 11.38a. Each widget on the IPM dashboard presents the health status of healthy (0.7–1), sick (0.3–0.7), and dead (0–0.3).

The equipment view of the IPM dashboard that contains EHI, Module Health Indexes (MHIs), and RUL is depicted in Figure 11.38b, which shows the results of CVD 1-1 in Table 9.1(a) of Chapter 9 for illustration. The health statuses of device, module, and equipment are displayed in green, yellow, or red light with health index values. Observing Figure 11.38b and Table 9.1(a), it becomes clear why CVD 1-1 (Tube1) is dead with its EHI value being 0.21, which is because the DHI value of throttle valve in the vacuum module of CVD 1-1 is 0.21. In fact, each production tool on the management view can be clicked and drilled down to the equipment view independently for finding out the root cause of a dead status.

In addition, the IPM system has also been integrated with the MES system. Figure 11.39 shows a shop floor form on MES with the IPM system information integrated. A lot of rectangles in different sizes with different notations according to factory layout are displayed in Figure 11.39; each rectangle represents a production tool, and the notation of a rectangle represents the current state of the specific production tool reported via equipment communication. The notation definitions of all the equipment states are shown at the top-left corner of Figure 11.39. To represent the status of EHI, each production tool has an icon of different shapes on the right; the definitions of the shapes are depicted at the top-left corner. Users can click the icon of a specific EHI status to link to the equipment view of the IPM dashboard in the IPM client. The integration between the MES and IPM systems is essential for prompt response to troubleshooting.

Figure 11.40 shows the sequence diagram of the interfaces between the MES and IPM systems. Each IPM server issues the status of equipment being either sick or dead to the IPM manager; then, the IPM manager sends summarized information of equipment health statuses to MES via web services. After receiving the message from the IPM manager, MES displays health statuses of equipment on the shop floor form of MES as shown in Figure 11.39. After clicking one of the EHI

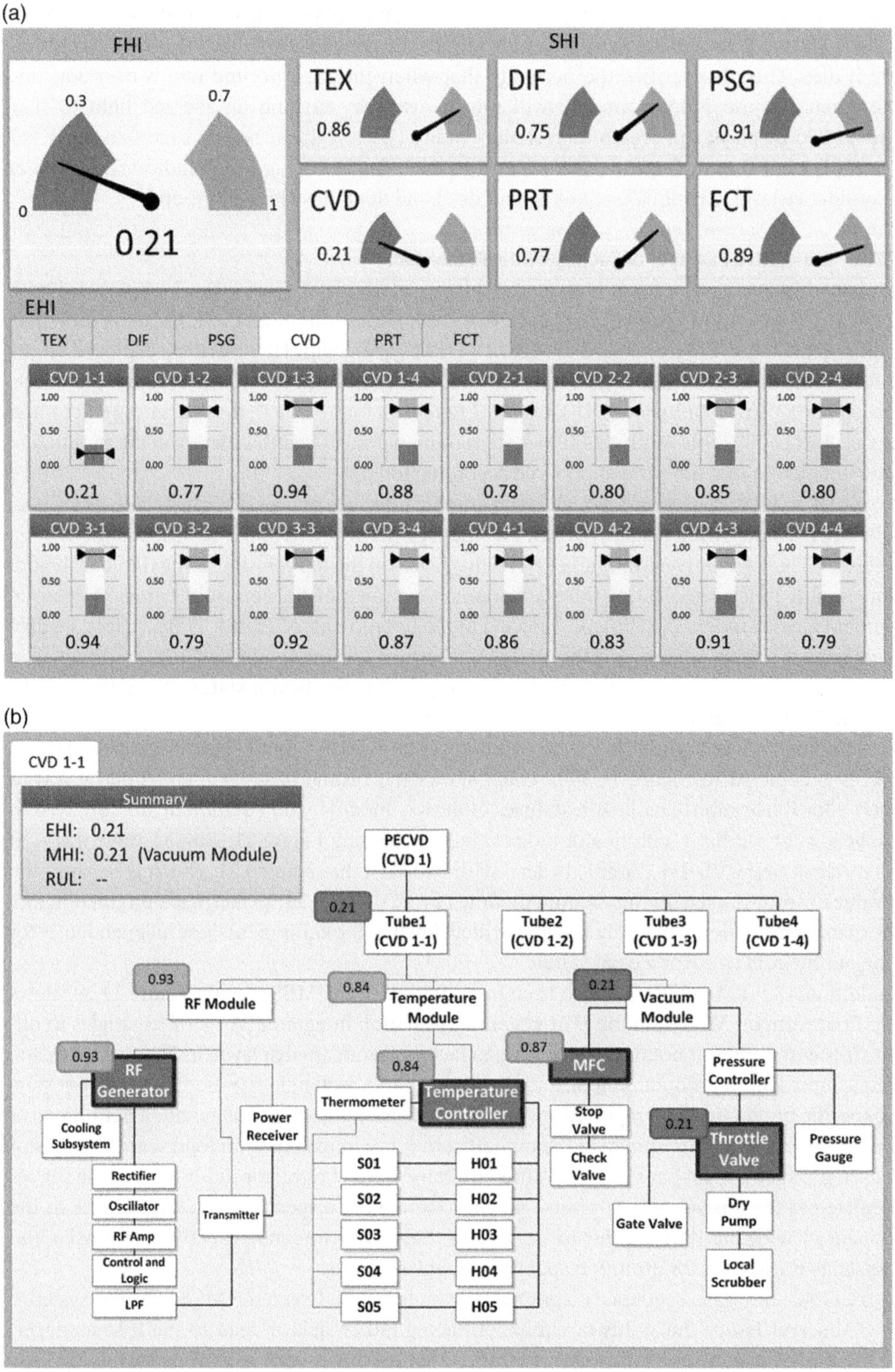

Figure 11.38 IPM dashboard. (a) Management view. (b) Equipment view. *Source:* Reproduced with permission from Ref. [3]; © 2017 JCIE.

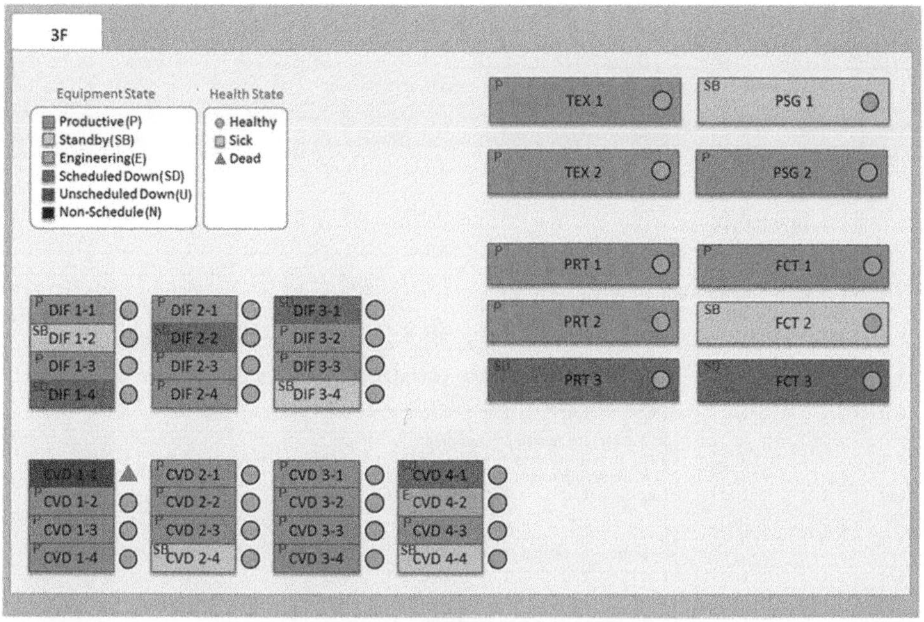

Figure 11.39 Equipment status dashboard in MES. *Source:* Reproduced with permission from Ref. [3]; © 2017 JCIE.

status icons, MES sends the selected equipment ID to the IPM client. Then, the IPM client requests and gets the related process data according to the selected equipment ID. After that, the IPM client displays the management view of the IPM dashboard to users.

11.3.4 Summary

The solar system is an option to solve the global warming situation encountered today. The solar system is composed of solar panels and inverters, and each solar panel contains many solar cells. Solar-cell is the most important part to generate electricity from sunlight. Currently, massive production with stable quality and low cost are the issues that solar-cell manufacturing factories need to resolve. To handle the challenge mentioned above, the implementations of T2T control with AVM and the factory-wide IPM system framework were demonstrated in this solar-cell application case.

11.4 Application Case III: Semiconductor Industry

AVM, IPM, and IYM have been deployed in the semiconductor industry. The deployment of AVM is illustrated first followed by the presentations of the IPM and IYM deployments.

11.4.1 AVM Deployment Example in the Semiconductor Industry

The authors would like to thank Taiwan Semiconductor Manufacturing Company (tsmc), Ltd. for providing the raw data used in this AVM application case.

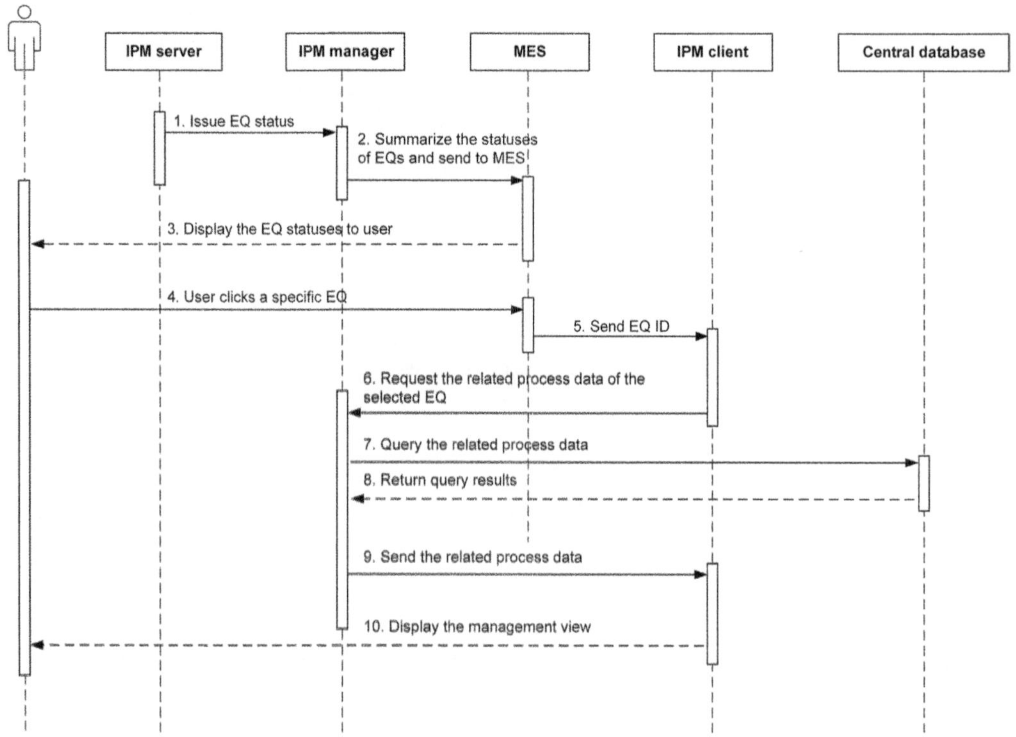

Figure 11.40 Sequence diagram showing the interfaces between the MES and IPM systems. *Source:* Reproduced with permission from Ref. [3]; © 2017 JCIE.

11.4.1.1 AVM Deployment Example of the Etching Process

The etching process of semiconductor manufacturing is adopted as an example to illustrate the functions of RI, GSI, and dual-phase schemes. The VM mean values of CD are selected for demonstration in this example. According to the physical properties of the etching process tools, 66 key process parameters, such as electrostatic chuck (ESC) leakage current, chuck backside pressure, throttle valve position, RF power, RF voltage, chamber temperature, etc., are selected for building the VM models. 178 sets of corresponding metrology and process data are collected in this example. The first 99 sets are applied to establish the VM models, whereas the remaining 79 sets are used to evaluate the conjecture accuracy of (a) without dual-phase scheme (free run) and (b) with dual-phase scheme as shown in Figure 11.41. The case of free-run test means that the BPNN and PLS VM conjecture models will not be refreshed during the entire test runs; while that of dual-phase test denotes that the VM conjecture models will be tuned/re-trained by the dual-phase scheme R2R.

Again, the BPNN and PLS are adopted as the designated algorithms for creating the VM conjecture models. Given the conjecture VM value $\hat{y}_i$, AM value y_i, target value y, and the sample size n, the conjecture accuracy is presented by the MAPE and maximum error (Max Error) denoted in (11.2) and (11.3), respectively. The closer the MAPE and Max Error are to zero, the better the conjecture accuracy is achieved. As shown in Figure 11.41, three preventive maintenances (PM_1, PM_2, & PM_3) occur in front of running numbers (RNs) 48, 51, and 52, respectively. Before the occurrence of PM_1 (i.e. RNs 1–47), the RI values of the free-run case (Figure 11.41a) and dual-phase case (Figure 11.41b) are all approaching to 1, which means that the reliability of VM results between the process data of RNs 1–47 of both cases is good. Also, before the occurrence of PM_1

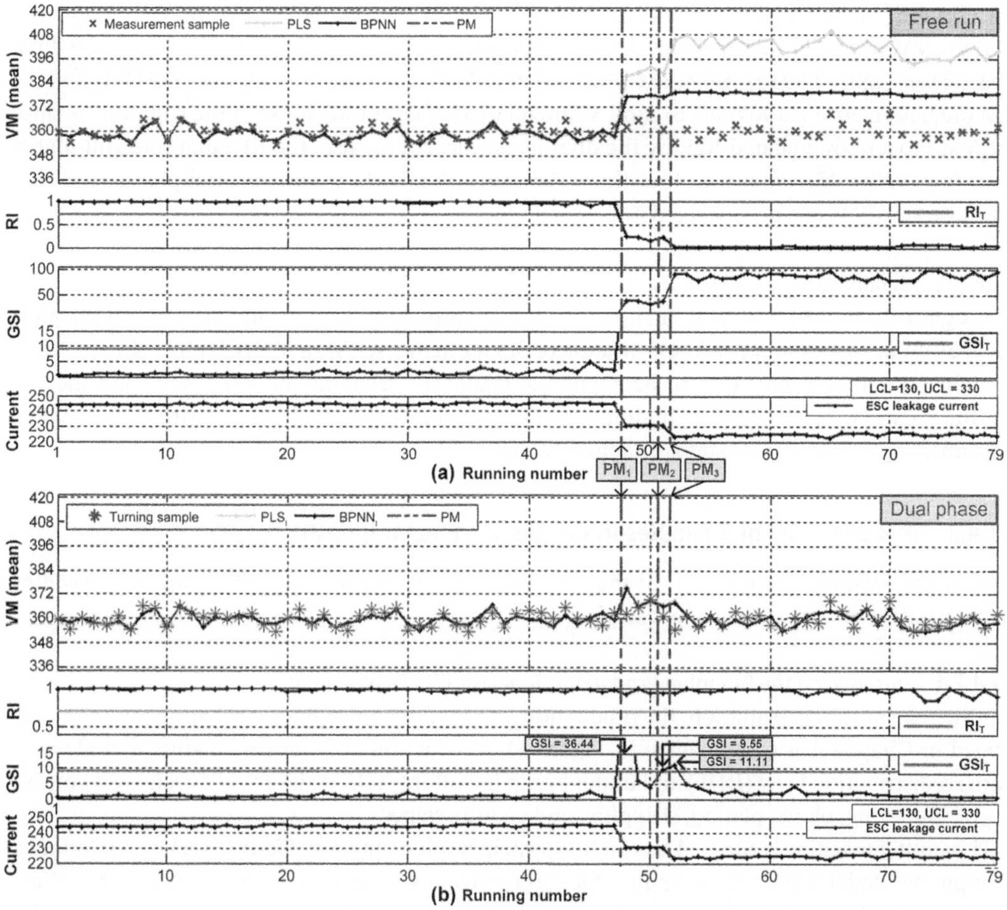

Figure 11.41 Illustrations of the functions of RI, GSI, and dual-phase schemes. (a) Without dual-phase scheme (free run). (b) With dual-phase scheme. *Source:* Reprinted with permission from Ref. [11]; © 2015 IJPR.

(i.e. RNs 1–47), the GSI values of the free-run case (Figure 11.41a) and dual-phase case (Figure 11.41b) are all approaching to 0, which means that the similarity between the process data of RNs 1–47 and the historical modeling data is good. As a result, the MAPEs of the free-run case (Figure 11.41a) and dual-phase case (Figure 11.41b) are almost the same. This implies that without manufacturing disturbances, the VM conjecturing algorithms, such as BPNN or PLS, can have acceptable accuracy even under the condition of free-run.

Encountering PM_1 just before RN 48, the GSI values of both the free-run and dual-phase cases at RN 48 go beyond the GSI_T, i.e. 9. After performing the ISI analysis, the key process parameter that causes major deviation is identified to be "ESC Leakage Current" which is shown at the bottoms of Figures 11.41a and b and exhibits a shift from value 245 to value 230 at RN 48. This process-data deviation due to PM_1 deflects the BPNN and PLS conjecturing results at RN 48 for the free-run case (see Figure 11.41a) as well as the phase-I BPNN ($BPNN_I$) and phase-I PLS (PLS_I) predicting results at RN 48 for the dual-phase case (see Figure 11.41b). As a result, the VM prediction errors at RN 48 of both free-run and dual-phase cases are enlarged. After RN 48, in the free-run case (Figure 11.41a), the GSI values stay beyond GSI_T and the VM prediction errors remain large for the rest of runs because the VM models are not refreshed; also, the RI values drop below RI_T (0.7), which means that the VM results of the free-run case after encountering PMs are not reliable. As for the dual-phase case (Figure 11.41b), only the GSI values of RNs 48, 51, and 52 are higher than GSI_T and the VM prediction errors of RNs 48, 51, and 52 are relatively larger than those of the rest of runs since the VM models are updated R2R by the dual-phase algorithm; also, the RI values are still higher than RI_T (0.7), which indicates that the VM results of RNs 48–79 of the dual-phase case after encountering PMs can be recovered and are reliable.

11.4.1.2 Summary

The significance of RI, GSI, and dual-phase algorithms of AVM is illustrated in this AVM deployment example of the etching process of semiconductor manufacturing. This example demonstrates that the RI and GSI schemes can be applied to gauge the reliability of VM results and the VM accuracy can be recovered and maintained by applying the dual-phase algorithm even when encountering unpredictable status changes (SCs).

11.4.2 IPM Deployment Examples in the Semiconductor Industry

The authors would like to thank Advanced Semiconductor Engineering (ASE), Inc. for providing the raw data used in this IPM application case.

11.4.2.1 Introducing the Bumping Production Tools for IPM Deployment

For the sake of easy explanation, the production line of the semiconductor bumping process mentioned in Chapter 1 is shown again in Figure 11.42. The bumping process mainly includes two parts: redistribution layer (RDL) and under bump metallurgy (UBM) layer with each layer consisting of sputtering, photo, plating, stripping, and etching stages.

The sputter machine is adopted for the IPM system implementation. Figure 11.43 is the Common Equipment Model (CEM) of the Sputter Machine, which is composed by chambers of three different functions: Degas, inductively coupled plasma (ICP), and physical vapor deposition (PVD). The Turbo Pumps in ICP chambers are chosen as the TDs for implementing IPM by applying the IPM system implementation architecture shown in Figure 9.15.

Based on domain knowledge, the current of the Turbo Pump is adopted as the TD's aging feature (Y_T).

11.4.2.2 Illustrative Example

The TSP algorithm is applied to perform RUL prediction. One hundred and ninety-seven (197) TD data are adopted to illustrate the TSP procedure depicted in Figure 9.11, and the results are shown below.

Step 1: Input TD's aging feature ($Y_T = \{y_1, y_2, \ldots, y_{197}, \ldots\}$) into the DHI module and RUL predictive module.

Step 2: When DHI < 0.7, start TSP model creation.

Step 3: Thirty samples ($Y_M = \{y_1, y_2, \ldots, y_{30}\}$) before the sick state (DHI ≥ 0.7) are adopted for TSP modeling.

Step 4: The variates are found to get larger gradually along with time.

Step 5: Let Y_T go through Log transformation to make variates smaller.

Step 6: Execute the ADF test on Y_M to ensure Y_M be in the stationary state.

Step 7: As it is stationary, there is no need to make the difference to Y_M.

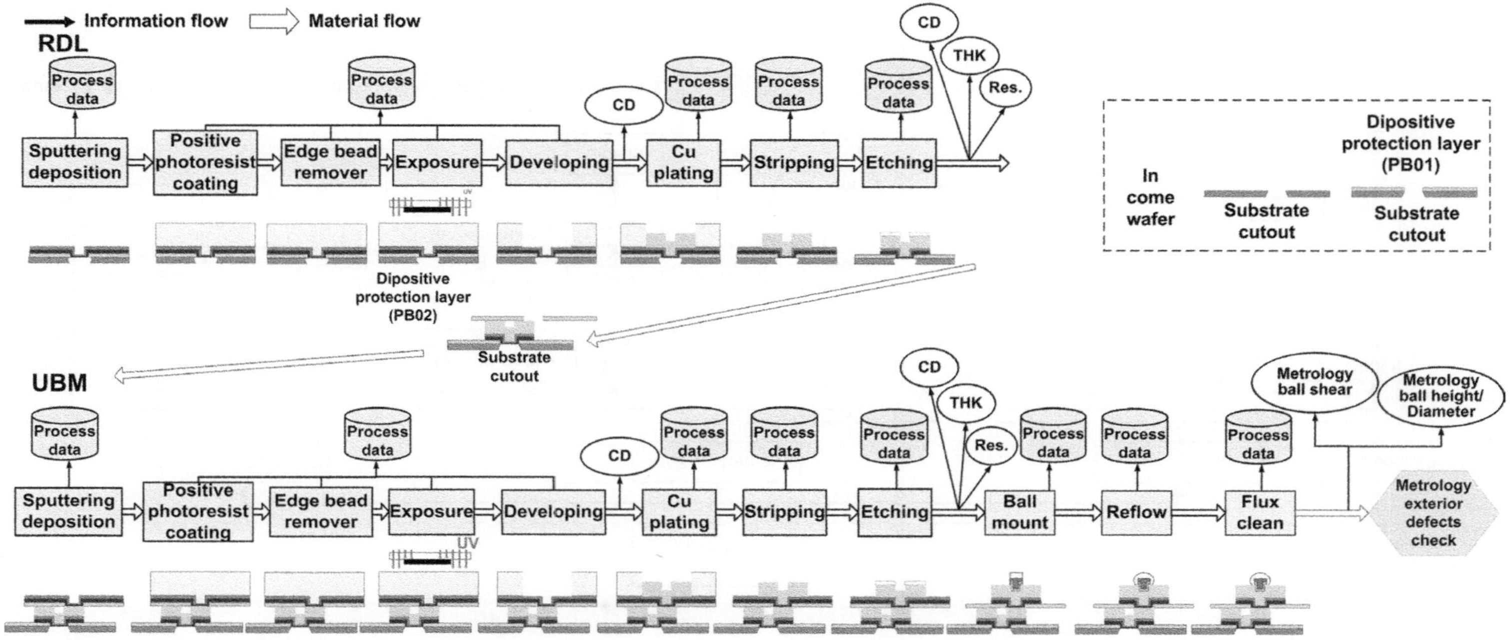

Figure 11.42 Production line of the bumping process.

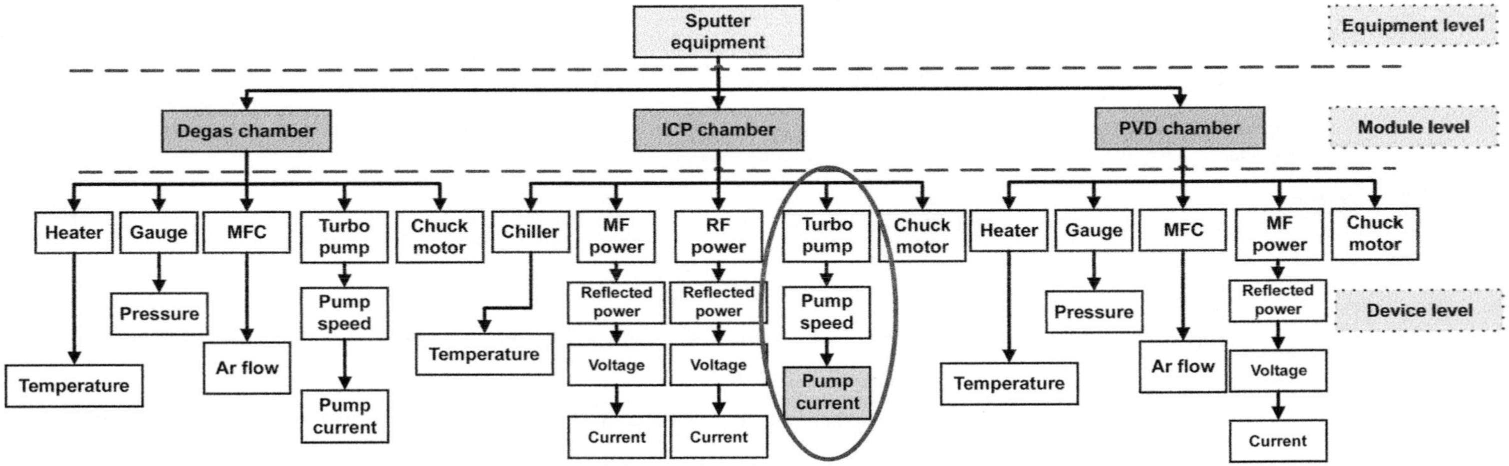

Figure 11.43 Common equipment model of the Sputter equipment.

Table 11.5 ACF and PACF.

	t-1	*t*-2	*t*-3	*t*-4	*t*-5	*t*-6	*t*-7	*t*-8	*t*-9
AR (PACF)	0.058	−0.127	−0.226	0.101	−0.158	0.001	−0.156	−0.022	0.058
MA (ACF)	0.042	−0.094	−0.014	0.014	−0.052	−0.012	−0.013	−0.002	0.042

Table 11.6 Significance of predictors.

Predictor	Coefficient	*s.e.*	Significance
AR(1)	0.56427	0.7142	Yes
MA(1)	−0.59594	0.6902	Yes

Step 8: Pick out the most correlated span from AR via (9.16) and MA via (9.15), the correlations between AR and MA are the degradations on spans and shown in Table 11.5, thus $A_{\text{ACF}} = 9$ via (9.24) and $B_{\text{PACF}} = 5$ via (9.25).

Step 9: If the spans of Y_M are correlated (reject H_0), jump to Step 11.

Step 11: Find out the most correlated span using Step 8 to establish 9×5 kinds of ARIMA combinations.

Step 12: Calculate the BIC of 9×5 ARIMA model combinations.

Step 13: ARIMA (1, 1, 1) is derived to be the combination with the lowest BIC, thus it is chosen as the best model temporarily.

Step 14: As shown in Table 11.6, MA(1) coefficient ($|-0.59594| < 1.96 \times 0.6902$) and AR(1) coefficient ($|0.56427| < 1.96 \times 0.7142$) are both significant predictors. Therefore, no predictors will be excluded.

Step 15: Recalculate ARIMA(1, 1, 1) coefficient after excluding the insignificant predicator and obtain $\hat{y}_t = 0.56427 y_{t-1} - 0.59594 \varepsilon_{t-1}$.

Step 16: Check the Y_M residual to see if the result is the white noise.

Step 17: Make sure ARIMA(1, 1, 1) is the final model and $\hat{y}_t = 0.56427 y_{t-1} - 0.59594 \varepsilon_{t-1}$, the model generated by TSP can then be added for prediction.

Set the BT of the PreAM module as 20 spans (each span is half an hour in this example) as it is necessary to contact the pump supplier prior to turbo pump changing. Figure 11.44a shows the aging feature prediction trend of the current, Y_T, with * being the actual aging feature of the current, and the blue curve (TSP-estimated line) being the predicted value generated by ARIMA (1, 1, 1) of TSP. Figure 11.44b presents the RUL prediction, with the blue curve (TSP-estimated line) being the RUL estimated by ARIMA (1, 1, 1) of TSP. Observing the prediction trend in Figure 11.44b, when it is stable during Spans 100–170, TSP will predict RULs steadily following the stable trend. The estimated RULs are around 30 spans. After entering the rapid decline period (Spans 180–197), TSP correctly predicts the trend of aging features and sends alarms to inform the users for scheduling a maintenance when the PreAM module displays blue light (TSP-estimated line) twice in consecution after Span 180.

Figure 11.44c displays the DCI values; before Span 182, DCI GO ($\text{DCI} < \text{DCI}_\text{T}$), but after Span 182, DCI NG ($\text{DCI} \geq \text{DCI}_\text{T}$), which indicates that under 95% confidence level, the current TD status is near dead and demands close attention.

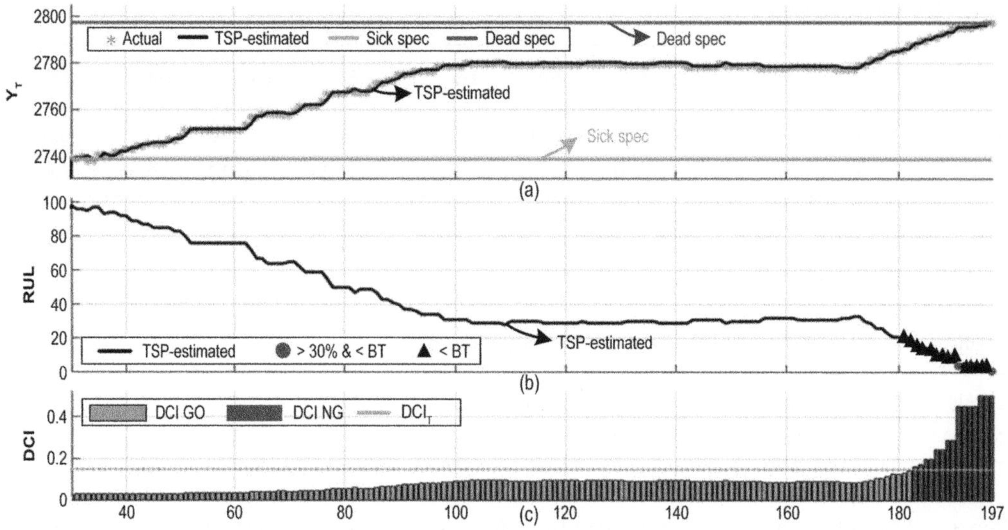

Figure 11.44 Turbo Pump RUL predictive results of the TSP algorithm. (a) Aging-feature prediction; (b) RUL prediction; (c) DCI analysis.

11.4.2.3 Summary

The significance of the TSP algorithm is illustrated in this IPM deployment example of the Turbo Pumps in the sputter equipment. This example demonstrates that the TSP algorithm can be applied to predict the RUL of a Turbo Pump so as to prevent unscheduled downs.

11.4.3 IYM Deployment Examples in the Semiconductor Industry

The authors would like to thank ASE, Inc. for providing the raw data used in this IYM application case.

11.4.3.1 Introducing the Bumping Process of Semiconductor Manufacturing for IYM Deployment

By applying the iFA system platform of Figure 10.3, the IYM system is deployed to the bumping process as shown in Figure 11.42.

11.4.3.2 Illustrative Example

Dealing with the interaction effect that causes a yield loss of the bumping process by applying the Interaction-Effect Search Algorithm (IESA) module in the KSA scheme of the IYM system is adopted as the illustrative example. The actual production data are collected from the UBM layer of the bumping process. IESA contains two phases. In the first phase, the IESA module is applied to identify the interaction-effect key variable; while the second phase focuses on determining the threshold value of the key variable that causes the interaction effect.

Phase I

The yield test in this phase is the resistance test. All lots in the resistance layer are included for analysis ($n = 431$). According to expert experience, high resistance usually happens in the Sputter Stage (SS) of the UBM layer due to the processing queuing time (Q-time) from the previous stages

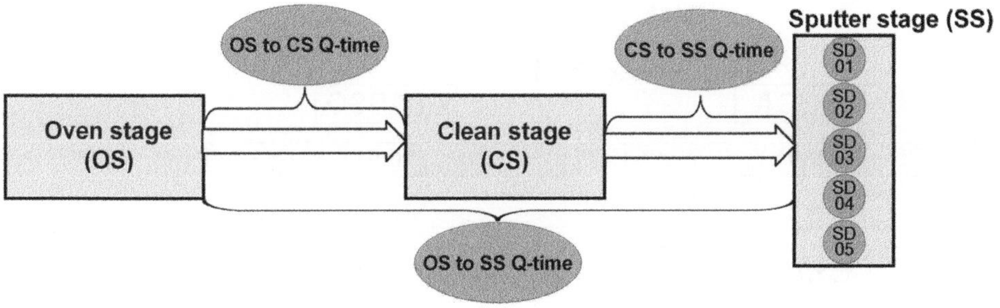

Figure 11.45 Illustration of UBM bumping process variables (5 device variables and 3 Q-time variables). *Source:* Reproduced with permission from Ref. [12]; © 2018 IEEE.

to the UBM layer SS. Note that, Q-time is a continuous variable. The inputs of this example have five UBM sputter devices and three Q-times from the previous stages to the UBM SS as shown in Figure 11.45. The three Q-times are Oven Stage (OS) to Clean Stage (CS) Q-time, OS-to-SS Q-time, and CS-to-SS Q-time. In summary, there are totally 5 device variables and 3 physical variables ($p = 5+3 = 8$) in this example.

To begin with, the original KSA scheme is applied with $\mathbf{X_R}$, $\mathbf{X_P}$ (Q-times), and $\mathbf{Y}$ of this illustrative UBM bumping process serving as the inputs. The outputs of KSA without IESA are shown in Figure 11.46a with the Top 1 root-cause device being Sputter Device 01 (SD 01). Then, the same input data are fed into the IESA module for interaction-effect analysis. The IESA Stage I result indicates that the Top 1 root-cause device goes to the interaction between SD 01 and OS-to-SS Q-time while SD 01, the original top 1 device, becomes the second place, as shown in Figure 11.46b.

Observing Figure 11.47, in the Regression Tree building process of the 431 samples, the parent node is SD 01 and the F-value is 29.8045. With the parent node being the decision variable, samples are segmented into two sub-nodes. The samples go through SD 01 will be gathered into the right sub-node and the ones that do not pass through the parent node go into the left sub-node. Continue to search for the most effective variable in the left sub-node denoted as Sputter Device 02 (SD 02), F-value $= 23.9989 > 1$ under all 431 samples testing; and F-value $= 75.655 > 1$ when it does not go through SD 01. This means that the value is significant in statistics but not important under the interaction-effect as Interaction F-value is smaller than 1, SD 02 is thus not marked as interactive with SD 01. Next, search for the most effective variable in the right sub-node, which is OS-to-SS Q-time; the F-value under all 431 samples is $2 > 1$, and after spilt F-value is $3.7579 > 1$, which means that the value is significant in statistics. Moreover, the interaction F-value with the parent node is $35.081 > 1$, this indicates that when a lot passes through SD 01, the length of OS-to-SS Q-time has significant impact on the high resistance, as shown in Figure 11.47. Hence, the ranking of the Regression Tree is as follows: (i) SD 01 and OS-to-SS Q-time interaction variable with F-value being 35.081; (ii) SD 01 with F-value being 29.8045; (iii) SD 02 with F-value being 23.9989.

On the other hand, TPOGA interaction-effect search framework (TPOGA IESF) uses the Top N ($N = 3$) of the KSA result to generate the interaction variables with strong-heredity influence which are [SD 01 × SD 02, SD 01 × OS-to-SS Q-time, SD 02 × OS-to-SS Q-time], respectively. The newly generated variables are combined into the matrix of the original ones as depicted in Figure 11.48, with $p = 8+3$ and $n = 431$. Then, the TPOGA searches in the combined matrix again to derive the result in Figure 11.46b. Since the results from Regression Tree and TPOGA are almost the same with the RI_I value being 0.954 (>0.7), it can be inferred that the results shown in Figure 11.46b are reliable.

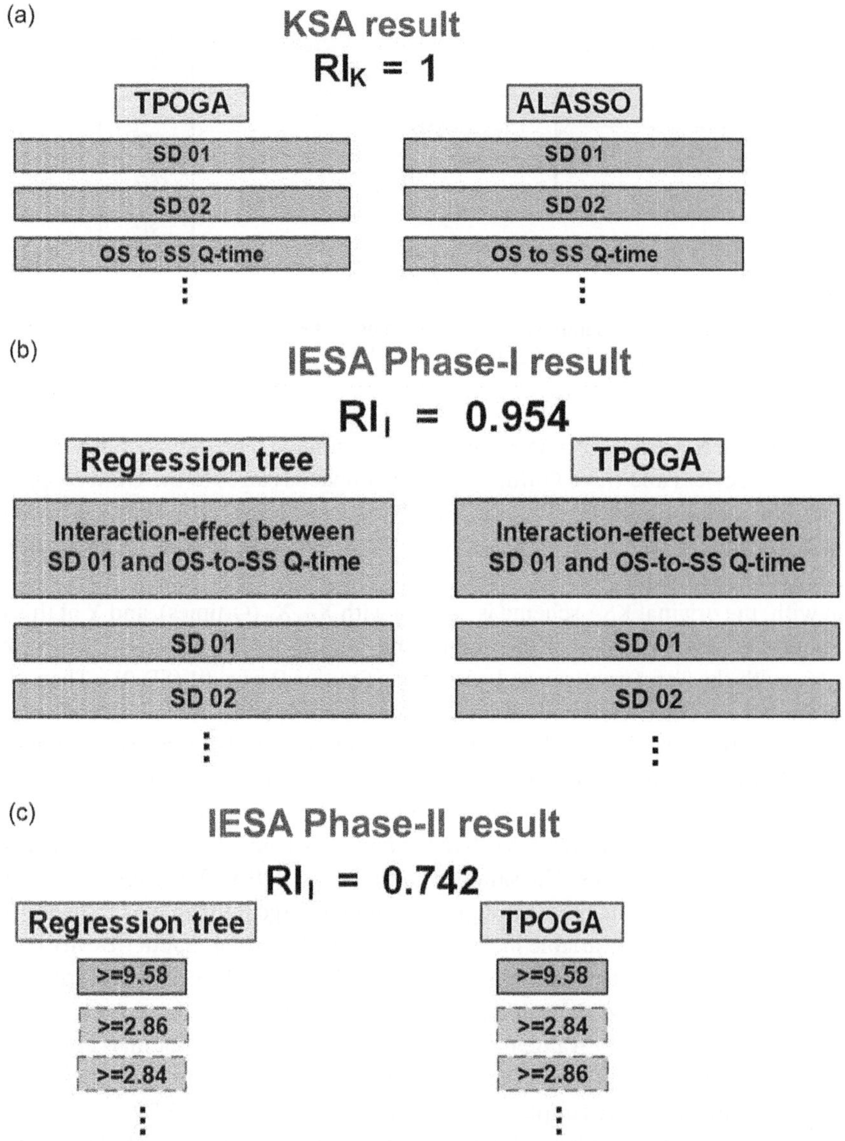

Figure 11.46 Analysis results with/without IESA. (a) Without IESA analysis. (b) With Phase-I IESA analysis. (c) With Phase-II IESA analysis. *Source:* Reproduced with permission from Ref. [12]; © 2018 IEEE.

Phase II

In Phase I, it reveals that when a lot goes through the OS-to-SS Q-time and then SD 01 will have significant impact on the high resistance after the IESA Phase I analysis. Then, the approach to determine the threshold value of OS-to-SS Q-time that causes the interaction effect is illustrated below.

To find out the Q-time that affects the high resistance, the variable conversion criteria in (10.47) should be performed to input $\mathbf{Y}$ and $\mathbf{X}_T$ for the IESA Phase II analysis. The samples passing through SD 01 include 96 lots ($n = 96$), which means there are 96 Q-time values going through SD 01. The size of binary matrix after conversion is 96×96 by (10.47) as shown in Figure 11.49. After

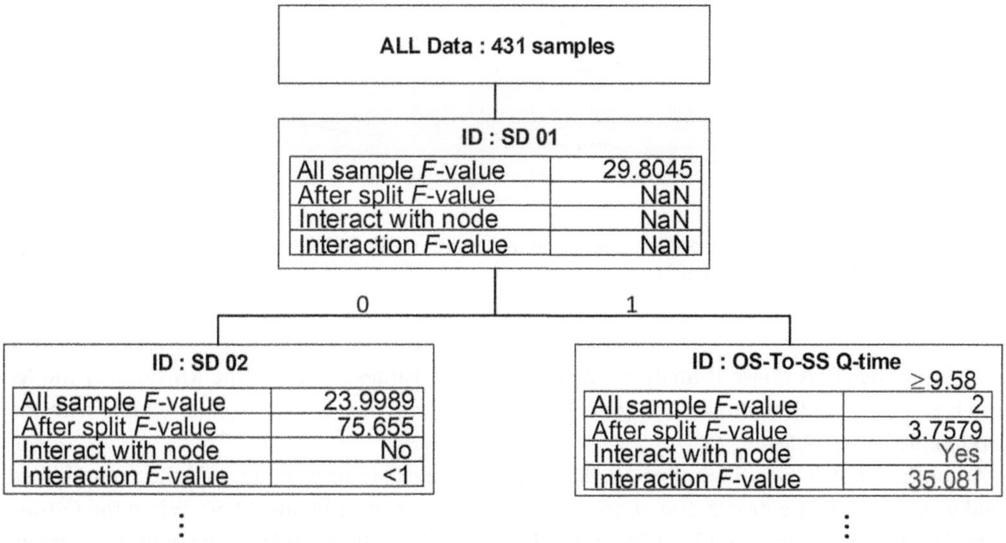

Figure 11.47 IESA Regression Tree analysis results. *Source:* Reproduced with permission from Ref. [12]; © 2018 IEEE.

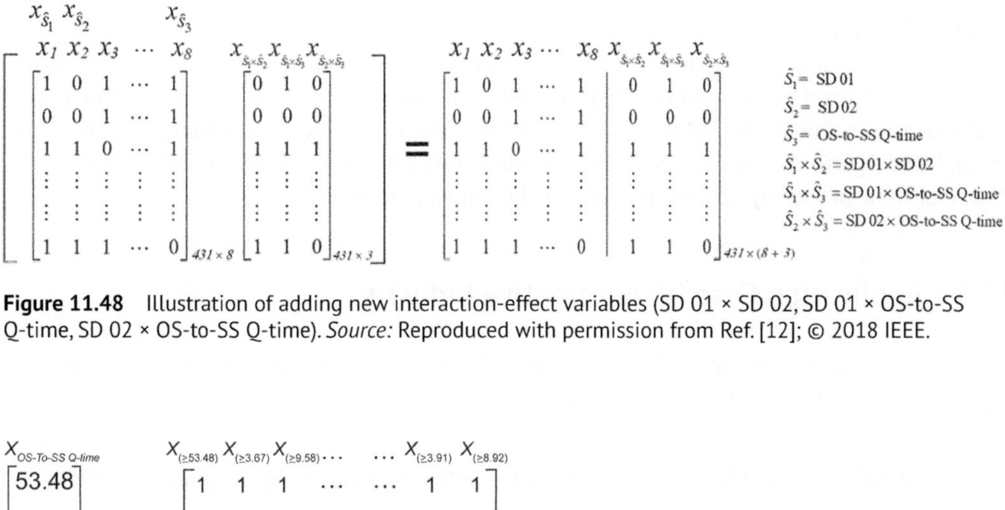

Figure 11.48 Illustration of adding new interaction-effect variables (SD 01 × SD 02, SD 01 × OS-to-SS Q-time, SD 02 × OS-to-SS Q-time). *Source:* Reproduced with permission from Ref. [12]; © 2018 IEEE.

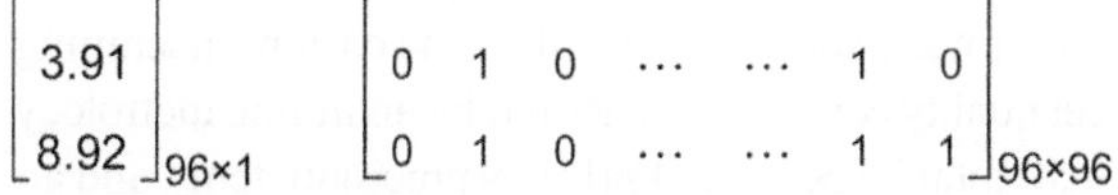

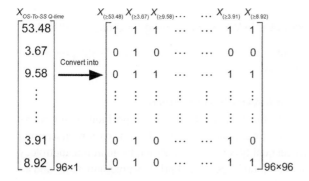

Figure 11.49 Conversion of the OS-to-SS Q-time variable into binary form. *Source:* Reproduced with permission from Ref. [12]; © 2018 IEEE.

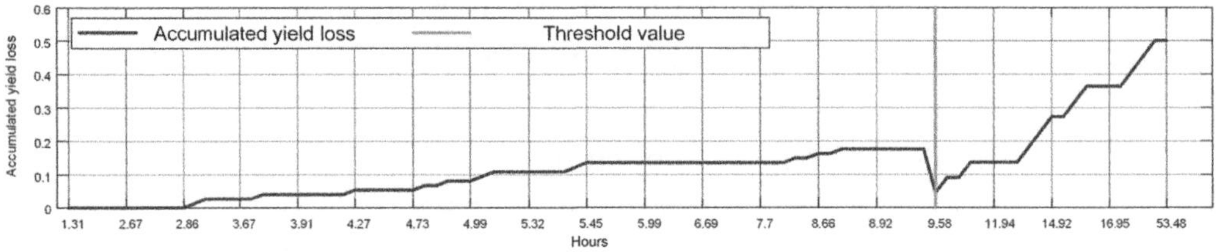

Figure 11.50 Trend chart of the accumulated yield loss vs. OS-to-SS Q-time. *Source:* Reproduced with permission from Ref. [12]; © 2018 IEEE.

conducting the IESA Phase II analysis on all the Q-time values of SD 01, the RI_I value of the $\mathbf{X}_T$ search is 0.742 (>0.7) as shown in Figure 11.46c. Therefore, the search result is reliable with the Top 1 value being ≥9.58.

To confirm that "OS-to-SS Q-time ≥9.58" is the root cause, Q-time values and the accumulated yield loss changing are drawn and verified as the chart shown in Figure 11.50. When the Q-time from OS-to-SS are segmented by 9.58 hours, there are 74 lots below 9.58 hours, with high resistance occurring in 13 lots out of 74, the abnormality rate is 17%. And, there are 22 lots above 9.58 hours, including 11 lots of high resistance, the abnormality rate is 50%. When "Q-time ≥ 9.58 hours" happens, the high resistance rate increases significantly as shown in Figure 11.50. Thus, it implies that the root cause of yield loss due to high resistance is "OS-to-SS Q-time ≥ 9.58."

11.4.3.3 Summary

The original KSA scheme may not correctly discover the root causes of a yield loss with the existence of a significant interaction effect. To remedy this flaw, the IESA module is proposed. The results shown in the illustrated example indicate that the IESA module can indeed improve the original KSA's performance for identifying the interaction-effect key variable and determining the threshold value of the key variable that causes the interaction effect.

11.5 Application Case IV: Automotive Industry

The authors would like to thank FEMCO Machine Tools Manufacturer in Taiwan, ROC for providing the raw data used in this AMCoT and AVM application cases.

11.5.1 AMCoT and AVM Deployment Examples in Wheel Machining Automation (WMA)

The AMCoT platform and its embedded AVM module have been deployed in the WMA production line. The deployment examples are presented below.

11.5.1.1 Integrating GED-plus-AVM (GAVM) into WMA for Total Inspection

As the requirements of highly securing mass-production increase, the machine-tool industry nowadays also begins to pursue the goal of full automation for total inspection [13, 14]. In the current machining scenario of a manufacturing cell for WMA, wheel quality can only be checked by an in-line metrology (ILM) equipment for total inspection equipment for total inspection of primary precision items, and an OMM (off-machine measuring) tool for sampling inspection of secondary precision items.

Proper embedded-sensors are prerequisite for collecting essential real-time process data and capturing any machining variance from machine tools [13, 14]. Three challenges of applying VM to machine tools addressed in [14] had been resolved. Sufficient process data are not only collected from sensors, but also can be (i) segmented to extract the essential parts from the original numerical control (NC) file for synchronizing various machining quality/metrology data, (ii) filtered for improving S/N (signal to noise) ratios, and (iii) transformed into essential features for building VM models. The corresponding solutions of those three challenges were implemented in Generic Embedded Device (GED) [14].

Figure 11.51 shows the architecture of GAVM and the ordinary manufacturing cell for WMA. The WMA cell layout includes two lathes (Lathe 1 and Lathe 2), one drilling machine (Drill), and one OMM tool for sampling inspection of all precision items. Moreover, one robot is placed in the middle of the cell for material handling and two buffers are installed for input/output purposes. GED [16] is installed at the up-right corner, it collects the machining parameters and process data from Lathe 1, Lathe 2, and Drill as well as gathers metrology data from OMM.

In addition, as depicted in Figure 11.51, two more devices: wheel-ID Reader and Laser Marker are also added into the manufacturing cell. Before Lathe 1, each wheel would be engraved a unique ID string by Laser Maker, then Reader is applied for work-in-process (WIP) tracking. After finishing all the machining operations, Laser Marker will again engrave all the essential production information and VM results on each wheel. In fact, WIP tracking is indispensable for Industry 4.0.

Actually, the case of integrating GAVM into WMA is a miniature of Industry 4.0, where GED is the prototype of an Internet of things (IoT) agent and AVM is an example of cyber-physical system (CPS). In fact, the functions of GED can be enhanced such that it can serve as a CPA. And then, by

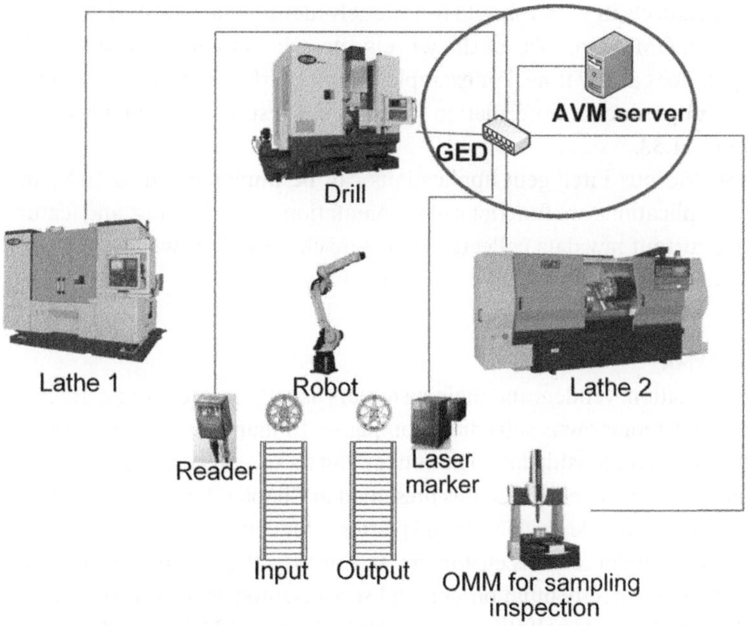

Figure 11.51 Integrating GAVM into WMA. *Source:* Reprinted with permission from Ref. [15]; © 2016 IEEE.

referring to Chapter 7, the so-called AMCoT for wheel production industries is developed based on the miniature GAVM system. Finally, AVM is adopted as a cyber system of AMCoT such that the goal of Zero Defects (ZD) can be achieved.

Enhancing GED to Become CPA

Originally, the functions of GED focus on data collection and implementation of pluggable applications. As shown on the left side of the GAVM architecture in Figure 11.52, machining parameters of the controller and process data of the machine tool are collected and sent to GED via Interface Box; also, metrology data from an OMM are collected and sent to GED. Various pluggable functions, such as segmentation, data cleaning, and feature extraction, are packed as pluggable application modules (PAM) and plugged into GED. For the purpose of illustration, the WMA system shown in Figure 11.51 is adopted as an example for the following explanations.

To reach the goal of Industry 4.0, the functions of 3C, including computation, communication, and control, should be implemented into GED so as to become a CPA for achieving "man-machine collaboration." Thus, the difference of CPA and GED lies not only in the control of interaction with physical objects and cyber systems, but also in the communication among the cyber systems (e.g. AMCoT), the physical objects (e.g. machine tools), and human operators. In this study, CPA serves as an important role of the CPS in (i) performing data collection as well as communicating with physical objects, cyber systems, and human operators, (ii) identifying all the physical objects, and (iii) supporting intelligent applications. More details are described below.

- **Data Collection and Communication**. Data collection from all the physical objects (such as machine tools and OMM) is the fundamental function of CPA. Communication among physical objects, cyber systems, and human operators can enable reporting and decision-making of CPS. A typical example of communication in the WMA case is to report the VM results of all the wheels for accomplishing the goal of ZD.
- **Identification**. All physical objects in WMA should be uniquely identifiable. Physical objects of WMA include machine-tools, OMMs, robots, all the wheels, etc. CPA should know where the object is and what the object does at any time. For example, a laser marker is applied for engraving a unique ID or any useful production information (such as VM results) on each wheel by a QR code as shown in Figure 11.53.
- **Intelligent Applications**. Various intelligent applications can be implemented as PAM and plugged into CPA. Typical applications are functions of segmentation, data cleaning, and feature extraction of vibration and current raw data collected from a machine tool, as well as tool wear prediction and various predictive maintenance (PdM) applications.

11.5.1.2 Applying AMCoT to WMA

Due to marketing internationalization, venders and their customers usually spread around different sites/countries; or one vender/customer owns subsidiary companies in many places. For example, Figure 11.54 shows a one-to-many relationship among a vender and its several customers; the platform to build this relationship is the so-called AMCoT as presented in Chapter 7. To efficiently manage them in different sites, AMCoT provides a cloud-based platform for connecting and sharing all information of things among the vender and his customers. In this way, the vender can create an attractive after-sales service such as directly build on-demand services/models in AMCoT, fan out real-time models over AMCoT and monitor all the machine-tools through AMCoT for decreasing maintenance costs. To take the functionality of AVM model-refreshing proposed in Chapter 8 as an example, this functionality can be enhanced and expanded under this AMCoT platform because each customer does not need to build its own AVM models but just downloads all the preliminary

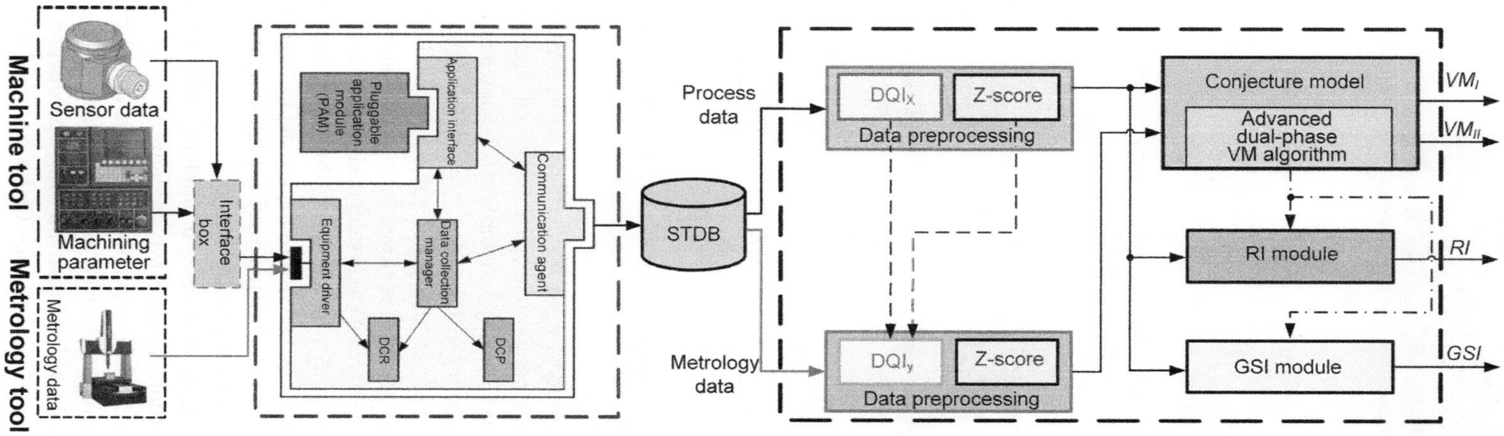

Figure 11.52 GAVM architecture for machine tools. *Source:* Reprinted with permission from Ref. [15]; © 2016 IEEE.

Figure 11.53 A unique QR-code-identification engraved on the mounting-surface of a wheel. *Source:* Reprinted with permission from Ref. [15]; © 2016 IEEE.

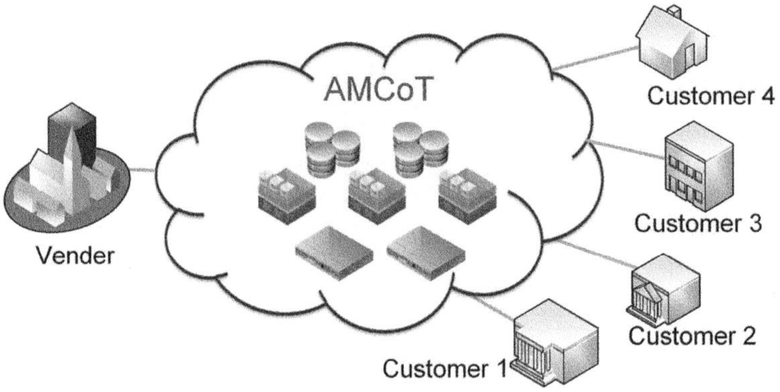

Figure 11.54 One-to-many relationship among a vender and its customers via AMCoT. *Source:* Reprinted with permission from Ref. [15]; © 2016 IEEE.

AVM models (provided by the vender) from AMCoT. As such, AMCoT provides a platform for bridging the technology supports among vender, customers, and manufacturing tools.

The architecture of the existing GAVM system for WMA is shown in Figure 11.55. As the clouds framing, excluding GED, all components of the AVM system including AVM servers, VM manager, Central database, Model creation server, and Simple Object Access Protocol (SOAP) web-services will be implemented in a private cloud-computing environment with a two-part scenario according to the architecture of the cloud-based AVM system designed in [17]. The first part is to virtualize all components of the AVM system while keeping the implementation of each component unchanged; the second part is to design an extra server to act as the system's operational core for hosting and performing the designed functional mechanisms so as to reduce the efforts of migrating the original AVM system to the cloud environment.

Expanding Figure 11.54, the detailed drawing of applying AMCoT into the WMA vender site (Vender) and customer sites (Customers) is shown in Figure 11.56. The complete infrastructure of AMCoT includes a cloud portion containing various cyber systems and one Vender site that owns a CPA and a WMA demonstrative line; as well as several Customers' sites. Each Customer possesses many WMA production lines. For brevity, a WMA production line is denoted as a Cell. Hence, each Customer owns many Cells and each Cell is equipped with a CPA for connecting to AMCoT and for deploying intelligent PAM functions which have real-time and on-line prediction/

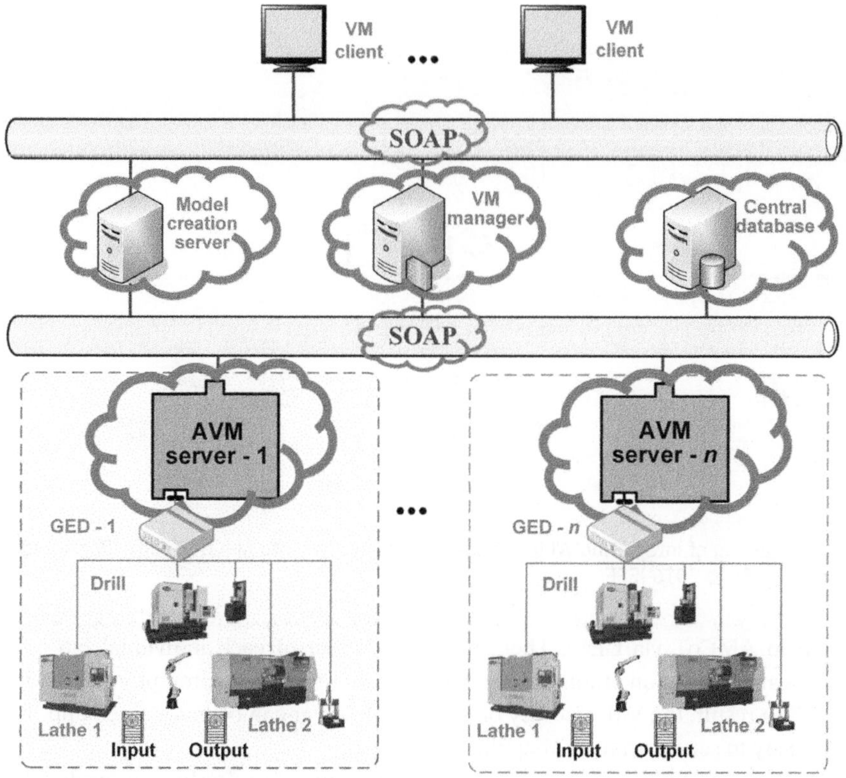

Figure 11.55 Architecture of the existing GAVM system. *Source:* Reprinted with permission from Ref. [15]; © 2016 IEEE.

monitoring capabilities. The intelligent PAM functions may include tool-wear prediction, collision detection, PdM, etc. The cyber systems developed in the cloud may contain the cloud-based AVM system, tool-life management, collision detection management, IPM service, etc.

11.5.1.3 Applying AVM in AMCoT to WMA

As shown at the bottom of Figure 11.56, two WMA factories for manufacturing alloy wheels are connected with AMCoT. The factory on the left side is the Vender of selling WMA-related machine-tools. Usually, the Vender owns one standard WMA cell for demonstration purposes. The factory on the right side is Customer 1 that manufactures massive wheels. The mass-production Customer 1 owns many identical WMA Cells for manufacturing wheels simultaneously. However, to manufacture various types of wheels, the machine-tools layouts, wheel materials, tool types, and NC paths are decided upon Customer's demands. In the following example, production conditions of Custom 1 are somewhat different from Vender even though the same-type of machine tools are utilized. The example demonstrates global cyber-physical interactions among Vender, AMCoT, and Cell 1 of Customer 1 by applying AVM models refreshing.

The AVM System shown in AMCoT of Figure 11.56 is composed of a Model-Creation Server, a VM Manager, several VM Clients, and many AVM Servers as depicted in Figure 11.55. The Model-Creation Server will generate the first set of VM models of a certain machine tool type in the standard WMA Cell of the Vender. Then, the VM Manager in the AVM System of AMCoT can fan out the first generated set of VM models to all the AVM Servers of the same tool type in the other

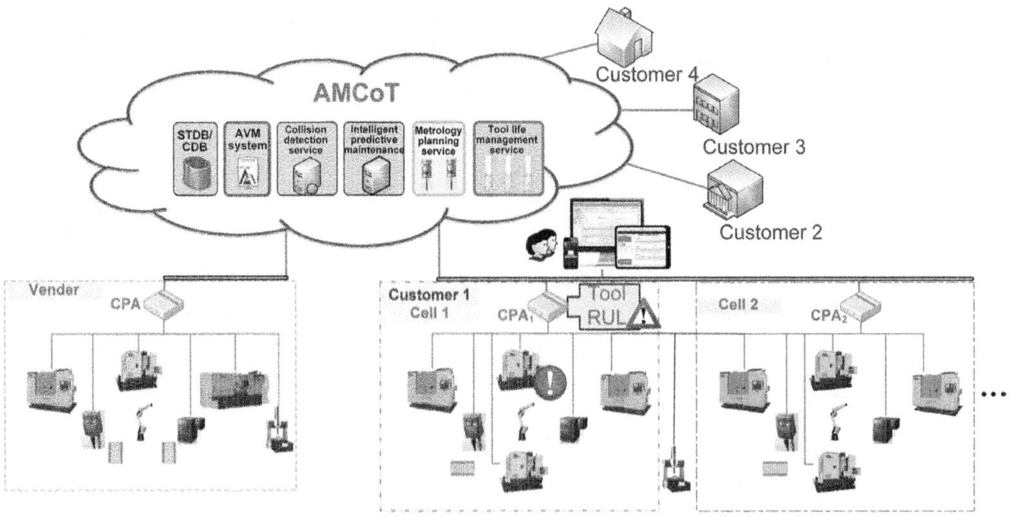

Figure 11.56 Detailed drawing of integrating WMA's vender and customers into AMCoT. *Source:* Reprinted with permission from Ref. [15]; © 2016 IEEE.

Cells that connected to AMCoT via CPAs. Then, the AVM Server of each individual fan-out-accepter of each Cell can perform an automatic model refreshing process to promptly refresh its own VM model set. As a result, the VM accuracy of each AVM Server can be maintained and the AVM Server is then ready to serve various VM applications.

Necessity for checking the conditions of a successful and complete refreshing procedure is explained below. The physical properties of two different machine tools are less likely to be the same. When conducting automatic fanning out, from Tool A to Tool B for example, the prediction accuracy of the fanned-out model will not be as good as that of the model created with the original data of Tool B. Therefore, automatic model refreshing should be applied to recover the prediction accuracy of the fanned-out model.

During the refreshing stage, the AVM Server will apply the advanced dual-phase VM algorithm to automatically refresh the old models in Tool B. As such, the Tool B models, originally copied from Tool A, can gradually retain its freshness and prediction accuracy.

In other words, the AVM's Model-Creation Server in the cloud will build the first set of AVM models according to Vender's machine-tool type in the demonstrative Cell. Then, the VM Manager in the cloud can fan out the first-set models to all Cells with the same machine-tool type via CPAs and AMCoT. In fact, the functionality of model-refreshing can be regarded as a series of interactions between the cyber (AVM) system in the cloud and physical objects (machine tools) in the Cells.

The BPNN is adopted as the algorithms of the VM models. The central-hole-diameter (CHD) of a wheel is selected as the key precision item for VM to demonstrate the model-refreshing effectiveness and adaptability of AMCoT. The AC4CH aluminum casting alloy wheels are machined in both Vender and Cell 1 of Customer 1, while the wheel materials, cutting tools, precision items, and machine-tool layouts are somewhat different between both sites. One accelerometer and three current sensors are installed in each machine tool for collecting the vibration and current signals. After segmenting signals and cleaning data, CPA extracts a so-called concise expert knowledge set (EK_C) of signal features (SFs) from the collected signals to capture real machining statuses. In EK_C, the six vibration SFs include maximum, root mean square (RMS), average, skewness, kurtosis, and

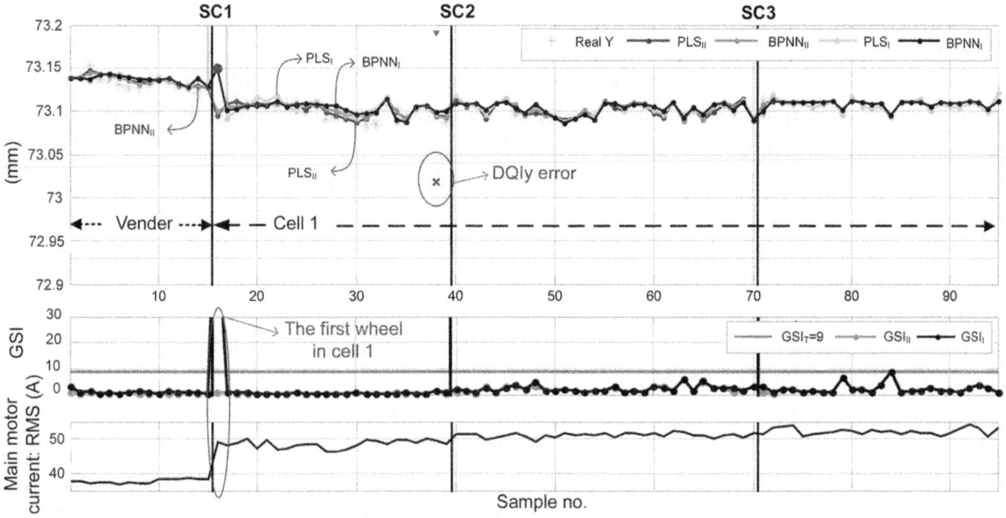

Figure 11.57 Global cyber-physical interactions (AVM models refreshing). LCL, lower control limit (72.93 mm); UCL, upper control limit (73.15 mm). *Source:* Reprinted with permission from Ref. [15]; © 2016 IEEE.

standard deviation; while the four current SFs are RMS, mean, maximum, and crest factor. Therefore, the total SF number of EK_C is 18 for one machine tool.

The CHD model-refreshing results of VM are displayed in Figure 11.57. Thirty (30) historical samples are collected from Vender and sent to AMCoT for creating the first set of AVM models. This set of models is, then, downloaded to Vender for testing and verification. Observing Samples 1–15, they are manufactured at Vender's site and their VM accuracy is verified as good with the maximal error being less than 0.02 mm. The Vender's CHD is close to 73.13 mm.

Afterwards, the same set of AVM models is downloaded to Cell 1. Eighty (80) samples, from Samples 16 to 95, are tested in Cell 1 for three (3) days. Three s (SC1–SC3) occur prior to Samples 16, 40, and 71, respectively. Note that, the first SC occurs at the beginning of the very first wheel machined by the machine-tool of Cell 1; the other two SCs result from tool idling for not operating during the evening at Cell 1.

During Day 1 (between SC1 and SC2), the CHD of every wheel is not only predicted by AVM but also measured by OMM for the purpose of verifying the accuracy of AVM. Observing Sample 16, the first sample exhibits a significant difference between the modeling samples and Sample 16 because the AVM's GSI value of that sample is much higher than its threshold (9). As stated in Chapter 6, the GSI is defined as the degree of similarity between the set of process data currently inputted and all the historical sets of process data used in the conjecture model. This high GSI value is due to the changes of main-motor-current RMS value at the bottom of Figure 11.57, with Vender using 38A to drive the main motor and Cell 1 using 49A. As a result, an obvious 0.02–0.03 mm of physical machining difference exists between Vender and Cell 1.

Apparently, $BPNN_I$ accuracy of Sample 16 is rather unreliable because the first set of AVM models is created from Vender without refreshing. However, after including the first physical sample of Cell 1 into the cyber AVM models in AMCoT for refreshing, the value of $BPNN_{II}$ at Sample 16 is almost the same as that of the real measurement value. The improved VM phase-II accuracy shows that cyber AVM models in AMCoT can be refreshed by a real measurement value collected from a physical measurement device of Cell 1. This is a typical example of cyber-physical interaction.

Subsequently, by applying the refreshed AVM models, the $BPNN_I$ accuracy of Sample 17 is more acceptable. Note that, a DQI_y (metrology-data-quality) [18] error is detected successfully at Sample 38, which indicates a metrology error resulted from erroneous measurement. Of course, the measurement value of Sample 38 should not be adopted for refreshing the AVM models. Except Samples 16 and 38, the VM accuracy of all the other samples is within 0.02 mm that is quite acceptable. As such, the current set of AVM models is verified and ready to be applied for production. Then, during Day 2 (between SC2 and SC3), only Sample 40 is needed to be measured by OMM for refreshing the AVM models just after SC2; the CHD values of all the other samples are provided by AVM. Similarly, during Day 3 (after SC3) only Samples 71 and 72 are measured by OMM for refreshing the AVM models; the CHD values of all the other samples are generated by AVM. In fact, only Sample 71 is required to be measured for refreshing the AVM models; the other measurement of Sample 72 is redundant for refreshing purposes.

In conclusion, during Days 2 and 3, Samples 40–95 are manufactured; while the CHD values of only Samples 40, 71, and 72 are measured by OMM, and those of all the other 53 samples are provided by AVM to achieve total inspection. Moreover, if there are 30 Cells in Customer 1's factory, then the same approach can be applied to all the 30 Cells in parallel and simultaneously. Therefore, the goal of total inspection that leads to ZD can be accomplished in Customer 1.

11.5.1.4 Summary

The AMCoT platform together with AVM is demonstrated above for the applications of WMA to accomplish the goal of ZD.

11.5.2 Mass Customization (MC) Example for WMA

One of the core values of Industry 4.0 targets to integrate people's demand into manufacturing for enhanced products, systems, and services for a wider variety of increasingly personalized customization of products. Thus, Industry 4.0 advances the traditional manufacturing techniques from mass production towards mass customization (MC). Take WMA as an example, to meet MC expectations, manufacturers should offer customized wheels at a large scale with low cost, short lead-time, and high quality. Thus, WMA cells with MC capability are requested to be designed to have high degree of quick responsiveness to accurately react in case of machining-condition changes for manufacturing different wheel types.

11.5.2.1 Requirements of MC Production for WMA

Like MC production in any other industries, wheel manufacturers have no choice but to face the requirement of quickly adapting to satisfy various needs of different customers. For a WMA cell, the MC production, or the so-called mixed-wheels manufacturing, means various customized-wheels are queueing on the production line for being batch-machined in the same WMA cell through switching customized machining conditions.

To meet MC expectations, manufacturers should offer customized wheels at a large scale with low cost, short lead-time, and high quality. Thus, WMA cells are requested to be designed to have high degree of quick responsiveness to accurately react in case of machining-condition changes for manufacturing different wheel types, as well as flexibility to promise benefits ranging from high utilization to large-scale volume of productivity.

To meet the requirements of MC production for WMA, we propose to apply the AVM system together with the so-called Target-Value-Adjustment (TVA) scheme. The TVA scheme is designed to enhance AVM's adaptive customization capability for automatically and rapidly accomplishing the goals of MC production.

11.5.2.2 Considerations for Applying AVM in MC-Production of WMA

SCs defined in [19] may result from maintenance operation, tool repair, recipe adjustment, or tool idle for a long period of time. The model-refreshing function of AVM has the ability to update the latest machining condition into VM models after encountering an SC. Originally, the design concept of the model refreshing focuses on "a stable machining environment for various mass productions" under the conditions of (i) good quality of real measurement of products, and (ii) fixed key signal-features (SFs) during a model refreshing process. With these two essential prerequisites, model refreshing of AVM can be rapidly completed within three samples/wheels to adapt any change(s) resulting from different SCs even if the VM models are initially adopted from other AVM servers of WMA cells.

MC-production in a WMA cell will generate many machining-condition adjustments. These machining-condition adjustments can be treated as a different kind of SCs. As such, from the perspective of AVM, MC production will cause a series of SCs. Every time there comes a new customized-wheel order, not only machine tools have to be manually re-adjusted, but also measurement devices (e.g. coordinate-measuring machine (CMM) or automated optical inspection (AOI)) have to be manually re-calibrated for new wheel-types. As a result, frequent SCs resulting from MC will make machining adjustment and measurement quality hard to maintain due to many human operations. Moreover, target values of inspection items may vary with SCs for different wheel types as well. This target-value change may not be reflected by SFs adopted in the VM models. Therefore, VM accuracies are greatly impacted by un-reliable measurements and target-value changes, which result in incorrect cause-and-effect relationship between SFs and the corresponding target value of the inspection item after each SC. To apply the AVM system in MC production of WMA, two issues should be carefully considered: (i) metrology quality evaluation and (ii) target value adjustment. These two issues will be resolved by the proposed TVA scheme as explained below.

11.5.2.3 The AVM-plus-Target-Value-Adjustment (TVA) Scheme for MC

The metrology values of modeling and running samples of the TVA scheme are shown in Figure 11.58. The metrology values of the modeling-samples will be adjusted by the TVA scheme and then be utilized to build various VM models; while the running-samples are newly manufacturing samples of the latest wheel-type. An SC exists in front of the running samples.

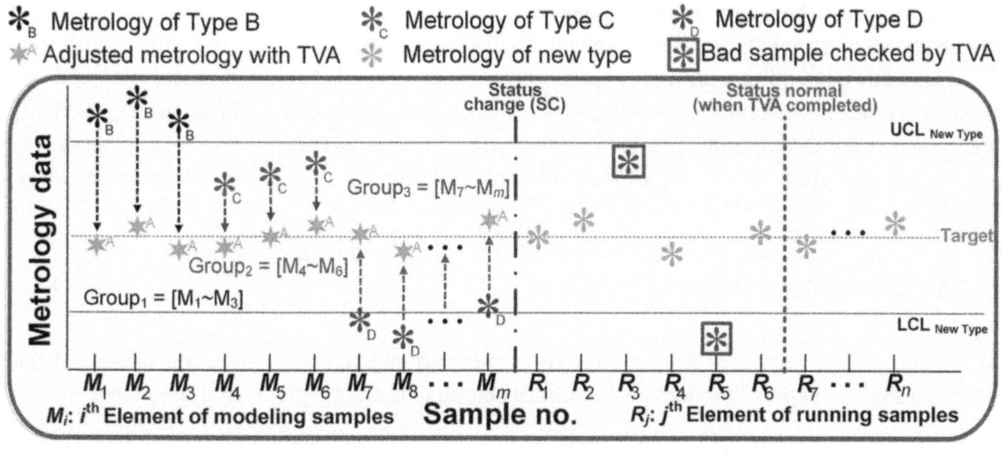

Figure 11.58 Operating scenarios of modeling and running samples of the TVA scheme. *Source:* Reprinted with permission from Ref. [20]; © 2017 IEEE.

The operating principles of the TVA scheme mainly include running-samples' quality check, modeling-samples' grouping, and modeling-samples' target value adjustments. A set of determinative samples (**DS**) is defined in TVA for computing target-value automatically. The size of **DS** is called $SIZE_{DS}$ and may vary from 0 to the threshold of $SIZE_{DS}$ denoted as T_{DS}. T_{DS} is defined as the needed amount of samples and is utilized to determine the number of iterations of the TVA scheme, which is about 3–5, suggested by expert knowledge. The target value of running samples is decided by the mean metrology value of **DS** as in (11.19), where y_j is the jth element of AM values, n = size of **DS**.

$$Target\ value = \bar{y}_{DS} = \frac{\sum_1^n y_j}{n}. \tag{11.19}$$

For machine tool industry, after wheel-type change, not only the features of the production tools alter, but also the metrology equipment needs to be recalibrated to ensure the metrology accuracy. Nevertheless, even after metrology equipment adjustment, the first several wheels after the wheel-type change may still be unstable. Unfortunately, the DQI_y functional module of the AVM system presented in Chapter 8 is not ready yet because of the metrology data collected are insufficient for model refreshing during this period. Therefore, in order to guarantee the metrology quality after the occurrence of an SC, a simplified quality check is performed first.

The quality check is utilized to compute a difference delta value (δ) between the neighboring samples and check if the δ is greater than the threshold value of δ (δ_T) or not, where δ_T is computed with specification tolerance and the tolerance ratio (α) is defined as in (11.20).

$$\delta_T = \alpha \left| UCL_{New \cdot Type} - LCL_{New \cdot Type} \right|. \tag{11.20}$$

The default value α is set to be 0.2. The specification has UCL and LCL that can be set with new type of samples (denoted as — in Figure 11.58). When a new metrology sample is checked as normal, it is added into DS; on the other hand, if the new sample is abnormal, it is denoted as ⊞ (such as R_3 and R_5 in Figure 11.58), and is not allowed to be added into DS.

As mentioned before, it will take too much time and cost to collect enough number of a specific type of samples for building the VM models. Therefore, to create models by utilizing multiple types of samples can reduce the time required for the model-creation process. Yet insufficient cause-and-effect relationships between the metrology values and their corresponding SFs resulting from different types of samples should be resolved. To distinguish different types of samples with various target values, grouping within the modeling samples is needed. As shown on the left-hand side of Figure 11.58, the modeling samples contain three metrology types of different target values, Type B ($*_B$), Type C ($*_C$) and Type D ($*_D$); they are sorted in ascending order to become $\mathbf{y}_M$ by applying (11.21), where y_i is the ith element of modeling samples, $m = SIZE_{y_M}$ = size of modeling samples.

$$\mathbf{y}_M = \left[sort\left(Model\ y_i\right)_{\text{Ascending}} \right], i = 1, 2, \ldots, m. \tag{11.21}$$

After that, a delta vector ($\boldsymbol{\delta}$) composed of differences among all the neighboring values of $\mathbf{y}_M$ is utilized to check whether the metrology samples belong to the same group or not, where

$$\delta_i = \mathbf{y}_{M(i+1)} - \mathbf{y}_{M(i)}, i = 1, 2, \ldots, m-1. \tag{11.22}$$

Set a δ_T value to search for the breakpoints of all $\mathbf{y}_M$ elements. Find the breakpoint vector (**BV**) with any of ($\boldsymbol{\delta}$) elements being greater than δ_T. As shown on the left-hand side of Figure 11.58, $\mathbf{BV} = [3, 5]$. Then, combine 0, **BV** and SIZE_{y_M} to become **BV′** as in (11.23).

$$\mathbf{BV}' = \left[0, \mathbf{BV}, \text{SIZE}_{y_M}\right]. \tag{11.23}$$

Group the modeling samples by **BV′** to obtain $\mathbf{G}_p$ as in (11.24), and calculate the mean of each group, denoted as $\bar{y}_{G_p}$. For example, as shown on the left of Figure 11.58, all modeling samples are sorted into $p = 3$ groups denoted as $\mathbf{G}_1 = [M_1 \sim M_3]$, $\mathbf{G}_2 = [M_4 \sim M_6]$, and $\mathbf{G}_3 = [M_7 \sim M_m]$, respectively, where M_i represents the ith element of modeling samples and $\text{SIZE}_{BV'}$ is the size of breakpoint vector.

$$\mathbf{G}_p = \left[\mathbf{y}_{M\left(\mathbf{BV}'_{(p)}+1\right)} \text{ to } \mathbf{y}_{M\left(\mathbf{BV}'_{(p+1)}\right)}\right], p = 1, 2, \ldots, \text{SIZE}_{\mathbf{BV}'} - 1. \tag{11.24}$$

Calculate $\bar{y}_{G_p}$ and apply (11.19) to obtain the new target value, $\bar{y}_{DS}$, then adopt $y^A_{G_p}$ as in (11.25) to adjust each element in $\mathbf{G}_p$ to become the newly adjusted modeling samples.

$$y^A_{G_p} = \left(\textit{Each Element in } \mathbf{G}_p - \bar{y}_{G_p}\right) + \bar{y}_{DS}, p = 1, 2, \ldots, \text{SIZE}_{BV'} - 1. \tag{11.25}$$

Consequently, metrology $*_B$, $*_C$, and $*_D$ are adjusted to become $\star^A$ by applying (11.25) as displayed in Figure 11.58.

After all the modeling samples have been adjusted by the new target value, revert SC (as ▬ · ▬ in Figure 11.58) into Status Normal (as ▪▪▪▪ in Figure 11.58), and finally delete all the elements in **DS** and reset SIZE_{DS} to 0. This concludes the process of the TVA scheme.

The results of TVA obtained from (11.25) provide all the samples in the VM models with correct cause-and-effect relationship for the newly-encountered wheel type. Therefore, these modeling samples after TVA treatment can then be applied to build the proper VM models of the newly-encountered wheel type.

The flowchart of the TVA scheme based on the scenarios mentioned above is depicted in Figure 11.59 and is explained as follows. Before running the operational flow of the TVA scheme shown in Figure 11.59, all the elements in **DS** should be cleared and the values of SIZE_{DS} count is reset to zero. Also, the initial status is set to be normal. Steps 3–5 are about the quality check of the new samples, Steps 6–11 are for grouping modeling samples, and Step 12 is concerning the final target value adjust of all the modeling samples. Detailed explanations are stated below.

Step 1: Obtain a metrology sample (y) and its corresponding process data (**X**).
Step 2: If an SC occurs, go to Step 3 for executing TVA; otherwise, jump to Step 13 to skip the TVA process.
Step 3: Conduct metrology-sample quality check. If the metrology quality is good, go to Step 3.1 to add this good sample into **DS** and increase SIZE_{DS} by one. On the contrary, if the metrology quality is poor, skip Step 3.1 to avoid this sample being added into **DS**.
Step 4: Check if SIZE_{DS} is smaller than or equal to T_{DS}; if not, TVA has finished all the procedures, then execute Step 4.1 to reset status to normal and jump to Step 13 to finish TVA; if yes, then TVA is still needed, go to Step 5.
Step 5: Compute the target value $\bar{y}_{DS}$, as in (11.19).
Step 6: Sort the modeling metrology samples in ascending order to get y_M as in (11.21).
Step 7: Compute the delta vector ($\boldsymbol{\delta}$) of y_M as in (11.22).

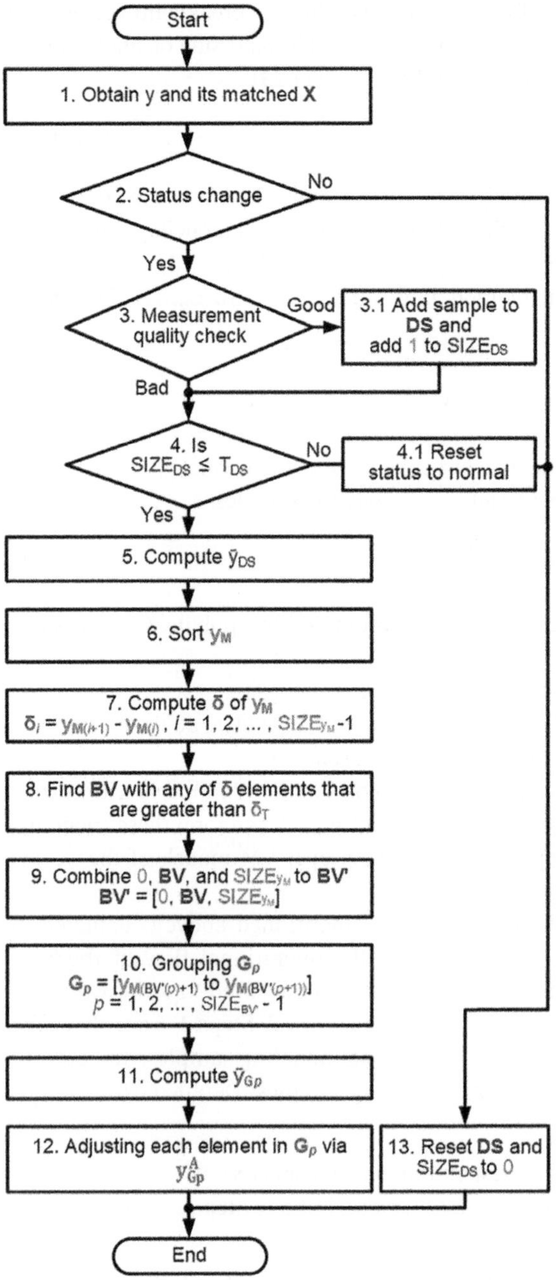

Figure 11.59 Flowchart of the TVA scheme. *Source:* Reprinted with permission from Ref. [20]; © 2017 IEEE.

Step 8: Find (**BV**) with any of **δ** elements that are greater than δ_T.

Step 9: Combine 0, **BV**, and SIZE_{y_M} to become **BV′**, to record the sorting indexes among groups as in (11.23).

Step 10: Group the modeling samples by **BV′** to obtain $\mathbf{G}_p$ as in (11.24).

Step 11: Compute the mean value of metrology data of each group, $\bar{y}_{G_p}$.

Step 12: Adjust each element in $\mathbf{G}_p$ via $y^A_{G_p}$ as in (11.25).

Step 13: Delete all the elements in **DS** and reset SIZE_{DS} to 0.

11.5.2.4 AVM-plus-TVA Deployment Example for WMA

This example adopts the wheel machining process to demonstrate that the MC issue with three different types of wheels is resolved by applying the AVM-plus-TVA scheme.

To demonstrate flexibility and effectiveness of the TVA scheme, three types of AC4CH aluminum casting alloy wheels are batch machined in a MC WMA cell. The central-hole diameter (CHD) of a wheel machined by a lathe is taken as the inspection item for this example. For the purpose of illustration, AOI is adopted in this example to provide the real metrology for accuracy check.

One accelerometer and three current sensors are installed in this lathe machine for collecting the vibration and alternating current signals. After segmentation, data cleaning, and feature extraction processes, six vibration signals including maximal magnitude, RMS, average, skewness, kurtosis, and standard derivation values; and four current signals involving: RMS, average, maximal magnitude, and crest factor values are selected as SFs. As a result, the total number of SFs is 18.

The BPNN is adopted as the algorithms of the VM model. Given the predictive VM value $\hat{y}_i$, AM value y_i, and the sample size n, the predictive accuracy in machine-tool industry is presented by the mean absolute error (MAE) denoted in (11.26).

$$\text{MAE} = \frac{\sum_{i=1}^{n}\left|\left(\hat{y}_i - y_i\right)\right|}{n} \times 100\%. \tag{11.26}$$

Totally 115 wheel samples (including 58 Type A, 33 Type B, and 24 Type C in sequence) are utilized in this example. The first 25 Type A wheels, which were machined several days ago, are adopted as the initial modeling samples. Prior to mass production, the AOI is calibrated to measure Type A wheels. Then, 33 Type A wheels are processed. After that, the AOI is re-calibrated to gauge Type B wheels followed by machining 33 Type B wheels. And then, the AOI is re-calibrated again to inspect Type C wheels. Finally, 24 Type C wheels are manufactured. According to the scenario mentioned above, three SCs occur before Samples 1, 34, and 67 as shown in Figure 11.60. All the SCs result in AOI calibrations; therefore, the TVA scheme (with $\text{SIZE}_{\text{DS}} = 3$) is applied to sustain the accuracy of VM results. To illustrate the effectiveness of TVA, the VM results without and with the TVA scheme are compared and depicted in Figure 11.60.

Note that, after encountering an SC, the metrology values of the first three samples are adopted to adjust the target value by TVA since SIZE_{DS} is set to be 3. As for the without-TVA Case, it still needs the first three metrology samples to refresh the VM models by the original AVM scheme. Then, the metrology values of one-out-of-ten wheels (such as Samples 14, 24, etc., as shown in Figure 11.60) are utilized for refreshing the VM models. The abovementioned scenario emulates the real mass-production environment. The rest nine-out-of-ten wheels' real metrology values measured by AOI are used to evaluate the VM accuracy.

As shown in Figure 11.60, the Type-A MAE of the without-TVA case is 0.0057 mm that is acceptable due to the fact that the original VM models are built by adopting Type A samples. However, after the wheels are changed from Type A (CHD = 83.10 mm) to Type B (CHD = 64.05 mm), the Type-B VM results of the without-TVA case go wild, which is undesirable. Then, the wheels are changed to Type C with CHD being 64.10 mm, which has slightly difference in comparison with that of Type B. Nevertheless, the Type-C VM results of the without-TVA case is still unacceptable.

Observing the with-TVA case in Figure 11.60, after encountering an SC before Sample 1, two good-metrology-quality samples are added into DS. Then, Sample 3's AOI metrology quality is bad, which is caught by TVA. As a result, Sample 3 cannot be added into DS; then, Sample 4 is included

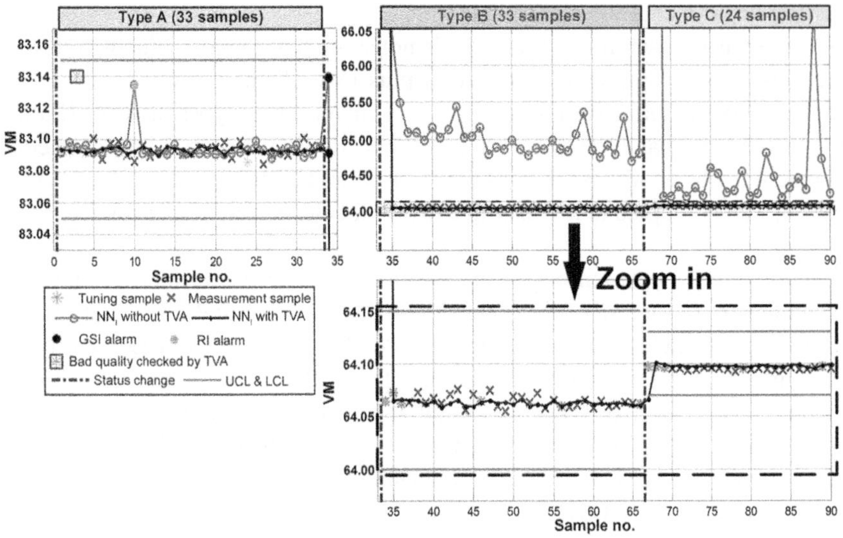

Figure 11.60 VM results with and without the TVA scheme. *Source:* Reprinted with permission from Ref. [20]; © 2017 IEEE.

into DS instead. And then, only one-out-of-ten samples are adopted to refresh the VM models. The Type-A MAE of the with-TVA case is 0.0034 mm which is better than that of the without-TVA case (0.0057 mm). After that, the wheels are changed to Type B such that another SC is encountered before Sample 34. Again, three good-metrology-samples are adopted into DS; and then, one-out-of-ten samples are adopted to refresh the VM models. As far as the VM accuracy is concerned, only the accuracy of Sample 34 (which is the first sample after the SC) is no good due to the first wheel changed from Type A to Type B. The VM results of all the other samples of Type B are close to the AOI metrology values with MAE value being 0.0040 mm. Finally, the wheels are changed to Type C such that another SC is encountered before Sample 67. The VM performance of Type C is as good as that of Type B.

11.5.2.5 Summary

The AVM-plus-TVA scheme is adopted in this paper to accommodate the MC production of a WMA cell. The TVA scheme is designed to enhance AVM's adaptive customization capability for automatically and rapidly accomplishing the MC goals. However, the purpose of the TVA scheme is not to replace the key features related to MC but to rapidly compensate the target-value gaps among modeling samples and running samples after SCs when the key features that reflect the target-value differences are not available. The illustrative examples reveal that the AVM-plus-TVA scheme is feasible and promising for meeting the requirements of MC production.

11.6 Application Case V: Aerospace Industry

The authors would like to thank Aerospace Industrial Development Corporation in Taiwan, ROC for providing the raw data used in this AVM application case.

11.6.1 Introducing the Engine-Case (EC) Manufacturing Process

Industry 4.0 creates a smart factory by applying novel technologies including IoT, CPSs, cloud-manufacturing, and big data analytics [15]. A smart factory will make a great contribution to the improvement of manufacturing efficiency and automation, which enables machine tools to make intelligent decisions themselves through extracting knowledge from big data by machine learning, artificial intelligence, or other high-powered analytics to reduce cost, improve efficiency, and even make the mass-customization possible [20]. The concept of smart factory can also benefit the machinery industry. For example, the techniques of machining condition monitoring [21, 22] have been extensively studied such that machine-health-status diagnosis, cutting-tool RUL prediction, etc., are advancing toward a more mature level than before.

However, the high barrier of entering the level of smart factory prevents this state-of-the-art technology from being easily adopted to the manufacturing of aerospace components. There are still significant challenges for aerospace components with complex machining conditions [23].

Compared to other components in machinery industries, the manufacturing of aerospace components possesses characteristics of complex operations, large variety, small batch, and frequent changes of production status [24]. Therefore, high-efficient communication and collaboration relationship among manufacturing departments are required so as to shorten lead time and reduce cost.

For manufacturing a turbine engine-case (EC), complex machining operations including rough and finish machining will be continuously processed on various types of lathes and milling machine-tools. Meanwhile, the machining quality should strictly be within the tolerable specification. Among several perspectives of challenges as mentioned above, this paper focuses on the most critical issues of real-time and on-line monitoring machining quality for aerospace components.

11.6.1.1 Manufacturing Processes of an EC

The key indexes of machining conventional components (such as car wheels, etc.) are short cycle time and high yield, which could be achieved by sampling inspections. However, as far as machining an engine case (EC) is concerned, several hundred operations are required to complete one cycle of an EC from a raw material to the finished product [25]. For example, main milling operations of an EC include various types of flange holes, base holes, location holes, blind holes, counterbores, scallop pockets, and slots. The total number of machining operations of an EC might take up more than 100 hours of cycle time. All machining precisions of each operation must be inspected for safety considerations and thus this results in a long-awaiting quality-inspection cycle time that raises production cost.

The purpose of the flange face of an EC is to serve as the contact surface to tightly link various types of cases among intermediate-pressure and high-pressure compressors for necessary connections. There are up to two hundred flange holes spreading on the end-face (flange-face) and they are one of the important machining-features whose quality needs to be ensured.

11.6.1.2 Inspection Processes of the Flange Holes

In the aerospace-components manufacturing industry, a quality inspection report which provides all machining precisions of an EC is essential for identifying whether the EC meets the specification or not [23, 26]. The upper part of Figure 11.61 shows various quality inspection methods for the flange precisions.

Conventionally, off-machine measurement (OMM) and on-machine probing (OMP) are two main measuring methods for flange-quality inspection. CMM can provide a more comprehensive quality inspection report than OMP does, including all types of precisions manufacturers need.

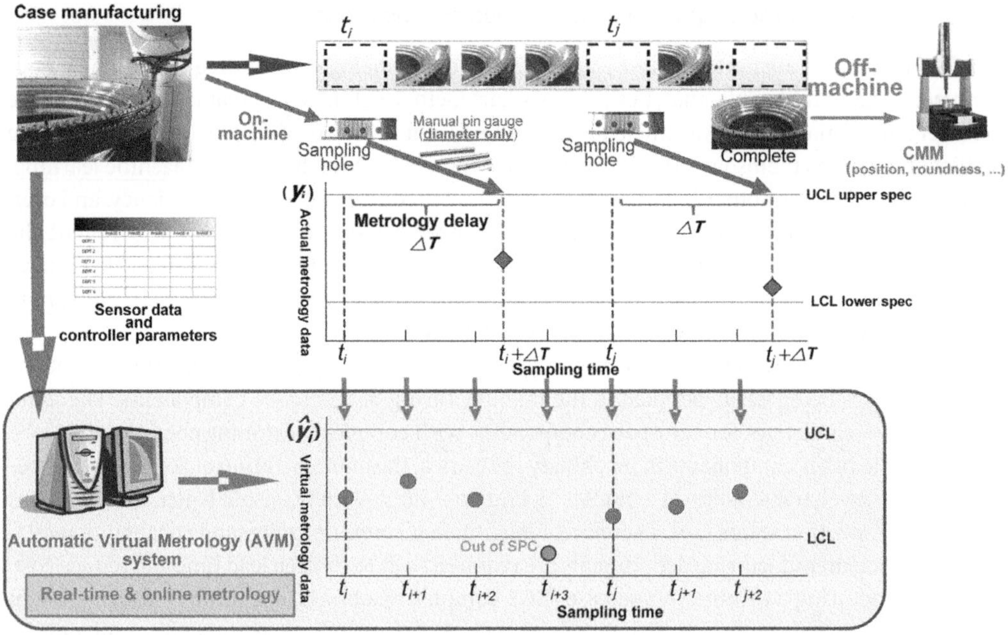

Figure 11.61 Comparison of on-machine probing (OMP), CMM, and VM. *Source:* Reprinted with permission from Ref. [27]; © 2018 IEEE.

However, the metrology delay is usually more than several days. On the other hand, OMP can save this metrology delay by installing a probe into the spindle to achieve on-line measurement; however, frequent switching processes between the probe and the cutting tools are time-consuming due to the fact that hundreds of machining features have to be inspected. Also, when OMP is applied, the manufacturing process needs to pause, and that will prolong the cycle time.

To save the cycle time, OMP is commonly replaced by the go/no-go pin gauge. However, only sampling flange-holes are chosen for manual gauging according to this rule. The operator inspects the Diameter of the first three flange-holes while changing to a new drill-cutter, and then it turns into the one-out-of-ten sampling inspection. Other critical measurement items, such as position error (Position) which is highly related to the EC deformation, can only be measured by CMM. On the other hand, although a probe or a pin gauge enables OMP to perform inspections on-line, the additional operations will also lead to extra cycle time.

In brief, cycle-time loss and/or metrology delay are unavoidable whether CMM, OMP, or a pin gauge is applied. Manufacturers need to balance appropriately between manufacturing cycle time and quality inspection. Due to the extremely high production cost and flight safety concern, aerospace-components manufacturers are more demanding than any other machinery industries for having an efficient and cost-effective machining-monitoring system [24, 28].

11.6.1.3 Literature Reviews

Literature reviews of machining operations of aerospace components are described below. Marinescu and Axinte [23] presented a sensor-fusion-based supervision system to monitor the milling quality and safety of aerospace components. Tangjitsitcharoen et al. [26] built a multiple-regression model to predict the in-process surface roughness during the ball-end milling operation of aerospace components. Griffin [29] used the genetic programming technique with neural networks (NN) algorithm to predict two complex geometrical machining features of aerospace components: grinding and hole-making operations. Gao et al. [25] optimized machining parameters to accumulate machining experience, avoid the repeated failures, and improve the machining quality for complicated aerospace components processing. Liu et al. [30] proposed a support-vector-machine-based system to avoid complex operations from resulting in the deformation of the thin-walled components in the aerospace industry.

To resolve the cycle-time-loss, metrology-delay, and deformation problems mentioned above, in this application case, AVM together with the Deformation Fusion (DF) scheme is applied to the EC manufacturing. The so-called DF scheme will estimate the EC deformation that is adopted as one of the key features for Position prediction of the AVM system.

By integrating AVM into the EC production line, as shown on the lower part of Figure 11.61, once the process data of manufacturing an EC are corrected completely and sent to the AVM system, which then conjectures the manufacturing quality of that EC on-line and in real-time.

11.6.2 Integrating GAVM into EC Manufacturing for Total Inspection

To apply AVM, the machine tool for EC manufacturing needs to be integrated with the so-called GAVM [14]. The architecture of GAVM, which is composed of a GED and an AVM Server [14, 19], is shown in Figure 11.52. GED plays the role of a bridge between the machine tool and the AVM server to fulfill the data-collection and data-preprocessing mission. All of the segmented sensor data, machining parameters, and metrology data will be gathered by GED. Then, the functions of data segmentation, data cleaning, and SFs extraction are also performed in GED [14]. After that, machining parameters, SFs, and metrology data will be stored into the standard-database (STDB) such that the AVM server can access the corresponding process data and metrology data from STDB to conduct real-time and on-line VM calculations.

The preparatory work of performing AVM is to build VM models, which include DQI_X and DQI_y, RI, and GSI as well as conjecture models; all the functions mentioned above are described in Chapter 8. The conjecture models can be created by adopting prediction algorithms, such as the BPNN and PLS. The cause-effect relationship between process data and their corresponding metrology data shall be strictly validated and matched during all VM-models creation scenarios.

11.6.2.1 Considerations of Applying AVM in EC Manufacturing

Due to the complexity of EC operations, GAVM applied into the EC manufacturing is somewhat different from the conventional machine industry. According to the machining characteristics of an EC manufacturing, two main challenges, long metrology-delay and EC deformation, are considered and discussed below.

The model-refreshing function in the advanced dual-phase algorithm provides an innovative solution to guarantee the efficiency and correctness of the AVM server by updating real measurement data into the AVM models. However, for EC manufacturing, it is difficult for the AVM server to update its models in time due to the fact that GED needs to wait for acquiring real measurement data from CMM after completing the machining. Owing to this long metrology delay, all VM prediction values may only be calculated in a free-run manner. Although real measurement data may be acquired from OMP techniques to reduce the waiting time, the switching time between cutting-tool and probe will prolong the manufacturing cycle time. As such, how to sustain the VM free-run prediction accuracy is one of the challenges.

The other challenge is that an EC would probably become deformed [30]. Any changes in the shape or size of an EC may happen due to an applied force, or a change in temperature during complex machining processes. That is, theoretically an EC might be deformed from a cylinder with a perfect circle on end-face to an ellipse in practice. Without the deformation information, the AVM models would just presuppose the end-face of the EC is a circle. For the diameter (Diameter) item, the machining precision can still be predicted by the sensor signals with acceptable accuracy. However, the other key measuring item: Position has a strong relationship with the EC deformation that is not sensed in the current machining process. Without the deformation information, the VM accuracy of Position is not acceptable.

These two challenges can be resolved by the proposed DF scheme. The flange-holes machining process on the EC end-face is adopted as an illustrative example presented below.

11.6.3 The DF Scheme for Estimating the Flange Deformation of an EC

Nowadays, for a fully automated manufacturing purpose, the touch-probe triggered system is widely used to measure the specific coordinates of the initial component during the set-up phase, such as volumetric-error compensation, zero location set-up, center-coordinate location, and ball-bar-test. The probe starts to execute the abovementioned initial scenarios at the beginning period rather than during machining operations. Therefore, the probe may also be applied to sense the real deformation information of the EC during the set-up phase such that the added process time of collecting deformation information will not contribute to the actual machining cycle time. In the following, the DF scheme is proposed to estimate the deformation information without adding massive machining cycle time.

11.6.3.1 Probing Scenario

Observing Figure 11.62, a probing procedure is applied to obtain a set of discrete real coordinates (RC), which represent a 2-D plane measurement of the real shape of a deformed end-face. The number of probing s is suggested to be at least 3 for constructing a basic circle. The larger the s is, the more accurate the EC shape measurement would be. As shown in (11.27), $\boldsymbol{RC}$ aggregates a combination of S pairs of discrete RC in total, and the sth pair of RC is denoted as (x_{Rs}, y_{Rs}) in (11.27), where x_{Rs}, y_{Rs} represent RC relative to the original positions in the x–y plane with $s = 1, 2, \ldots, S$, respectively.

$$\boldsymbol{RC} = \{(x_{R1}, y_{R1}), (x_{R2}, y_{R2}), \ldots, (x_{RS}, y_{RS})\}. \tag{11.27}$$

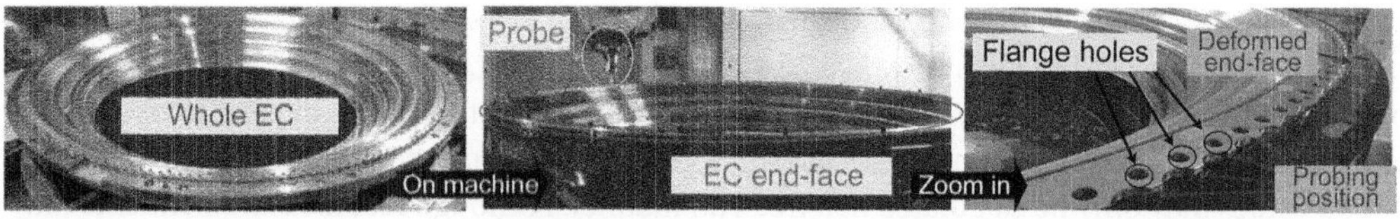

Figure 11.62 Using a probe to touch the outside of an EC end-face. *Source:* Reprinted with permission from Ref. [27]; © 2018 IEEE.

11.6.3.2 Ellipse-like Deformation of an EC

The Position trends (curves with irregular oscillations) of 10 ECs are shown in Figure 11.63, where the vertical axis and horizontal axis are the real Position and flange-hole order number, respectively. The curve-fitting results made by the least square method approximate the position trends to the ones that are ellipse-like. Thus, for approximation, any deformed EC would be assumed to be of an ellipse shape, which is denoted by a set of five parameters $\{a, b, h, k, \theta\}$ as defined below. In addition, several uncertain factors (such as the ones illustrated within dashed frames in EC1 and EC7 of Figure 11.63) which cannot be precisely estimated by generic algorithm (GA) in local regions still exist. These uncertain factors will be handled by interpolation fitting (IF) that will be explained in Section 11.6.3.3.

Definition of {a, b, h, k, θ}

As shown in Figure 11.64, the circle is the top view of the end-face of an ideal EC with the radius length r, which is measured from the Origin(0, 0); while the ellipse is the top view of the end-face of an actual EC with the major-axis length r_a, the minor-axis length r_b, and the actually deformed radius (r_D) at the rotation degree θ. Moreover, suppose that the EC surface is completely flat so that all coordinates exactly lie on the x–y plane. A set of five parameters $\{a, b, h, k, \theta\}$ are utilized to describe the deformation of an EC located at the real position as below.

The first two parameters (a, b) are defined to determine the deformation ratio. Parameter a represents the ratio of r_a to r ($a = r_a/r$); and Parameter b stands for the ratio of r_b to r ($b = r_b/r$). Component deformation is regarded as one of the major factors that degrades machining quality. Depending on the deformation degree, Parameters a and b may be greater or less than 1.

The parameters (h, k) are the offset displacements of the center of the end-face of an EC measured from the Origin(0, 0).

The last parameter (θ) is defined as the rotation angle of a point on the end-face of an EC measured from the major axis of the EC end-face.

Relationship between EC Deformation Amount and Approximate Position on the End-Face

Observing Figure 11.64, a basic ellipse expressed in (11.28) is used to describe a shape of the actual deformed end-face.

$$\left(\frac{x-h}{r_a}\right)^2 + \left(\frac{y-k}{r_b}\right)^2 = 1. \tag{11.28}$$

As shown in Figure 11.64, the relation between actually deformed position (D) and its corresponding nondeformed ideal position (I) of flange holes are described below. Point I is the sth ideal position for machining the flange hole at θ, with Position I: $(x_I, y_I) = (r\cos\theta, r\sin\theta)$; while Position D: (x_D, y_D) lying on the real EC end-face (assumed to be an ellipse) is the corresponding real machined position. Hence, the real position error is related to the distance between Positions I and D. However, this distance is not easy to calculate. To simplify the computation, instead of Position D, an approximate Position A is proposed as explained below. Figure 11.65 depicts the sth approximate Position A for the probing position (dots on the ellipse) with $h = 0$, $k = 0$, & $\theta = 0$. In other words, Position A is regarded as an approximate deformed Position D when $h = 0$, $k = 0$, & $\theta = 0$.

Let $\omega = 360/S$ be the included angle between any two adjacent probing-positions, the approximate angle of the sth approximate position can be defined as

$$\varphi_s = \theta + \omega(s-1), \text{where } s = 1, 2, \ldots, S. \tag{11.29}$$

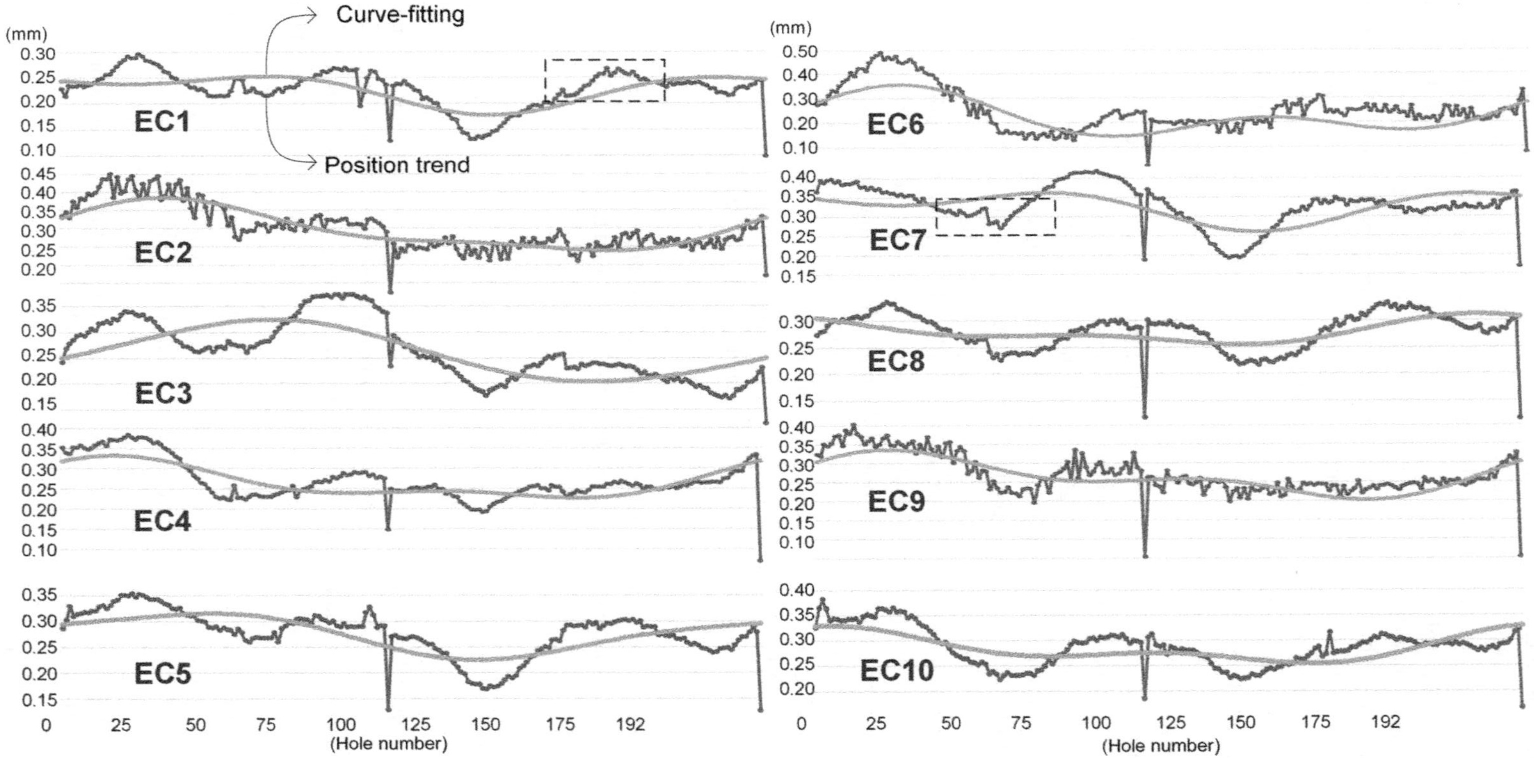

Figure 11.63 Position trends and their curve-fitting results of 10 ECs. *Source:* Reprinted with permission from Ref. [27]; © 2018 IEEE.

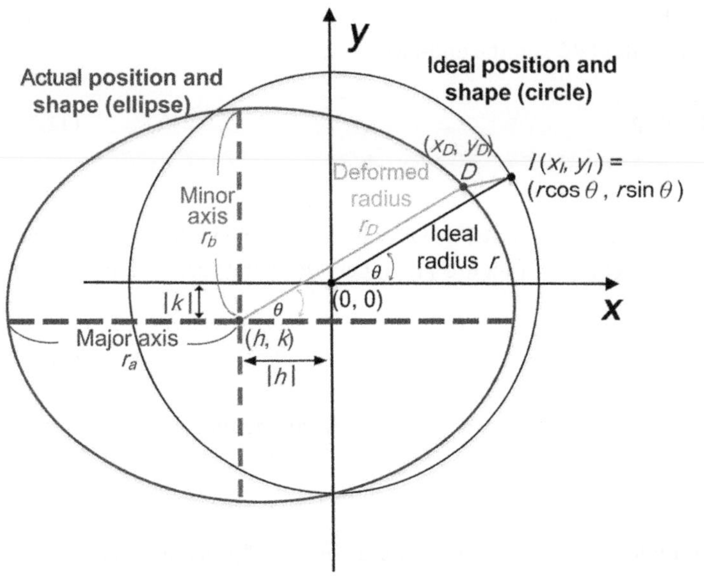

Figure 11.64 Actual deformed position (D) and ideal position (I) of an EC. *Source:* Reprinted with permission from Ref. [27]; © 2018 IEEE.

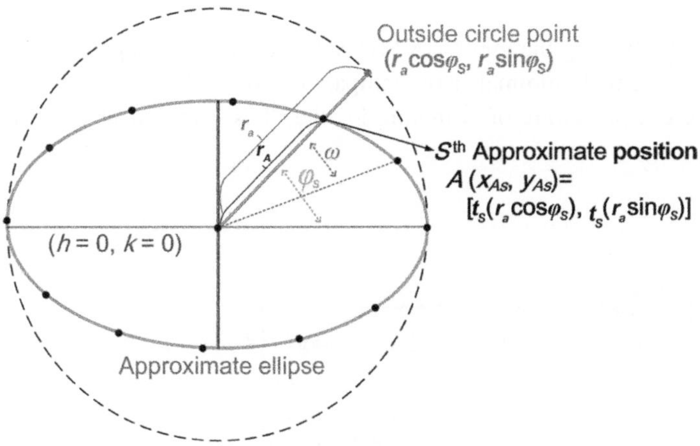

Figure 11.65 Approximate machining position (A) on a deformed EC. *Source:* Reprinted with permission from Ref. [27]; © 2018 IEEE.

Observing Figure 11.65, one large circle with radius r_a is drawn outside the ellipse for helping to understand the relationship between r_A and r_a. The corresponding coordinate of outside-circle point at angle φ_s is ($r_a\cos\varphi_s$, $r_a\sin\varphi_s$). Thus, define a parameter t_s being the ratio of r_A to r_a, then the sth coordinate of the approximate Position A: (x_{As}, y_{As}) can be expressed as in (11.29), where t_s ranges from 0 to 1 and $0° \leq \varphi_s \leq 360°$

$$AC: \begin{cases} x_{As} = h + t_s\left(r_a \cos\varphi_s\right). \\ y_{As} = k + t_s\left(r_a \sin\varphi_s\right). \end{cases}, s = 1, 2, \ldots, S. \tag{11.30}$$

Substitute (11.30) into (11.28), then t_s can be derived as in (11.31), which will be used to explicitly identify all the approximate coordinates (***AC***) on the end-face.

$$t_s = \sqrt{1/\left(\cos^2\varphi_s + \left(r_a/r_b\right)^2 \sin^2\varphi_s\right)}. \tag{11.31}$$

11.6.3.3 Position Error

Prior to define the position error, each element in ***AC*** needs to be transformed into its corresponding approximate radius, r_A, which is depicted in Figure 11.65. Let **AR** be the vector of all approximate radii as

$$\mathbf{AR} = \left[r_{A1}, r_{A2}, \ldots, r_{AS}\right], \tag{11.32}$$

with $r_{As} = \sqrt{\left(x_{As}\right)^2 + \left(y_{As}\right)^2}$. Then, the estimated Position Errors ($\mathbf{P_E}$) are defined as

$$\mathbf{P_E} = 2\left[r_{A1} - r, r_{A2} - r, \ldots, r_{AS} - r\right]. \tag{11.33}$$

where r stands for the default radius r. Eventually, AVM would consider $\mathbf{P_E}$ as one of key SFs for Position prediction.

Three approaches denoted GA, IF, and DF are adopted in this application case to obtain $\mathbf{P_E}$. These approaches are described below.

Genetic Algorithm

With all the elements of RC in (11.27) being the input, GA [31] can generate the five parameters that represent an approximated ellipse with minimal errors compared to ***RC*** of the deformed end-face. Figure 11.66 illustrates the GA procedure of searching for the fittest solution of the five parameters as follows.

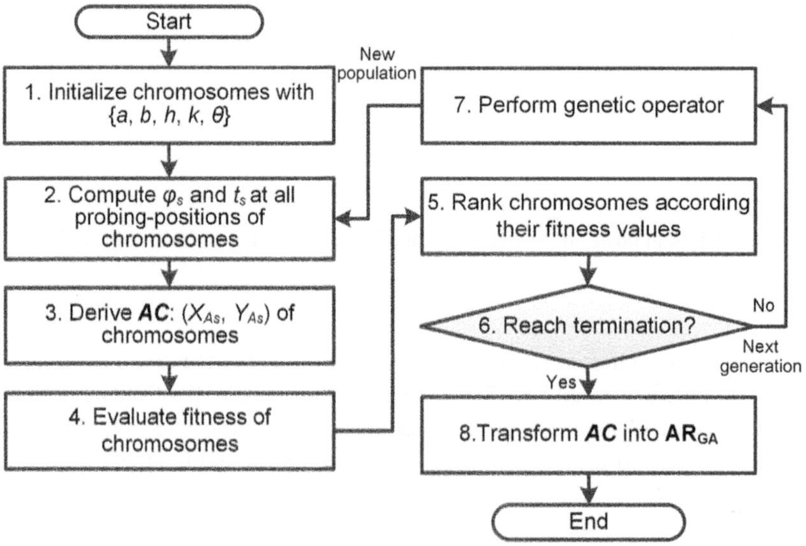

Figure 11.66 Flowchart of generating the fittest ellipse and AC via GA. *Source:* Reprinted with permission from Ref. [27]; © 2018 IEEE.

Step 1: Initialize each chromosome with $\{a, b, h, k, \theta\}$ by defining reasonable ranges of the five parameters and assigning nonduplicated values in a random manner. The population and generation number are set to be 50 and 200, respectively.

Step 2: Compute φ_s and t_s of each chromosome and let $s = 1, 2, \ldots, S$.

Step 3: Derive $\boldsymbol{AC}$ in (11.30) via φ_s and t_s of each chromosome.

Step 4: Evaluate fitness value of each chromosome. The fitness function defined in (11.34) minimizes the total errors between $\boldsymbol{RC}$ and $\boldsymbol{AC}$ from the first to the Sth probing position, and the smaller the better, because $\boldsymbol{AC}$ is expected to be as close to $\boldsymbol{RC}$ as possible.

$$\text{Fitness} = sum\left[\sqrt{\left(x_{Rs} - x_{As}\right)^2 + \left(y_{Rs} - y_{As}\right)^2}\right], \text{for } s = 1, 2, \ldots, S. \tag{11.34}$$

Step 5: Rank chromosomes according to their fitness values. Chromosomes with smaller fitness values are of higher priority to be selected into the mating pool.

Step 6: Go to Step 8 to stop the evolution while reaching the maximal limit or when no better solutions are satisfied for several generations; else, jump to Step 7.

Step 7: Perform the genetic operator which includes crossover, mutation, and selection operators. The given crossover rate and mutation rate are 0.9 and 0.1, respectively. The roulette-wheel selection is utilized to allow better chromosomes to pass on their genes to the next generation in the mating pool.

Step 8: Transform each element in $\boldsymbol{AC}$ into its corresponding r_A in $\mathbf{AR}$ as defined in (11.32). Define the number of the actual manufacturing-positions as M and let $S = M$ to derive all the M $\boldsymbol{AC}$-pairs on the deformed end-face via (11.30). These M $\boldsymbol{AC}$-pairs are then denoted as $\mathbf{AR_{GA}}$ to represent the $\mathbf{AR}$ vector generated from GA for all the actual manufacturing-positions.

Interpolation Fitting

As depicted in Figure 11.66, several information related to the deformation of an EC can be extracted from the five parameters, including setup position, approximate ellipse-shape, $\mathbf{AR_{GA}}$, etc. However, except ellipse-shape deformation, there are still some uncertain factors, as shown within the dashed frames in EC1 and EC7 of Figure 11.63, which cannot be precisely estimated by GA and would affect the severe deformation in local regions of the EC end-face. Thus, to emulate a more close-to-reality deformation distribution, the IF with a cubic-spline function [32] is also adopted to describe the detailed deformation in local regions.

Prior to perform IF, each element in $\boldsymbol{RC}$ as in (11.27) needs to be transformed into its corresponding radius, r_D, with

$$r_{Ds} = \sqrt{\left(x_{Rs}\right)^2 + \left(y_{Rs}\right)^2}, \tag{11.35}$$

for $s = 1, 2, \ldots, S$. Let $\mathbf{DR_{RC}}$ denote the vector of all r_{Ds} as

$$\mathbf{DR_{RC}} = \left[r_{D_1}, r_{D_2}, \ldots, r_{D_S}\right]. \tag{11.36}$$

As such, $\mathbf{DR_{RC}}$ is regarded as known nodes for the purpose of interpolating. Then, IF rearranges the serial numbers of r_D in $\mathbf{DR_{RC}}$ into their original manufacturing position ($\mathbf{OP}$) sequence spreading among 1~M. Denote the ratio of M to the number of probing-positions as "Magnification", $\mathbf{OP}$ is then expressed as

$$\mathbf{OP} = \left[\left(op_1\right), \left(op_2\right), \ldots, \left(op_S\right)\right], \tag{11.37}$$

with

$$op_s = (s-1) \times \text{Magnification} + 1. \tag{11.38}$$

Based upon the sth op_s of **OP**, r_{Ds} in $\mathbf{AR_{RC}}$ is assigned to the op_s position in the size-M $\mathbf{AR_{IF}}$, while the unknown radii between two consecutive $(r_D)_{op_s}$ and $(r_D)_{op_s+1}$ will be generated by IF. As such, the vector of **AR** generated from IF for all the actual manufacturing-positions denoted as $\mathbf{AR_{IF}}$ expresses as

$$\mathbf{AR_{IF}} = \begin{cases} r_{A(op_s)\sim(op_1)} = \text{Spline}\left[(r_D)_{op_S}, (r_D)_{op_1}\right], \text{if } s = S. \\ r_{A(op_s)\sim(op_{s+1})} = \text{Spline}\left[(r_D)_{op_s}, (r_D)_{op_{s+1}}\right], \text{else.} \end{cases} \tag{11.39}$$

Note that, when $s = S$, for any r_A beyond the Sth r_D: $(r_D)_{op_S}$, IF will use the first r_D: $(r_D)_{op_1}$ as an end-point to complete the interpolation during the last interval between $(r_D)_{op_S}$ and $(r_D)_{op_1}$.

Deformation Fusion

$\mathbf{AR_{IF}}$ is expected to generate more accurate deformation shape happening around local regions than those of $\mathbf{AR_{GA}}$. On the other hand, the advantage of $\mathbf{AR_{GA}}$ is that it possesses a more extensive deformed information than $\mathbf{AR_{IF}}$.

In order to take advantage of both $\mathbf{AR_{GA}}$ and $\mathbf{AR_{IF}}$, the so-called DF scheme is proposed. The DF scheme combines both $\mathbf{AR_{IF}}$ and $\mathbf{AR_{GA}}$ together to produce the fusion **AR** denoted $\mathbf{AR_{Fusion}}$ as follows

$$\mathbf{AR_{Fusion}} = f\mathbf{AR_{IF}} + (1-f)\mathbf{AR_{GA}}. \tag{11.40}$$

where f is the weighting coefficient and ranges from 0 to 1. The value of f depends on the degree of EC deformation, which varies from the residual stress of semi-finished EC after pre-machining. In general, the more EC deforms locally, the larger f is recommended. In this case, the value of f is assigned to be 0.5 that is derived from the degree of local deformation of those 10 ECs as shown in Figure 11.63.

11.6.3.4 Integrating the On-Line Probing, the DF Scheme, and the AVM Prediction

The flowchart of integrating on-line probing, DF scheme, and AVM prediction is depicted in Figure 11.67. Detailed explanations are stated below.

Step 1: Use a probe to touch S sampling points on the exterior/surface area of the deformed end-face for collecting ***RC***.

Step 2: Input ***RC*** into GA to obtain $\mathbf{AR_{GA}}$.

Step 3: Generate $\mathbf{AR_{IF}}$ based on $\mathbf{DR_{RC}}$ by using IF.

Step 4: Compute $\mathbf{AR_{Fusion}}$ by the linear combination of $\mathbf{AR_{IF}}$ and $\mathbf{AR_{GA}}$ so as to obtain $\mathbf{P_E}$.

Step 5: Combine $\mathbf{P_E}$ with other SFs to serve as the inputs of AVM for on-line and real-time flange-hole Position prediction.

11.6.4 Illustrative Examples

To validate the accuracy and effectiveness of the AVM-plus-DF scheme, a commonly used material, superalloy, for the aerospace turbine EC with a radius r = 727.75 mm of the flange-face is utilized as illustrative examples in this application case. A five-axes ($x/y/z/a/b$) milling machine is equipped with four drill-cutters to bore 192 holes (M = 192) on the flange-face of an EC. Each

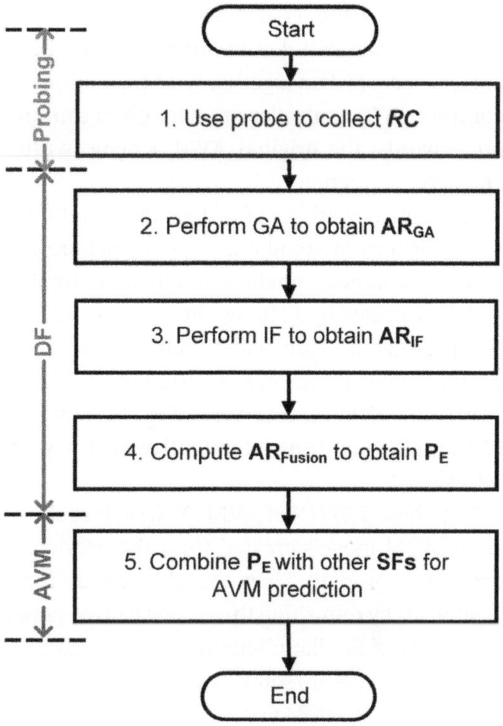

Figure 11.67 Flowchart of integrating the on-line probing, the DF scheme, and the AVM prediction. *Source:* Reprinted with permission from Ref. [27]; © 2018 IEEE.

flange hole requires two critical measurement/inspection items: Diameter and Position. Thus, the first example illustrates typical AVM predictions for all the Diameters of 192 flange holes. The second example demonstrates a complete scenario of performing the on-line probing, the DF scheme, and the AVM predictions for all the Positions of 192 flange holes.

The multi-sensor fusion-based technique is used to combine sensory data of vibration and current signals. Three accelerometers are installed on the spindle for collecting vibration signals of $x/y/z$ axes; and six current sensors are implemented on servomotors in the electric cabinet for gathering current signals of $x/y/z/a/b$ axes and spindle. The rotary b-axis controls the spindle swiveling to a degree of vertical direction that parallels to the z-axis-operation drilling. The rotary c-axis rotates the EC to the degree of any two adjacent flange-hole with $\omega = 360°/192 = 1.875°$ for the spindle drilling purpose before each flange hole is drilled. The five parameters generated from GA are $\{a = 1.00003, b = 1.00000, h = -0.0004\,\text{mm}, k = -0.0015\,\text{mm}, \theta = 1.07°\}$ which reveal that the center-position setup is under the acceptable range (less than 0.01 mm); while the $a = 1.00003$ ($r_a = r \times a = 727.7718\,\text{mm}$) and $\theta = 1.07°$ show a certain degree of deformation that might result in poor Position-quality.

The BPNN of the AVM model contains three layers of input nodes, five hidden nodes, and two output nodes for VM values of Position and Diameter. With the uth predictive VM value being VM_u, AM value being AM_u, and the sampling size being v, the predictive accuracy is evaluated by the MAE as

$$\text{MAE} = \frac{\sum_{u=1}^{v} \left|(\text{VM}_u - \text{AM}_u)\right|}{v} \times 100\%. \tag{11.41}$$

The closer the MAE is to zero, the better the conjecture accuracy is achieved.

11.6.4.1 Diameter Prediction

Due to the fact that deformation will not affect the quality of Diameter, real-time vibration, and current sensors data together with machining parameters collected from the computer-numerical-control (CNC) controller are enough to capture all machining statuses during manufacturing. In other words, the original AVM scheme without the DF scheme is proper enough to perform Diameter prediction.

Standard deviations of *x*/*y*/*z*-axes vibrations as well as maximal magnitudes and root-mean-square values of spindle/*x*/*y*/*z* currents are selected as the concise SFs for building the BPNN model of Diameter prediction. The total number of SFs is 11 for each flange-hole sample, and the BPNN is created by 32 flange-hole samples collected from the previous finished ECs. Observing the AVM results in Figure 11.68, four SCs (SC_1–SC_4) resulted from drill-cutters changes occur before testing sample numbers 1, 47, 91, and 137, respectively. All the SCs lead to on-line probe measurements and AVM models refreshing. This means that after encountering an SC, the probing values of the following three consecutive samples are collected and inputted to the AVM models for refreshing.

The phase-I BPNN of AVM (NN_I) values shown at Samples 1–3 are obtained by applying the first set of AVM models created from the previous EC without model refreshing, and the prediction errors are not small. Then, the advanced dual-phase algorithm of the AVM system recovers the VM accuracies by refreshing three consecutive samples. As a result, the prediction accuracy of Samples 4–48 regains. Similar phenomena occur again at SC_2–SC_4, which indicate that the performances of new drill-cutters are somehow different from that of the old one. In brief, the NN_I MAE among 192 flange holes is 0.0044 mm, which is a free-run result excluding the first three calibration-samples in four SCs. Compared with Diameter tolerance: 0.36 mm (UCL: 9.16 mm; LCL:8.80 mm), the original AVM scheme is feasible enough to provide a reliable Diameter VM prediction.

11.6.4.2 Position Prediction

On the other hand, unlike Diameter, the original AVM scheme is not accurate enough for Position prediction due to the fact that Position has a strong correlation [33] with the deformation-distribution on the end-face. To evaluate the effectiveness of the DF scheme, the AVM performance on Position will have four cases for comparison: (i) without DF (without $\mathbf{P_E}$), (ii) with $\mathbf{P_E}$ generated from $\mathbf{AR_{GA}}$, (iii) with $\mathbf{P_E}$ generated from $\mathbf{AR_{IF}}$, and (iv) with $\mathbf{P_E}$ generated from $\mathbf{AR_{Fusion}}$. Note that, the number of probing, *S*, is set to be 32. It takes about five to eight seconds for a probe to detect the outside shape at one position, and around three to four minutes for all the 32 points. The scheme used in Case 1 is the original AVM without DF, while Case 2, Case 3, and Case 4 utilize the AVM-plus-DF scheme with the SFs of the original AVM plus $\mathbf{P_E}$ from $\mathbf{AR_{GA}}$, $\mathbf{AR_{IF}}$, and $\mathbf{AR_{Fusion}}$, respectively.

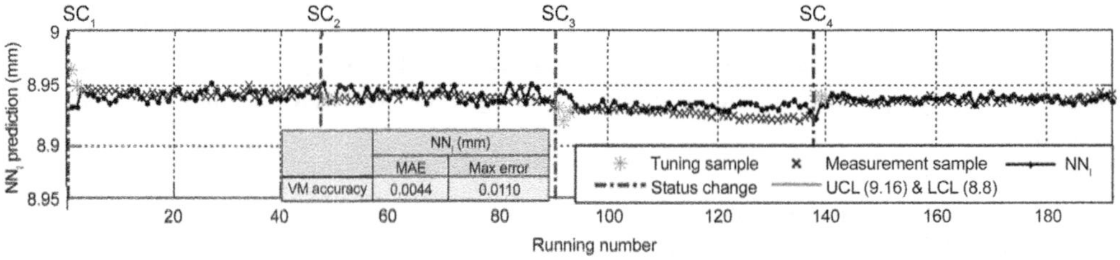

Figure 11.68 AVM results for diameter prediction. *Source:* Reprinted with permission from Ref. [27]; © 2018 IEEE.

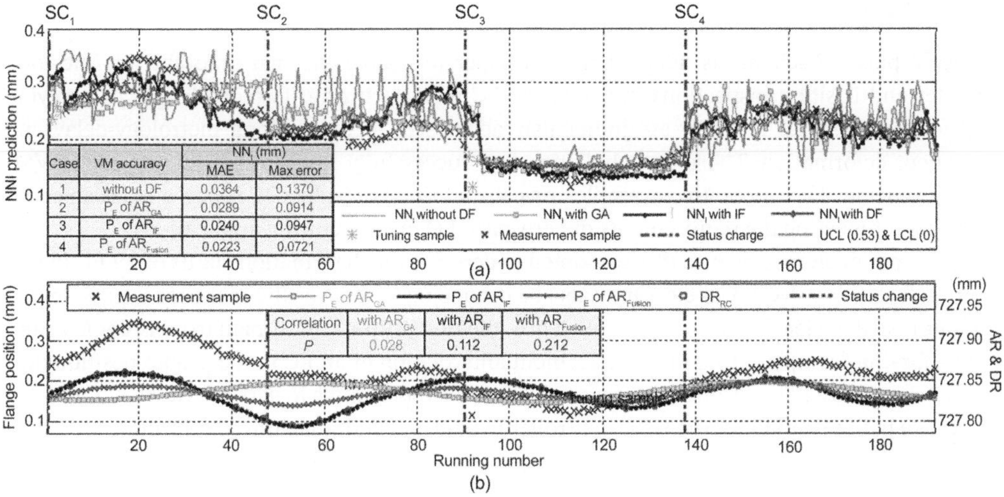

Figure 11.69 Position prediction. (a) VM results of four cases: (1) without DF (2) with P_E generated from AR_{GA} (3) with PE generated from AR_{IF}, and (4) with PE generated from AR_{Fusion} for Position. (b) The corresponding ARs of Case 2, Case 3, and Case 4. *Source:* Reprinted with permission from Ref. [27]; © 2018 IEEE.

The real metrology and VM results of the four cases are depicted in Figure 11.69a. Notice that the real measurement values vividly show the trend of deformation. The NN_I MAE accuracy of Case 1 (without DF) is 0.0364 mm that is not acceptable. Observing Figure 11.69a, the Case-1 VM prediction trend loses the prediction ability to real metrology due to the fact that the Case-1 VM models do not possess the key SF generated from the DF scheme. After being refreshed with three consecutive samples, the VM accuracy is still unacceptable.

As shown in Figure 11.69a, both the MAEs in Case 2 (considering $\mathbf{P_E}$ generated from $\mathbf{AR_{GA}}$) and Case 3 (considering $\mathbf{P_E}$ generated from $\mathbf{AR_{IF}}$) are better than that of Case 1 with 0.0289 and 0.0240 mm, respectively.

Observing the fitting results of various **AR** shown in Figure 11.69b, the square-dotted line of $\mathbf{AR_{GA}}$ makes NN_I of Case 2 gradually vary with real metrology especially between SC_3 and SC_4. This indicates that deformation issue may be handled by considering the $\mathbf{P_E}$ generated from $\mathbf{AR_{GA}}$ being one of the key SFs.

However, some special occasions such as insufficient sampling points or intense deformations may mislead $\mathbf{P_E}$. When this happens, the DF scheme must utilize the IF method to resolve the problem, and this is Case 3 (considering $\mathbf{P_E}$ generated from $\mathbf{AR_{IF}}$). The circle-dotted line in Figure 11.69b represents $\mathbf{AR_{IF}}$ that is interpolated from the larger circle dots of $\mathbf{DR_{RC}}$. Obviously, $\mathbf{AR_{IF}}$ shows more sensitive to reflect deformation during the local regions than that of $\mathbf{AR_{GA}}$, which shows too smooth fitting-trend between SC_1 and SC_3.

$\mathbf{P_E}$ of Case 4 (considering $\mathbf{P_E}$ generated from $\mathbf{AR_{Fusion}}$) inherits the advantages of both $\mathbf{AR_{GA}}$ and $\mathbf{AR_{IF}}$ with $f = 0.5$. As shown in Figure 11.69b, the diamond-dotted line of $\mathbf{AR_{Fusion}}$ possesses the best correlation ($\rho = 0.212$) with measurement samples of Position in comparison with those of $\mathbf{AR_{GA}}$ and $\mathbf{AR_{IF}}$. As a result, Case 4, which also skips the counts of errors of each first three samples in four SCs, NN_I MAE can be significantly improved to 0.0223 mm that is the best among the four cases. Moreover, the inspection time of four drill-cutters is also reduced by 20%, from originally 2660 seconds for Diameter only of sampling flange-holes to 2080 seconds for all measurement items (including Diameter and Position) of all flange-holes.

11.6.5 Summary

The AVM-plus-DF scheme is adopted in this paper to enable the real-time and on-line total Diameter and Position inspections in aerospace EC manufacturing. Two challenges of applying AVM to the aerospace industry are judiciously addressed, including long metrology-delay and component deformation. These two challenges are successfully resolved by the proposed AVM-plus-DF scheme. The illustrative examples reveal that the AVM-plus-DF scheme is feasible and promising for meeting the requirements of aerospace EC manufacturing applications.

The five parameters $\{a, b, h, k, \theta\}$ are adopted to describe the deformation status of an EC in this paper. In fact, those five parameters may also be promising to serve as the real-time inputs for further automated feedback-control purposes. Thus, the first future work is plan to consider h, k, and θ for calibrating/compensating the displacement and rotation of an EC to the proper position during the set-up phase. The second future work is then to adopt a and b to perform the recommended ellipse-like path planning for the CNC controller to adjust default machining coordinates.

11.7 Application Case VI: Chemical Industry

The authors would like to thank Formosa Plastics Corporation in Taiwan, ROC for providing the raw data used in this application case.

11.7.1 Introducing the Carbon-Fiber Manufacturing Process

Carbon fiber is produced after the oxidization, carbonization, and starching of its main raw material polyacrylonitrile (PAN). Carbon fiber is a composite material with good physical properties (good heat resistance, high strength, and anti-corrosion) that has been widely used in aerospace, sports equipment, medical equipment, and wind power industries.

The manufacturing process of carbon fiber is of continuous production type, which can be divided into precursor (PR) stage and Carbon-Fiber stage with PR stage in the front. In the process of the Carbon-Fiber stage as shown at the upper left part of Figure 11.70, firstly, the products of the PR stage are yarned with the weft machines into a desired number of spins. Then, all the spins are proceeded through the stations of oxidation, pre-carbonization, carbonization, surface treatment, and drying in the Carbon-Fiber stage. Finally, the Carbon-Fiber stage products are grouped into spins at each individual spinning positions through the crimping machine. Among them, the temperature of oxidation, pre-carbonization, and carbonization ranges from 100 °C to a maximum of 3000 °C. Oxidation is mainly applied to soften PR with heat, and keep the original yarn a linear state by the tension of the production machine tool. The internal structure will also be transformed into a more stable hexagonal array during this heating process. The process of pre-carbonization and carbonization are mainly utilized to oxidize the PR so as to convert it into the carbon-fiber product through high temperature. After carbonization, the PR enters surface treatment to attach a protective layer to the surface of the carbon fiber to avoid surface damage from other processes in the future [35, 36].

The current carbon-fiber process-quality inspection method is displayed at the upper right portion of Figure 11.70: only a small percentage (e.g. 6%) of the spins of the daily production are randomly sampled to be inspected, and the final 60 m of the sampled spins are examined with destructive tests. Then, nearly 24 hours are spent to inspect the quality items, including sizing percentage, tensile strength, and tensile modulus. If the results of the 6% sampling inspections are

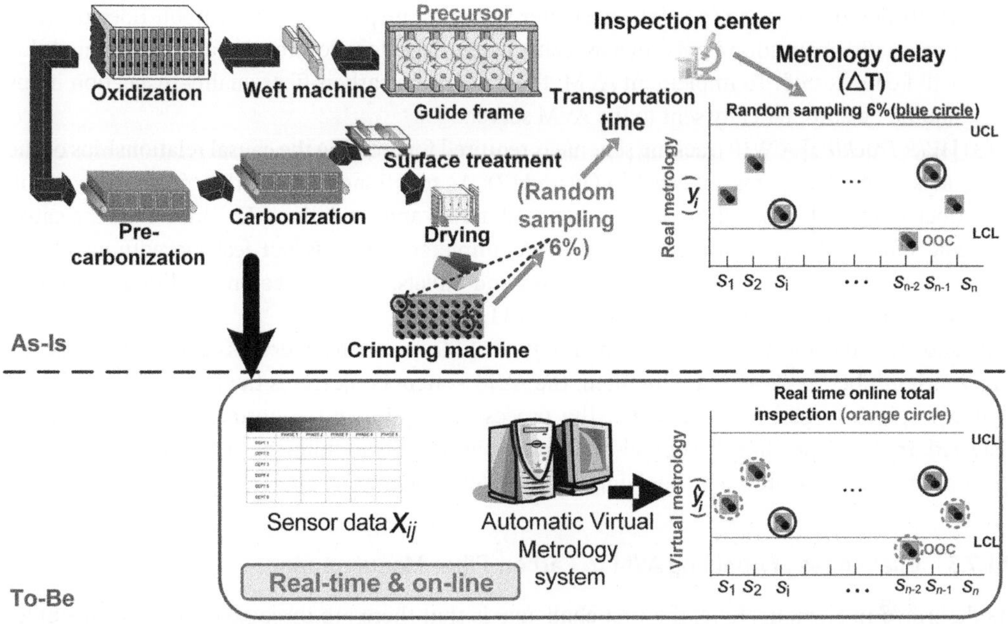

Figure 11.70 Comparison of off-line measurement and virtual metrology. *Source:* Reprinted with permission from Ref. [34]; © 2019 IEEE.

good, then the remaining 94% will be delivered to the customers without inspection. However, those remaining 94% spins may still contain defective ones. One way to detect defective spins is to increase the sampling rate. Nevertheless, due to the fact of destructive inspection, increasing sampling rate is not a feasible solution. The best way to resolve the problem mentioned above is to apply the technology of AVM presented in Chapter 8 with its capability of converting sampling inspection with metrology delay into real-time and on-line total inspection.

Generally, in the common production process, the identification (ID) of each workpiece is usually defined by a bar code or radio frequency identification (RFID), and the workpieces will be scanned at certain stages and/or stations for the purpose of WIP tracking. With a unique ID, various parameters of each workpiece during all processes can be collected and stored for intelligent manufacturing applications such as AVM, intelligent yield enhancement, and engineering data analysis [37–39]. However, this scenario cannot be applied to the continuous roll-to-roll carbon-fiber production. The reason is that the product unit or batch number of a workpiece of continuous production cannot be defined, thus it is unlikely to correctly track the WIP.

11.7.2 Three Preconditions of Applying AVM

There are three preconditions of applying AVM.

(1) [*Cause Collection*] All the factors that would affect quality inspection items are equipped with sensors for online and real-time data collection. The factors in the carbon-fiber stage include the key parameters and processing conditions of the oxidation, pre-carbonization, carbonization, surface treatment, and drying stations.

(2) [*Effect Collection*] Equipped with the actual measurement instruments (for constructing or refreshing the AVM models), and the AM values can be sent to the AVM server through

electronic transferring (e.g. internet) after sampling inspection. After carbon-fiber processing, quality inspection items such as sizing percentage, tensile strength, and tensile modulus will be measured. To implement AVM, the data of the carbon-fiber quality-inspection items should be automatically sent to the AVM server.

(3) [*WIP Tracking*] A WIP tracking scheme is required for creating the causal relationships of the cause and effect items collected in (1) and (2). As mentioned earlier, the continuous carbon fiber roll-to-roll production processes cannot define and track the WIP. Thus, even the cause and effect collections are completed, AVM implementation is not feasible without a WIP tracking scheme to match all these causes and effects. These three preconditions of implementing AVM will be elaborated in Section 11.7.3.

As shown in the lower (To-Be) portion of Figure 11.70, once sensor/process data of manufacturing a carbon-fiber workpiece are inputted, the AVM system conjectures the processing quality of that workpiece in real time such that the process or equipment abnormalities can be timely detected. In short, the objective of real-time and online quality monitoring of all the workpieces of carbon fiber can be achieved by applying the AVM technology.

11.7.3 Challenges of Applying AVM to Carbon-Fiber Manufacturing

As described in Section 11.7.2, the first challenge is that there are three preconditions of implementing AVM including Cause Collection, Effect Collection, and WIP Tracking. Moreover, the second challenge is described below.

The current carbon-fiber raw material is about 3000 m, and the sampling length of carbon-fiber quality inspection is the last 60 m. In other words, the length of a workpiece for AVM prediction should be defined as 60 m. There are 300 spinning positions and 365 sensors in a typical carbon-fiber manufacturing process. The sampling rate of the sensor data collection is one per 30 seconds, as such, if production rate is 6 m per minute, a workpiece (60 m) would contain 20 samples in each sensor. As there are 365 sensors in total, the total amount of the samples of all sensors would be 7300 ($20 \times 365 = 7300$) in a workpiece. A spin of carbon-fiber final product contains 50 ($3000/60 = 50$) workpieces, and there are 300 spinning positions. Therefore, the total amount of carbon-fiber samples of a lot with 300 spins would be 109 500 000 ($50 \times 300 \times 7300 = 109\,500\,000$).

As the carbon-fiber production is a roll-to-roll process, with long processing time and large amount of collected parameters, the total amount of workpieces would reach 85 GB in five days. With these, it is not easy to connect all the Cause and Effect data of the workpieces for the AVM system to conduct quality inspections to achieve ZD of all the workpieces. Adding the fact that the factory site and the cloud site shown in Figure 11.74 may locate at different places, big-data-collection and cloud-manufacturing technologies are required for deploying AVM.

To tackle the two challenges of applying AVM to carbon-fiber manufacturing, this work proposes the following solutions.

11.7.3.1 CPA+AVM (CPAVM) Scheme for Carbon-Fiber Manufacturing

The so-called CPA+AVM = CPAVM scheme is proposed and shown in Figure 11.71 to take care of the three pre-conditions of deploying AVM. The Cause, such as the sensor data and machining parameters, serves as inputs of the AVM server for predicting processing accuracy as indicated in the upper-left portion of Figure 11.71; while the Effect, the metrology data shown in the lower-left portion of Figure 11.71, is collected to serve as the other input of AVM for tuning the prediction model.

As for the WIP Tracking, the so-called Production-Data-Traceback (PDT) mechanism is proposed and illustrated in Figure 11.72. This work applies Meter Counter to trace the specific

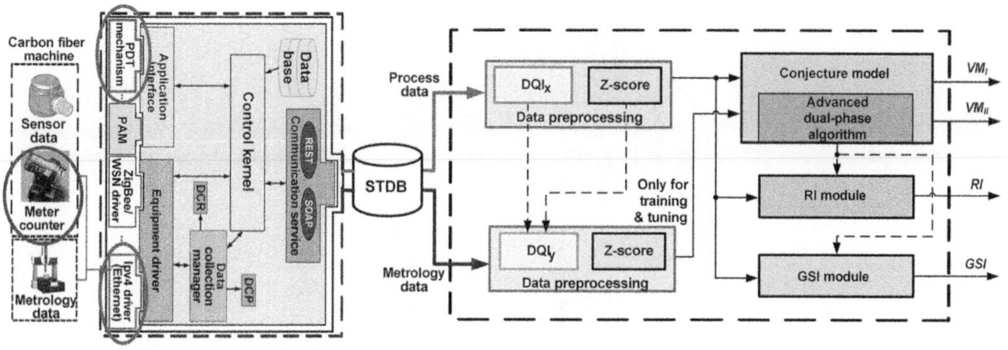

Figure 11.71 CPAVM scheme. *Source:* Reprinted with permission from Ref. [34]; © 2019 IEEE.

workpiece position in meters of carbon fiber during processing. In addition to name each workpiece, it is necessary to accurately record the time that a specific workpiece passes at each processing area and measure the distances of all the sensors to the inspection spot. The distance, denoted as SD, of each sensor to the Meter Counter as well as that of Meter Counter to each spin, denoted as FSD, can be measured. For instance, the distance of Oxidation Sensor1 to the Meter Counter is 950 m ($SD_1 = 950$); and that of the Meter Counter to Spin1 is 14 m ($FSD_1 = 14$). Due to the roll-to-roll characteristics of the carbon-fiber process, if the process completion time of a workpiece in Spin1 can be acquired, then the timing of that workpiece passing through Sensor1 can be estimated by conversing the rolling speed and the traveling distance so as to obtain the corresponding Sensor1 value of this workpiece.

Process data of a specific workpiece need to be traced back with its ID due to the property of continuous production. After finishing processing the products, the time of completion is used to refer to the timing of passing through each station. Then all the process data are collected and stored as the inputs of CPAVM. As such, the PDT mechanism is developed in this work for effective WIP tracking and data collection.

Generally, there are about 365 sensors and 300 spins in the carbon-fiber manufacturing process. The formula of finding the completion time at the ith spinning position of sensor j, denoted as SR_{ij} in the PDT mechanism, is illustrated below:

$$SR_{ij} = ST_j - \frac{SD_i + FSD_j}{RS} \tag{11.42}$$

where

ST_i	process completion time at the ith spinning position;
SD_j	distance of sensor j;
FSD_i	distance between meter counter and the ith spinning position;
RS	rolling rate (meter/minute);
with	$i = 1, 2, \ldots, 300$;
	$j = 1, 2, \ldots, 365$.

After obtaining SR_{ij}, the workpiece WIP tracking of the carbon-fiber process can be achieved.

As shown in the middle of Figure 11.71, the components of the CPA presented in Section 7.3 include Equipment Driver, Application Interface (AI), PAM, Data Collection Manager (DCM),

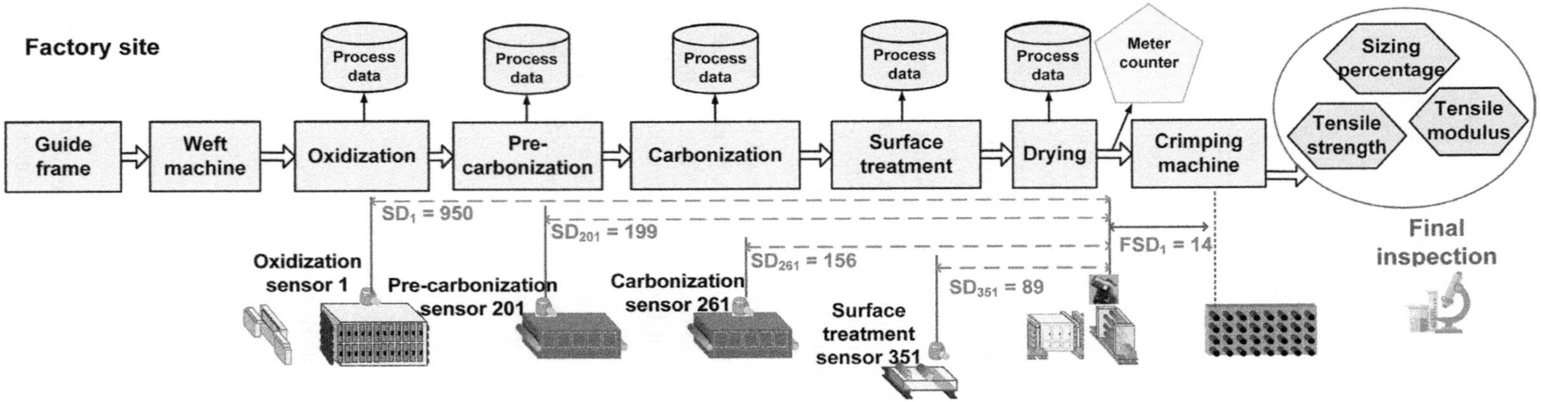

Figure 11.72 Illustration of the production-data-traceback (PDT) mechanism. *Source:* Reprinted with permission from Ref. [34]; © 2019 IEEE.

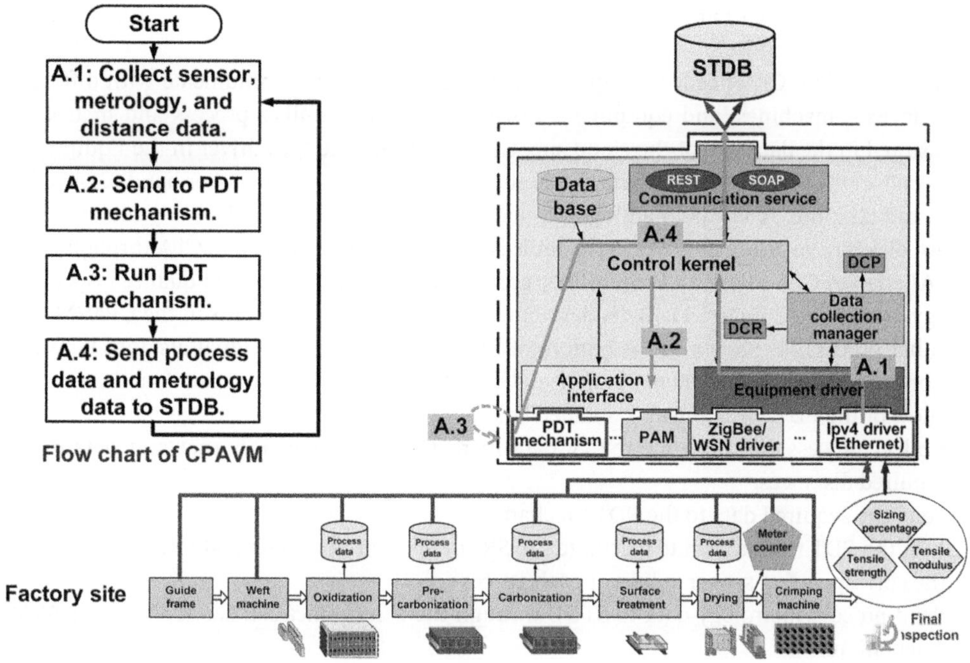

Figure 11.73 Information flow of the PDT mechanism. *Source:* Reprinted with permission from Ref. [34]; © 2019 IEEE.

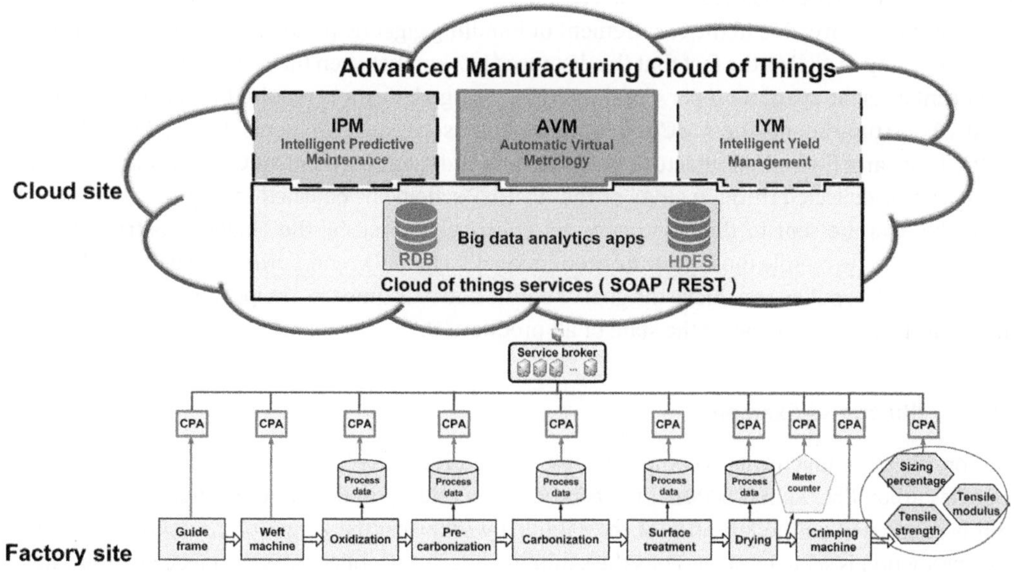

Figure 11.74 AMCoT for carbon-fiber manufacturing. *Source:* Reprinted with permission from Ref. [34]; © 2019 IEEE.

Data Collection Plan (DCP) and Data Collection Report (DCR), Database, CPA Control Kernel, and Communication Service.

For data collection, CPA needs the Equipment Driver in order to communicate with different sensors, devices, machines, and equipment. Currently, the carbon-fiber process and metrology data are stored in the database of the cloud site. CPA can utilize the Ipv4 driver in the Equipment Driver to collect the required sensor data and Meter Counter values. On the other hand, CPA relies on AI to add all kinds of PAMs in a plug-and-play manner for users to replace any module they need with different requirements. As such, embedding the PDT mechanism to CPA through PAM plug-in AI enables CPA to collect carbon-fiber process parameters (Cause) and quality inspection metrology data (Effect). Figure 11.73 displays how the PDT mechanism works in the CPAVM and carbon-fiber process. As shown in the bottom of Figure 11.73, each individual distance between the sensor of each process to Meter Counter and each Spin position are measured, and the information flow of the PDT mechanism is detailed below.

A.1: Collect process data and metrology data in real time through Ethernet and obtain the required distances.

A.2: Send the acquired data to the PDT mechanism.

A.3: Run the PDT mechanism to calculate the SR_{ij} of all metrology data and refer the values to the corresponding sensor values of each process.

A.4: Send process/metrology data to STDB through the communication service for later AVM calculation.

With the development of the PDT mechanism and its integration into CPA, the architecture of CPAVM is developed so as to fulfill the three preconditions of applying AVM into the carbon-fiber processes.

11.7.3.2 AMCoT for Carbon-Fiber Manufacturing

As mentioned above, due to the requirement of handling gigantic production data of carbon-fiber manufacturing as well as the fact that the distance (147 km) between the factory site and cloud site (data center) is far apart, a proper cloud manufacturing platform is required to meet the requirements. This study adopts the AMCoT platform [40] to solve such problems. Figure 11.74 displays AMCoT for carbon-fiber manufacturing. All process and metrology data of Carbon-fiber manufacturing can be collected through CPAs of the AMCoT scheme as depicted in Figure 11.73 A1–A4. Then the data are sent to the cloud site and effectively stored by the Hadoop Distributed File System (HDFS) to handle those gigantic production data for AVM computing. As such, the AMCoT platform can be applied to achieve the goal of online and real-time total inspection on each carbon-fiber workpiece so as to realize the state of all product ZD.

11.7.4 Illustrative Example

Carbon-fiber manufacturing accomplished by AVM and AMCoT is illustrated below. The length of a single carbon fiber spin is 3000 m and its production rate is 6 m per minute. Since quality inspection is conducted on the last 60 m of the sampled workpiece, the length of a workpiece for real-time prediction is set to be 60 m. Thus the sampling interval of collecting workpiece process data is 10 minutes.

Referring to Table 11.7, there are 365 process parameters in total, including 200 Process Oxidation, denoted as Sensor1–200; 60 Process Pre_carbonization, denoted as Sensor201–260; 90 Process Carbonization, denoted as Sensor261–350; and 15 Process Surface_Treatment, denoted as Sensor351~365.

Table 11.7 WIP tracking of Spin-1 sizing percentage with the PDT mechanism.

		Distance	Sampling time	
Station	**Sensor ID**	$SD_i + FSD_{kj}$	SR_{ij}	**Value**
OXIDATION	Sensor1	964	2018/3/29 12:19:51	127
OXIDATION	Sensor2	950	2018/3/29 12:22:08	124
OXIDATION	...	...	...	...
PRE_CARBONIZATION	Sensor201	213	2018/3/29 14:25:02	64
PRE_CARBONIZATION	Sensor202	205	2018/3/29 14:26:27	51
PRE_CARBONIZATION	...	...	...	...
CARBONIZATION	Sensor261	170	2018/3/29 14:32:15	450
CARBONIZATION	Sensor262	166	2018/3/29 14:32:51	1045
CARBONIZATION	...	...	...	...
SURFACE_TANK	Sensor351	103	2018/3/29 14:45:17	158
SURFACE_ TANK	Sensor352	92	2018/3/29 14:43:19	58
SURFACE_ TANK	...	...	...	...

Source: Reprinted with permission from Ref. [34]; © 2019 IEEE.

11.7.4.1 Production Data Traceback (PDT) Mechanism for Work-in-Process (WIP) Tracking

Distances from each sensor to Meter Counter and from Meter Counter to each Spin (e.g. FSD_1, SD_1, SD_{201},. . .), spinning speed and the sampling time of the product are previously measured and shown in Figure 11.72. Then, the PDT mechanism (11.42) is applied to trace back the process data of each workpiece. As such, the Cause and Effect data required for AVM are collected. Spin1 is taken as the WIP tracking example with the Sizing-Percentage value being 23.6. Each process data item can then be estimated by (11.42). Take Sensor1 of Spin1 for example, the distance between Sensor1 and Meter Counter is 950 m ($SD_1 = 950$), the distance of Meter Counter to Spin1 is 14 m ($FSD_1 = 14$), process completion time at Spin1 position (ST_1) is 2018/03/29 15:00:00, and the spinning speed is 6 m minute^{-1} (RS), then

$$SR_{11} = ST_1 - \frac{SD_1 + FSD_1}{RS}$$

$$\longrightarrow SR_{11} = ST_1 - \frac{950 + 14}{6}$$

$$\longrightarrow SR_{11} = 2018/3/29\,12:19:51.$$

Therefore, as shown in Row 1 of Table 11.7, Spin1 at 2018/03/29 15:00:00 passed through Sensor1 at 2018/03/29 12:19:51 with value 127. In this way, WIP tracking can be accomplished and the processing conditions are acquired. By the same token, the process data of other sensors are calculated and filled in Table 11.7. The CPAVM scheme in Figure 11.71 is then applied to deliver all the data computed by PDT in Table 11.7 to STDB through CPA for the AVM server to conduct inspection item prediction.

11.7.4.2 AVM for Carbon-Fiber Manufacturing

The AMCoT platform in Figure 11.74 is adopted to solve the issue of different locations of factory site and cloud site (data center) to effectively manage carbon-fiber manufacturing for AVM real-time prediction.

Carbon-fiber production data and inspection results collected by CPAVM are used to conduct AVM accuracy verification on the paired-up carbon-fiber data. Workpiece, the unit for AVM prediction, should be defined and encoded before AVM accuracy verification. The coding logic of the workpiece ID is "Year Month Day Hour Minute:Spin_MeterRange." For example, ID 201806100019:001_60 indicates that this workpiece of Spin1 is between 0 and 60 m at 00:19, June 10, 2018. By utilizing (11.42), the Cause-and-Effect data of each workpiece required by AVM can be obtained on-line and in real-time.

The main prediction algorithms of the AVM models are the BPNN and PLS. Set the VM value as $\hat{y}_i$, AM value as y_i, and the sample size as n. The MAE depicted in (11.26) is utilized here to assess the prediction accuracy.

Firstly, off-line AVM accuracy verification is executed. The total collected verification data count is 125; among them, 60 are used for model building and 65 for testing. After discussing with the domain expert, 17 signals such as the temperature, transfer speed, tension and slurry concentration of the oxidation, pre-carbonization, and carbonization stations are taken as model building parameters.

The AVM results of Sizing Percentage are shown in Figure 11.75, where the solid line tagged with "NN" represents the prediction results of BPNN, solid line tagged with PLS displays those of PLS, and the line of asterisks indicate the AM results. MAE are BPNN: 1.410%, PLS: 1.438%, respectively. As seen from Figure 11.75, even when the Sizing Percentages change drastically among samples 1–10, the prediction results still catch up the trend of AM, which indicates good conjecture quality.

After that, online and real-time total inspection of carbon-fiber manufacturing is to be carried out. The online results are displayed in Figure 11.76. The upper part of Figure 11.76 is the online AVM Sizing Percentage results of all spins, with the vertical axis being Sizing Percentage and the horizontal axis being the IDs of workpieces. For concise display, the interval is set as three zones, and "Year" from the ID encoding is deleted and listed from left to right according to time series. For instance, the squared 06201217:165_3000 represents the Sizing Percentage prediction result of Spin 165 at 2940–3000 m at 2018/06/20 12:17. The Sizing Percentages of all spins such as Spins 001, 004, 126, 080, 119. . .are displayed from left to right in Figure 11.76.

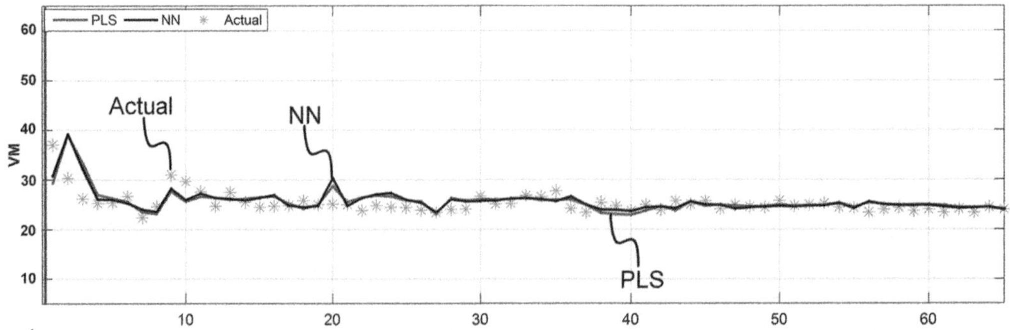

Figure 11.75 AVM results of sizing percentage. *Source:* Reprinted with permission from Ref. [34]; © 2019 IEEE.

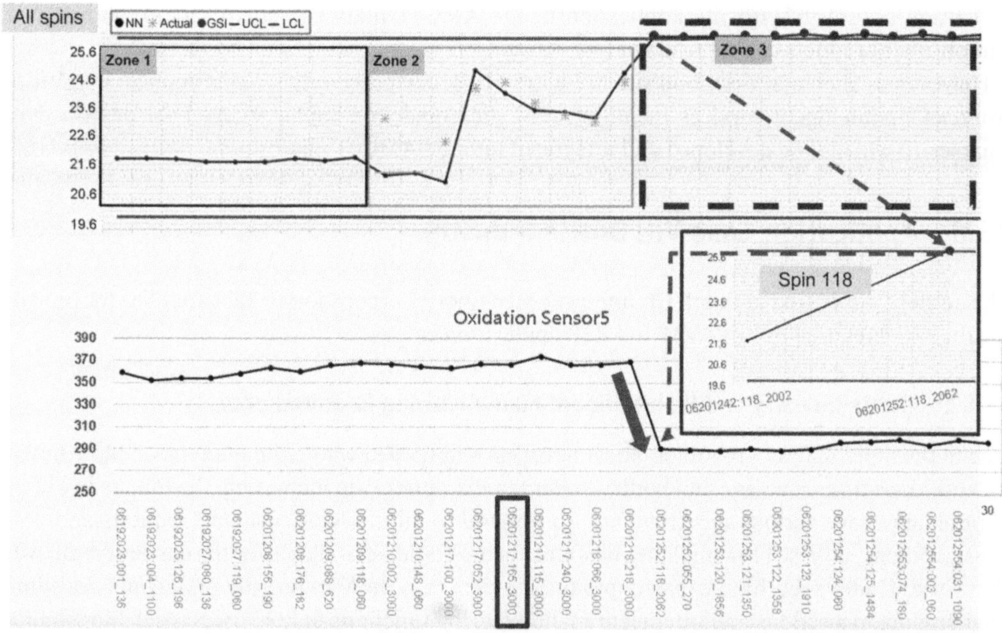

Figure 11.76 Carbon-fiber manufacturing on-line display results. *Source:* Reprinted with permission from Ref. [34]; © 2019 IEEE.

In Zone 1, since there are no AM values to those spins, only AVM is available for real-time quality monitoring. Spins in Zone 2 are with AM values, and the results match well with the trend and accuracy of the AVM prediction. GSI [5] abnormalities occur in Zone 3, with the first abnormality falls on Spin 118. As shown in the figure of Sizing Percentage of Spin 118 in the middle of Figure 11.76, abnormality did not occur at 1942–2002 m at 2018/06/20 12:42 but appears at 2002–2062 m. As such, the root cause of this abnormality should be discovered and fixed. When GSI abnormality occurs, ISI analysis [5] can be adopted to find out the process parameter with the biggest deviation, which is the root cause. Referring to the bottom part of Figure 11.76, oxidation Sensor 5 is analyzed to be the most deviated parameter and it is the root cause of false Spin 118 prediction at 2002–2062 m at 2018/06/20 12:52 (06201252:118_2062). In summary, with AVM implementation, when carbon-fiber parameter deviation occurs, ISI analysis can be conducted to find out the problematic parameter for continuous improvement, so as to achieve the goal of ZD of carbon-fiber manufacturing.

11.7.5 Summary

Section 11.7 proposes a scheme for carbon-fiber manufacturing to achieve online and real-time total inspection. As the CPAVM illustrated in Section 11.7.3.1, through the Equipment Driver and the Database, sensor data and Meter Counter values can be acquired. The PDT mechanism is also proposed and packed as a PAM to be plugged into CPA and to be integrated into the CPAVM scheme for carbon-fiber manufacturing. In this way, the three preconditions of applying AVM to the carbon-fiber process can be fulfilled. Further, AMCoT is adopted to solve the issues of big data and different locations of factory site and cloud side (data center). The offline and online experiments in Section 11.7.4.2 signify the feasibility of CPAVM as well as good AVM prediction accuracy. With the application of the AMCoT platform, smart factory with ZD of carbon-fiber products can be realized.

However, as shown at the cloud site in Figure 11.74, currently only AVM is applied. For future work, other intelligent services including IPM presented in Chapter 9 and IYM described in Chapter 10 will also be developed and integrated into the AMCoT platform.

11.8 Application Case VII: Bottle Industry

The authors would like to thank ChumPower Machinery Corporation in Taiwan, ROC for providing the raw data used in this AVM and R2R application cases.

11.8.1 Bottle Industry and Its Intelligent Manufacturing Requirements

Due to the rapid market growth of global beverage and bottled water, the requests on blow-molding machines from beverage and bottled water manufacturers are increasing. Customers in different domains have various requirements. To respond to all kinds of customer requests and cost considerations, diverse blow-molding machine types should be available for the customers all over the world. Under such high-capacity production demands, how to provide online and real-time total product inspections and intelligent manufacturing functions to help users reach blow-molding process optimization with high process capability and nearly ZD is a big challenge.

To overcome the challenge mentioned above, this work adopts cloud-based AVM System and AVM-based cloud R2R control scheme. Big-data collection of process data and bottle quality metrology data are conducted for the AVM System to convert offline sampling inspection with metrology delay into online and real-time total inspection.

11.8.1.1 Introducing the Blow-Molding Manufacturing Process

Figure 11.77 shows the two-stage blow-molding machine structure that comprises of the following processes. First, put the preforms after cooling into the aligner feeder, and transfer these preforms to the heating box for warming through the conveyor. Then, the variable pitch and traverse mechanism are used to send the preheated preforms to the forming group for blow molding polyethylene terephthalate (PET) bottle containers of various shapes with high-pressure gas and stretch blowing. Finally, the PET bottles are sent to the next stage for offline inspection or online metrology to check their center-of-mass or thickness. Among them, the heating and blow molding processes are the key factors of affecting the product quality.

In view of the existing processes, the quality control of the two-stage blow molding machine and the optimization of the process quality parameters are the two issues to be addressed primarily. Then PET bottle inspection and measurement methods are required to evaluate the PET bottle quality. Finally, the cloud-based AVM System and R2R control scheme for the two-stage blow molding machine are developed to solve the PET quality monitoring and process optimization problems. The R2R control scheme adopts VM results and RI/GSI indexes generated by AVM for automatic recipe-adjustment calculation such that the center-of-mass of PET bottles can be adjusted to reach the target value accordingly.

11.8.2 Applying AVM to Blow Molding Manufacturing Process

The heating and blow-molding processes are considered the two key stages for the entire blowing manufacturing process. Observing the red-segments in Figure 11.77, temperature and pressure

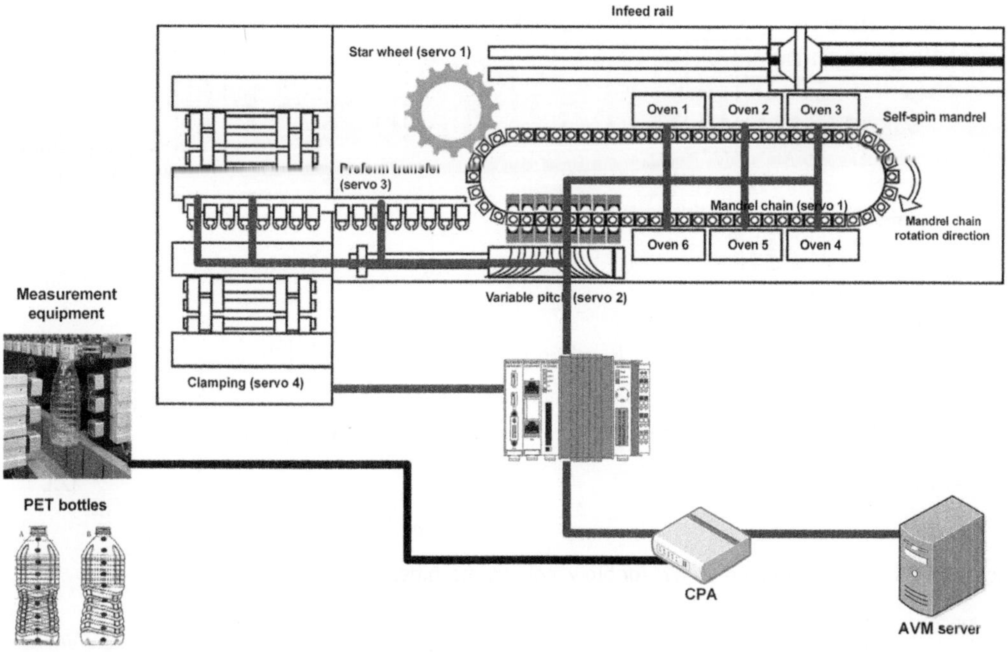

Figure 11.77 Two-stage PET stretch-blow molding machine.

sensor data of these two stages are collected via CPA to serve as the main process data of AVM. Besides the abovementioned temperature and pressure data, lever position of the machine is also adopted as the other process data of AVM. Moreover, as shown in the black segments of Figure 11.77, the measurement values of the thickness and/or center-of-mass of PET bottles are also collected via CPA to serve as the metrology data of AVM.

Applying the AVM technology in the PET blow molding machine can convert sampling inspection with metrology delay into real-time and on-line total inspection for monitoring the quality of thickness and/or center-of-mass of every PET bottle in the production line, as shown in Figure 11.78.

11.8.3 AVM-Based Run-to-Run (R2R) Control for Blow Molding Manufacturing Process

The architecture of the In-situ-Metrology based R2R control scheme is shown in Figure 11.79. The actual measurement values of every single lot should be collected in each machine. In other words, every machine demands a measuring equipment to collect actual measurement values for R2R control. All the process data are considered in the key feature identification experiment to search for the key adjustable R2R control factor. As a result, Preform-3 heating parameter is selected.

To reduce the cost of measuring equipment, this work develops the architecture of the AVM-based R2R control scheme as in Figure 11.80. The VM values generated by the AVM System can replace most of the AM values. As such, fewer measuring machines are required. Not only the VM values but also their accompanying RIs and GSIs are generated by the AVM System. The RI and GSI values are applied to judge that the associated VM values are qualified to be utilized for R2R control or not. The metrology values from regular sampling inspection can be feed-backed to the AVM

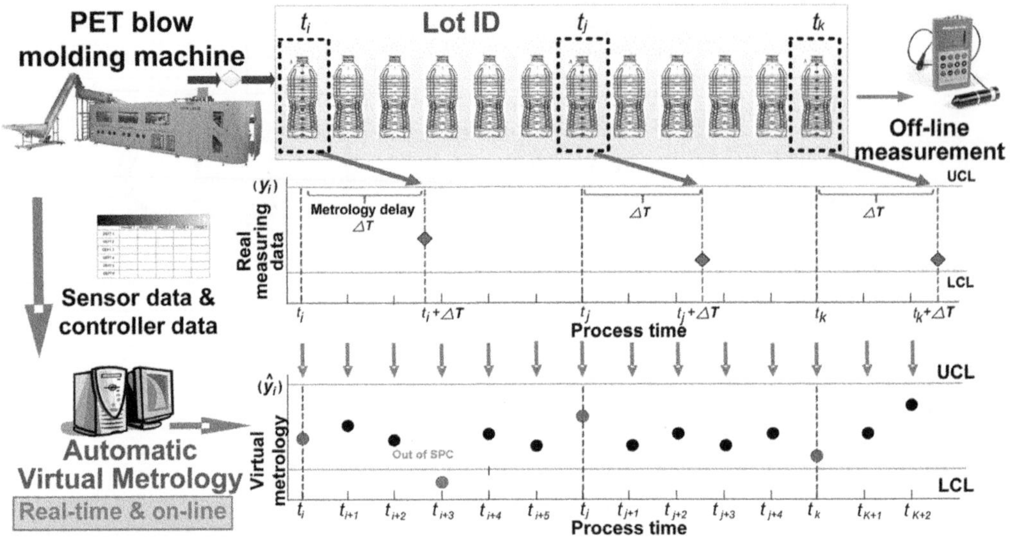

Figure 11.78 Implementation of AVM for blow molding machines.

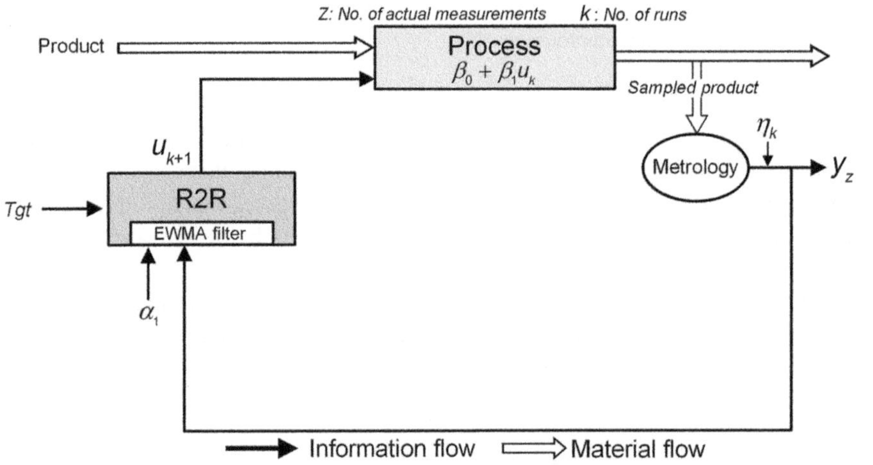

Figure 11.79 Architecture of IM-based R2R control. *Source:* Reprinted with permission from Ref. [41]; © 2013 IEEE.

System for model refreshing, and DQI_y algorithm of the AVM System can be further adopted to check the quality of the AM values. Unreliable AM values would be discarded or replaced by VM values instead. The architecture of the AVM-based R2R control scheme can thus be expanded into the scenarios containing multiple blow molding machines.

11.8.4 Illustrative Example

Figure 11.81 shows the cloud-based implementation of the AVM-based R2R control in the PET bottle blowing industry. Based on the AVM technology, the proposed R2R control scheme provides an affordable online and real-time quality control architecture, which can save the cost of

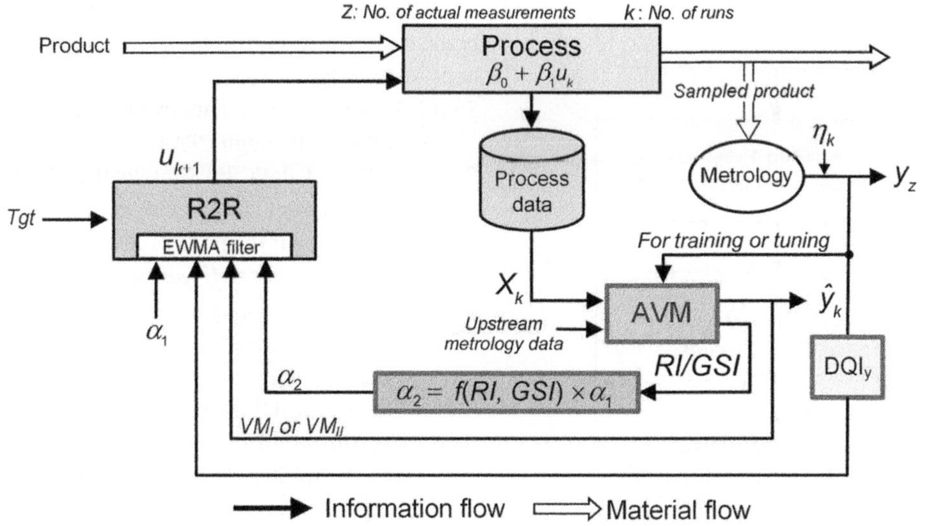

Figure 11.80 Architecture of AVM-based R2R control. *Source:* Reprinted with permission from Ref. [41]; © 2013 IEEE.

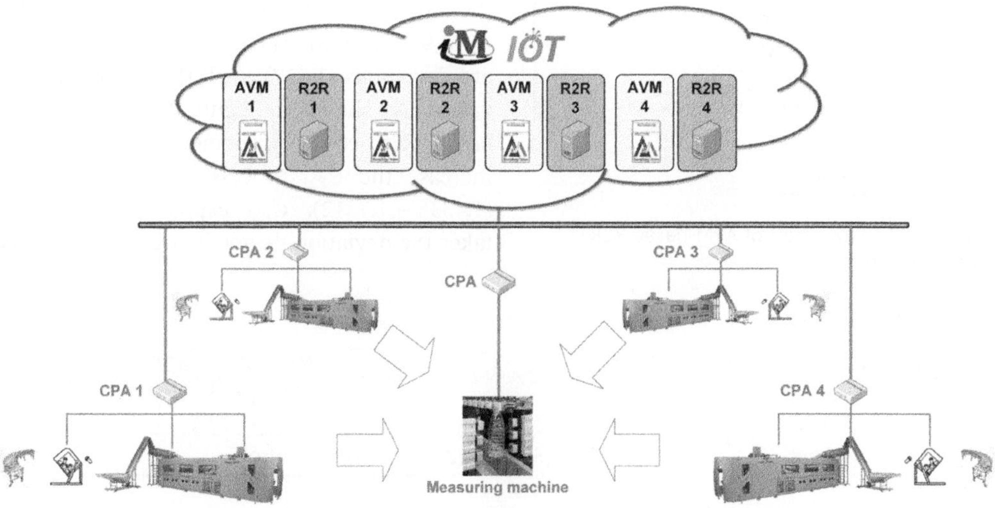

Figure 11.81 AVM-based R2R control implementation in multiple machines.

measuring instruments, improve overall process capability, and provide an economic and feasible solution for multiple machines to share the same measuring instrument.

As shown in the scenario of Figure 11.81, four blow molding machines share one measuring equipment and each machine produces 8 bottles per lot simultaneously. The sampling rate for AVM model refreshing is 1 per 4 lots. In other words, 1 out of 32 bottles that a single machine produces is adopted for sampling inspection. After the R2R controller receives the VM values from AVM, quality control of the center-of-mass of PET bottles can be conducted on these four blow molding machines.

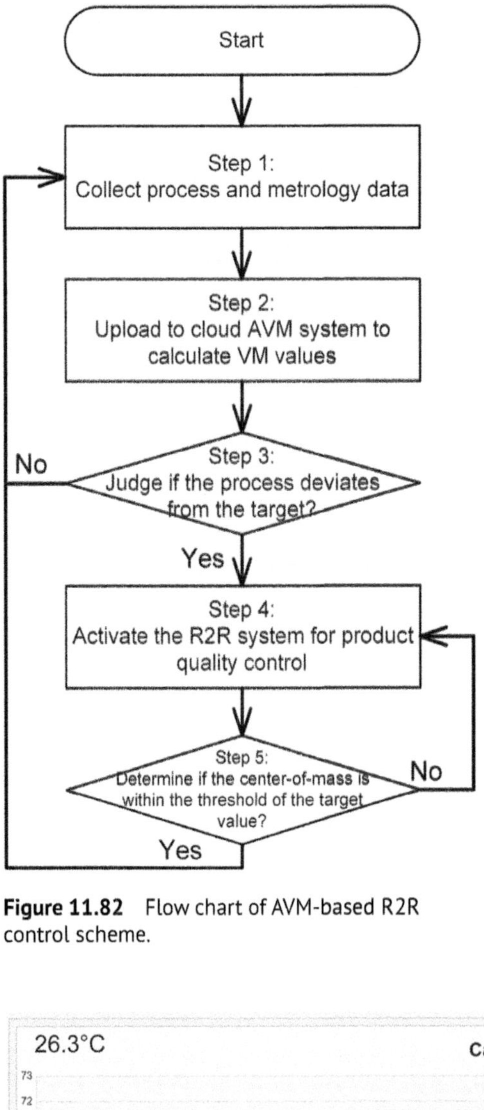

Figure 11.82 Flow chart of AVM-based R2R control scheme.

The flow chart of the AVM-based R2R control scheme is shown in Figure 11.82, and there are 5 steps as described below.

Step 1: Process data and metrology data are collected through CPAs.

Step 2: Collected data are sent to the cloud AVM to calculate the VM values.

Step 3: Determine if the process deviates from the target value. If yes, go to Step 4; otherwise, jump to Step 1.

Step 4: Activate the R2R scheme to conduct product quality control.

Step 5: Determine if the center-of-mass within the threshold of the target value. If yes, go to Step 1; otherwise, jump to Step 4.

The 1-out-of-1 lot experimental results are displayed in Figure 11.83. After the R2R scheme adjusts the VM values for 30 batches, the mean of center-of-mass is stabilized to reach the target value. The results also show that the AVM 95% MaxError (1.303) is less than 2 standard deviations (1.437) of all the real metrology values. As shown in Figure 11.84, the AVM-based R2R control scheme can successfully increase the C_{PM} process capability by 22% (1.4336→1.7512). C_{PM} expressed in (11.43) takes the deviation degree between the process

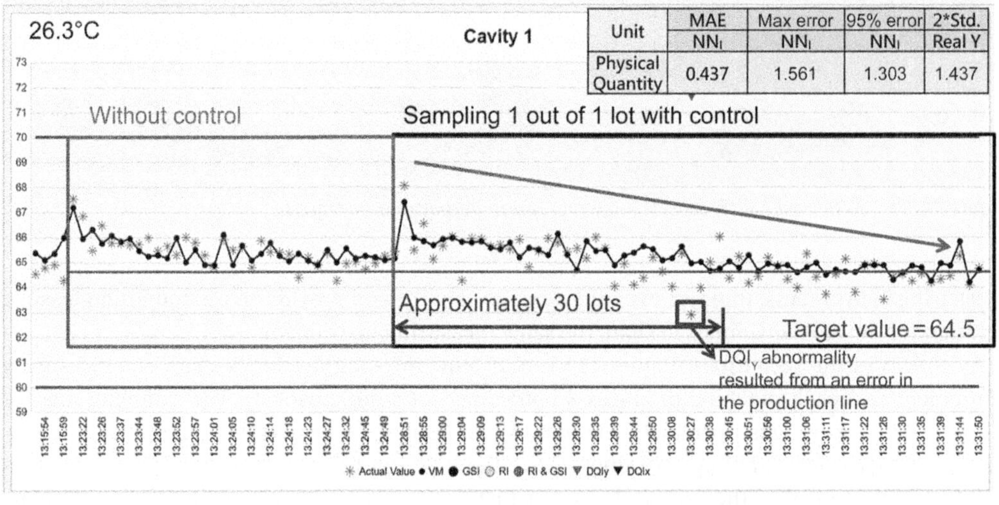

Figure 11.83 Experimental results of AVM-based R2R control for Case 1-out-of-1 lot.

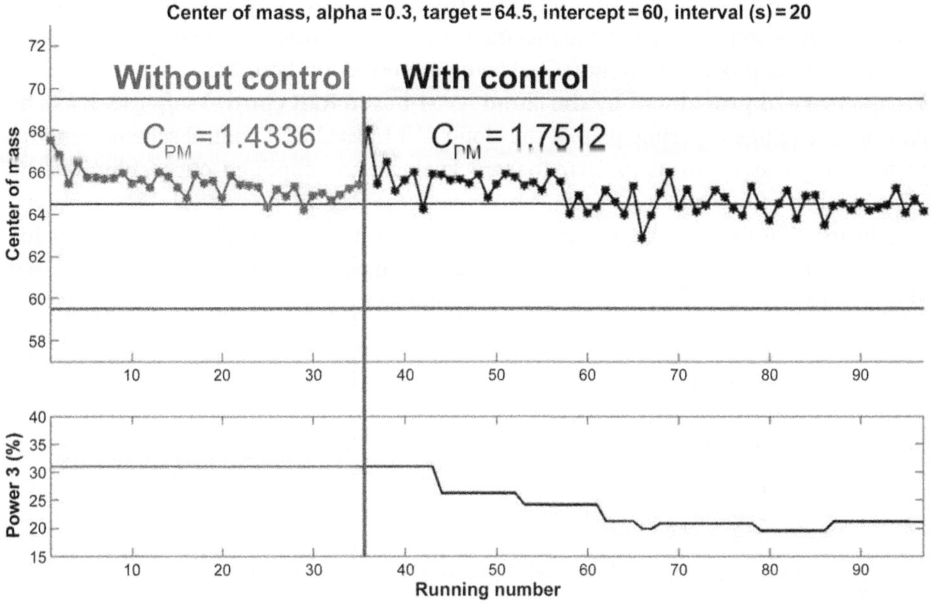

Figure 11.84 C_{PM} values of Case 1-out-of-1 lot.

mean and the target value into consideration. It emphasizes on the loss caused by the process mean deviating from the target value.

$$C_{PM} = \frac{USL - LSL}{6\sqrt{\sigma^2 + (\mu - T)^2}} \tag{11.43}$$

with

USL upper specification limits;

LSL lower specification limits;

σ^2 process variance;

μ process mean;

T target value of the process.

11.8.5 Summary

The implementation of the AVM-based cloud R2R control scheme proposed in this work can solve the issues of real-time production quality inspection and control in the blow molding industry for process improvement in three aspects:

- **Total quality inspection of the cloud AVM System.** Concerning the accuracy of the cloud AVM System, 95% MaxError of AVM results is less than 2 standard deviations of the real metrology values. As such, total inspection of the blowing production line can be achieved; also, abnormalities of production machines can be detected so as to reduce defective products or production waste.

- **Process capability improvement by the cloud IM-based R2R control scheme.** By adopting the IM-based R2R control scheme, product quality can be stabilized to approach to the target value, with the overall process capability (C_{PM}) being improved by about 22%.
- **Process capability improvement by the cloud AVM-based R2R control scheme.** It can be observed from the online experiments that the cloud AVM-based R2R control scheme can also stabilize the product quality to approach to the target value with sampling 1-out-of-1 lot, and the overall C_{PM} is improved by about 22%. It is also proved that multiple machines can share one measuring instrument to control and stabilize product quality to approach to the target value simultaneously. Further, simulation results show that one measuring instrument can be shared by up to seven production machines at its best performance.

Appendix 11.A – Abbreviation List

AC	Approximate Coordinates
ACF	Autocorrelation Function
ADF	Augmented Dickey-Fuller
ALASSO	Automated Least Absolute Shrinkage and Selection Operator
AI	Application Interface
AM	Actual Metrology
AMCoT	Advanced Manufacturing Cloud of Things
AOI	Automated Optical Inspection
APC	Advanced Process Control
AR	Approximate Radius
ARIMA	Autoregressive Integrated Moving Average Model
ASE	Advanced Semiconductor Engineering
AVM	Automatic Virtual Metrology
AVM_{G1}	AVM Server of Stage I for G1 Conjecturing
AVM_{G2}	AVM Server of Stage II for G2 Conjecturing
BEI	Baseline Error Index
BEI_T	Threshold of BEI
BIC	Bayesian Information Criterion
BPNN	Back-Propagation Neural Networks
$BPNN_I$	Phase-I BPNN
BT	Buffer Time
CD	Critical Dimension
CEM	Common Equipment Model
CF	Color Filter
CHD	Center-Hole Diameter

CMM	Coordinate-Measuring Machine
CMP	Chemical Mechanical Polishing
CNC	Computer Numerical Control
CPA	Cyber-Physical Agent
CPAVM	CPA+AVM
Cpk	Process Capability Index
CPS	Cyber-Physical System
CS	Clean Stage
CVD	Chemical Vapor Deposition
C&H	Concise and Healthy
DCI	Death Correlation Index
DCM	Data Collection Manager
DCP	Data Collection Plan
DCR	Data Collection Report
DF	Deformation Fusion
DHI	Device Health Index
DHI_T	Threshold of DHI
DQI_X	Data Quality Index of the Process Data
DQI_{XR}	Data Quality Index of X_R
DQI_y	Data Quality Index of the Metrology Data
DR	Deformed Radius
DS	Determinative Samples
EC	Engine Case
ECF	Exponential Curve Fitting
EHI	Equipment Health Index
EK_C	Concise Expert Knowledge Set
ESC	Electrostatic Chuck
EWMA	Exponentially Weighted Moving Average
FDC	Fault Detection and Classification
FHI	Factory Health Index
GA	Generic Algorithm
GAVM	An Integration system of GED and AVM
GED	Generic Embedded Device
GSI	Global Similarity Index
GSI_I	Global Similarity Index of Phase I

GSI_{II}	Global Similarity Index of Phase II
GSI_T	Threshold of GSI
HIH	Health Index Hierarchy
ICP	Inductively Coupled Plasma
ID	Identification
IESA	Interaction-Effect Search Algorithm
IESF	Interaction-Effect Search Framework
IF	Interpolation Fitting
iFA	Intelligent Factory Automation
ILM	In-line Metrology
IoT	Internet of Things
IPM	Intelligent Predictive Maintenance
ISI	Individual Similarity Index
ISI_B	Baseline Individual Similarity Index
ISI_{B_T}	Threshold of ISI_B
ITO	Indium Tin Oxide
IYM	Intelligent Yield Management
KSA	Key-variable Search Algorithm
KIS	Keep-Important-Sample
LCD	Liquid Crystal Display
LCL	Lower Control Limit
LCM	Liquid Crystal Module
LSL	Lower Specification Limit
MA	Moving Average Model
MAE	Mean Absolute Error
MAPE	Mean Absolute Percentage Error
$MAPE_P$	Process Mean-Absolute-Percentage Error with Respect to the Target Value
MC	Mass Customization
MHI	Module Health Index
MES	Manufacturing Execution System
MR	Multi-Regression
NC	Numerical Control
NG	No Go
NN	Neural Networks
NN_I	MAPE of Phase-I NN-based VM value

NN_{II}	MAPE of Phase-II NN-based VM value
OEE	Overall Equipment Effectiveness
OMM	Off-Machine Measuring
OMP	On-Machine Probing
OOC	Out-of-Control
OP	Original Manufacturing Position
OS	Oven Stage
PACF	Partial Auto Correlation Function
PAM	Pluggable Application Module
PAN	Polyacrylonitrile
PdM	Predictive Maintenance
PDT	Production-Data-Traceback
PECVD	Plasma-Enhanced Chemical Vapor Deposition
PEP	Photo Engraving Processes
PLS	Partial Least Squares
PLS_I	Phase-I Partial-Least-Square-based VM
PM	Preventive Maintenance
PVD	Physical Vapor Deposition
PR	Precursor
PreAM	Pre-Alarm Module
PS	Photo Space
RC	Real Coordinates
RDL	Redistribution Layer
RF	Radio- Frequency
RFID	Radio Frequency Identification
RI	Reliance Index
RI_I	Phases-I RI Value
RI_{II}	Phases-II RI Value
RI_T	RI Threshold Value
RMS	Root Mean Square
RNs	Running Numbers
RUL	Remaining- Useful- Life
SC	Status Change
SD	Sputter Device
SFs	Signal Features

SHI	Station Health Index
SiN	Silicon Nitride
SS	Sputter Stage
$SIZE_{DS}$	The Size of DS
SOAP	Simple Object Access Protocol
SPC	Statistical Process Control
STDB	Standard Database
T2T	Tube to Tube
TD	Target Device
T_{DS}	Threshold of $SIZE_{DS}$
TFT	Thin Film Transistor
TFT-LCD	Thin Film Transistor-Liquid Crystal Display
TPOGA	Triple- Phase Orthogonal Greedy Algorithm
TPOGA IESF	TPOGA Interaction-Effect Search Framework
TSP	Time-Series-Prediction
TVA	Target-Value-Adjustment
UB	Upper Bound
UBM	Under Bump Metallurgy
UCL	Upper Control Limit
USL	Upper Spec Limit
VM	Virtual Metrology
VM_I	Phase-I VM Value
VM_{II}	Phase-II VM Value
VMS	Virtual Metrology System
WIP	Work-in-Process
WMA	Wheel Machining Automation
X	Process Data
X_P	Process Data
X_R	Production Routes

Appendix 11.B – List of Symbols in Equations

$\widehat{G1}$	stage-I VM output
$\tilde{G}2$	stage-II metrology data
G	stage-II process

$\hat{y}_i$	conjecture VM value
y_i	target value
n	sample size
$\mathbf{Y}$	yield test result
$\boldsymbol{y}$	inline data
OS_l	original score
FS_l	final score with $l = 1, 2,\ldots, 10$ being the pick order
FS_{Oi}	final score of O_i
FS_{Lj}	final score of L_j
O_i	ith pick variable of TPOGA, $i = 1, 2, 3,\ldots, 10$
L_j	jth pick variable of ALASSO, $j = 1, 2, 3,\ldots, 10$
y_j	jth element of actual metrology values
α	EWMA coefficient
α_1	EWMA coefficient when AM is utilized
α_2	EWMA coefficient when VM is utilized
$\tilde{\eta}_{k+1}$	estimated model offset or disturbance of the (k+1)th run
A	gain parameter estimated for the system
μ_k	control action taken for run k
σ_y	standard deviation of y
$\sigma_{\hat{y}}$	standard deviation of $\hat{y}$
y	AM data
$\hat{y}$	VM data
$Z_{\hat{y}N_i}$	statistical distribution of the VM value built by NN
$Z_{\hat{y}_{r_i}}$	statistical distribution of the reference prediction value built by MR
Z_λ	standardized λth set process data
R^{-1}	inverse matrix of correlation coefficient
p	number of process parameters
σ	standard deviation
y_k	process output measured from AM
η_k	model offset or disturbance of the kth run
Tgt	target value
Z_{y_i}	standardized AM value
$Z_{\hat{y}_{r_i}}$	standardized MR prediction value

ΔT	amount of cycle time reduction
T_P	PECVD process time
T_A	APC+AVM calculation time
T_S	manual recipe-inspection time
T_C	manual recipe-calculation time
T_M	manual recipe-modification time
y_T	actual value of TD
$\bar{y}_T$	mean of y_T
$\hat{y}_B$	baseline predictive value of TD
y_E	absolute value of "$\hat{y}_B$ minus y_T"
y_{E_S}	sick value of y_E
$\bar{y}_E$	mean of y_E
σ_{y_E}	standard deviation of y_E
y_{E_B}	begin value of y_E
y_{E_D}	dead value of y_E
k_i	ith sample number
$\hat{y}_{E_i}$	ith y_E predictive value corresponding to k_i
y_{E_i}	ith y_E actual value corresponding to k_i
k_B	Bth sample number corresponding to y_{E_B}
k_S	Sth sample number corresponding to y_{E_S}
A	interception value of the ECF formula
C	slope value of the ECF formula
$\hat{y}_{E_i\_UB_{S+j}}$	predictive $S+j$th UB value
$\hat{y}_{E_i\_LB_{S+j}}$	predictive $S+j$th LB value
$\hat{k}_{D\_RUL}$	D_RULth sample number corresponding to y_{E_D}
Δ_{S+j}	95% PI value corresponding to the $S+j$th sample
$t_{\frac{\alpha}{2}}(k_S - k_B + 1 - p)$	a t-distribution with $k_S - k_B + 1 - p$ degrees of freedom with p being the number of parameters and $a = 0.05$
$\hat{k}_{D\_UB}$	D_UBth sample number corresponding to y_{E_D}
$\hat{k}_{D\_LB}$	D_LBth sample number corresponding to y_{E_D}.
k_i	ith sample number
$\hat{y}_{E_i}$	ith y_E predictive value corresponding to k_i
Y_T	TD's aging feature
y_t	TD's aging feature value at time t

$\hat{y}_t$	predicted TD's aging feature value at time t
ε_{t-j}	white noise error terms at time $t-j, j = 1, 2, \ldots, q$
H_0	null hypothesis
H_{a}	alternative hypothesis
Y_{M}	model matrix of TD's aging feature
A_{ACF}	most y_{t-1}-related ACF time
B_{PACF}	most y_{t-1}-related PACF time
M	sample size of Y_{M}
y_{death}	death value of aging feature
$\hat{y}_{t+n}$	predicted TD's aging feature value at time $t+n$
$\hat{y}_{\mathrm{death}}$	predictive death value of aging feature
δ	delta value
δ_{T}	threshold value of δ
$*$	abnormal new sample
$*_{\mathrm{B}}, *_{\mathrm{C}}, *_{\mathrm{D}}$	three metrology types of different target values, type B, type C, and type D
$\mathbf{y}_{\boldsymbol{M}}$	modeling samples sorted in ascending order
$\mathrm{SIZE}_{\mathrm{y}_M}$	size of modeling samples
BV	breakpoint vector
BV′	derived by combining 0, **BV**, and $\mathrm{SIZE}_{\mathrm{y}_M}$
$\mathbf{G}_{\boldsymbol{p}}$	modeling samples grouped by **BV′**
$\overline{\mathrm{y}}_{G_p}$	mean of each group
M_i	ith element of modeling samples
$\mathrm{SIZE}_{\mathbf{BV}'}$	size of breakpoint vector
$\bar{y}_{DS}$	new target value
$\mathrm{y}^{\mathrm{A}}_{\mathrm{G}_p}$	used to adjust each element in $\mathbf{G}_{\boldsymbol{p}}$ to become the newly adjusted modeling samples
$\star^{\mathrm{A}}$	derived by adjusting metrology $*_{\mathrm{B}}$, $*_{\mathrm{C}}$, and $*_{\mathrm{D}}$
x_{Rs}, y_{Rs}	real coordinates relative to the original positions in the x–y plane with $s = 1, 2, \ldots, S$, respectively
S	the number of probing
a	the ratio of r_a to r ($a = r_a/r$)
b	the ratio of r_b to r ($b = r_b/r$)
(h, k)	the offset displacements of the center of the end-face of an EC measured from the origin(0, 0)
θ	rotation angle of a point on the end-face of an EC measured from the major axis of the EC end-face

r	default radius
ω	included angle between any two adjacent probing-positions
φ_s	approximate angle of the sth approximate position
r_A	approximate radius
r_a	major-axis length of an ellipse
r_b	minor-axis length of an ellipse
$\mathbf{P_E}$	Position errors
t_s	ratio of r_A to r_a
r_D	corresponding radius
r_{Ds}	sth corresponding radius
$\mathbf{AR}$	vector of all approximate radii
$\mathbf{AR_{IF}}$	vector of **AR** generated from IF for all the actual manufacturing positions
$\mathbf{AR_{GA}}$	**AR** vector generated from GA for all the actual manufacturing-positions
$\mathbf{AR_{Fusion}}$	fusion **AR** generated by combining both $\mathbf{AR_{IF}}$ and $\mathbf{AR_{GA}}$
$\mathbf{DR_{RC}}$	known nodes for the purpose of interpolating
$\mathbf{AR_{RC}}$	**AR** vector generated from RC for all the actual manufacturing-positions
f	weighting coefficient ranging from 0 to 1
VM_u	uth predictive VM value
AM_u	uth actual metrology value
SR_{ij}	completion time at the ith spinning position of sensor j
ST_i	process completion time at the ith spinning position
SD_j	distance of sensor j
FSD_i	distance between meter counter and the ith spinning position
RS	rolling rate (m/minute)
SD_1	distance between sensor1 and meter counter
FSD_1	distance of meter counter to spin1
ST_1	process completion time at spin1 position
RS	spinning speed

References

1 Cheng, F.T., Kao, C.A., Chen, C.F. et al. (2015). Tutorial on applying the VM technology for TFT-LCD manufacturing. *IEEE Transactions on Semiconductor Manufacturing* 28 (1): 55–69. https://doi.org/10.1109/TSM.2014.2380433.

2 Cheng, F.T., Hsieh, Y.S., Zheng, J.W. et al. (2017). A scheme of high-dimensional key-variable search algorithms for yield improvement. *IEEE Robotics and Automation Letters* 2 (1): 179–186. https://doi.org/10.1109/LRA.2016.2584143.

3 Chiu, Y.C., Cheng, F.T., and Huang, H.C. (2017). Developing a factory-wide intelligent predictive maintenance system based on Industry 4.0. *Journal of the Chinese Institute of Engineers* 40 (7): 562–571. https://doi.org/10.1080/02533839.2017.1362357.

4 Kao, C.A., Cheng, F.T., and Wu, W.-M. (2011). Preliminary study of run-to-run control utilizing virtual metrology with reliance index. In: *Proceedings of the IEEE International Conference on Automation Science and Engineering, Trieste, Italy (24–27 August 2011).* USA: IEEE.

5 Cheng, F.T., Chen, Y.T., Su, Y.C. et al. (2008). Evaluating reliance level of a virtual metrology system. *IEEE Transactions on Semiconductor Manufacturing* 21 (1): 92–103. https://doi.org/10.1109/TSM.2007.914373.

6 Cheng, F.T. and Chiu, Y.C. (2013). Applying the automatic virtual metrology system to obtain tube-to-tube control in a PECVD tool. *IIE Transactions* 45 (6): 670–681. https://doi.org/10.1080/0740817X.2012.725507.

7 Castillo, E.D. and Rajagopal, R. (2002). A multivariate double EWMA process adjustment scheme for drifting processes. *IIE Transactions* 34 (12): 1055–1068. https://doi.org/10.1080/07408170208928934.

8 Su, Y.C., Lin, T.H., Cheng, F.T. et al. (2008). Accuracy and real-time considerations for implementing various virtual metrology algorithms. *IEEE Transactions on Semiconductor Manufacturing* 21 (3): 426–434. https://doi.org/10.1109/TSM.2008.2001219.

9 Hsieh, Y.S., Cheng, F.T., Huang, H.C. et al. (2013). VM-based baseline predictive maintenance scheme. *IEEE Transactions on Semiconductor Manufacturing* 26 (1): 132–144. https://doi.org/10.1109/TSM.2012.2218837.

10 Lin, C.Y., Hsieh, Y.M., Cheng, F.T. et al. (2019). Time series prediction algorithm for intelligent predictive maintenance. *IEEE Robotics and Automation Letters* 4 (3): 2807–2814. https://doi.org/10.1109/LRA.2019.2918684.

11 Cheng, F.T., Chen, C.F., Hsieh, Y.S. et al. (2015). Intelligent sampling decision scheme based on the AVM system. *International Journal of Production Research* 53 (7): 2073–2088. https://doi.org/10.1080/00207543.2014.955924.

12 Lin, C.Y., Hsieh, Y.M., Cheng, F.T. et al. (2018). Interaction-effect search algorithm for the KSA scheme. *IEEE Robotics and Automation Letters* 3 (4): 2778–2785. https://doi.org/10.1109/LRA.2018.2838323.

13 Tieng, H., Yang, H.C., Hung, M.H. et al. (2013). *A novel virtual metrology scheme for predicting machining precision of machine tools. Proceedings of the 2013 IEEE International Conference on Robotics and Automation (ICRA 2013), Karlsruhe, Germany (May 6–10 2013).* USA: IEEE.

14 Yang, H.C., Tieng, H., and Cheng, F.T. (2015). Total precision inspection of machine tools with virtual metrology. *Journal of the Chinese Institute of Engineers* 39 (2): 1–15. https://doi.org/10.1080/02533839.2015.1091279.

15 Cheng, F.T., Tieng, H., Yang, H.C. et al. (2016). Industry 4.1 for wheel machining automation. *IEEE Robotics and Automation Letters* 1 (1): 332–339. https://doi.org/10.1109/LRA.2016.2517208.

16 Su, Y.C., Cheng, F.T., and Hung, M.H. et al. (2006). Intelligent prognostics system design and implementation. *IEEE Transactions on Semiconductor Manufacturing* 19 (2): 195–207. https://doi.org/10.1109/TSM.2006.873512.

17 Huang, H.C., Lin, Y.C., Hung, M.H. et al. (2015). Development of cloud-based automatic virtual metrology system for semiconductor industry. *Robotics and Computer-Integrated Manufacturing* 34: 30–43. https://doi.org/10.1016/j.rcim.2015.01.005.

18 Huang, Y.T. and Cheng, F.T. (2011). Automatic data quality evaluation for the AVM system. *IEEE Transactions on Semiconductor Manufacturing* 24 (3): 445–454. https://doi.org/10.1109/TSM.2011.2154910.

19 Yang, H.C., Tieng, H., and Cheng, F.T. (2015). Automatic virtual metrology for wheel machining automation. *International Journal of Production Research* 54 (21): 6367–6377. https://doi.org/10.1080/00207543.2015.1109724.

20 Tieng, H., Chen, C.F., Cheng, F.T. et al. (2017). Automatic virtual metrology and target value adjustment for mass customization. *IEEE Robotics and Automation Letters* 2 (2): 546–553. https://doi.org/10.1109/LRA.2016.2645507.

21 Abellan-Nebot, J.V. and Subirón, F.R. (2010). A review of machining monitoring systems based on artificial intelligence process models. *The International Journal of Advanced Manufacturing Technology* 47: 237–257. https://doi.org/10.1007/s00170-009-2191-8.

22 Teti, R., Jemielniak, K., O'Donnell, G. et al. (2010). Advanced monitoring of machining operations. *CIRP Annuals Manufacturing Technology* 59 (2): 717–739. https://doi.org/10.1016/j.cirp.2010.05.010.

23 Marinescu, I. and Axinte, D.A. (2011). An automated monitoring solution for avoiding an increased number of surface anomalies during milling of aerospace alloys. *International Journal of Machine Tools and Manufacture* 51 (4): 349–357. https://doi.org/10.1016/j.ijmachtools.2010.10.005.

24 Chu, W., Li, Y., Liu, C. et al. (2016). Collaborative manufacturing of aircraft structural parts based on machining features and software agents. *The International Journal of Advanced Manufacturing Technology* 87: 1421–1434. https://doi.org/10.1007/s00170-013-4976-z.

25 Gao, X., Mou, W., and Peng, Y. (2016). An intelligent process planning method based on feature-based history machining data for aircraft structural parts. *Procedia CIRP* 56: 585–589. https://doi.org/10.1016/j.procir.2016.10.115.

26 Tangjitsitcharoen, S., Thesniyom, P., and Ratanakuakangwan, S. (2017). Prediction of surface roughness in ball-end milling process by utilizing dynamic cutting force ratio. *Journal of Intelligent Manufacturing* 28: 13–21. https://doi.org/10.1007/s10845-014-0958-8.

27 Tieng, H., Tsai, T.H., Chen, C.F. et al. (2018). Automatic virtual metrology and deformation fusion scheme for engine-case manufacturing. *IEEE Robotics and Automation Letters* 3 (2): 934–941. https://doi.org/10.1109/LRA.2018.2792690.

28 Wang, S.M., Chen, Y.S., Lee, C.Y. et al. (2016). Methods of in-process on-machine auto-inspection of dimensional error and auto-compensation of tool wear for precision turning. *Applied Sciences* 6: 107. https://doi.org/10.3390/app6040107.

29 Griffin, J.M. (2015). The prediction of profile deviations from multi process machining of complex geometrical features using combined evolutionary and neural network algorithms with embedded simulation. *Journal of Intelligent Manufacturing* 29: 1171–1189. https://doi.org/10.1007/s10845-015-1165-y.

30 Liu, C., Li, Y., Zhou, G. et al. (2016). A sensor fusion and support vector machine based approach for recognition of complex machining conditions. *Journal of Intelligent Manufacturing* 29: 1739–1752. https://doi.org/10.1007/s10845-016-1209-y.

31 Malhotra, R., Singh, N., and Singh, Y. (2011). Genetic algorithms: concepts, design for optimization of process controllers. *Computer and Information Science* 4 (2): 39–54. https://doi.org/10.5539/cis.v4n2p39.

32 Prenter, P.M. (2013). *Splines and variational methods*. New York: Dover Publications.

33 Sedgwick, P. (2012). Pearson's correlation coefficient. *British Medical Journal* 345 (7) https://doi.org/10.1136/bmj.e4483.

34 Hsieh, Y.M., Lin, C.Y., Yang, Y.R. et al. (2019). Automatic virtual metrology for carbon fiber manufacturing. *IEEE Robotics and Automation Letters* 4 (3): 2730–2737. https://doi.org/10.1109/LRA.2019.2917384.

35 Buckley, J.D. and Edie, D.D. (1993). *Carbon-carbon materials and composites*. Amsterdam: Elsevier Inc.

36 Newcomb, B.A. (2016). Processing, structure, *and properties of carbon fibers. Composites Part A: Applied Science and Manufacturing* 91 (1): 262–282. https://doi.org/10.1016/j.compositesa.2016.10.018.

37 Jin, W.J. and Ye, W.H. (2006). Study on the barcode based WIP tracking and management system under the lean production mode. *Machine Building & Automation* 1: 34.

38 Huang, G.Q., Zhang, Y.F., and Jiang, P.-Y. (2007). RFID-based wireless manufacturing for real-time management of job shop WIP inventories. *The International Journal of Advanced Manufacturing Technology* 36: 752–764. https://doi.org/10.1007/s00170-006-0897-4.

39 Yin, C., Gao, Q., and Tian, J. (2012). *WIP tracking and monitoring system based on RFID. Proceedings of 2012 Third International Conference on Intelligent Control and Information Processing 365–368, Dalian, China (15–17 July 2012).* USA: IEEE.

40 Lin, Y.C., Hung, M.H., Huang, H.C. et al. (2017). Development of advanced manufacturing cloud of things (AMCoT)-a smart manufacturing platform. *IEEE Robotics and Automation Letters* 2 (3): 1809–1816. https://doi.org/10.1109/LRA.2017.2706859.

41 Kao, C.A., Cheng, F.T., Wu, W.M. et al. (2013). Run-to-run control utilizing virtual metrology with reliance index. *IEEE Transactions on Semiconductor Manufacturing* 26 (1): 69–81. https://doi.org/10.1109/TSM.2012.2228243.

Index

Industry 4.1: Intelligent Manufacturing with Zero Defects, First Edition. Edited by Fan-Tien Cheng.

Printed and bound by CPI Group (UK) Ltd, Croydon, CR0 4YY
16/04/2025
14658423-0001